METHODS IN MOLECULAR BIOLOGY™

Series Editor
John M. Walker
School of Life Sciences
University of Hertfordshire
Hatfield, Hertfordshire, AL10 9AB, UK

For other titles published in this series, go to
www.springer.com/series/7651

METHODS IN MOLECULAR BIOLOGY™

Micro and Nano Technologies in Bioanalysis

Methods and Protocols

Edited by

James Weifu Lee and Robert S. Foote

Oak Ridge National Laboratory, Oak Ridge, TN, USA

Humana Press

Editors
James Weifu Lee
Oak Ridge National Laboratory
Oak Ridge
TN, USA

Robert S. Foote
Oak Ridge National Laboratory
Oak Ridge
TN, USA

ISSN: 1064-3745
ISBN: 978-1-61779-487-2 e-ISBN: 978-1-59745-483-4
DOI: 10.1007/978-1-59745-483-4
Springer Dordrecht Heidelberg London New York

Printed on acid-free paper

Springer is part of Springer Science+Business Media (www.springer.com)

Preface

This book provides current information on the development of microfluidics, nanotechnologies, and physical science techniques for the separation, detection, manipulation, and analysis of biomolecules, and should be useful to a wide audience, including molecular and cell biologists, biochemists, microbiologists, geneticists, and medical researchers. Chapters cover a variety of topics and techniques ranging from lab-on-chip technologies and microfluidics-coupled mass spectrometry for separation and detection of biomolecules, including proteins and nucleic acids, to manipulating and probing biomolecules with nanopores, nanochannels, optical, and other physical means, with the possibility of isolation and analysis of individual biomolecules from a single cell, and to structural and functional analysis of biomolecules with liquid nuclear magnetic resonance, X-ray and neutron scattering techniques. The book presents emerging nanotechnologies including quantum dots and molecular fluorescence for imaging and tracking of biomolecules and nanotechnologies for biomolecular delivery, gene therapy, and gene-expression control. Each chapter describes a specific technology with its fundamental mechanism and practical applications for a particular subject area, so that a competent scientist who is unfamiliar with the technology can understand its capabilities and basic procedures. In many cases, a reader should be able to carry out the techniques successfully at the first attempt by simply following the detailed practical procedures (protocols) and/or information (including useful notes) provided in the book. For sophisticated technologies such as neutron scattering, the book describes their physical concepts and discusses the new opportunities that these new technologies may bring for both basic and applied research in the fields of molecular biology and biotechnology. This book consists of 41 chapters that are organized into four parts. The chapters were contributed by nearly 100 authors worldwide, who are among the world's prominent scientists in their fields.

The first half of the volume covers microfluidic and physical methods of bioanalysis. It consists of Part I on applications of microfluidics and nanopores in separation, manipulation, detection, and analysis of biomolecules, and Part II on technologies of physical science in detection and analysis of biomolecules. It contains valuable protocols on microfluidics and physical science-related technologies that may benefit the field of molecular biology. Chapter topics are briefly described below.

Part I consists of Chaps. 1–10: Chap. 1 describes a commercially available nanoflow analytical technology conducted on a microfabricated chip that allows for highly efficient HPLC separation and superior sensitivity for MS detection of complex proteomic mixtures; Chaps. 2–4 describe fabrication of nanofluidic channels for manipulation of DNA molecules, a single-molecule barcoding system using nanoslits for DNA analysis, and microfluidic devices with photodefinable pseudovalves for protein separation, respectively; Chap. 5 introduces specific antibody detection by using a microbead-based assay with quantum dot (QD) fluorescence on a microfluidic chip; Chap. 6 describes a biomolecular sample-focusing method based on a device design incorporating arrays of addressable on-chip microfabricated electrodes that can locally increase the concentration of DNA in solution by electrophoretically sweeping it along the length of a microchannel; Chap. 7 describes a solid-state nanopore technique for detecting individual biopolymers, and Chap. 8 reports a method of inserting and manipulating DNA in a nanopore with optical tweezers; Chaps. 9 and 10 describe techniques of forming an α-hemolysin nanopore for single-molecule analysis and for nanopore force spectroscopy of DNA duplexes.

Part II consists of Chaps. 11–22: Chap. 11 describes an electrochemical method for quantitative chemical analysis of neurotransmitter release from single cells; Chaps. 12–14 introduce techniques for trapping and detection of single molecules in water, ZnO nanorods as an intracellular sensor for pH measurements, and analysis of biomolecules using surface plasmons; Chap. 15 reports use of residual dipolar couplings in structural analysis of protein–ligand complexes by solution NMR spectroscopy; Chaps. 16 and 17 report Raman-assisted X-ray crystallography for the analysis of biomolecules and methods and software for diffuse X-ray scattering from protein crystals, and Chaps. 18–20 describe deuterium labeling for neutron structure–function–dynamics analysis, the basics and instrumentation of small-angle neutron scattering for molecular biology, and small-angle scattering and neutron contrast variation for studying biomolecular complexes, respectively; Chap. 21 describes the application of tandem mass spectrometry to identification of protein biomarkers of disease, and Chap. 22 describes the use of hyphenated MS techniques for comprehensive metabolome analysis.

The second half of the volume covers nanotechnologies for biosystems, and consists of Part III on applications of quantum dots and molecular fluorescence in detection, tracking, and imaging of biomolecules, and Part IV on nanotechnologies for biomolecular delivery, gene therapy, and expression control. It contains valuable information on nanoscience-empowered molecular biotechnologies.

Part III consists of Chaps. 23–32: Chaps. 23–25 describe multicolor detection of combed DNA molecules using quantum dots, quantum dot molecular beacons for DNA detection, and a gel electrophoretic blotting technique for identifying quantum dot–protein/protein–protein interactions; Chaps. 26 and 27 present techniques for in vivo imaging of quantum dots and efficient biolabels in cancer diagnostics, respectively; Chap. 28 describes monitoring and affinity purification of proteins using dual tags with tetracysteine motifs, and Chap. 29 reports use of genomic DNA as a reference in DNA microarray analyses; Chap. 30 describes single-molecule imaging of fluorescent proteins expressed in living cells; Chap. 31 describes micropositron emission tomography (PET), single-photon emission computed tomography (SPECT), and near-infrared (NIR) fluorescence imaging of biomolecules in vivo, which could lead to a number of exciting possibilities for biomedical applications, including early detection, treatment monitoring, and drug development; Chap. 32 reports a revolutionary photo-based imaging technology: the ultrahigh resolution imaging of biomolecules by fluorescence photoactivation localization microscopy (FPALM) that can now image molecular distributions in fixed and living cells with measured resolution better than 30 nm, which likely represents a breakthrough technology that has now shattered the classic limit of light microscopy resolution associated with the wavelength-dependent light diffraction barrier, thought to be unbreakable for more than 100 years.

In Part IV, Chaps. 33–41 describe nanotechnologies with potential biomedical applications. Specifically, Chap. 33 describes real-time imaging of gene delivery and expression with DNA nanoparticle technologies and Chap. 34 reports nanoparticle-mediated gene delivery. Chapters 35 and 36 describe magnetic nanoparticles for local drug delivery using magnetic implants and functionalized magnetic nanoparticles as an in vivo delivery system, and Chap. 37 reports formulation/preparation of functionalized nanoparticles for in vivo targeted drug delivery; Chap. 38 reports detection of mRNA in single living cells using atomic force microscopy nanoprobes; Chap. 39 describes a gene transfer technique through reverse transfection using gold nanoparticles; Chap. 40 presents custom-designed molecular scissors for site-specific manipulation of the plant and mammalian genomes, and Chap. 41 describes a technique for determining DNA sequence specificity of natural and artificial transcription factors by cognate site identifier analysis, both of which could lead to modern applications in molecular biology and biomedicine.

Oak Ridge, TN

James Weifu Lee
Robert S. Foote

Acknowledgments

The editors, James Weifu Lee and Robert S. Foote, thank the nearly 100 authors throughout the world for their contributions and collaboration on this book project. The editing work of this volume was accomplished using significant amounts of the editors' spare time including their family time. Therefore, the editors also wish to thank their respective families: the Lee family and the Foote family, for their wonderful support and understanding.

Contents

Contributors

SAFAA AL-HILLI • *Department of Physics, Gothenburg University, Gothenburg, Sweden*

ASEEM Z. ANSARI • *Department of Biochemistry, and the Genome Center, University of Wisconsin-Madison, Madison, WI, USA*

JOERG BEWERSDORF • *Institute for Molecular Biophysics, The Jackson Laboratory, Bar Harbor, ME, USA*

DOMINIQUE BOURGEOIS • *Institut de Biologie Structurale Jean-Pierre Ebel, Grenoble, France*

NATHANIEL C. CADY • *College of Nanoscale Science and Engineering, University at Albany, Albany, NY, USA*

PHILIPPE CARPENTIER • *Institut de Biologie Structurale Jean-Pierre Ebel, Grenoble, France*

SRINIVASAN CHANDRASEGARAN • *Department of Environmental Health Sciences, Johns Hopkins University, Baltimore, MD, USA*

XIAOYUAN CHEN • *Department of Radiology and Bio-X Program, Stanford University, Stanford, CA, USA*

AURÉLIEN CRUT • *Laboratoire Kastler Brossel, Département de Physique, Ecole Normale Supérieure, Paris, France*

B. DANIELSSON • *Department of Pure and Applied Biochemistry, Lund University, Lund, Sweden*

CEES DEKKER • *Kavli Institute of Nanoscience, Delft University of Technology, Delft, The Netherlands*

NYNKE H. DEKKER • *Kavli Institute of Nanoscience, Delft University of Technology, Delft, The Netherlands*

EVE DE ROSNY • *Institut de Biologie Structurale Jean-Pierre Ebel, Grenoble, France*

PIERRE DESBIOLLES • *Laboratoire Kastler Brossel, Département de Physique, Ecole Normale Supérieure, Paris, France*

CHRISTOPHE ESCUDÉ • *Muséum National d'Histoire Naturelle, Paris, France*

ANDREW G. EWING • *Department of Chemistry, The Pennsylvania State University, University Park, PA, USA, Department of Chemistry, Göteborg University, Göteborg, Sweden*

Z. HUGH FAN • *Department of Mechanical and Aerospace Engineering, Department of Biomedical Engineering, University of Florida, Gainesville, FL, USA*

PATRICIA M.A. FARIAS • *Department of Biophysics and Radiobiology, Federal University of Pernambuco, Cidade Universitária, Recife, PE, Brazil*

OMID C. FAROKHZAD • *Harvard-MIT Center for Cancer Nanotechnology Excellence, Massachusetts Institute of Technology, Cambridge, MA, USA, Laboratory of Nanomedicine and Biomaterials, Brigham and Women's Hospital, Boston, MA, USA*

RODRIGO FERNÁNDEZ-PACHECO • *Instituto Universitario de Investigación en Nanociencia de Aragón (INA), Universidad de Zaragoza, Zaragoza, Spain*

ADRIANA FONTES • *Department of Biophysics and Radiobiology, Federal University of Pernambuco, Cidade Universitária, Recife, PE, Brazil*

SATOSHI FUJITA • *Research Institute for Cell Engineering, National Institute of Advanced Industrial Science and Technology, Tokyo, Japan*

BÉNÉDICTE GÉRON-LANDRE • *Muséum National d'Histoire Naturelle, Paris, France*

RICHARD J. GIANNONE • *Biosciences Division, Oak Ridge National Laboratory, Oak Ridge, TN, USA, Graduate School of Genome Science and Technology, University of Tennessee-Oak Ridge National Laboratory, Knoxville, TN, USA*

JENE A. GOLOVCHENKO • *Department of Physics, Division of Engineering and Applied Sciences, Harvard University, Cambridge, MA, USA*

TRAVIS J. GOULD • *Department of Physics and Astronomy, and Institute for Molecular Biophysics, University of Maine, Orono, ME, USA*

FRANK GU • *Harvard-MIT Center for Cancer Nanotechnology Excellence, Massachusetts Institute of Technology, Cambridge, MA, USA*

Laboratory of Nanomedicine and Biomaterials, Brigham and Women's Hospital, Boston, MA, USA

MUDALIGE GUNEWARDENE • *Department of Physics and Astronomy, and Institute for Molecular Biophysics, University of Maine, Orono, ME, USA*

TAKAHIRO HATANAKA • *Mitsubishi Kagaku Institute of Life Sciences, Tokyo, Japan*

MICHAEL L. HEIEN • *Department of Chemistry, The Pennsylvania State University, University Park, PA, USA*

WILLIAM T. HELLER • *Center for Structural Molecular Biology, Oak Ridge National Laboratory, Oak Ridge, TN, USA*

SAMUEL T. HESS • *Department of Physics and Astronomy, and Institute for Molecular Biophysics, University of Maine, Orono, ME, USA*

KAYO HIBINO • *Cellular Informatics Laboratory, RIKEN, Wako, Japan*

MICHIO HIROSHIMA • *Cellular Informatics Laboratory, RIKEN, Wako, Japan*

M. RICARDO IBARRA • *Instituto Universitario de Investigación en Nanociencia de Aragón (INA), Universidad de Zaragoza, Zaragoza, Spain, Instituto de Ciencia de Materiales de Aragón (ICMA), Universidad de Zaragoza-CSIC, Zaragoza, Spain*

YUKO ICHIYANAGI • *Department of Physics, Graduate School of Engineering, Yokohama National University, Yokohama, Japan*

ATSUSHI IKAI • *Department of Life Science, Graduate School of Bioscience and Biotechnology, Tokyo Institute of Technology, Yokohama, Japan*

NITIN U. JAIN • *Biochemistry, Cellular and Molecular Biology Department, University of Tennessee, Knoxville, TN, USA*

NAHID N. JETHA • *Department of Physics and Astronomy, University of British Columbia, Vancouver, Canada*

SHA JIN • *DNA Resource Center, University of Arkansas, Fayetteville, AR, USA*

Kyubong Jo • *Department of Chemistry, University of Wisconsin, Madison, WI, USA, Department of Chemistry & Interdisciplinary Program of Integrated Biotechnology, Sogang University, Seoul, Korea*

Véronique Josserand • *ANIMAGE, CERMEP, Lyon, France INSERM U823, Institut Albert Bonniot, La Tronche, France*

Karthikeyan Kandavelou • *Department of Environmental Health Sciences, Johns Hopkins University, Baltimore, MD, USA*

Gergely Katona • *Institut de Biologie Structurale Jean-Pierre Ebel, Grenoble, France*

Ulrich. F. Keyser • *Cavendish Laboratory, University of Cambridge, UK*

Hak-Sung Kim • *Department of Biological Sciences, KAIST, Daejeon, Korea*

Michael Kinter • *Free Radical Biology and Aging Research Program, Oklahoma Medical Research Foundation, Oklahoma City, Oklahoma, USA*

Robert Langer • *Harvard-MIT Center for Cancer Nanotechnology Excellence, Massachusetts Institute of Technology, Cambridge, MA, USA*

John C. Leach • *Biomedical Engineering Program, College of Engineering, University of Arkansas, Fayetteville, AR, USA*

Dohoon Lee • *Environment and Energy Division, Korea Institute of Industrial Technology, Cheonan, Korea*

Jiali Li • *Department of Physics, University of Arkansas, Fayetteville, AR, USA*

Zi-Bo Li • *Department of Radiology and Bio-X Program, Stanford University, Stanford, CA, USA*

Kenneth C. Littrell • *Neutron Scattering Sciences Division, Oak Ridge National Laboratory, Oak Ridge, TN, USA*

Hong Yan Liu • *Department of Biomedical Engineering, Oregon Health & Science University, Portland, OR, USA*

Yie Liu • *Biosciences Division, Oak Ridge National Laboratory, Oak Ridge, TN, USA*

Andre Marziali • *Department of Physics and Astronomy, University of British Columbia, Vancouver, Canada*

Michael D. Mason • *Department of Chemical and Biological Engineering, and Institute for Molecular Biophysics, University of Maine, Orono, ME, USA*

Flora Meilleur • *Department of Molecular & Structural Biochemistry, North Carolina State University, Raleigh, NC, USA, Spallation Neutron Source, Oak Ridge National Laboratory, Oak Ridge, TN, USA*

Jun Miyake • *Research Institute for Cell Engineering, National Institute of Advanced Industrial Science and Technology, Tokyo, Japan*

Masato Miyake • *Research Institute for Cell Engineering, National Institute of Advanced Industrial Science and Technology, Tokyo, Japan*

Shinji Moritake • *Department of Physics, Graduate School of Engineering, Yokohama National University, Yokohama, Japan*

DEAN A.A. MYLES • *Center for Structural Molecular Biology, Oak Ridge National Laboratory, Oak Ridge, TN, USA*

HANBEN NIU • *Institute of Optoelectronics, Shenzhen University, Shenzhen, China*

O. NUR • *Department of Science and Technology, Campus Norrköping, Linköping University, Norrköping, Sweden*

TOSHIYA OSADA • *Department of Life Science, Graduate School of Bioscience and Biotechnology, Tokyo Institute of Technology, Yokohama, Japan*

MARY S. OZERS • *Department of Biochemistry, University of Wisconsin-Madison, Madison, WI, USA*

K. RISVEDEN • *Department of Pure and Applied Biochemistry, Lund University, Lund, Sweden*

YASUSHI SAKO • *Cellular Informatics Laboratory, RIKEN, Wako, Japan*

BEATE S. SANTOS • *Department of Pharmaceutical Sciences, Federal University of Pernambuco, Cidade Universitária, Recife, PE, Brazil*

TIMOTHY M. SCHRAMM • *Laboratory for Molecular and Computational Genomics, Department of Chemistry, Laboratory of Genetics, and Biotechnology Center, University of Wisconsin, Madison, WI, USA*

DAVID C. SCHWARTZ • *Laboratory for Molecular and Computational Genomics, Laboratory of Genetics, Department of Chemistry and Biotechnology Center, University of Wisconsin, Madison, WI, USA*

MITSUTOSHI SETOU • *Mitsubishi Kagaku Institute of Life Sciences, Tokyo, Japan, National Institute for Physiological Sciences, National Institute of Natural Sciences, Aichi, Japan*

FAISAL A. SHAIKH • *Artie McFerrin Department of Chemical Engineering, Texas A&M University, College Station, TX, USA*

WENCHAO SUN • *Department of Pediatrics, Case Western Reserve University, Cleveland, OH, USA*

SHU TAIRA • *Mitsubishi Kagaku Institute of Life Sciences, Tokyo, Japan*

MASAHIRO TAKAHASHI • *Cellular Informatics Laboratory, RIKEN, Wako, Japan*

ISABELLE TEXIER • *Micro-Technologies for Biology and Healthcare Department, CEA Grenoble, Grenoble, France*

VLADIMIR V. TOLSTIKOV • *University of California Davis Genome Center, Davis, CA, USA*

JILL TREWHELLA • *School of Molecular and Microbial Biosciences, University of Sydney, Sydney, NSW, Australia*

EIICHIRO UCHIMURA • *Research Institute for Cell Engineering, National Institute of Advanced Industrial Science and Technology, Tokyo, Japan*

HIRONORI UEHARA • *Department of Ophthalmology & Visual Science, University of Utah, Salt Lake City, UT, USA*

VICTOR M. UGAZ • *Artie McFerrin Department of Chemical Engineering, Texas A&M University, College Station, TX, USA*

J. Gabriel Valdivia • *Instituto Universitario de Investigación en Nanociencia de Aragón (INA), Universidad de Zaragoza, Zaragoza, Spain, Hospital Clínico Universitario "Lozano Blesa", Zaragoza, Spain*

Tom van de Goor • *Agilent Technologies, Waldbronn, Germany*

J. van der Does • *Kavli Institute of Nanoscience, Delft University of Technology, Delft, The Netherlands*

Martin Vollmer • *Agilent Technologies, Waldbronn, Germany*

Tania Q. Vu • *Department of Biomedical Engineering, Oregon Health and Science University, Portland, OR, USA*

Michael E. Wall • *Computer, Computational, and Statistical Sciences Division, Bioscience Division, and Center for Nonlinear Studies, Los Alamos National Laboratory, Los Alamos, NM, USA*

Kai-Ge Wang • *Institute of Photonics and Photonic Technology, Northwest University, Xi'an, China, Institute of Optoelectronics, Shenzhen University, Shenzhen, China*

Yisong Wang • *Biosciences Division, Oak Ridge National Laboratory, Oak Ridge, TN, USA*

Christopher L. Warren • *Department of Biochemistry, University of Wisconsin-Madison and VistaMotif LLC, Madison, WI, USA*

Kevin L. Weiss • *Center for Structural Molecular Biology, Oak Ridge National Laboratory, Oak Ridge, TN, USA*

Andrew E. Whitten • *Bragg Institute, Australian Nuclear Science and Technology Organisation, Lucas Heights, NSW, Australia*

Matthew Wiggin • *Department of Physics and Astronomy, and Department of Biochemistry, University of British Columbia, Vancouver, Canada*

M. Willander • *Department of Science and Technology, Linköping University, Campus Norrköping, Norrköping, Sweden, Department of Physics, Gothenburg University, Gothenburg, Sweden*

Shigeru Yamada • *Research Institute for Cell Engineering, National Institute of Advanced Industrial Science and Technology, Tokyo, Japan*

Yunfeng Yang • *Biosciences Division, Oak Ridge National Laboratory, Oak Ridge, TN, USA*

Kaiming Ye • *Biomedical Engineering Program, College of Engineering, University of Arkansas, Fayetteville, AR, USA*

Euisik Yoon • *Department of Electrical Engineering and Computer Science, University of Michigan, Ann Arbor, MI, USA*

Kwang-Seok Yun • *Department of Electronic Engineering, Sogang University, Seoul, Korea*

Assem G. Ziady • *Department of Pediatrics, Case Western Reserve University, Cleveland, OH, USA*

Part I

Applications of Microfluidics and Nanopores in Separation, Detection, Manipulation, and Analysis of Biomolecules

Chapter 1

HPLC-Chip/MS Technology in Proteomic Profiling

Martin Vollmer and Tom van de Goor

Summary

HPLC-chip/MS is a novel nanoflow analytical technology conducted on a microfabricated chip that allows for highly efficient HPLC separation and superior sensitive MS detection of complex proteomic mixtures. This is possible through on-chip preconcentration and separation with fluidic connection made automatically in a leak-tight fashion. Minimum precolumn and postcolumn peak dispersion and uncompromised ease of use result in compounds eluting in bands of only a few nanoliters. The chip is fabricated out of bio-inert polyimide-containing channels and integrated chip structures, such as an electrospray emitter, columns, and frits manufactured by laser ablation technology. Meanwhile, a variety of HPLC-chips differing in design and stationary phase are commercially available, which provide a comprehensive solution for applications in proteomics, glycomics, biomarker, and pharmaceutical discovery. The HPLC-chip can also be easily integrated into a multidimensional separation workflow where different orthogonal separation techniques are combined to solve a highly complex separation problems. In this chapter, we describe in detail the methodological chip usage and functionality and its application in the elucidation of the protein profile of human nucleoli.

Key words: HPLC-chip/MS, Nanoflow LC/MS, Multidimensional separation, Proteomics, Nucleolus

1. Introduction

To comprehensively elucidate a complex proteome, such as that of a cell organelle, it is necessary to combine different orthogonal separation techniques. In the past, numerous techniques have been combined that exploit different chemical and physicochemical properties of the protein and peptide analytes (for recent reviews *see* **refs.** *1,2*). Liquid-based techniques, in contrast to gel-based approaches, bear the advantage that the analytes always stay in the liquid phase. This avoids labor-intense and error-prone

James Weifu Lee and Robert S. Foote (eds.), *Micro and Nano Technologies in Bioanalysis*,
Methods in Molecular Biology, vol. 544
DOI 10.1007/978-1-59745-483-4_1,

extraction processes, and the final separation step can easily be directly coupled to nano electrospray, which allows for highly sensitive MS detection.

In the following study, we report the two-dimensional separation of the human nucleolus proteome, where strong cation-exchange chromatography was conducted offline in the first separation dimension, while reversed phase based HPLC-chip/MS was chosen for the last separation step. This separation scheme is similar to the Mudpit approach *(3)*. However, because the first and second dimensions are separated, enhanced flexibility concerning loading capacity and solvent compatibility is achieved.

Proteins are isolated and digested by standard procedures. Tryptic peptides are then fractionated according to their charge by using strong cation-exchange chromatography. The fractions are further separated on an HPLC-chip/MS system that contains an enrichment column for sample cleanup and concentration, and a reversed phase separation column. Because the chip separation column and the electrospray emitter are both integrated on a single chip, no peak broadening occurs after the peptide analytes elute off the column and enter the chip electrospray tip. This finally results in small peak volumes and hence superior sensitivity, especially for low abundant protein species from the investigated proteome. The method described in the following section can be applied and adapted for any multidimensional proteomic workflow.

1.1. Functionality of HPLC-Chip/MS

Proteomic studies usually face the dilemma that, on the one hand, sample size is limited and, on the other hand, high sensitivity and a wide dynamic range are required to identify and quantitate peptides and corresponding proteins comprehensively. High sensitivity in combination with ESI MS is best achieved by lowering the overall HPLC flow rate to a few hundreds of nanoliters or less. This results in a decrease of the dimension of the Taylor cone and of the size of the formed droplets in the ESI spray chamber, such that, due to the higher surface tension, the overall ionization efficiency is increased *(4)*.

The drawback of nano-HPLC/MS is the occurrence of small leaks and blockages that are difficult to trace under extremely low flow conditions. Dead volumes before the separation column affect the composition of the LC gradient and the analyte elution time. Dead volumes downstream of the separation column lead to significant peak broadening and result in loss of sensitivity. Therefore, chip-based separation devices have been introduced recently that integrate the nanoflow separation and the electrospray process, such that error-prone connections susceptible to introducing dead volumes are avoided. An overview of chip-based formats used in combination with electrospray MS was recently published *(5)*.

Although most of the chip formats that include separation and electrospray in a single device are still academic research tools, the Agilent HPLC-chip/MS system was introduced commercially in 2005. The system and the corresponding chips and their fabrication process have been described in detail in several reports *(6,7)*.

In short, the HPLC-chip is fabricated from layers of bio-inert polyimide that are first laser-ablated to form the microfluidic channels, fluidic inlet ports, column chambers, frits, and electrospray emitter. Different layers are then attached to each other by heat vacuum lamination followed by deposition of electrical contacts by metallization. The column channels of the chips are packed with standard silica-based reversed phase particles (Zorbax 300 SB-C18, 5 µm, Agilent Technologies, Waldbronn, Germany).

The chip is automatically inserted into the HPLC-chip/MS interface by a software command and clamped in a leak-tight fashion between a valve stator that bears the inlet connections of the fluidic transfer capillaries and a ceramic nano rotary valve.

HPLC-chips are available with packed separation channels of 50 µm (D) × 75 µm (W) in a length of 43 mm or 150 mm. Separation of the analytes is usually performed at 200–600 nL/min with the aid of a nanopump. A second column serves as enrichment and sample clean-up column and is available at volumes of 40 nL and 160 nL, depending on the loading capacity and complexity of the sample that is required for the specific separation workflow **(Fig. 1)**. The sample is loaded first onto the enrichment column using a capillary pump at flow rates of 4 µL/min. While salts are washed off, peptides and proteins are retained. A nano rotary valve operates directly on the surface of the chip and is turned 60 degrees to switch the enrichment column to

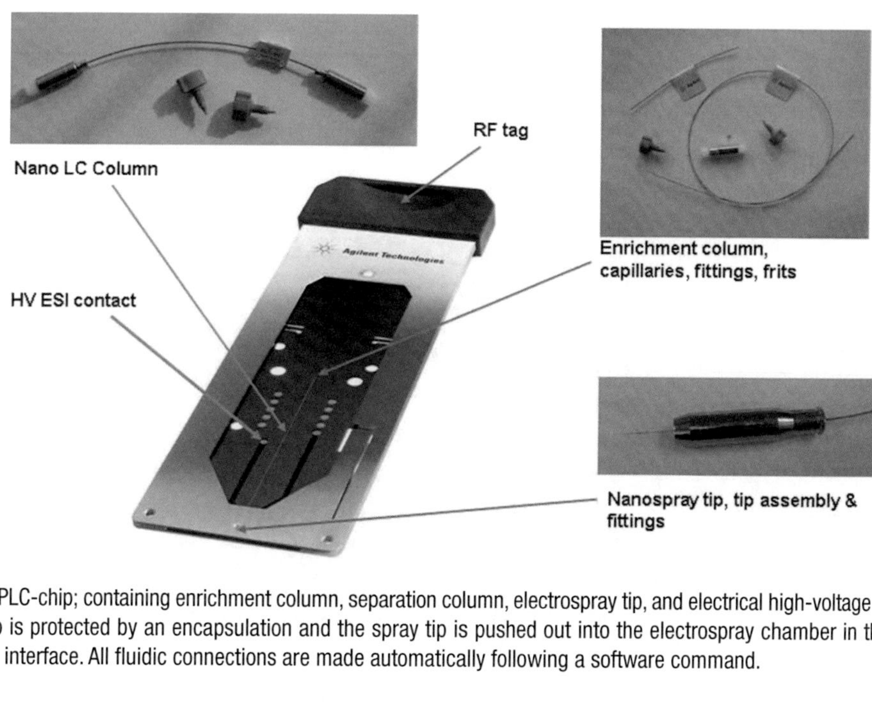

Fig. 1. HPLC-chip; containing enrichment column, separation column, electrospray tip, and electrical high-voltage contacts. The chip is protected by an encapsulation and the spray tip is pushed out into the electrospray chamber in the HPLC-chip/MS interface. All fluidic connections are made automatically following a software command.

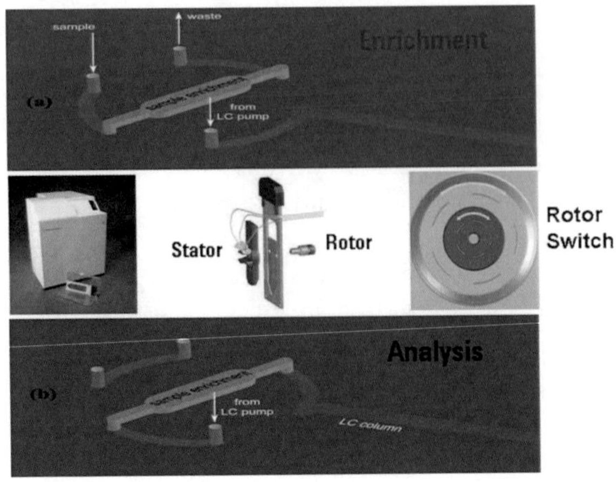

Fig. 2. Sample loading and analysis: The sample is loaded onto the enrichment column at 4 μL/min using the LC capillary pump. Salts and contaminants are flushed into waste while the valve is in the enrichment position (*upper panel*). The nano rotary valve, which operates directly on the surface of the chip clamped between the rotor and stator (*intermediate panel*), is then switched into the analytical flow path. The sample is then desorbed by opposite flow from the enrichment column using a nanopump at 300 nL/min and transferred to the analytical column where the sample is separated using a gradient of increasing organic concentration (*lower panel*).

the same flow path as the separation column. Peptides elute off the enrichment column by a solvent gradient delivered from the nanopump and are transferred to the separation column (**Fig. 2**). Analytes eluting from the separation channel travel 2 mm further into an 8 μm ID emitter channel to exit finally through the outlet hole of the electrospray tip into the ionization chamber. The chip can be used with backpressures of up to 150 bars. Typical chip lifetime exceeds 200 working hours.

The column material of the chip is retained by narrowing the chip channel on the column outlet side and a filter layer is attached at the column inlet after the packing process. HPLC-chip/MS has been successfully applied for biomarker discovery and several proteomic and glycomic research studies *(8–13)*. Using the described method, we were able to identify 2,024 unique peptides with high confidence that corresponded to 206 nucleolar proteins *(11)*.

2. Materials

2.1. Nucleolus Protein Extraction

1. Eagles minimum essential medium (Sigma Aldrich, St. Louis, MO) supplemented with 5% calf serum (Eurobio, Les Ulis, France).
2. Washing buffer, cold phosphate-buffered saline, pH 7.4.

3. Hypotonic cell buffer: 10 mM Tris-HCl, pH 7.4, 10 mM NaCl, 1 mM $MgCl_2$.
4. Nonidet P-40 for cell lysis (at 0.3% final concentration) (Roche Applied Science, Mannheim, Germany).
5. Resuspension of nuclei in 0.25 M sucrose, 10 mM $MgCl_2$.
6. Purification of nuclei and sonicated nucleoli fraction through 0.88 M sucrose, 0.05 mM $MgCl_2$.
7. Resuspension of purified nuclei and nucleoli in 0.34 M sucrose, 0.05 mM $MgCl_2$.
8. Resuspension of purified nucleoli for protein extraction in 0.34 M sucrose, 0.05 M $MgCl_2$, 0.2 M magnesium acetate, addition of two volumes of glacial acetic acid for nucleic acid precipitation.
9. Dialysis in 1 M acetic acid.

2.2. Digestion and Alkylation of Nucleolar Proteins

1. Coomassie Plus Protein Assay Kit (Pierce, Rockford, IL) for protein concentration determination *(14)*.
2. 50 mM Ammonium bicarbonate for protein resuspension.
3. 100 mM DTT stock, working solution, 1 mM DTT for denaturation, 1 M urea (*see* **Note 1**).
4. 10 mM iodoacetamide (Sigma-Aldrich) for alkylation from 100 mM stock.
5. 10 mM DTT for quenching.
6. TPCK trypsin (Pierce) (1 mg/mL stock, frozen, *see* **Note 2**).
7. 10% Formic acid to stop enzymatic digestion.
8. 0.1% Formic acid for resuspension of lyophilized digest.

2.3. Strong Cation Exchange Chromatography

1. Mobile phase A: 0.1% formic acid, 5% acetonitrile (HPLC grade, Merck, Darmstadt, Germany). Mobile phase B: 0.1% formic acid, 500 mM KCl, 5% acetonitrile.
2. Separation column: Agilent BioSCX series 2, 50 mm L × 0.8 mm, ID (Agilent Technologies, Waldbronn, Germany, *see* **Note 3**).
3. HPLC configuration: Agilent 1200 Capillary LC system containing micro degasser, capillary pump, diode array detector (equipped with a 300-nL flow cell), thermostated μ-well plate sampler and thermostated μ-fraction collector (*see* **Note 4**).
4. 96-conical well plates (Eppendorf, Hamburg, Germany).

2.4. Reversed Phase Separation on the HPLC-Chip with Online MS

1. Mobile phase A: 0.1% formic acid. Mobile phase B: 0.1% formic acid, 99.9% acetonitrile.
2. HPLC-chip, containing a 160-nL high-capacity enrichment column and a 150-mm separation column, packed with Zorbax SB-300 C18, 5-μm particles (*see* **Note 5**).

3. HPLC configuration: Agilent 1200 LC system consisting of micro degasser, capillary pump, nanopump, thermostated µ-well plate sampler, HPLC-chip/MS (cube) interface.
4. LC/MSD ion trap XCT ultra (Agilent, *see* **Note 6**).

2.5. Data Analysis

1. Spectrum Mill MS Proteomic workbench (Agilent) installed on a dual Xenon 2.4-GHz computer.
2. IPI human database (http://www.ebi.ac.uk/IPI).
3. Swiss-Prot database (http://www.expasy.org/ch2d).

3. Methods

To achieve a comprehensive profile of a cellular/subcellular proteome, it is important to optimize the workflow of a multidimensional separation such that sufficient protein digest is loaded onto the first separation column without compromising significantly the separation efficiency by overloading the separation column. To achieve the required peak capacity *(15)*, it is important to collect a sufficient number of fractions, which are further processed by the second orthogonal separation step. However, there is always a tradeoff in the number of collected fractions because a huge number of fractions elongates the total analysis time significantly.

For very complex proteomes, it is therefore advisable to use a prefractionation technique that either removes highly abundant proteins (such as for serum or cerebrospinal fluid [CSF] *(16)*) or to use a technique that decreases the complexity without reducing the overall information content of the sample, e.g., by specific enrichment for certain amino acid-containing peptides. For the described workflow, 50 µg of total protein digest is used in the first dimension. Because 24 fractions are collected, every fraction contains an average of 2.3 µg of digest. The collection time of the fractions is optimized by varying the collection time to ensure that the total amount of peptide in the different fractions is approximately evenly distributed *(11)*.

3.1. Protein Isolation

1. HeLa Cells are grown in Eagle's minimum medium containing 5% fetal calf serum at 37°C under 5% CO_2 atmosphere to 80% confluence.
2. Cells on ice are washed with cold phosphate-buffered saline and scraped off using a Teflon cell scraper.
3. Cells are resuspended in 12–15 volumes hypotonic buffer and incubated on ice for 40 min.

4. Cell lysis *(17)* is initiated by addition of 0.3% Nonidet P-40; homogenization is best performed using a Dounce homogenizer.
5. Nuclei are obtained in a pellet by centrifugation at 1,300×*g* using a Heraeus benchtop centrifuge, after resuspension of the homogenate in ten volumes of 0.25 M sucrose, 10 mM $MgCl_2$. The supernatant containing the cytoplasm is discarded (*see* **Note 7**).
6. Nuclei are further purified at 1,300×*g* for 10 min through a 0.88 M sucrose, 0.05 M $MgCl_2$ layer.
7. Nucleoli are obtained from nuclei by resuspension in ten volumes 0.34 M sucrose followed by five 30-s sonications on ice (*see* **Note 8**).
8. Nucleoli can be separated from the remaining fraction by centrifugation at 2,000×*g* for 20 min through a 0.88 M sucrose, 0.05 mM $MgCl_2$ layer.
9. The supernatant is discarded and purified nucleoli are resuspended in 0.34 M sucrose containing 0.05 mM $MgCl_2$.
10. Protein extraction and nucleic acid removal is performed according to Madjar et al. *(18)* by addition of magnesium acetate to a final concentration of 0.2 M, followed by the addition of two volumes of glacial acetic acid. The solution is then incubated at 4°C for 1 h. Precipitated nucleic acids are removed by centrifugation at 13,000×*g*. The supernatant is collected and stored at 4°C and the precipitate is extracted a second time to achieve an increased yield of nucleolar proteins. The obtained supernatants are combined and dialyzed against 500 volumes of 1 M acetic acid.

3.2. Digestion and Alkylation of Nucleolar Proteins

1. The protein concentration of the dialyzed nucleoli protein solution can be determined by using the Coomassie Plus Protein Assay Kit according to the assay method described by the manufacturer. Aliquots of 50 µg protein are then evaporated to dryness using an Eppendorf SpeedVac and stored at −18°C until further analysis.
2. Aliquots of 50 µg protein are resuspended in 50 mM ammonium bicarbonate, 1 M urea, 1 mM DTT for 1 h at 37°C to denature and reduce the proteins.
3. Alkylation is performed by the addition of iodoacetamide to a final concentration of 10 mM, followed by incubation for 30 min at room temperature in the dark, followed by the addition of 10 mM DTT to quench the alkylation reaction (*see* **Note 9**).
4. TPCK–trypsin is added in an enzyme/substrate ratio of 1:30. The protein solution is then incubated for 15 h at 37°C in a rotary shaker at 100 rpm.

5. The enzymatic reaction is stopped by the addition of 10% formic acid until a pH of 3.0 is reached (*see* **Note 10**).
6. The digest is evaporated to dryness using a SpeedVac and the resulting peptide pellet can be stored in a freezer at –18°C until further processing.
7. Directly before HPLC analysis, the pellet should be resuspended in 20 μL of mobile phase A of the strong cation exchange chromatography (5% acetonitrile, 0.1% formic acid) and stored in the thermostated μ-well plate sampler of the Agilent 1200 Capillary LC system at 6°C.

3.3. Strong Cation Exchange Chromatography

1. The BioSCX series 2 column (50 × 0.8 mm) has to be conditioned with 500 mM KCl, 5% acetonitrile, 0.1% formic acid, followed by flushing the column with mobile phase A until baseline stability. Detection is performed by using an Agilent 1200 diode array detector tuned at a wavelength of 222 nm (*see* **Note 11**).
2. The complete sample is loaded onto the column by flushing mobile phase A at a flow rate of 20 μL/min across the cation exchanger.
3. The separation gradient can be performed under the following conditions: 0 min: 0% B; 5 min: 0% B; 8 min: 10% B; 18 min: 15% B; 29 min: 70% B; 32 min: 100% B; 38 min: 100% B (*see* **Note 12**) by applying a continuous gradient.
4. Reconditioning of the column should be done for at least 15 min with 100% mobile phase A before the next run.
5. Different fraction collection times of 0.5 min, 1 min, 2 min, and 4 min are useful to distribute the total amount of peptides in the fraction more evenly (illustrated in **Fig. 3**, *see* **Note 13**). This results in fractions between 10 and 80 μL, which are collected in 96-well plates with conical wells of 100 μL volume.
6. Well plates are directly transferred to the second separation dimension performed on the HPLC-chip and stored in the respective HPLC μ-well plate sampler cooled to 6°C.
7. Special precaution has to be taken with the HPLC equipment when using high-concentration salt solutions (*see* **Note 14**).

3.4. Reversed Phase Separation on the HPLC-Chip with Online MS

1. 50% of the fraction is injected and loaded onto the 160-nL enrichment column of the HPLC-chip at a constant flow rate of the loading pump of 4 μL/min. The chip user interface is set automatically to enrichment during this process if the "injection flush volume" feature is used in the Agilent ChemStation. An injection flush volume of 5–8 μL is recommended to make sure that all salts are flushed off before switching the enrichment column into the nanoflow path. The chip is operated in backward flush mode to achieve optimum separation efficiency (*see* **Note 15**).

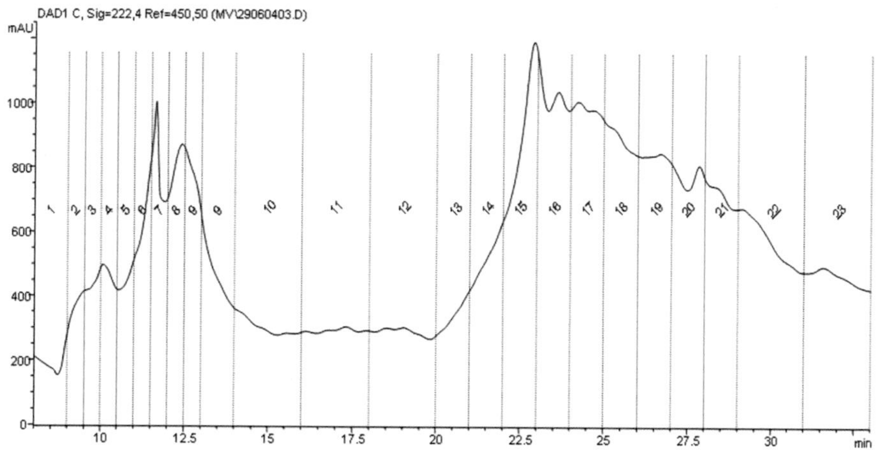

Fig. 3. UV trace of the first separation dimension on an Agilent BioSCX series 2 strong cation exchanger (0.8 mm × 50 mm). The signal was recorded at a wavelength of 222 nm. Collected fractions are indicated by *dashed lines*. Fractions were taken with different collection times depending on the peptide concentration of the individual fractions.

2. After loading and desalting of the sample on the chip enrichment column, the latter is switched on-line by the use of the HPLC-chip interface nano rotary valve with the analytical chip column (packed with the same material as the enrichment column). The sample is eluted in backward flush from the enrichment column and transferred to the analytical column at a flow rate of 300 nL/min by a nanopump with increasing percentage of mobile phase B.
3. A continuous linear gradient is used for separation. Gradient conditions are: 0 min: 2% mobile phase B; 10 min: 2% B; 12 min: 18% B; 42 min: 55% B; 45 min: 75% B; 48 min: 5% B; followed by a post time of 6 min for column re-equilibration (*see* **Note 16**).
4. Data-dependent MS acquisition is performed on the Agilent LC/MSD Trap XCT with the following MS conditions: drying gas for solvent dissolvation at 4 L nitrogen/min and 325°C; MS capillary voltage: 1,800 V (*see* **Note 17**); skimmer 1: 30 V; capillary exit: 75 V; and trap drive: 85. For each precursor ion, two averages are taken. The maximum accumulation time for ions in the trap is 150 ms, with a maximum target of 125,000. MS scan range is selected in a mass-over-charge ratio range of 300–2,000; "Ultra scan," the fastest scanning mode of the machine, is selected for detection (*see* **Note 18**).
5. Fragmentation conditions for MS/MS: A maximum of three parent ions is selected in each MS/MS cycle for fragmentation. Fragmentation amplitude for peptide fragmentation: 1.25 V; SmartFrag: on, 30–200%. Spectra are actively excluded for fragmentation after four recorded spectra for 2 min to allow the detection of less-abundant coeluting compounds; doubly

charged ions are preferred for selection and further fragmentation; and the MS/MS scan range is between 100 and 1,800 m/z for fragment ions.

3.5. Data Analysis and Processing

1. Database searches are performed against the IPI human database and on specific nucleus or nucleolus localization in the Swiss-Prot database by applying distinct auto validation criteria in the Agilent Spectrum Mill software (Rev. A03.02.), installed on a dual Xenon 2.4 GHz computer.
2. The following auto validation criteria are used to validate the identified proteins and peptides: Minimum score for proteins is 13. Minimum scores for spectra resulting from fragmentation of 1+, 2+, 3+, and 4+ parent ions are: 8, 7, 9, and 9, respectively, with a scored peak intensity (SPI, the percentage of total peak intensity that is assigned to particular ion types) value of at least 70%. Additionally, for 2+ with a SPI greater than 90%, the minimum score is 6.
3. To minimize the number of false-positive hits, all MS/MS spectra should also be searched against the reversed entries of the IPI human database. Only spectra with a reversed score that are at least two scoring units smaller than the real score should be taken into account for the auto validation. After the auto validation, only the subset of already identified proteins is used to search the IPI database again, also allowing nonspecific digestion. In this case, the peptide criteria can be lowered to a score greater than 6 and an SPI greater than 50%.

4. Notes

1. DTT stock should be freshly prepared when used. An appropriate amount of urea is weighed before use and doubly distilled water is added to make up the final concentration of 1 M urea, 1 mM DTT. The solution should not be heated above 37°C to avoid carbamylation *(19)*. Iodoacetamide should be handled with gloves because of its toxic properties.
2. Trypsin stock should be stored in small aliquots in an acidic buffer like acetate buffer and stored at –18°C to prevent autodigestion and loss of activity. Freeze-thaw cycles should be minimized.
3. A 0.3 mm × 35 mm BioSCX series 2 column is also commercially available, however, the loading capacity of this column is between 5 and 10 μg, whereas, for the 0.8 mm × 50 mm column, the loading capacity was determined to be between 50 and 100 μg. Recommended flow rates are 5 μL/min and 20 μL/min, respectively.

4. Thermostats should be set at 6°C to prevent degradation of the sample. The 96-well plate should be sealed using a plate sealer if samples need to be stored after the first separation dimension.
5. Alternatively, an HPLC-chip is available that contains a 40-nL enrichment column. This chip can be run in forward and backward flush mode. If high loading capacity is required, such as for a complex proteome sample, the 160-nL enrichment column chip is the preferred choice.
6. The Agilent HPLC-chip/MS system can also be equipped with a LC/MSD trap XCT and XCT plus mass spectrometer. However, these instruments have slower scan rates and slower electronic signal processing. Therefore, the use of the XCT ultra leads to a higher number of identified compounds in highly complex samples.
7. It is crucial to strictly follow the described centrifugation speed and sucrose concentrations to obtain a good separation of cellular components.
8. Longer sonication intervals should be avoided to keep the temperature of the suspension low.
9. Quenching of the reaction is recommended to prevent alkylation of the trypsin.
10. Acidification of the digest supports evaporation of the solution in the SpeedVac.
11. For proper function of the BioSCX column, the following conditioning steps at 20 μL/min are recommended: 5% acetonitrile, 0.03% formic acid for 10 min followed by 5% acetonitrile, 0.03% formic acid, 500 mM KCl for 15 min and 5% acetonitrile, 0.03% formic acid for at least 20 min or until the baseline is flat again at a wavelength of 220 nm UV detection.
12. A flatter gradient with a slower increase of %B/min is recommended with samples that are more complex. This, however, increases the number of fractions and the total analysis time.
13. The UV signal gives a rough estimate of the overall peptide concentration in the corresponding fraction. Performing a pre-run under identical separation conditions with a low amount of sample might give a good estimate of how to adjust the collection time of the individual fractions.
14. The system needs to be flushed thoroughly for at least 2 h immediately after analysis to prevent precipitation of salts, which can cause blockage or leakage in the LC system and to avoid corrosion of stainless steel components. Organic solvents must not be used as long as salt solutions are still in the system because they cause precipitation of crystalline

salts that might damage the system. The column should be stored in a refrigerator after use. It is recommended to use an inline filter in the loading path of the system to prevent clogging of the SCX columns by sample debris and to extend the lifetime of the column. Samples should always be centrifuged after resuspension to prevent residual sample debris from clogging the column.

15. In general, an HPLC chip can be operated in forward and backward flush mode. In forward flush mode, the loading capillary enters the chip stator side of the rotary valve at port 6, whereas the waste capillary exits at port 5 (**Fig. 2**). For backward flush mode, the loading capillary is connected to port 5 whereas the waste capillary leaves on port number 6. For chips with enrichment columns larger than 40 nL, backward flush mode is recommended to preserve narrow peak width.
16. Reconditioning time can be shortened by 1–2 min if higher primary flows are selected for the LC pumps. However, this increases solvent consumption.
17. Capillary voltage might be variable from setup to setup and is usually in the range of 1,750–1,950 V at starting conditions. After a few hours of operation, the voltage should be increased by 50–100 V to guarantee stable spray performance over a long period of time.
18. As an alternative, the operation mode standard enhanced can be used. This results in slower scanning and data processing. If higher mass resolution is required, the standard enhanced setting should be used.

References

1. Issaq, J. H., Chan, K. C., Janini, G. M., Conrads, T. P., and Veenstra, T. D. (2005). Multidimensional separation of peptides for effective proteomic analysis. *J. Chromatogr. B* **817**, 35–47
2. Jandera, P. (2006). Column selectivity for two-dimensional liquid chromatography. *J. Sep. Sci.* **29**, 1763–1783
3. Wolters, D. A., Washburn M. P., and Yates, J. R. (2006). An automated multidimensional protein identification technology for shotgun proteomics. *Anal. Chem.* **73**, 5683–5690
4. Wilm, M. and Mann M. (1996). Analytical properties of the nanoelectrospray ion source. *Anal Chem.*, **68**, 1–8
5. Koster, S. and Verpoorte, E. (2007). A decade of microfluidic analysis coupled with electrospray mass spectrometry: an overview. *Lab Chip*, 7, 1394–1412
6. Yin, H., Killeen, K., Brennen, R., Sobek, D., Werlich, M., and van de Goor T. (2005). Microfluidic chip for peptide analysis with an integrated HPLC column, sample enrichment column and nanoelectrospray tip. *Anal. Chem.* 77, 527–533
7. Yin, H. and Killeen, K. (2007). The fundamental aspects and applications of Agilent HPLC-Chip. *J. Sep. Sci.* **30**, 1427–1434
8. Fortier, M.-H., Bonneil, E., Goodley, P., and Thibault, P. (2005). Integrated microfluidic device for mass spectrometry-based proteomics and its application to biomarker discovery programs. *Anal. Chem.* 77, 1631–1640
9. Hardouin, J., Duchateau, M., Caron-Joubert, R., and Caron, M. (2006). Usefulness of an integrated microfluidic device (HPLC-Chip-MS) to enhance confidence in protein

identification by proteomics. *Rapid Commun. Mass Spectrom.* **20**, 3236–3244

10. Ninonuevo, M. R., Park, Y., Yin, H., Zhang, J., Ward, R. E., Clowers, B. H. et al. (2006). A strategy for annotating the human milk glycome. *J. Agric. Food Chem.* **54**, 7471–7480
11. Vollmer, M., Hoerth, P., Rozing, G., Coute, Y., Grimm, R., Hochstrasser, D., and Sanchez, J.-C. (2006). Multi-dimensional HPLC/MS of the nucleolar proteome using HPLC-chip/MS. *J. Sep. Sci.* **29**, 499–509
12. Hoerth, P., Miller, C.A., Preckel, T., and Wenz, C. (2006). Efficient fractionation and improved protein identification by peptide OFFGEL electrophoresis. *Mol. Cell. Proteomics* 5.**10**, 1968–1974
13. Staes, A., Timmerman, E., Van Damme, J., Helsens, K., Vandekerckhove, J., Vollmer, M., and Gevaert, K. (2007). Assessing a novel microfluidic interface for shotgun proteome analyses. *J. Sep. Sci.* **30**, 1468–1476
14. Bradford, M. M. (1976). A rapid and sensitive method for the quantitation of microgram quantities of protein utilizing the principle of protein-dye binding. *Anal. Biochem.* **72**, 248–254
15. Giddings, J. C. (1987). Concepts and comparisons in multidimensional chromatography. *J. High Res. Chromatogr.* **10**, 319–323
16. Zolotarjova, N., Martosella, J., Nicol, B., Bayley J.,. and Boyes, B. (2005). Differences among techniques for high-abundant protein removal depletion. *Proteomics* **5**, 3004–3013
17. Scherl, A., Coute, Y., Deon, C., Calle, A., Karine, K., Sanchez, J.-C. et al. (2002). Functional proteomic analysis of human nucleolus. *Mol. Biol. Cell.* **13**, 4100–4109
18. Madjar, J.-J., Arpin, M., Buisson, M., Reboud, J. P. (1979). Spot position of rat ribosomal proteins by four different two-dimensional electrophoreses in polyacrylamide gel. *Mol. Gen. Genet.* **171**, 121–134
19. Lippincott, J. and Apostol I. (1999). Carbamylation of cysteine: a potential artifact in peptide mapping of hemoglobins in the presence of urea. *Anal. Biochem.* **267**, 57–64

Chapter 2

Nanofluidic Channel Fabrication and Manipulation of DNA Molecules

Kai-Ge Wang and Hanben Niu

Summary

Confining DNA molecules in a nanofluidic channel, particularly in channels with cross sections comparable to the persistence length of the DNA molecule (about 50 nm), allows the discovery of new biophysical phenomena. This sub-100 nm nanofluidic channel can be used as a novel platform to study and analyze the static as well as the dynamic properties of single DNA molecules, and can be integrated into a biochip to investigate the interactions between protein and DNA molecules. For instance, nanofluidic channel arrays that have widths of approximately 40 nm, depths of 60 nm, and lengths of 50 μm are created rapidly and exactly by a focused-ion beam milling instrument on a silicon nitride film; and the open channels are sealed with anodic bonding technology. Subsequently, lambda phage DNA (λ-DNA; stained with the fluorescent dye, YOYO-1) molecules are introduced into these nanoconduits by capillary force. The movements of the DNA molecules, e.g. stretching, recoiling, and transporting along channels, are studied with fluorescence microscopy.

Key words: Nanofluidic channels, Nanopore, Focused-ion beam, DNA molecules, Fluorescence microscopy

1. Introduction

Recently, with the advancements of nanotechnologies, many scientists in different research areas, including both fundamental studies and applied techniques, have focused their attention on the fabrication of nanofluidic devices and the applications in the research of single biomolecules, such as DNA and protein molecules ***(1–3)***. Nanofluidic channels with critical dimensions comparable to the size of molecules provide new possibilities for direct observation, manipulation, and analysis of single biomolecules, and

James Weifu Lee and Robert S. Foote (eds.), *Micro and Nano Technologies in Bioanalysis*,
Methods in Molecular Biology, vol. 544
DOI 10.1007/978-1-59745-483-4_2, © Humana Press, a part of Springer Science+Business Media, LLC 2009

provide a novel technological platform as an ultrasensitive and high-resolution sensor for studying single DNA molecules.

The nanofluidic channel is defined as a channel with at least one cross-section dimension (depth or diameter) in the nanometer range (one-dimension or two-dimension nanochannel) ***(4)***. In particular, a nanopore is thought of as a channel with all three dimensions in the nanoscale range, so that the work done with a nanopore falls under the realm of nanofluidics. The potential application of nanopores as detectors for ultrafast genome sequencing is its most attractive application ***(5–7)***. Nanofluidics has great benefits for bioscience studies ***(8)***, the practical applications of nanofluidics are improvements to the state of the art of DNA separation and sequencing providing significant reductions in both time and cost. The small dimensions of the nanoscale structure reduce processing times and the amount of reagents necessary for assay, substantially reducing costs.

At present, different approaches have been undertaken to successfully fabricate nanoscale structures. In general, nanofabrication methods can be divided roughly into two groups ***(9)***: top-down and bottom-up methods. Top-down methods start with patterns made on a large scale and reduce their lateral dimensions before forming nanostructures; these methods are mainly adopted by physics scientists. On the other hand, bottom-up methods begin with atoms or molecules to build nanostructures, in some cases, through smart use of self-organization.

Top-down methods can be classified into two categories, that is, optical masked lithography and optical maskless lithography. The focused-ion beam (FIB) milling tool is a maskless lithography technique that can image features on a lithographic surface directly ***(10)***. This technique has the advantages of facility and celerity; the patterns created are smooth and can be easily controlled and faithfully reproduced for different applications.

Micro-scale and submicro-scale fluidic channel arrays have been used for studying single biomolecules for many years ***(11)***. However, applications with sub-100 nm fluidic channel arrays in biomolecular studies are rarely reported. The limitation is partially caused by the difficulty of fabrication process and metrology ***(12)***. In addition, the properties of DNA molecules confined in these fluidic channels and the dynamics of DNA movements under this condition are not well known.

Although there are many possible applications of nanofluidic channels for DNA study, in this chapter, we focus on the manipulation of single DNA molecules, e.g., stretching, recoiling, and transporting. We describe the fabrication of open nanofluidic channel arrays (40 nm width, 60 nm depth) in silicon nitride (Si_3N_4) membrane surfaces using the FIB milling technique and other nanofabrication techniques. Next, we describe the sealing of these channels with Pyrex glass by the anodic bonding technique; followed by a description of nanoscale channel arrays used to study the properties of single DNA molecules with the help of fluorescence microscopy. DNA molecules (e.g., λ-phage DNA molecules) stained with the fluorescent dye YOYO-1 can be driven to stretch and transport along these open nanoconduits by capillary force, and also to recoil in the enclosed nanoconduits under the force of the electrode field. Because the dimension of the channel is approximately the natural-state DNA molecule persistence length (~50 nm) in aqueous buffer, this nanostructured channel can provide an essential new method for detecting and analyzing single DNA molecules. Such nanoconduits can be used as one component of a "lab-on-a-chip" in the manipulating single biomolecules. These nanochannel systems are also expected to find significant applications in medical diagnostic systems.

2. Materials

2.1. Substrates for Nanofluidic Channels

2.1.1. Substrate Silicon Wafers

1. The substrate silicon wafers are 3-inch or 4-inch diameter, *n*-type, 390-μm or 525-μm thick, double-sided mirror-polished, <100> single crystal oriented standard bare wafers.
2. Preclean the wafers with a H_2SO_4-H_2O_2 (10:1, v/v) mixture at 120°C for 20 min followed by buffered HF (BHF; NH_4F:HF = 7:1, v/v) for 2 min at room temperature to remove surface organics and metals.
3. Rinse with doubly deionized water (Millipore S.A., Molsheim, France).
4. Dry with pure nitrogen gas.

2.1.2. Encapsulating Pyrex Glasses

1. Pyrex 7740 borosilicate glass (3-inch or 4-inch; 600-μm thick, Corning Inc., Corning, NY), matched with the substrate silicon wafer used. The surface roughness of glass is less than 1 nm.
2. Preclean the glass with a standard solution of H_2SO_4-H_2O_2 at 120°C for 20 min, and then dip into a solution of buffered HF (BHF; NH_4F:HF = 7:1, v/v) for 2 min at room temperature to remove surface organics and metals.
3. Rinse with doubly deionized water (Millipore S.A.).
4. Dry with a stream of pure nitrogen.

2.2. Biophysics Experimental Buffers

1. Tris-EDTA: 10 mM Tris base, 1 mM EDTA, pH 8.0. All buffers are made with 18.2 M water purified through the Milli-Q water Purification System (Millipore).
2. TBE: 45 mM Tris base, 1 mM EDTA, 45 mM boric acid.

2.3. Biomolecule Sample

1. λ-Phage DNA molecules (Sino-American Biotechnology Company, Beijing, China), stored in alcohol at –20°C. The final DNA concentration is 1 μg/mL in buffer containing 10 mM Tris-HCl, 10 mM NaCl, and 1 mM EDTA (Sigma, St. Louis, MO, USA), pH 8.0 (*see* **Note 1**).
2. DNA (1 ng/μL) is stained with 0.25 μM fluorescent dye YOYO-1 (Molecular Probes, Carlsbad, CA, USA) at a ratio of ten base pairs per dye molecule (bp/dye = 10:1), mixing DNA complex molecules with a specific volume of freshly prepared 0.1 μM dye solution (10 mM Tris, 1 mM EDTA buffer, pH 8.0) (*see* **Note 2**).

3. Methods

Nanofluidics fabrication and applications have now attracted great enthusiasm because of their brilliant prospects. Nanochannel fabrication techniques should be cost-effective and one should be able to control the channel dimensions precisely. With the rapid improvements of nanotechnological manufacturing, four methods are now (normally) used for fabrication of nanofluidics channels, e.g., bulk nanomachining and wafer bonding ***(13)***, surface nanomachining ***(14)***, buried channel technology ***(15)***, and nanoimprinting lithography ***(16)***. In general, the bulk nanomachining technique is the preferred approach for nanoscale fabrication. Among the nanomachining techniques, the FIB milling technique has many advantages ***(17)***—it is an extremely versatile technique for making arbitrary micro- and nano-structures with no essentially required preprocessing or postprocessing.

Nanofluidics channels can act as a novel basis for more precisely controlling the behaviors of single DNA molecules when the diameter of the channel is comparable to or less than the persistence length of the DNA molecule ***(18)***. At nanoscale dimensions, different biophysical phenomena start to dominate, this leads to new scientific insights and applications ***(19)***. We can use these nanoscale structures (open nanofluidic channels and enclosed nanofluidic channels) to manipulate, detect, and analyze individual biological molecules, and can also carry out individual molecular reactions within these nanofluidic environments while electric fields are used to drive flow, move analytes, and separate ionic species.

3.1. Fabrication of Nanofluidic Channels

3.1.1. Creating Free-standing Si_3N_4 Crystal Membranes

1. 500-nm thick lower stress (~200 MPa tensile) Si_3N_4 films are deposited on both sides of the prepared bare silicon wafer by standard low-pressure chemical vapor deposition (LPCVD, M80100, Sevenstar Electronics Co., Beijing, China). The working condition are: temperature, 800°C; pressure, 200 mTorr; gas, $SiCl_2H_2$ and NH_3.

2. Approximately 100-μm thick photoresist (ARN7500, GermanTech Co., Beijing, China) is spun onto the front side of the silicon wafer with the spin coater (KW-4A, XiaMen Chemat Scientific Instrument Company, XiaMen, China) at the speed of 5,000 rpm.
3. Bake the wafer at 140°C for approximately 30 min and then store at room temperature in a dust-free environment.
4. A standard photolithography process is used to pattern an appropriate square (~1,200 × 1,200 μm^2) in the photoresist layer, that is, the same square pattern is exposed on the Si_3N_4 surface.
5. The reactive ion etching (RIE, Plasmalab 80Plus RIE, Oxford Instruments Co., Abingdon, UK) is used to open the hole in the Si_3N_4 membrane with a SF_6/O_2 (1:4, v/v) gas mixture for 2 min, working conditions: RF, 100 W; pressure, 110 mTorr; temperature, 100°C.
6. The residual photoresist on the front surface is removed using oxygen plasma with an O_2/CF_4 gas mixture (CF_4 is ~20% in total gas mixture volume); RF power, 60 W; pressure, 135 mTorr; temperature, 120°C.
7. The wafer is immersed into 40% (m/v) potassium hydroxide KOH(aq) at 60°C for ~10 min to create a 100 × 100-μm^2 free-standing Si_3N_4 membrane (*see* **Note 3**).

3.1.2. Fabricating Nanofluidic Channels

1. Vent the FIB (DB235, FEI Company, Hillsboro, OR, USA) system to mount the silicon wafer with a free-standing Si_3N_4 membrane sample carefully and tightly and then pump down the system. When vacuum is reached, switch on the beam. Carefully move the sample in the Z-direction to get closer to the working distance.
2. On the backside of the free-standing Si_3N_4 membrane, a standard FIB milling technique is used to fabricate nanoscale fluidic channel arrays. The model of FIB drilled is single-pass with a 30 keV Ga^+ ion beam. The initial incident ion beam full-width half-maximum (WHFM) diameter is 20 nm.
3. Choose appropriate working conditions to control the channels' depth and width, where ion beam current, overlap, and dwell time are 10 pA, 50%, and 0.3 μs, respectively (*see* **Note 4**).
4. Deposit platinum in the two reservoirs as the electrode.
5. A wafer bonder (EV501, EV Group, St. Florian, Austria) is used to bond the Pyrex glass to the substrate wafer. The voltage applied on the glass wafer is negative with respect to that of the silicon wafer. The bonding process is approximately 30 min at 350°C with an applied voltage of 800 V. A sketching image of nanofluidic channel fabrication is shown in **Fig. 1**. Some typical nanochannel arrays created are shown in **Fig. 2**.

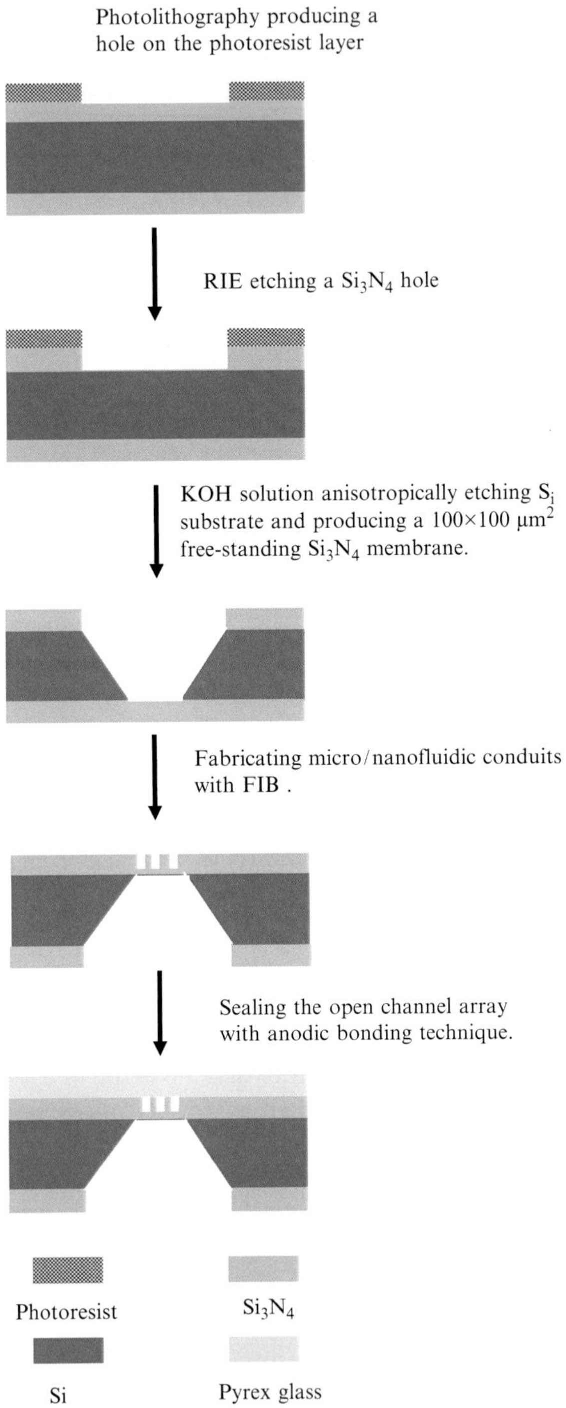

Fig. 1. Schematic drawing of fabricating process for nanofluidic channels.

Fig. 2. SEM images of nanofluidic channel arrays at the center of the free-standing Si_3N_4 membrane.

3.2. Manipulation of DNA Molecules

3.2.1. Preparing Biomolecule Samples

1. Incubate the DNA/YOYO-1 mixture solution in a dark room for ~30 min. In all experiments, the DNA base pair-to-dye ratio is kept at 10:1 (bp/dye = 10).
2. Dilute the DNA/YOYO-1 complex solution to 6.5 pM in a 50 mM Bis-Tris buffer (pH 7.5, Sigma).
3. Admix the buffer with 5% (v/v) β-mercaptoethanol (Sigma) as an antiphotobleaching agent and 2.5% (w/w) poly (*n*-vinylpyrrolidone) (PVP, Sigma) to reduce both electroosmotic flow and nonspecific binding of DNA to channel walls.

3.2.2. Manipulating DNA Molecules in the Nanofluidic Channels

1. Carefully place the nanofluidics channel system on the luggage carrier.
2. With a syringe, transfer the DNA/YOYO-1 complex solution into one reservoir of the open fluidic channel system with a Digital Precision Microliter Pipette (Gilson S.A.S., Roissy en France, France). The solution migrates into and is transported along the open channels by capillary action as soon as it arrives at the channel entrance (*see* **Note 5**).
3. With a syringe, transfer the DNA/YOYO-1 complex solution into one reservoir of the enclosed nanochannel system; the solution is loaded into the nanochannels via capillary action and then transported through the nanochannels by using an applied electrical field with platinum electrodes inserted into the reservoirs.
4. A 5-V bias is applied to drive a DNA molecule from the reservoir into a nanochannel (*see* **Note 6**).
5. Switch off the bias field before the DNA molecule has completely entered the nanochannel. As a result, DNA molecule is driven back to the reservoir and can be observed to both recoil and unstretch simultaneously.
6. The DNA complex molecule is then driven entirely into the nanochannel, and the bias field is switched off (*see* **Note 7**).
7. A DNA molecule is electrophoretically driven from the reservoir into a nanochannel. Upon fully entering a nanochannel, the molecule begins to relax and finally reaches its equilibrium extension length inside the channel (*see* **Note 8**).
8. After the molecule has completely relaxed to its equilibrium length inside the nanochannel, it is electrophoretically driven to the exit of the channels. Once the tip of the molecule is straddling the interface, the voltage is turned off and the molecule is observed to completely recoil from the nanochannel (*see* **Note 9**).

3.2.3. Single Molecular Optical Imaging

1. The fluorescently stained DNA/YOYO-1 complex molecules are observed using an inverted optical microscope (1X-70, Olympus, Tokyo, Japan) by epifluorescence with a 20× objective.
2. A 100-W mercury lamp is used in combination with a U-MWB excitation cube (BP450–480, dm500, BA515) for light-induced fluorescence illumination (*see* **Note 10**).
3. Fluorescence light from the complex molecules is detected by a cooled charge-coupled device (CCD) camera (1,300 × 1,300 pixels, 12-bit digitization; Cool SNAP-HQ, Roper Scientific, Inc., Tucson, AZ, USA). MetaMorph software (Universal Imaging Corporation, West Chester, PA, USA) is used for the system control, data acquisition, and data processing.

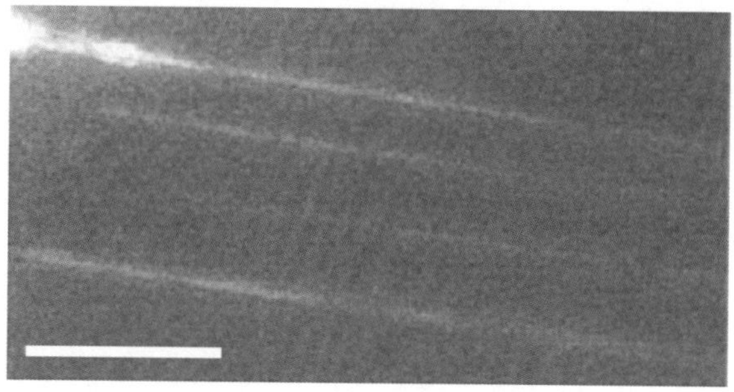

Fig. 3. A typical fluorescence image of stained λ-phage DNA inside fluidic channels. Scale bar, 10 μm.

The CCD acquisition time is 3 s (*see* **Note 11**). **Figure 3** shows a typical fluorescence image of the λ-phage DNA molecules passing along the open fluidic channels.

4. Notes

1. Lambda-phage DNA is a linear double-stranded helix that contains 48.502 kbp, its molecular mass is ~30.6 MDa, and its contour length is ~16.2 μm. It is widely used in life sciences.
2. When the mixing ratio (dye molecules per base pair) is below 1:8, the predominant binding mode of YOYO-1 on DNA is bis-intercalation; when the mixing ratio is above 1:8, groove association (external binding) with DNA begins to contribute significantly.
3. When the wafer is immersed into the KOH solution, the silicon is etched at 54.7-degree angles relative to the surface normal. This anisotropic etching creates a free-standing 100 × 100 μm^2 Si_3N_4 membrane on the backside of the wafer.
4. These open nanochannels can be made with different shapes, e.g., uniform-linear or curvilinear. All of these nanochannels are combined by two bigger containers, which act as the solution reservoirs. The linear nanofluidic channel can be created down to 40-nm width and 60-nm depth. The channel lengths are 50 μm, and the distance between two channels is 5 μm.
5. λ-phage DNA molecules can be observed stretched and threaded along these open nanochannels. DNA molecules can be moved along the channels, although they move only a short distance, not through the whole conduit. In addition, it can be seen that not all channels are filled with DNA

molecules, i.e., there is only buffer liquid within some conduits.

6. This bias resulted in E = 100 V/cm in the nanochannels, which is large enough to drive DNA molecules. DNA molecules carry negative charges, which prevent them from adhering to the nanofluidic channel walls, which are also negatively charged. This electrostatic repulsion effectively prevents the nonspecific binding of biomolecules to the nanofluidic channel surface.
7. Once the DNA molecule has contracted, it is slowly driven back down the nanochannel until a small portion of the molecule has reached the reservoir. At this point, the field is switched off and the molecule is observed to undergo a pure recoil process.
8. Stretching is caused by the electric force pulling the molecules into the nanochannel against a resistance at the entrance. The resistance at the entrance is probably caused by the entropic interface force and friction for molecules encountering the entrance edges.
9. Because molecules are allowed to reach equilibrium before beginning to recoil, this process is driven purely by the entropic recoil force and unaffected by elastic restoration.
10. YOYO-1 has an excitation maximum at 491 nm and an emission maximum at 509 nm; that is, YOYO-1 molecules emit green fluorescence under the excitation of blue light.
11. YOYO-1 molecules bind strongly to doubled-strand DNA molecules and the fluorescence quantum yields of the bound dyes are very high. The amount of intercalated dye is proportional to the length of the molecule, therefore, measuring the total fluorescent intensity from a single molecule gives a direct measurement of its length.

Acknowledgments

This work is supported by grants from the National Natural Science Foundation of China (No. 60771048, No. 60025516, and No. 10334100), and the Major Project of National Science Foundation of China (No. 60138010), and partly supported by National Center for Nanoscience and Technology, China.

References

1. Tegenfeldt, J. O., Prinz, C., Cao, H., Huang, R. L., Austin, R. H., Chou, S. Y., Cox, E. C., Sturm, J. C, (2004) Micro- and nanofluidics for DNA analy. *Anal. Bioanal. Chem.* **378**, 1678–1692
2. van der Heyden, F. H. J., Stein, D., Dekker, C. (2005) Streaming currents in a single nanofluidic channel. *Phys. Rev. Lett.* **95**, 116104
3. Baldessari, F. and Santiago, J. G. (2006) Electrophoresis in nanochannels: brief review and speculation. *J. Nanobiotechnology.* **4**, 12–16
4. Eijkel, J. C. T. and van den Berg, A. (2005) Nanofluidics: what is it and what can we expect from it? *Microfluid Nanofluids.* **1**, 249–267
5. Henrickson, S. E., Misakian, M., Robertson, B., Kasianowicz, J. J. (2000) Driven DNA transport into an asymmetric nanometer-scale pore. *Phys. Rev. Lett.* **85**, 3057–3060
6. Li, J., Stein, D., McMullan, C., Branton, D., Aziz, M. J., Golovchenko, J. A. (2001) Ion-beam sculpting at nanometre length scales. *Nature.* **412**, 166–169
7. Dekker, C. (2007) Solid state nanopores. *Nat. Nanotechnol* **2**, 209–215
8. Lin, Y., Huang, M., Chang, H. (2005) Nanomaterials and chip-based nanostructures for capillary electrophoretic separations for DNA. *Electrophoresis.* **26**, 320–330
9. Mijatovic, D., Eijkel, J. C. T., van den Berg, A. (2005) Technologies for nanofluidic systems: top-down vs. bottom-up. *Lab. Chip.* **5**, 492–500
10. Biance, A. L., Gierak, J., Bourhis, E., Madouri, A., Lafosse, X., Patriarche, G., Oukhaled, G., Ulysse, C., Galas, J. C., Chen, Y., Auvray, L. (2006) Focused ion beam sculpted membranes for nanoscience tooling. *Microelectro. Eng.* **83**, 1474–1477
11. Squires, T. M. and Quake, S. R. (2005) Microfluidics: fluid physics at the nanoliter scale. *Rev. Mod. Phys.* 77, 977–1025
12. Stein, D., van der Heyden, F. H. J., Koopmans, W. J. A., Dekker, C. (2006) Pressure-driven transport of confined DNA polymers in fluidic channels. *Proc. Natl. Acad. Sci. U.S.A.* **103**, 15853–15858
13. Cao, H., Yu, Z. N., Wang, J., Tegenfeldt, J. O., Austin, R. H., Chen, E., Wu, W., Chou, S. Y. (2002) Fabrication of 10 nm enclosed nanofluidic channels. *Appl. Phys. Lett.* **81**,171–176
14. Ahpan, H., Mondin, G., Hegelbach, N. G., de Roij, N. F., Staufer, U. (2006) Filling kinetics of liquids in nanochannels as narrow as 27 nm by capillary force. *J. Colliod Interface Sci.* **293**, 151–157
15. de Boer, M. J., Tjerkstra, R. W., Berenschot, J. W., Jansen, H. V., Burger, G. J., Gardeniers, J. G. E., Elwenspoek, M., van den Berg, A. (2000) Micromachining of buried micro channels in silicon. *J. Microelectromech. Syst.* **9**, 94–103
16. Guo, L. J., Cheng, X., Chou, C. (2004) Fabrication of size controllable nanofluidics channels by nanoimprinting and its applications for DNA stretching. *Nano. Lett.* **4**, 49–73
17. Yanagi, H. and Kwawi, Y. (2004) Organic field effect transistor with narrow channel fabricated using focused ion beam. *J. Appl. Phys.* **43**, L1575–L1577
18. Craighead, H. G. (2000) Nanoelectromechanical systems. *Science.* 290, 1532–1535
19. Mannion, J. T., Reccius, C. H., Cross, J. D., Craighead, H. G. (2006) Conformational analysis of single DNA molecules undergoing entropically induced motion in nanochannels. *Biophys. J.* **90**, 4538–4546

Chapter 3

A Single-Molecule Barcoding System using Nanoslits for DNA Analysis: Nanocoding

Kyubong Jo, Timothy M. Schramm, and David C. Schwartz

Summary

Single DNA molecule approaches are playing an increasingly central role in the analytical genomic sciences because single molecule techniques intrinsically provide *individualized* measurements of selected molecules, free from the constraints of bulk techniques, which blindly average noise and mask the presence of minor analyte components. Accordingly, a principal challenge that must be addressed by all single molecule approaches aimed at genome analysis is how to immobilize and manipulate DNA molecules for measurements that foster construction of large, biologically relevant data sets. For meeting this challenge, this chapter discusses an integrated approach for microfabricated and nanofabricated devices for the manipulation of elongated DNA molecules within nanoscale geometries. Ideally, large DNA coils stretch via nanoconfinement when channel dimensions are within tens of nanometers. Importantly, stretched, often immobilized, DNA molecules spanning hundreds of kilobase pairs are required by all analytical platforms working with large genomic substrates because imaging techniques acquire sequence information from molecules that normally exist in free solution as unrevealing random coils resembling floppy balls of yarn. However, nanoscale devices fabricated with sufficiently small dimensions fostering molecular stretching make these devices impractical because of the requirement of exotic fabrication technologies, costly materials, and poor operational efficiencies. In this chapter, such problems are addressed by discussion of a new approach to DNA presentation and analysis that establishes scaleable nanoconfinement conditions through reduction of ionic strength; stiffening DNA molecules thus enabling their arraying for analysis using easily fabricated devices that can also be mass produced. This new approach to DNA nanoconfinement is complemented by the development of a novel labeling scheme for reliable marking of individual molecules with fluorochrome labels, creating molecular barcodes, which are efficiently read using fluorescence resonance energy transfer techniques for minimizing noise from unincorporated labels. As such, our integrative approach for the realization of genomic analysis through nanoconfinement, named nanocoding, was demonstrated through the barcoding and mapping of bacterial artificial chromosomal molecules, thereby providing the basis for a high-throughput platform competent for whole genome investigations.

Key words: DNA labeling, Genomics, Nanofabrication, Polymer confinement, Low ionic strength, FRET, Nicking enzyme, Physical mapping

James Weifu Lee and Robert S. Foote (eds.), *Micro and Nano Technologies in Bioanalysis*,
Methods in Molecular Biology, vol. 544
DOI: 10.1007/978-1-59745-483-4_3, © Humana Press, a part of Springer Science+Business Media, LLC 2009

1. Introduction

The explosion of single molecule approaches and microfluidics offers promising routes for effectively dealing with real-world biological applications requiring large data sets. Advances in this direction have stemmed from the appreciation of polymer behavior exhibited within typical microfluidic devices that have laid the basis for the development of practical approaches for molecular presentation. However, taking new theoretical insights forward for the development of practical genome analysis systems built around microfluidic and nanofluidic devices has proven difficult. Consider that most previously described microfluidic devices lack functionalities required for the large-scale manipulation of very large DNA molecules typically extracted from cells as genomic substrates. In addition, few devices have been specifically designed for operation within an integrated system—a necessary step if any device is to be used for meaningful application as a platform genomic analysis.

In this regard, Optical Mapping *(1)* is an unique high-throughput system using microfluidic devices designed to manipulate ensembles of very large genomic DNA molecules with sequence-specific decoration *(2, 3)*. With the Optical Mapping system, a solution of DNA molecules flows in microfluidic channels via capillary force, elongating then depositing individual molecules in the same orientation on a positively charged surface via electrostatic interactions, creating massive single molecule arrays. These electrostatic interactions are strong enough that the molecules are held to the surface in an elongated fashion yet remain viable biochemical substrates *(4, 5)*. After individual DNA molecules are presented as arrays, restriction enzyme action recognizes and cleaves specific sequences along the DNA backbones for generating discrete DNA restriction fragments that remain ordered on the surface. Subsequent staining with fluorochrome dyes that bind DNA enable fluorescence microscopy to rapidly image and analyze molecules using a fully automated system. Image analysis software identifies DNA molecule backbones, determines the size of each daughter restriction fragment, and generates one physical map (ordered restriction map) per molecule. Consensus maps are constructed from many single molecule maps using dedicated algorithms *(6, 7)* for subsequent biological analysis, such as useful scaffolds for guiding sequence assembly *(8–11)*, comparative genomic studies *(12)*, and the discovery of genomic structural alterations or "differences" involving kilobase- to megabase-sized changes in human genomes *(3, 6)*.

Direct analysis of individual DNA molecules for genome analysis has been realized through the development of Optical Mapping. Nevertheless, a principal challenge faced using single molecule approaches is still the immobilization of a large number of samples, yet maintenance of competent biochemical activity. For example, fixation of elongated DNA molecules on positively charged surfaces suffers from a range of shortcomings that molecules in free solution can obviate. In this regard, the optimum approach for substrate immobilization is no immobilization; the best chemical linker is no chemical linker. As such, entropic confinement techniques offer a uniquely powerful route to purely physical means for immobilization, which can be realized through the nanofabrication of features that leverage the wonderfully facile entropic properties of large coils such as DNA *(13–15)*. The "stiffness" of double-stranded DNA and the long polymer length provide ample opportunities through entropic confinement techniques *(16)*.

This chapter describes a single-molecule system using nanoconfinement for genomic analysis using disposable nanoscale silicone rubber devices and chemistries that reproducibly elongate large DNA molecules by 60% of their polymer contour length in concert with a single DNA molecule labeling scheme possessing low-noise characteristics. This approach facilitates the development of a truly high-throughput system for genomic analysis.

2. Materials

2.1. Fabrication of Master Wafer

1. Chrome mask array pattern of 750 nm × 5 mm (Center for Nanotechnology of the University of Wisconsin-Madison, Madison, WI).
2. Photoresist SU-8 2000.5 for nanoslits and SU-8 2005 for microchannel overlay (Microchem, Newton, MA).
3. Reactive ion etching instrument (Unaxis 790 RIE, Unaxis Wafer Processing, St. Petersburg, FL).
4. Piranha: 80% H_2SO_4 and 20% H_2O_2. Extremely corrosive. Wear proper eye/hand/body protection.
5. Alpha step 200 profilometer (KLA-Tencor, San Jose, CA).

2.2. PDMS Nanoslit Preparation

1. Polydimethylsiloxane (PDMS, Sylgard 184, Dow Corning, Midland, MI).
2. Oxygen plasma chamber (Technics Plasma GmbH 440, Technics Plasma GmbH, Florence, KY).
3. Ethylenediamine tetraacetic acid (EDTA, 0.5 M) adjusted to pH 8.5.

2.3. Clean Glass Preparation

1. Coverslips (22 mm × 22 mm) (Fischer Scientific, Pittsburgh, PA).
2. Nano-Strip (sulfuric acid and hydrogen peroxide) (Cyantek Corp., Fremont, CA). Extremely corrosive. Wear proper eye/hand/body protection.

3. Concentrated hydrochloric acid (12 M). Wear proper eye/hand/body protection.

2.4. DNA Barcoding

All concentrations listed are final; stocks that are more concentrated need to be made and diluted in the reaction solution to final concentration.

1. Bacterial artificial chromosomes (BAC) 79, 150, 614 from *E. coli* K-12; any BAC clone will suffice.
2. NEBuffer 2: 50 mM NaCl, 10 mM Tris-HCl, 10 mM $MgCl_2$, 1 mM dithiothreitol, pH 7.9 (New England Biolabs, Ipswich, MA).
3. NEBuffer 4: 50 mM potassium acetate, 20 mM Tris-acetate, 10 mM magnesium acetate, 1 mM dithiothreitol, pH 7.9 (New England Biolabs).
4. Restriction enzymes FseI and SpeI (New England Biolabs).
5. T4 DNA ligase (New England Biolabs).
6. 1 mM ATP to be used as a cofactor for T4 DNA ligase (Sigma-Aldrich).
7. Nicking enzyme: Nb.BbvCI (New England Biolabs).
8. Deoxyribonucleotides (dNTP): dATP, dCTP, dGTP, dTTP (New England Biolabs). 20 μM solutions of each were made in Tris-EDTA buffer (TE).
9. Alexa Fluor 647-aha-dCTP and Alexa Fluor 647-aha-dUTP (Invitrogen) were prepared as 2 μM solutions in TE. Protect from light.
10. *E. coli* DNA polymerase I, endonuclease-free grade (Roche Applied Sciences, Indianapolis, IN).
11. Dideoxyribonucleotides (ddNTP): ddATP, ddCTP, ddGTP, ddTTP (Amersham Biosciences, Piscataway, NJ) prepared as 0.2 μM each in TE.
12. Proteinase K (Bioline, Taunton, MA).
13. *n*-Lauroyl sarcosine (Sigma-Aldrich, St. Louis, MO).
14. Micro dispodialyzer (Spectrum Laboratories, Rancho Dominguez, CA).

2.5. DNA Sample Preparation in Low Ionic Strength and Loading

1. DNA of bacteriophage λ (48.5 kbp) from New England Biolabs. Stock concentration of 500 ng/μL, final concentration 1 ng/μL.
2. DNA of bacteriophage T4 (166 kbp) from Waco Chemicals USA, which is the vendor of Nippon Gene, Japan. Stock concentration of 390 ng/μL, diluted to final concentration of 0.78 ng/μL.
3. YOYO-1 from Invitrogen, Inc. (Eugene, OR). Final concentration of 0.25 μM from stock, which is 1 mM in DMSO. Protect from light.

4. β-mercaptoethanol as an antibleaching agent from Sigma-Aldrich.
5. Tris-EDTA buffer (1× TE): 10 mM Tris-HCl and 1 mM EDTA, pH 8.0. Tris EDTA buffer is made as follows: Tris base and EDTA acid are dissolved together, and titrated to pH 8.0 with HCl (*see* **Note 1**).
6. POP6 (Applied Biosystems, Foster City, CA).

2.6. Microscopy and Image Processing

1. Argon ion laser (488 nm; Spectra Physics 2017, Spectra Physics Laser Inc., Irvine, CA).
2. Inverted Microscope (Zeiss 135M equipped with a 63× Zeiss Plan-Neofluor oil immersion objective, Carl Zeiss Inc., Jena, Germany) *(2)*.
3. Charge-coupled device digital camera (CCD, Hamamatsu ORCA-ER, 1344 × 1024 pixels, 12-bit digitization, Hamamatsu Photonics Inc., Hamamatsu, Japan) and Cooke PixelFly CCD cameras (1,376 × 1,040, 12-bit, Applied Scientific Instrumentation, Eugene, OR).
4. Emission filters for the green channel (XF3086) and for the red channel (XF3076) (Omega Optical, Inc., Brattleboro, VT).
5. All software programs from image collection to image analysis are written in our laboratory. For microscopy and image collection, programs are written in Borland C++ Builder 6.0 (Cupertino, CA) and, for image analysis, programs are written in C++ using GTK and GNOME library in the Linux system.

3. Methods

3.1. Fabrication of Master Wafer

1. Prepare a chrome mask of 750 nm × 5-mm array fabricated by e-beam lithography.
2. Spin coat a negative photoresist (SU-8 2000.5) onto a silicon wafer.
3. Illuminate the silicon wafer through the chrome mask to create arrays of 1-μm-wide, 5-mm-long slits (*see* **Note 2**).
4. Etch the wafer 100 nm deep using a reactive ion etching (RIE) machine with CF_4 at 10 mTorr for 8 min (*see* **Note 3**).
5. Clean the etched wafer using piranha solution to lift off the photoresist layer (SU-8 2000.5).
6. Measure the height of the nanoslits (100-nm high × 1-μm wide) using an alpha step profilometer and measure the width under a scanning electron microscope (*see* **Fig. 1d**).
7. Overlay a microchannel array (3-μm high, 100-μm wide, and 10-mm long) on the nanopatterned wafer using the negative

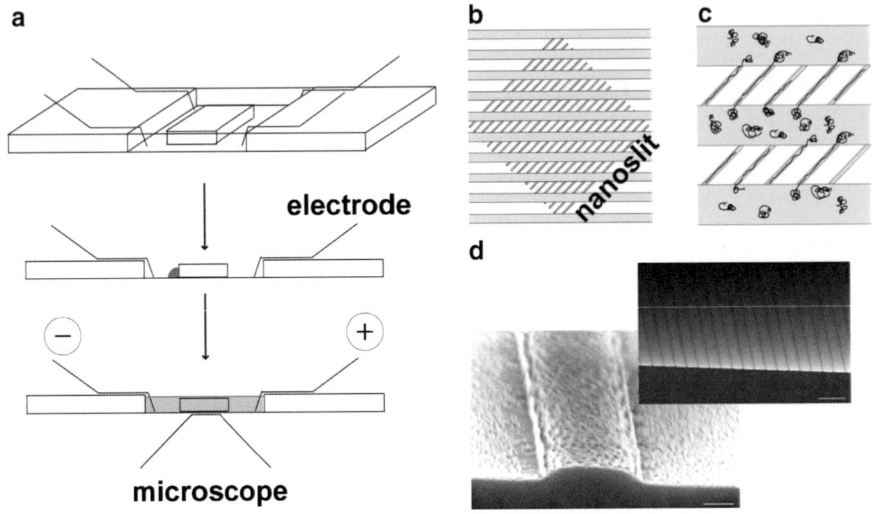

Fig. 1. The micro nanoslit device design and loading scheme. **(a)** For microscopy, a small chamber is fashioned from a Plexiglas™ slide (25.4 × 76.2 mm) with a rectangular opening to which a glass coverslip window (18 × 18 mm) is affixed with wax. The PDMS device is adhered to the coverslip window within the chamber. Pipetting applies DNA solution to microchannels, loading into the device by capillary action and then a buffer solution is added for electrokinetic loading. **(b)** Illustration (*top view*) shows nanoslits (diagonal; 100-nm high × 1,000-nm wide) overlaid with microchannels (horizontal; 3-μm high × 100-μm wide). **(c)** Cartoon depicts relaxed and elongated DNA molecules as occurring during electrokinetic loading within the microchannels and nanoslits, respectively. **(d)** Scanning electron micrograph of the silicon master shows a single nanoslit mold feature (bar = 300 nm); inset image shows many such nanoslit features spaced 4-μm apart (center-to-center; bar = 10 μm) (Reproduced from ref. *(16)*. Copyright 2007 National Academy of Sciences, USA.

photoresist SU-8 2005 in the second cycle of photolithography (*see* **Fig.1b**). The mask for the second cycle is a transparency film drawn in AutoCAD 2002 (http://www.autodesk.com/autocad).

8. Perform vapor deposition of tridecafluoro-1,1,2,2-tetrahydro octyl-tricholoro silane for silanization of the patterned wafer to promote PDMS releasing *(17)*. Place a patterned wafer in a Petri dish, add a drop of coating chemical at the corner of the Petri dish, close, and allow for vapor to deposit for an hour.

3.2. PDMS Nanoslit Preparation

1. Mix PDMS prepolymer with catalyst in a 10:1 ratio for 10 min.
2. Pour PDMS onto the silicon wafer master contained in a Petri dish.
3. Cure PDMS at 65°C for longer than 24 h, and peel it from the master wafer (*see* **Note 4**).
4. Perform oxygen plasma treatment in the Technics Plasma GmbH 440 to make the PDMS surface hydrophilic (O_2 pressure, ~0.67 millibars; load coil power, 100 W; 36 s).
5. Store plasma-treated devices in high-purity water for 24 h (*see* **Note 5**).

6. Sonicate the PDMS devices in a 50-mL conical tube filled with 0.5 M EDTA (pH 8.5) for 15 min to extract platinum (II) ions (*see* **Note 6**).
7. Sonicate the PDMS devices thoroughly in a 50-mL conical tube filled with high-purity water three times for 15 min each and store in high-purity water.
8. Dry the PDMS devices before use.

3.3. Clean Glass Preparation

Cleaning glass surfaces follows the cleaning procedure for Optical Mapping surfaces *(3, 18, 19)* (*see* **Note** 7). Heating concentrated acids can be dangerous and great care should be taken when executing this procedure. Proper eye/hand/skin protection must be used and the procedure must be done in a fume hood to avoid noxious vapors.

1. Fit cover slips (22 × 22 mm) in a Teflon rack and wrap with Teflon tape to hold them securely.
2. Set Teflon racks in a Pyrex glass cylinder (*see* **Note** 7).
3. Heat coverslips in Nano-Strip for 50 min after reaching 70°C. Allow for the cylinders to cool before draining the Nano-Strip.
4. Rinse the cylinder meticulously six times with high-purity and dust-free water.
5. Pour hydrochloric acid into the cylinder; make sure to cover the top of the Teflon rack by a few inches because some volume will be lost during boiling.
6. Boil in hydrochloric acid solution for 6 h once the liquid reaches 104°C to impart a uniform hydrolysis of the glass surface.
7. Rinse extensively with high-purity water to a neutral pH.
8. Remove the cover slips from the Teflon racks one at a time, and rinse them three times in absolute ethanol.
9. Store the clean coverslips under absolute ethanol in polypropylene containers at room temperature.

3.4. DNA Barcoding

All concentrations are final concentrations.

1. Linearize DNA molecules using a one-cut restriction enzyme if they are circular. FseI was used to linearize BAC79 and BAC150; SpeI for BAC614 (*see* **Note 8**).
2. Add 2 U T4 DNA ligase at 16°C for 2 h in 17.5 μL of NEBuffer 2 or 4 to attenuate indigenous nicks (*see* **Note 9**).
3. Inactivate T4 DNA ligase at 65°C for 10 min.
4. To the DNA solution, add 10 U *E. coli* DNA polymerase I and 0.2 μM ddNTPs at 37°C for 30 min in 40 μL in NEBuffer 2 or 4 to block remaining nicks (*see* **Notes 10** and **11**).

5. Add the labeling reaction mix of 20 U Nb.BbvCI, 2 μM Alexa Fluor 647-aha-dCTP, 2 μM Alexa Fluor 647-aha-dUTP, 20 μM dATP, 20 μM dGTP, 1 μM dCTP, and 1 μM dTTP (*see* **Note 12**)."
6. Incubate the mix for 30 min at 37°C.
7. Stop the reaction by adding 20 μM EDTA, pH 8.5.
8. Digest enzymes by adding proteinase K to a final concentration of 100 ng/μL and *n*-lauroyl sarcosine to 0.1%, w/v and incubating for 3 h at 50°C.
9. Adjust buffer conditions by simple dilution with water (~2,000–4,000 times) or dialysis against 500 mL of 100 μM Tris and 10 μM EDTA (pH 8.0; 0.01× TE) buffer solution overnight with micro dispodialyzer at 4°C.

3.5. DNA Sample Preparation in Low Ionic Strength and Loading

Figure 2 shows images of electrokinetically loaded λ DNA (48.5 kbp), T4 DNA (166 kbp), and fragments of *E. coli* genomic DNA. To form these elongated single DNA molecules, a molecule undergoes electrophoresis within a microchannel toward a nanoslit entrance, where it proceeds to enter, and then stretch. Low ionic-strength solution facilitates the stretching of DNA molecules within nanoslits.

1. Prepare low ionic-strength buffer by adding high-purity water. An example is demonstrated in **Fig. 3**, where 1× TE buffer is diluted by 5-, 10-, 15-, 20-, 50-, or 100-fold with water. To the dilutions of TE, add intercalating dye YOYO-1 (0.25 μM, final), antibleaching agent β-mercaptoethanol (4%, v/v, $HSCH_2CH_2OH$, final), and POP6 (0.1%, w/v, final) for suppressing electroendoosmosis *(15)* into a final volume of 1 mL.

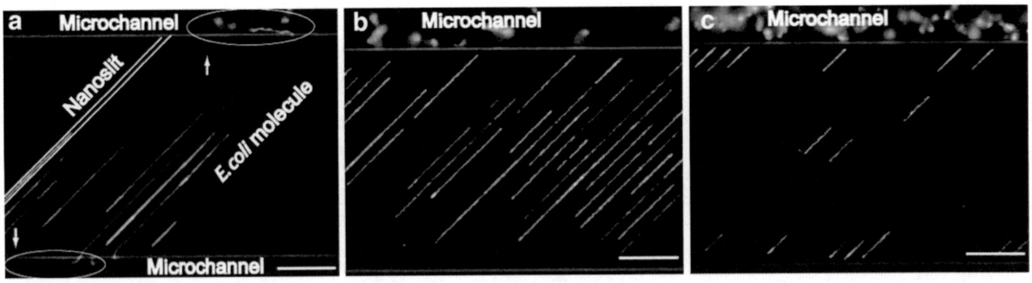

Fig. 2. Gallery of fluorescence micrographs shows stretched and relaxed DNA molecules within the nanoslit device after electrokinetic loading. Relaxed molecules within the microchannel regions appear as diffuse, partly out of focus, fluorescent balls, whereas stretched molecules appear as long linear objects. (**a**) A large *E. coli* DNA molecule spans across the 105-μm-long nanoslit (0.01× TE buffer) showing relaxed ends (*circled*) within abutting microchannels. (**b**) T4 DNA (166 kbp) molecules in 0.05× TE buffer. (**c**) λ DNA (48.5 kbp) molecule in 0.01× TE buffer. Scale bars = 20 μm (Reproduced from ref. *(16)*. Copyright 2007 National Academy of Sciences, USA.

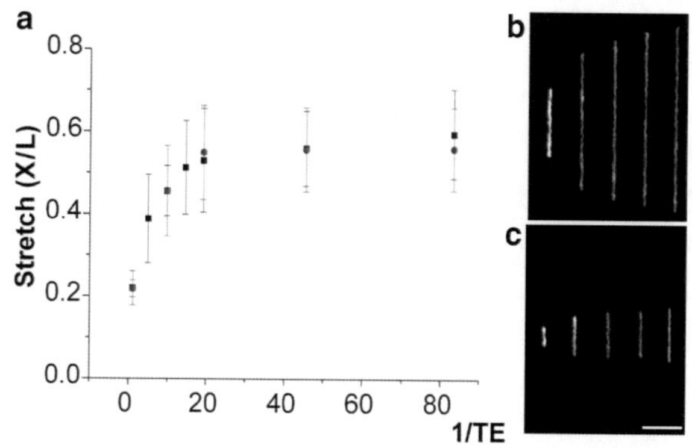

Fig. 3. DNA stretch varies with diluted TE concentration of the λ DNA (*open square*) and T4 DNA (*filled circle*). (**a**) Ionic strength varied through dilutions of Tris-EDTA buffer (1× TE: 10 mM Tris-base, 1 mM EDTA, pH 8.0). The dilution factors are 1.0, 5.0, 9.8, 14.6, 19.3, 45.5, and 83.5 of 1× TE (ionic strength = 8.5 mM). The stretch is defined by apparent length (*X*) divided by the polymer contour length (*L*) of YOYO-1-stained DNA. Each data point represents measurements from 50 to 300 molecules and error bars show standard deviations on the means. (**b**, **c**) Fluorescence micrographs (a combination of five separate experiments) show T4 DNA (166 kbp) (**b**) and λ DNA (48.5 kbp) (**c**) at five different TE dilutions: 1.0, 9.8, 19.3, 45.5, and 83.5 (dilution factors). Scale bar = 10 μm (Reproduced from ref. *(16)*. Copyright 2007 National Academy of Sciences, USA.

2. Add DNA sample. In an example in **Fig. 3**, we add 2 μL of DNA in 1× TE into 1 mL buffer described in **step 1** (*see* **Note 13**).
3. Affix a glass coverslip (22 × 22 mm) with wax on a rectangular opening (18 × 18 mm) of a Plexiglas slide (25.4 × 76.2 mm) (*see* **Fig. 1a**).
4. Place a PDMS device on this glass window. Press gently to seal the PDMS to the clean glass but take care to not collapse the small features.
5. Load DNA sample into microchannels by capillary loading (*see* **Fig. 1a**).
6. Fill the reservoir with the same ionic strength buffer of the DNA sample for electrokinetic loading via the indicated electrodes in **Fig. 1a**.
7. Place the Plexiglas slide on the table of the fluorescence microscope and focus on the nanoslits.
8. Apply an electric potential (70 V) to transport relaxed DNA coils to nanoslit entrances for subsequent elongation (*see* **Note 14**).
9. After turning off the electric field, wait for several minutes for DNA to reach the steady state of relaxation within nanoslits (*see* **Note 15**).

3.6. Microscopy and Image Processing

1. Use microscopes equipped with two CCD cameras, that have two filtering optics for green and red colors, respectively (*see* **Subheading 2.6**). The green channel acquires images of DNA backbone stained with YOYO-1 (491 nm, absorption; 509 nm, emission), and the red channel acquires images of sequence-specific decorations of Alexa Fluor 647 (650 nm absorption; 665 nm emission) punctuates via fluorescence resonance energy transfer (FRET) (*see* **Note 16**).
2. Flatten images by image-processing software (*see* **Note 17**).
3. Identify DNA molecules in images by connecting neighboring pixels with fluorescence intensities above a threshold value.
4. Overlap the two corresponding images.
5. Determine molecular size based on integrated fluorescence intensities as well as the end-to-end length.
6. Determine corresponding punctate positions within a molecule using integrated fluorescence intensity profiles and unity based mapping (*see* **Note 18**).

4. Notes

1. In this case, the ionic strength of 1× TE (10 mM Tris, 1 mM EDTA) is 8.4 mM. On the other hand, typical 1× TE buffer usually starts from EDTA sodium salts and Tris-HCl titrated with NaOH. In this case, the ionic strength is 13.7 mM.
2. Although the width of a nanoslit pattern in the chrome mask is 750 nm, the width of a nanoslit template in a wafer is wider than 750 nm because of light diffraction. The width of a nanoslit can be optimized by adjusting exposure time.
3. A master wafer of nanoslits is dry-etched according to **ref.** *(20)* instead of a photoresist template built on the wafer like **ref.** *(17)*. We notice a gradual erosion of the SU-8 photoresist pattern (100-nm height) with the repetition of replica molding; in contrast, an etched silicon wafer master has a higher mechanical stability.
4. Long curing times are critical for stable PDMS nanostructures. With short curing times, 100-nm high PDMS nanoslits often collapse and disappear.
5. PDMS surfaces have a highly polar surface immediately after plasma treatment. If DNA solution is loaded in this reactive PDMS device, a significant amount of DNA molecules will affix to the PDMS device.
6. Without EDTA treatment, fluorescence intensity decreases as DNA molecules travel inside nanochannels. This problem

may be caused by platinum ions, the catalyst for PDMS polymerization, which may cause cyanine dye (YOYO-1) to dissociate from DNA backbones *(21, 22)*. Thus, YOYO-1-stained DNA molecules become dimmer either because the local Pt^{2+} concentration may be high within a nanoslit or because the Pt^{2+} effect is accumulated as DNA traverses a nanoslit. To resolve this issue, sonicating a PDMS device submerged in an EDTA buffer solution extracts platinum ions from PDMS.

7. Commercial coverslips have coating materials that have been added by the glass manufacturer. Thus, the acid cleaning procedure removes coating materials on the glass surface. For safety and cleanliness, a self-contained acid boiling system was built. The main components of this system are made from Pyrex glass cylinders, Teflon tubing, and Teflon sealing "O" rings, all of which are resistant to strong acids. Vacuum grease is not used to seal any joints. Custom Teflon racks were also designed to hold the cover slips securely during the cleaning process.
8. All steps are performed in test tubes and then loaded into the nanoslit device after dilution or dialysis, because the shortage of biochemically meaningful salt concentrations in the nanoslits obviously causes problems for most DNA modification enzymes used for genome analysis.
9. Because preexisting DNA nicks would produce spurious signals, such sites are repaired or disabled using T4 ligase or polymerase incorporation of dideoxyribonucleotides (ddNTPs) before labeling.
10. Because nick-translation efficiently incorporates fluorochrome-labeled nucleotides, crossing of mobilized nick sites on complementary DNA strands occurs, producing double-strand breaks; this is reduced by the continued presence of ddNTPs in the labeling reaction mix, thus, limiting the number of nucleotides incorporated per nick site through chain termination.
11. The termination of polymerase action by ddNTPs controls the size of fluorescent punctuates, which would otherwise expand into each other if nucleotide incorporation was unchecked, thus, diminishing the number of discrete markers.
12. A nicking enzyme (Nb.BbvCI; GC^TGAGG) cleaves only cognate sites on single strands of double-stranded molecules made detectable by nick translation using fluorochrome-labeled nucleotides *(23–25)*.
13. Low ionic strength increases DNA intrachain electrostatic repulsion. DNA molecules could be enlarged in terms of physical measures of size that would hinge on a polymer's persistence length—a characteristic proportional to the

"stiffness" of a chain reflecting the relative directional orientation of several infinitesimal segments. DNA molecules are stretched up to 60% of their polymer contour length in disposable PDMS devices having 100 × 1,000-nm channels under low ionic-strength conditions (*see* **Fig. 3**); remarkably, these results are comparable to what was previously obtained under standard buffer conditions using 30 × 40-nm channels fabricated on fused silica substrates using nanoimprint or electron beam lithography *(14)*.

14. DNA molecules electrokinetically progress through the microchannels as relaxed coils until approaching a nanoslit entrance. There, as one end of a molecule enters the nanoslit, the DNA molecule transiently elongates.
15. DNA molecules expectedly enter nanoslits with different conformations and reach equilibrated forms via different relaxation processes, such as dynamic shrinking and unfolding (elongation).
16. Because labeled DNA molecules are globally stained with the intercalating dye, YOYO-1, we reasoned that fluorescence resonance energy transfer (FRET) would operate between YOYO-1 (FRET donor) and the sequence-specifically placed AlexaFluor-647 labeled nucleotides (FRET acceptor) because they are intimately situated within the same DNA backbone and spectrally compatible. **Figure 5** shows FRET detection of barcode features and their spacing (in kilobases) using integrated fluorescence intensity measurements of the YOYO-1 signals between them.
17. Laser light has a Gaussian shape of light intensity, which generates varying fluorescent intensities for an equivalent amount of fluorochrome. For quantitative analysis, this nonhomogenous fluorescent response should be computationally corrected *(2)*.
18. A "unity-based" approach *(1, 5, 26)* uses integrated fluorescence intensity or apparent length for estimation of restriction fragment masses, with the assumption that molecules are not broken. Such measurements were performed on a per fragment basis then normalized by total fluorescence intensity (or size) of the entire molecule, so that the apportionment of fragment fluorescence intensities (or sizes) sums to 1.0. The prevailing assumption generally holds for relatively small DNA molecules (<180 kb) *(27, 28)*. Final maps are then constructed by averaging fluorescence intensity measurements of like restriction fragments over a number of molecules, allowing estimation of precision expressed as a standard deviation. Here, we adapt this approach by measuring the integrated fluorescence intensity of intervals, demarcated by fluorescent punctates (*see* **Fig. 4**).

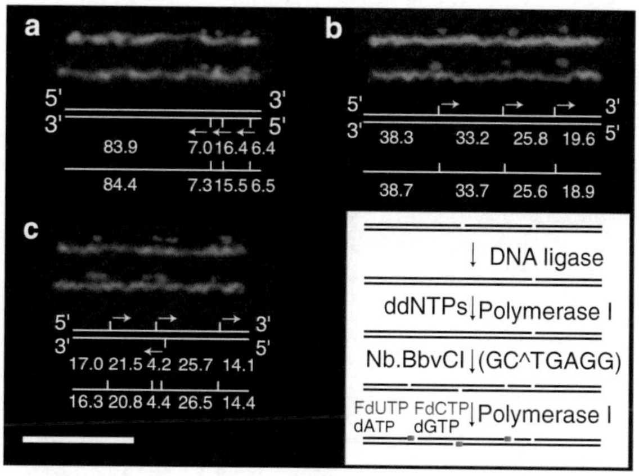

Fig. 4. Molecular barcoding scheme and image gallery of molecules "barcoded" by a nicking enzyme, followed by fluorochrome labeling via nick translation and detection by fluorescence resonance energy transfer (FRET). (*Inset*) First, DNA ligation and nick translation with ddNTPs eliminates or incapacitates inherent nicks on molecules to be barcoded; second, Nb.BbvCI places site-specific nicks on these molecules; and lastly, incorporation of fluorochrome-labeled dNTPs (AlexaFluor 647-aha-dCTP, UTP) nick sites by *E. coli* DNA polymerase I. The fluorescence micrographs of labeled DNA molecules compared with the expected labeling pattern derived from the sequence. (**a**) BAC79 (113.7 kb), (**b**) BAC150 (116.8 kb), (**c**) BAC614 (82.5 kb) with two molecules corresponding in silico map shown per sample. *Arrows* indicate the direction of nick translation on a given strand. Scale bar = 10 μm (Reproduced from ref.*(16)*. Copyright 2007 National Academy of Sciences, USA.

Acknowledgments

The authors thank Dr. Dalia M. Dhingra for assisting in the development DNA barcoding, Prof. Theo Odijk for his theory development of DNA elongation within nanoslits, Prof. Juan J. de Pablo and Prof. Michael D. Graham for their advice, and Dr. Guy Plunkett III for bacterial artificial chromosomes and their sequence data. This work was supported by National Institutes of Health Grant 5R01HG000225 and National Science Foundation Grant NSEC DMR-0425880.

References

1. Schwartz, D. C., Li, X., Hernandez, L. I., Ramnarain, S. P., Huff, E. J., & Wang, Y. K. (1993). *Science* **262**, 110–114
2. Dimalanta, E. T., Lim, A., Runnheim, R., Lamers, C., Churas, C., Forrest, D. K., de Pablo, J. J., Graham, M. D., Coppersmith, S. N., Goldstein, S., et al. (2004). *Anal. Chem.* **76**, 5293–5301
3. Zhou, S., Herschleb, J., & Schwartz, D. C. (2007). in New Methods for DNA Sequencing ed. Mitchelson, K. R. (Elsevier Scientific Publishers, Amsterdam, Netherland), pp. 265–300
4. Cai, W., Aburatani, H., Stanton, V. P., Jr., Housman, D. E., Wang, Y. K., & Schwartz, D. C. (1995). *Proc. Natl. Acad. Sci. U.S.A.* **92**, 5164–5168
5. Meng, X., Benson, K., Chada, K., Huff, E. J., & Schwartz, D. C. (1995). *Nat. Genet.* **9**, 432–438

6. Valouev, A., Schwartz, D. C., Zhou, S., & Waterman, M. S. (2006). *Proc. Natl. Acad. Sci. U. S. A.* **103**, 15770–15775
7. Valouev, A., Zhang, Y., Waterman, M. S., & Waterman, M. S. (2006). *Bioinformatics* **22**, 1217–1224
8. Lin, J., Qi, R., Aston, C., Jing, J., Anantharaman, T. S., Mishra, B., White, O., Daly, M. J., Minton, K. W., Venter, J. C., et al. (1999). *Science* **285**, 1558–1562
9. Armbrust, E. V., Berges, J. A., Bowler, C., Green, B. R., Martinez, D., Putnam, N. H., Zhou, S. G., Allen, A. E., Apt, K. E., Bechner, M., et al. (2004). *Science* **306**, 79–86
10. Lai, Z., Jing, J., Aston, C., Clarke, V., Apodaca, J., Dimalanta, E. T., Carucci, D. J., Gardner, M. J., Mishra, B., Anantharaman, T. S., et al. (1999). *Nat. Genet.* **23**, 309–313
11. Perna, N. T., Plunkett, G., III, Burland, V., Mau, B., Glasner, J. D., Rose, D. J., Mayhew, G. F., Evans, P. S., Gregor, J., Kirkpatrick, H. A., et al. (2001). *Nature* **409**, 529–533
12. Zhou, S., Kile, A., Bechner, M., Place, M., Kvikstad, E., Deng, W., Wei, J., Severin, J., Runnheim, R., Churas, C., et al. (2004). *J. Bacteriol.* **186**, 7773–7782
13. Cao, H., Yu, Z. N., Wang, J., Tegenfeldt, J. O., Austin, R. H., Chen, E., Wu, W., & Chou, S. Y. (2002). *Appl. Phys. Lett.* **81**, 174–176
14. Reisner, W., Morton, K. J., Riehn, R., Wang, Y. M., Yu, Z. N., Rosen, M., Sturm, J. C., Chou, S. Y., Frey, E., & Austin, R. H. (2005). *Phys. Rev. Lett.* **94**, 196101
15. Tegenfeldt, J. O., Prinz, C., Cao, H., Chou, S., Reisner, W. W., Riehn, R., Wang, Y. M., Cox, E. C., Sturm, J. C., Silberzan, P., et al. (2004). *Proc. Natl. Acad. Sci. U.S.A.* **101**, 10979–10983
16. Jo, K., Dhingra, D. M., Odijk, T., de Pablo, J. J., Graham, M. D., Runnheim, R., Forrest, D., & Schwartz, D. C. (2007). *Proc. Natl. Acad. Sci. U.S.A.* **104**, 2673–2678
17. Duffy, D. C., McDonald, J. C., Schueller, O. J. A., & Whitesides, G. M. (1998). *Anal. Chem.* **70**, 4974–4984
18. Zhou, S., Deng, W., Anantharaman, T. S., Lim, A., Dimalanta, E. T., Wang, J., Wu, T., Chunhong, T., Creighton, R., Kile, A., et al. (2002). *Appl. Environ. Microbiol.* **68**, 6321–6331
19. Reed, J., Singer, E., Kresbach, G., & Schwartz, D. C. (1998). *Anal. Biochem.* **259**, 80–88
20. Effenhauser, C. S., Bruin, G. J. M., Paulus, A., & Ehrat, M. (1997). *Anal. Chem.* **69**, 3451–3457
21. Markstrom, M., Cole, K. D., & Akerman, B. (2002). *J. Phys. Chem. B* **106**, 2349–2356
22. Eriksson, M., Karlsson, H. J., Westman, G., & Akerman, B. (2003). *Nucleic Acids Res.* **31**, 6235–6242
23. Heiter, D. F., Lunnen, K. D., & Wilson, G. G. (2005). *J. Mol. Biol.* **348**, 631–640
24. Xu, S. Y., Zhu, Z., Zhang, P., Chan, S. H., Samuelson, J. C., Xiao, J., Ingalls, D., & Wilson, G. G. (2007). *Nucleic Acids Res.* **35(14)**, 4608–4618
25. Xiao, M., Phong, A., Ha, C., Chan, T. F., Cai, D., Leung, L., Wan, E., Kistler, A. L., DeRisi, J. L., Selvin, P. R., et al. (2007). *Nucleic Acids Res.* **35**, e16
26. Jing, J., Reed, J., Huang, J., Hu, X., Clarke, V., Edington, J., Housman, D., Anantharaman, T. S., Huff, E. J., Mishra, B., et al. (1998). *Proc. Natl. Acad. Sci. U.S.A.* **95**, 8046–8051
27. Cai, W., Aburatani, H., Stanton, V. P., Jr., Housman, D. E., Wang, Y. K., & Schwartz, D. C. (1995). *Proc. Natl. Acad. Sci. U.S.A.* **92**, 5164–5168
28. Cai, W., Jing, J., Irvin, B., Ohler, L., Rose, E., Shizuya, H., Kim, U. J., Simon, M., Anantharaman, T., Mishra, B., et al. (1998). *Proc. Natl. Acad. Sci. U.S.A.* **95**, 3390–3395

Chapter 4

Microfluidic Devices with Photodefinable Pseudo-valves for Protein Separation

Z. Hugh Fan

Summary

Plastic microfluidic devices are fabricated with an array of pseudo-valves for two-dimensional (2D) protein separation. The devices are made by compression molding; the mold is created by electroplating on a glass master fabricated by photolithography. Each device consists of one channel for isoelectric focusing (IEF) and multiple parallel channels for polyacrylamide gel electrophoresis (PAGE). The IEF channel (first dimension) is orthogonal to the PAGE channels (second dimension). Microfluidic pseudo-valves are created at the intersections of orthogonal channels by photodefinable, in situ gel polymerization. These valves enable the introduction of two types of separation media into orthogonal channels for performing 2D protein separation in the device. The presence of the pseudo-valves prevents one separation medium from being contaminated by the other medium, although proteins are allowed to transfer from the first to the second dimension under an electric field. Two-dimensional protein separation is achieved in less than 10 min, an improvement of two orders of magnitude compared with the conventional 2D gel electrophoresis using an IEF strip and a PAGE slab.

Key words: Microfluidics, Valves, Isoelectric focusing, Gel electrophoresis, Proteins, Plastics, Photo-initiated polymerization

1. Introduction

Proteomics is emerging as an important tool in modern drug discoveries and medical diagnostics *(1)*. One of the popular techniques for proteomics studies is two-dimensional (2D) gel electrophoresis (2DGE) *(2)*. 2DGE consists of a first dimensional separation (isoelectric focusing [IEF]) and a second dimensional separation (polyacrylamide gel electrophoresis [PAGE]). The key advantage of 2DGE is its enormous separation resolution *(2)*. Hundreds of protein spots are detected in a single 2D gel image

James Weifu Lee and Robert S. Foote (eds.), *Micro and Nano Technologies in Bioanalysis*,
Methods in Molecular Biology, vol. 544
DOI 10.1007/978-1-59745-483-4_4,

called a 2D map. The major limitations of 2DGE include poor reproducibility and the time-consuming process *(3)*.

To address the limitations, efforts have been reported to develop 2D electrophoresis devices that have the potential to replace conventional 2DGE *(3–6)*. In these devices, one channel is designed for IEF and 10–100 parallel channels orthogonal to the IEF channel are created for PAGE.

This chapter focuses on a 2D electrophoresis device with a two-layer structure *(6)*. The device is made from cyclic olefin copolymers, which offer a number of benefits including increased solvent resistance, higher optical clarity, and reduced absorption of moisture *(7, 8)*. One of the key elements in the device is microfluidic pseudo-valves that separate the first from the second dimension. In situ, photo-initiated gel polymerization is used to define precisely the boundary of the second dimension. The pseudo-valves significantly reduce cross-contamination between the two dimensions. The detailed protocol of the device fabrication, valve operation, and protein separation are discussed.

2. Materials

2.1. Device Fabrication

1. Computer-aided design (CAD) software for device design, e.g., AutoCAD (Autodesk, San Rafael, CA).
2. Photolithography and cleanroom facility.
3. Microscope slides from Fisher Scientific (Atlanta, GA).
4. Cyclic olefin copolymer (COC) resins (Zeonor® 1020, Louisville, KY) and Topas® 8007 films from Ticona (Florence, KY).
5. A hydraulic press (Carver, Wabash, IN).
6. A computer numerically controlled (CNC) milling machine (e.g., Flashcut 2100; Menlo Park, CA).
7. A laminator (Catena 35, GBC, Northbrook, IL).

2.2. Microfluidic Pseudo-valves

1. Acrylamide monomer (acrylamide/*N*, *N*′-methylenebisacrylamide, 19:1 ratio, 40%) from Sigma-Aldrich (St. Louis, MO). Tris-HCl (1 M, pH 8.0) and tricine from Fisher Scientific.
2. 1-hydroxycyclohexyl-phenylketone (HCPK) from Sigma-Aldrich. HCPK of 100 mM is prepared in propanol. The solution for making photodefinable gel valves consists of 10%T acrylamide/bis-acrylamide, 40 mM Tris-HCl, 40 mM tricine, and 13 mM HCPK.
3. A chrome mask (or transparency mask) with the desired pattern for blocking light.

4. A UV source with collimated light (Oriel Instruments, Stratford, CT).
5. A vacuum pump from Fisher Scientific for priming the device and removing unpolymerized reagents.

2.3. Protein Separation

1. Acrylamide monomer (acrylamide/N, N′-methylenebisacrylamide, 19:1 ratio, 40%) from Sigma-Aldrich. The monomer solution for PAGE consists of 10% acrylamide/bis-acrylamide, 40 mM Tris-HCl, 40 mM tricine, and 13 mM HCPK.
2. Hydroxypropyl cellulose (HPC, MW 80,000) from Sigma-Aldrich, and glycerol, acetic acid, ethanolamine, and sodium dodecyl sulfate (SDS) from Fisher Scientific.
3. Carrier ampholytes (e.g., pH 3–10 and 4–6) from Bio-Rad Laboratories (Hercules, CA). The IEF medium consists of 2% carrier ampholytes, 8% glycerol, 2.3% HPC, and proteins of interest *(9)*.
4. Fluorescein-EX protein labeling kit from Invitrogen Molecular Probes (Eugene, OR).

3. Methods

Microfluidic devices have been studied for a number of applications, including DNA analysis, protein separations, and cell manipulation. They are fabricated from silicon, glass, or plastics. Plastics include polymethylmethacrylate (PMMA), polycarbonate, polyester, fluorinated ethylene propylene, poly(ethylene terephthalate), and elastomers such as poly(dimethyl siloxane) (PDMS), although this chapter focuses on cyclic olefin copolymers (COC).

Microfluidic valves are often required to isolate one region from other regions of an integrated device, in which several elements are designed for different functions. When valves are absent, reagents in different regions flow into each other without control due to diffusion, convection, and other mechanisms. In the 2D electrophoresis device, valves are needed to prevent possible contamination and interference between two separation media. However, fabrication of reliable valves in a microfluidic device is very challenging *(10, 11)*. Photodefinable pseudo-valves are created to address this need *(6, 12)*.

3.1. Device Fabrication

1. Design microfluidic devices using AutoCAD. The layout of one device used in the author's lab is shown in **Fig. 1** as an example (*see* **Note 2**). The CAD file is sent to a vendor (Photo Sciences, Torrance, CA) and a chrome photomask is created.

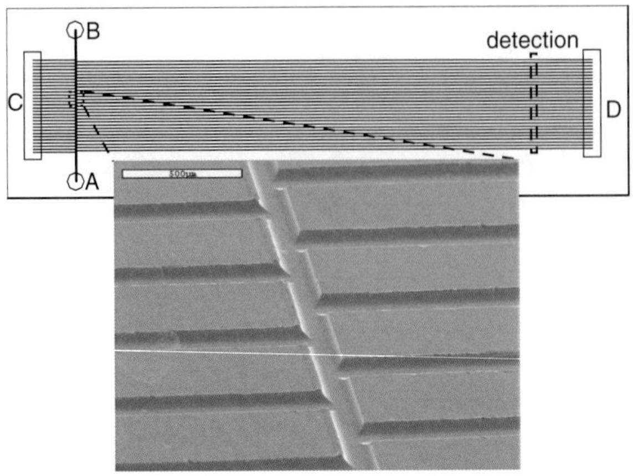

Fig. 1. Layout of a microfluidic device for two-dimensional protein separation. Channel AB is for IEF and channels CD for PAGE. The detection region is indicated by dashed lines. An exploded view of the channel intersections is shown in the scanning electron micrograph (SEM) in the inset with a 500-μm scale bar. The two sections of channels CD are not aligned with each other, allowing the protein transfer from the first to the second dimension. All channels are 40 μm deep and 110 μm wide.

2. Perform photolithography on glass microscope slides. Deposition of chromium and gold on glass slides is done using an electron-beam process. Photolithography is performed using a mask aligner. The pattern on the chrome photomask is transferred into the glass slides by chemical etch. The glass slides are etched using a mixture containing 20% HF/14% HNO3/66% H_2O; the depth of the resultant channels is determined by the etching time (*see* **Note 3**).
3. Generate a metal mold (called the E-form) by an electroplating process (Optical ElectroForming, Clearwater, FL). The glass slide with best etching quality is sent to the vendor (*see* **Note 4**). After smoothing the back of the E-form, it is then used for producing plastic parts.
4. Produce plastic substrates using compression molding with a hydraulic press (*see* **Note 5**). Plastic resins are placed on the E-form and then covered with a flat plate. This assembly is placed between platens of the press that is heated to the predetermined temperature (*see* **Note 6**) *(8)*. After warming resin for 5 min, platens are compressed using a predetermined force (*see* **Note 6**). After holding the force for 5 min, the force is released and the assembly is removed from the hydraulic press. While the assembly is cooling down, the plastic part is removed from the E-form.
5. Trim the plastic parts and create reservoirs at the end of channels. A CNC milling machine is used to drill through the substrate

at the ends of the channels to create 2-mm-diameter wells A and B and 2 mm × 12 mm slots C and D, as shown schematically in **Fig. 1**. The CNC milling machine is also used to trim the edges of the plastic substrate to a size of 1 × 3 in. House air, directed through a 12-in. flexible hose (Loc-line, Lake Oswego, OR) with a 1/16 diameter nozzle, is used to remove the debris from the area of the cutting tool.

6. Laminate the plastic substrate with a cover film to complete the device fabrication. Before lamination, both film and plastic substrate are washed in a 1% detergent (Alconox, White Plains, NY) solution in an ultrasonic bath for 5–10 min. They are then rinsed three times with Nanopure water, followed by air-drying in a laminar hood. The film and substrate are then sandwiched between two layers of a 0.1-mm-thick Mylar® film (Hydrofarm, Petaluma, CA), and then run through the laminator at the predetermined temperature (*see* **Notes 7** and **8**).

3.2. Microfluidic Pseudo-valves

1. Fill all channels in the device with the solution consisting of HCPK and acrylamide monomers (*see* **Notes 9** and **10**). The procedure starts by cleaning all channels with purified water. The monomer solution is then introduced into slot C and wells A and B (**Fig. 1**). Vacuum is then applied to slot D to introduce the solution into all channels (*see* **Notes 11** and **12**).
2. Place the mask on the device to define the location of pseudo-valves. To make gel valves at precise locations, the line in the mask is aligned with channel AB in the device (*see* **Notes 13** and **14**).
3. Expose the device with the monomer solution to UV light for 2–5 min. Photo-initiated polymerization takes place in all channels CD because they are exposed to light, whereas the solution in channel AB does not polymerize because UV light is blocked by the mask. Nonpolymerized solution in channel AB is removed by vacuum and replaced with water.
4. Ensure complete polymerization of the monomer solution everywhere in the device. After channel AB is replaced with water, the mask is removed. The whole device is then exposed again to UV light for 5 min to ensure complete polymerization everywhere (*see* **Note 15**).

3.3. Protein Separation

1. Label proteins with a fluorescent dye if the detection system is based on fluorescence detection. For instance, the Fluorescein-EX protein labeling kit is used to label nonfluorescent proteins (*see* **Note 16**).
2. Fill in the IEF separation medium in channel AB. The IEF medium consists of 2% carrier ampholytes, 8% glycerol, 2.3% HPC, and proteins of interest *(9)*.

3. Add 10 mM acetic acid (anolytes) and 10 mM ethanolamine (catholytes) to wells A and B, respectively, and add 20 mM Tris-HCl buffer with 10% SDS buffer solution to slots D and C.
4. Position the device appropriately in the apparatus (*see* **Note 17**). Platinum wire electrodes are put in wells A and B, and platinum foil strip electrodes are placed in slots C and D.
5. Apply a voltage across channel AB to implement IEF (*see* **Notes 18** and **19**).
6. Apply a voltage across channels CD to implement PAGE (*see* **Notes 18**, **20**, and **21**). Separated proteins are detected by using the laser-induced fluorescence (LIF) imaging system described previously (*see* **Note 22**) *(13)*.
7. Construct 2D protein separation map by processing acquired CCD images (*see* **Note 23**).

4. Notes

1. All solutions (except where specified otherwise) are prepared using water purified from a Barnstead Nanopure Water System (Model: D11911, Dubuque, IA).
2. Device design is critical to the separation resolution. The closer the PAGE channels are, the higher IEF resolution one gets when the focused proteins in the first dimension are transferred into the second dimension. However, the narrow space between two adjacent channels will make it challenging to bond the substrate layer with the cover film.
3. Glass microscope slides should be annealed to remove the internal stress. Unannealed glass slides could lead to many defects after chemical etching, as discussed in the literature *(14)*.
4. The glass slide with best etching quality should be used to create the molding die. Any defect in the glass slide will be reproduced in the E-form, and then transferred into the plastic parts. The feature dimensions such as channel width and depth should be measured before sending the glass slide out because it is likely to be broken during its separation from the E-form.
5. The topology of the E-form is exactly the opposite of the pattern on the glass slide. For instance, a channel in the glass slide becomes a ridge in the E-form. The topology of the plastic substrate is again the negative image of the E-form. Therefore, the plastic substrate has the exactly same surface feature as the glass slide. Thousands of substrates can be made from a single E-form, making the device fabrication reproducible and cost-effective because of mass production.

6. The conditions used in the hydraulic press are experimentally determined *(8)*. The desired temperature at the platens depends on the glass transition temperature of the plastic resins used. The compression force also has an effect on the fidelity of the feature size in the plastic substrate to the E-form.
7. During lamination of the plastic substrate and the cover film, two dummy layers of Mylar® film are used to sandwich the assembly. Without these dummy layers, the textures on the rubble rollers of the laminator will be transferred onto the device. These textures will result in a rough surface, affecting optic detection. The Mylar layers are removed after lamination is completed.
8. The conditions used for lamination are determined experimentally *(8)*. The lamination temperature depends on the glass transition temperature of the plastic resin and the film used. Too high a temperature will cause the distortion of channels and other microfeatures, whereas too low a temperature will not bond the plastic substrate and the cover film. The pressure used for lamination also has an effect on the lamination.
9. Before filling a solution into a microfluidic device, all channels should be rinsed with purified water. The device should be examined under a microscope to ensure that there are no particles or fibers in the channels. These foreign materials could block the flow in microchannels.
10. The monomer solution of acrylamide and HCPK should be prepared fresh, mixed well, and kept from exposure to room light. Because HCPK is light sensitive, samples exposed to light will not flow properly into the device and polymerize fully.
11. After the device is filled with the monomer solution, it should be checked under a microscope to ensure that no bubbles exist in the channels. Bubbles will break the electrical connection during electrophoresis. Bubbles may be removed by applying continuous vacuum from one well while the other wells are filled with solutions.
12. After the device is filled with the monomer solution, the solution in the wells (A and B) and slots (C and D) should be removed. If they are not replaced, they will also polymerize, resulting in no room for introducing separation buffers later.
13. When placing the mask on the device, the plastic device should be upside down so that the cover film is on the top. The thin cover film allows less optical diffraction than the thick substrate when light passes through it during UV exposure.
14. The alignment between the mask and device is critical; it should be carried out under a microscope. In addition, the dimension of the mask line is slightly larger than the channel

width, because light-initiated radicals diffuse to monomers nearby and form a polymerized gel *(12)*.

15. Complete polymerization everywhere in the device is essential considering the fact that the collimated UV light covers a circle with a diameter of ~25 mm. The regions at the D end of channels are not exposed during the first exposure when it is focused on channel AB. In addition, the duration of the first exposure tends to be short to avoid possible radical diffusion near channel AB *(12)*. After the monomer solution in channel AB is removed, additional exposure is used to generate enough radicals to complete polymerization of the monomer solution everywhere.
16. The current detection scheme requires proteins to be labeled by a fluorescent dye. However, labeling alters the electrophoretic behavior of proteins *(15)*. To address this challenge, differential gel electrophoresis (DIGE) has been developed for proteomics study *(16, 17)*. It is accomplished by fluorescently labeling experimental and control samples with different dyes (e.g., Cy3 and Cy5). Two samples are combined and run in the same gel, thereby canceling out the variation caused by dye derivatization.
17. The device must be placed in the detection system properly. The laser line beam should focus in the gel in channels and the fluorescence emission should also be in the focus of the CCD lens. XYZ-transition stages in the detection system facilitate the alignment.
18. One high-voltage power supply (Glassman High Voltage Inc., High Bridge, NJ) is used for each dimension, and the power supplies are controlled by a computer using software written in Labview (National Instrument, Austin, TX).
19. IEF separation time depends on the voltage applied, device layout, and proteins used. The laser line can be focused on the IEF channel to monitor the dynamic process as demonstrated elsewhere *(9)*. It takes 2–5 min to complete IEF based on the images obtained *(9)*.
20. When a voltage is applied to the second dimension, SDS at the C end is electrokinetically pumped into the device, and then interacts with the focused proteins as reported *(4)*. These proteins are then transported into one corresponding CD channel or two adjacent channels depending on the location of the focused proteins and the size of the protein band *(6)*.
21. PAGE separation time depends on the electric field applied and proteins used. The PAGE separation time for four proteins is ~5 min, as demonstrated *(6)*. Therefore, the total analysis time is less than 10 min for IEF and PAGE. Compared with a typical analysis time of 1–2 days for conventional

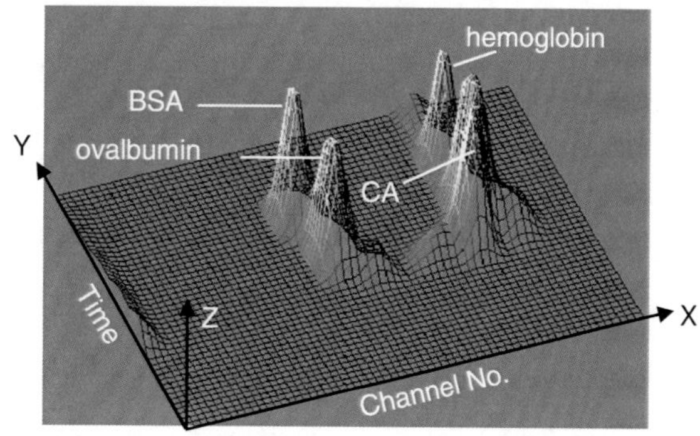

Fig. 2. Two-dimensional separation of bovine serum albumin (BSA), ovalbumin, hemoglobin, and carbonic anhydrase (CA) in the microfluidic device.

2D gel electrophoresis using an IEF strip and a PAGE slab, the microfluidic device shows an improvement of approximately two orders of magnitude.

22. The LIF imaging system *(13)* can detect all PAGE channels simultaneously. In brief, light from an argon ion laser is directed by a series of mirrors and goes through a beam expander (Newport HB-20X), which expands it into a column of light. The laser light is then focused by a cylindrical lens to form a line. This laser line is used to illuminate simultaneously all channels CD, as indicated by the detection region in **Fig. 1**. The fluorescence emission is collected by a CCD camera (Hamamatsu C4742-80-12AG) after passing through band-pass filters (Chroma Technology HQ585/40). The data collection frequency is 2 Hz.
23. The 2D image is obtained by processing the images acquired by the CCD camera over a period of time. Software ImageJ from the National Institute of Health (http://rsb.info.nih.gov/ij) is used. The map consists of the channel number in the X axis (related to the p*I* value), the migration time in the Y axis (related to the molecular weight), and the fluorescence signal in the Z axis (related to the protein concentration), as shown in **Fig. 2** *(6)*.

Acknowledgement

This work is supported by grants from the Army Research Office (48461-LS, 52924-LS-II) and the startup fund from the University of Florida. The author thanks his students and postdoctoral fellows, including Champak Das, Jiyou Zheng, Alexander Stoyanov, Carl Fredrickson, Zheng Xia, and Andrew Simon, for their work cited in this chapter.

References

1. Petricoin, E. F., Belluco, C., Araujo, R. P., Liotta, L. A. (2006). The blood peptidome: a higher dimension of information content for cancer biomarker discovery. *Nat. Rev. Cancer* **6**, 961–967
2. Service, R. F. (2001). Gold rush - High-speed biologists search for gold in proteins. *Science* **294**, 2074–2077
3. Chen, X. X., Wu, H. K., Mao, C. D., Whitesides, G. M. (2002). A prototype two-dimensional capillary electrophoresis system fabricated in poly(dimethylsiloxane). *Anal. Chem.* **74**, 1772–1778
4. Li, Y., Buch, J. S., Rosenberger, F., DeVoe, D. L., Lee, C. S. (2004). Integration of isoelectric focusing with parallel sodium dodecyl sulfate gel electrophoresis for multidimensional protein separations in a plastic microfludic network. *Anal. Chem.* **76**, 742–748
5. Tsai, S. W., Loughran, M., Karube, I. (2004). Development of a microchip for 2-dimensional capillary electrophoresis. *J. Micromech. Microeng.* **14**, 1693–1699
6. Das, C., Zhang, J., Denslow, N. D., Fan, Z. H. (2007). Integration of isoelectric focusing with multi-channel gel electrophoresis by using microfluidic pseudo-valves. *Lab Chip* 7, 1806– 1812
7. Koh, C. G., Tan, W., Zhao, M. Q., Ricco, A. J., Fan, Z. H. (2003). Integrating polymerase chain reaction, valving, and electrophoresis in a plastic device for bacterial detection. *Anal. Chem.* **75**, 4591–4598
8. Fredrickson, C. K., Xia, Z., Das, C., Ferguson, R., Tavares, F. T., Fan, Z. H. (2006). Effects of fabrication process parameters on the properties of cyclic olefin copolymer microfluidic devices. *J. Microelectromech. Syst.* **15**, 1060–1068
9. Das, C., Fan, Z. H. (2006). Effects of separation length and voltage on isoelectric focusing in a plastic microfluidic device. *Electrophoresis* **27**, 3619–3626
10. Anderson, R. C., Bogdan, G. J., Puski, A., Su, X. (1998). Genetic analysis systems: improvements and methods. In: *Solid-State Sensor and Actuator Workshop*, Transducer Research Foundation: Hilton Head Island, SC, pp. 7–10
11. Lagally, E. T., Medintz, I., Mathies, R. A. (2001). Single-molecule DNA amplification and analysis in an integrated microfluidic device. *Anal. Chem.* **73**, 565–570
12. Das, C., Fredrickson, C. K., Xia, Z., Fan, Z. H. (2007). Device fabrication and integration with photodefinable microvalves for protein separation. *Sens. Actuators A Phys.* **134**, 271–277
13. Das, C., Xia, Z., Stoyanov, A., Fan, Z. H. (2005). A laser-induced fluorescence imaging system for isoelectric focusing. *Instrum. Sci. Technol.* **33**, 379–389
14. Fan, Z. H., Harrison, D. J. (1994). Micromachining of capillary electrophoresis injectors and separators on glass chips and evaluation of flow at capillary intersections. *Anal. Chem.* **66**, 177–184
15. Rabilloud, T. (2000). Detecting proteins separated by 2-D gel electrophoresis. *Anal. Chem.* **72**, 48A–55A
16. Unlu, M., Morgan, M. E., Minden, J. S. (1997). Difference gel electrophoresis: a single gel method for detecting changes in protein extracts. *Electrophoresis* **18**, 2071–2077
17. Hoorn, E. J., Hoffert, J. D., Knepper, M. A. (2006). The application of DIGE-based proteomics to renal physiology. *Nephron Physiol.* **104**, 61–72

Chapter 5

Microfluidic Chips Designed for Measuring Biomolecules Through a Microbead-Based Quantum Dot Fluorescence Assay

Kwang-Seok Yun, Dohoon Lee, Hak-Sung Kim, and Euisik Yoon

Summary

This chapter introduces the demonstration of specific antibody detection by using a microbead-based assay with quantum dot (QD) fluorescence on a polydimethylsiloxane (PDMS) microfluidic chip. The microfluidic chip is designed to isolate a single microbead where the binding reaction of antibodies occurs on the surface. The microfluidic chip is fabricated on a glass substrate using a transparent silicone elastomer, PDMS, for easy access of monitoring and flexible gate operations to capture the single microbead. For antibody detection, a sequence of functionalized assays has been performed in the fabricated chip, including the capturing of microbeads, antibody injection into a microchamber, quantum dot injection, and fluorescence detection. Various concentrations of human IgG antibodies have been introduced to bind to a single microbead captured and isolated inside a designated microchamber in a small volume of 75 pL. Fluorescence detection is monitored using a CCD camera after the second binding with the QDs conjugated with anti-human IgG. In this experiment, a human IgG antibody concentration below 0.1 μg/mL has been successfully detected.

Key words: Microbeads, Quantum dot, Microfluidics, Biomolecules, MEMS, Single microbead, Human IgG

1. Introduction

The demand for microfluidics is increasing rapidly in the fields of miniaturized chemical analysis systems, micro total analysis systems (μ-TAS), embedded medical devices, microdosing systems, and miniaturized production systems. Current micromachining and microelectromechanical systems (MEMS) technologies

James Weifu Lee and Robert S. Foote (eds.), *Micro and Nano Technologies in Bioanalysis*,
Methods in Molecular Biology, vol. 544
DOI 10.1007/978-1-59745-483-4_5, © Humana Press, a part of Springer Science+Business Media, LLC 2009

have made it possible to implement the microfluidic functions on microchips, which are generally referred to as microfluidic chips. One of the most interesting applications of microfluidic chips is found in the detection of biomolecules because the analysis using microfluidic chips generally provides low cost, high throughput, fast analysis, and high sensitivity, so that it can be eventually implemented in micro portable systems *(1, 2)*.

Microfluidic chips are classified into two types according to their method of handling fluids on the chip. First, in droplet-based microfluidic systems, all of the fluids, including analytes, reagents, and buffer solution, are manipulated in a droplet form, being separated from each other by air or oil *(3, 4)*. This system can be imagined as a miniaturized method of the conventional analysis procedure normally done in a biological laboratory where the solutions are prepared, processed, and analyzed in glassware or microtubes. In the droplet-based microfluidic chip, a small droplet (normally from several tens of nanoliters to several microliters) substitutes for the whole solution in a glassware container *(3)*. Second, in continuous microfluidic systems, the fluids are manipulated in continuous streams. The different solutions are not isolated physically but make direct contact with each other, which causes unwanted mixing of solutions at the boundary. At the microscale, however, the mixing effects are generally negligible because the fluid stream occurs as laminar flow where little chaotic mixing is observed *(5, 6)*.

Bioanalytical methods using microbeads have been reported by many research groups because of the ease of modifying microbead surfaces for specific binding and of manipulation of the microbeads inside the microfluidic channels. The manipulations of microbeads have been achieved both in droplet-based microfluidic systems *(7)* and in continuous microfluidic systems *(8–15)*. The recent advancement of microfluidic technology has introduced a few methods of manipulating a single microbead on the microfluidic chip *(16–18)*. Yun et al. have reported on a passive manipulation of single microbeads in the microfluidic chip based on continuous microfluidics. A microbead can be passively manipulated in a fluid stream and positioned in a predetermined target microwell *(17)*. Medoro et al. used an active control method to manipulate cells (or microbeads) by using dielectrophoresis *(16, 18)*. This active control method has an advantage of providing high flexibility in bead/cell manipulation. However, it requires a complicated layout of control signals in chip implementation. This chapter describes a microfluidic chip for bioassay that allows the manipulation of microbeads down to single-bead resolution by using simple pneumatic control of a micro gate, without using complicated active manipulations.

1.1. Bioassay Using Quantum Dot Fluorescence

Quantum dots (QDs) are tiny nanocrystals that glow when stimulated by an external source such as ultraviolet (UV) light. Their size is determined by the number of atoms included in the QDs and determines their specific physical and optical properties, such as the color of light emitted. The possible applications of QDs include medical applications, lighting such as high-resolution television screens or flat panel displays, and quantum computers of the future. Especially, in optical detection for bioanalytical assays, QDs have many advantages as fluorescent reporters compared with conventional organic dyes.

The QDs provide detection results that are more sensitive because of the brighter luminescence resulting from their high extinction coefficients and quantum yields. The emission spectra of QDs can be modulated according to particle size. Therefore, simultaneous excitation of QDs in different sizes can be used for identifying various biomolecules with different emission colors, owing to their broad excitation spectra. In addition, the narrow emission bandwidth reduces interference in the detection spectrum. Because of their low sensitivity to photobleaching, QDs are optically more stable in long-term measurements. In addition, QDs can be well used in in vivo measurements because they can be easily transferred into cells after specific chemical modifications of the surfaces of the QDs. The toxicity to cells is known to be insignificant if the cells are exposed to QDs in appropriate duration *(19–21)*.

With these advantages, QDs have been widely used for the detection of biomolecules *(15, 22–32)*. QD-based Western blot technology was introduced by Bakalova et al. *(25, 26)*. Sun et al. used QDs modified with anti-human IgG to detect human IgG antibodies on protein microarrays *(27)*, reporting a detection limit of 2 μg/mL using a laser confocal scanner. Goldman et al. reported the detection of protein toxins (staphylococcal enterotoxin B, cholera toxin) by using QD-antibody conjugates *(28, 29)*. They detected the toxins coated on microtiter plates, and the lowest concentration of toxins that gave detectable signals over the background was approximately 15 ng/mL. In most cases, the detection of molecules has been performed on solid substrates such as well plates or glass plates after fixing the molecules on the surface.

This chapter introduces a QD-based bioassay using microbeads in a microfluidic chip. Use of the microfluidic chip is expected to result in high sensitivity because the efficiency of the chemical reaction will be improved by the continuous supply of reactants on the surface of the detection region—the surface of microbeads in this experiment. In the following sections, the sensitivity of single microbead-based bioassays using QD fluorescence on a microfluidic chip is discussed.

2. Materials

2.1. Chip Fabrication

1. *Substrate for mold formation.* Silicon wafers (Siltron, Gumi-si, South Korea), 4-in. diameter, single-side polished, test grade, 500-μm thick.
2. *Glass substrate.* Glass microscope slides (1 in. × 3 in.) (Corning, Corning, NY) are used as the substrate layers of the microfluidic chips.
3. *Mold structures.* SU-8 2005 and SU-8 2015 (MicroChem Corp., Newton, MA) for thin and thick structures, respectively. Store at a temperature of 4–21°C. Warm to room temperature before spin coating.
4. *Development.* SU-8 developer (MicroChem Corp.). Store at room temperature.
5. *Polydimethylsiloxane (PDMS).* Sylgard 184 (Dow Corning, Midland, MI). Store at room temperature.
6. *Surface modification.* The surface of mold structures are self-assembled monolayer (SAM)-coated with tridecafluoro-(1,1,2,2-tetrahydrooctyl)-1-trichlorosilane (United Chemical Technologies, Inc., Bristol, PA). Toxic and irritating gas.

2.2. Microbead-Based Assay

1. *Buffer.* Phosphate-buffered saline (PBS), pH 7.4. Composition: 137 mM NaCl, 2.7 mM KCl, 10 mM Na_2HPO_4, 2 mM KH_2PO_4. Sterilize by filtration (0.45 μm). Store at room temperature.
2. *Surface coating of microchannel.* 0.5% bovine serum albumin (BSA) in PBS. Store at 4°C.
3. *Reactive biotin.* EZ-Link® sulfo-NHS-LC-biotin (Pierce Biotechnology, Inc., Rockford, IL). Store desiccated at −20°C.
4. *Column.* Excess biotins are removed using D-Salt™ dextran desalting column (MW cutoff = 5 kDa, Pierce Biotechnology Inc.)
5. *Modified microbeads.* ProActive® streptavidin-coated microspheres (Bangs Laboratories, Inc., Fishers, IN), 10 μm in diameter. Store at 4°C.
6. *Quantum dot.* Qdot® 605 goat $F(ab')_2$ anti-human IgG conjugate (H + L) (Qdot, Invitrogen, Carlsbad, CA). Emission at 605 nm. Store at 4°C.

3. Methods

3.1. Fabrication of Microfluidic Chips

3.1.1. Design

Figure 1 shows the schematic view and operation of the proposed microfluidic chip for single microbead bioassay by exploiting QDs for enhanced fluorescence markers. The chip is composed of a microchamber in which a single microbead can be isolated by gate operation. The two gates located on both sides of a microchamber are operated by pneumatic pressure applied in each gate control channel. These gates are designed to be partially open with a small gap at the initial phase when there is no pressure applied to the gate control channel. The gap is small enough to block the microbead but large enough to allow liquid or reagent to flow. When negative pneumatic pressure (or vacuum pressure) is applied to the control channel, the gate is fully opened and microbeads can be introduced into the microchamber. Initially the inlet gate is closed (or partially open) to prohibit any microbeads from entering into the microchamber. To capture a single microbead, the diluted solution containing a small number of microbeads is introduced through the inlet **(Fig. 1a)**. Next, the inlet gate is opened by applying negative (or vacuum) pressure to the upper gate control channel **(Fig. 1b)**. After a microbead enters the microchamber, the gate is closed to capture the microbead. Finally, the microbeads remaining outside of the chamber are washed away by reversing the flow direction from outlet to inlet **(Fig. 1c)**.

3.1.2. Fabrication

The microfluidic chip has been implemented using two polydimethylsiloxane (PDMS) layers and a glass substrate that are bonded together. **Figure 2** shows the fabrication procedure,

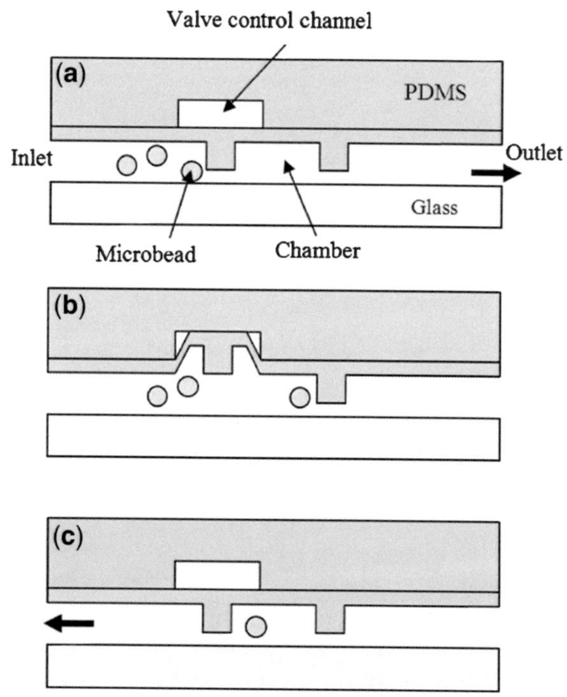

Fig. 1. Schematic view of each stage for the capturing of a single microbead. **(a)** Introduction of the beads; **(b)** valve open; and **(c)** valve closed/single bead capturing. (Reproduced from ref. *32* with permission from IOP).

which is composed of fabrication of the upper PDMS structure, fabrication of the lower PDMS layer, and the assembly process.

1. Fabrication of upper PDMS layer
   (a) Formation of PDMS replica molds **(Fig. 2a)**: A photosensitive epoxy, SU-8 (MicroChem Corp.), is patterned on silicon substrate to be used as a replica mold for PDMS. Next, the self-assembled monolayer (SAM) of tridecafluoro-(1,1,2,2-tetrahydrooctyl)-1-trichlorosilane (United Chemical Technology, Inc.) is coated on both the silicon and SU-8 surfaces for easy peeling of the PDMS layer.
   (b) Application of PDMS **(Fig. 2b)**: The PDMS prepolymer (Sylgard 184, Dow Corning) is poured onto the mold structure prepared in step *(1)* (*see* **Note 1**).
   (c) Peeling off PDMS **(Fig. 2c)**: The PDMS is cured at 80°C for 1 h and peeled off the substrate mold (*see* **Note 2**).
2. Fabrication of lower PDMS layer
   (a) Formation of PDMS replica molds **(Fig. 2d)**: SU-8 is coated with a thickness of 5 μm and exposed to ultraviolet light (UV) to define the gate structure. Without developing the first-coated SU-8 layer, the second SU-8 layer (thickness of 30 μm) is spin coated and exposed to UV to define the channel regions. Then, the unexposed region of SU-8 is developed by forming a PDMS mold for the channel and gate structures.
   (b) Application of PDMS **(Fig. 2e)**: The thin PDMS prepolymer is spin coated with a thickness of 60 μm on the fabricated SU-8 mold, followed by curing in an oven.
3. Assembly
   (a) Bonding of upper and lower PDMS layers **(Fig. 2f)**: The fabricated upper and lower PDMS layers are bonded together after surface modification by using oxygen plasma (*see* **Note 3**).
   (b) Peeling off PDMS **(Fig. 2g)**: The bonded PDMS layer is peeled off the silicon substrate, and the access holes for sample inlet, outlet, and vacuum control are formed by manual punching.
   (c) Bonding of PDMS structure and glass substrate **(Fig. 2h)**: The completed PDMS structure is bonded with the glass substrate after surface treatment by oxygen plasma.

Figure 3 shows the fabricated microfluidic chip and its magnified view of a microchamber. The size of a microchamber is 50 μm × 50 μm and the height is 30 μm. The microchamber is located between two gate valves. The initial gap of the gate valve is 5 μm to block the microbeads (10 μm in diameter) while maintaining the flow of liquid. Platinum electrodes (bright areas in magnified view) are

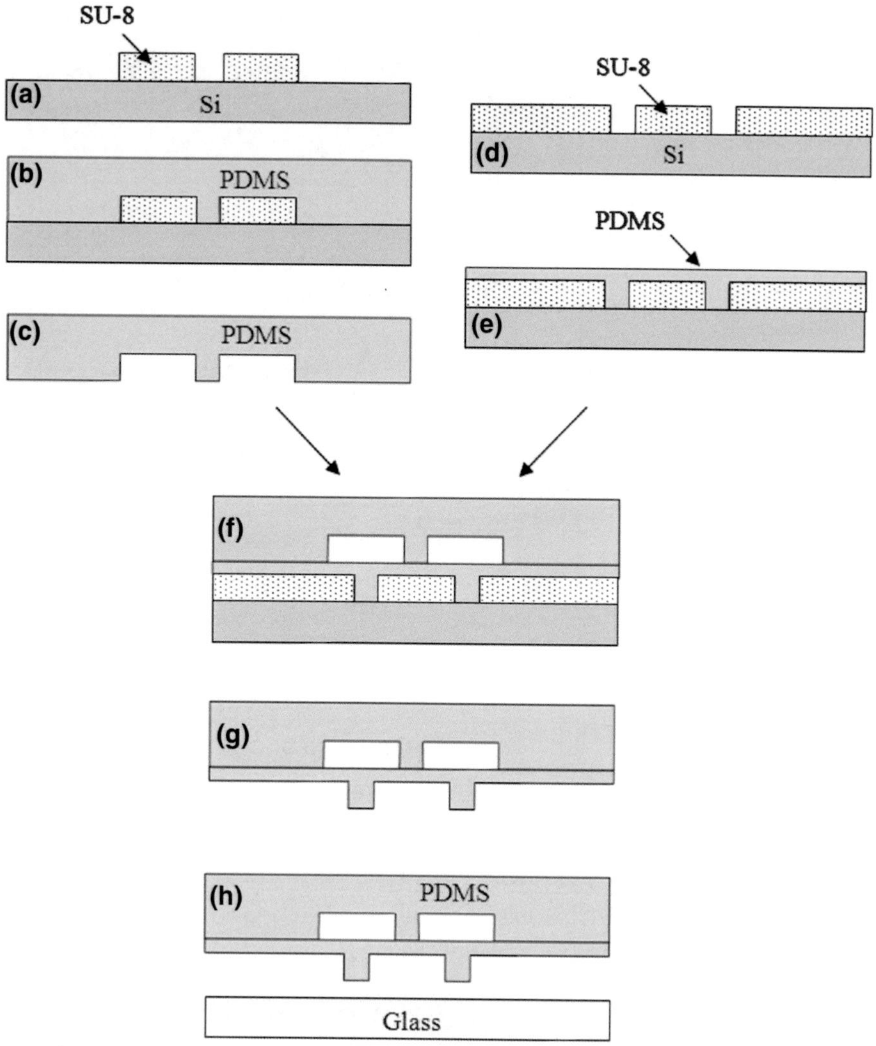

Fig. 2. Fabrication process. **(a)** SU-8 mold patterning for top PDMS plate; **(b)** PDMS pouring and curing; **(c)** peel off from mold; **(d)** SU-8 mold patterning for bottom PDMS membrane; **(e)** PDMS spin coating and curing; **(f)** bonding of top and bottom PDMS plates; **(g)** bonded PDMS layer peeled off from substrate; and **(h)** bonding with glass substrate. (Reproduced from ref. *32* with permission from IOP).

formed on the bottom surface of the microchannel and the microchamber for future applications.

3.2. Assay Using Single Microbeads on Microfluidic Chips

3.2.1. Preparation

The fabricated microfluidic chip is cleaned with ethanol, sterilized in a commercial autoclave, and fully dried in the oven (70°C) overnight (*see* **Note 4**). The surface of the microchannel is coated with 0.5% bovine serum albumin (BSA) in phosphate-buffered saline (PBS) for 2 h at room temperature to prevent nonspecific binding of proteins.

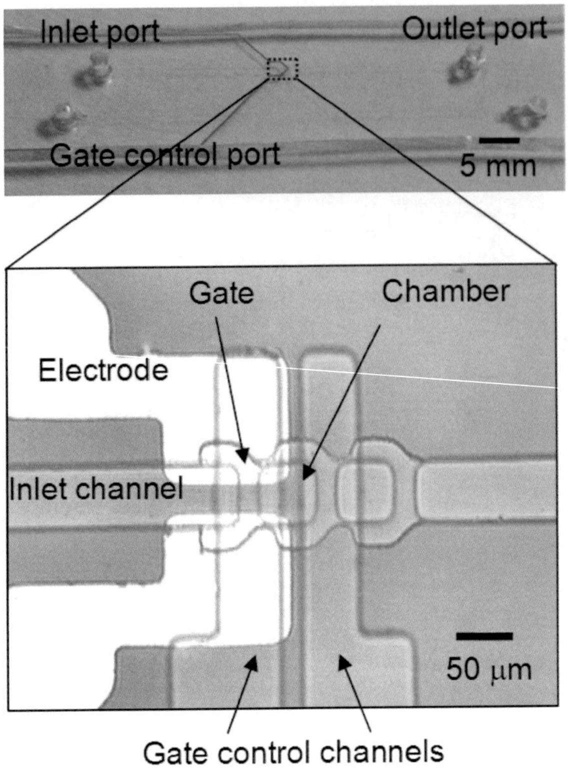

Fig. 3. Photograph of the fabricated microfluidic chip and magnified view. (Reproduced from ref. *32* with permission from IOP).

In this experiment, human IgG antibody is used to demonstrate the feasibility of QD fluorescent detection using single microbeads. The microbeads in this experiment are polystyrene microbeads (10 μm in diameter) coated with streptavidin (ProActive® streptavidin-coated microspheres, Bangs Laboratories, Inc.). For specific binding of human IgG, the surface of the microbead is coated with biotinylated protein A. The detailed procedures are as follows: sulfo-NHS-LC-biotin (Pierce Biotechnology Inc.) is added to protein A solution (10 mg/mL in PBS, 100 μL) at 12× molar excess. After 30 min reaction at room temperature, excess biotins are removed using a desalting column (Pierce Biotechnology Inc.). The biotin/protein molar ratio is determined to be ~4 by 4′-hydroxyazobenzene-2-carboxylic acid (HABA) assay (*see* **Note 5**). The biotinylated protein (b-protein A, ~56 μM) is divided into small aliquots and stored at –70°C until use. Before the immobilization of protein A, the microbeads are washed with PBS three times. The b-protein A (6 μM in PBS with 0.01% BSA and 0.05% Tween 20, 100 μL) is mixed with microbeads (1 mg) and allowed to bind to the surface of the microbeads by avidin-biotin interaction (overnight at 4°C) (*see* **Note 6**). After the binding reaction, the modified microbeads are washed with PBS three times, resuspended in PBS (with 0.01% BSA and 0.05% Tween 20), and stored at 4°C (*see* **Note 7**).

The procedure for capturing a single microbead is as follows (refer to **Fig. 1**):

(a) The solution containing microbeads is deposited in the inlet port using a pipette (*see* **Notes 8** and **9**).

(b) The introduced microbeads are transferred to the inlet gate by applying suction at the outlet port using a microsyringe **(Fig. 1a)** (*see* **Notes 10** and **11**).

(c) When a microbead arrives at the inlet gate, the gate is opened to allow the microbead to pass through and then closed immediately, while keeping the outlet gate closed. This results in capturing and isolating a single microbead inside the microchamber **(Fig. 1b)** (*see* **Note 12**).

(d) After trapping the single microbead, the rest of the microbeads loaded in the microchannel are washed out by the reverse flow of PBS from the outlet to the inlet, as explained in the previous section **(Fig. 1c)** (*see* **Note 13**).

3.2.2. Detection of Protein

Figure 4 shows the experimental protocol for antibody detection using a microbead loaded in the microfluidic chip. All of the liquids, including PBS, antibody solution, and QD-conjugated secondary antibody solution, are introduced into the microfluidic chip through a silicone tube connected to the inlet port. To maintain a steady flow at a very low flow rate, all of the media and reagents are supplied at constant pressure by applying the same height difference of liquid levels between the inlet and outlet ports (*see* **Note 14**).

1. Experimental procedures

(a) Introduction of protein **(Fig. 4a)**: After a single microbead is isolated inside the microchamber, human IgG antibody is introduced through the inlet. To examine the detection limit of the single-bead assay, the experiment is conducted on five different microfluidic chips. Five different concentrations of human IgG are injected into each chamber for 30 min. The injected concentrations are 0 μg/mL, 0.01 μg/mL, 0.1 μg/mL, 1 μg/mL, and 10 μg/mL, respectively. The injected antibody is attached to the surface of the microbead, which is coated with biotinylated protein A.

(b) Introduction of QDs **(Fig. 4b)**: First the microchamber is washed with PBS for 30 min. Then, a 5 nM solution of anti-human IgG conjugated with CdSe/ZnS QDs ($\lambda_{emission}$ = ~605 nm, Qdot Corp.) is injected into each microchamber for 30 min (5 nM is the concentration of QDs).

(c) Washing and detection **(Fig. 4c)**: The channels and microchambers are completely washed with PBS solution. The

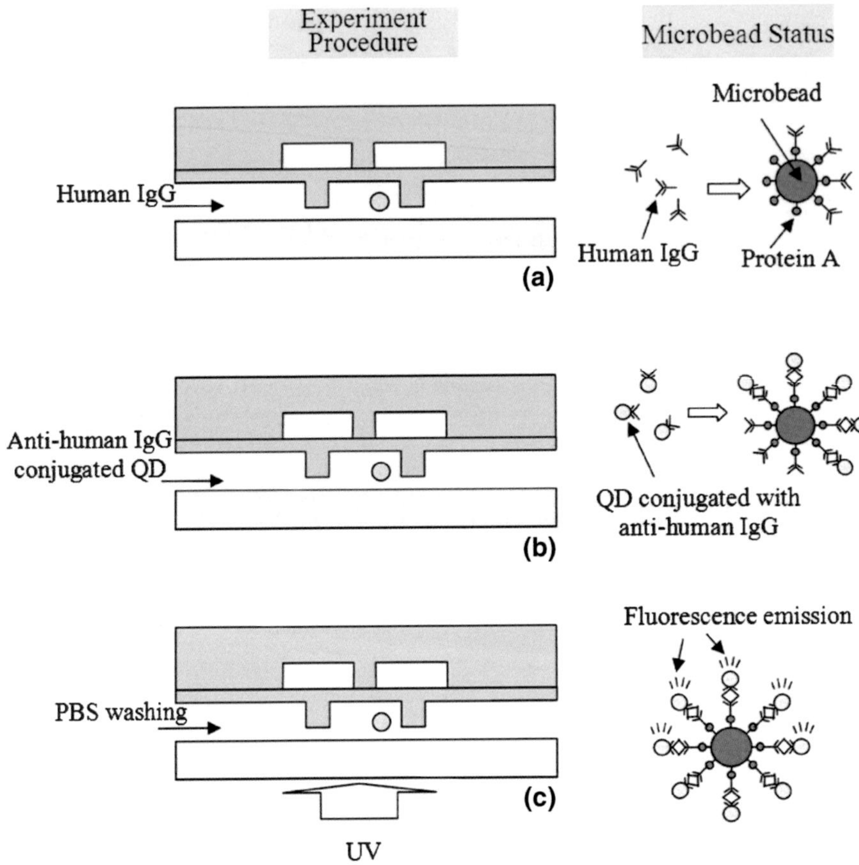

Fig. 4. The experimental procedure for microbead assay in a microchamber. **(a)** Human IgG injection after bead capturing; **(b)** anti-human IgG-conjugated QD injection after washing; and **(c)** washing with PBS and fluorescence detection. (Reproduced from ref.*32* with permission from IOP).

experimental results of this microbead-based assay are monitored using a fluorescence microscope (IX71, Olympus Co., Tokyo, Japan) with specialized filter set (XF304-2, Omega Optical, Inc., Brattleboro, VT) (*see* **Note 15**). The fluorescence images from microbeads are captured using a CCD camera and the average brightness is extracted using a graphics software (Photoshop, Adobe Systems, San Jose, CA).

3.3. Experimental Results

Figure 5 shows the experimental results presenting both the bright-field and the fluorescent images of the microchamber with a single microbead. The images are captured after the injection of human IgG followed by consequent introduction of QDs conjugated with anti-human IgG. The bright-field images show the successful isolation of a single microbead in each microchamber (**Fig. 5a**). Although purified distilled water is used for the experiment, unknown substances sometimes flow into the microfluidic

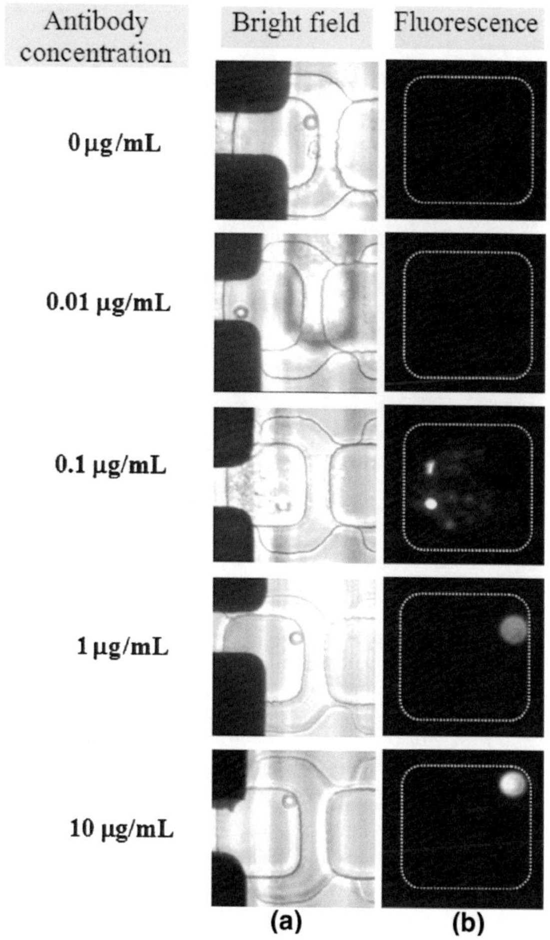

Fig. 5. Experimental results. **(a)** Bright field images of microchamber with a single microbead; and **(b)** fluorescence image according to each microbead in **(a)**. (Reproduced from ref.*32* with permission from IOP).

chip and obstruct the fluid stream, as shown in the 0.1-µg/mL case in this figure.

Figure 5b shows the experimental results as fluorescent images for various concentrations of antibody. The fluorescent image of the antibody concentration of 0 µg/mL verifies that the nonspecific binding of anti-human IgG (conjugated with QDs) to the bare microbead is negligible. The fluorescence signal can be detected when the antibody concentration becomes higher than 0.1 µg/mL and the intensity increases at higher concentrations. The normalized average intensities are plotted in **Fig. 6** as a function of antibody concentration. The deviation from the linear relationship at the concentration of 0.1 µg/mL may result from the reduced binding of antibody caused by the reduced flow rate of the reagent in the contaminated microchannel. The detection limit of this experiment is much lower than that of the previous

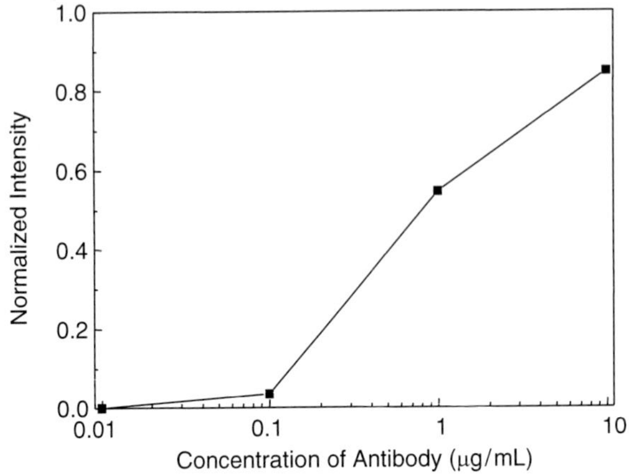

Fig. 6. Measured fluorescence intensity for various antibody concentrations. (Reproduced from ref. *32* with permission from IOP).

report, which was reviewed in **Subheading 1**. The detection limit in this experiment, 0.1 μg/mL, is approximately one order of magnitude lower than that reported in **ref.** *(27)*, in which human IgG antibodies are detected down to 2 μg/mL on a glass plate. In the microfluidic chip-based assay, binding efficiency can be improved because the reagents are continuously transferred to the surface of the microbeads, although this requires further experiments and analyses to be confirmed quantitatively. In addition, the commercial QD-conjugated anti-human IgG used in this experiment may have contributed to the increased sensitivity relative to that obtained using the QD reagents synthesized in **ref.** *(27)*.

4. Notes

1. In this step, after application of the PDMS solution onto the mold structure, the inherent bubbles are removed in a vacuum chamber.
2. The PDMS curing time can be reduced at higher temperature, but low curing temperature (<80°C) is recommended to prevent the thermal deformation of the mold structure made of SU-8.
3. To obtain the best bonding property, bonding should be completed in 5 min after surface modification.
4. The autoclave step is performed for 15 min at 120°C and 15 psi and the temperature of the drying oven is 70°C. These high-temperature sterilization and drying procedures do not disrupt the bonding between the PDMS layers and glass.

5. The assay was performed by following instructions provided by Pierce Biotechnology Inc. Generally, the addition of 12× molar excess sulfo-NHS-LC-biotin results in ~4 mol of conjugated biotin per mole of protein. Underderivatization or overderivatization can cause poor binding of biotinylated protein A to the microbead surface or a decrease in capture performance of the protein A.
6. Vigorous mechanical mixing with a stir bar should be avoided in this step, because it can damage the proteins. Mild mixing using a rocking plate is recommended.
7. The modified microbeads can be stored under these conditions for 2–3 weeks without degrading performance.
8. Before introducing microbeads into the microfluidic chip, the channel should be first filled with phosphate-buffered saline (PBS) to prevent the trapping of air in the microchannel.
9. To prevent a large number of microbeads from entering and clogging the microfluidic channel, diluted solution (~10^4/mL) is used in the experiment.
10. This step, as well as the following steps, should be done under microscope observation.
11. To transfer the microbeads to an inlet gate, the suction operation at the outlet port should be used instead of pushing the fluid to the inlet port. The pushing operation can generate pressure buildup inside the microchannel and opens the gates. This may result in failure of capturing microbeads.
12. The gate manipulation is manually performed with a syringe connected to the gate control channel through a silicone tube.
13. Again, in this step, suction at the inlet port is used.
14. The reservoir containing the injection liquid is placed approximately 40 cm higher than the outlet, which creates a constant pressure of about 4 kPa, making a continuous flow rate of 2 μL/min in the microfluidic chip.
15. The wavelength of excitation light was 470 nm and the emission wavelength from QDs was approximately 605 nm.

Acknowledgments

The authors acknowledge the financial support from the Intelligent Microsystems Program (IMP) of KIST under the "21C Frontier R&D program" and a research grant from Sogang University in 2007.

References

1. Manz, A. and Bekker, H. (1998). Microsystem Technology in Chemistry and Life Science (Springer Topics in Current Chemistry, Vol. 194), Springer, Berlin.
2. Hong, J. W. and Quake, S. R. (2003). Integrated nanoliter systems. *Nat. Biotechnol.* **21**, 1179.
3. Cho, S. K., Moon, H. and Kim, C.-J. (2003). Creating, transporting, cutting, and merging liquid droplets by electrowetting-based actuation for digital microfluidic circuits. *J. Microelectromechanical Systems* **12**, 70–80.
4. Wheeler, A. R., Kim, C.-J., Loo, J. A. and Garrell, R. L. (2004). Electrowetting-based microfluidics for analysis of peptides and proteins by matrix-assisted laser desorption/ionization mass spectrometry. *Anal. Chem.* **76**, 4833–4838.
5. Thorsen, T., Maerkl, S. J. and Quake, S. R. (2002). Microfluidic large scale integration. *Science* **298**, 580–584.
6. Koch, M., Evans, A. and Brunnschweiler, A. (ed.) (2000). Microfluidic Technology and Applications. RSP, Baldock, Hertfordshire, England, pp. 227–237.
7. Shah, G. J., Pierstorff, E., Ho, D. and Kim, C.-J. (2007). Meniscus-assisted magnetic bead trapping on ewod-based digital microfluidics for specific protein localization. *Solid-State Sensors, Actuators and Microsystems Conference (TRANSDUCERS) 2007*, 707–710.
8. Lettieri, G.-L., Dodge, A., Boer, G., de Rooij, N. F. and Verpoorte, E. (2003). A novel microfluidic concept for bioanalysis using freely moving beads trapped in recirculating flows. *Lab Chip* **3**, 34– 39.
9. Verpoorte, E. (2003). Beads and chips: new recipes for analysis. *Lab Chip* **3**, 60N-68N.
10. Choi, J.-W., Oh, K. W., Thomas, J. H., Heineman, W. R., Brian Halsall, H., Nevin, J. H., Helmicki, A. J., Henderson, H. T., Ahn, C. H. (2002). An integrated microfluidic biochemical detection system for protein analysis with magnetic bead-based sampling capabilities. *Lab Chip* **2**, 27–30.
11. Sato, K., Tokeshi, M., Kimura, H., Kitamori, T (2001). Determination of carcinoembryonic antigen in human sera by integrated bead-bed immunoasay in a microchip for cancer diagnosis. *Anal. Chem.* **73**, 1213–1218.
12. Thomas, J. H., Kim, S. K., Hesketh, P. J., Halsall, H. B. and Heineman, W. R. (2004). Bead-based electrochemical immunoassay for bacteriophage MS2. *Anal. Chem.* **76**, 2700–2707.
13. Buranda, T., Huang, J., Perez-Luna, V. H., Schreyer, B., Sklar, L. A. and Lopez, G. P. (2002). Biomolecular recognition on well-characterized beads packed in microfluidic channels. *Anal. Chem.* **74**, 1149–1156.
14. Ali, M. F., Kirby, R., Goodey, A. P., Rodriguez, M. D., Ellington, A. D., Neikirk, D. P. and McDevitt, J. T. (2003). DNA hybridization and discrimination of single-nucleotide mismatches using chip-based microbead arrays. *Anal. Chem.* **75**, 4732–4739.
15. Gao, X. and Nie, S. (2004). Quantum dot-encoded mesoporous beads with high brightness and uniformity: rapid readout using flow cytometry. *Anal. Chem.* **76**, 2406–2410.
16. Medoro, G., Manaresi, N., Leonardi, A., Altomare, L., Tartagni, M. and Guerrieri, R. (2003). A lab-on-a-chip for cell detection and manipulation. *IEEE Sensors J.* **3**, 317–24.
17. Yun, K.-S. and Yoon, E, (2005). Micro/nanofluidic device for single-cell-based assay. *Biomed. Microdevices* 7, 35–40.
18. Kim, B.-G., Yun, K.-S. and Yoon, E. (2005). Active positioning control of single cell/microbead in a micro-well array chip by dielectrophoresis. *Technical Digest of IEEE Int. Conf. on MEMS*, pp. 702–705.
19. Hoshino, A., Fujioka, K., Oku, T., Suga, M., Sasaki, Y. F., Ohta, T., Yasuhara, M., Suzuki, K. and Yamamoto, K. (2004). Physicochemical properties and cellular toxicity of nanocrystal quantum dots depend on their surface modification. *Nano Lett.* **4**, 2163–2169.
20. Chen, F., and Gerion, D. (2004). Fluorescent CdSe/ZnS nanocrystal-peptide conjugates for long-term, nontoxic imaging and nuclear targeting in living cells. *Nano Lett.* **4**, 1827–1832.
21. Derfus, A. M., Chan, W. C. W. and Bhatia, S. N. (2004). Probing the cytotoxicity of semiconductor quantum dots. *Nano Lett.* **4**, 11–18.
22. Bruchez, M., Moronne, M., Gin, P., Weiss, S. and Alivisatos, A. P. (1998). Semiconductor nanocrystals as fluorescent biological labels. *Science* **281**, 2013–2016.
23. Chan, C. W. and Nie, S. (1998). Quantum dot bioconjugates for ultrasensitive nonisotopic detection. *Science* **281**, 2016–2018.
24. Yeh, H. C., Simone, E., Zhang, C. and Wang, T.-H. (2004). Single bio-molecule detection with quantum dots in a microchannel. *IEEE Int. Conf. on MEMS*, pp. 371–374.
25. Bakalova, R., Zhelev, Z., Ohba, H. and Baba, Y. (2005). Quantum dot-based western blot technology for ultrasensitive detection of tracer proteins. *J. Am. Chem. Soc.* **127**, 9328–9329
26. Zhelev, Z., Bakalova, R., Ohba, H., Imai, Y. and Baba, Y. (2006). Uncoated, broad fluorescent,

and size-homogeneous CdSe quantum dots for bioanalyses. *Anal. Chem.* **78**, 321–330.

27. Sun, B., Xie, W., Yi, G., Chen, D., Zhou, Y. and Cheng, J. (2001). Microminiaturized immunoassays using quantum dots as fluorescent label by laser confocal scanning fluorescence detection. *J. Immunol. Methods* **129**, 85–89.
28. Goldman, E. R., Balighian, E. D., Mattoussi, H., Kuno, M. K., Mauro, J. M., Tran, P. T. and Anderson, G. P. (2002). Avidin: a natural bridge for quantum dot antibody conjugates. *J. Am. Chem. Soc.* **124**, 6378–6382.
29. Goldman, E. R., Anderson, G. P., Tran, P. T., Mattoussi, H., Charles, P. T. and Mauro, J. M. (2002). Conjugation of luminescent quantum dots with antibodies using an engineered adaptor protein to provide new reagents for fluoroimmunoassays. *Anal. Chem.* **74**, 841–847.
30. Goldman, E. R., Clapp, A. R., Anderson, G. P., Uyeda, H. T., Mauro, J. M., Medintz, I. L. and Mattoussi, H. (2004). Multiplexed toxin analysis using four colors of quantum dot fluororeagents. *Anal. Chem.* **76**, 684–688.
31. Yun, K.-S., Lee, D., Kim, M. S., Kim, H.-S., Lee, G. M. and Yoon, E. (2004). High-throughput bio-molecule detection using microbead-based assay with quantum dot fluorescence in a microfluidic chip. *Proceedings of the International Conference on Miniaturized Systems for Chemistry and Life Sciences (Micro TAS'04)*, pp. 222–224.
32. Yun, K.-S., Lee, D., Kim, H.-S. and Yoon, E. (2006). A microfluidic chip for measurement of bio molecules using microbead-based quantum dot fluorescence assay. *Meas. Sci. Technol.* **17**, 3178–3183.

Chapter 6

DNA Focusing Using Microfabricated Electrode Arrays

Faisal A. Shaikh and Victor M. Ugaz

Summary

Focusing methods are a key component in many miniaturized DNA analysis systems because they enable dilute samples to be concentrated to detectable levels while being simultaneously confined within a specified volume inside the microchannel. In this chapter, we describe a focusing method based on a device design incorporating arrays of addressable on-chip microfabricated electrodes that can locally increase the concentration of DNA in solution by electrophoretically sweeping it along the length of a microchannel. By applying a low voltage (~1–2 V) between successive pairs of neighboring electrodes, the intrinsically negatively charged DNA fragments are induced to migrate toward and collect at each anode, thereby allowing the quantity of accumulated DNA to be precisely metered. We have characterized the kinetics of this process, and found the response to be robust over a range of different sample compositions and buffer environments.

Key words: Focusing, Injection, Concentration, Purification, DNA, Microfabrication, Microfluidics

1. Introduction

Miniaturized systems continue to be developed that are capable of performing a variety of increasingly sophisticated bioanalysis assays. However, as these devices become smaller, the correspondingly minute sample and reagent quantities become more challenging to detect and manipulate. In the case of DNA, for example, concentration enhancement has been traditionally accomplished by either physical processes (e.g., centrifugation) or chemical amplification (e.g., via polymerase chain reaction [PCR]). However, these processes are often difficult to miniaturize and interface with other components in an integrated lab-on-a-chip format. Consequently, many bioassays critically depend on the ability to

James Weifu Lee and Robert S. Foote (eds.), *Micro and Nano Technologies in Bioanalysis*, Methods in Molecular Biology, vol. 544
DOI 10.1007/978-1-59745-483-4_6, © Humana Press, a part of Springer Science+Business Media, LLC 2009

focus an analyte by *concentrating* it to a detectable level while simultaneously *confining* it within a well-defined spatial zone.

Sample focusing is particularly important in the design of miniaturized gel electrophoresis systems for performing size-selective fractionation of charged biomolecules (DNA, proteins, etc.). Injection of a nonconcentrated and unfocused sample zone not only requires a long separation distance to distinguish each component, but the corresponding signal from each species may fall below the detectable range as the zones spread by diffusion. Injection of a concentrated and focused sample zone, on the other hand, allows each component to be detected in a considerably shorter separation distance.

A variety of electrokinetic concentrators has been developed based on an arrangement wherein two microchannels intersect to form in a "T"-shaped geometry *(1)*. In this configuration, analytes are electrokinetically transported through an injection channel until they reach an intersection with the analysis channel. At this point, the voltage is switched such that only the sample volume confined within this intersection point is injected into the analysis channel. Most of these approaches use off-chip electrodes with potentials in the 1–2 kV range. Subsequent refinements include the cross *(2, 3)* and double-T *(4)* designs that offer greater control over the size of the compacted zone. In addition to geometry, the electric field in the vicinity of the channel intersection can also be manipulated to further define the size and shape of the compacted zone *(5)*. Further variations on the basic "T" configuration have also been demonstrated, including the incorporation of porous membranes *(6, 7)*, thin film electrodes *(8)*, optical gating *(9, 10)*, and locally narrow channel geometries *(11)*. Electrodes positioned within a microchannel have been used to focus and extract DNA samples *(12–15)*, and stacking at the interface between the sample and a background buffer has been performed in T-injector geometries *(16–20)*. Other techniques using bulk flow *(21)*, capillary *(22)*, diffusion-based *(23)*, pressure-driven *(24)*, hydrodynamic *(25)*, and dielectrophoretic trapping *(26–28)* processes have also been demonstrated. The use of nanocapillary array interconnects *(29)* and nanoporous membranes *(30–32)* to concentrate DNA have also been explored. Despite these advances, it remains challenging to achieve simultaneous concentration and focusing of analytes in a microchannel environment.

In the technique described here, we take a different approach that involves using arrays of addressable microfabricated electrodes positioned along the floor of a microchannel to locally increase the concentration of DNA in solution by electrophoretically sweeping it from one electrode to the next *(33)*. The use of very low potentials and currents (e.g., equivalent to a single AA-size battery) combined with the cost benefits of photolithographic fabrication makes this technology broadly applicable as a highly efficient mechanism to achieve sample focusing, as well as to "digitally" collect and meter precise quantities of charged biomolecules in miniaturized bioanalysis systems.

2. Materials and Equipment

2.1. Reagents

Electrode capture experiments were carried out using a 100-bp double-stranded DNA ladder as a test sample (Bio-Rad Laboratories, Inc., Hercules, CA) (*see* **Note 1**). The DNA ladder was suspended in Tris-borate-EDTA (TBE) (Bio-Rad Laboratories, Inc.) and histidine (Sigma-Aldrich Co., St. Louis, MO) buffers. An intercalating dye (YOYO-1; Invitrogen/Molecular Probes, Carlsbad, CA) was used to fluorescently label the DNA, and β-mercaptoethanol (BME) (Sigma-Aldrich Co.) was added to inhibit photobleaching (intensity measurements were calibrated to account for residual photobleaching). Fluorescently labeled DNA samples were prepared by adding the as-supplied DNA ladder solution to 1 mM YOYO-1 intercalating dye in a 2:1 v/v ratio. After incubation for several minutes at room temperature, the suspension buffer was extracted using a vacuum centrifuge evaporator. The sample was then resuspended in the appropriate buffer system, and the desired amount of BME was added.

2.2. Imaging and Detection

The electrode capture process was imaged using an Axioskop-2 Plus microscope (Carl Zeiss, Thornwood, NY) with an Orca-ER digital CCD camera (Hamamatsu, Hamamatsu City, Japan). Openlab software (Improvision, Inc., Waltham, MA) was used to acquire the intensity data from the fluorescence measurements. For flow visualization studies, a dilution of 1-μm diameter carboxylate-modified polystyrene microspheres (FluoSpheres®, Invitrogen/Molecular Probes) was added to serve as tracers.

2.3. Microdevice Fabrication

The electrode array chips consist of three components: a silicon device (~7.5 mm × 35 mm), a glass microchannel (~3.9 mm × 33.5 mm), and a printed circuit (PC) board (~4 cm × 6.4 cm), as shown in **Fig. 1a**. Silicon wafers (500-μm thick, 150-mm diameter, ‹100›, with a 5,000-Å-thick oxide layer [University Wafer, South Boston, MA]) were cleaned using a reactive ion etcher, spin coated with hexamethyldisilazane (J.T. Baker, Phillipsburg, NJ) and positive photoresist SPR 220-7.0 (Rohm and Haas Electronic Materials LLC, Philadelphia, PA), patterned, and developed using MF-319 developer (Rohm and Haas Electronic Materials LLC). Gold electrodes were fabricated by depositing a 500-Å layer of chromium followed by a 1,000-Å layer of gold using a thermal evaporator. Glass microchannels were fabricated by depositing a 600-Å chromium layer followed by a 4,000-Å gold layer on glass wafers (borofloat, 500-μm thick, 150-mm diameter [Precision Glass & Optics, Santa Ana, CA]) that had been cleaned using a reactive ion etcher. The wafers were then spin coated with positive photoresist SPR 220-7.0, patterned, developed using MF-319, and hard baked. After immersion for 10 min each in gold etchant (Transene Co., Danvers, MA) and chromium etchant (Cyantek Inc., Fremont, CA), the exposed glass was then etched using a freshly prepared 7:3 v/v solution of hydrofluoric and nitric acids (etch rate ~5 μm/min) to a channel depth of 45 μm (width = 275 μm). After dicing, fluidic access holes were drilled using an electrochemical discharge process. Silicon devices were mounted on PC boards and wire bonded to provide electrical connections. The wire bonds were encapsulated with epoxy for protection. Glass microchannels were then bonded to the silicon device using a UV curable optical adhesive (SK-9 Lens Bond; Summers Laboratories, Collegeville, PA).

3. Methods

3.1. Microdevice Operation

After injecting a DNA sample into a microchannel **(Fig. 1a)**, the concentration process begins by applying a 1 V DC potential between a pair of electrodes in the array (*see* **Note 2**). Under these conditions, a sufficient electrostatic driving force is developed to direct DNA migration while maintaining a voltage below the threshold for bubble formation arising from hydrolysis reactions at the electrode surfaces *(34, 35)*. The DNA present between these two electrodes, being intrinsically polyanionic, migrates toward the anode in response to the electric field and becomes trapped or "captured" there **(Fig. 1b)**. The incremental increase in concentration is defined by the quantity of DNA initially contained between a given electrode pair, and can be precisely controlled by the interelectrode spacing and channel cross-section dimensions.

Now, if a further increase in DNA concentration is desired, the anode can be switched to the next electrode in the array **(Fig. 1d)** so that the captured DNA is released and begins to migrate toward the new anode. Once the released DNA is completely swept away from the surface of the second electrode, the potential applied to the first electrode is switched to the second electrode so that it becomes the cathode **(Fig. 1e)**. As this sequential capture–release process is repeated, the collected DNA experiences a "snowballing" effect whereby enrichment occurs in discrete intervals as a consequence of migration from one electrode to the next. In addition to an increase in concentration,

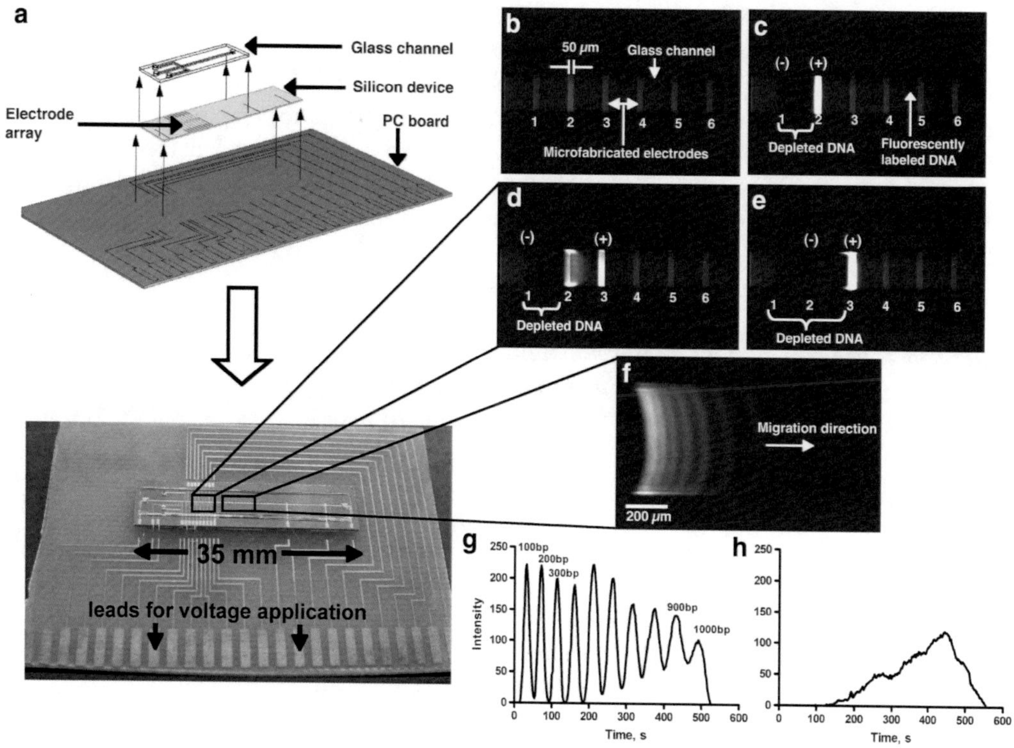

Fig. 1. (**a**) Schematic illustration of microdevice construction and assembly. (**b**) Image of the electrode array within the device (channel dimensions are 275-μm wide by 45-μm tall, electrodes are 50-μm wide with a 225-μm edge-to-edge spacing). A 100-bp dsDNA ladder (12 μg/mL in 1× TBE buffer with 10%, v/v BME) fluorescently labeled with YOYO-1 intercalating dye is loaded inside the microchannel. (**c**) A potential of 1 V is applied to the first two electrodes in the array, resulting in migration toward the anode (electrode #2) of DNA initially between the two electrodes, after which it becomes "captured" at the anode (migration is from left to right). (**d**) The anodic potential is switched to electrode #3 while the cathodic potential is switched to electrode #2, resulting in release of the captured DNA and migration toward electrode #3. (**e**) DNA initially between electrodes #1 and #3 becomes captured at electrode #3. This process is repeated until a desired concentration is achieved. (**f**) Electrophoretic separation of a 100-bp dsDNA ladder sample initially at 6 μg/mL (1× TBE buffer with 10%, v/v BME) after concentration and injection into a gel matrix photopolymerized inside the microchannel immediately downstream of the final electrode in the capture array (5% T crosslinked polyacrylamide gel, $E = 23$ V/cm). (**g**) All fragments in the DNA ladder are resolved in a separation length of 3 mm. (**h**) Without focusing and concentration before injection into the gel, fragments are unresolvable at the same separation length. The total area under the peaks in (**g**) is approximately double that in (**h**), as expected because the quantity of DNA injected in the focused/concentrated separation of (**g**) was nearly double that in the unfocused separation of (**h**) (based on the initial sample concentration and volume contained between electrodes in each case). Reproduced with permission from **ref.** *33*; copyright 2006, The National Academy of Sciences of the USA.

the captured sample zone becomes focused as it assumes a size and shape nearly identical to that of the anode (**Fig. 1c**).

After a sufficient number of capture–release cycles have been performed to raise the concentration to a desired level, the resulting highly focused and concentrated sample can be dispensed for subsequent analysis. The utility of this scheme is illustrated by using it to achieve enhanced sensitivity in microchip-based gel electrophoresis of DNA at concentrations that are initially below a detectable level (**Fig. 1f**). After completing the sequential capture–release process, the sample is injected into the gel by switching the applied potential to a set of on-chip electrodes spanning the analysis channel. This focusing process allows all bands in a fluorescently labeled double-stranded DNA ladder to be resolved within a distance of 3 mm from the gel interface (**Fig. 1g**), whereas the fragments are not easily resolvable in an identical sample that was not prefocused and concentrated by the capture–release process (**Fig. 1h**).

3.2. Simplified Model

A basic physical understanding of the electrode capture process can be obtained by constructing a simplified theoretical framework subject to the following approximations: *(1)* electric field is unaffected by the charge on the DNA and its movement in solution, *(2)* reaction and generation of chemical entities are negligible, *(3)* electroosmosis (if present) can be represented by an equivalent current flux in the direction of the electroosmotic flow, and *(4)* conductivity, charge, temperature, and mobility values remain constant. The fundamental physics associated with ionic species transport can then be modeled by coupling the electrokinetic flow to the electric field generated in the microchannel.

3.2.1. Electrokinetic Flow

The time derivative of the species concentration is related to the divergence of the total species flux due to diffusion and electromigration. The migration term is proportional to the charge, mobility, and electric field intensity at that location and the diffusive flux depends on the diffusion coefficient and the concentration gradient.

$$\frac{\partial c_i}{\partial t} + \nabla \cdot (-D_i \nabla c_i - z_i \mu_i F c_i \nabla V) = 0. \tag{1}$$

Here, c_i = ionic species concentration (mol/m^3), D_i = diffusivity (m^2/s), z_i = ionic charge (C/mol), μ_i = ion mobility (m^2/V s), F = Faraday constant = 96,487 C/mol, and V = potential (V).

3.2.2. Electric Field

The conservation law (**Eq. 1**) is then coupled to the electric field profile, where the electric field is represented as a gradient of the potential function.

$$E = -\nabla V. \tag{2}$$

The current density in the conductive media can be expressed as

$$J = \sigma E + J^e. \tag{3}$$

Here, J = current density, σ = conductivity, and J^e = externally generated current density. The static form of this equation of continuity, along with **Eq. 2**, gives

$$\nabla \cdot J = -\nabla \cdot (\sigma \nabla V - J^{e}) = 0. \qquad (4)$$

We have applied this model to the single electrode capture process using a two dimensional (2-D) finite element approach with the software package COMSOL Multiphysics (formerly FEMLAB). The domain geometry is defined to represent a 2-D side-view profile of a pair of capture–release electrodes. In this view, the electrodes are located on the bottom edge (i.e., at the floor) of the rectangular section of the microchannel **(Fig. 2a)**, with the DNA suspended in buffer in the enclosed channel above it. Zero species flux and electrically insulating boundary conditions are applied at all surfaces except the two electrodes, where a 1 V potential difference is imposed. Species mobilities are expressed as $\mu_i = D_i/RT$. The relatively high characteristic values of DNA mobility, however, introduce convergence problems because fluxes become dominated by migration (convective flux). This is dealt with by introducing an artificial diffusion component (streamline, Petrov–Galerkin: compensated, tuning parameter = 0.5) to stabilize the solution. Characteristic electrochemical data are obtained from literature *(12, 36, 37)* **(Table 1)**.

Despite its relative simplicity, this model is capable of capturing many features of the electrode compaction process **(Fig. 2b–d)**. As a potential is applied, the DNA migrates toward the anode, following the field lines that are established between electrodes. The enrichment in DNA concentration at the anode agrees with corresponding observations of increased fluorescence intensity, and the DNA concentration can be quantified to obtain kinetic

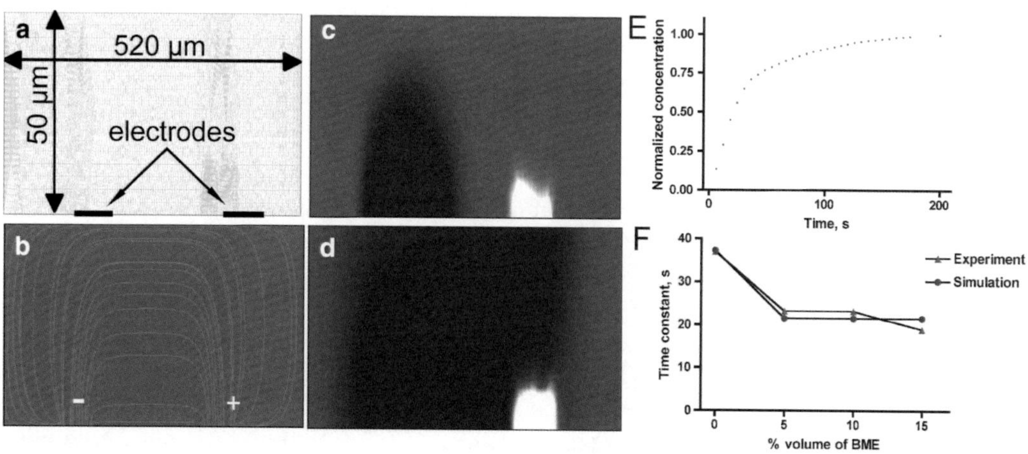

Fig. 2. Electrode capture process simulated for DNA initially at 12 μg/mL in 1× TBE buffer under a 1 V applied potential showing: (**a**) mesh generated within the geometry, and (**b**) electric field lines in the geometry. Simulation results depicting the migration and capture of DNA after elapsed times of (**c**) 10 s and (**d**) 200 s. (**e**) Simulated transient enrichment in DNA concentration at the anode. (**f**) Comparison of simulated capture time constants with corresponding experimental results (parameter values of $f_1 = 0.168$ and $f_2 = 2.3$ were used here).

Table 1
Characteristic species parameter values used in simulation studies

Sample component	Mobility (m^2/V s)	Diffusivity (m^2/s)	Concentration (g/L)
Tris	2.90×10^{-8}	8.00×10^{-10}	15.75
Borate	3.90×10^{-9}	1.00×10^{-10}	2.78
DNA	1.50×10^{-8}	5.00×10^{-12}	0.012

parameters, as is done experimentally. Quantitative kinetic predictions, however, require consideration of the interplay between the parameters in **Table 1**. From the experimentally obtained values of the time constant, we notice that the kinetics are accelerated (higher time constants) at lower TBE concentrations (up to 1×). The addition of BME further accelerates the capture. In case of TBE, the mobility of DNA exhibits a dependence on buffer concentration *(38)*. To incorporate this behavior into the model, we define an effective mobility for the DNA within terms of two adjustable parameters, f_1 and f_2; where f_1 accounts for the change in mobility with TBE concentration and f_2 for the change observed with addition of BME ($\mu_{eff} = \mu f_1 f_2$). The values of the parameters are determined from experimental data, yielding $f_1 = 0.073x^2 - 0.371x + 0.466$ and $f_2 = -7.7848x^2 + 32.892x - 22.812$, where x = TBE concentration (viz., 1, 2, 3 for 1×, 2×, 3× buffers, respectively). The ability of this relatively simple model to capture the essential features (both qualitatively and quantitatively) associated with the electrode capture process provides further evidence that electrophoretic transport of DNA is the dominant process **(Fig. 2e, f)**.

3.3. On-Chip Buffer Exchange

In addition to DNA focusing and metering, the on-chip electrode array format allows additional functionality not possible with conventional electrokinetic approaches. For example, this scheme makes it possible to completely replace the sample buffer by immobilizing a DNA plug on the surface of an electrode while simultaneously flowing a different desired buffer over it **(Fig. 3)**. The flow displaces the original sample buffer so that the DNA can then be resuspended in the new buffer environment when the applied voltage is switched off. This "buffer exchange" capability is useful, for example, in applications like post-PCR analysis where subsequent reaction steps may require much different buffering conditions. By applying a superimposed bulk hydrodynamic flow whose strength does not exceed the electrostatic forces confining the captured DNA at the anode, a new buffer solution can be

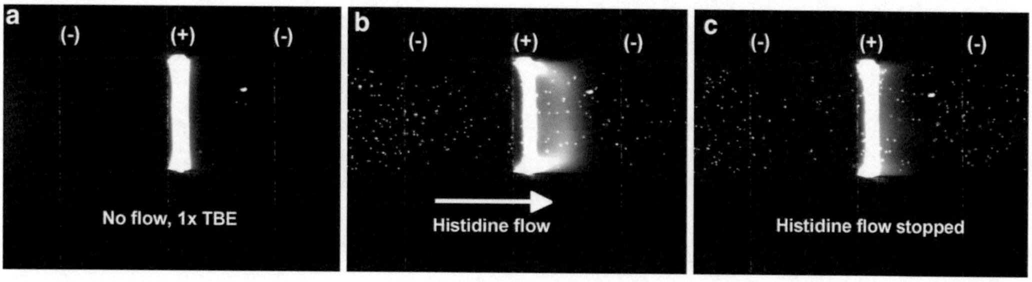

Fig. 3. Illustration of a buffer-exchange process performed by superimposing a bulk hydrodynamic flow over a captured DNA sample. (**a**) DNA sample initially at 12 μg/mL in 1× TBE buffer with 10%, v/v BME is captured at the central anode by dual focusing. (**b**) A hydrodynamic flow is introduced by placing a drop of 50 mM histidine (no BME added) labeled with carboxylated polystyrene microsphere tracers at the microchannel inlet, generating velocities ranging from 11 to 18 μm/s (flow direction is from left to right). Depending on the flow velocity, the captured DNA is partially swept downstream from the electrode surface but remains confined between cathodes (2.4 V potential). (**c**) Captured DNA returns to the anode when the flow stops. The sample can then be released and resuspended in the new buffer environment when the potential is switched off.

introduced into the microchannel without allowing the DNA to be swept away by the flow.

3.4. Conclusion

Sample preconcentration and focusing components are likely to become increasingly important as microfluidic systems become smaller and more complex. The use of microfabricated electrode arrays located inside a microchannel offers enhanced flexibility relative to conventional electrokinetic approaches that use off-chip electrodes. We demonstrate this in a design that allows DNA to be simultaneously concentrated and focused using very low potentials and currents (e.g., equivalent to a single AA-size battery). In addition to simultaneous concentration and focusing, this system can also be used to precisely meter and temporarily localize DNA at any electrode in the array. When coupled with a bulk pressure-driven flow, this scheme can also be used to collect and resuspend a DNA sample in a new buffer environment.

4. Notes

1. We have also tested the electrode-mediated focusing process with longer double-stranded DNA (dsDNA) fragments (e.g., λ-DNA) and found it to perform as robustly and reproducibly as with the shorter 100-bp ladder samples described above. This scheme can also be used to collect and focus single-stranded DNA fragments, but the process displays somewhat slower kinetics compared with dsDNA.

2. The likelihood of bubble formation caused by electrolysis increases greatly as the potential applied between electrodes increases beyond ~1.5 V, with higher potentials resulting in more rapid bubble formation. This bubble formation is undesirable because it will completely block the microchannel. Thus, the device design is dictated by the requirement to establish locally high electric fields at potentials below the threshold for bubble formation.

Acknowledgments

We thank Prof. Mark A. Burns for helpful discussions and assistance with fabricating some of the microelectrode array chips. This work was supported in part under a grant from the National Science Foundation (CTS-0554108).

References

1. Manz, A., Harrison, D.J., Verpoorte, E.M.J., Fettinger, J.C., Paulus, A., Ludi, H., Widmer, H.M. (1992). Planar chips technology for miniaturization and integration of separation techniques into monitoring systems – capillary electrophoresis on a chip. *J. Chromatogr.* **593**, 253–258.
2. Harrison, D.J., Fluri, K., Seiler, K., Fan, Z.H., Effenhauser, C.S., Manz, A. (1993). Micromachining a miniaturized capillary electrophoresis-based chemical-analysis system on a chip. *Science* **261**, 895–897.
3. Harrison, D.J., Manz, A., Fan, Z.H., Ludi, H., Widmer, H.M. (1992). Capillary electrophoresis and sample injection systems integrated on a planar glass chip. *Anal. Chem.* **64**, 1926–1932.
4. Effenhauser, C.S., Manz, A., Widmer, H.M. (1993). Glass chips for high-speed capillary electrophoresis separations with submicrometer plate heights. *Anal. Chem.* **65**, 2637–2642.
5. Jacobson, S.C., Hergenroder, R., Koutny, L.B., Warmack, R.J., Ramsey, J.M. (1994). Effects of injection schemes and column geometry on the performance of microchip electrophoresis devices. *Anal. Chem.* **66**, 1107–1113.
6. Khandurina, J., Jacobson, S.C., Waters, L.C., Foote, R.S., Ramsey, J.M. (1999). Microfabricated porous membrane structure for sample concentration and electrophoretic analysis. *Anal. Chem.* **71**, 1815–1819.
7. Song, S., Singh, A.K., Kirby, B.J. (2004). Electrophoretic concentration of proteins at laser-patterned nanoporous membranes in microchips. *Anal. Chem.* **76**, 4589–4592.
8. Lin, Y.C., Ho, H.C., Tseng, C.K., Hou, S.Q. (2001). A poly-methylmethacrylate electrophoresis microchip with sample preconcentrator. *J. Micromech. Microeng.* **11**, 189–194.
9. Lapos, J.A., Ewing, A.G. (2000). Injection of fluorescently labeled analytes into microfabricated chips using optically gated electrophoresis. *Anal. Chem.* **72**, 4598–4602.
10. Roddy, E.S., Lapos, J.A., Ewing, A.G. (2003). Rapid serial analysis of multiple oligonucleotide samples using optically gated injection. *J. Chromatogr. A* **1004**, 217–224.
11. Zhang, C.X., Manz, A. (2001). Narrow sample channel injectors for capillary electrophoresis on microchips. *Anal. Chem.* **73**, 2656–2662.
12. Brahmasandra, S.N., Ugaz, V.M., Burke, D.T., Mastrangelo, C.H., Burns, M.A. (2001). Electrophoresis in microfabricated devices using photopolymerized polyacrylamide gels and electrode-defined sample injection. *Electrophoresis* **22**, 300–311.
13. Inoue, A., Ito, T., Makino, K., Hosokawa, K., Maeda, M. (2007). I-shaped microchannel array chip for parallel electrophoretic analyses. *Anal. Chem.* **79**, 2168–2173.
14. Lin, R., Burke, D.T., Burns, M.A. (2003). Selective extraction of size-fractionated DNA

samples in microfabricated electrophoresis devices. *J. Chromatogr. A* **1010**, 255–268.

15. Lin, R., Burke, D.T., Burns, M.A. (2005). Addressable electric fields for size-fractionated sample extraction in microfluidic devices. *Anal. Chem.* 77, 4338–4347.
16. Jacobson, S.C., Ramsey, J.M. (1995). Microchip electrophoresis with sample stacking. *Electrophoresis* **16**, 481–486.
17. Jung, B., Bharadwaj, R., Santiago, J.G. (2003). Thousandfold signal increase using field-amplified sample stacking for on-chip electrophoresis. *Electrophoresis* **24**, 3476–3483.
18. Kim, D.K., Kang, S.H. (2005). On-channel base stacking in microchip capillary gel electrophoresis for high-sensitivity DNA fragment analysis. *J. Chromatogr. A* **1064**, 121–127.
19. Kurnik, R.T., Boone, T.D., Nguyen, U., Ricco, A.J., Williams, S.J. (2003). Use of floating electrodes in transient isotachophoresis to increase the sensitivity of detection. *Lab Chip* **3**, 86–92.
20. Vazquez, M., McKinley, G., Mitnik, L., Desmarais, S., Matsudaira, P., Ehrlich, D. (2002). Electrophoresis using ultra-high voltages. *J. Chromatogr. B* **779**, 163–171.
21. Huang, X.J., Pu, Q.S., Fang, Z.L. (2001). Capillary electrophoresis system with flow injection sample introduction and chemiluminescence detection on a chip platform. *Analyst* **126**, 281–284.
22. Smith, E.M., Xu, H.W., Ewing, A.G. (2001). DNA separations in microfabricated devices with automated capillary sample introduction. *Electrophoresis* **22**, 363–370.
23. Slentz, B.E., Penner, N.A., Regnier, F. (2002). Sampling BIAS at channel junctions in gated flow injection on chips. *Anal. Chem.* **74**, 4835–4840.
24. Lee, N.Y., Yamada, M., Seki, M. (2004). Pressure-driven sample injection with quantitative liquid dispensing for on-chip electrophoresis. *Anal. Sci.* **20**, 483–487.
25. Backofen, U., Matysik, F.M., Lunte, C.E. (2002). A chip-based electrophoresis system with electrochemical detection and hydrodynamic injection. *Anal. Chem.* **74**, 4054–4059.
26. Asbury, C.L., Diercks, A.H., van den Engh, G. (2002). Trapping of DNA by dielectrophoresis. *Electrophoresis* **23**, 2658–2666.
27. Asbury, C.L., van den Engh, G. (1998). Trapping of DNA in nonuniform oscillating electric fields. *Biophys. J.* **74**, 1024–1030.
28. Chou, C.-F., Tegenfeldt, J.O., Bakajin, O., Chan, S.S., Cox, E.C., Darnton, N., Duke, T., Austin, R.H. (2002). Electrodeless dielectrophoresis of single- and double-stranded DNA. *Biophys. J.* **83**, 2170–2179.
29. Cannon, D.M., Kuo, T.C., Bohn, P.W., Sweedler, J.V. (2003). Nanocapillary array interconnects for gated analyte injections and electrophoretic separations in multilayer microfluidic architectures. *Anal. Chem.* **75**, 2224–2230.
30. Dai, J.H., Ito, T., Sun, L., Crooks, R.M. (2003). Electrokinetic trapping and concentration enrichment of DNA in a microfluidic channel. *J. Am. Chem. Soc.* **125**, 13026–13027.
31. Dhopeshwarkar, R., Sun, L., Crooks, R.M. (2005). Electrokinetic concentration enrichment within a microfluidic device using a hydrogel microplug. *Lab Chip* **5**, 1148–1154.
32. Wang, Y.-C., Stevens, A.L., Han, J. (2005). Million-fold preconcentration of proteins and peptides by nanofluidic filter. *Anal. Chem.* 77, 4293–4299.
33. Shaikh, F.A., Ugaz, V.M. (2006). Collection, focusing, and metering of DNA in microchannels using addressable electrode arrays for portable low-power bioanalysis. *Proc. Natl Acad. Sci. U. S. A.* **103**, 4825–4830.
34. Heller, M.J. (2002). DNA microarray technology: devices, systems, and applications. *Annu. Rev. Biomed. Eng.* **4**, 129–153.
35. Heller, M.J., Forster, A.H., Tu, E. (2000). Active microelectronic chip devices which utilize controlled electrophoretic fields for multiplex DNA hybridization and other genomic applications. *Electrophoresis* **21**, 157–164.
36. Beckers, J.L. (2000). Window optimization in isotachophoresis superimposed on capillary zone electrophoresis. *Electrophoresis* **21**, 2788–2796.
37. Stellwagen, N.C., Gelfi, C., Righetti, P.G. (1997). The free solution mobility of DNA. *Biopolymers* **42**, 687–703.
38. Stellwagen, E., Stellwagen, N.C. (2003). Probing the electrostatic shielding of DNA with capillary electrophoresis. *Biophys. J.* **84**, 1855–1866.

Chapter 7

Solid-State Nanopore for Detecting Individual Biopolymers

Jiali Li and Jene A. Golovchenko

Summary

Solid-state nanopores have been fabricated and used to characterize single DNA and protein molecules. Here we describe the details on how these nanopores were fabricated and characterized, the nanopore sensing system setup, and protocols of using these nanopores to characterize DNA and protein molecules.

Key words: Solid-state nanopore, Ionic current blockage, DNA, Protein

1. Introduction

1.1. Background

Detecting the change in electrical conductance or in resistance of a small aperture in ionic solution containing particles has been used to count and size particles *(1, 2)* for decades. Starting in the 1990s, protein pores or ion channels suspended in lipid membrane have been used to characterize single DNA and RNA molecules *(3–5)*. Remarkable results have been published from studying the translocation of these biopolymers through protein pores *(6–10)*. Recently, solid-state nanometer size pores have been successfully fabricated from materials such as silicon nitride and silicon dioxide, and these synthetic nanopores have been used to study biopolymers such as single DNA *(11–17)* and protein molecules *(18, 19)*.

1.2. Ionic Current Measurement

As illustrated in **Fig. 1**, the main component of a solid-state nanopore sensing system is a single 2- to 20-nm-diameter pore in a silicon nitride membrane that separates two chambers connected electrically only by the electrolyte solution inside the nanopore. The electrical behavior of a nanopore in an electrolyte solution

James Weifu Lee and Robert S. Foote (eds.), *Micro and Nano Technologies in Bioanalysis*,
Methods in Molecular Biology, vol. 544
DOI 10.1007/978-1-59745-483-4_7, © Humana Press, a part of Springer Science+Business Media, LLC 2009

approximately obeys Ohms law. When a voltage is applied and no biopolymers are presented in the chambers, a stable ionic current, the open pore current I_0, will be observed (**Fig. 1a, c**):

$$I_0 = \frac{VA_p}{\rho H_{eff}}. \quad (1)$$

Here ρ is the resistivity of the electrolyte solution, V is the applied voltage across the membrane, H_{eff} is the effective membrane thickness, and A_p is the mean cross-section area of a nanopore. When charged biopolymers are added to the cis chamber near the nanopore, the molecules will be captured by an electric field of the right polarity, and forced to pass through the nanopore to the other side (trans chamber) of the membrane. In the simple case of high ionic strength, a biopolymer inside the nanopore (**Fig. 1b**) simply blocks part of the ionic flow in the nanopore, causing a detectable ionic current blockade (**Fig. 1c**). This transient signal depends on the properties of the nanopore, electrolyte solution, and the passing molecule.

1.3. Properties of a Biopolymer Can Be Measured from a Current Blockage Signal

Three basic parameters can be extracted from a current blockage event like that shown in **Fig. 1c**: the average blockade current I_b, the translocation time t_d, and the integrated area of an event A_{ecd}. There is no universal equation to describe the amplitude change of this current blockage as a function of time $I_b(t)$. Approximate analytical solutions can be obtained in the limit of a globular shape molecule much shorter than the nanopore thickness, H_{eff}, or a chain-like polymer much longer than H_{eff}.

In the first case, when the length of a biopolymer $l_m < H_{eff}$, by neglecting the interaction between the biopolymer and the nanopore (free translocation), Ohm's law and the work by DeBlois and Bean *(2)* estimates the parameters I_b, t_d, and A_{ecd} in our study of protein *(18)*:

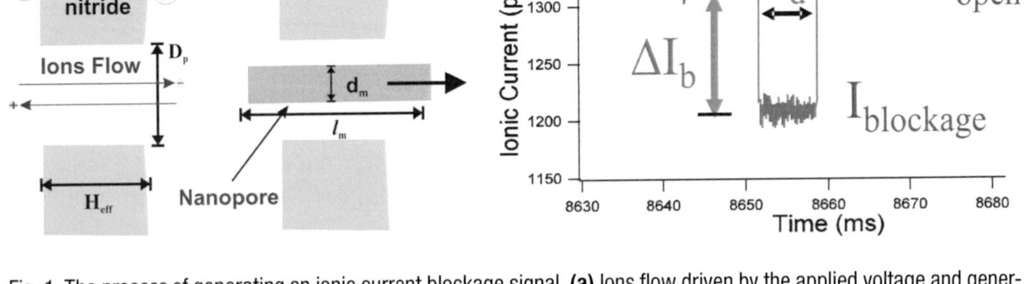

Fig. 1. The process of generating an ionic current blockage signal. **(a)** Ions flow driven by the applied voltage and generate an open pore current I_{open}. **(b)** A nanopore is partially occupied by a long chain shape negatively charged biopolymer. **(c)** A current blockage signal generated due to the translocation of a DNA molecule through a small nanopore.

$$\Delta I_b = \frac{VA_m l_m}{\rho H_{eff}^2}[1 + f(A_m, A_p, l_m, H_{eff})], \tag{2}$$

$$\Delta I_b = \frac{VA_m}{\rho H_{eff}} \quad \text{for } l_m \gg H_{eff}, \tag{3}$$

$$t_d \propto \frac{\eta H_{eff}}{QV} M, \tag{4}$$

$$A_{ecd} = \int_{event} \Delta I_b(t)dt \propto M \Delta I_b / Q. \tag{5}$$

Here, A_m is the mean cross-section area of the biopolymer, M is the molecular weight of a biopolymer, Q is its total effective electrical charge, and η is the solution viscosity. $f(A_m,A_p,l_m,H_{eff})$ is a correction factor that depends on the relative values associated with the dimensions of a biopolymer and a nanopore. The estimated value of H_{eff} is approximately 10 nm in the nanopores we fabricated. According to **Eq. 2**, in the same nanopore measuring system in which H_{eff}, V, and ρ are kept constants, I_b is directly proportional to the product of $A_m l_m$ (the volume) of a biopolymer, thus a change of the size of biopolymer could be measured by the change of I_b. According to **Eq. 4**, a change in the charge Q can be measured by the change in t_d. A_{ecd} (event charge deficit [ecd], in the unit of kilo electrons [ke] of charge) represents the net excluded charges caused by an ionic current blockade event. A_{ecd} is determined primarily by the identity of the molecule, its charge, geometry, applied voltage, and ionic solution parameters *(16, 20)*.

1.4. An Example of a Nanopore Measurement and Its Application on Determining DNA Conformation and Size

Our results showed that a nanopore assay can simultaneously evaluate a DNA molecule's conformation and its number of bases *(21)*. The conformation of a DNA molecule can be evaluated by the blockage amplitude I_b. The number of bases or the size of a DNA molecule can be estimated from the integrated area A_{ecd} of an event. In this example, a linear dsDNA ladder (contains 2,027-, 2,322-, 4,361-, 6,557-, 9,416-, and 23,130-bp dsDNA) was used as a marker (New England Biolabs, Ipswitch, MA). A known-size circular dsDNA pBR322 (4.4 kbp) was to be evaluated regarding the distribution of its supercoiling. An "unknown" DNA sample "X" was to be evaluated regarding its conformation and size.

First, the DNA ladder was added to the negatively biased cis chamber and ~7,000 events were recorded. The chambers were washed, the pBR322 was added, and ~7,000 events were recorded. The chamber was again flushed, the unknown sample "X" was added, and ~9,000 events were recorded. The plot of number density versus I_b (vertical) and t_d (horizontal) in **Fig. 2a** shows the linear ~2.17 kbp (a mixture of 2,027 kbp and 2,322 kbp), 4.36k, 6.56k, and 9.42k events from the ladder DNA,

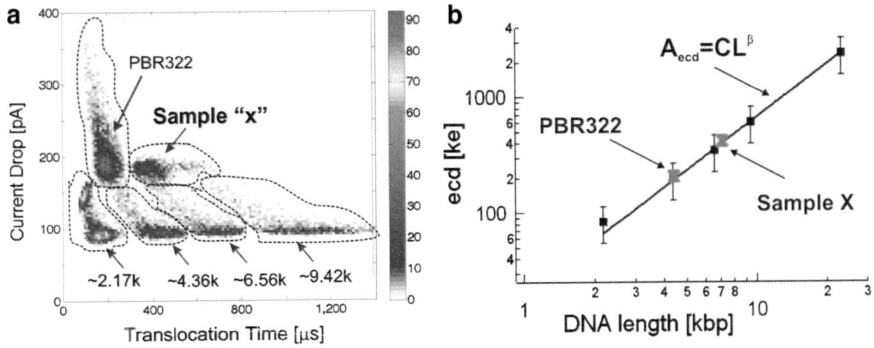

Fig. 2. **(a)** Event density plot versus t_d and I_b for a sample contains a mixture of a linear dsDNA ladder (contains 2,027-, 2,322-, 4,361-, 6,557-, and 9,416-bp dsDNA), a supercoiled circular dsDNA pBR322 (4.4 kbp), and an "unknown" sample measured with a 12-nm pore. The experiment was performed in a 1.6 M buffered KCl solution containing 20% glycerol at pH 7. The applied voltage across the electrodes was 120 mV. **(b)** Length determination of a DNA sample "X" from the nanopore experiment. A_{ecd} of the DNA ladder (*filled square*) and sample X (*X*). The *solid curve* is a fit of $A_{ecd} = CL\beta$ for the DNA ladder.

the 4.4-kbp circular DNA, and the unknown sample "X." From the I_b (doubled comparing with linear dsDNA) and the shape of each DNA shown in the density plot, the 4.4-kbp PBR322 DNA contains circular relaxed (the main peak, ~40% events), and supercoiled (the tail, 60% events) form of DNA. The unknown "X" contains circular relaxed DNA only.

To estimate the length of the unknown "X," we extracted the integrated areas of events or A_{ecd} as shown in **Fig. 2b**. Fitting the A_{ecd} data calculated from the DNA ladder measurement to $A_{ecd} = CL^{\beta}$, $C = 26.4 \pm 0.7$ and $\beta = 1.40 \pm 0.01$ were obtained. The corresponding A_{ecd} peak value for the unknown sample "X" is $A^{X}_{ecd} = 392 \pm 47$ ke, the calculated DNA size from this relation, $L^{X} = \left(A^{X}_{ecd} / 26.4\right)^{1/1.40}$ is $L^{X} = 6.9 \pm 0.6$ kbp (it was a circular relaxed 7.2-kbp dsDNA).

This example demonstrates that a nanopore measurement can determine the conformation of a DNA molecule accurately. The measurement can also estimate the conformation distribution and the size of a biopolymer at the single molecule level.

2. Materials

1. Silicon wafers (CZ-silicon wafers, Allegro Microsystems Inc., Worcester, MA, with diameter 100 mm [4 in.], type N/phosphorous, thickness of the wafers 380 μm, Resistivity: 1–10 Ω cm).

2. Ag/AgCl electrodes (1 mm, World Precision Instruments, Sarasota, FL) were bleached (Ultra Clorox regular bleach) before use.
3. Sylgard 184 Silicon Elastomer (Dow Corning Corporation, Midland, MI) was used to make cis and trans chambers shown in **Fig. 7**.
4. Polyethylene tubing (Intramedic Clay Adams Brand), Luer lok syringes (Becton-Dickinson, Franklin Lakes, NJ), and Luer valves (World Precision Instruments) were used to connect the fluidic system shown in **Fig. 7**.

3. Methods

3.1. Fabrication of Solid-State Nanopore Chips

There are several methods to fabricate solid-state nanopores, here we only introduce two methods: by feedback-controlled low energy (0.5–5.0 keV) noble gas ion beam sculpting (IBS), and by high-energy (200–300 keV) electron beam illumination.

3.1.1. Feedback-Controlled Ion Beam Sculpting Method

Three major steps are involved in this method: (A1) Fabrication of a freestanding silicon nitride membrane supported by a silicon substrate as illustrated in **Fig. 3a**. (A2) Fabrication of a ~100-nm diameter hole or cavity in the freestanding silicon nitride membrane as illustrated in **Fig. 3b**. (A3) Low-energy IBS of the ~100-nm hole or cavity to a desired-size nanopore as illustrated in **Fig. 3c, d**.

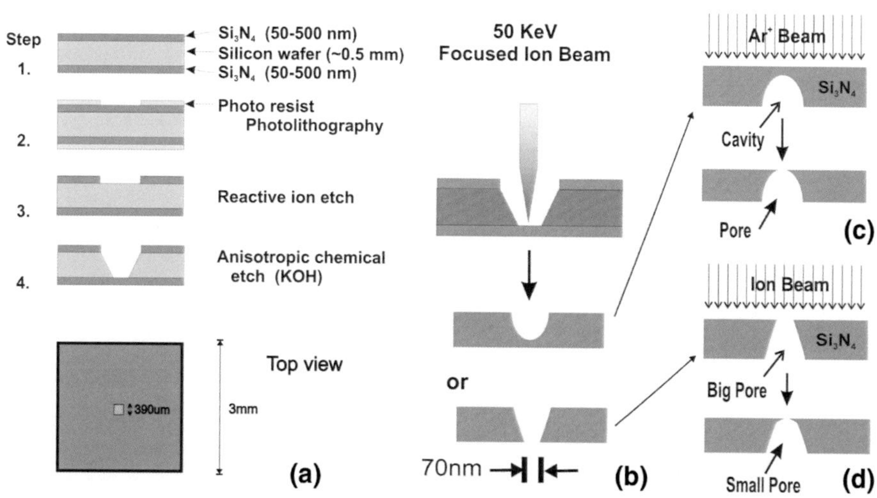

Fig. 3. Fabrication steps for solid-state nanopore chips **(a)**. FIB milling **(b)**. Ion beam sculpting nanopores by keV noble gas ions by sputter removing materials from the flat side of a cavity **(c)**, and by closing a larger pore **(d)**.

Major Steps A1 to A3

A1. Fabrication of a Freestanding Silicon Nitride Membrane Supported by a Silicon Substrate

Double-sided polished silicon wafers coated with low stress (~200 MPa tensile) silicon nitride membranes of 50–500 nm in thickness can be custom made at the Cornell Nanofabrication Center (CNF). **Figure 3a** shows the fabrication steps to create the freestanding membrane. Photolithography and wet chemical (KOH, 30% by volume) etching of silicon are used. The dimension of the freestanding silicon nitride membranes is ~25 μm × 25 μm supported by 3 mm × 3 mm (or 4 mm × 6 mm) silicon substrate (100).

A2. Fabrication of a ~100-nm Diameter Hole or Cavity in the Freestanding Silicon Nitride Membrane

The initial 0.1-μm-diameter holes or ball-shaped cavities (**Fig. 3**, bottom) in the membrane's center can be milled using a focused ion beam (FIB) machine. In our work, a focused 50-keV Ga^+ ion beam (Micrion 9500) was used. The initial 0.1-μm-diameter hole or cavity has also been made by electron beam lithography at the CNF. Transmission electron microscope (TEM) images are taken at the end of this step to verify and measure the size of the FIB holes or cavities.

A3. Feedback-Controlled Low-Energy IBS

Here, exposure of silicon nitride films to a feedback-controlled low-energy ion beam provides an approach to fabricating and fine-tuning the size of nanopores by controlling the lateral transport of matter across the surface. This ion beam sculpting [IBS] technique has demonstrated that tunable single nanometer-sized pores can be fabricated in silicon nitride and silicon dioxide membranes *(11, 22, 23)*.

There are two ways to make a nanopore by feedback-controlled IBS, as illustrated in **Fig. 3c, d** *(11)*. The first is to shrink a larger pore by ion beam-induced lateral mass transport on the membrane surface **(Fig. 3d)**. The second way is to remove membrane material by ion beam sputtering **(Fig. 3c**) layer by layer from the flat side of the membrane containing a cavity on its opposite side. The flat surface will ultimately intercept the bottom of the bowl-shaped cavity, forming a sharp-edged nanopore. In both cases, the ion beam exposure is extinguished when the ion current transmitted through the pore is appropriate for the desired pore size.

The Apparatus of Feedback-Controlled Low-Energy Ion Beam Sculpting

Figure 4a is a schematic drawing of the feedback-controlled IBS system. The main components of this system are housed in a high vacuum chamber pumped to 10^{-9} mbar by a turbo molecular pump. The system includes a differentially pumped ion gun that generates an ion beam of energy 0.5–5.0 keV; a focusing einzel lens and 60° electrostatic deflection system; an electron gun to neutralize insulating surface (SiN) charging effects; a "Channeltron" electron multiplier-style single-ion detector to measure the ions transmitted through the forming nanopore. The temperature of the sample holder is monitored with a resistor temperature detector, and adjusted with liquid or gas nitrogen and a resistance heater.

This system not only uses keV ions as a nanopore sculpting tool, the number of ions transmitted per second through the pore is directly proportional to the area of the pore, thus the system also measures the area (the size) of the forming pore. The transmitted ions are refocused, energy analyzed, and measured by a single-ion detector, which converts every detected ion into an electronic pulse. The electronic pulses are then counted by a counter and read by a PC Labview program that also controls a voltage that can deflect the ion beam away from the sample. When the desired pore size is reached, the Labview program will terminate the IBS process. **Figure 4d** illustrates the IBS process of shrinking a larger ~100-nm FIB hole (**Fig. 4b**) to a ~4-nm-diameter pore (**Fig. 4c**).

If we adjust the IBS parameters so that the ion sputtering removing material process dominates, we can fabricate nanopores from a bowl-shaped cavity as illustrated in **Fig.3c**. **Figure 5** shows a real example of a 10-nm pore fabricated by this method.

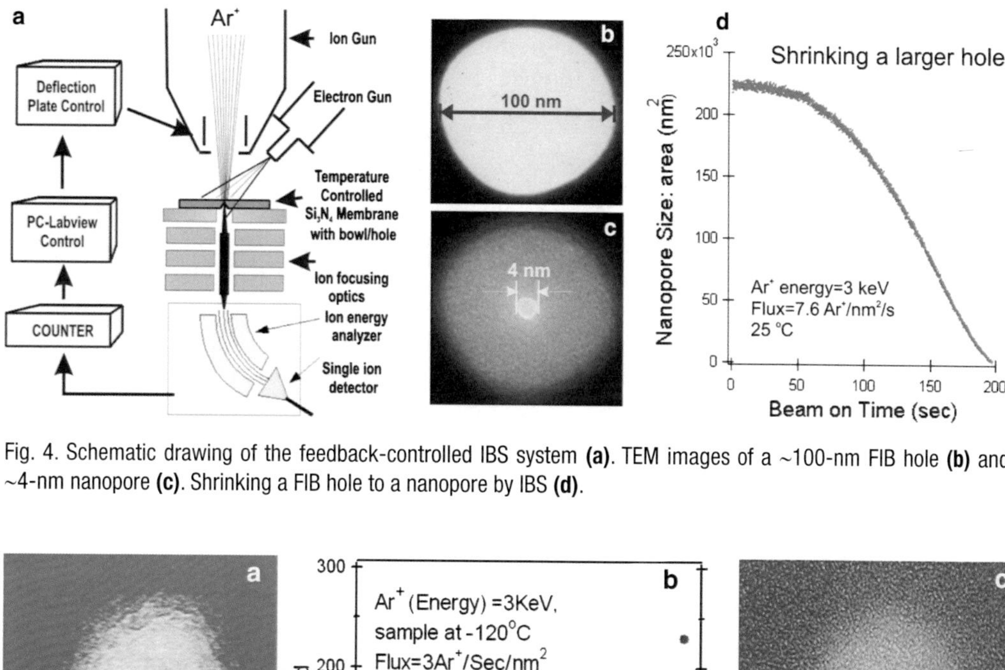

Fig. 4. Schematic drawing of the feedback-controlled IBS system **(a)**. TEM images of a ~100-nm FIB hole **(b)** and a ~4-nm nanopore **(c)**. Shrinking a FIB hole to a nanopore by IBS **(d)**.

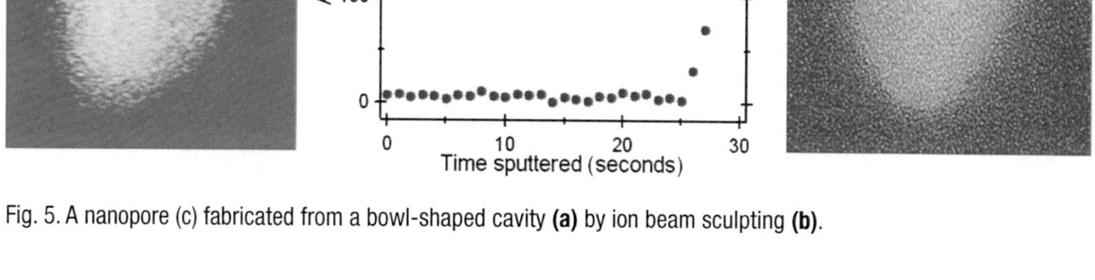

Fig. 5. A nanopore (c) fabricated from a bowl-shaped cavity **(a)** by ion beam sculpting **(b)**.

In summary, feedback-controlled IBS technique is capable of fabricating silicon nitride nanopores with desired diameters (D_p > ~1 nm). The estimated length of the nanopores (H_{eff}) is between 5 and 15 nm. However, the length or thickness of nanopores fabricated by this technique is not well controlled.

Major Steps of Making Nanopores by IBS

1. Align the nanopore chip (with a FIB hole in it, for example) to the sample holder.
2. Load the nanopore chip into the main chamber.
3. Calculate the ion beam flux (in ions nm^{-2} s^{-1}) using the initial ion beam counts divided by the area of the FIB hole, and then estimate the final ion beam count threshold for a targeted size nanopore.
4. Starting the IBS, the process will be automatically terminated when the ion beam counting rate passes the threshold set.

3.1.2. High-Energy Electron Beam Illumination

Dekker and co-workers at the Delft University of Technology have developed another method for producing SiO_2 and SiN pores by high-energy (200–300 keV) electron beam illumination in a TEM *(15, 16)*. Using semiconductor processing techniques, e-beam lithography, reactive-ion etching of SiO_2 mask layers, and anisotropic KOH etching of Si, they produced pyramidal 20 × 20-nm and larger pores in a 40-nm-thick membrane. They used the electron beam in a TEM to shrink the larger 20-nm pores to small ones. Because, in the TEM approach, the electron beam is used to image the pore normally, the shrinking process can be observed in real-time. Using a high-energy focused electron beam in a TEM, nanopores could also be drilled in freestanding <10-nm-thick SiO_2 membranes *(13)*.

3.2. Characterization of Solid-State Nanopores: TEM and AFM

Several nanopore properties are important for biopolymer analysis. These properties are the geometry of the nanopore (diameter, length, and the profiles of these parameters), the chemical properties, and the charging state of the nanopore surface. Characterization of these properties is challenging because of the small scale involved. A TEM can measure the smallest diameter of a nanopore accurately, atomic force microscopy (AFM) can be used to characterize the surface properties of a nanopore, and an *I–V* measurement can be used to probe the nanopore's electrical properties. The details of these characterization tools are described below.

3.2.1. Transmission Electron Microscope

The size and shape of a nanopore with sub-nanometer resolution can be measured by TEM. TEM has been the best tool for nanopore size characterization. Standard TEM images can characterize two-dimensional nanopore profiles (**Figs. 4c** and **5c**). Recently, tomographic techniques have been applied to extract three-dimensional information *(24)*.

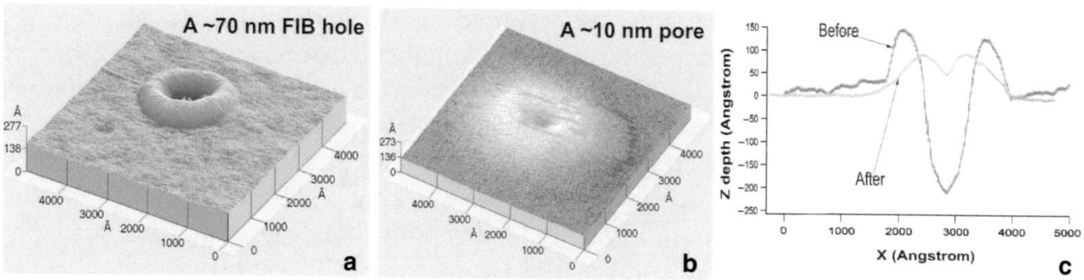

Fig. 6. AFM images of a FIB hole **(a)** and a nanopore made by IBS. **(c)** The cross-section profiles of **(a)** and **(b)**.

3.2.2. Atomic Force Microscopy

AFM has been used to characterize nanopore surface profiles *(24–27)*. **Figure 6** shows AFM images for a FIB-made ~70-nm hole **(Fig. 6a)**, a 10-nm nanopore made by IBS, and the cross-section profiles of the two images **(Fig. 6c)**. AFM images provide surface roughness and shape, however, limited by the lateral resolution, this technique is not widely used for measuring the size of small nanopores.

3.3. Materials Sciences of Nanopore Formations

3.3.1. Ibs

The demonstrated ability of the IBS method to continuously monitor changing dimensions in the nanometer scale while varying experimental parameters provides an unusual opportunity to test microscopic models to account for observed materials phenomena. Competing processes are probably at work. One is responsible for opening the pore, probably driven by ion sputter erosion of the pore edge, this process dominates at low temperatures and high fluxes. The other process is responsible for closing the pore, probably driven by ion beam-induced mass transport on the surface.

When an ion beam is directed to a material surface, a number of processes are understood to occur. In one such process, a sputtering process, atomic-scale erosion occurs at the material surface, removing atoms from the surface for incident ions. Because of this phenomenon, as material is removed from a solid-state surface, e.g., a Si_3N_4 surface, which has been processed to contain a bowl-shaped cavity on its opposite surface, as shown in **Fig. 3c**, the flat surface will ultimately intercept the bottom of the bowl-shaped cavity, forming a nanopore. Production of a nanopore requires knowledge of precisely when to stop the ion sputtering erosion process. Because we do not have a built-in reference point on the size of the nanopore being made, the absolute size of the nanopores could be difficult to anticipate by this method.

It is preferred that the apparatus be operated to control a number of processing parameters, those that can create favorable conditions for the flow of matter to a developing nanopore. These parameters are sample temperature, ion beam duty cycle (defined as the time of an ion beam "on" the sample divided by the total time of the beam "on" plus "off" the sample for pulsed beams), and the instantaneous ion beam flux, F (in ions nm^{-2} s^{-1}), when the beam is directed to the eroding material. Because the initial FIB hole can serve as a reference point for determining the size of a nanopore, the shrinking hole method (**Fig. 3d**) has better size control for making nanopores compared with the opening a cavity method (**Fig. 3c**).

Two different views can explain the motion of matter necessary to account for the pore closing phenomenon. In the first, a very thin (~5 nm) stressed viscous surface layer may be created by the energy and matter deposited by the ion beam. An enhanced collective motion, driven by a reduced viscosity and/or enhanced stress caused by implantation effects or surface tension, causes the layer to relax. Alternatively, we can account for our observations with a model in which incident ions both create and annihilate excess, independent, and mobile surface "adatoms" (e.g., atoms or molecular clusters) that can diffuse to the pore *(11, 28)*.

3.3.2. High-Energy Electron Beam Illumination

The underlying mechanism for the shrinking process by electron beam illumination appears to differ from the IBS. The authors believe that the electron beam effectively melts the SiO_2, and surface tension causes the molten SiO_2 to flow to minimize its free energy. A model was also proposed by Storm et al. *(15)* describing the contraction and expansion kinetics based on the competition of surface tension-driven contraction with electron beam-induced sputtering.

3.4. Nanopore Biopolymer Sensing Apparatus

3.4.1. Sample Holders, Electrodes, and Fluidic System

An example of the nanopore sensing system is shown in **Fig. 7**. In our design, the nanopore chip is sandwiched between two polydimethylsiloxane (PDMS) chambers. Using the solid-state nanopore chip as the main sensing component, we have designed and constructed a nanopore sensing system as illustrated in **Fig. 7**. This sensing system has the capability of changing sample solution with a microfluidic system (manually or computer controlled), and varying the solution temperature from 0°C to 100°C by a thermoelectric cooling-heating system. The PDMS chambers are made by casting PDMS into metal molds. Both chambers are held in place separately. The measuring system shown in **Fig. 7** can be assembled under an optical microscope. Each chamber is equipped with a Ag/AgCl electrode to which a voltage bias is applied during experiments. The bottom of the cis chamber has a 50- to 100-μm funnel aperture that sits atop the freestanding membrane window with a nanopore in it. It is aligned with the help of the optical microscope. The whole apparatus is placed in a Faraday cage on top of a vibration isolation table to reduce noise.

3.4.2. Electronics and Data Recording

Ionic current through a nanopore is generated and recorded by an Axopatch 200B patch clamp system (Molecular Devices, Sunnyvale, CA) with the amplifier set in resistive feedback mode. The 10- or 100-kHz low-pass Bessel filter in the Axopatch 200B was selected for measurements. Current blockage events are recorded in event-driven mode with the pretrigger length set for 1 ms.

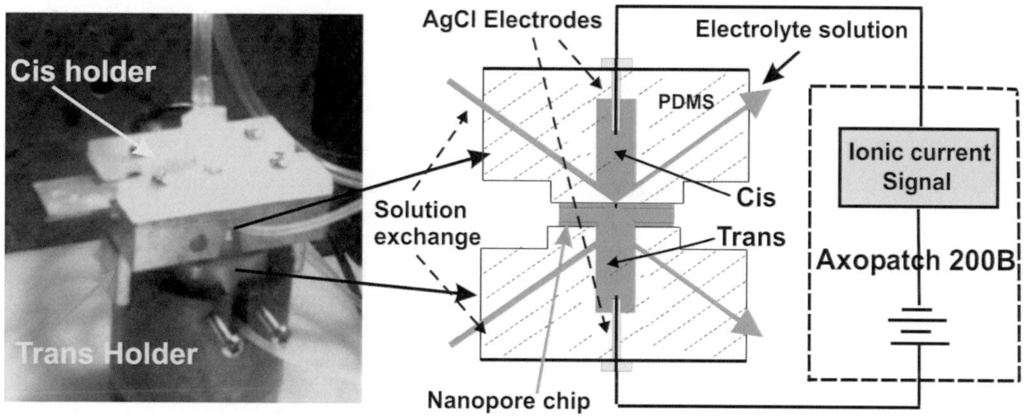

Fig. 7. Schematic of a solid-state nanopore detection system for charged biopolymers.

4. Notes

Recommended procedures for a nanopore measurement:

Nanopore Sensing System Preparations

1. Clean the nanopore chip by soaking the whole chip in acetone, isopropanol, and deionized (DI) water for 10 min each. Clean the PDMS chambers with isopropanol, followed by DI water, and blow dry the chambers with clean air or nitrogen. Prepare the Ag/AgCl electrodes by bleaching them in regular bleach solution (Clorox, for example) for approximately 10 min.
2. Align the nanopore chip with the cis and trans chambers as shown in **Fig. 7b**. New PDMS chambers are used for each set of experiments.
3. Fill the cis and trans chambers with a standard 1 M KCl TE (10 mM Tris and 1 mM EDTA at pH 7) buffer solution for initial nanopore conductivity testing. The solution should be degassed before use.
4. Apply a biased voltage V (120 mV usually) to the electrodes. If the open pore current is stable, is close to the value estimated from **Eq. 1**, and the ionic current noise root mean square (RMS) is below 10 pA, then the system is ready for measurements.

Sample Preparation and Translocation Event Recording

DNA Molecules

1. Depending on the size of DNA molecules, approximately ~10 nM final concentration of a DNA in the cis chamber is usually recommended for a translocation experiment. The concentration can be lower for long and higher for short molecules.
2. For one DNA sample, the recommended number of recording events is approximately 3,000–5,000.
3. For a series of DNA samples, after the measurement of one sample, the chambers should be extensively flushed with fresh

electrolyte solution to remove residual DNA. The chambers are considered clean when no translocation events are observed during approximately 10 min.

Protein Molecules

Protein molecules could be positively, neutral, or negatively charged depending on the solution pH values. Both positive and negative polarities are applied to test the electrical charge of an unknown protein. Different voltage amplitudes are also applied to find the appropriate voltage for the measurements.

Data Analysis

Here we describe the data analysis routines to extract basic parameters from the recorded translocation data.

1. Baseline correction is usually applied to the raw data. To correct for baseline drift in the open pore current during an experiment, the average current amplitude of the beginning 70% of the 1-ms pretrigger point was used to define the open pore current (I_0), or baseline, for each event.
2. Classification and sorting of translocation events are effected after baseline correction. The start of a translocation event was defined as one that caused the nanopore current to drop monotonically below two thresholds; the end of a translocation event was signaled by the current trace climbing monotonically back to the open pore current past both of these thresholds. All of these events have to be classified and sorted first to discard the short time spikes and extremely long events. We discard the current blockage pulses that are either very short (less than the time constant of the electrical measuring system, 30 μs for a 10-kHz low-pass filter, for example) or extremely long in time (molecules stuck in the nanopore).
3. Calculate parameters I_b, t_d, and A_{ecd}. The arithmetic mean of the current blockage value was calculated from the current range between the downward and upward crossings of the second threshold. The event time duration, t_d, was defined by the time between the current drop across the first threshold (~40 pA below the baseline) and its rise through the first threshold to the open channel current. The first threshold value was set to be beyond the baseline noise level and approximately the half peak height of the I_b. The second threshold was set to be at 70% of the maximum I_b values. The event integral, A_{ecd}, was defined as the integral of obstructed ionic current over the duration of an event.
4. Further data analysis routines can be applied to detect features such as DNA folding states, supercoiling distribution, hybridization, and protein secondary structures.

Acknowledgments

The authors acknowledge financial support for this work from the Defense Advanced Research Projects Agency (DARPA) (N65236-98-1-5407), the National Science Foundation (DMR 0073590), National Science Foundation (NSF)/Materials Research Science and Engineering Center (MRSEC) grant 0080054, ABI-PT06, and National Institutes of Health (NIH) grants 1R21HG003290 and 5 R01 HG003703.

References

1. Coulter WH (1953) Means for Counting Particles Suspended in a Fluid. U.S. Patent Number 2656508.
2. DeBlois RW, Bean CP (1970). Counting and sizing of submicron particles by the resistive pulse technique. *Rev. Sci. Instrum.* **41**: 909–916
3. Bezrukov SM, Vodyanoy I, Parasegian VA (1994). Counting polymers moving through a single ion channel. *Nature* **370**: 279–281
4. Kasianowicz JJ, et al. (1996). Characterization of individual polynucleotide molecules using a membrane channel. *Proc. Natl Acad. Sci. U. S. A.* **93**: 13770–13773
5. Bezrukov SM (2000). Ion channels as molecular coulter counters to probe metabolite transport. *J. Membr. Biol.* **174**: 1–13
6. Henrickson SE, et al. (2000). Driven DNA transport into an asymmetric nanometer-scale pore. *Phys. Rev. Lett.* **85**: 3057–3060
7. Meller A, Nivon, L., Branton, D. (2001). Voltage-driven DNA translocations through a nanopore. *Phys. Rev. Lett.* **86**: 3435–3438
8. Meller A, et al. (2000). Rapid nanopore discrimination between single polynucleotide molecules. *Proc. Natl Acad. Sci. U. S. A.* 97: 1079–1084
9. Movileanu L, et al. (2005). Interactions of peptides with a protein pore. *Biophys. J.* **89**: 1030–1045
10. Sutherland TC, et al. (2004). Structure of peptides investigated by nanopore analysis. *Nano Lett.* **4**: 1273–1277
11. Li J, et al. (2001). Ion-beam sculpting at nanometre length scales. *Nature* **412**: 166–169
12. Li J, et al. (2003). DNA molecules and configurations in a solid-state nanopore microscope. *Nat. Mater.* **2**: 611–615
13. Heng JB, et al. (2004). Sizing DNA using a nanometer-diameter pore. *Biophys. J.* **87**: 2905–2911
14. Fologea D, et al. (2005). Slowing DNA translocation in a solid state nanopore. *Nano Lett.* **5**: 1734–1737
15. Storm AJ, et al. (2003). Fabrication of solid-state nanopores with single-nanometre precision. *Nat. Mater.* **2**: 537–540
16. Storm AJ, et al. (2005). Translocation of double-stranded DNA through a silicon oxide nanopore. *Phys. Rev. E* **71**: 051903
17. Chang H, et al. (2004). DNA-mediated fluctuations in ionic current through silicon oxide nanopore channels. *Nano Lett.* **4**: 1551–1556
18. Fologea D, et al. (2007). Electrical characterization of protein molecules in a solid-state nanopore. *Appl. Phys. Lett.* **91**: 053901
19. Han A, et al. (2006). Sensing protein molecules using nanofabricated pores. *Appl. Phys. Lett.* **88**: 093901–093903
20. Fologea D, et al. (2005). Detecting single stranded DNA with a solid state nanopore. *Nano Lett.* **5**: 1905–1909
21. Fologea D, et al. (2007). DNA conformation and base number simultaneously determined in a nanopore. *Electrophoresis* **28**: 3168–3192
22. Stein D, Li J, Golovchenko JA (2002). Ion-beam sculpting time scales. *Phys. Rev. Lett.* **89**: 276106
23. Stein DM, et al. (2004). A feeback-controlled ion beam sculpting apparatus. *Rev. Sci. Instrum.* **75**: 900–905
24. Kim MJ, et al. (2007). Characteristics of solid-state nanometre pores fabricated using a transmission electron microscope. *Nanotechnology* **18**: 205302
25. Chen P, et al. (2004). Atomic layer deposition to fine-tune the surface properties and diameters of fabricated nanopores. *Nano Lett.* **4**: 1333–1337
26. King GM, Golovchenko JA (2005). Probing nanotube-nanopore interactions. *Phys. Rev. Lett.* **95**: 216103
27. Mitsui T, et al. (2006). Nanoscale volcanoes: accretion of matter at ion-sculpted nanopores. *Phys. Rev. Lett.* **96**: 036102
28. Cai Q, et al. (2006) Nanopore sculpting with noble gas ions. *J. Appl. Phys.* **100**: 024914

Chapter 8

Inserting and Manipulating DNA in a Nanopore with Optical Tweezers

U. F. Keyser, J. van der Does, C. Dekker, and N. H. Dekker

Summary

The translocation of small molecules and polymers is an integral process for the functioning of living cells. Many of the basic physical, chemical, and biological interactions have not yet been studied because they are not directly experimentally accessible. We have shown that a combination of optical tweezers, single solid-state nanopores, and electrophysiological ionic current detection enable deeper insight into the behavior of polymers in confinement. Here we describe the experimental procedures that are necessary to manipulate single biopolymers in a single nanopore, not only by electrical fields, but also through mechanical forces using optical tweezers.

Key words: Nanopore, Optical tweezers, DNA translocation, Biopolymers, Polymer transport, Single-molecule sensors, Single-channel recording

1. Introduction

The fabrication of the first single solid-state nanopore in an insulating membrane *(1)* pioneered the development of new techniques for nanopore fabrication and for single-molecule detection in aqueous solutions *(2)*. The ionic current flowing through a nanopore proves to be a useful tool for the label-free detection of biologically relevant polymers, ranging from DNA to proteins.

Nanopores in solid-state membranes are more robust than their biological counterparts, e.g., those found in cells. Solid-state nanopores are easily tailored in size and length to match the required characteristics for an experiment *(3)*. One intriguing idea is to use solid-state nanopores as model systems to gain

James Weifu Lee and Robert S. Foote (eds.), *Micro and Nano Technologies in Bioanalysis*,
Methods in Molecular Biology, vol. 544
DOI: 10.1007/978-1-59745-483-4_8, © Humana Press, a part of Springer Science+Business Media, LLC 2009

a deeper understanding of polymer transport in living organisms. Prominent examples for these fundamental processes are the gene transfer between bacteria and transport of RNA and proteins through the nuclear membrane.

Here we explain the experimental techniques for mechanical manipulation of a single polymer in a solid-state nanopore *(4)*. This enables technological advances as well as new insights into basic problems in chemical physics of polymers. This novel single-molecule technique combines the high force resolution of optical tweezers with the means to detect local structures on DNA *(5)*. In a recent theoretical study, this technique was proposed for unraveling the structure of RNA molecules *(6)*. The possibility to slow down or even reverse the translocation of DNA through nanopores with optical tweezers holds potential for the detection of the primary sequence of DNA.

Nanopores are also promising building blocks for future lab-on-a-chip technologies. Together with the integration of optical techniques into microfluidic chips in combination with automation, the detection and identification of biomolecules by measuring both the ionic current and the force seems feasible.

Finally, the combination of solid-state nanopores with their biological counterparts like α-hemolysin from *Staphylococcus aureus* would circumvent common problems like the long-term stability of lipid membranes, while providing control over the nanopore shape on the single-atom level.

In the following paragraphs, we describe the experimental procedures to insert a single DNA molecule into a solid-state nanopore. Special emphasis will be given to the microfluidic cell design and suggestions for how to solve the problems we encountered during the measurements.

2. Materials

2.1. Fabrication of Solid-State Nanopores

The fabrication of solid-sate nanopores is briefly described here (excellent descriptions can be found in the literature, see, e.g., **ref.** *(7)*). In brief, a 700-nm, free-standing membrane is produced using standard semiconductor technology. Part of this membrane is thinned down by chemical etching, which leads to a 20-nm thin round silicon nitride membrane with a diameter of 5 μm. This design was chosen to make the membranes resistant to mechanical stress. The round membrane yields a diffraction pattern that is used to monitor the distance between the nanopore and optical trap (*see* 3.6).

Coating the SiN-membrane on both sides with sputtered silicon oxide (thickness 10–20 nm) facilitates wetting and reduces sticking of colloidal particles coated with proteins. This empirical finding proved to be essential for carrying out successful experiments.

The membranes are mounted into a transmission electron microscope (TEM). The TEM is used to drill nanopores by focusing a 200- to 300-kV electron beam onto the membrane. Nanopores with <1-nm diameter can be directly drilled. The TEM shrinking method developed by Dekker and coworkers in Delft *(3)* allows for tailoring of the nanopores to the desired diameter and shape.

The nanopores are usually used within a few days of fabrication. If this is not the case, they are stored in a desiccator under vacuum. Before usage, they are treated with an oxygen plasma for 20 s to remove carbon contamination and facilitate wetting.

2.2. Optical Tweezers

The optical tweezers setup is also extensively described in the literature *(8)*. We briefly summarize the most important details, necessary for understanding the functions and proper operations (*see* **Note 1**).

The setup consists of a custom-built inverted optical microscope with a high numerical aperture water immersion objective (UPLSAPO, 60×, NA = 1.2, water immersion, Olympus, Tokyo, Japan) combined with a high-resolution patch-clamp amplifier (Axopatch 200B, Molecular Devices, Sunnyvale, CA). The water immersion objective was chosen because it is possible to trap colloidal particles up to 200 μm above the cover slip. This allows a relatively simple design of the sample cell that will be described in the next section.

The optical trap is created by overfilling the back aperture of the objective with an infrared laser (λ = 1,064 nm; P = 1.5 W; linearly polarized; CrystaLaser Inc., Reno, NV). In the focal spot, dielectric particles are confined in all three dimensions. Near the focal region, the trapping potential is harmonic, thus the restoring force F exerted on the bead is directly proportional to the distance from the center of the trap, $F= -k\,X$, where k is the stiffness of the trapping potential. The system represents a simple Hooke spring in all three directions. In the most general case, all three spring constants k_x, k_y, and k_z have to be characterized independently.

We use the same objective for monitoring the motion of the trapped colloid. A second, red laser (λ = 635 nm, P = 5 mW, linearly polarized, Coherent, Santa Clara, CA) is used to illuminate the bead. Decoupling of the trapping and detection lasers has certain advantages that are discussed in the literature *(9)*. The red laser is coupled into the beam path with a dichroic mirror DM1 **(Fig. 1)**that transmits in the visible range and reflects infrared (IR) light. For relative alignment to the IR laser path, its angle and position are adjusted by the mirrors M_{red}**(Fig. 1)**. The red laser is partly reflected by the trapped particle, collected by the trapping objective, and focused onto a quadrant photo detector. Spatial filtering is achieved by placing a pinhole in front of the detector. Alignment of the pinhole relative to the optical axis is crucial and has to be checked from time to time. The intervals between realignments depend on the stability of the temperature, the mechanical parts, and the pointing stability of the red laser.

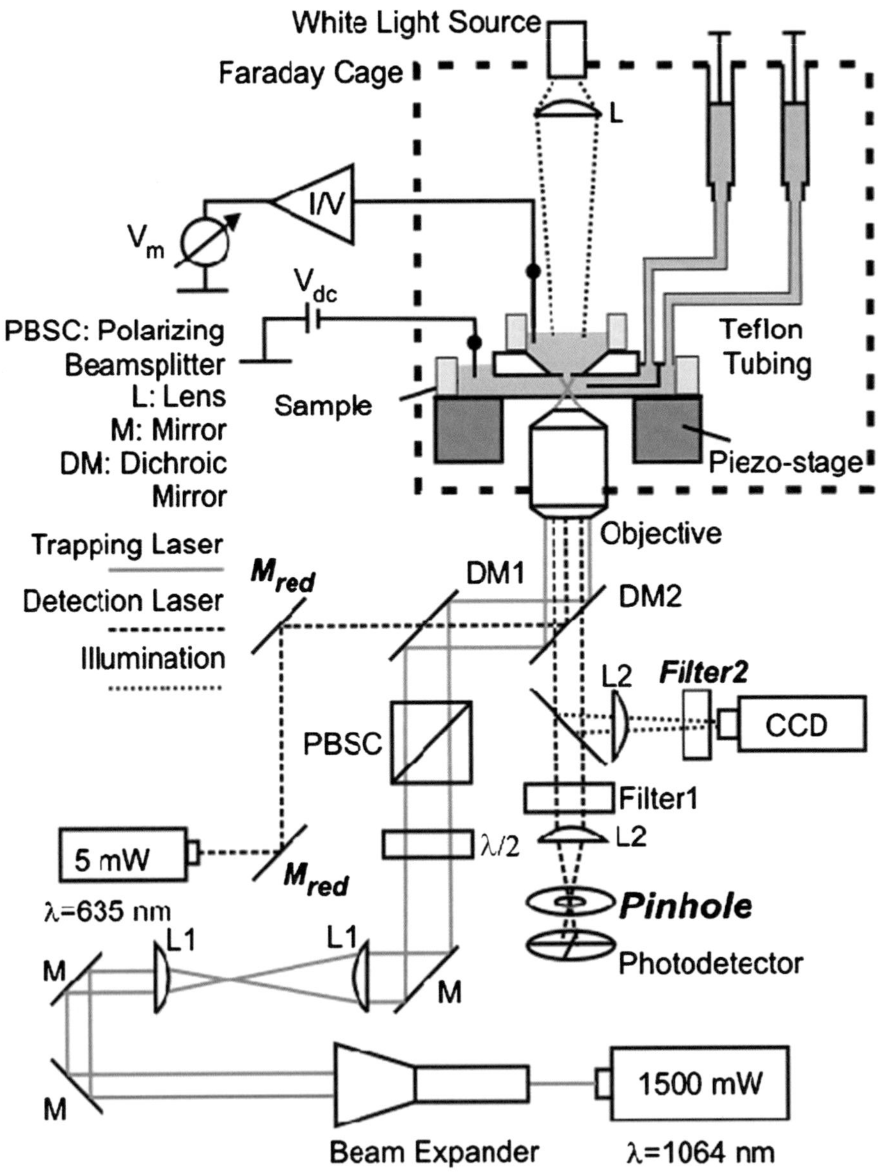

Fig. 1. Schematic of the optical tweezers setup. Two lasers (infrared and red) are coupled collinear and concentric into an objective. The objective, the head-stage amplifier, and the nanopore fluid cell are enclosed by a Faraday cage. The main elements, crucial for alignment during normal operations or maintenance, are shown in bold: two mirrors M_{red}, pinhole, and Filter2.

Alignment is preferably done in two steps to reach an optimal result as rapidly as possible. First, the IR laser path is aligned so that this trapping laser is coupled into the center of the objective. Afterward, the red laser is coupled into the IR laser path at DM1 (**Fig. 1**). A laser power meter (Melles Griot, Carlsbad, CA, USA) and a beam profiler (TaperCamD, GenTec, Quebec, Canada) can be used to check for correct alignment behind every optical element. By inserting irises at regular intervals along the beam path, errors in the alignment can be easily identified later.

The biggest advantage of this single objective-based optical tweezers design is the ability to build a fluid cell with any size, thickness, and geometry on top of the setup. This simplifies working with solid-state nanopore chips in combination with high NA objectives.

The relative positions of the nanopore and optical trap are controlled by a three-axis piezoelectric stage (Physik Instrumente, Karlsruhe, Germany). A small Faraday cage is built around the piezoelectric stage and the microscope objective. The Faraday cage should be large enough to include all liquid handling systems to avoid electrical interference. It also contains the head stage of the patch-clamp amplifier (Axopatch 200B, Axon, Molecular Devices).

2.3. PDMS Sample Cell

The sample cell is one of the most crucial parts of the setup (**Fig. 2**). We assemble it out of four parts. A baseplate is used as a carrier for the other parts. This allows assembling of the fluid cell containing the nanopore before mounting it on the optical tweezers. The design also facilitates testing of the nanopore and the search for leaks. A glass cover slide is used as an optical window for the microscope objective. On top, a thin layer of polydimethylsiloxane (PDMS) is used to establish a seal and create the channels for flushing in the electrolyte solutions or beads coated with DNA. A covalent bonding between the PDMS and the glass helps prevent leaks (**Fig. 2a**). If both the PDMS and cover slide are clean, good seals are possible, allowing reuse of the PDMS layer with different cover slides. The nanopore chip is pressed onto the PDMS layer by an O-ring with a diameter of 2 mm that is held in place by a Perspex block. We apply a very thin layer of vacuum grease between the nanopore chip and the PDMS to get a good seal. The Perspex block contains the liquid connections to pump the solutions into the PDMS channel next to the objective and the top of the nanopore chip (**Fig. 2a**). The alignment of all of the parts can be easily done by hand.

The configuration of the assembled flow cell is shown in **Fig. 2b**. The beads are flushed in beneath the nanopore, which is pressed on the PDMS layer by the O-ring in the Perspex block.

For making the PDMS layer, one has several options. The simplest option is to make a mould from aluminum that can be

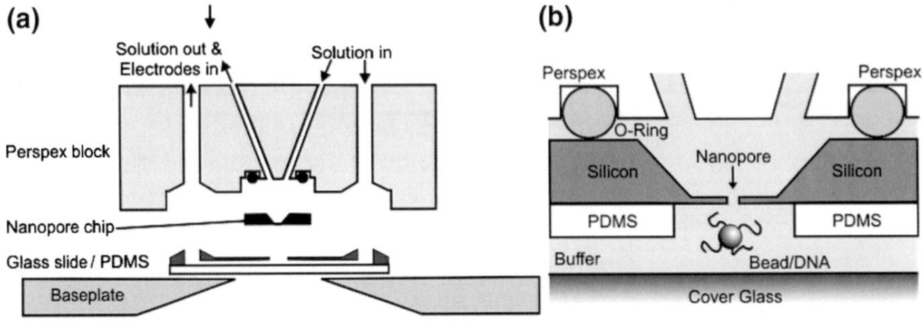

Fig. 2. Nanopore fluidic cell. (**a**) Shown are the main parts before assembly, base plate, glass slide with PDMS layer, and Perspex block, from bottom to top. The fluid inlets and outlets are shown, as well as the points where the electrodes are inserted. (**b**) The outline of the center part of the flow cell is depicted. The hole is cut into the PDMS. The diameter should be as small as possible. Please note that both figures are not drawn to scale. (The second inlet for the bottom channel, as shown in **Fig. 1**, is not depicted here for clarity. A sketch of the PDMS channels is published in **ref. (*8*)**).

reused. Modern machining tools allow accuracies of better than 0.01 mm, which is by far enough for channels with an average height of 100 μm and a total thickness of the PDMS of 150 μm. However, it is necessary to make sure that all surfaces are extremely flat to ensure a good seal. The more elaborate possibility is the use of SU8 lithography on silicon wafers. This must be weighed against the increased difficulty of obtaining usable flow cells if there is no proper equipment for photolithography available in the laboratory.

The PDMS layers are made according to the following protocol:

1. Mix PDMS 1:10 (curing agent: polymer base solution) (Sylgard 184, Dow Corning, Midland, MI).
2. It is very important to degas the PDMS before curing. This can be done with either a vacuum pump that is available in most laboratories, or with a centrifuge in combination with an ultrasonic bath.
3. Check if the mould is clean from dust particles or PDMS from previous use. Clean it if necessary.
4. After pouring the PDMS onto the mould, cure for ~10 min at 150°C.
5. Clean cover slides in an ultrasonic bath in acetone, and then in isopropanol (10 min each). Make sure the slides are well separated to ensure cleaning on both sides. Blow dry immediately before mounting in an oxygen plasma chamber (SPI Plasma-Prep II) (*see* **step** 7).
6. Check the PDMS for holes caused by air inclusions. If there are no holes in the PDMS, continue to **step** 7.

7. Treat the cover slide and the PDMS with oxygen plasma for 30 s.
8. Cut a hole into the PDMS after the film is mounted on the cover slide. This allows liquid to touch the surface of the nanopore chip. Use an injection needle that is sharpened with fine sand paper. Make sure the needle is clean by flushing it with acetone/isopropanol, wiping, and blowing it dry with nitrogen gas to remove debris from the sharpening step (*see* **Note 2**). The position of the hole can be marked with a holder, e.g., a small aluminum block where the cover slide fits into a pit with the center indicated by a small indentation. The usual diameter is >0.5 mm, depending on the injection needle. This allows easy alignment by hand.

2.4. Electrodes

Silver/silver chloride electrodes are light sensitive. To avoid the usage of these electrodes, we use platinum wires that are connected to the headstage and immersed in 1 mM potassium-ferri/ferrocyanide (Sigma Aldrich, St. Louis, MO) with 1 M KCl background solution. The solution should be stored in the dark in a refrigerator to avoid degradation. The platinum electrode configuration with the salt bridges has the advantage that no silver wire has to be chlorinated, reducing the light sensitivity compared with silver/silver chloride electrodes *(10)*. This is important because there is intense laser irradiation from the infrared laser, which can cause electrical interference. Use small salt bridges made from agar gel with 1 M KCl to connect the sample cell with the headstage (*see* **Note 3**).

2.5. Salt Bridges

1. Heat the agar (1% in 1 M KCl) to 100°C in a microwave. Aliquot this solution into Eppendorf tubes with volumes of 1–2 mL. Let the agar harden for later use and store in the refrigerator.
2. Cut thin Teflon tubing (outer diameter <1 mm, inner diameter as large as possible) into short pieces of 10 cm in length.
3. For preparation of the salt bridge, take one agar tube and heat in a standard heating block for tubes to 100°C.
4. Using a syringe, pull the agar slowly into the Teflon tubing and afterward harden the agar by rinsing with deionized water.
5. Store the salt bridge in 1 M KCl solution in the refrigerator.

2.6. Buffers

The electrolyte solutions are made from 3 M KCl stock solutions. For the stock, potassium chloride (Sigma Aldrich ≥99.0%) is dissolved in deionized water. We use Tris-EDTA buffer solution (Sigma Aldrich) at pH 8.0 and add it to all solutions. All liquids are stored at 4°C. All solutions, excluding the ones containing DNA or beads, are filtered using 20-nm filters (Anotop

25, Whatman, Maidstone, UK) immediately before use in the setup to remove any bacteria or dust particles. The solutions are subsequently sterilized in an autoclave.

2.7. DNA and Primers

We used λ-phage DNA (Promega, Madison, WI) for the experiments. DNA primers were purchased from Biolegio, Nijmegen, The Netherlands.

3. Methods

3.1. Biotinylated DNA

The procedure to attach biotin to one end of the DNA consists basically of the hybridization and subsequent ligation of a specially designed primer to the *cos* site of the lambda DNA.
The primer sequence is: P-5′-GGGCGGCGACCTT-3′-B.

This primer is phosphorylated (P) at the 5′ end and contains a single biotin label (B) at the 3′ end.

1. First, we hybridize the primer to the *cos* site of the lambda molecule, using a 100 times molar excess of the primer. For the hybridization, use the following amounts, with concentration given in parentheses:
2. 50 μL lambda DNA (0.5 μg/μL).
3. 15 μL T4 ligase buffer (10× buffer).
4. 2 μL primer (100 μM).
5. 78 μL milliQ water.
6. Use the following hybridization program (in a polymerase chain reaction [PCR] machine) with a slow cool-down program:

   65°C for 1 h, followed by 5 min at each of the following temperatures: 63.8°C–62.5°C–61.3°C–60°C–58.8°C–57.5°C–56.3°C–55°C–53.8°C–52.5°C–51.3°C–50°C–48.8°C–47.5°C–46.3°C–45°C–43.8°C–42.5°C–41.3°C–40°C–38.8°C–37.5°C–36.3°C–35°C–33.8°C–32.5°C–31.3°C–30°C–28.8°C–27.5°C–26.3°C–25°C.
7. Cool to 4°C following the last step of the program.
8. For ligation, add 5 μL of T4 DNA ligase (New England Biolabs, Ipswitch, MA) after running the hybridization program.
9. Ligate overnight at 16°C.
10. Following the ligation, perform a phenol-chloroform extraction followed by an EtOH precipitation to remove the remaining enzymes and primers. The DNA is then dissolved

in 10 mM Tris-EDTA buffer at pH 8.0 and stored at −20°C. The typical concentration is 100–200 ng/μL.

3.2. Binding Multiple DNA Strands to Beads

1. Use 5 μL of bead suspension (e.g., 2.1-μm diameter, streptavidin coated, Polysciences, Eppelheim, Germany) to bind 5 μL of biotinylated λ-DNA (48.5 kb) at a concentration of 1 nM.
2. For washing the beads, add 100 μL of buffer (100 mM KCl, 10 mM Tris-EDTA) to beads in a 1.5-mL centrifuge tube.
3. Spin down the beads with a simple laboratory centrifuge and carefully remove the supernatant with a pipette.
4. Repeat the washing step.
5. Resuspend the beads in 5 μL of binding buffer (2 M KCl, 10 mM Tris-EDTA).
6. Add 5 μL of DNA at a concentration of 1 nM in 10 mM Tris-EDTA buffer.
7. Incubate for 15 min at room temperature. Carefully shake the vial from time to time.
8. Repeat the washing in 100 mM KCl, 10 mM Tris-EDTA.
9. Resuspend the DNA/bead constructs in measurement buffer to the desired concentration. This depends on the layout of your flow cell.
10. The concentrated DNA/beads solution can be used for 1–2 days if stored at 4°C (*see* **Note 4**).
11. Adding DNA to the beads enhances the friction coefficient (*see* **Notes 5**–7).

3.3. Assembling Flow Cell

1. Put the PDMS layer/cover slide on the base plate (**Fig. 2**). Apply a tiny amount of vacuum grease to the PDMS layer and mount the nanopore chip with the membrane side toward the PDMS. Press lightly to distribute the grease and do not allow contact of the membrane with the vacuum grease.
2. Hold the baseplate assembly against a light source over your head. Check if the hole in PDMS and the membrane are concentrically aligned. The membrane containing the nanopore is transparent, so a bright spot is visible when held against a light source.
3. Mount the Perspex block (**Fig. 2**) to complete the cell. M3 screws (not shown) are used to attach the Perspex block to the baseplate. Very carefully tighten the screws by hand to avoid breaking the cover slide. Because only slight overpressure is used to flush the cell, moderate tightening is sufficient to prevent leaking.

4. Fill the flow cell on both the top and bottom using a pipette. Flush slowly using a pipette to avoid creating air bubbles or breaking the membrane. Check for air bubbles before mounting the assembled cell.
5. Mount onto the optical tweezers setup.
6. Important: test all electrical connections, especially the salt bridges and the platinum electrodes.
7. Connect the tubing for flushing the cell.

3.4. Characterizing Nanopores

3.4.1. Current Voltage Characteristics

It is necessary to check the current-voltage characteristics before any experiments involving DNA translocations (*see* **Note 8**). Nanopores should exhibit a linear current voltage characteristic in the range of ±500 mV. Testing this before every experiment can save a lot of time in identifying problematic nanopores. Check if the resistance R of the nanopore differs more than a factor of two from the anticipated resistance from the basic formula for cylindrical nanopores:

$$R = \rho\left(\frac{h}{\pi r^2} + \frac{1}{2r}\right).$$

Here, r is the radius of the nanopore inferred from the TEM measurements, ρ the resistivity of the salt solution (depending on the concentration), and h is the nominal thickness of the membrane (in our case, 20–60 nm depending on the membrane, i.e., plain SiN or coated with SiO_x). Experience shows that, if the resistance happens to be higher than the expected value, there usually is a problem with the nanopore and it should be cleaned again by oxygen plasma.

3.4.2. Localizing Nanopores

The exact position of a nanopore can be easily determined by using IR laser-induced heating. The buffer absorbs part of the laser light and thus the conductivity of the aqueous salt solution increases. By scanning the laser over the membrane and simultaneously measuring the ionic current, a single nanopore is easily localized *(10)*. This is a straightforward process that can also be used to determine the spatial extension of the laser focus *(10)*. Furthermore, the laser scanning can be also used to assess the quality of the nanopores *(11)*. A typical measurement is shown in **Fig. 3a**. Depicted is a two-dimensional map of the ionic conductance of a nanopore as a function of the x and y position of the laser; high conductance regions appear as black and low conductance regions appear as white. The nanopore is in the center of the black region, which gives direct feedback about the nanopore position on the membrane.

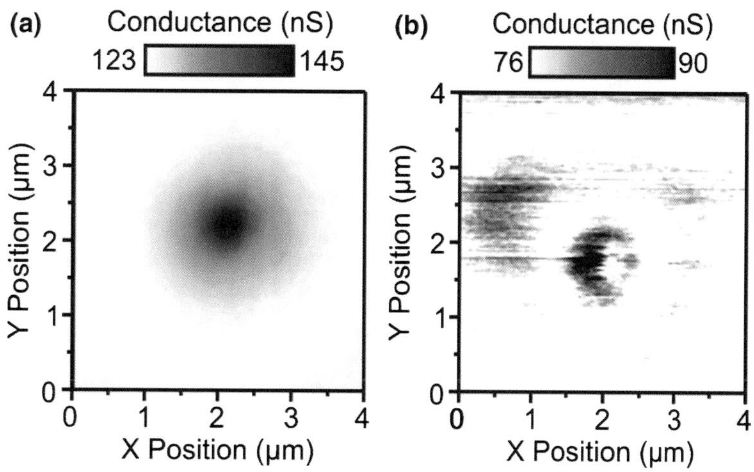

Fig. 3. (**a**) Ionic conductance as a function of laser position. The diameter of the nanopore was 11 nm. *Black* denotes high conductance and *white* low conductance. (**b**) The same measurement as (**a**) for a nanopore (10-nm diameter) with a nanobubble. The whole conductance map looks distorted and the peak at the nanopore position shows a clear substructure, indicating a bubble. Both nanopores were immersed in 1 M KCl at pH 8.0.

3.4.3. Noise in Nanopores

For a successful experiment, the number of DNA molecules inserted in the nanopore should be known at all times. This is only possible if the signal-to-noise ratio is sufficient to detect a single DNA molecule. Three major factors influence the ionic current measurements and thus the noise properties of the system, namely the electrodes, flow cell, and the nanopore itself. The electrodes were described earlier. Reducing their capacitances and access resistances is possible but not necessary because noise is determined mainly by the nanopore chip. By using a smaller flow cell with low contact resistance and small capacitance, the high-frequency noise can be reduced. The main capacitance, however, is caused by the silicon chip. The usually conductive silicon is covered by a thin insulating layer of SiN. This leads to a huge capacitance that can reach up to several 100 pF depending on the area over which the salt solution is touching the membrane. Reduction of this area is possible by adjusting the design of the flow cell (*see* **Fig. 2b**). This can be accomplished by using soft lithography techniques in PDMS. However, the main problems often arise from the solid-state nanopores. In particular, problems with wetting of nanopores were identified as sources of excessive f^{-a} (with $1<a<2$) noise in the low-frequency regime *(11)*. However, when using the two-dimensional scanning of nanopores through the laser focus, nanopores that are not properly functioning can be easily identified *(11)*. This is shown in **Fig. 3b**, in which a two-dimensional current map of a partially blocked nanopore (diameter 10 nm,

1 M KCl) is depicted. The entire scan depicts instabilities and the peak at the nanopore position exhibits a prominent substructure. This is obvious when comparing this measurement with a stable measurement depicted in **Fig. 3a**. Such unstable nanopore characteristics make controlled insertions of DNA impossible. The nanopore has to be exchanged for a new one (*see* **Notes 8–10**).

3.5. Trapping Beads in the Optical Trap

To flush solutions into the flow cell, we use 1-mL syringes that are pushed by manual micrometer screws. This has the advantage that there is no back flow after the flow is stopped because water is incompressible in this low pressure range. Of course, it is crucial to avoid any bubbles in the syringe, tubing, and the fluid cell itself to ensure proper operation (*see* **Note 10**).

1. For trapping a bead, first position the optical trap in the middle between the cover clip and the membrane containing the nanopore.
2. Start by flushing a small amount of buffer into the cell by using a syringe (**Fig. 1**).
3. Push buffer containing the bead/DNA solution (second syringe in **Fig. 1**) into the cell until you see beads flowing through the field of view.
4. Wait until one bead is trapped. This can take up to a few minutes depending on the concentration used.
5. Move the trap close to a surface (membrane or cover slip), stay typically a few microns away from the surface to avoid losing the bead because of sticking. This step avoids losing the bead because of the flushing of other buffer to get rid of the remaining beads in **step 6**.
6. Flush the rest of the beads away by flowing clean buffer solution through, using the syringe with buffer **(Fig. 1)**.
7. Characterize the trapped bead by looking at the power spectral densities in all three directions of the optical trap. Distortions of the spectrum point to problems of the alignment. This should be double checked before starting the measurements.
8. Bring the bead close to the nanopore and start inserting DNA into the nanopore.

3.6. Detecting the Distance Between Bead and Nanopore

1. Focus the laser on the nanopore by measuring the ionic current (at 50 mV) and changing the laser position. Maximize the current until the nanopore is in focus without a trapped bead.
2. Using the CCD camera, record images of the diffraction pattern caused by the nanopore at increasing distances between the focal point of the laser and the nanopore. Three examples

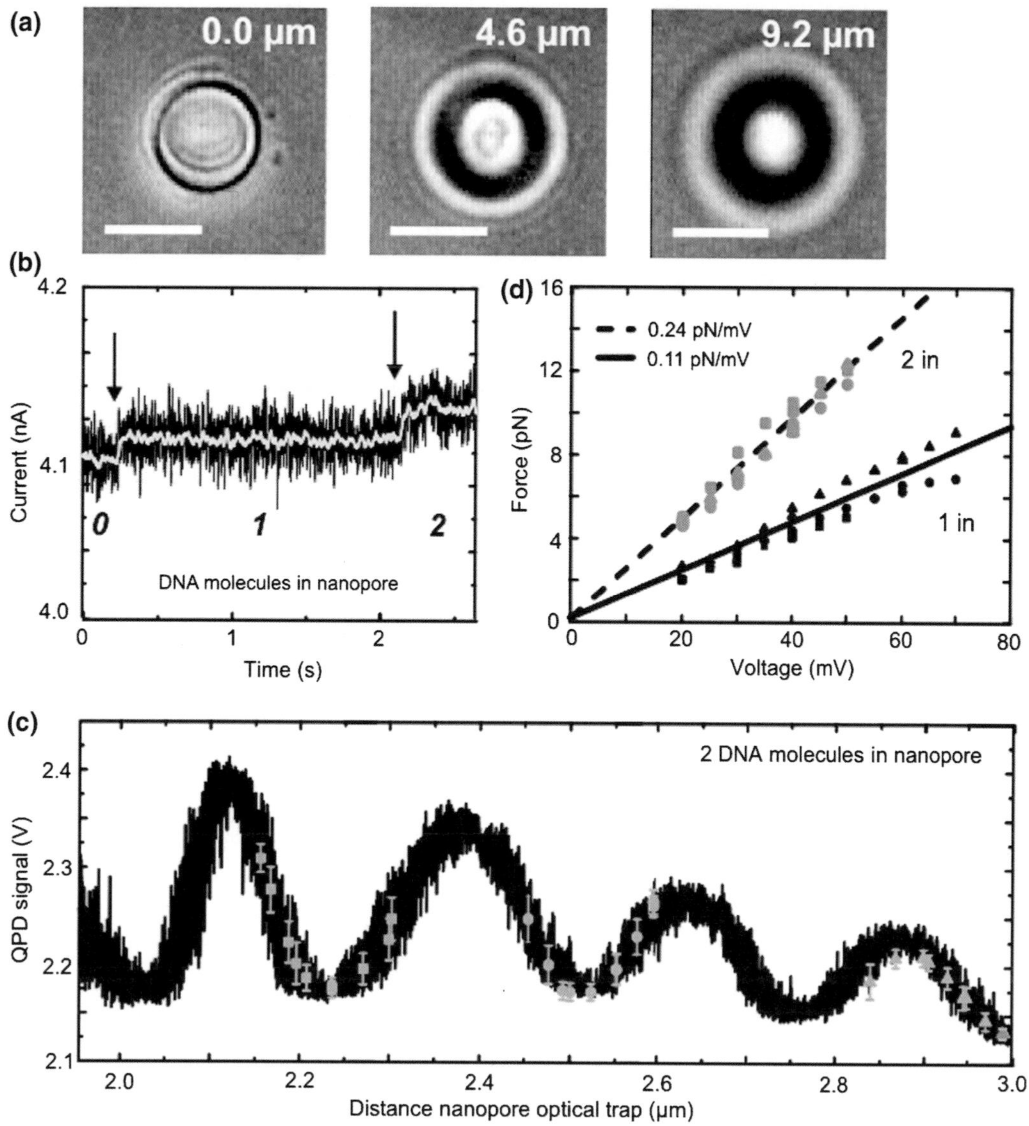

Fig. 4. Force measurements on DNA in a 80-nm nanopore. (**a**) Video images of the membrane containing the nanopore. The diffraction pattern depends on the distance between focus and membrane. The scale bar is 5 μm. (**b**) Ionic current as a function of time showing the stepwise increase in signal indicating insertion of a single DNA molecule in the nanopore (indicated by *arrows*). The *black* data is filtered at 1 kHz. The *light grey* data is filtered by a moving average over 20 points. (**c**) Calibration curve (*black*) of the total intensity on the quadrant photo detector (QPD) as a function of distance between the nanopore and optical trap. *Points* indicate force measurements with the two DNA molecules from (**a**) in the nanopore. (**d**) Comparison of force for one and two molecules in this nanopore. The force doubles when two DNA molecules are inserted. All measurements were done in 33 mM KCl, pH 8.0.

are shown in **Fig. 4a**. Use a step size of 300 nm and 31 steps to get a calibration up to 9 μm from the nanopore. This is similar to the process used, e.g., in magnetic tweezers; details can be found in the literature *(12)*. This technique works perfectly if the diameter of the membrane is at least a factor of two larger than the bead diameter.

3. Return the laser focal point to the nanopore. Increase the distance between the focal point and the nanopore to 2, 4, and 6 μm, respectively. At each position, check that an accurate position measurement can be determined from the diffraction pattern as explained in the literature *(12)*. In brief, the positions of the diffraction rings are compared with the recorded diffraction pattern (*see* **step 2**). Intermediate position values are interpolated.

3.7. Getting DNA into the Nanopore

1. Depending on the length of the DNA, insertion of DNA is most likely if the distance between the bead surface and the nanopore is smaller than the radius of gyration of the DNA. Move the surface of the bead coated with DNA to within 1 μm of the nanopore if λ-DNA is used.
2. Apply a voltage with polarity such that DNA is pulled into the nanopore, i.e., positive voltage at the top chamber in **Fig. 1**.
3. Monitor the ionic current.
4. When the current changes, one or more DNA molecules are inserted into the nanopore. The current change indicates the number of DNA molecules in the nanopore *(5)*. The ionic current can increase or decrease depending on salt concentration *(5)*. Typical data for DNA entering a nanopore at 33 mM KCl is shown in **Fig. 4b**. DNA entering the nanopore can be clearly observed.
5. If too many molecules are trapped, reverse the voltage to drive the DNA out of the nanopore. Increase the distance between bead and nanopore and/or reduce the voltage to avoid multiple captures. This is especially important when using large nanopores with diameter >20 nm, as can be observed by the short interval between captures of less than 2 s observed in **Fig. 4b**.
6. After successful capture, increase the distance between nanopore and bead to more than four times the radius of gyration of the DNA to avoid further captures of DNA (*see* **Note 11**).
7. Start experiments.

3.8. Force Measurements

1. Reduce the voltage to 10–20 mV to keep the DNA in the nanopore.
2. Increase the distance between the optical trap and the nanopore using the piezoelectric stage and monitor the total power at the quadrant photo detector. This will lead to data similar to the black curve in **Fig. 4c**.
3. Increase the voltage stepwise to 100 mV and record the power at the detector for several seconds.

4. Repeat the experiment at the same position at least twice. Always return to the starting voltage to check for sticking of the DNA to the nanopore wall or some other surface (*see* **Note 12**).
5. Average the detected power at every voltage step and compare this with the measured retraction curve that is shown in **Fig. 4c** by the black line. Comparison of the averaged signals with the calibration curves (points in **Fig. 4c**) returns the position and force values at this voltage. The result for two DNA molecules in 33 mM KCl is shown in **Fig. 4d**. The scatter of the points reflects the error in the position detection. The dashed line is a fit through the data taken at three different distances between nanopore and optical trap. For comparison, data for one DNA molecule in the same nanopore is also shown in **Fig. 4c**. The gradient or force/voltage (0.24 ± 0.01 pN/mV) on two DNA molecules is twice as high as for a single DNA molecule (0.11 ± 0.01 pN/mV) within experimental errors (*see* **Note 13**).

3.9. Conclusions

With the above-described experimental procedures, single DNA molecules can be readily inserted into solid-state nanopores. All of the common pitfalls we encountered are described in great detail in the above text. For an experienced master student, it is possible to mount two nanopores per day on the setup, including all cleaning steps, provided the nanopores are already available. It is also worth mentioning that we had experiments where a single nanopore worked for several days in a row. Reusing nanopores is possible after thorough cleaning. This is always determined by how clean the solutions are. In conclusion, we hope that this chapter is helpful in inserting single molecules in nanopores and enables exciting new experiments in the coming years.

4. Notes

In this section, we review the most common problems encountered during the experiments, with their respective diagnoses and solutions.

1. Bad alignment of the optical trap can be easily identified by observing the power spectral density of the bead as a function of x-, y-, and/or z-position. If it is not of Lorentzian form (dropping off with $1/f^2$ at frequencies above f_c), something is wrong. Start by checking alignment of the red laser relative to the IR laser. Trap a bead, remove the laser line filter (Filter 2, **Fig. 1**) in front of the camera, and check if the bead is

uniformly illuminated by the red laser. If this is not the case, realign the red laser with the two mirrors (M_{red}, **Fig. 1**). Next, it is necessary to check the alignment of the pinhole relative to the red laser. Do this also with a trapped bead. Afterward, look at the power spectrum. If it is normal, start to measure. Otherwise, test if the bead is the problem. Trap a new bead and look at the spectrum again. If the problem persists, repeat the red laser/pinhole alignment steps. When you are sure that this is not the problem, realign the IR laser. Important: only move the mirrors of the IR path (labeled M, **Fig. 1**) if you are absolutely sure that there is indeed a problem.

2. Clean PDMS layers are crucial. Cleaning in methanol and with ultrasound removes any remaining particles from previous experiments.

3. No ionic current through nanopore: If there are problems with the electrical signal, first test the electrodes. Start with the platinum wires and salt bridges. Next, check for bubbles in the PDMS or Perspex channels. If the wires do not work, clean them again. If the salt bridges have high resistance, cut off short pieces on both ends with a sharp scalpel or use a new salt bridge. Bubbles can normally be seen by eye and flushed away. This has to be done carefully to avoid breaking the membrane containing the nanopore.

4. To ensure long-term stability of the DNA, only use autoclaved pipette tips and wear gloves at all times. Contamination of the DNA stock can lead to rapid degradation. When the DNA is stored at 4°C, it is possible to use the same DNA for more than half a year with the same results.

5. A large amount of DNA on the bead will increase the friction coefficient of the bead. Because the trap stiffness is not influenced by the DNA on the bead, a reduction of the corner frequency *(9)* should be observed. The corner frequency in the power spectrum for a bead with a lot of DNA is usually half compared with a bead of similar size without DNA.

6. If a fluorescence setup is available, there is a convenient way to see DNA on the beads. Stain the DNA with YoYo-1 (Invitrogen, Carlsbad, CA) after performing the binding procedure and observe the beads with a fluorescent microscope to determine if DNA is bound. This can avoid much wasted time in case the binding does not work. It can also be used for fast determination of whether the DNA stock is still intact.

7. The exact concentration of beads and DNA varies from batch to batch. Thus, the parameters have to be slightly adjusted from time to time.

8. Only after excluding the problems described in **Note 3** should possible problems with the nanopore be considered.

9. Usually, in our experiments, a nanopore that shows high $1/f$ noise, randomly changing conductance levels or total blockades of ionic currents should be replaced and the experiment should be restarted. Trying to insert DNA controllably in such a situation is not possible.
10. Bubble formation in the buffer can be prevented by degassing before the experiments. This can be done by placing the solution into a vacuum flask and removing the gas with a vacuum pump. Ultrasound also helps to remove dissolved gases. It is best to combine both procedures.
11. If DNA seems not to enter the nanopore, first check the power spectrum of the bead in the trap. If f_c of the trapped bead is close to f_c of beads without DNA, find a bead with more DNA (lower f_c). Afterward, increase the voltage and decrease the distance. If the surface is nonsticky and the bead is densely coated with DNA, the bead can be pressed against the membrane without getting stuck. If this does not help and all of the beads have a high f_c, test the DNA (*see* **Notes 5** and **6**). If the DNA is okay, the next most likely explanation is that the nanopore has a high surface charge. Use larger nanopores and high salt concentrations to get DNA into the nanopore.
12. Sticking of the DNA/beads to the membranes is one of the most serious problems because under normal circumstances this implies that the experiment is over. The odds are minimized by using glass surfaces (sputtered SiO_x). There is also the possibility of adding a tiny amount of Tween to the solutions. This tends to help.
13. Check the alignment of the IR and red laser on a regular basis. Otherwise, results from the force measurements can be totally incorrect.
14. Avoid flushing the solutions with too much force because this can lead to rupture of the DNA *(13)* on the beads or breaking of the membranes *(14)*.

Acknowledgments

We thank Peter Veenhuizen, Stijn van Dorp, Bernard Koeleman, Ya-Hui Chen, and Suzanne Hage for making the biotinylated lambda-DNA, and Bernadette Quinn for help with electrochemical questions. Ralph Smeets, Diego Krapf, and Meng-Yue Wu fabricated the nanopores. Stijn van Dorp and Bernard Koeleman are especially acknowledged for obtaining some of the data presented here. Financial support of FOM and NWO is gratefully acknowledged.

References

1. Li, J., Stein, D., McMullan, C., Branton, D., Aziz, M. J., and Golovchenko, J. A. (2001). Ion-beam sculpting at nanometre length scales. *Nature* **412**, 166–169
2. Dekker, C. (2007). Solid-state nanopores. *Nature Nanotechnology* **2**, 209–215
3. Storm, A. J., Chen, J. H., Ling, X. S., Zandbergen, H. W., and Dekker, C. (2003). Fabrication of solid-state nanopores with single-nanometre precision. *Nature Materials* **2**, 537–540
4. Keyser, U. F., Koeleman, B. N., van Dorp, S., Krapf, D., Smeets, R. M. M., Lemay, S. G., Dekker, N. H., and Dekker, C. (2006). Direct force measurements on DNA in a solid-state nanopore. *Nature Physics* **2**, 473–477
5. Smeets, R. M. M., Keyser, U. F., Krapf, D., Wu, M. Y., Dekker, N. H., and Dekker, C. (2006). Salt dependence of ion transport and DNA translocation through solid-state nanopores. *Nano Letters* **6**, 89–95
6. Gerland, U., Bundschuh, R., and Hwa, T. (2004). Translocation of structured polynucleotides through nanopores. *Physical Biology* **1**, 19–26
7. Krapf, D., Wu, M. Y., Smeets, R. M. M., Zandbergen, H. W., Dekker, C., and Lemay, S. G. (2006). Fabrication and characterization of nanopore-based electrodes with radii down to 2 nm. *Nano Letters* **6**, 105–109
8. Keyser, U. F., van der Does, J., Dekker, C., and Dekker, N. H. (2006). Optical tweezers for force measurements on DNA in nanopores. *Review of Scientific Instruments* 77, 105105
9. Visscher, K. and Block, S. M. (1998). Versatile optical traps with feedback control. *Molecular Motors and the Cytoskeleton, Part B* **298**, 460–489
10. Keyser, U. F., Krapf, D., Koeleman, B. N., Smeets, R. M. M., Dekker, N. H., and Dekker, C. (2005). Nanopore tomography of a laser focus. *Nano Letters* **5**, 2253–2256
11. Smeets, R. M. M., Keyser, U. F., Dekker, N. H., and Dekker, C. (2006). Nanobubbles in solid-state nanopores. *Physical Review Letters* **97**, 088101
12. Strick, T. R., Allemand, J. F., Bensimon, D., Bensimon, A., Croquette, V. (1996). The elasticity of a single supercoiled DNA molecule. *Science* **271**, 1835–1837
13. Davison, P. F. (1959). The effect of hydrodynamic shear on the deoxyribonucleic acid from T2 and T4 bacteriophages. Proceedings of the national Academy of Sciences USA **45**, 1560–1568
14. Tong, H. D., Jansen, H. V., Gadgil, V. J., Bostan, C. G., Berenschot, E. van Rijn, C. J. M., and Elwenspoek, M. (2004). Silicon Nitride Nanosieve Membrane. Nano Letters **4**, 283–287

Chapter 9

Forming an α-Hemolysin Nanopore for Single-Molecule Analysis

Nahid N. Jetha, Matthew Wiggin, and Andre Marziali

Summary

Nanopore analysis of single molecules can be performed by measuring the modulation in ionic current passing through the nanopore while an individual biomolecule such as DNA or RNA is resident in, translocating through, or otherwise interacting with the pore. The corresponding current signature has been shown to reveal properties of the biomolecule and information on its interactions with the pore. The α-hemolysin nanopore remains the pore of choice, particularly for single-molecule analysis of nucleic acids, because of its internal dimensions, hydrophilicity, and low-noise characteristics. In this chapter we present a detailed protocol for forming a robust α-hemolysin nanopore (or multiple nanopores) for single-molecule analysis.

Key words: Nanopore, α-Hemolysin, Lipid bilayer, Single-molecule analysis

1. Introduction

Nanopore single-molecule analysis is based on the electrical detection of individual molecules, electrophoretically driven into a nanometer-scale pore, by measurement of the modulation of ionic current as each biomolecule enters, resides in, and exits the pore. Of the available organic pores, α-hemolysin (α-HL), a heptameric protein that self-assembles as a pore in a lipid bilayer *(1)*, is the most commonly used for studies of nucleic acids. Its dimensions (in particular its 1.5-nm-diameter limiting aperture) *(2)*, internal hydrophilicity, and low-noise characteristics make it ideally suited for this purpose.

James Weifu Lee and Robert S. Foote (eds.), *Micro and Nano Technologies in Bioanalysis*,
Methods in Molecular Biology, vol. 544
DOI 10.1007/978-1-59745-483-4_9, © Humana Press, a part of Springer Science+Business Media, LLC 2009

Single-molecule analysis using nanopores was first introduced in 1996 *(3)*. In this seminal paper, Kasianowicz et al. detected individual strands of RNA and DNA by the transient blockage of ionic current as they were driven through an α-HL nanopore. The authors confirmed the relationship between molecule translocation through the pore and current blockage events, and showed a correlation between blockage event times and the molecule length. Nanopore current was later shown to display a high degree of sensitivity to the nature of the molecule blocking the pore, as homopolymers that differ only in sequence were distinguished from each other based on the nucleic acid–α-HL pore interactions that ensue during translocation *(4)*.

Since then, there has been an explosion of studies and techniques developed using nanopores. These include single-molecule force spectroscopy *(5)*, studies of nucleic acid dynamics *(6, 7)*, progress towards single-molecule sequencing *(8, 9)*, single-molecule chemistry *(10, 11)*, protein dynamics *(12)*, and unfolding studies *(13)*. A number of excellent reviews are available on nanopore techniques for DNA analysis *(14–17)*. A force spectroscopy technique using multiple α-HL nanopores incorporated into a single lipid bilayer has also been developed for rapid single-nucleotide polymorphism detection *(18)* (refer to the Chapter "Nanopore Force Spectroscopy on DNA Duplexes").

In this chapter we present the protocol for forming single or multiple α-HL nanopores. The ease with which an α-HL nanopore can be formed is strongly correlated with the quality of the lipid bilayer in which it incorporates. We therefore also present considerable detail on methods for forming robust lipid bilayers.

2. Materials

2.1. Aperture Construction in PTFE/FEP Tubing (Forming the U-Tube)

1. Constantan® thermocouple wire, 0.001-in. diameter (OMEGA Engineering Inc., Stamford, CT).
2. Polytetrafluoroethylene (PTFE)/fluorinated ethylene-propylene (FEP) dual-shrink tube 0.036-in. inner diameter (ID) (Small Parts Inc., Miramar, FL).
3. Instant adhesive (Loctite® 401).
4. 24 AWG solid wire (insulation removed).
5. Varitemp® heat gun (MASTER® Appliance, Racine, WI).

2.2. Silver Chloride (Ag/AgCl) Electrodes

1. Silver wire (Alfa Aesar, Ward Hill, MA), 1-mm (0.04-in.) diameter, 99.9% purity.

2. 0.06-in. ID and 0.130-in. ID PTFE/FEP dual-shrink tube (Small Parts Inc.).
3. Bleach (6% sodium hypochlorite).
4. Connector pins appropriate to the patch clamp amplifier head stage (e.g., MaxBit® 1-mm diameter, male, crimp pins if using the Axopatch 200B, Molecular Devices Corp., Sunnyvale, CA). Refer to **Subheading 2.5** regarding the patch clamp amplifier.
5. 24 AWG solid wire.
6. Varitemp® heat gun (MASTER® Appliance).
7. Imperial® wetordry® sanding paper (3M Inc., St. Paul, MN).

2.3. Black Lipid Bilayer Formation

1. KCl buffer: 1 M potassium chloride (KCl), 10 mM HEPES, and 1 mM EDTA, pH 8.0. Filter through a 0.1-μm vacuum filter (e.g., Stericup® filter units, Millipore Corp., Billerica, MA). Adjust the pH using 2 M potassium hydroxide (KOH). Store at room temperature.
2. 10 mg/mL solution of 1,2-diphytanoyl-*sn*-glycero-3-phosphocholine, powdered form (Avanti Polar Lipids, Alabaster, AL) in chloroform. Store lyophilized lipid and lipid chloroform solution at −20°C (*see* **Note 1**).
3. Single bristle lipid brush **(Fig. 1)** made from a 10-μL pipette tip, a paint brush bristle, and a small stick from a cotton swab (*see* **Note 2**).
4. Varitemp® heat gun (MASTER® Appliance).
5. Three NORM-JECT® syringe connectors **(Fig. 2)**. Using the heat gun, shrink 0.130-in. ID PTFE/FEP dual-shrink tube (Small Parts Inc.) around stainless steel hypodermic needles (with plastic hubs), 14 and 15 gauge (Small Parts Inc.) (*see* **Note 3**).
6. 1-Hexadecene (Sigma-Aldrich, St. Louis, MO). Store at room temperature.
7. Concentrated nitric acid (HNO_3).
8. Hexane. Store at room temperature.
9. Ethanol. Store at room temperature.
10. 50-μL Glass capillary pipettes (e.g., Drummond® Wiretrol micropipettes, VWR International, West Chester, PA).
11. Oil-less vacuum pump (24-in. Hg, McMaster-Carr®, Cleveland, OH) and vacuum chamber (VWR).
12. Stirring hot plate (Corning®, Corning, NY).
13. Polyvinyl chloride (PVC) gloves (Winter Monkey Grip®).
14. 6 M Sodium hydroxide (NaOH) solution.
15. Safety glasses.

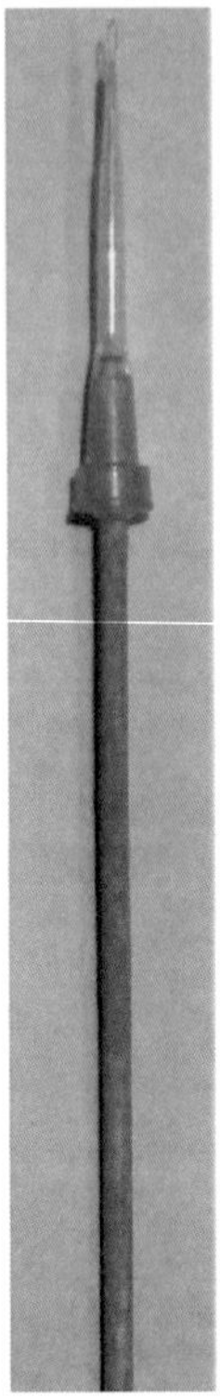

Fig. 1. Lipid brush. A single paint brush formed into a loop extends out the end of the pipette tip.

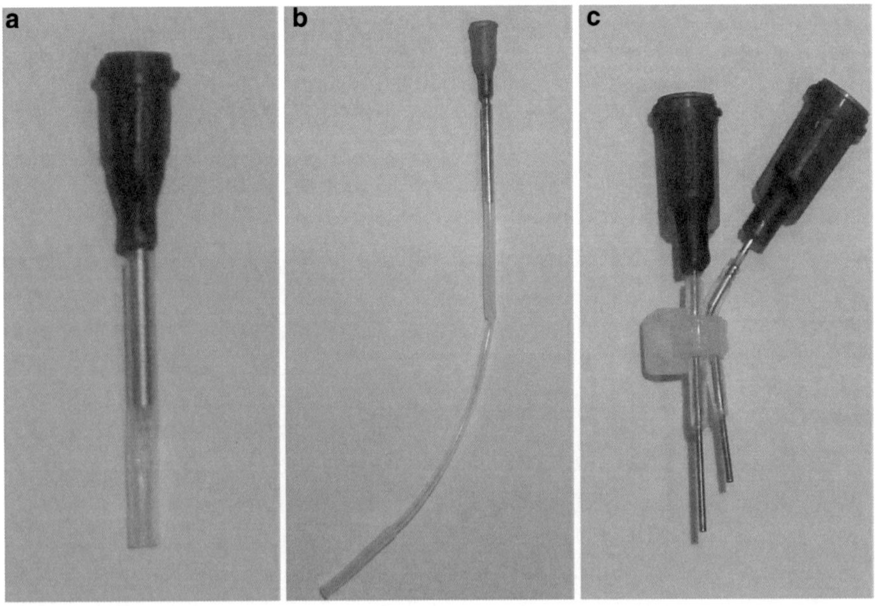

Fig. 2. The syringe connectors. **(a)** Connector for flushing solutions through the U-tube (**step 4** of **section 3.3**). **(b)** Two of this type of connector should be made. One used for pushing air through the U-tube and one for pushing KCl buffer (refer to **section 3.3.**). **(c)** Tygon®-syringe connector for flushing out and replacing KCl buffer (**step 6** of **section 3.4**).

2.4. α-Hemolysin Pore (or Multiple Pore) Formation

1. α-HL from *Staphylococcus aureus* (Sigma-Aldrich). Follow MSDS precautions when handling α-HL, which is highly toxic. Mix lyophilized α-HL powder with a 1:1 glycerol to water solution at 1 mg/mL. Store powdered α-HL and solution at −20°C. The solution should be kept for a maximum of 3 months.
2. Venturi pump with a 0.25-in. ID Tygon® hose attached.
3. Tygon®-syringe connector **(Fig. 2c)**. The connector is made by bending the steel of a 21-gauge hypodermic needle by approximately 30° and zip tying it to a second 21-gauge hypodermic needle (the bottom of the bent hypodermic needle should not be flush with that of the straight needle). It is helpful to heat-shrink some PTFE/FEP tubing around a portion of each needle to improve the tightness of the tying.

2.5. Experimental Apparatus

1. PTFE cell **(Fig. 3)**.
2. Aluminum cell housing **(Fig. 4)** to secure the PTFE cell, Ag/AgCl electrodes, and light source. The housing should be grounded to reduce electrical noise.
3. Vibration isolation table for mechanical noise reduction. Light load air suspension tables are adequate (e.g., LW series tables, Newport Corp., Irvine, CA).
4. Microscope light source. (e.g., a high intensity white LED and DC battery or a fiber-Lite high intensity illuminator, model 170D, with flexible fiber optic cable, Dolan-Jenner Industries Inc., Boxborough, MA).
5. Stereomicroscope with swingarm stand, 40× magnification (e.g., Leica S6E, Leica Microsystems, Ontario, Canada).
6. Two-channel oscilloscope, 100-MHz bandwidth, 1.25 GS/s (giga-samples per second) (e.g., TDS 3012B, Tektronix, Richardson, TX).
7. At least one faraday cage over the cell housing to eliminate external electromagnetic fields.
8. Thermoelectric module (peltier) and temperature controller (TE Technology Inc., Traverse City, MI) for performing temperature-dependence studies (*see* **Note 4**).
9. Low-noise, patch clamp amplifier (e.g., Axopatch 200B, Molecular Devices). The signal from the patch clamp amplifier may need to be grounded, which can be accomplished by connecting the signal ground to the vibration isolation table.

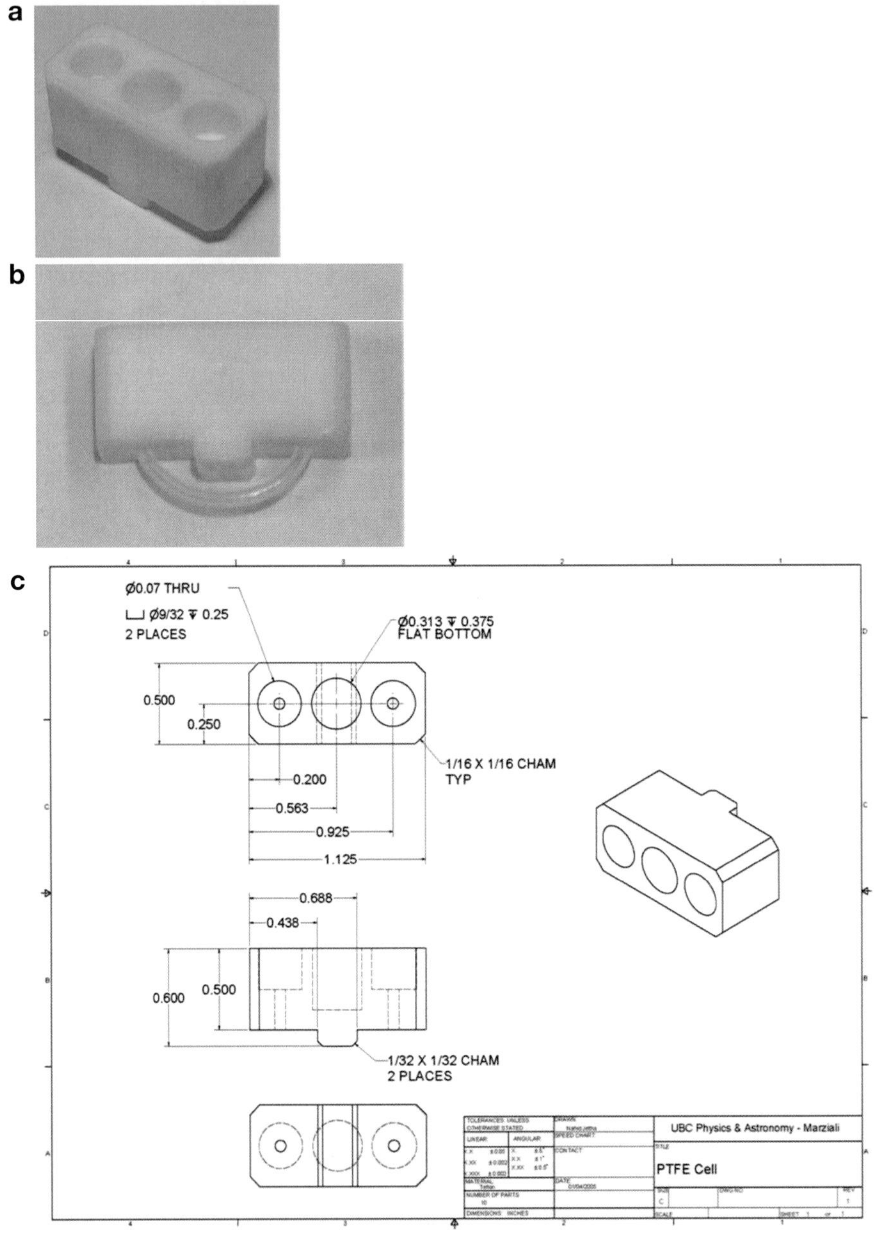

Fig. 3. **(a)** PTFE cell. **(b)** Front view of the PTFE cell with the U-tube in place. **(c)** Mechanical drawing of the PTFE cell.

3. Methods

3.1. Aperture Construction in PTFE/FEP Tubing (Forming the U-Tube)

1. Wrap a piece of the thermocouple wire around the solid wire several times and glue using instant adhesive. One to two inches of thermocouple wire should extend past the end of the solid wire.

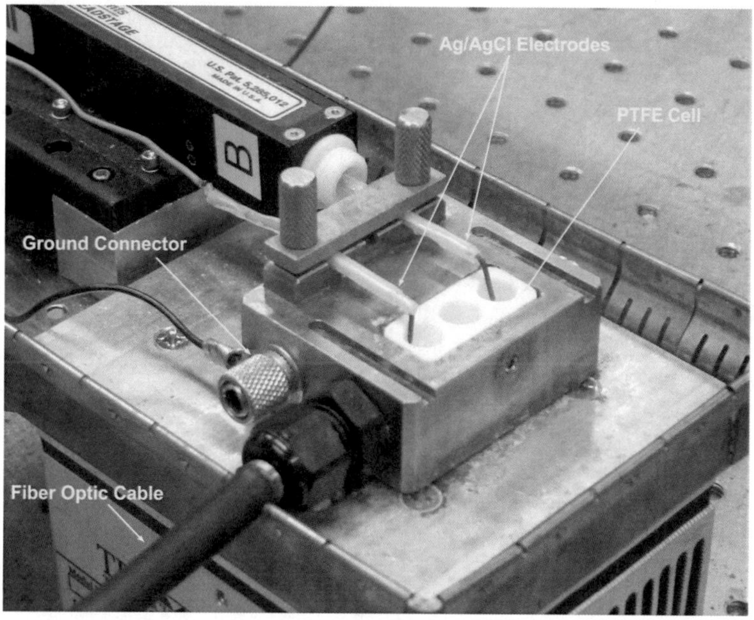

Fig. 4. The experimental apparatus with components in place. The aluminum cell housing has been gold plated to remove the oxide layer and improve the attenuation of electromagnetic noise. A first faraday cage is placed overtop of the electrodes and PTFE cell (positioned by the grooves in the aluminum cell housing). A second faraday cage encloses the entire system.

2. Position the solid wire inside a 6-in.-long piece of 0.036-in. ID PTFE/FEP tubing such that only a small portion of the thermocouple wire protrudes from the end.
3. Using the heat gun, shrink ~0.25 in. of tubing around the thermocouple wire until the tubing closes completely.
4. Use a new razor blade to cut the tubing near the end of the closed portion, preserving ~1/16 in. of the closed portion on the end of the tubing (to ensure that the aperture is 0.001 in. in diameter).
5. Gently pull the solid wire from the back end of the tubing. Be careful not to break the thermocouple wire in the tubing. To check that the aperture has been created properly, push water through the back end of the tubing using a NORM-JECT® syringe (with the appropriate syringe connector, **Fig. 2a**). A very fine stream of water will spray from a good aperture.
6. Insert the tubing into the cell and bend it into the U-shaped formation as shown in **Fig. 3** (*see* **Note 5**).

3.2. Silver Chloride (Ag/AgCl) Electrode Fabrication

1. Insulate a 1-in.-long section of 1-mm-diameter silver wire by heat-shrinking 0.06-in. ID PTFE/FEP using the heat gun. Leave approximately 0.25 in. of silver overhanging on each end (the silver wire should be approximately 1.50-in. long). Repeat this step using the 0.130-in. ID PTFE/FEP tubing.

2. For the power electrode **(Fig. 5a)**, solder one end of the silver wire to a crimp pin, then heat-shrink a piece of PTFE/FEP tubing around the soldered portion of the pin.
3. For the ground electrode **(Fig. 5b)**, solder an approximately 8-in.-long piece of solid wire to one end of the silver wire, then heat-shrink a piece of PTFE/FEP tubing around the soldered portion of the wire. Solder the other end of the solid wire to a crimp pin and heat-shrink a piece of PTFE/FEP tubing around the soldered portion of the pin.
4. Bend the free end of the silver wire by 90° (for both electrodes) and immerse the electrodes into bleach solution overnight to create a layer of AgCl on the electrode surface. This step should be repeated before every experiment.

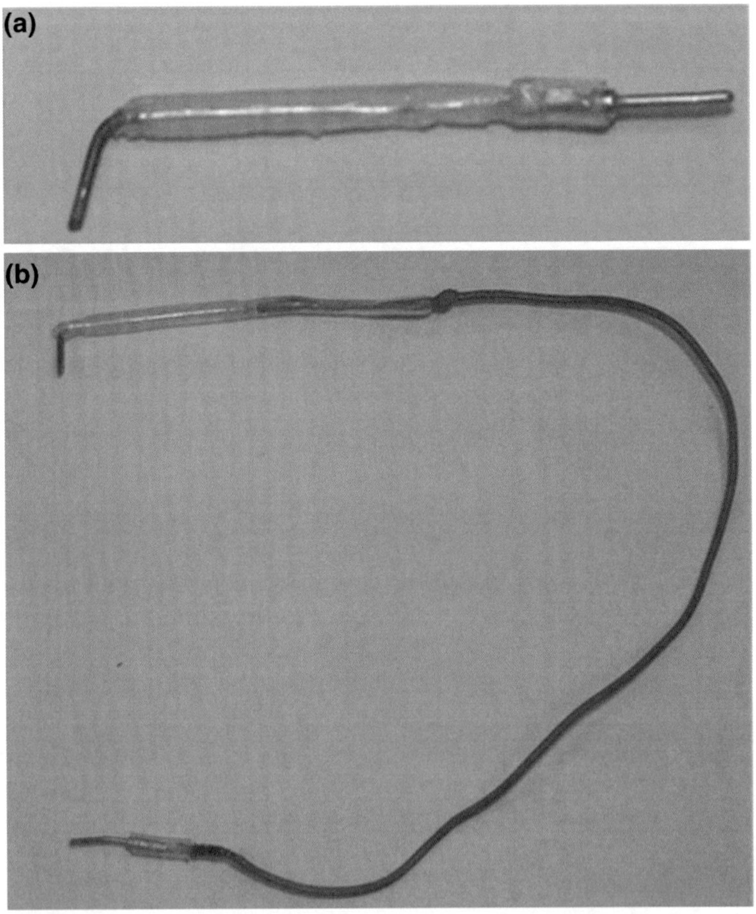

Fig. 5. **(a)** The power electrode positioned on the *trans* side (the non-aperture side) of the PTFE cell. **(b)** The ground electrode positioned on the *cis* side (aperture side).

3.3. Forming a Black Lipid Bilayer

1. The day before forming a nanopore, set some lipid out to dry in test tubes. Clean three test tubes by blowing pressurized air in them to remove any dust that may have settled on the bottom of the tube. Transfer 50 μL of 10 mg/mL 1,2-diphytanoyl-*sn*-glycero-3-phosphocholine/chloroform solution into each test tube (using a glass capillary pipette). Let the lipid dry overnight in a fume hood (*see* **Note 6**).
2. The day of the experiment, turn on the patch clamp amplifier. Allow it to warm up for 1 h before beginning to record data.
3. Perform this step in a fume hood; use PVC gloves, a lab coat, and safety glasses. Immerse a PTFE cell in a solution of ~15% nitric acid (v/v) in dH_2O. Bring the solution to a boil and then dump it into a basin (prefilled with ~0.5 L of water; neutralize with 6 M NaOH). Repeat this boiling procedure two more times with deionized water only.
4. Sequentially flush the U-tube of the cell with ~200 μL of each of the following solutions: dH_2O, ethanol, and hexane. To flush the tubing, load a NORM-JECT® syringe (fitted with the appropriate connector, **Fig. 2a**) with the corresponding solution and push the solution through the tubing from the *trans* side (i.e., the non-aperture end). A thin stream of liquid should spray from the aperture. To prevent introducing air bubbles, remove any air trapped at the forefront of the syringe before connecting it (*see* **Note 7**).
5. Push any remaining solution out of the tube with air (using a dry NORM-JECT® syringe connector) and place the cell in the vacuum chamber. Pump vacuum for approximately 5 min, or until the remaining solution inside the U-tube has evaporated (*see* **Note 8**).
6. Using a glass pipette, aliquot 250 μL of hexane into one of the test tubes of dried lipid to dissolve the lipid. Cap the tube to prevent evaporation.
7. Remove the silver chloride electrodes from the bleach solution and rinse them with deionized water.
8. Clean the lipid brush with an ethanol soaked KimWipe®. Allow the brush to dry.
9. Securely fasten the PTFE cell into the aluminum cell housing (**Fig. 4**). With the aid of the stereomicroscope and illuminator, pipette 1.5 μL of the hexane-lipid solution (from **step 6**) onto the aperture of the U-tube. This coats the surface and inner walls of the tube with a lipid monolayer. Repeat this once more after the hexane has evaporated.
10. Using a NORM-JECT® syringe and syringe connector (**Fig. 2b**), push air through the U-tube from the *trans* side to remove

any lipid solution inside the tube. Note the location of efflux of the hexane-lipid solution because this is the location of the aperture. Once the tube is visually dry, place the aluminum cell housing (with the PTFE cell securely fastened) into the vacuum chamber and pump vacuum for approximately 5 min (*see* **Note 8**).

11. Wear safety glasses for this step. Aliquot 3–4 μL of 1-hexadecene to another test tube of dried lipid. Allow the 1-hexadecene to coat and soften the lipid for approximately 20–30 s. Invert the test tube; gently and repeatedly tap the base of the tube to remove the excess 1-hexadecene from the lipid. Be careful not to shatter the test tube. Store the tube in the inverted position to prevent the 1-hexadecene from collecting at the bottom (*see* **Note 9**).
12. Install the electrodes **(Fig. 4)** and connect them to the head stage of the patch clamp amplifier (the *cis* side—i.e., the aperture side—is ground). Connect the aluminum cell housing to ground (through a connection to the vibration isolation table). Set the output gain on the patch clamp amplifier to 1. Ensure that the applied voltage is 0, the pipette capacitance compensation is minimized, and that the whole cell capacitance compensation and the series resistance compensation are off (these are the parameters for the Axopatch 200B). The patch clamp amplifier should be set to low-pass filter at 10 kHz (if performing nanopore force spectroscopy, refer to the Chapter "Nanopore Force Spectroscopy on DNA Duplexes").
13. Using a NORM-JECT® syringe and syringe connector **(Fig. 2b)**, gently push the KCl buffer or analyte solution (e.g., DNA solution in 1 M KCl, if performing nanopore force spectroscopy, refer to the Chapter "Nanopore Force Spectroscopy on DNA Duplexes") through the U-tube from the *trans* side (be wary of air bubbles and ensure that none are introduced into the tubing) (*see* **Note** 7).
14. Fill the *cis* and *trans* sides of the PTFE cell with KCl buffer. Fill both the wells such that the electrodes from each particular side are fully immersed in the solvent.
15. Apply a 5-mV square wave potential using the patch clamp amplifier (Seal Test toggle switch on the Axopatch 200B). Current through an open tube will appear as a square wave; if the tube contains an obstruction (e.g., an air bubble), there will be a zero DC bias with spikes during potential steps, due to the capacitance of the obstruction (*see* **Note** 7).
16. Adjust the offset potential to zero the current (Pipette Offset knob of the Axopatch 200B). Adjust the scale (and the trigger) of the oscilloscope to view a single cycle of the seal test.

Monitor the current for a few minutes; significant, continual current drift indicates cell leaks or depletion of the AgCl layer on the electrodes (*see* **Note 10**).

17. Using the lipid brush, transfer a small amount of lipid from the lipid/1-hexadecene test tube to the *cis* side of the U-tube (i.e., the end containing the aperture). When collecting the lipid, keep the test tube in the inverted position at all times. When transferring the lipid, transfer it onto the edge of the U-tube (away from the aperture). After transferring the lipid, wipe the bristle with a KimWipe® (*see* **Note 11**).
18. Using the lipid brush, mix the lipid for approximately 20–30 s until it becomes noticeably less viscous. Move the lipid to the edge of the U-tube, and clean excess lipid from the aperture using the lipid brush. It is important not to excessively mix the lipid because it will become too soft for practical use (*see* **Note 12**).
19. Using a NORM-JECT® syringe and syringe connector, push a small amount of KCl buffer (or analyte solution, if performing nanopore force spectroscopy, refer to the Chapter "Nanopore Force Spectroscopy on DNA Duplexes") through the U-tube from the *trans* side to clear the inside of lipid. If it is cleared then the capacitive spike should disappear (*see* **Note 13**).
20. Form a lipid bilayer. Using a 20-μL micropipettor, blow an air bubble across the aperture and suck it back up. If a bilayer is formed, a capacitive spike will appear on the oscilloscope.
21. Check for leaks by applying a 100-mV potential across the bilayer. Ionic current of <1 pA indicates a good bilayer. Current higher than this indicates leaky bilayers, or leaks in the cell itself (*see* **Note 14**).
22. Check the mechanical properties of the bilayer by applying an increasing potential until the bilayer breaks, causing the current to instantaneously saturate. A good bilayer will burst at a voltage of 450–550 mV. Bilayers bursting at <300 mV are generally unstable and should not be used (*see* **Note 15**). Membranes bursting at >600 mV occur when a thick lipid blockage occludes the tube; such a blockage is not a bilayer and cannot be used to form a nanopore (*see* **Note 16**). Reform a bilayer and repeat this step a couple of more times to ensure that the bilayers consistently burst between 450 and 550 mV.

3.4. α-Hemolysin Pore (or Multiple Pore) Formation

1. Prepare a NORM-JECT® syringe by filling it with KCl buffer.
2. For forming a single nanopore, prepare a diluted solution of α-toxin by transferring 3 μL of the α-HL solution (the α-HL dissolved in glycerol/water) with 50 μL of KCl buffer. For forming multiple nanopores, mix 3 μL of the α-HL solution with 20 μL of KCl.

3. Form a lipid bilayer (refer to **step 20** of **Subheading 3.3**) and add 3 μL of the diluted α-toxin solution to the *cis* side of the PTFE cell (the side containing the aperture of the U-tube). Add the α-toxin away from the aperture, and mix it into the KCl buffer by refluxing with the micropipette (*see* **Note 17**).
4. Set the potential to 100 mV and wait for channel insertion. An α-HL nanopore will conduct a current of 90–100 pA at 100 mV and a current of –70 to –79 pA at –100 mV (open channel current values for a single pore in 1 M KCl). Check both polarities of the potential before proceeding (it is possible for a secondary structure of the protein to incorporate into the lipid bilayer in which case the conducted current will reside outside of the above ranges. In this case, burst the bilayer and reform it).
5. For forming multiple nanopores, wait until 50–150 nanopores incorporate into the lipid bilayer before proceeding (the conducted current should be approximately 50–150 times the single-channel values, respectively).
6. Flush out and replace the KCl buffer on the *cis* side of the PTFE cell. Attach the syringe from **step 1** to the longer needle of the Tygon®-syringe connector; squirt some liquid out to remove any air bubbles. Connect the other needle to the hose of the venturi pump, and turn the pump on. Insert the longer needle into the *cis* well and inject the clean KCl solution while simultaneously removing excess liquid through the shorter needle connected to the venturi pump. Make sure that the liquid level on the *cis* side of the PTFE cell does not drop below the top surface of the U-tube, otherwise the pore(s) will disappear. Perform this step quickly to prevent further pores from incorporating into the bilayer. Repeat this step five times (*see* **Note 18**).
7. Turn off the 100-mV applied potential. Zero the current if required (*see* **Note 19**).

4. Notes

1. Avoid transferring chloroform (and hexane) with plastic, which can contaminate the solutions and affect the quality of the lipid. Wherever possible, use glass pipettes or capillary pipettes.
2. To make the brush, form a U-shaped loop with a single bristle from a Sable artist's paint brush, and insert both free ends of the bristle into a 10-μL pipette tip, leaving a loop ~2-mm long. Glue the loop in place using instant adhesive.

3. 0.06-in. ID PTFE/FEP dual-shrink tube may also be used depending on the outer diameter of the hypodermic needles. When heat-shrinking the tubing, the goal is to shrink the suspended portion such that when placed around the native 0.036-in. ID PTFE/FEP dual-shrink tube it forms a snug fit.
4. Because the experiments are sensitive to electrical and mechanical noise, the peltier device should be liquid cooled as opposed to air cooled by a fan. Liquid hard-drive coolers that pump liquid through a water block (to which the peltier should be fastened) is a good noninvasive cooling option.
5. It can be difficult to insert the tubing into the through hole of the PTFE cell because the fit is tight. Heat-shrink ~1½ in. at the non-aperture end of the PTFE/FEP tubing and insert that end ("guide end") into the through hole. Use pliers to pull on the guide end to pull the rest of the tubing through. Once the tube is in place, cut the excess tube from the non-aperture end.
6. Lipid takes at least 4–5 h to dry once transferred. Lipid oxidizes over time; do not keep dried lipid for more than 2 days.
7. If air bubbles do get trapped inside the tube and are obstructing the flow of liquid, push solution through the *cis* side of the tubing (i.e., the aperture end). If doing this, it is crucial to remove any air trapped at the forefront of the syringe. After clearing the obstruction, return to pushing solution through the *trans* side. If the flow remains impeded after numerous attempts, restart with a different PTFE cell.
8. The syringe connector used to push air through the U-tube should always remain dry (i.e., for pushing solution through the U-tube use a different connector).
9. Extending this time beyond 30 s will likely result in over-softening of the lipid.
10. Depleted electrodes are white or have white patches present, in which case they need to be replated in bleach. If this is not the case, then gently disassemble the PTFE cell (without emptying it) and check for leaks or spills; wipe any KCl solution from the cell or apparatus and reassemble. In general, the current will drift slightly over time, in which case it should be re-zeroed. The offset current should be re-zeroed periodically, e.g., every 10 min during experiments.
11. Avoid contact between the harvested lipid and any 1-hexadecene residing on the walls of the test tube. Excess 1-hexadecene may over-soften the lipid.
12. When moving the lipid back to the edge of the U-tube, leave a very small layer of lipid near the aperture to act as a reservoir for bilayer formation.

13. If the lipid clears from the aperture but spontaneously reenters, it may imply that the lipid is too soft. Spontaneous clogging of the aperture may be remedied by pushing solution through the tubing from the *trans* side with one hand and using the lipid brush with the other to push lipid away from the aperture. Refer to **Note 15** for a possible remedy to the excess softening of the lipid.
14. If a substantial leakage current is observed, remix the lipid (refer to **step 18** of **Subheading 3.3**). If this does not solve the problem, then check for leaks or spills (*see* **Note 10**).
15. If the bilayer bursts at potentials less than 300 mV, it can imply one of two things. Either the lipid has not been mixed well enough, in which case, remix the lipid (refer to **step 18** of **Subheading 3.3**). The alternative is that the lipid is too soft (which may be a product of over-mixing or too much 1-hexadecene). If this is the case, transfer another fraction of lipid (**step 17** from **Subheading 3.3**; alternatively, aliquot 1-hexadecene into a third test tube of dry lipid and transfer this lipid to the *cis* side of the U-tube), which can be used as a mop to collect the soft lipid that surrounds the aperture, replacing it with fresh, unmixed lipid. This new lipid need not be mixed thoroughly as required in **step 18** of **Subheading 3.3**.
16. Clear the U-tube of lipid (refer to **step 19** of **Subheading 3.3**).
17. Do not add the α-toxin too close to the aperture to prevent runaway nanopore formation. If this occurs, then flush out and replace the KCl buffer on the *cis* side of the cell (refer to **step 6** of **Subheading 3.4**). If, after 10 min, no nanopores have formed, add another 2 μL of the diluted α-toxin solution to the *cis* side of the cell and mix appropriately. Burst and reform the bilayer.
18. It is a good idea to have the Tygon®-syringe connector inserted into the Tygon® tubing of the venturi pump and attached to the NORM-JECT® syringe before reaching this step (**step 6** of **Subheading 3.4**). Once a pore incorporates in the bilayer (e.g., in the case when making a single nanopore) there is limited time to flush out and replace the KCl solution before a second pore forms. It is important to avoid spilling any solution around the cell, because this creates leaks from the *cis* to the *trans* chamber.
19. When cleaning up (at the end of performing any experiments), empty the *cis, trans*, and U-tube contents of the PTFE cell into the basin **(step 3**, **Subheading 3.3)** and place the cell into a separate container than the clean PTFE cells (cleaned in the manner described in **step 3**, **Subheading 3.3**). Clean the electrodes by lightly sanding the surface with the wetordry® sanding paper, and then re-immerse the electrodes in bleach.

Acknowledgments

We thank members of the University of British Columbia Applied Biophysics Laboratory, in particular, Vincent Tabard-Cossa and Dhruti Trivedi. Many thanks to Jonathan Nakane, Mark Akeson, and Carolina Tropini for their important contributions to this protocol. This work was supported by the US National Institutes of Health (NIH). Financial support for Matthew Wiggin was provided by the Natural Sciences and Engineering Research Council of Canada (NSERC).

References

1. Bhakdi, S., and Tranum-Jensen, J. (1991). Alpha-toxin of Staphylococcus aureus. *Microbiol. Mol. Biol. Rev.* **55**, 733–751.
2. Song, L.Z., Hobaugh, M.R., Shustak, C., Cheley, S., Bayley, H., and Gouaux, J.E. (1996). Structure of staphylococcal alpha-hemolysin, a heptameric transmembrane pore. *Science* **274**, 1859–1866.
3. Kasianowicz, J.J., Brandin, E., Branton, D., and Deamer, D.W. (1996). Characterization of individual polynucleotide molecules using a membrane channel. *Proc. Natl Acad. Sci. U. S. A.* **93**, 13770–13773.
4. Akeson, M., Branton, D., Kasianowicz, J.J., Brandin, E., and Deamer, D.W. (1999). Microsecond time-scale discrimination among polycytidylic acid, polyadenylic acid, and polyuridylic acid as homopolymers or as segments within single RNA molecules. *Biophys. J.* 77, 3227–3233.
5. Nakane, J., Wiggin, M., and Marziali, A. (2004). A nanosensor for transmembrane capture and identification of single nucleic acid molecules. *Biophys. J.* **87**, 615–621.
6. Howorka, S., Movileanu, L., Braha, O., and Bayley, H. (2001). Kinetics of duplex for individual DNA strands within a single protein nanopore. *Proc. Natl Acad. Sci. U. S. A.* **98**, 12997–13001.
7. Sauer-Budge, A.F., Nyamwanda, J.A., Lubensky, D.K., and Branton, D. (2003). Unzipping kinetics of double-stranded DNA in a nanopore. *Phys. Rev. Lett.* **90**, 238101-1–238101-4.
8. Ashkenasy, N., Sanchez-Quesada, J., Bayley, H., and Ghadiri, M.R. (2005). Recognizing a single base in an individual DNA strand: a step toward DNA sequencing in nanopores. *Angew. Chem. Int. Ed.* **44**, 1401–1404.
9. Astier, Y., Braha, O., and Bayley, H. (2006). Toward single molecule DNA sequencing: direct identification of ribonucleoside and deoxyribonucleoside 5′-monophosphates by using an engineered protein nanopore equipped with a molecular adapter. *J. Am. Chem. Soc.* **128**, 1705–1710.
10. Shin, S., Luchian, T., Cheley, S., Braha, O., and Bayley, H. (2002). Kinetics of a reversible covalent-bond-forming reaction observed at the single molecule level. *Angew. Chem. Int. Ed.* **41**, 3707–3709.
11. Luchian, T., Shin, S., and Bayley, H. (2003). Kinetics of a three-step reaction observed at the single-molecule level. *Angew. Chem. Int. Ed.* **42**, 1925–1929.
12. Jung, Y., Bayley, H., and Movileanu, L. (2006). Temperature-responsive protein pores. *J. Am. Chem. Soc.* **128**, 15332–15340.
13. Oukhaled, G., Mathe, J., Biance, A.L., Bacri, L., Betton, J.M., Lairez, D., Pelta, J., and Auvray, L. (2007). Unfolding of proteins and long transient conformations detected by single nanopore recording. *Phys. Rev. Lett.* **98**, 158101-1–158101-4.
14. Healy, K. (2007). Nanopore-based single-molecule DNA analysis. *Nanomedicine* **2**, 459–481.
15. Nakane, J.J., Akeson, M., and Marziali, A. (2003). Nanopore sensors for nucleic acid analysis. *J. Phys.: Condens. Matter.* **15**, 1365–1393.
16. Rhee, M., and Burns, M.A. (2006). Nanopore sequencing technology: research trends and applications. *Trends Biotechnol.* 24, 580–586.
17. Dekker, C. (2007). Solid-state nanopores. *Nat. Nanotechnol.* **2**, 209–215.
18. Tropini C., and Marziali A. (2007). Multi-nanopore force spectroscopy for DNA analysis. *Biophys. J.* **92**, 1632–1637.

Chapter 10

Nanopore Force Spectroscopy on DNA Duplexes

Nahid N. Jetha, Matthew Wiggin, and Andre Marziali

Summary

Force spectroscopy can be applied using nanopores to study charged molecules such as nucleic acids. This technique can be used to study the binding energy of a DNA duplex by threading an anchored single-stranded DNA (ssDNA) probe molecule through a nanopore (having a diameter large enough to accommodate only a single strand) and allowing target DNA on the backside of the pore to hybridize to the probe. Electric potential can be used to apply a force to the charged ssDNA in a direction tending to translocate the duplex through the pore. If the pore is only large enough to accept ssDNA, the duplex must dissociate for the probe to escape the pore. The dissociation time of the duplex can therefore be measured under applied force, and (provided that enough dissociation events have been recorded) a characteristic time scale for dissociation can be determined. In this chapter, we present a detailed protocol for performing nanopore force spectroscopy on DNA duplexes using one or more α-hemolysin nanopores. We present the details of the measurement of the duplex survival probability under force, and show that dissociation time scales for duplexes that are perfectly complimentary differ by greater than approximately two orders of magnitude from those containing a single sequence mismatch, offering opportunities for sequence detection.

Key words: Nanopore, Force spectroscopy, Genotyping, SNP detection

1. Introduction

The dissociation energy of a DNA duplex is largely determined by its length, sequence, and any mismatches between the two strands. Single-molecule force spectroscopy *(1)* using nanopores (nanopore force spectroscopy *(2, 3)*) provides an excellent means to study the energy landscapes that govern DNA duplex stability, and thus to uncover sequence mismatches in duplexes of known length. Briefly, this is done by observing the effects of a range of

James Weifu Lee and Robert S. Foote (eds.), *Micro and Nano Technologies in Bioanalysis*,
Methods in Molecular Biology, vol. 544
DOI 10.1007/978-1-59745-483-4_10, © Humana Press, a part of Springer Science+Business Media, LLC 2009

forces on dissociation kinetics of a duplex. Nanopores provide a means of electrically applying such a force, and of simultaneously detecting dissociation one molecule at a time.

Nanopore force spectroscopy has been used extensively *(2–8)* with application to nucleic acids, including in studying the effects of sequence length and homology (i.e., the presence or absence of mismatches) on duplex stability. Because mismatches reduce the dissociation energy of a duplex, nanopore force spectroscopy can be used for genotyping tests (e.g., single-nucleotide polymorphism [SNP] detection) *(3)*. High sensitivity and label-free detection may enable the technique to operate faster and cheaper than conventional genotyping techniques, providing an avenue for the introduction of genomic analysis to personalized health care. A modification of the process has been used to study DNA-protein interactions *(7)*.

In a nanopore force spectroscopy experiment, a DNA molecule containing a double-stranded region and a single-stranded overhang is driven through a nanopore by electrostatic forces. The pore is large enough to accommodate only a single strand, forcing the DNA duplex to dissociate if the single-stranded region is pulled through the pore with sufficient force. The electrostatic force is chosen such that it lowers, but does not entirely remove the energy barrier for dissociation: an additional contribution from thermal energy is required. The force-modified thermal dissociation time is recorded, and the process is repeated many times at each of a number of experimental conditions (force, temperature, etc.). Kinetic parameters (i.e., characteristic dissociation time, or dissociation rate) under each set of conditions are estimated by fitting the distribution of dissociation times; such analysis is complicated by interactions between DNA and the nanopore *(9)*. For genotyping applications, qualitative comparison of characteristic dissociation times can be sufficient to distinguish between perfectly complementary duplexes and duplexes containing mismatches. If measurement of the absolute height of the energy barrier is desired, experiments must be performed at a number of different temperatures.

In this chapter, we present the details for performing single-molecule force spectroscopy on DNA duplexes using an α-hemolysin (α-HL) nanopore, and an extension of this technique that uses many nanopores in parallel (multi-nanopore force spectroscopy). The method described here assumes that the reader is capable of forming an α-HL nanopore as outlined in the Chapter "Forming an alpha-hemolysin nanopore for single-molecule analysis." A considerable portion of this chapter describes the details of the data acquisition (DAQ) and analysis software that we have developed, and the method by which we analyze the data to distinguish between DNA duplexes that are perfectly complementary and those that contain one or more mismatches.

The protocol and molecules described here, which follows Nakane et al. *(2)*, allows label-free sequence detection in a DNA sample, as well as keeping probe molecules separate from an unlabeled sample (target molecules) on opposite sides of the pore. A single-stranded DNA (ssDNA) probe bound to avidin at one end is placed on the *cis* side of the pore, with the target molecule placed on the *trans* side. In experiments, the probe is driven into the nanopore by an applied electric potential, allowing for duplex formation on the *trans* side of the pore (**Fig. 2**). Upon formation of the DNA duplex, the potential is reversed, causing duplex dissociation. Other groups have used DNA hairpins *(5, 8)* or have tethered ssDNA to the pore itself *(10)*; the principles of operation in such cases are similar, although some details of the protocol will necessarily be different.

2. Materials

The following assumes the use of the materials for forming an α-HL nanopore as outlined in the Chapter "Forming an alpha-hemolysin nanopore for single-molecule analysis".

2.1. Nanopore Force Spectroscopy

The following are details regarding the design and concentration of oligonucleotides used in a typical nanopore force spectroscopy experiment. Proper design of the ssDNA probe is especially important. During experiments, the probe is threaded through the pore, anchored by avidin on the *cis* side to prevent translocation, and forms a duplex with the analyte on the *trans* side.

1. Probe (Integrated DNA Technologies, Coralville, IA):
   i. The probe should be biotinylated at the 3′ end; this is subsequently bound to avidin.
   ii. A 30-nucleotide linker sequence is required to extend through to the *trans* side when a probe molecule is captured in the pore. Check that the linker does not contain significant homology with the target or form a secondary structure.
   iii. The active sequence (complementary to the target) should be at the 5′ end of the probe DNA sequence. Mismatches between the probe and target sequences should be as close to the middle of the active sequence as possible, because they have the greatest effect on the binding energy of the duplex. In **refs 2** and **3**, the active portion of the probe was 14-nucleotides long.
   iv. As an example, the probe used in **ref 3** was 5′-CCACCAACCAAACC$(dA)_{30}$-3′-biotin with avidin bound on the 3′ end. (*see* **Note 1**).

2. Target (Integrated DNA Technologies):
   i. The target sequence must be single stranded (to allow it to hybridize to the probe).
3. 100 μM Avidin (Invitrogen, Carlsbad, CA) in 1× phosphate-buffered saline (PBS) (e.g., 6.6 mg in 1 mL of 1× PBS). The avidin solution along with the powdered avidin should be stored at –20°C.
4. 100 μM solution of probe in TE buffer (10 mM Tris, 1 mM EDTA, pH 8.0). The probe solution should be stored at –20°C.
5. 100 μM solution of target in TE buffer (10 mM Tris, 1 mM EDTA, pH 8.0). The target solution should be stored at –20°C.
6. Working probe solution:
   i. 6 parts 1 M KCl, 10 mM HEPES, 1 mM EDTA, pH 8.0.
   ii. 2 parts 2 M KCl, 20 mM HEPES, pH 8.0.
   iii. 1 part 100 μM probe solution.
   iv. 1 part 100 μM avidin solution.Final probe concentration: 10 μM. Working probe solution should be stored at –20°C (*see* **Note 2**).
7. Working analyte solution:
   i. 248 parts 1 M KCl, 10 mM HEPES, 1 mM EDTA, pH 8.0.
   ii. 1 part 2 M KCl, 20 mM HEPES, 2 mM EDTA, pH 8.0.
   iii. 1 part 100 μM target solution.

Final target concentration: 0.5 μM. The working target solution should be stored at –20°C (*see* **Note 3**).

2.2. Control, Acquisition, and Analysis Software and Hardware

It is possible to purchase packages of computer hardware and software for DAQ and analysis from electrophysiology instrument manufacturers. A less expensive alternative is to purchase a patch clamp amplifier, a multipurpose DAQ card, and write custom DAQ and analysis software. Specifications discussed below are appropriate for either option.

2.2.1. Hardware

1. Low-noise, patch clamp amplifier (e.g., Axopatch 200B, Molecular Devices, Sunnyvale, CA).
2. DAQ card with the following features: at least two analog inputs and one analog output, minimum 8-bit sampling resolution, and maximum sampling rate of at least 250 KS/s on two input channels simultaneously (e.g., NI-PCI-6040E, National Instruments, Austin, TX).
3. Shielded I/O connector block for connecting the DAQ card to the patch clamp amplifier (e.g., National Instruments SCB-68) and shielded cable (National Instruments).

4. Vibration isolation table. Mechanical vibrations can cause significant noise during experiments. To minimize noise from mechanical sources, the nanopore cell should be mounted on an isolation table (e.g., Newport Instruments LW Series, Newport Instruments, Irvine, CA).

2.2.2. Acquisition Software

The main purpose of the DAQ software is to control the electrostatic potential across the nanopore while recording the current and any events that ensue resulting from entry and exit of probes from the pore. The software functions as a state machine using the current to monitor the blocked or open state of the pore (i.e., states in which a probe is present or absent from the pore, respectively), and responding by changing the applied potential. For example, probe capture decreases the ionic current by a factor of nearly 4. The acquisition software senses this change and executes a user-defined protocol, e.g., holding the probe to allow probe-target hybridization, and subsequently reducing the potential to the negative dissociation potential (**Fig. 2**). The software then waits in this new state to observe the decrease in impedance that accompanies dissociation and probe exit. Concurrently, the acquisition software records current and voltage information via the DAQ card for later analysis.

We have written our software in LabVIEW 8.0 (National Instruments); prewritten functions for data acquisition and DAQ card interfacing make this relatively straightforward.

Acquisition software requirements:

1. Recording of the voltage and current data at a rate of at least twice the patch clamp amplifier's filter frequency. Higher sampling rates are preferred.
2. The capacity to output data (i.e., corresponding to the applied potential) to a single channel on the DAQ card.
3. The capability of creating user-defined states that place conditions for state switching (in response to the ionic current, voltage, and/or time in a particular state).

2.2.3. Single-Nanopore Analysis Software

The following is a description of the analysis software that we have developed to analyze the current signature recorded from a single-nanopore force spectroscopy experiment. Its main purpose is to scan the data generated by the acquisition software, and extract dissociation time information from probe-analyte binding events. The statistical spread in dissociation times over a large number of events is later used to calculate the survival probability of the probe in the pore (i.e., the probability that the probe will remain in the pore after an elapsed period of time). In the case of nanopore force spectroscopy, the survival probability of the probe in the pore is calculated based only on events in which the probe was bound to an analyte molecule such that probe escape required dissociation from the analyte.

1. The software reads the potential and current from the data file, and identifies dissociation events, recording relevant information, such as duration, for each event.
2. The analysis software functions as a state machine. A series of expected voltage and current states are defined by the user to match potential and current states that occur during DAQ (refer to **Subheading 3.2.1**).
3. State switching decisions are triggered by:
   i. Voltage and/or current entering into a specified range.
   ii. Voltage and/or current going outside a specified range.
   iii. "Timing out," i.e., remaining in the current state for greater than a specified time.

   The identity of the new state depends on which of i, ii, or iii occurs.
4. Event timing is triggered by states defined by the user. Timing starts when the potential reaches the dissociation potential, and ends when the ionic current increases in magnitude from the blocked to the open value upon duplex dissociation and probe exit.
5. The following event properties are recorded by the software (in tabulated form):
   i. The event duration in seconds.
   ii. The average applied voltage.
   iii. The start time of the event (in seconds) relative to the beginning of the acquired data file.
   iv. The calculated survival probability of the probe in the pore (*see* **Note 4**).
   v. A Boolean flag indicating an event that was terminated before probe-analyte dissociation.
   vi. The pathname to the data file from which the event was read.
6. The software includes a means of viewing the events with the event start and end times marked, which is useful for debugging analysis.

2.2.4. Multi-nanopore Analysis Software

The following is a description of our custom-made software to analyze voltage and current signatures acquired from a multi-nanopore force spectroscopy experiment. In these experiments, many pores exist in the membrane. As probes dissociate from duplexes over many pores, the current through the membrane returns asymptotically to the open state (i.e., the number of pores times the open channel current through a single pore). Each event recorded in this case consists of synchronized dissociation over many pores. To reconstruct the survival probability of the probes in the pores (analogous to the single-nanopore case, refer to **Subheading 2.2.3**), many such events are averaged together; the resulting normalized current trace is a measure of the survival probability of the probe-analyte duplexes over time under a specific applied force. The multi-nanopore analysis software consists of two programs and is designed to determine the normalized current trace during dissociation.

1. The first program (the multi-pore raw analysis program) functions as a state machine where a series of voltage states are user-defined, which are equivalent to those that are cycled through during a dissociation cycle (refer to **Subheading 3.2.2**).
2. The program reads the acquired data file, and cycles through each defined state depending on the voltage at a particular time.
3. Threshold voltage levels and recording states are defined that signal the start and end of the events to be recorded. State switching is determined by the voltage, time elapsed in the current state, or a combination of the two. State switching decisions are triggered by achieving a voltage within a specified range; having a voltage that exits a specified range; or by timing out of the current state. The new state is determined by which condition is met.
4. The multi-pore raw analysis program is designed to calculate and write to a first text file (in tabulated form) specific properties of an event as defined by the threshold voltage levels, which are the:
   i. Start time of the event (in seconds) relative to the beginning of the acquired data file.
   ii. Event time (i.e., the length of time of the event in seconds).
   iii. The pathname to the data file from which the event was read.
5. After successful writing of the first text file, a multi-pore event viewer program (the second program) is run that reads the start time and end time (calculated) from the text file for each event with the corresponding data file for the event, and writes to a second text file (in tabulated form) the following properties of the event, *at each particular moment in time* (as defined by the sampling rate upon acquisition of the data):
   i. The time elapsed from the start of the event.
   ii. The average voltage between all the events.
   iii. The instantaneous current for each recorded event.
   iv. The instantaneous current averaged over all the events.

3. Methods

3.1. Configuring the Hardware

Refer to **Fig. 1**.

1. Connect the DAQ device (e.g., The NI-PCI-6040E) to the shielded I/O connector block with the shielded cable.
2. Connect a BNC T-connector to the output voltage and one to the output current on the patch clamp amplifier. Connect one channel from the oscilloscope to the output voltage and another channel to the output current (*see* **Note 5**).
3. Using BNC cable, connect the output voltage and output current from the patch clamp amplifier to the I/O connector block in the following manner. Each BNC cable has two wires, one power wire and one ground wire. The power line and ground line from each cable should be connected to separate analog input channels on the I/O connector block (*see* **Note 5**).
4. Connect one of the analog output channels on the I/O connector block to the patch clamp amplifier. The patch clamp amplifier should have some connection that enables it to be controlled externally (*see* **Note 5**).
5. The signal from the patch clamp amplifier may need to be grounded. This can be accomplished by connecting the signal ground to a common ground.

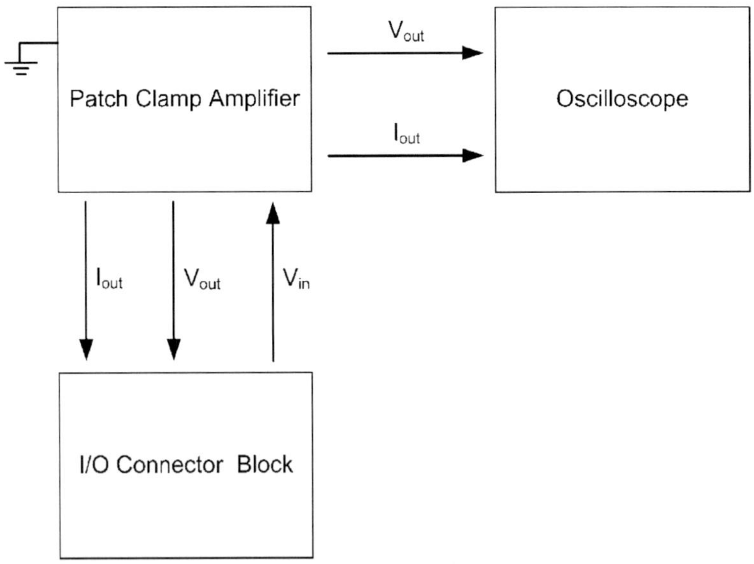

Fig. 1. Block diagram of the hardware configuration to enable external control through the DAQ card. The I/O connector block is connected (via the shielded cable) to the DAQ card.

3.2. Nanopore Force Spectroscopy

This section describes the protocol for setting up and running a nanopore force spectroscopy experiment. Note that different protocols are listed below for single pore **(Subheading 3.2.1)** and multi-nanopore **(Subheading 3.2.2)** experiments.

A series of nanopore force spectroscopy experiments (i.e., dissociation cycles) should be performed, which differ from each other by the magnitude of the dissociation potential that is applied (i.e., the potential that is applied to force probe-analyte dissociation on the *trans* side).

The optimal dissociation potential for our probe-analyte ranges between −50 and −90 mV. Use 5- to 10-mV increments between experiments. Each recorded experiment at a given dissociation potential should have at least 500 dissociation events to extract meaningful statistics (which, in the case of multi-nanopore force spectroscopy, can be achieved in a matter of minutes depending on the number of pores) (*see* **Note 6**).

1. Form a nanopore according to the instructions in the Chapter "Forming an alpha-hemolysin nanopore for single molecule analysis". Note that the target molecule must be placed on the *trans* side of the cell before nanopore formation. Flush the *cis* side of the cell with new solvent (1 M KCl). Add 10 μM working probe solution to the *cis* side of the PTFE cell (15 μL for a 250 μL *cis* chamber). Be careful not to introduce any air bubbles.
2. Cover the aluminum cell housing with the faraday cage(s), be careful not to introduce any vibrations during this step because vibrations can cause the bilayer to burst.
3. In the case of single-nanopore force spectroscopy, set the output gain on the patch clamp amplifier to 10. For multi-nanopore force spectroscopy, set the output gain to 0.5.
4. With the seal test on, adjust the capacitance compensation on the patch clamp amplifier. This is done by adjusting the pipette and whole-cell capacitance compensation parameters (parameters as defined for the Axopatch 200B) such that the capacitive spike (which should appear on the oscilloscope with the seal test on) is minimized.
5. Turn the seal test off, and set the patch clamp amplifier to external command (i.e., enabling it to be controlled by an external source). Zero the current if required.

3.2.1. Voltage States for Single-Nanopore Force Spectroscopy

Figure 2 shows a single event from a nanopore force spectroscopy experiment. Three main steps occur during each event:

1. A high *trans* side positive potential (200 mV in **Fig. 2**) is applied. A probe is captured (decreasing the current) and held, giving time to allow probe-target hybridization.

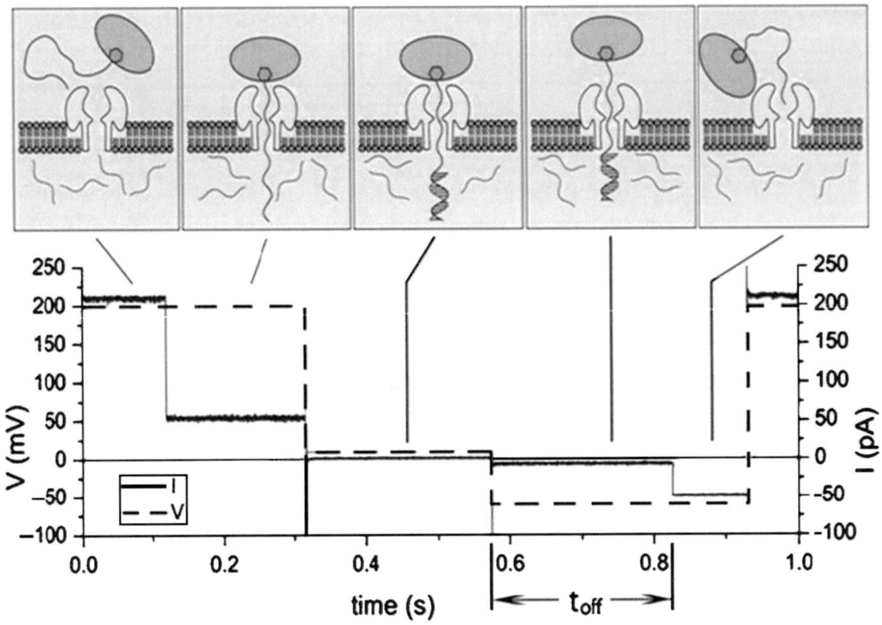

Fig. 2. A voltage and current trace from a single-molecule event. Applied potential is shown as a *dashed line*; recorded ionic current is shown as a *solid line*. Drawings above depict the state of the pore *(2)*.

2. Probe-target hybridization is tested by reducing the potential to a low, *trans* positive value (10 mV). At this potential, probes not bound to target rapidly escape from the pore. Target-bound probes are trapped in the pore by the duplex region.
3. The potential is reduced to a *trans* negative dissociation potential. The probe exits the pore after probe-target dissociation and is detected by a sudden increase in the magnitude of the current. The process is then repeated.

Set up the acquisition software to proceed through the following voltage states:

i. 0 mV reset potential for 0.1 s to record the 0 mV current between events, which is indicative of the magnitude of the offset potential (if any) being applied (not shown in **Fig. 2**).

ii. Apply a capture voltage of 200 mV. A sudden change in impedance from 1 to ~4 GΩ indicates probe capture.

iii. Hold the potential at 200 mV for 1 s (~100 ms is shown in **Fig. 2**, however we find that 1 s is optimal to allow a duplex to form). This is to provide the probe with enough time to bind to an analyte on the *trans* side.

iv. Reduce the voltage to 10 mV ("check voltage") for 0.25 s. This is to allow any unbound probes to thermally escape from the pore via the *cis* side (*see* **Note 7**).

v. Apply the dissociation potential (–60 mV in **Fig. 2**). Monitor the impedance for a reduction to ~1.5 GΩ then proceed to **step vi**. If after 10 s the probe has not escaped, manually force the probe out of the pore by applying a large negative potential. This can be done with a separate state, or manually by user control of the patch clamp amplifier (*see* **Note 8**).

vi. Once the probe has escaped, hold the potential at the dissociation voltage for 0.1 s and then return to the reset potential (voltage state i).

3.2.2. Voltage States for Multi-nanopore Force Spectroscopy

Multi-nanopore experiments proceed similarly to single-pore experiments with one important difference: the acquisition software cycles through a fixed set of potentials instead of state-switching in response to current feedback. These potentials include the following four main steps (refer to **Fig. 3**):

1. A large *trans* positive capture potential (200 mV in **Fig. 3**) is applied to trap probe molecules into the pore, causing an asymptotic decrease in the current as probes enter pores. Probes are held in place, allowing them to form probe-target duplexes on the *trans* side of the pore.

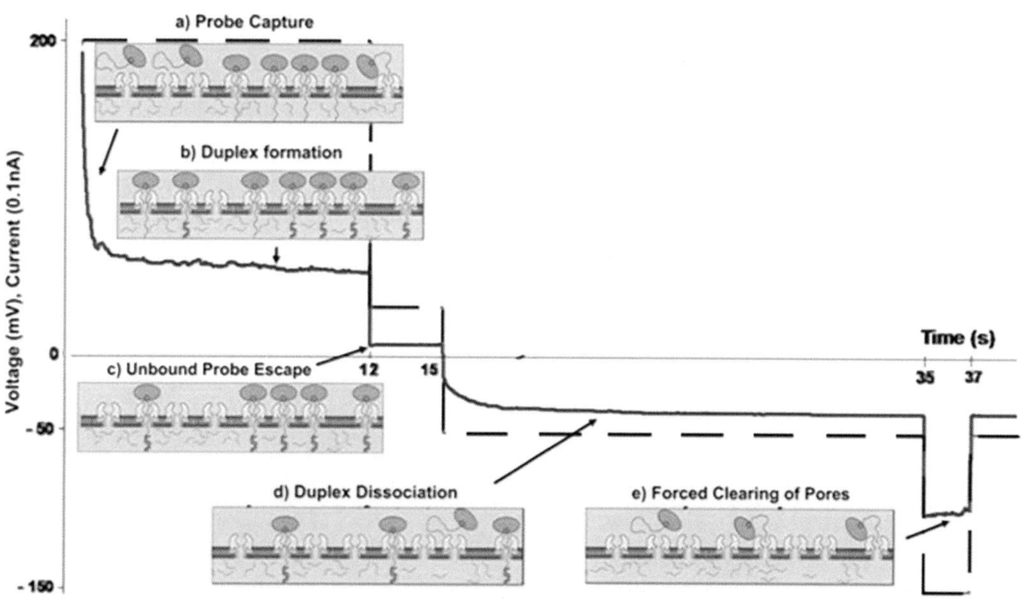

Fig. 3. Sample current (*solid line*) and voltage (*dashed line*) trace from probe capture to forced clearing of the pores during a multi-nanopore dissociation cycle. As probes are captured in the pores, the current begins to decay. After application of the capture voltage (for 10 s), a check voltage is applied to allow any unbound probes (those that did not form a duplex) to escape from the pore. The dissociation potential is then applied, forcing the probes out of the pore by shearing the formed duplex on the *trans* side. After 10 s of the dissociation potential, a –150 mV reset potential is then applied to ensure that all pores are cleared of the probe. Note that **steps i** and **vi** (subheading 3.2.2., item 4) are omitted from the figure *(3)*.

2. A reduced "check" voltage is then applied (after an elapsed period of time at the capture potential) allowing unbound probes to thermally escape from the pore.
3. The potential is reduced to a *trans* negative dissociation potential. Probe-target duplexes dissociate, causing an asymptotic increase in current magnitude.
4. The system is then "reset" by applying a very large *trans* negative potential to force any remaining probes out of the pores.

Set the following voltage states on the acquisition software:

i. 0 mV potential for 0.1 s to record the 0 mV current between events, which is indicative of the magnitude of the offset potential (if any) being applied (not shown in **Fig. 3**).

ii. Apply a capture voltage of 200 mV for 10 s to allow (for an appropriate probe concentration) most of the pores to fill with probe molecules.

iii. Reduce the voltage to 10 mV (check voltage) for 2 s (probe escape step in **Fig. 3**). This is to allow any unbound probes to escape from the pore via the *cis* side (*see* **Note 7**).

iv. Reduce the voltage to the dissociation potential for 10 s.

v. Increase the negative potential to –150 mV (reset potential) for 0.1 s to force any remaining bound probes to dissociate.

vi. Apply a 10 mV potential for 0.1 s to open any pores that may have gated upon application of the large negative potential in state v (not shown in **Fig. 3**).

vii. Apply the dissociation potential for 0.1 s before returning to state i. This is to record the open channel current at the dissociation potential, which will be required during the data analysis (**Subheading 3.3.2**).

3.3. Data Analysis

The dissociation time for a single probe-analyte duplex is the time between application of the dissociation potential and probe-target duplex dissociation (t_{off} in **Fig. 2**). The aim of raw data analysis is to determine the distribution of dissociation times for probe-target duplexes in the pore. We characterize this distribution by calculating the survival probability—the probability that the probe-target duplex will survive for at least time t after the application of the dissociation potential. The process of determining event times and survival probabilities proceeds differently in single-nanopore versus multi-nanopore experiments, but the resulting survival probability curves are effectively identical.

3.3.1. Single-Nanopore Data Analysis

In a single-nanopore force spectroscopy experiment, the analysis software scans the recorded voltage and current data, and looks for dissociation events. An event begins at the moment the dissociation potential is applied to the probe-analyte duplex, and ends when the duplex dissociates and the probe exits from the pore. The software determines the event time (i.e., the length of time, in seconds, of the event), and can be configured to record other data such as the average current during the event. When an entire set of event times has been determined, the software assigns survival probabilities to each of the times. Data such as event time, survival probability, and other relevant data is then recorded to a file for further analysis (refer to **Subheading 2.2.3**).

The analysis software only records single-molecule events that it identifies as probe-target dissociation rather than events that block the pore by other mechanisms (e.g., events in which the probe is not threaded through the pore correctly). This is done by recording only those events that follow a tightly controlled sequence of potential and current signatures, beginning with probe capture and ending after the pore clears upon probe exit. Current signatures are a sensitive indicator of the orientation and physical configuration of molecules in the pore *(11, 12)*. Events deviating from these expected signatures are discarded. The only exception to this rule occurs for events that are manually terminated after long times by changing the potential: these events are not timed, but are counted when determining the total number of events for survival probability calculations.

The current states to be defined in the analysis program are defined in terms of the expected I/I_0. value for the given state (i.e., the ratio of the current in that specific state to the open channel current measured repeatedly throughout the experiment). This rescaling simplifies data processing because most variations in current (caused by structural variation between pores, or drift caused by evaporation of KCl buffer during an experiment) affect I and I_0 proportionally. Values included here are for probe capture potentials ($V_{capture}$) of 200 mV and serve as an example (these values should be adjusted as necessary):

1. Voltage and current states as defined in the analysis software:
   i. Capture state, open channel range; 204 mV $\geq V_{capture} \geq$ 196 mV; $1.05 \geq I/I_0 \geq 0.95$; 1.05 is used as an upper bound to account for any statistical variation in the open channel current between events. I_0 is the average open channel current.
   ii. Capture state, blocked channel range; 204 mV $\geq V_{capture} \geq$ 196 mV; $0.27 \geq I/I_0 \geq 0.17$.
   iii. Analyte binding check state, blocked channel range 14 mV $\geq V_{check} \geq$ 6 mV; $0.03 \geq I/I_0 \geq -0.01$. Ignore I/I_0 until

the capacitive transient caused by the voltage transition decays.

iv. Dissociation state, blocked channel range. Example for –50 mV dissociation potential: –46 mV ≥ V_{FS} ≥ –54 mV; 0.0 ≥ I/I_0 ≥ –0.15. Ignore I/I_0 until the capacitive transient caused by the voltage transition decays.

v. Dissociation state, open channel. –46 mV ≥ V_{FS} ≥ –54 mV; –0.17 ≥ I/I_0 ≥ –0.19

(*see* **Note 9**).

2. The total number of events is needed for determining the probe survival probability. Therefore, events too short or too long to be timed accurately must be recorded, and flagged as either short or long, i.e., events where the channel was blocked during the check step (meaning a duplex was formed) but resulted in an open channel reading immediately during dissociation (or during the capacitative current spike) is listed as a zero-length event.

3. After all events have been recorded, they are sorted by event time (shortest to longest) and the survival probability is then calculated (using the index number) by:

$$P_{\text{survival}}(t) = 1 - \frac{\text{Event\#}}{\text{Total\#of events}},$$

where the first event (i.e., the event with the shortest time) is event number 0.

3.3.2. Multi-nanopore Data Analysis

Multi-nanopore methods involve many synchronous single-molecule events in a single dissociation cycle, which (because of synchronized application of the dissociation potential over all the pores) is equivalent to, but faster than, performing the same number of events with a single nanopore. The survival probability at time *t* is determined from the sum of the ionic current through each of the blocked and open pores in parallel:

$$P_{\text{survival}}(t) = 1 - \frac{I(t) - I_{\text{start}}}{I_{\text{end}} - I_{\text{start}}}, \quad (1)$$

where $I(t)$ is the current (through all pores) during the dissociation phase (**step iv, item 4** of **Subheading 3.2.2**). I_{start} is the current at the onset of the dissociation phase, and I_{end} is the open channel current at the dissociation potential. $I_{\text{end}}-I_{\text{start}}$ is the difference in ionic current resulting from all probes-target duplexes blocking pores at the beginning of the experiment; $I(t)-I_{\text{start}}$ is the difference in current from probe-target duplexes that remain at time *t*. The ratio of the two, therefore, serves as an estimate of the fraction of pores that remain blocked by probe-analyte duplexes at time *t*. I_{start} must be calculated because its value is lost in the capacitive current spike that ensues upon changing the voltage to initiate the dissociation phase. The following is a description of how to calculate the survival probability of the probe in the pore (and thus the survival probability for the duplex under force) during the dissociation of the probes from the analytes, using the multi-nanopore force spectroscopy scheme.

1. Run both the multi-pore raw analysis and event viewer programs, recording the dissociation events to determine the average current (between all the events) at each particular moment in time (I(t)).
   Steps in the raw analysis state machine:
   i. Capture state; 204 mV ≥ $V_{capture}$ ≥ 196 mV.
   ii. Analyte binding check state 14 mV ≥ V_{check} ≥ 6 mV.
   iii. Dissociation state −46 mV ≥ V_{FS} ≥ −54 mV.
   iv. Forced clearing of any probes remaining in the pore (−150 mV).
   v. Pore relaxation state (0 mV).
   vi. Open channel impedance check state −46 mV ≥ V_{FS} ≥ −54 mV.
   The tabulated text file should be imported into MS Excel or another spreadsheet application that is capable of basic operations such as averaging, plotting, etc.
2. Run both the multi-pore raw analysis and event viewer programs, recording the open channel current at the dissociation potential (state vii, **item 4** of **Subheading 3.2.2**). Calculate the average of the average current between all events (property iv of **item 5** in **Subheading 2.2.4**). This value is I_{end} (*see* **Note 10**).
3. I_{start} can be calculated from the *IV*-curves for α-HL *(3)*, the number of pores, and the current (through all the pores) just before the dissociation state (state iv, **item 4** of **Subheading 3.2.2**). The *IV*-curves are empirical expressions for the current through a single α-HL nanopore, immersed in 1 M potassium chloride (KCl) solution, during both the open channel and blocked channel (i.e., probe present in the pore) states. The *IV*-curves are:
   Open channel current (pA, V in mV):

$$V > 0{:}I = 0.97V,$$
$$V < 0{:}I = 0.74V.$$

   Blocked channel current (in pA, by probe for $V > 0$, probe and analyte for $V < 0$):
   $V > 0{:}I = 0.0006V^2 + 0.0953V,$
   $V < 0{:}I = -0.0003V^2 + 0.0645V.$
   (*see* **Note 11**).
4. Run both the multi-pore raw analysis and event viewer programs, recording the check state (state iii, **item 4** of **Subheading 3.2.2**). This will determine the current just before

the dissociation state ($I_{capture}$). The number of pores (N) can be approximately determined by dividing the open channel current (at some voltage, typically 100 mV) by the current through a single pore at the same voltage.

5. Determining I_{start} involves the following calculations:
   i. Determine the number of blocked pores (N_b) before the dissociation state (state iv, **item 4** of **Subheading 3.2.2**).

$$N_b = \frac{I_{capture} - 0.97V_{check}N}{(0.0006V_{check}^2 + 0.0953V_{check}) - 0.97V_{check}},$$

where V_{check} is the voltage (in mV) during the check state (state iii, **item 4** of **Subheading 3.2.2**).

   ii. Calculate I_{start},

$$I_{start} = (-0.0003V_{dis}^2 + 0.0645V_{dis})N_b + (0.74V_{dis})(N - N_b),$$

where V_{dis} is the dissociation potential (in mV - state iv, **item 4** of **Subheading 3.2.2**).

6. Using $I(t)$, I_{start}, and I_{end}, the survival probability of the probe in the pore as a function of time can be determined (and the curve plotted) by using **Eq. 1**.

3.4. Interpreting Results

Physical characteristics of the DNA duplex energy landscape are estimated from the manner in which dissociation time scales vary with experimental parameters such as applied electrostatic force and temperature. Interpretation of nanopore force spectroscopy results therefore consists of determining time scales from dissociation time distributions, and using these time scales to estimate physical parameters such as binding energies and energy barrier width along the reaction coordinate. Extraction of time scales from dissociation time distributions is complicated by weak, variable energy, bonds that appear to form between the probe DNA and the pore *(9)*. It is easiest to discuss the analysis in the case of a deterministic (i.e., nonvariable) energy barrier, and then discuss modifications required where the energy barrier is variable. Dissociation times for a deterministic energy barrier follow an exponential distribution of the form:

$$P_{survival}(t) = e^{-t/\tau},$$

where τ is the characteristic dissociation time (i.e., the inverse of dissociation rate). The dissociation time scale is related to the energy barrier and applied force by Kramer's rule:

$$\tau = \tau_D \, e^{\frac{E_b - F\Delta x_{barrier}}{k_b T}}$$

where E_b is the height of the energy barrier in the absence of force, F is the applied electrostatic force, $\Delta x_{barrier}$ is the distance through which the force acts before the duplex dissociates, and τ_D is the diffusive relaxation time *(1)*.

When the probe forms bonds with the nanopore, E_b increases over its value for duplex binding alone. These bonds are weak, meaning that they have a minor impact on estimation of E_b ($\sim 2k_bT$) *(9)*. However, because the probe-pore interaction energy varies from event to event, there is a distribution of characteristic dissociation times. This, in turn yields a non-exponential survival probability (**Fig. 4**, *see* **Note 12**). The goal of this step in the analysis is therefore to determine the dominant dissociation time scale by fitting survival probability data to an appropriate function. We have found that a stretched exponential fit yields good results without the need for lengthy numerical analysis, and we therefore apply it here to extract time scales:

$$P_{survival}(t) \cong e^{-(t/\tau)^{\alpha}}$$

where α, a stretch exponent, varies between 1, for purely exponential behavior (i.e., a deterministic energy barrier), and 0. The dominant time scale is estimated as τ^{α}. This time scale still varies with force according to a slightly modified version of Kramer's rule *(13)*. Analyte molecules differing in sequence by a single nucleotide hybridized to the same probe sequence yield measurably

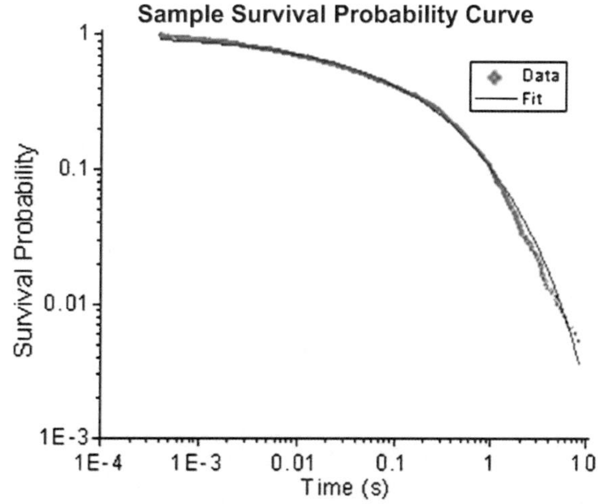

Fig. 4. A sample survival probability curve of the probe in pore as a function of time during dissociation of the DNA duplex. The fit is a stretched exponential fit.

different duplex-lifetime versus force characteristics, and can be discriminated using this sensor if enough events are detected. For most molecules, the dissociation time scale is approximately exponential with applied force, but only at low applied force. At larger forces, the plots for all molecules converge as the energy barrier becomes so small that probe escape from the pore becomes a diffusive rather than a barrier-crossing process. For genotyping analyses, qualitative comparison of characteristic dissociation time versus the applied potential is sufficient to distinguish between perfectly complementary probe-target duplexes and duplexes containing one or more mismatches. Determining energy barrier parameters such as E_b or Δx requires fitting time scale-temperature data, or time scale-force data.

The following is a description of how to extract the appropriate time constant from the survival probability curves. Methods for extracting energy barrier widths and heights from time scale data are described below. This following analysis is similar for both single nanopore and multi-nanopore force spectroscopy data.

1. Perform a least-squares fitting of the survival probability curve using a stretched exponential function. Single exponential behavior is not typically exhibited because of variable probe-pore interactions *(9)*.

$$P_{\text{survival}}(t) \cong e^{-(t/\tau)^{\alpha}},$$

where α and τ are free parameters (**Fig. 4**).

2. The time constant (τ) will obey Kramer's rule for escape over an energy barrier with the barrier height discounted by the applied force, given by

$$\tau^{\alpha} = \left(\tau_{\mathrm{D}}^{\alpha}\, e^{\frac{E_b}{k_b T}}\right)\left(e^{-\frac{NzeV\Delta x}{\Delta l k_b T}}\right) \qquad (2)$$

where τ^{α} is the characteristic dissociation time, $\tau_{\mathrm{D}}^{\alpha}$ is the diffusive relaxation time, E_b is the native energy barrier, k_b is the Boltzmann constant, T is the temperature, N is the number of nucleotides in the influence of the electric field (~25), z is the effective charge per nucleotide (~0.2), e is the elementary charge, V is the voltage, Δx is the location of the energy barrier maximum along the pulling coordinate (i.e., the displacement required to break the duplex apart), and Δl is the distance over which the electrostatic potential drop occurs (10 nm - the length of the pore).

The first term (in **Eq. 2**) is the energetic contribution from the probe-analyte duplex, while the second term is the energetic contribution from the applied force.

3. Plot the time constant (τ^{α}) versus voltage (V) on a semi-log plot (**Fig. 5**). Do a least-squares linear fit to $\ln(\tau^{\alpha})$ versus V of the form:

$$\ln(\tau^{\alpha}) = b - mV$$

(*see* **Note 13**).

4. Taking the ln of **Eq. 2**, and setting the potential term = mV, the location of the energy barrier maximum, Δx can be estimated:

$$\Delta x = m\frac{\Delta l k_{\mathrm{b}} T}{Nze}$$

5. The intercept of the plot is an estimate of the characteristic time scale at zero force, τ_0^{α}

$$\tau_0^{\alpha} = \mathrm{e}^{b}$$

6. The intrinsic attempt rate is estimated from the intercept:

$$\tau_{\mathrm{D}}^{\alpha} = \mathrm{e}^{a}$$

7. The energy barrier is estimated from the slope, using the estimate of Δx calculated above:

$$E_{\mathrm{b}} = k_{\mathrm{b}} n - \frac{Nze\Delta x}{\Delta l} V$$

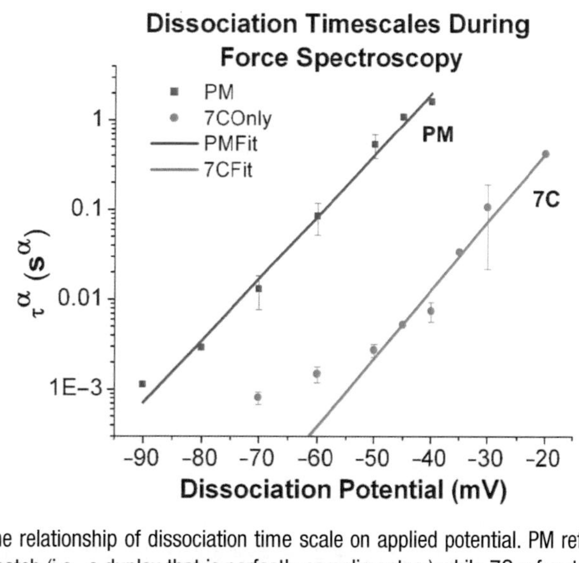

Fig. 5. The relationship of dissociation time scale on applied potential. PM refers to the perfect match (i.e., a duplex that is perfectly complimentary) while 7C refers to a duplex containing a SNP at the seventh nucleotide position from the 3′ end of the target strand (for a 14-nucleotide target molecule, this corresponds to the middle of the strand). The data is well approximated by Kramer's law (**Eq. 2**) with the barrier height for dissociation discounted by the applied force. The time scale for dissociation of the PM is approximately two orders of magnitude larger than that for the 7C (at lower negative potentials).

4. Notes

1. Secondary structure and DNA hybridization energies can be estimated using a program such as Mfold: http://frontend.bioinfo.rpi.edu/applications/mfold/.
2. Suggested volumes are:
   i. 60 μL of 1 M KCl, 10 mM HEPES, 1 mM EDTA, pH 8.0.
   ii. 20 μL of 2 M KCl, 20 mM HEPES, pH 8.0.
   iii. 10 μL of 100 μM probe solution.
   iv. 10 μL of 100 μM avidin solution.

   The 100 μM probe solution should be thawed relatively infrequently. It is a good idea to make a few aliquots of the 10 μM nanopore probe solution that can be cycled through between experiments.
3. Suggested volumes are:
   i. 24.8 mL of 1 M KCl, 10 mM HEPES, 1 mM EDTA, pH 8.0.
   ii. 100 μL of 2 M KCl, 20 mM HEPES, 2 mM EDTA, pH 8.0.
   iii. 100 μL of 100 μM probe solution.

Avoid repeated freezing and thawing. 1.0-mL aliquots of the 0.5-μM target solution are appropriate for the apparatus and methods described here.

4. To determine the survival probability of the probe in the pore, the analysis software will first order all the events in a data file by increasing event time (i.e., shortest first and longest last). The survival probability for a particular event time is given by:

$$P_{\text{survival}}(t) = 1 - \frac{\text{Event \#}}{\text{Total \# of events}},$$

   where the first event is event 0. To calculate the survival probability, the total number of events must be accurately counted. If a long event is terminated by changing the potential, it must still be recorded, but flagged to indicate that it is not accurately timed.
5. Label the cables (e.g., V_{out}, I_{out}, and V_{in}) for future reference.
6. Split up any given experiment (e.g., the –50 mV experiment) into a few data files so that between files the current can be re-zeroed (it will tend to drift slightly during the experiment).
7. The magnitude of the check voltage and the amount of time it is applied for should be optimized for the probe length and sequence. The magnitude of the check voltage is such that once applied, any unbound probes will thermally escape from

the nanopore within a few milliseconds. The values given above are for the probe sequence outlined in **Subheading 2.1, item 1, iv**.

8. Be careful when manually forcing the probe out of the pore. Any potential that is applied manually through the patch clamp amplifier will be in addition to the potential being applied by the acquisition software. If the combined voltage is greater than ±200 mV, the bilayer will be at risk of bursting.
9. In defining the states in the analysis program, be sure to take into account the capacitative spike in the current that will appear in the data upon transitions between voltages. This can be done by including dummy states between transition states. Check that events are not being systematically discarded or included improperly, as this can result in the distortion of the survival probability curves (e.g., systematically discarding long events). Typical I and I_0 values, for different voltages, can be calculated using the α-HL *IV*-curves (refer to **Subheading 3.3.2**).
10. When calculating the average do not include values of the current that correspond to the capacitive spike (i.e., only use values representative of the true open channel current after the capacitive spike has decayed).
11. The calculated value for I_{start} is approximate, and in particular assumes that all the pores have incorporated in the bilayer in the forward direction, and that no pore has incorporated as one of the possible secondary structures. It is important to note that the given *IV*-curves are for the case of a 1 M KCl buffer. Although the open channel current scales linearly with salt concentration (roughly), the blocked channel current does not.
12. A complete discussion of the manner in which energy barrier variation affects survival probability distributions is beyond the scope of this chapter. Refer to *(9)* for more details.
13. Performing the fit on the logarithmically transformed function has two benefits. First, it makes the fit a simple linear fit. Second, and much more importantly, the residuals in this fit are weighted similarly for all data points. This is in contrast to the exponential fit, where residuals on large points tend to dominate the overall error term.

Acknowledgements

We thank members of the University of British Columbia Applied Biophysics Laboratory, in particular Vincent Tabard-Cossa, and

Dhruti Trivedi. Many thanks to Jonathan Nakane and Carolina Tropini for their important contributions to the development of this protocol. This work was supported by the US National Institutes of Health (NIH). Financial support for Matthew Wiggin was provided by the Natural Sciences and Engineering Research Council of Canada (NSERC).

References

1. Evans, E. (2001). Probing the relation between force lifetime and chemistry in single molecular bonds. *Annu. Rev. Biophys. Biomol. Struct.* **30**, 105–128.
2. Nakane, J., Wiggin, M., and Marziali, A. (2004). A nanosensor for transmembrane capture and identification of single nucleic acid molecules. *Biophys. J.* **87**, 615–621.
3. Tropini, C., and Marziali, A. (2007). Multi-nanopore force spectroscopy for DNA analysis. *Biophys. J.* **92**, 1632–1637.
4. Sauer-Budge, A.F., et al. (2003). Unzipping kinetics of double-stranded DNA in a nanopore. *Phys. Rev. Lett.* **90**, 238101-1–238101-4.
5. Dudko, O.K., et al. (2007). Extracting kinetics from single-molecule force spectroscopy: Nanopore unzipping of DNA hairpins. *Biophys. J.* **92**, 4188–4195.
6. Chen, P., and Li C.M. (2007). Nanopore unstacking of single-stranded DNA helices. *Small* **3**, 1204–1208.
7. Hornblower, B., et al. (2007). Single-molecule analysis of DNA-protein complexes using nanopores. *Nat. Methods* **4**, 315–317.
8. Mathé, J., et al. (2004). Nanopore unzipping of individual DNA hairpin molecules. *Biophys. J.* **87**, 3205–3212.
9. Wiggin, M., et. al. Nonexponential kinetics of DNA escape from alpha-hemolysin nanopores. *Biophys. J.* **95**, 5317–5323.
10. Howorka, S., et al. (2001). Sequence-specific detection of individual DNA strands using engineered nanopores. *Nat. Biotechnol.* **19**, 636–639.
11. DeGuzman, V., et al. (2006). Sequence-dependent gating of an ion channel by DNA hairpin molecules. *Nucleic Acids Res.* **34**, 6425–6437.
12. Vercoutere, W.A., et al. (2003). Discrimination among individual Watson-Crick base pairs at the termini of single DNA hairpin molecules. *Nucleic Acids Res.* **31**, 1311–1318.
13. Metzler, R., and Klafter, J. (2000). The random walk's guide to anomalous diffusion: a fractional dynamics approach. *Phys. Rep.* 339, 1–77.

Part II

Technologies of Physical Science and Chemistry in Detection and Analysis of Biomolecules

Chapter 11

Quantitative Chemical Analysis of Single Cells

Michael L. Heien and Andrew G. Ewing

Summary

Exocytosis, the fusion of intracellular vesicles with the membrane and subsequent release of vesicular contents, is important in intercellular communication. The release event is a rapid process (milliseconds), hence detection of released chemicals requires a detection scheme that is both sensitive and has rapid temporal dynamics. Electrochemistry at carbon-fiber microelectrodes allows time-resolved exocytosis of electroactive catecholamines to be observed at very low levels. When coupled with constant-potential amperometry, the number of molecules released and the kinetics of quantal release can be determined. The rapid response time (milliseconds) of microelectrodes makes them well suited for monitoring the dynamic process of exocytosis.

Key words: Amperometric detection, Electrochemistry, Oxidation, Exocytosis, Microelectrode, Monoamines, Quantal size, PC12 cell

1. Introduction

Exocytosis, the process by which a cell directs secretory vesicles to the cell membrane and releases the vesicles contents into the extracellular space, is common to many types of cells. The concept of exocytosis has grown from anatomical, biochemical, and physiological evidence *(1)*. However, direct measurements of exocytosis allow both the number of molecules released and the kinetics of quantal release to be elucidated.

The first electrochemical measurements of exocytosis determined release of catecholamines from large dense-core vesicles from bovine chromaffin cells *(2)*. Electrochemical measurements of exocytosis have also been made at mast cells *(3)*, pancreatic β-cells *(4)*, and rat pheochromocytoma (PC12) cells *(5)* and

James Weifu Lee and Robert S. Foote (eds.), *Micro and Nano Technologies in Bioanalysis*,
Methods in Molecular Biology, vol. 544
DOI 10.1007/978-1-59745-483-4_11,

dopaminergic neurons *(6, 7)*. The PC12 clonal cell line, established from a transplantable rat adrenal pheochromocytoma, is particularly useful as a model for sympathetic neurons and their development because of its response to nerve growth factor. Like sympathetic neurons in primary cell culture, PC12 cells synthesize, store, release, and take up dopamine and norepinephrine.

To monitor single-vesicle exocytosis, constant-potential amperometry is performed at carbon-fiber microelectrodes. Briefly, an electrode is firmly placed adjacent to a cell, and a potential sufficient to oxidize dopamine is applied to the electrode. This setup is presented in **Fig. 1**. A carbon-fiber microelectrode is located near the cell surface and a stimulating pipette is positioned so that chemicals may be applied to the cell. After application of a secretagogue, vesicles fuse with the plasma membrane and release their contents. If these vesicles are located under the electrode surface, their contents are quantitatively oxidized, and the resultant current is measured and plotted versus time.

During each fusion event, a spike is generated in the current trace. Several important parameters relating to the vesicles can be obtained by examining the trace **(Fig. 2)**. Individual spikes can be integrated, measuring the total charge (Q) in one event. By integrating the current, the total charge present in one event can be determined ($Q = \int idt$). Using Faraday's Law this reveals the amount of neurotransmitter in each vesicle. The total charge is related to the amount of neurotransmitter present by **Eq. 1**:

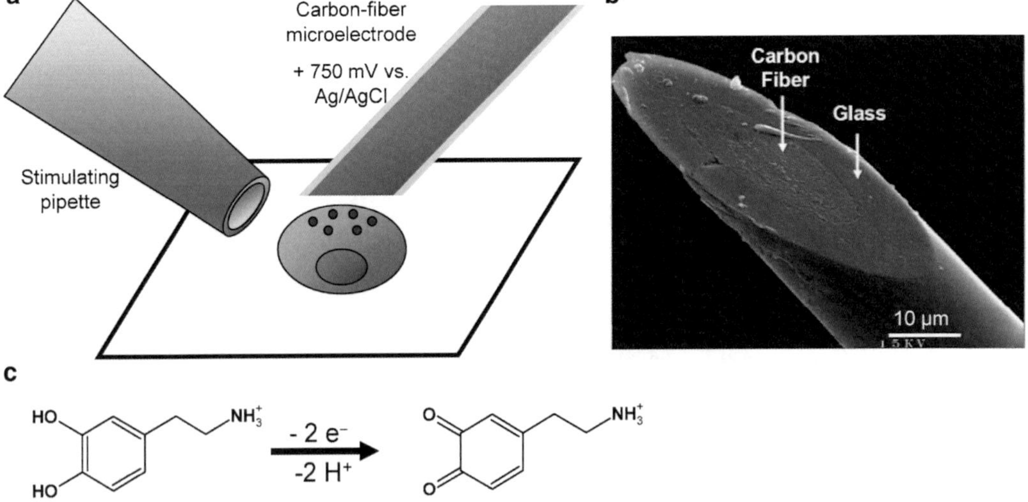

Fig. 1. A typical experimental setup for amperometry at a single cell. A carbon-fiber microelectrode is positioned above the cell with a stimulating pipette nearby **(a)**. A scanning electron micrograph (SEM) of a carbon-fiber microelectrode can be used to determine the active area of the probe **(b)**. A potential (+750 mV) sufficient to oxidize dopamine is applied to the electrode causing dopamine to be converted to dopamine-ortho-quinone **(c)**.

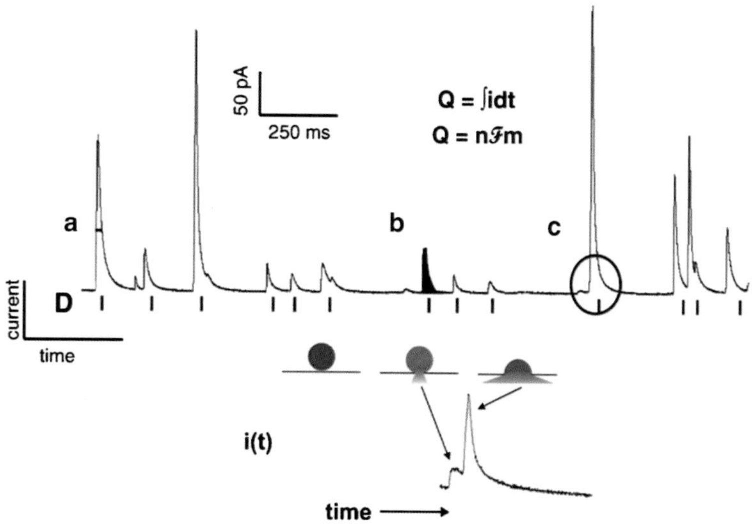

Fig. 2. A representative section of an amperometric trace measured during stimulated exocytosis of epinephrine from a bovine adrenal medullary chromaffin cell. Important features for quantitative analysis include **(a)** spike half-width ($t_{1/2}$), **(b)** spike area (Q), **(c)** foot events, and **(d)** spike frequency (f). The spike half-width is related to the kinetics of the release event. The area under a spike is calculated by integrating the curve, and can be directly related to the number of molecules released, using Faraday's Law, as shown. The inset demonstrates the transition from a fusion-competent state to full fusion with the intermediate formation of a fusion pore that is related to the foot. The spike frequency provides an indication of the mechanisms controlling the initiation and regulation of the release events (reproduced from ref. *(14)* with permission from Elsevier).

$$Q = nFN, \tag{1}$$

where Q is the total charge measured, n is the number of electrons per mole, F is Faraday's constant, and N is the number of moles.

The half-width ($t_{1/2}$) of each event is a measure of the rate of extrusion of each vesicle. The frequency of spikes can easily be measured by digitizing each event and counting the number of events over a specified amount of time. This may reveal changes in the release pattern of a cell or changes in the number of vesicles readily available for release *(8, 9)*. Also shown in **Fig. 2** is the presence of a prespike feature in the amperometric response, known as a foot. These events are caused by an impediment during the formation of a fusion pore, which allows neurotransmitters to escape through the fusion pore before full fusion. The fusion pore also seems to be able to "flicker" open and closed, leading to a different type of amperometry. Altered release via the fusion pore might lead to alternative mechanisms in which a presynaptic cell can regulate the amount of neurotransmitter released and allow for rapid recycling of vesicles *(7, 10)*.

This review provides a detailed procedure to perform electrochemical measurements at single cells. Central to making these measurements is the fabrication of carbon-fiber microelectrodes.

The procedure for fabricating these electrodes and characterizing them is described. The measurements at single cells are made, and the data are analyzed.

2. Materials

2.1. Chemicals

1. Norit A activated carbon (ICN, Costa Mesa, CA).
2. Electrolyte solution (4 M potassium acetate, 150 mM potassium chloride) in distilled water.
3. Dopamine stock solution (10 mM dopamine in 0.1 N perchloric acid) is prepared. This solution can be stored at 4°C for 2 weeks.

2.2. Cell Culture

1. Cell media consists of phenol red-free RPMI-1640 media (Gibco, Grand Island, NY) supplemented with 10% heat-inactivated equine serum, 5% fetal bovine serum (Hyclone Laboratories, Logan, UT), and 0.4% penicillin streptomycin solution (Sigma Chemical Co., St. Louis, MO).
2. Cell culture dishes (60 mm), coated with mouse collagen (type IV, 0.5 $\mu g/cm^2$, Collaborative Biomedical Products, Bedford, MA) are used to culture cells.
3. Physiological saline consisting of 4.2 mM KCl, 150 mM NaCl, 2 mM $CaCl_2$, 0.7 mM $MgCl_2$, 1 mM NaH_2PO4, and 10 mM HEPES is prepared in distilled water and adjusted to pH 7.4.
4. Physiological saline with elevated K^+ consisting of 80 mM KCl, 50 mM NaCl, 2 mM $CaCl_2$, 0.7 mM $MgCl_2$, 1 mM NaH_2PO4, and 10 mM HEPES is prepared in distilled water and adjusted to pH 7.4.

2.3. Electrode Fabrication

1. Carbon fibers are available in sizes from 5 to 33 μm. Five-micron carbon fibers (T-40 Thornel, Amoco Performance Products, Greenville, SC) are widely used.
2. Glass capillaries (1.2 mm × 0.68 mm ID, 10 cm, A-M Systems, Carlsborg, WA).
3. Micropipette puller (P-97 Flaming Brown, Sutter Instrument Company, Novato, CA).
4. Epoxy (Epon 828 with 14 wt% *m*-phenylenediamine, Miller-Stephenson Chemical Co., Danbury, CT).
5. Microelectrode beveller (BV-10, Sutter Instrument Company).
6. Silver wire (0.5-mm diameter, A-M Systems).

2.4. Electrochemical Measurements

1. Current amplifier (Axopatch 200B, Molecular Devices, Sunnyvale, CA).
2. Microinjector for application of substances to cells (FemtoJet Microinjection System, Eppendorf, Westbury, NY).
3. A Ag/AgCl reference electrode (BAS, West Lafayette, IN) is used for all electrochemical measurements.
4. An inverted microscope (Olympus IX70, Center Valley, PA) is used to visualize cells and electrodes during an experiment.
5. Micromanipulators (Narashigi MHW-3, Tokyo, Japan) are used to position the electrode and stimulating pipette near cells.

2.5. Data Analysis

1. Mini Analysis (Synaptosoft, Fort Lee, NJ) is used to determine spike parameters such as amplitude, area, half-width, and frequency.
2. Histograms and statistical analysis are performed in Excel (Microsoft, Redmond, WA) and GraphPad Prism (GraphPad Software, Inc., San Diego, CA).

3. Methods

3.1. Preparations of samples

1. A standard solution of dopamine is prepared (10 μM in physiological saline solution). The dopamine stock solution is diluted to give the desired concentration. It should be used within 2 h of manufacture.

3.2. Cell Culture

1. Cells are grown on mouse collagen (type IV, 0.5 íg/cm^2, Collaborative Biomedical Products)-coated cell culture dishes (60 mm) and are subcultured every 7–9 days.
2. Medium is replaced every 2–3 days throughout the lifetime of cultures. Cells are maintained in a 7% CO_2, 100% humidity atmosphere at 37°C.

3.3. Electrode Fabrication

1. Carbon-fiber microelectrodes are central to electrochemical measurements of neurotransmitters, and their production has been previously described *(11)*. Fabrication of microelectrodes is a learned skill, and a novice may become proficient quickly (*see* **Note 1**). Individual fibers must be isolated. Bundles of carbon fibers are placed on a well-lit white surface (typically a sheet of paper). Individual carbon fibers are then teased apart from the bundle, aided by the use of fingers, tweezers, or paper to secure the bundle. The isolated fiber is then placed on the surface.

2. The fiber is inserted into a capillary. One end of the carbon fiber is secured with tape or a finger. The capillary is connected to a vacuum hose, and the carbon fiber is drawn into the capillary. The vacuum hose is then removed.
3. The electrode is pulled in a capillary puller. The heater and pull settings may have to be adjusted to yield a satisfactory electrode (*see* **Note 2**).
4. The electrode is secured on a glass slide and placed on a microscope. Using a scalpel, the electrode is cut flush with the tip, near the region where the glass meets the carbon fiber.
5. Epoxy is mixed and heated to approximately 90°C in a water bath. Electrodes are dipped in the epoxy for 30–60 s and removed. The electrodes are cured in an oven (150°C for 2 h, 100°C for 24 h).
6. The electrodes are microscopically polished using a microelectrode beveller at the desired angle (typically 45°, *see* **Note 3**), yielding an elliptical surface. Distilled water is placed on the polishing wheel, and the electrode is lowered until it makes contact with the polishing wheel. It is important to polish the electrode until the desired geometry is obtained.
7. Before use, electrodes are soaked in 2-propanol purified with Norit A activated carbon (ICN) for at least 10 min *(12)*, then backfilled with electrolyte solution, and a silver wire is inserted for electrical contact.
8. A stimulating pipette is fabricated by pulling a glass capillary to a narrow opening (~10 μm).

3.4. Electrochemical Measurements

3.4.1. Electrode Screening

1. Before making measurements, the electrodes are tested to ensure ideal behavior. This is typically done by collecting a steady-state cyclic voltammogram *(13)*. The electrode is placed in a solution of an electroactive substance, the potential of the electrode is ramped, and the resultant current is measured. This is done to ensure the electrode displays low noise and has a good seal. The procedure for this is outlined below.
2. A standard solution (500 μM dopamine) is prepared by diluting the dopamine stock solution in HEPES buffer solution, and then it is deoxygenated by bubbling argon or dry nitrogen through it.
3. The carbon-fiber microelectrode (working electrode) and a Ag/AgCl reference electrode are placed in the solution and connected to the current amplifier, taking care to ensure the reference electrode is connected first.
4. The potential of the working electrode is then ramped from a resting potential (typically –200 mV) to a potential sufficient to oxidize dopamine (typically +800 mV) and back to –200

mV again. The current is measured, and plotted versus the potential, to create a cyclic voltammogram.

5. Electrodes with poor responses are discarded, whereas electrodes with ideal responses are kept for use.
6. A stimulating pipette is filled with elevated potassium solution to simulate cells.

3.4.2. Measurements from Single Cells

1. The culture media used to grow and support cells in culture is quickly replaced with physiological saline. These cells are placed on an inverted microscope for experiments. Cells are visually inspected to ensure they appear healthy.
2. A Ag/AgCl reference electrode is then placed in the solution, and the working electrode is positioned near the cells in the bath, ensuring the sensing surface is parallel to the bottom of the dish.
3. The stimulating pipette is then placed in the solution, and positioned near the cells in the bath, opposite the working electrode. It is connected to a microinjector, and the pressure is adjusted to yield satisfactory application of substances.
4. To carry out amperometry, a potential sufficient to oxidize dopamine (+750 mV) is applied to the electrode. A bandpass of 2 kHz is typically used for signal collection to capture the rapid temporal dynamics of vesicle fusion (*see* **Note 4**). A stimulating pipette can be used to test the working electrode's response (*see* **Note 5**).
5. The working electrode is then positioned over the cell, and carefully lowered onto the cell surface until the cell surface is seen to slightly distort. This ensures the electrode traps a small volume of solution next to the cell surface, meaning the contents of a vesicle that fuses with the membrane will be completely oxidized by the electrode.
6. The recording is started; the current is measured and stored on a personal computer. After a baseline is obtained (typically 10 s) the cell is stimulated with elevated potassium, and the resultant spikes are recorded.

3.5. Data Analysis

1. Data is read into MiniAnalysis software for spike quantification and analysis.
2. The data is digitally filtered to reduce noise and allow quantification of the desired parameters. Different filter values may alter the temporal dynamics revealed in the data **(Fig. 3)**. A smaller filter frequency reduces the noise, allowing more peaks to be identified; however, the temporal dynamics observed may be altered (*see* **Note 6**).

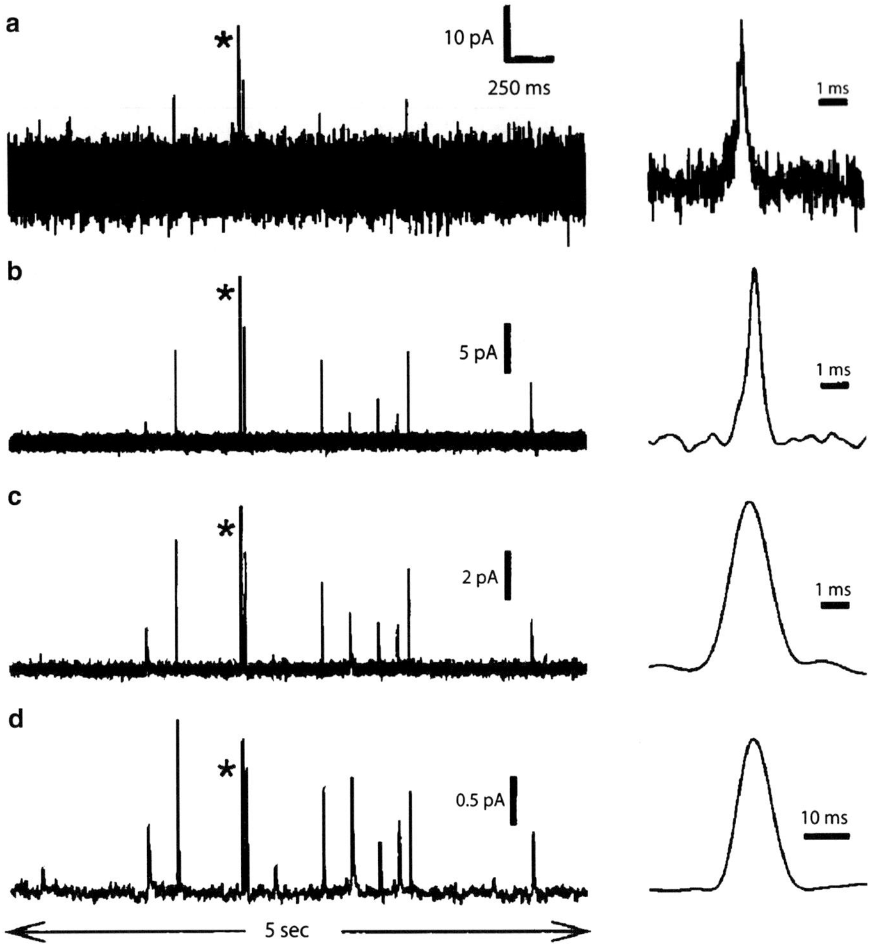

Fig. 3. Effects of low-pass filtering on amperometric spikes. A 5-s segment of amperometric data collected from a DA cell is presented at four low-pass filter rates: 10 kHz (analog four-pole Bessel), 1 kHz (digital eight-pole Bessel), 200 Hz (digital eight-pole Bessel), and 40 Hz (digital eight-pole Bessel) (**a**–**d**, respectively). The spike marked with an *asterisk* is shown on an expanded time scale to the right of each trace. Note that, for each spike, the amplitude decreases while the full width at half-height increases with each filter rate (reproduced from ref.*(6)* with permission from the American Chemical Society).

3. Individual spike parameters (charge, amplitude, and half-width) are then evaluated.
4. The data can be averaged and statistically compared across cells and populations.

4. Notes

1. To facilitate mass production, 25–40 electrodes are typically made in an "assembly line" fashion. Many variations on the procedure described exist; steps can be modified for individuals.

2. To verify a good seal is obtained, the electrode may be inspected under a microscope. Electrodes with breaks or cracks are discarded. A good seal will have the glass forming a seal with the carbon fiber, and will appear black under a microscope; however, the glass should not be pulled so thin that the electrode is weak or brittle.
3. The exposed area of the electrode can be varied by polishing at different angles. The "top" of the electrode is also marked so that the electrode can be positioned on top of cells with the beveled surface on the cell.
4. If the Axopatch 200B is being used for signal collection, the instrument is set up in the "Voltage-Clamp" mode, with the "Holding Command" set to +750 mV. The noise present is checked to ensure satisfactory recordings can be made. Typically, an I_{RMS} of less than 1.5 pA is expected. Larger noise values will interfere with recordings and data analysis. If more noise is present, it may have to be eliminated before experiments.
5. The stimulating pipette may be filled with a dopamine solution (typically 100 μM, although different concentrations will work) to test the working electrode response, or to obtain optimal parameters for application of the stimulating pipette's contents. The stimulating pipette is positioned near the working electrode's surface, and the current is recorded while the solution is applied. The current should increase sharply when the application begins, then level off, and finally slowly decrease when the application is ended.
6. In some cases, distortion from filtering might not be a problem. Peak areas are relatively unaffected by filtering; however, the half-with and amplitude of signal peaks are distorted by filtering. The amount of data pretreatment should be selected based on the experiment.

Acknowledgments

This work was supported by funding from the National Institutes of Health (NIH). A.G.E. is supported by a Marie Curie Chair from the European Union 6th Framework.

References

1. Heuser, J. E. (1989). Review of electron microscopic evidence favouring vesicle exocytosis as the structural basis for quantal release during synaptic transmission. *Q J Exp Physiol.* **74**, 1051–69
2. Wightman, R. M., Jankowski, J. A., Kennedy, R. T., Kawagoe, K. T., Schroeder, T. J., Leszczyszyn, D. J., Near, J. A., Diliberto, E. J., Jr., and Viveros, O. H. (1991). Temporally resolved catecholamine spikes correspond to

single vesicle release from individual chromaffin cells. *Proc Natl Acad Sci U S A*. **88**, 10754–8

3. Pihel, K., Travis, E. R., Borges, R., and Wightman, R. M. (1996). Exocytotic release from individual granules exhibits similar properties at mast and chromaffin cells. *Biophys J*. **71**, 1633–40
4. Kennedy, R. T., Huang, L., Atkinson, M. A., and Dush, P. (1993). Amperometric monitoring of chemical secretions from individual pancreatic beta-cells. *Anal Chem*. **65**, 1882–7
5. Zerby, S. E., and Ewing, A. G. (1996). Electrochemical monitoring of individual exocytotic events from the varicosities of differentiated PC12 cells. *Brain Res*. **712**, 1–10
6. Hochstetler, S. E., Puopolo, M., Gustincich, S., Raviola, E., and Wightman, R. M. (2000). Real-time amperometric measurements of zeptomole quantities of dopamine released from neurons. *Anal Chem*. **72**, 489–96
7. Staal, R. G., Mosharov, E. V., and Sulzer, D. (2004). Dopamine neurons release transmitter via a flickering fusion pore. *Nat Neurosci*. **7**, 341–6
8. Rettig, J., and Neher, E. (2002). Emerging roles of presynaptic proteins in Ca++-triggered exocytosis. *Science*. **298**, 781–5
9. Rizzoli, S. O., and Betz, W. J. (2005). Synaptic vesicle pools. *Nat Rev Neurosci*. **6**, 57–69
10. Sulzer, D., and Pothos, E. N. (2000). Regulation of quantal size by presynaptic mechanisms. *Rev Neurosci*. **11**, 159–212
11. Cahill, P. S., Walker, Q. D., Finnegan, J. M., Mickelson, G. E., Travis, E. R., and Wightman, R. M. (1996). Microelectrodes for the measurement of catecholamines in biological systems. *Anal Chem*. **68**, 3180–86
12. Bath, B. D., Michael, D. J., Trafton, B. J., Joseph, J. D., Runnels, P. L., and Wightman, R. M. (2000). Subsecond adsorption and desorption of dopamine at carbon-fiber microelectrodes. *Anal Chem*. **72**, 5994–6002
13. Wightman, R. M. (1981). Microvoltammetric electrodes. *Anal Chem*. **53**, 1125–34
14. Haynes, C. L., Siff, L. N., and Wightman, R. M. (2007). Temperature-dependent differences between readily releasable and reserve pool vesicles in chromaffin cells. *Biochim Biophys Acta*. **1773**, 728–35

Chapter 12

Trapping and Detection of Single Molecules in Water

M. Willander, K. Risveden, B. Danielsson, and O. Nur

Summary

An innovative nanoprobe-based device that can measure and adjust the pH, can mimic biochemistry, can create microscale vortices in water, and can be used to trap single molecules is presented. Because the analytes in question to trap and detect are small in dimensions, we start by presenting scaling issues and challenging limitations for miniaturized chemical nanosensors. Advantages of using nanoprobes e.g., isolated nanowires, as the components in chemical sensing are discussed. How the observation of the physical property can beneficially change with isomorphic scaling is highlighted. Some of the technology-related constrains are presented for specific sensors. Solutions to overcome such problems are also given. Different aspects, e.g., sample size and sensitivity, for chemical sensing at the nanoscale are highlighted.

Key words: Trapping single molecules, Scaling, Miniaturized Sensors, Nanoelectronics, pH, Nanowires, Biosensors, Chemical Sensors

1. Introduction

Traditional chemical and biological analytical techniques used in various fields involve reactions that take place in solutions on addition of reagents or other bioreactive species. In addition, systems that do not involve reagents are more common and of interest. In these systems, the reagents are already immobilized. In some systems, these reactions take place at an electrode and they are commonly called *sensors (1)*. By definition, sensors are devices that are composed of an analyte-selective interface, which is connected to or in proximity to a transducer. The transduction mechanism relies on the interaction between the surface and the analyte directly or through mediators *(2)*. The analyte selective

James Weifu Lee and Robert S. Foote (eds.), *Micro and Nano Technologies in Bioanalysis*,
Methods in Molecular Biology, vol. 544
DOI 10.1007/978-1-59745-483-4_12, © Humana Press, a part of Springer Science+Business Media, LLC 2009

interface can be a gas, a bioactive substance, a protein such as an enzyme or an antibody, a microorganism, etc. These interfaces can be very capable of recognizing, sensing, and regulating sensitivity and specificity with regard to the analyte. The transducer converts the biochemical signal into an electric signal. Finally, an amplification of the sensed signal may be needed. This implies that the sensor system is mainly composed of three subunits. Nevertheless, the sensor is sometimes referred to as the input transducer; the word transducer is derived from the Latin verb *traduco*, which means a device that transfers energy from one system to another in the same or another form. Biochemical sensors are often simple and can offer real-time analysis of human body analytes. They represent a broad area of emerging technologies ideally suited for human health care analysis. The human being's natural sensing system is a notoriously poor gauge even to approximately measure simple things like weight, length, distance, humidity, etc. Using measurement tools to help us to obtain accurate values, in general, is as old as the existence of human beings. Noah when building his ark used the cubit (the length of the arm from the tip of the middle finger to the elbow) as a standard to measure length. Today sensor science and technology require a multidisciplinary environment where biology, chemistry, physics, electronics, and technology work hand in hand to achieve the ultimate goal of a time domain, small-size, selective, and sensitive sensor. The modern sensor concept dates back to 1956, when Clark published a paper on the oxygen electrode *(3)*. Since then, sensor science and technology have evolved dramatically and become multidisciplinary. In its modern concept, which begins in 1962, a biosensor is based on the fact that enzymes can be immobilized at an electrochemical detector to form enzymatic detectors that can be used for sensing *(4)*. It is worth mentioning that bioactivity is characterized to be of a chemical nature, as will be elaborated in the next section. The main biosensor categories are divided into optical, calorimetric, piezoelectric, and electrochemical biosensors. Calorimetric or thermal biosensors, such as the enzyme thermistor developed at Lund University, have proven very versatile and scalable, and have unsurpassed operational stability *(5)*. Electrochemical biosensors respond to electron transfer, electron consumption, or electron generation during a chem/bio-interaction process. This class of sensors is of major importance and they are more flexible in terms of miniaturization (scaling down) than most other biosensors. They are further divided into conductometric, potentiometric, and amperometric devices. In conductometric sensors, the change of conductance between two metal electrodes caused by the biological reaction is measured *(6)*, whereas, in potentiometric sensors, the relative potential change is measured at a reference electrode *(7)* due to accumulation of charge (electrons) with no current conduction. Amperometric biosensors are based on the current change caused by electron transfer in the chemical reactions at the electrodes at a certain applied voltage. The principle of operation of this class of sensors is of great importance to miniaturized future sensors.

There is an increasing need for selective, sensitive, time domain chemical sensors for physiological environments. This is driven by human health care and the need for new drug discovery. Almost all chemical and biochemical reactions involve a process where the acidity (pH) is subjected to relatively small changes, sometimes only momentarily. When considering real physiological mediums, the problem becomes more complicated because the pH changes have to be detected in volumes that are relatively quite small. This obviously implies that the new needed sensors have to also be small in dimension. In general, when objects are scaled down isomorphically (i.e., all dimensions are scaled uniformly), the change in length, area, and volume ratios increases as we scale down, and this renders surface effects to be significant. This alters the relative influence of the different physical effects in question in an unexpected way. If the object (e.g., analyte) in question shrinks to the same length scale as the boundary layer (effect), being thermal, diffusion, optical, etc., continuum theories break down and the laws of micro scaling no longer apply. This fact makes such scaled objects possess unusual properties that can be further engineered depending on the way these objects are scaled and/or arranged. Fortunately, sometimes these properties are beneficial to our nanosensors purpose, as is illustrated below. For the analyte in question, the total sample size needed for the detection is determined by the analyte concentration. The analyte concentration is in fact out of our control, i.e., sometimes we need to detect mediums with relatively very low analyte concentrations. One important property of scaled objects (sensors) of particular interest is the sensitivity of these scaled sensors. Obviously, a sensor with a wide dynamic range for detection sensitivity is indeed an aim of the scientific community. Before proceeding, we define the sensors of interest here to be those called electrochemical sensors.

In this chapter, we first present sensor domains and a brief summary of the history of using silicon in sensor technology. Because of its importance to our human health, the emphasis will be on sensors of a chemical nature because they are vital to physiological investigations. Different aspects of scaling issues and challenging limitations are briefly discussed. The nanoscale water transistor developed recently is then presented as a multifunction platform (Lab on a Chip).

1.1. Sensor Domain and Historical Background

We first need to define the different signal domains that distinguish between different types of sensors. In general, sensors are devices that make use of one or several of many different effects. Some documented effects dates back to as early as the 18th century (thermocoupling effect). Most of the actual effects belong to six main domains *(8)*. These domains are radiant, mechanical, thermal, electrical, magnetic, and chemical.

Silicon is the dominant semiconductor material in use today, and has been explored in sensor technology both for conventional sensors as well as for state-of-the-art of nanosensors. Below, we briefly mention the history of using Si as a material for sensors. Sensors have been fabricated using many material technologies, ranging from piezoelectric quartz crystals and compound semiconductors to metals. However, the real advances in sensor technology occurred when pure germanium and, later, silicon became available. Although the availability of these two elementary semiconductors was not aimed at sensor technology, the impact cannot be neglected. An elegant review article that deals with many issues in the history of using Si in sensor technology can be found in reference *(9)*. To our knowledge, the first two Si-based sensors were based on the piezoresistive effect (the electrical to mechanical or the opposite transfer domain) and were demonstrated more than four decades ago *(10,11)*. The first was demonstrated in 1954 *(10)* and the second was published in 1970 *(11)*, although related activities started in 1965. Moreover, the emergence of silicon dioxide use and various processing techniques, especially lithography, have made Si in its pure crystalline state with controlled doping the best material choice for sensors. Nevertheless, because silicon dioxide is one of the reasons for the domination of silicon technology today, ironically silicon dioxide still plays an indirect important role in today's nanosensors. The first two pioneering chemical sensors were the ion-sensitive field effect transistor (ISFET) developed by Bergveld *(12)* and the Pd-gate metal oxide field effect transistor (MOSFET) invented by Lundström et al. *(13)*. Pd-gate MOSFET (Pd-MOSFET) operation is restricted to gaseous medium and relies on the catalytic ability of the Pd to decompose hydrogen molecules into hydrogen atoms. Nevertheless, the first "enzyme transistor" was made with a Pd-MOSFET that also showed some ammonia sensitivity *(14)*. By using hydrophobic gas-permeable membranes (e.g., Gore-Tex) as the interface between the aqueous medium and the gas phase, interesting sensor constructions could, however, be made, especially with the use of noble metals other than Pd. Sensitive ammonia detectors were made with iridium as the catalytic metal and combined with ammonia-producing enzymes to form excellent biosensors for clinically important analytes such as urea and creatinine *(15)*. On the other hand, the ISFET is suitable for operating in electrolyte medium. This has led to its usefulness for many interesting experiments for physiological investigations. This property has forced the ISFET to be the building block of the modern knowledge we have today in chemical nanosensors.

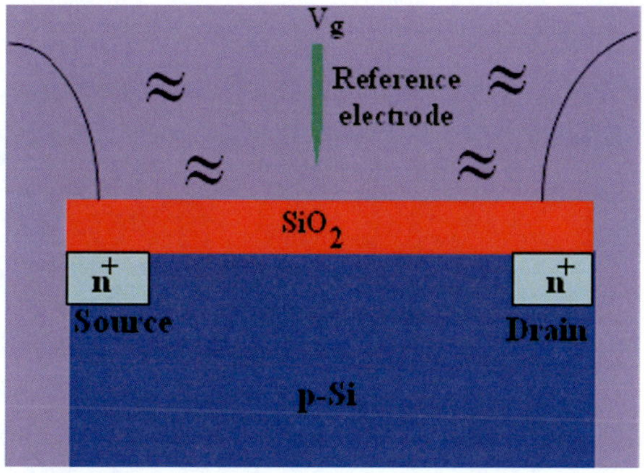

Fig. 1. Schematic diagram showing the ISFET (as an example of a conventional planar sensor) in its simplest configuration, with the metal gate replaced by a small cage to host the electrolyte.

Figure 1 shows a schematic diagram of the ISFET in its simplest form, with the gate metal replaced by a cage for hosting the electrolyte.

The need for sensors to be smaller is in fact because of two reasons, the first is the so-called nonlinear effect when using an isolated nano-electrode for sensing and the second is due to fundamental reasons (both are discussed below). To mention some examples of the state of the art of modern sensors, we choose two recently developed nanosensors, both based on Si technology. The first is the Si nanowire-based field effect transistor (SiNW-FET) *(16)*. This represents the first NANO-FET-based chemical nanosensor. It operates on a functionalized site binding approach. Using this SiNW-FET, selective sensitive time-domain sensors were demonstrated. The second example is, however, different, it is the first wet nanoscale transistor operating in the bipolar mode (presented in a separate section below) *(17)*. It relies on the control and sensitive detection of the acidity (pH) of water or any other compatible electrolyte. It can be used to mimic conditions of chemical reactions taking place in living cells as well as detecting and manipulating single molecules in a very convenient way *(18–20)*. **Figure 2** shows the outer microparts of the recently developed nanoscale wet transistor. Here the nanoscale active parts are placed within the small dark middle part in the figure. The nanoscale active parts are shown in **Fig. 3c–f**. The role of the larger parts is to adjust and control the pH.

1.2. Scaling Issues and Challenging Limitations

The ISFET has been a very interesting and useful device for sensor technology and has been a fertile tool for scientific knowledge; however, for multiple reasons, but mainly physical limitations, it is not the most suitable tool for applications that involve low concentration and single cell or molecule detection.

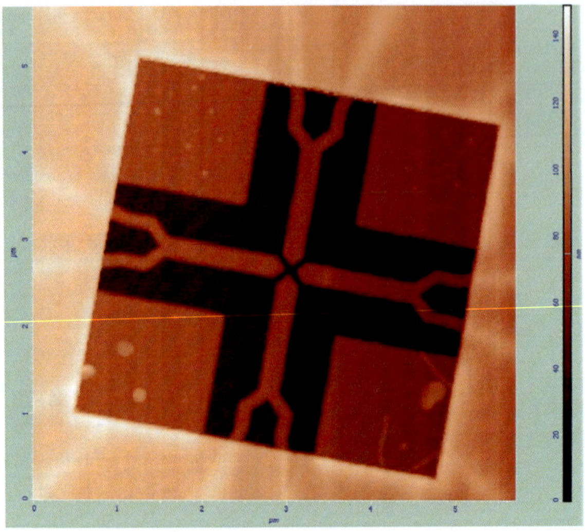

Fig. 2. Atomic force microscopy (AFM) image of the nanoscale active parts of the "wet" transistor/RISFET chip. The inner electrodes in this AFM image are separated with a 200-nm gap.

The "nonclassical" nanostructure-based sensors are more suitable for selective, sensitive, and time domain chemical sensing for many applications that involve low concentration and single cell or molecule detection.

In this section, we briefly mention some important facts and challenges regarding scaling issues, i.e., reducing the size of the sensors to smaller dimensions (specifically to the nanometer regime). Nevertheless, because of the fundamental importance of scaling issues, the present section is rather comprehensive and it discusses different aspects of scaling. When a system is reduced isomorphically in size, i.e., scaled down with all dimensions of the system decreased uniformly (isomorphic scaling), the change in length, area, and volume ratios alters the relative influence of the physical effect in question in an unexpected way *(21)*. The famous Russian nesting doll represents an illustrative example of isomorphic scaling **(Fig. 4)**. Note that the smallest doll has the largest surface area-to-volume (S/V) ratio. In other words, if the object (analyte) under focus shrinks down to the same length scale as the boundary layer (effect), being of thermal, diffusion, optical, etc., continuum theories break down and the laws of macro scaling no longer apply. When an object shrinks down, the S/V ratio is larger. This renders the surface forces and effects more important and dominating than other forces. Generally speaking, as objects decrease in size, force scaling follows the following: forces are scaled for those with a lower power of the linear dimensions in a

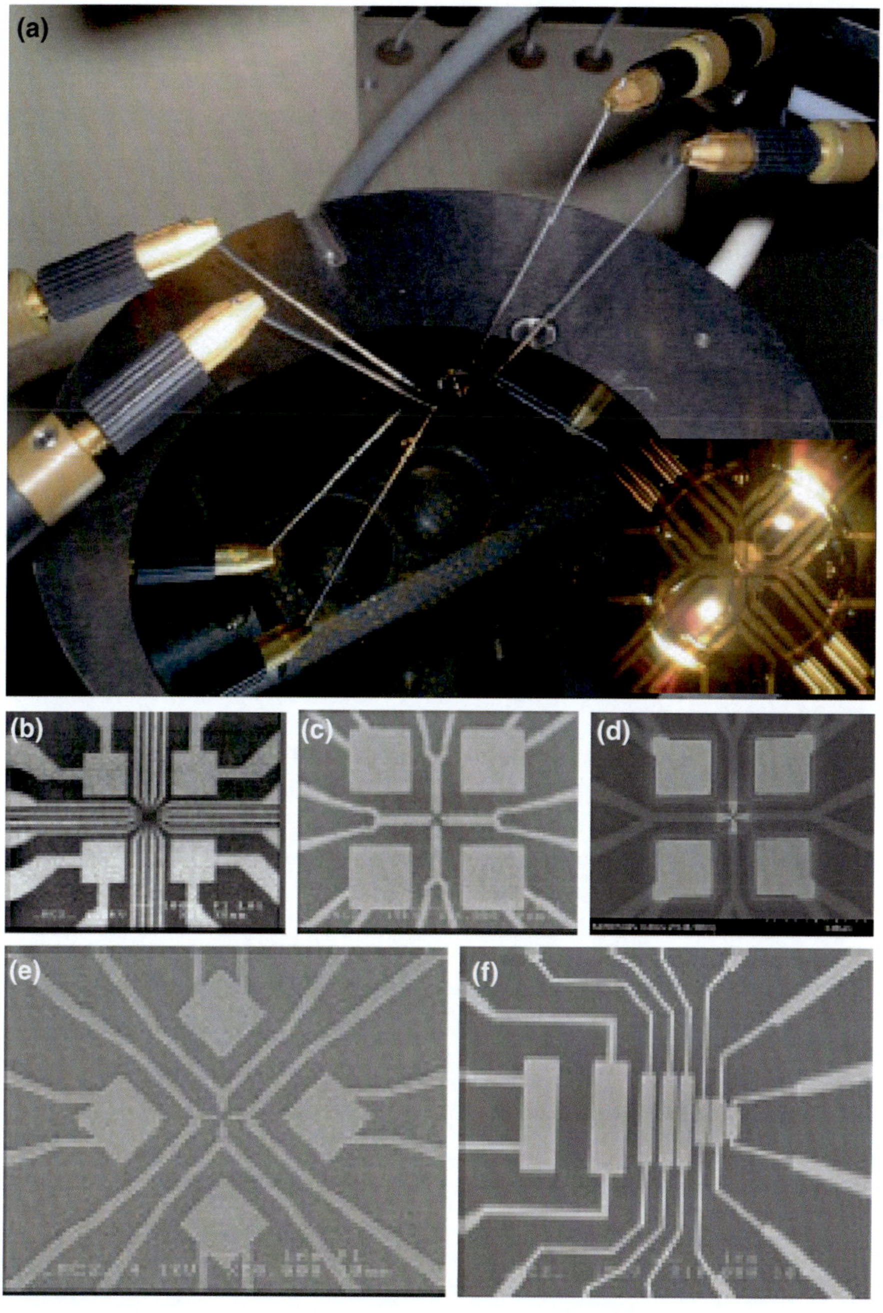

Fig. 3. (**a**) The sensor nanotransistor under measurement (the insert is a colored micrograph of the enlarged middle part showing the large pH electrodes); (**b**) scanning electron microscopy (SEM) image showing the large pH electrode (base); (**c**) SEM image of the investigated small pH electrodes; (**d**) the small pH electrode (base) and the small sensing electrodes (emitter-collector) with windows exposing the active device areas; (**e**) another design of the small electrode; and (**f**) a special design used for time-dependant electrical characterization.

Fig. 4. The famous Russian nesting dolls are an illustrative example of isomorphic scaling. The largest doll has the smallest surface to volume ratio, whereas the smallest doll has the largest ratio.

way that keeps them dominant over those with a higher power. A simple example of this is that, when considering an object with a mass *m*, when shrinking the dimensions, the surface tension will dominate over the gravitational force. Another example is that, for an object in an electromagnetic field, the electrostatic forces will gain over the magnetic forces if the object shrinks *(22)*. The later example has important consequences and is also one of the reasons that necessitated the need for nanosensors for detection of a small low-concentration analyte (e.g., single-molecule detection). Below we discuss the influence of the miniaturization (reduction in size) on different aspects, namely, technological challenges of nanosensors, the influence of the sample size, and the effect of the sensor size on the sensor detection sensitivity. We start by defining the sensor detection sensitivity (η) as the amount of change in the sensor's output, I_{out}, in response to a change in the sensor's input, I_{in}, over the entire sensor range (i.e., $\eta = dI_{out}/dI_{in}$). Obviously, a sensor with a wide dynamic range for the detection sensitivity is an aim of the scientific community.

1.3. Technological Challenges

In connection with technological challenges, we consider an example where liquid ejection is needed. In a real man-made sensor device, a small wet sensor must be embedded in a fluidic circuit to provide an environment for the sensor to survive (e.g., a cell). This is mentioned to emphasize why such an example is illustrative. We consider scaling down the liquid contactless ejection isomorphically, e.g., a water droplet is to be ejected, without contact, from a nozzle with different volumes. Nowadays, volumes from 10 μL down to 25 nL can routinely be ejected without contact through a commercial 8-channel dispenser assembly. Such an ejection can be made and observed, e.g., the event can be photographed. However, scaling down to smaller droplets (such as 10-nL water droplets) will lead to evaporation of the ejected droplet before observation of the event (e.g., taking a picture of the 10-nL droplet being ejected) *(22,23)*. This simple example implies that an engineering solution for such a situation is challenging. Different approaches for such problem can be adopted, e.g., mixing the liquid with a compatible low vapor-pressure solvent (i.e., stirring). The wet environment we mentioned above (as the environment that enables the sensor to survive) can be a cell, for example, as demonstrated in **ref.** *(24)*. Here, a sophisticated experiment demonstrated that a subtle change in the concentration of ions and small molecules (≤20 nm in size) embedded inside a cell can be detected. The cell here represents the environment that kept the molecule wet. This simple example of the challenge of small (a few nanoliters of volume) aqueous matter ejection suggests that, probably, mature future chemical nanosensors will relay on engineered bacteria, proteins, etc. In addition, self-assembling to house the sensing molecules is expected to be importantant. Self-assembling is a general term in our opinion. Conventionally, it means an engineered substrate and/or a substrate achieved by other standard processing techniques. However, we use the term self-assembling here with an insight toward non-conventional scenarios, e.g., site binding is a form of the non-conventional self-assembling scenarios intended above, but not achieved by conventional processing techniques. Naturally, this is enforced by the small size of the analyte to be sensed. Another technological appealing issue is the sensor arrays. Arraying is one way of providing fast, efficient, and consistent analysis for accurate comparative investigations. Moreover, sensor arrays with reduced signal to noise (S/N) ratio ($\approx n^{+1/2}$ for n connected identical sensors), and having enhanced selectivity/sensitivity are required. Many difficulties in achieving these requirements arise when scaling down to the sensor nanoregime. The final issue discussed before concluding this section is sensor storage and diffusive mixing. This is important and connected to the first discussed issue in this section, i.e., the aqueous wet environment challenges. Clearly, shelf-life storage of different sensor reagents is limited, and this implies that storage in a limited chemical environment should be followed by rehydration using the appropriate buffer before sensor usage. In addition, small amounts of reagents that make the sensor cocktail should be prepared according to diffusive mixing just before the sensor usage. Using this strategy increases the sensor lifetime and also opens new sensor opportunities. Hence, diffusivity is important. Considering a spherical molecule, the diffusion coefficient, D (m^2/s) is given by:

$$D = \frac{kT}{6\pi\rho r}, \tag{1}$$

where k is the Boltzmann constant (1.38×10^{-23} mJ/K), T the absolute temperature (K), ρ is the absolute viscosity (in kg/m/s), and r is the molecule hydrodynamic radius *(21)*. A molecule with molecular weight between 500 and 100 Da will have a diffusion coefficient of approximately 5×10^{-5} cm²/s. Considering the random walk equation, the diffusion length x of a molecule in a solution is given by:

$$x = \sqrt{2Dt}\ , \tag{2}$$

where t is the time required for the molecule to diffuse over a distance x. This equation implies that the diffusion of a molecule in a bulk solution over 10 μm is a million times faster than the diffusion over 1 cm. Now considering a specific volume of a liquid (for simplicity, we take cubic volume), we consider diffusion of a molecule (contained in a fixed solution mixed with the liquid volume in question) from one side of the cube to the opposite side. We consider here the case of saline solution in a bucket *(25,26)*. Assuming a diffusion constant D of the order of 10^{-5} cm²/s, and considering two volumes of 1 μL and 1 fL (10^{-15} L), the corresponding cube length for the molecule to diffuse across are 1 mm and 1 μm, respectively. The diffusion time t to cross a distance of 1 mm is 500 s whereas it is only 0.5 ms to diffuse across 1 μm. It is worth mentioning that in this solution reagent mixing example, the number of molecules present in the two volumes (taking a concentration of 1 μM solution) are 6×10^{11} and 600 molecules for the two solution volumes considered, respectively. From the preceding mixing example, it is evident that, although mixing is mediated by diffusion, it is fast at micro levels and this allows for reaction times dictated by inherent kinetics rather than the time it takes for the reactant species to meet in mixed solutions. This takes us back to the fact mentioned above; namely that as sensors are scaled, the physical effect in question will often be altered in an unexpected way. In nature, only the smallest species made of one or fewer cells relay purely according to diffusion, but other species made of many cells reply on other routes for transport, e.g., hearts, blood vessels, lungs, digestive system, etc. Nevertheless, from the discussion above, aqueous mixed volumes for chemical sensors should be approximately 1 mm³ or lower if a fast sensor response is to be achieved.

1.4. Sample Size Effect

Moving from technological challenges illustrated by the above example, we proceed to the effect of the sample analyte size. Miniaturization is a mixed issue for both the sensor and the analyte. The sample volume (V) required for detecting a given analyte concentration is given by *(27)*:

$$V = \frac{1}{\phi N_A C_i}, \tag{3}$$

where is the sensor efficiency (between 1 and 0), N_A is Avogadro's number (6.02×10^{23} mol^{-1}), and C_i is the concentration of analyte i (in mol/L). This equation clearly indicates that the analyte concentration fundamentally determines the sample volume. It has been noted that numerous chemicals and other biological species are routinely present with concentrations ranging from 100 to 10^7 units-copies/mL *(25)*. Consequently, large volumes are required to sense and detect almost most of the natural analyte concentrations; e.g., a relatively large volume of approximately 100 μL is required for accurate DNA assays. If the sensor sensitivity does not scale with its size, there is no benefit in future miniaturization. Nevertheless, as we mentioned in the paragraph above, new engineered materials are needed. Fortunately, such materials are available today and they help in shifting the detection for low analyte concentration and hence reduce the sample volume size required for sensing low concentrations. Such "exotic" materials have the property of producing enhanced fluorophores. It worth mentioning that colloidal particles, as well as nanocrystals or quantum dots, also play an important role because of their ability to approximate the ideal fluorophores, i.e., nonphotobleaching, narrow emission, and symmetric with multiple resolvable color that can be excited simultaneously using a single excitation wavelength. In addition, an important property of quantum dots is that their color, for both emission and absorption, can be tuned to any color by only changing their size. This will improve and push the sensitivity to a better level. However, with our current knowledge, even the most sensitive exotic particle will only slightly improve the sensitivity, not to the level we require *(21)*. On the other hand, the extreme case of detecting "single" molecules can be achieved either by trapping, which technically means that the concentration of the molecule is infinite, or by interrogating ultrasmall (≈femtoliter) volumes, as performed in fluorescence correlation spectroscopy *(28)*. Nevertheless, some experiments demonstrated that excellent sensitivity still could be achieved by using nanoprobes. Even a single ion placed near a single electron transistor (SET) can cause observable modulation of the current. This critical issue, namely, the miniaturized sensor sensitivity, is addressed in the next section of this chapter.

1.5. Sensitivity Issues

To discuss miniaturized sensor sensitivity, we choose a special type of sensor, namely, electrochemical sensors. We have chosen electrochemical sensors because they are more flexible to miniaturization. Electrochemical sensors are divided into conductometric, potentiometric, and amperometric sensors. It is important to mention that measuring a voltage in a potentiometric sensor, such as the ISFET or ISE (ion sensitive electrode), is scaling invariant; amperometric sensors, on the other hand, measure currents and they are affected by miniaturization. Most of the research efforts in miniaturization were focused on potentiometric sensors, although more benefit can be achieved from amperometric effect sensors, as is elaborated below and at the end of the next section.

Electrochemical reactions are governed by the electrode size with respect to the diffusion layer of the analyte to be recognized. If the diffusion layer of the analyte is of the order of the sensor electrode size, then the laws of classic macroscale electrochemistry breaks down *(21)*. This leaves us with unexpected effects; fortunately, some turn out to be beneficial to miniaturization. The total diffusion-limited current I_1 on a large substrate of an area A based on diffusion-limited current i_1 is given by:

$$i_1 = nFD_0 \frac{C_\infty^0}{\delta}, \qquad (4)$$

where n is the number of electrons, F is the Faraday constant, D_0 is the diffusion coefficient of the reactant species, δ is the diffusion layer thickness, and C is the concentration of the bulk of the solution. This implies that the total diffusion current is $I_1 = i_1 Ax$, with x being the diffusion length. If we reduce the size of the sensing electrodes to sizes comparable to the thickness of the diffusion layer, and keep them isolated, nonlinear diffusion caused by curvature effects "as the electrodes are isolated" should be considered. Analysis of such a situation showed that, as the nonlinear curvature effects become more and more pronounced, more diffusion takes place, i.e., diffusion occurs from all directions and ion collection increasingly persists, leading to more ion supply to the electrode, i.e., a beneficial unexpected effect. The diffusion layer thickness δ caused by linear effects is time dependant and given by *(29)*:

$$\delta = \left(\pi D_0 t\right)^{\frac{1}{2}}, \qquad (5)$$

Substituting this into the above expression for the total diffusion-limited current, namely I_1, we obtain the so-called Cottrell equation, which reads:

$$I_1 = nFAC_\infty^0 \left(\frac{C_0}{\pi t}\right)^{\frac{1}{2}}, \qquad (6)$$

This equation represents the current versus time for an electrode subjected to a potential step large enough to cause surface concentration of electro-active species to reach zero. This expression is appropriate regardless of the electrode geometry or the solution stirring conditions, as long as the diffusion layer thickness is much less than the hydrodynamic boundary layer thickness.

This is applicable to stirred aqueous solutions. For unstirred (i.e., pure) solutions, the thickness of the diffusion layer is not well defined and all types of disturbances can affect the transport. Hence, to prevent random connective motion from affecting the transport to and from the electrode, we want to keep the diffusion layer thickness smaller than the hydrodynamic boundary layer thickness and we want the hydrodynamic layer thickness to be regular. Ironically, stirring "mixing" can enable us to achieve this goal, a fact that was also necessary to avoid; is other un-wanted effects, e.g. to avoid fast evaporation of small volumes of ejected aqueous, mixing for long self storage, etc. Nonlinear diffusion at the edges of a microelectrode results in deviation from the simple Cottrell equation at longer collection times. The corrected equation will then read:

$$I_1 = nFAC_{\infty}^{0}\left(\frac{C_0}{\pi t}\right)^{\frac{1}{2}} + AnFD_0\frac{D_{\infty}^{0}}{r}. \qquad (7)$$

For longer times and small electrodes, the Equation 7 predicts that the correction term can become significant (note the electrode surface area A is divided by radius of the collected species). It is also important to note that the charge transfer is located at the outer edge of the electrode. This is actually a very favorable scaling. The correction term is proportional to A/r r^1, whereas the background current I_c (associated with the Helmholtz capacitance) is proportional to r^2, thus, the ratio of the Faradic current (charging current) to the background current should decrease as the electrode radius decreases. The fact to be emphasized here is that, with single small microelectrode, the analytical current is still small enough to easily be exploited.

We have presented several advantages that could be achieved by scaling amperometric sensors. However, we conclude by the following: (1) higher mass transfer rates at ultrasmall electrodes makes it possible to experiment with shorter time scale (faster dynamics), (2) an array of closely spaced ultrasmall electrodes can lead to collection of sufficient electrogenerated species with high efficiency if designed appropriately, and (3) the preceding discussion implies that electrochemical measurement (sensing) is possible even in highly resistive media.

1.6. Size and Sensitivity

The aim of this section is to elucidate the advantages of nanostructures in chemical sensing. In addition to the previously mentioned nonlinear effects (denoted as geometrical effects), other fundamental issues are highlighted.

The key issue is due to simple thermodynamic reasons. It is difficult to detect a single bioanalyte, because the net charge of the analyte is shielded by a double layer. However, there are two options to still get high sensitivity and be able to observe an immunological reaction. The first option is when working at low ionic conductivity. Here, we try to increase the Debye length as much as possible. In this case, we can measure the Donnan potentials. These small electrostatic surface potentials occur because of the formation of gradients of diffusible ions. The second option is when we deal with high ionic concentrations; in this case, the strong electrostatic potentials will be by far dominating the potential changes on the surface. Thus, the only way of observing an immunological reaction in this case is to make dynamic measurements. We can change the salt concentration of our solution and, for a short period of time, we can observe a transient signal caused by the rearrangement of the double layer. Why do nanowires make a difference when used as electrodes for sensing? The answer is that at low concentrations small surface potential changes become more and more visible with decreasing the nanowire dimensions (the nonlinear effects mentioned previously). The question is now the following: can we actually "observe" the net charge of our biomolecule, because the dimensions of the wire are smaller than the Debye length? There is no clear answer yet. However, from the experimental point of view, it seems that more than only the Donnan potentials or streaming potentials (electrokinetic) lead to the observed effects. If we have high salt concentrations, we end up with the same problem as for planar sensors.

2. Materials

The innovative electrochemical device used for the experiments presented here was processed using standard silicon technology. The materials used were as follows:

1. Low-doped oxidized Si 001 (resistivity 100 Ω-cm).
2. Gold was used for electrodes and electrical measurements pads.
3. PMMA electron beam resist was used for electrical isolation.

3. Methods

3.1. Device Fabrication Procedures

We used low-doped silicon as a starting substrate. A thermal high-quality oxide layer was first grown on the Si 001 wafers. The thickness of the oxide was 0.4 μm. We then combined optical and electron beam lithography to obtain the device structure and used metal electrodes composed of gold for both the inner nanoscale electrodes and for the final out large-scale electrodes. Before deposition of the gold, a thin chromium layer (~5 nm) was deposited to enhance the adhesion of gold to the silicon dioxide at the same lift-off step. The device is composed of five different processing lithography layers. The first was composed of depositing orientation marks (global and chip marks) for the electron beam lithography and, in addition, it contained a part of the connectors of the nano-electrodes. It is important to mention that the metal connectors from the nano-electrodes to the outer measurements pads were divided into three different parts: outer (thick ~400 nm), middle (thin ~150 nm), and inner (thin ~100 nm) parts. The middle part was the one deposited during the first lithography step (optical lithography). Such division is necessary as the pads have to be thick and the inner nano-electrodes have to be very thin. This is due to the fact that the measurements pads can not be less than 0.4 μm in thickness if stable measrements are to be performed, and the small nanoelectrodes has to be thin (less than 100 nm) due to processing requirement of the fact that high resolution electron beam resist has a thickness which is not larger than the 100 nm. After that, the first electron beam lithography layer was performed to deposit the smallest electrodes, i.e., the nano-electrodes. Then this was followed by an optical lithography step to deposit the out thick electrode part containing the measurements pads. We next used electron beam resist as the insulating layer. Two electron beam steps were next; both exposed parts of the covered surface. The covering was necessary so the aqueous liquids placed on top of the electrodes could be biased from different places with no electrical shorts (*see* **Fig. 3**). The need for two different steps was because some areas were in the nanometer size (sensing electrodes), whereas others were in excess of 100 μm × 100 μm, and we are restricted by the electron beam lithography machine, which performs either at high resolution/low speed or low resolution/high speed. The large windows for pads were made in the low-resolution mode whereas the nanometer-sized window was exposed by using the high-resolution mode. More information regarding the fabrication and measurements can be found in refs. *(17,18,20,30)* (also *see* **Note 1**).

3.2. A Nanoprobe Tool for Trapping Single Molecules

As mentioned above, when considering sensing single molecules, trapping is the ideal approach for sensing. Technically, this means that the concentration of the molecule is infinite. Because single molecules usually exist in aqueous mediums, a flexible device is needed for such a purpose. According to the previous section, both the analyte and the detection device are to be comparable in dimensions. Because molecules have dimensions of few nanometers, a device with these dimensions will be the ideal choice for developing a sensitive, selective platform for the trapping and detection of single molecules. Below, we introduce a recently developed nanoscale water transistor that can operate in aqueous mediums and perform multifunctions. **Figure 3** illustrates different parts of the nanoscale water transistor.

3.3. The Nanoscale Water Transistor

The basic operation of the sensor (transistor) is based on the variation of the pH of water by applying a voltage and independently measuring the current variation between the two nano-spaced electrodes. The pH control electrode plays the role of the base of the transistor, whereas the two nano-electrodes act as emitter and collector. When a voltage is applied between the emitter and base (V_{BE}), the simplest envisioned situation is that water dissociates into H_3O^+ and OH^-, consequently the pH of the water changes. In the presence of an electric field between the emitter and collector (V_{EC}), the dehydrated OH^- ions will be attracted to the collector, and a potential drop is established between the OH^- dehydrated hydroxyl ions and their images at the metal. Charge neutrality is postulated for the whole system and, therefore, OH^- dehydrated hydroxyl ions have to be compensated for by oppositely charged ions, namely the H^+ hydrated ions in the water. The hydroxyl ions are then located in a plane adjacent to the collector electrode called the inner Helmholtz plane (IHP). The hydrated ions (for simplicity, we consider H_3O^+) that diffuse from the bulk will be located at the so-called outer Helmholtz plane (OHP) *(31)*. These hydrated ions are then surrounded by water dipoles and will become nonconducting species. They cannot enter the outer layer, and, hence, a potential drop is then established between the OHP and the metal. This is the boundary condition at the metal/water electrodes emitter or collector electrodes. However, inside the bulk, the situation is different. The hydrated protons are envisioned in a number of ways. Namely, they exist as $H(H_2O)^+_n$ clusters. The most commonly existing configurations are Zundel cation ($n = 2$) and Eigen cation ($n = 4$) clusters. In the Zundel ion, the proton is placed between two water molecules, whereas, in the Eigen ion, a H_3O^+ ion is strongly bound to three water molecules. The proton kinetics in water as well as the absolute hydration free energy of H^+ ions in aqueous solutions is a topic of current research. For more information on these topics and on proton transport and kinetics in water, we refer the reader to refs. *(32–34)*. In addition, the protons have high mobility in presence of electric field and the ionic transport hops between water molecules in extended hydrogen-bonded structures (Grotthus mechanism) *(35)*. The total charge per unit area on the metal (σ) is given by the electronic charge multiplied by the difference between the number of anions and cations per unit area *(36)*. Using the geometrical capacitance between the metal and the OHP, and proceeding, we obtain:

$$I_{EC} = (W_E / L_{EC})\sigma\mu V_{EC}, \tag{8}$$

Here I_{EC} is the current between the emitter and collector, W_E is the emitter width, L_{EC} is the emitter to collector distance, μ is the mobility of the protonic species, and (σ) is the difference between the number of anions (dehydrated OH^-) and cations (hydrated H^+) per unit area. In deriving this expression, we only considered decomposition of water, and we have ignored other reactions that might exist *(37,38)*.

Figure 5 displays the measured *I–V* characteristics of two nano-electrodes 20-nm apart and for the choice of biasing V_{BE} between a large pH electrode **(Fig. 3b)** and a small pH electrode **(Fig. 3c)**. As observed in the figure, by applying V_{BE} with steps of 0.2 V, clear distinct I_{EC} currents with different threshold voltages were obtained. We then investigated the effect of changing the strength of the electric field, by applying V_{BE} between the two small pH electrodes shown in **Fig. 3c**. This device configuration will provide electric field strengths 1,500 times larger than the previous configuration **(Fig. 3)**.

We observed almost the same current range as shown in **Fig. 5**. The only qualitative difference was observed in the reverse direction (negative V_{EC}). For this negative V_{EC} bias, a different shift of the threshold voltage was observed compared with the previous

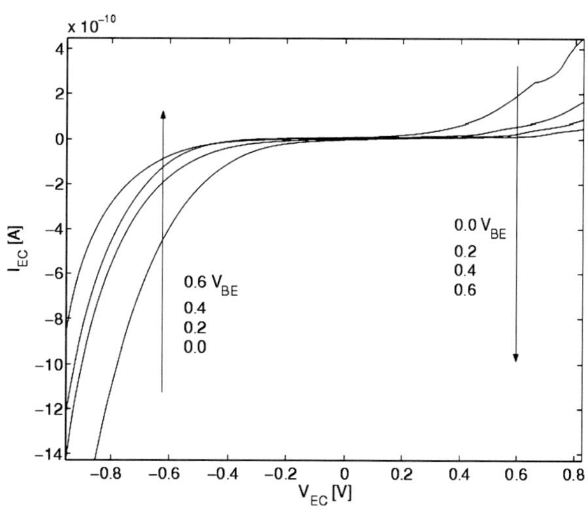

Fig. 5. The measured I_{EC}–V_{CE} of the nanoscale water-based transistor with emitter to base biasing of 0.6 V. The pH control biasing was used for a configuration that involved large pH electrodes **(Fig. 3b)** and the small pH electrodes shown in **Fig. 3c**. Reproduced from **ref.** *(17)* with permission from the American Institute of Physics.

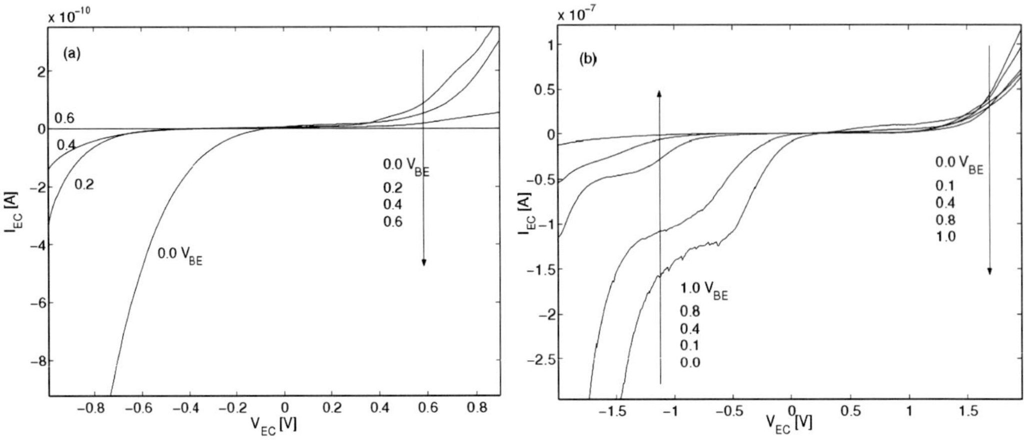

Fig. 6. *I-V* characteristics using two different designs of the small pH electrodes, (**a**) for the choice of pH electrodes shown in **Fig. 3c**, and (**b**) *I-V* obtained for the electrodes shown in **Fig. 3d**. Reproduced from **ref. (*17*)** with permission from the American Institute of Physics.

behavior, as shown in **Fig. 6a**. To study the effect of the position of the pH electrode (applied electric field with regard to the applied V_{EC}), we then compared the case of **Fig. 6a** with the case using the pH electrodes of **Fig. 3e**. Here, these small pH electrodes provide the same electric field as in the case of **Fig. 6a**, but they lie along the same line joining the emitter and collector (i.e., the electric field of the pH electrodes is along the same direction as the electric field between the emitter and collector). In addition, the pH electrodes in this case are much closer to the emitter-collector nano-gap active area. The measured *I-V* characteristics are shown in **Fig. 6b**. As clearly seen, the current is now in the range of 10^{-7} A. The magnitude of the measured I_{EC} is 1,000 times larger compared with the previous two cases shown in **Figs. 5** and **6a**. Beside the observed larger current, a large shift in the threshold voltage is detected in the reverse bias direction, whereas almost no shift is observed for the forward direction (*see* **Fig. 6b**). The reason for this is attributed to the fact that, in this case, the electric field inducing the pH variation lies along the same line as the electric field between the emitter and collector that is driving the current through the device. It is also observed that, in the case of **Fig. 6b**, saturation (plateau) starts to be pronounced for negative V_{EC} and at higher positive V_{BE} values. The origin of this is a matter of present investigation. The previous DC characteristics shown in **Figs. 5** and **6a, b**, imply that the electric field strength for the variation of the pH (i.e., V_{BE}) alone without considering the proximity of positioning of this field has a weak influence on the output characteristics. This also indicates that the device pH sensitivity can be designed according to the application in question. The demonstrated water-based nanoscale transistor presented here is of interest for many chemical and bioelectrical applications because of the biocompatibility of and the wide usage and presence of water in different living systems. In fact, it can be used to mimic conditions of living cells, and at the same time create microscale vortices in water *(39)*. This is in addition to the use as an efficient nanoprobe of trapping single molecules in aqueous solutions, as is demonstrated below..

3.4. Trapping Single Molecules

The ideal way of detecting single molecule is by trapping, because this implies that the molecule concentration is infinite at the trapping surface. Indeed, with the above sophisticated nanoscale water transistor, it was possible to trap and detect a single molecule *(20)*. By using a special design of the electrodes (sharp edges) of the above-mentioned nanoscale "wet" transistor and applying a radio frequency voltage between the two sharp nanoelectrodes, a strong electric field gradient is established. The use of alternating electric fields is routinely exploited for trapping biological molecules *(38)*. Depending on the temporal properties as well as the geometrical field distribution, microscopical particles (such as cells, etc.) can be moved, attracted, oriented, rotated, stretched, or trapped. This technique is supported by theories dealing with electrokinetic phenomena to help estimate the experimental conditions and confirm the observation. The technique used here is denoted as dielectrophoresis (DEP) and can lead to particle movement toward electrode edges (positive DEP) or repletion away from the electrode edge (negative DEP). Positive DEP was used and successfully trapped a single protein (R-phycoerythrin) in an aqueous solution. Using this technique, different types and sizes of molecules have been trapped in different experiments; for detailed information, we refer the reader to **ref.** *(21)* and references therein. **Figure 7a** displays the sharp electrode configuration used in the trapping experiment. In **Fig. 7b–d**, dielectrophoresis trapping of a single R-phycoerythrin protein is clearly seen because the molecule shows strong fluorescence in the cases. In **Fig. 7b**, in a protein solution of 0.6 nM and electrodes, before applying the AC voltage, no fluorescence is observed. After applying an electric field for 10 s, the molecule is trapped. In **Fig. 7d**, the scale is changed, but the experimental conditions are the same as those in **Fig. 7e**. **Figure 7e** presents results of a calculated electric field gradient. The color-coded projection shows the field gradient in the electrode plane. The upper gray enclosure defines the region where the electric field force gradient exceeds the molecular diffusion. This experiment was complemented by fluorescence correlation spectroscopy performed in the original solution (Zeiss Confocor 2). The results of this experiment indiçate that up to four molecules might be present in the solution. Nevertheless, the results are shown in **Fig. 8**.

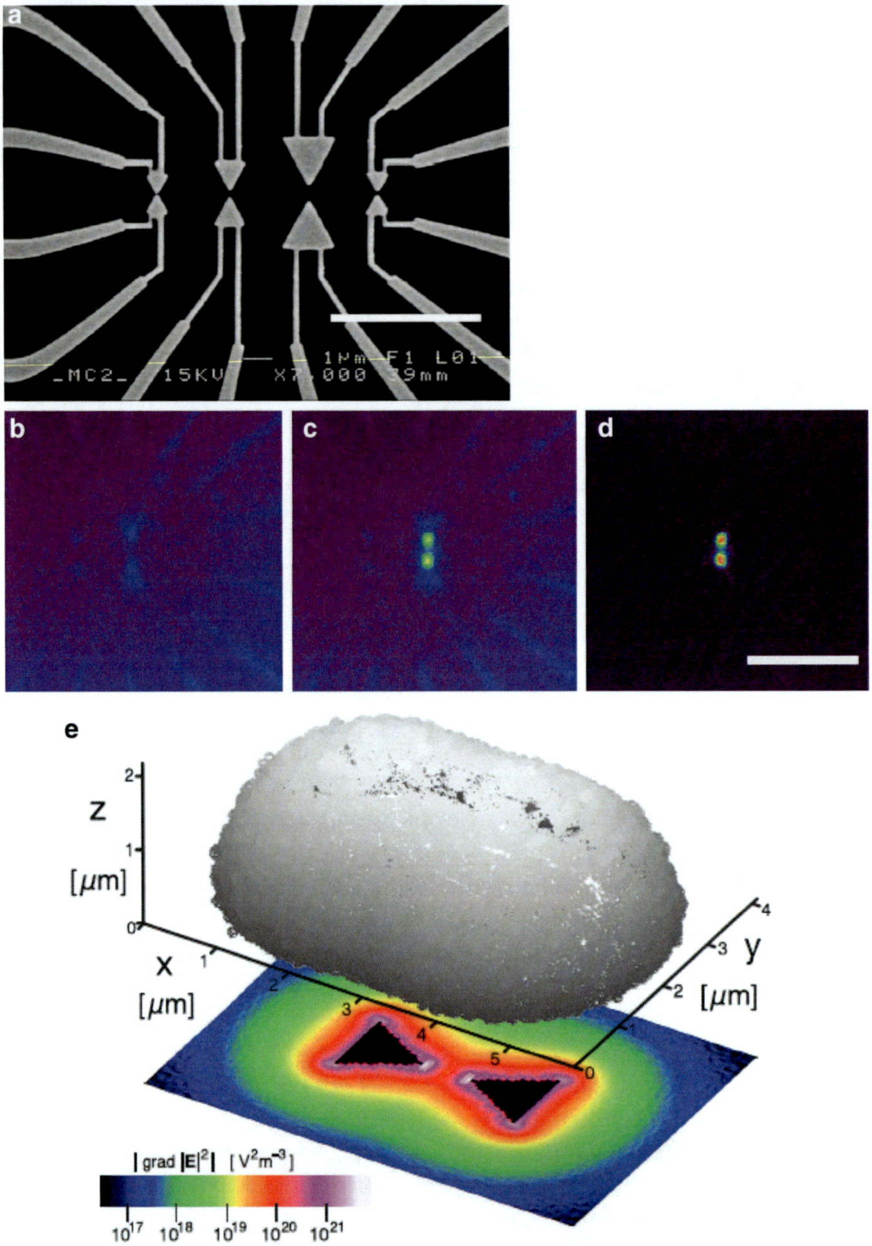

Fig. 7. (**a**) SEM showing the specially designed sharp edge electrodes, (**b**) Dielectrophoresis trapping of single R-phycoerythrin molecule between the two sharp electrodes shown in (**a**), here no field is applied and no florescence is observed. (**c**) and (**d**) field is applied and flourescence is observed. Finally in (**e**) calculated electric field gradient. Reproduced from **ref.** *(20)* with permission from the American Physical Society.

3.5. The Region Ion-Sensitive Field Effect Transistor

Recently, it was shown that the wet-transistor could be converted to a bioelectronic region ion-sensitive field effect transistor (RISFET) sensor for the detection of glucose *(30)*. In the RISFET system, the diffuse double layer above the SiO_2 surface in the vicinity of the sensing electrodes is modulated with the gate/source-drain potential. The low-profile sensing electrodes (58 ±

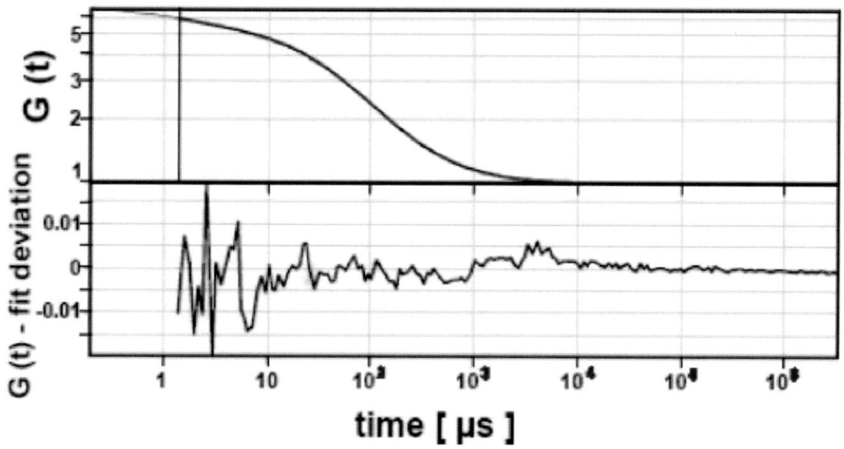

Fig. 8. Upper panel shows the florescence autocorrelation function of the R-phycoerythrin solution, and the lower panel shows the difference between the data and the numerical simulation assuming a single molecule is displayed. Reproduced from **ref.** *(20)* with permission from the American Physical Society.

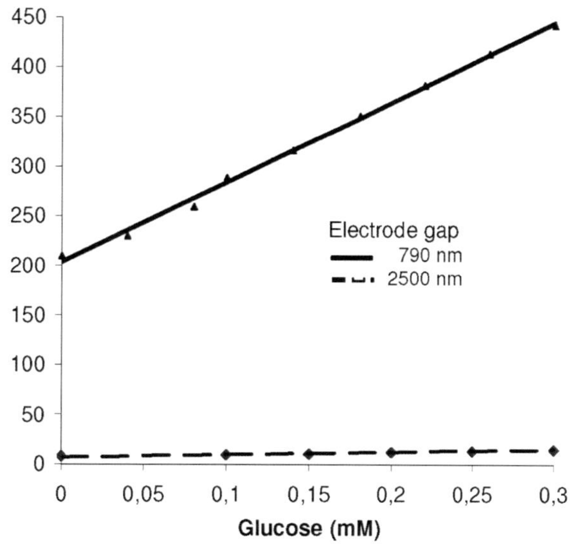

Fig. 9. The signal response current for different glucose concentrations measured with a 790-nm sensing electrode gap versus a 2,500-nm sensing electrode gap. Decreasing the size of the electrode gap enhances the sensitivity of RISFET sensor systems many fold.

2 nm high, with a median of 57.9 nm) can use this modulation to enhance the sensitivity of the sensor. Furthermore, it was shown that the field strength between the sensing electrodes had a great impact on the RISFET sensitivity. At the same applied source-drain voltage and gate potential, for two different sensing gaps, the field strength was increased approximately five times, going from a 2,500-nm gap to a 790-nm gap (*see* **Fig. 9**, $x = 0–0.3$ mM, $y = 0–450$ pA).

However, the sensitivity was increased 30 times when going from 28 to 830 pA/mM glucose. Decreasing the sensing electrode gap should enhance the sensitivity of the sensor because the field strength between the sensing electrodes would increase. Also decreasing the height of the sensing electrodes could potentially increase the sensitivity of the sensor even further, because this change would more effectively sense the modulation of the diffuse double layer. A possible way of achieving the first point is through conventional nanolithography. As for the second point, the height of the sensing electrodes could be decreased to a few hundred angstroms using conventional nanolithography, however, the introduction of conducting nanowires or nanotubes as sensing electrodes would certainly push the limits to an optimum somewhere in the vicinity of the first Helmholtz plane, as the coins are almost depleted for the favor of the desired counterions. Both suggestions are presently under investigation. The recent development of nanotrees opens up new types of nanobiosensors. **Figure 10** shows a future nanotree RISFET nanobiosensor system with gallium phosphide (GaP) nanotrees in the center of two platinum electrodes. The GaP nanotrees functions as supporting material to the catalytic enzymes of the sensor system.

3.6. Conclusion

As objects are isomorphically scaled down to the nanometer regime, some observations change in an unexpected way, sometimes this change is beneficial for the observation. Although many technological constrains appear when scaling, the advancement of nanotechnology provides possible solutions for such problems and constrains. An innovative example of a nanodevice

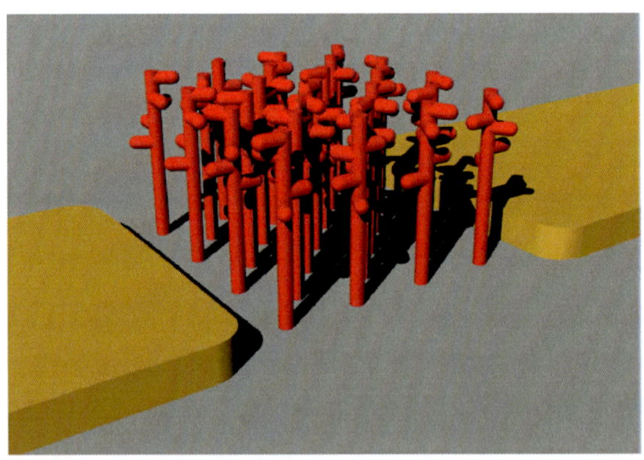

Fig. 10. A future nanotree RISFET biosensor system. Nanotrees are placed in between the sensing platinum electrodes. Enzymes are immobilized on the tops of the trunks and on the ends of the branches using thiol coupling. The substrate is measured through impedance measurements on low AC voltage in a RISFET sensor system. The impedance change of the liquid is caused by the enzymatic conversion of the substrates via the immobilized enzymes.

(multifunction platform) developed recently is highlighted as an example of achieving laboratory on a chip (lab on a chip). This nanodevice is the nanoscale water transistor, which measures and adjusts the pH value, can mimic biochemistry, can create microscale vortices in aqueous mediums, and can trap single molecules. Owing to the critical need of human health monitoring, much research is still needed to provide mature, reliable nanosensors.

4. Notes

1. The difficulty in fabricating such a device can arise from the fact that high- and low-resolution modes in electron beam lithography are combined. The metal contact from the measurement pads to the nanosensing electrode is composed of three different parts, each requiring a separate lithography step, two steps using low-resolution lithography and the final step using high-resolution lithography. In the low-resolution starting step, a thick metal layer, usually approximately 400 nm, is deposited and followed by a metal lift-off technique. The final high-resolution step is associated with thin metal lift-off and breakage at corners can lead to a nonworking device. To avoid this problem, the fabrication process is started by the middle metal layer lithography (low resolution) and a rather thin metal layer is deposited followed by lift-off. This is the middle layer. Together with this layer, alignment marks are deposited. This is followed by the other low-resolution lithography step, to form the metal line leading to the measurement pads. Here a thick metal is deposited. Finally, the high-resolution step is performed and a thin metal is deposited after lithography. In this way, metal breakage at corners is avoided and a reliable device consisting of nanoparts as well as microparts is achieved.

References

1. Wilson, G. S., Hu, Y. (2002). Enzyme based nano-sensors for in-vivo measurement. *Chem. Rev.* **100**, 2693–2704
2. Patolsky, F., Lieber, C. M. (2005). Nanowires nanosensors. *Materials Today* **8**, 20–28
3. Clark, Jr., L. C. (1956). Monitor and control of blood and tissue oxygen tensions. *Trans. Am. Artif. Inter. Organs* **2**, 41–57
4. Lyons, C., Clark, Jr., L. C. (1962). Electrode systems for continuous monitoring in cardiovascular surgery. *Ann. N. Y. Acad. Sci.* **102**, 29–45
5. Xie, B., Danielsson, B. (2007). Thermal biosensor and microbiosensor techniques. In: Handbook of Biosensors and Biochips (C. Lowe, et al. eds.). John Wiley & Sons, New York
6. Archer, M., Christopherson, M., Fauchet, P. M. (2004). Microporous silicon electrical sensor for DNA hybridization detection. *Biomed. Microdevices* **6**, 203–211

7. Koncki, R., Lenarczuk, T., Radomska, R., Glab, S. (2001). Optical biosensors based on Prussian blue films. *Analyst* **126**, 1080–1085
8. Middelhoek, S., Noorlag, D. J. (1982). Signal conversion in solid state transducers. *Sens. Actuators* **2**, 211–228
9. Middelhoek, S. (2000). Celebration of the tenth transducers conference: the past, present and future of transducer research and development. *Sens. Actuators A* **82**, 2–23
10. Smith, C. S. (1954). Piezoresistance effect in germanium and silicon. *Phys. Rev.* **94**, 42–49
11. Wise, K. D., Angle, P. D. (1970). An integrated-circuit approach to extra cellular microelectrodes. *IEEE Trans. Biomed. Eng.* **17**, 238–247
12. Bergveld, P. (1970). Development of an ion-sensitive solid-state device for neurophysiological measurements. *IEEE Trans. Biomed. Eng.* **17**, 70–71
13. Lundström, I., Shivaraman, M. S., Svensson, C. M. (1975). Hydrogen sensitive MOS structures. *J. Appl. Phys.* **46**, 3876–3881
14. Danielsson, B., Lundström, I., Mosbach, K., Stiblert, L. (1979). On new enzyme transducer; the enzyme transistor. *Anal. Lett.* **12**, 1189–1199
15. Winquist, F., Lundström, I., Danielsson, B. (1986). Determination of the creatinine by an ammonia sensitive semiconductor structure and immobilized enzymes. *Anal. Chem.* **58**, 145–148
16. Cui Y., Yei, Q., Lieber, C. M. (2001). Functional nanoscale electronic devices assembled using silicon nanowire building blocks. *Science* **291**, 851–853
17. Chiragwandi, Z., Nur, O., Willander, M., Calander, N. (2003). DC characteristics of a nanoscale water based transistor. *Appl. Phys. Lett.* **83**, 5310–5312
18. Chiragwandi, Z., Nur, O., Willander, M., Panas, I. (2005). Vortex rings in pure water under static external electric field. *Appl. Phys. Lett.* **87**, 153109–3
19. http://physicsweb.org/articles/news/9/10/7
20. Hözel, R., Calander, N., Chiragwandi, Z., Willander, M., Bier, F. (2005). Trapping single molecules by dielectrophoresis. *Phys. Rev. Lett.* **95**, 128102–4
21. Madou, M. J., Cubiccoiotti, R. (2003). Scaling issues in chemical and biological sensors. *Proc IEEE.* **91**, 830–38
22. Madou, M. J. (2002). Fundamentals of Microfabrication, 2nd edition. CRC, Boca Raton, FL
23. http://www.seyonic.com/flow.htm
24. Kopelman, R., Miller, M. T., Brause, M., Clark, H. A. (1999). Optochemical nanosensors for intracellular chemical measurement. *Proc. SPIE (Int. Soc. Opt. Eng.)* **3540**, 198–205
25. Harrision, D. J., Seilier, K., Manz, A., Fan, Z. (1992). Chemical analysis and electrophoresis systems integrated on glass and silicon chip. In: Technical Digest IEEE Solid State Sensors and Actuators Workshop, pp. 110–118
26. Brown, P. R., Grushka, E. (1993). Advances in Chromatography. Marcel Dekker, New York, pp. 50–51
27. Petersen, K., McMillan, W., Kovacs, G., Northrup, A., Christel, L. (1998). Toward next generation clinical diagnostic instruments: scaling and new processing paradigms. *Biomed. Microdevices* **1**, 43–49
28. Eigen, M., Rigler, R. (1994). Sorting single moleculels: application to diagnostics and evolutionary biotechnology. *Proc. Natl. Acad. Sci. U.S.A.* **91**, 5740–5747
29. Bausells, J., Carrabina, J., Errachid, A., Merlos, A. (1999). A simple REFET for pH detection in differential mode. *Sens. Actuators* **B57**, 56–62
30. Risveden, K., Pontén, J. F., Calander, N., Willander, M., Danielsson, B. (2007). The region ion sensitive field effect transistor, a novel bioelectronic nanosensor. *Biosens. Bioelectron.* **22**, 3105–3112
31. Mott, N. F., Twose, W. D. (1961). The theory of impurity conduction. *Adv. Phys.* **10**, 107–163
32. Walbran, S., Kornyshev, A. A., (2001). Proton transport in polarizable water. *J. Chem. Phys.* **114**, 10039–10048
33. Grabowski, P., Riccardio, D., Gomez, M. A., Asthagiri, D., Pratt, L. R. (2002). Quasi-chemical theory and the standard free energy of H+(aq). *J. Phys. Chem.* **106**, 9145–9148
34. Kornyshev, A. A., Kuznetov, A. M., Spohr, E., Ulstrup, J. (2003). Kinetics of proton transport in water. *J. Phys. Chem. B* **107**, 3351–3366
35. Pectina, O., Schmickler, W. (1998). A model for electrochemical proton transfer reactions. *Chem. Phys.* **228**, 265–277
36. Atkins, P., de Paula, J. (2002). Atkin's Physical Chemistry, 7th edition. Oxford University Press, New York,271
37. Kek, D., Bonanos, N., Mogensen, M., Pejovink, S. (2000). Effect of electrode material on the oxidation of H2 at the metal-Sr0.995Ce0.95Y0.05O2.970 interface. *Solid State Ionics* **131**, 249–259
38. Pohl, H. A. (1987). In: Dielectrophoresis. Cambridge University Press, Cambridge, England
39. http://spie.org/x8832.xml

Chapter 13

ZnO Nanorods as an Intracellular Sensor for pH Measurements

M. Willander and Safaa Al-Hilli

Summary

High-density ZnO nanorods of 60–80 nm in diameter and 500–700 nm in length grown on the silver-coated tip of a borosilicate glass capillary (0.7 μm in diameter) demonstrate a remarkable linear response to proton H_3O^+ concentrations in solution. These nanorods were used to create a highly sensitive pH sensor for monitoring in vivo biological process within single cells. The ZnO nanorods exhibit a pH-dependent electrochemical potential difference versus an Ag/AgCl microelectrode. The potential difference was linear over a large dynamic range (pH, 4–11) and had a sensitivity equal to 51.88 mV/pH at 22°C, which could be understood in terms of changes in surface charge during protonation and deprotonation. Vertically grown nanoelectrodes of this type can be smoothly and gently applied to penetrate a single living cell without causing cell apoptosis.

Key words: Intracellular pH, ZnO nanorods, Potentiometric measurement, pH of adipocyte or fat cell

1. Introduction

One-dimensional (1D) nanostructure research has elucidated many biomarkers that have the potential to greatly improve disease diagnosis *(1–3)*. pH sensor miniaturization is highly important because the large surface-to-volume ratio leads to a short diffusion distance of the analyte toward the electrode surface, thereby providing an improved signal-to-noise ratio, faster response time, enhanced analytical performance, and increased sensitivity *(4)*. These results enable the sensitive and rapid detection of biochemical and physiological processes, essential to basic biomedical research applications. However, our research is at a very primitive stage and many additional efforts are necessary to obtain reliable instrumentation for intracellular measurements.

James Weifu Lee and Robert S. Foote (eds.), *Micro and Nano Technologies in Bioanalysis*,
Methods in Molecular Biology, vol. 544
DOI 10.1007/978-1-59745-483-4_13, © Humana Press, a part of Springer Science+Business Media, LLC 2009

The sensor in this study was used to detect and monitor real changes in cell behavior using changes in the electrochemical potential at the single cell/ZnO nanorod surface interface in the intracellular microenvironment.

When a solid emerges in a polar solvent or an electrolyte, a surface charge will develop through one or more of the following mechanisms: preferential adsorption of ions; dissociation of surface charged species; isomorphic substitution of ions; accumulation or depletion of electrons at the surface; and physical adsorption of charged species onto the surface. The polar and nonpolar surface structures of ZnO nanorods are of interest in understanding the mechanism of interaction of these surfaces with the medium surrounding them. The sensing mechanism is the polarization-induced bound surface charge by interaction with the polar molecules in the liquids. Significant progress in understanding the surface properties of ZnO was achieved recently *(5–15)* stimulated by the importance of this material for a number of applications ranging from cosmetics and medicine to heterogeneous catalysis. ZnO crystallizes in the hexagonal wurtzite structure with lattice parameters (a = 3.25 Å and c = 5.206 Å). ZnO is a polar crystal whose polar axis is the c-axis, and it belongs to the $C_{6v}^{4} = p6_3\,mc$ space group.

A practical application of some metal oxides is associated with their pH sensitivity. These oxides (insoluble and stable in water) may function as metal oxide pH electrodes on electronically conducting substrates (not necessarily on the metal from which they are formed). This group of pH sensors is distinguished from metal/metal oxide pH-sensitive electrodes by a different kind of reaction determining their potential-pH response. In this case, a metal oxide/metal oxide (higher and lower valency) couple instead of a metal/metal oxide couple is involved in the pH-dependent equilibrium *(16)*. The oxide layer that covers the electronic conductor contains a mixture of both oxides, one with some oxygen deficiency.

In zero-current potentiometry, the relative size of the two electrodes is immaterial *(17)*. Acquiring useful information in this case requires only that the potential of the working electrode be measured against a well-defined and stable potential from a reference electrode. The conventional reference electrode consists of an electrode of the second type, such as the Ag/AgCl/KCl system. Any foreign potential inadvertently present within the measuring circuit can contaminate the information, as a result, it is mandatory that the reference electrode be placed as close to the working electrode as practically possible. A determination by direct potentiometric measurement is accomplished either by calibrating the electrode with solutions of known concentration or by using the techniques of standard addition or standard subtraction.

An advantage of ZnO nanorod sensors is their small size, which allows intracellular sensing of physiological and biological parameters in nanoenvironments, and a strong, stable, and reversible signal with respect to pH changes. The detection sensitivity of the pH sensor is achieved by monitoring minimal changes in electrochemical potential caused by binding of biomolecular species on the surfaces of the probe, owing to the high isoelectric point of the material comprising the sensor (in the case of ZnO, the isoelectric point is between 9 and 10) *(18)*.

1.1. Intracellular pH

The acid and base properties of electrolytes in living cells play an important role in any biological process, because the pH value is the most critical parameter in chemical and biochemical reactions. Intracellular pH has been studied extensively for longer than a century, using a wide range of techniques. These techniques have been subject to constant improvements, to the extent that useful measurements can now be made in the smallest of cells. Although intracellular pH changes were observed much earlier, the first measurements of what could loosely be described as intracellular pH were made around 1910 using cell extracts and platinum/hydrogen electrodes (for a review, *see* **ref.** *(19)*). High electrical resistance, however, presented a major problem for the miniaturization of metal minielectrodes. It was not until the 1950s that significant progress was made in producing microelectrodes that could be widely used. pH-sensitive glass, although discovered around 1900, became the material of choice. It has relatively low electrical resistance; no sensitivity to oxidizing and reducing agents, dissolved gasses, anions, or buffers; and it is stable and can produce a relatively rapid response. The electrode consisted of a portion of exposed pH-sensitive glass, which is approximately 500-μm long (*see* **Fig. 1**). Improvements in both the pH-sensitive glass and input impedance of modern electrometers allowed the exposed length of the pH-sensitive glass to be reduced to approximately 100 μm, but the use of these electrodes was restricted to giant cells. The first true pH-sensitive *micro*electrode was produced in 1974 by Thomas *(20)* (*see* **Fig. 2**). Thomas introduced a design wherein the exposed length of pH-sensitive glass (still at least 100

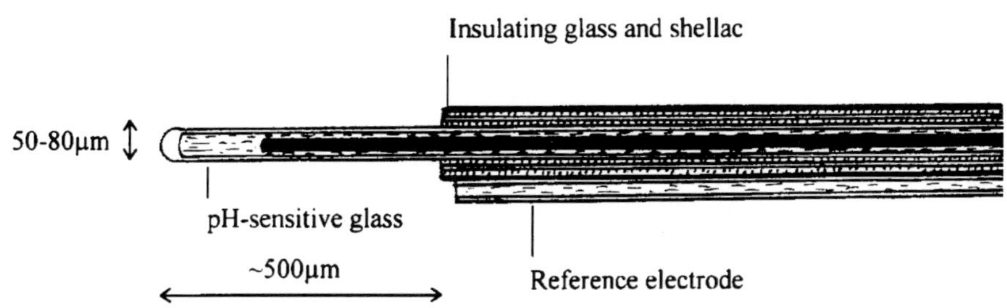

Fig. 1. Glass pH-sensitive electrode used by Caldwell *(19)* to measure intracellular pH in crab muscle fibers. Reproduced from ref. *(19)* with permission from PMC.

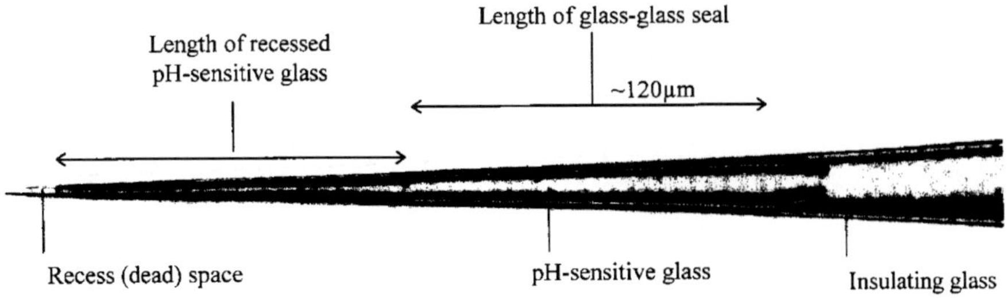

Fig. 2. Recessed-tip pH-sensitive microelectrode as used by Thomas *(20)* to measure intracellular pH in snail neurons. Reproduced from ref. *(20)* with permission from PMC.

μm in length) was recessed within the insulating glass such that the pH of the recessed space was measured. The tip diameters of the completed pH-sensitive microelectrodes were <1 μm. The result was that only the 1- to 2-μm tip had to be placed within the cell. Such microelectrodes require considerable skill to manufacture. The recessed-tip pH-sensitive microelectrode has been used to measure intracellular pH in numerous cell types (snail neurons, skeletal muscle, and cardiac muscle), and despite its relatively slow response, for those who can produce them, it remains the method of choice for measuring intracellular pH in cells approximately 100 μm in diameter. Even with such large cells, the requirement to place two microelectrodes into one cell can be difficult to fulfill, especially where cell boundaries are obscured. The problems of slow response, large tips, and difficult construction of recessed-tip pH-sensitive microelectrodes have been partially resolved by the discovery of submicron capillary tubes.

At intracellular pH, the ZnO nanorods are positively charged, which provides a suitable environment for the adsorption of low isoelectric point biological function groups (proteins and enzymes) and the retention of bioactivity. In this chapter, we focus on the fabrication of nanostructure ZnO nanorods for intracellular pH sensing. Our main effort has been directed toward the construction of tips capable of penetrating the cell membrane as well as optimization of the electrochemical potential properties. To demonstrate the electrode performance, we use it in biological media. Our results indicate that the electrode acts as an extremely sensitive intracellular pH sensor.

2. Materials

1. Standard buffers: potassium phthalate, pH 4.0; sodium potassium phosphate, pH 6.0; sodium phosphate/potassium phosphate, pH 7.0; hydrochloric acid-borate, pH 8.0; boric

acid-sodium-potassium borate, pH 9.0; and boric acid-sodium-potassium borate, pH 11. Store at room temperature.

2. Buffer solutions having pH values equal to 5.7, 6.1, 6.5, 6.9, 7.3, and 7.7 prepared from sodium phosphate monobasic dehydrate $NaH_2PO_4 \cdot 2H_2O$ and disodium hydrogen phosphate Na_2HPO_4. Store at room temperature.
3. Borosilicate glass capillaries: sterile Femtotip® II with a tip inner diameter of 0.5 μm, an outer diameter of 0.7 μm, and a length of 49 mm (Eppendorf AG, Hamburg, Germany).
4. 0.2 M HCl solution.
5. High-purity silver conductive paint (a division of HK Wentworth Ltd., Kingsburg Park, Midland Road Swadlincote, Derbyshire, DE 11 OAN, UK).
6. Zinc nitrate hexahydrate, reagent grade $[Zn(NO_3)_2 \cdot 6H_2O]$ and hexamethylenetetramine $[(CN_2)_6N_4]$.
7. Ag/AgCl reference electrode from Metrohm (no. 6.0733.100, Zofingen, Switzerland).
8. pH combined electrode from Metrohm (no. 6.0228.020).
9. Pt wire, 10-cm length, for use as anode electrode.
10. DC power supply for applying 1 V potential difference.
11. All electrochemical experiments were carried out using a Metrohm pH meter model 827 at room temperature(22 ± 2°C).
12. Single human adipocyte or fat cells.
13. Nikon inverted microscope for verification that individual cells are probed.
14. Glass slides with dimensions 5 cm × 4 cm × 0.177 mm.

3. Methods

The pH electrode behavior is a function of its extensive interior surface area, whereas that of ZnO nanorods is based on their exterior surface. Assembling many ZnO nanorods together yields the same benefits as the pH electrode (low volume and high reactivity). We use the tip of the electrode surface, which is an important component of electrochemical sensors, for detection. The tip contains hundreds of individual ZnO nanorod sensors. These sensors can be combined to produce a greater level of accuracy for a single proton or hydroxyl group. The pH response of ZnO polar and nonpolar surfaces has been explained in terms of the formation of hydroxyl groups that lead to a pH-dependent net surface charge with a resulting change in voltage at the electrode/liquid interface *(21–24)*. The principle of the ZnO nanorod electrochemical potential pH sensor shows that when a solid is submerged in a polar solvent or an electrolyte solution, a surface charge will develop. ZnO is an amphoteric oxide in which an electropositive metal atom gives the oxygen a sufficient negative charge to strip a proton from a neighboring H_3O^+. However, the metal ion must be electronegative enough to serve as an electron acceptor from a neighboring OH^-. Theoretically, the potential determining reaction of ZnO nanorods in aqueous solution can be represented by *(25)*:

$$ZnO + 2H^+ + 2e^- = Zn + H_2O. \tag{1}$$

The relation between the ZnO electrode potential difference E and the solution pH is then determined by the Nernst equation:

$$E = E^{o} - \frac{RT}{nF}\ln\left(\frac{f_{\mathrm{Production}}\left[\mathrm{Production}\right]}{f_{\mathrm{Reaction}}\left[\mathrm{Reaction}\right]}\right) + \frac{2.303RT}{nF}\log(a_{\mathrm{H}^+}) - E_{\mathrm{ref}}, \tag{2}$$

where E° is the standard electrode potential of the ZnO redox probe, F is the Faraday constant (96,500 C/mol), T is absolute temperature (298 K), R is the gas constant (8.314 J/mol/K), n is the number of electrons in the redox reaction, a_{H} is the concentration of protons, E_{ref} is the reference electrode potential, [production] and [Reaction] are the concentrations of species, and $f_{\mathrm{production}}$ and f_{Reaction} are the related activity coefficients. Thus, the plot of the measured open-circuit potential E versus pH shows a Nernstian curve with a theoretical limit of 59.15 mV/pH at 25°C.

3.1. ZnO Nanorods Electrode

A major challenge in producing electrodes for intracellular sensing is tip geometry. Intracellular electrodes must have extremely sharp tips (submicrometer dimensions) and they must be long (>10 μm in length). These characteristics are necessary for effective bending and penetration of the flexible cell membrane.

The intracellular pH measuring method uses two electrodes with ZnO nanorods serving as the intracellular working electrode and Ag/AgCl as the intracellular reference microelectrode (*see* **Fig. 3**). The electrochemical potential difference response recorded in this manner measures the electrochemical surface potential generated near electrodes that may cause voltage differences between the two electrodes. The specific design and fabrication details of the electrochemical potential electrodes are described in **ref.** *(26)*.

Ag/AgCl reference microelectrodes are commonly used to ensure high stability in intracellular probing devices. However, it was suggested that even the reliable Ag/AgCl electrode may fail to support a very high fidelity recording. This may be due

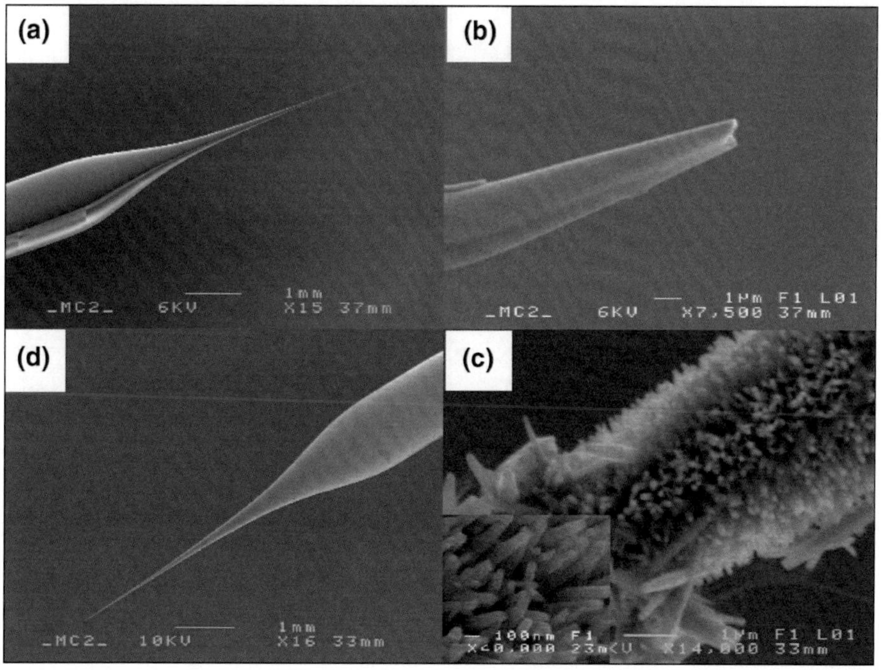

Fig. 3. Ag/AgCl reference microelectrode (**a**) and (**b**) sterile Femtotip II (Eppendorf AG) tip coated with 100-nm silver at different magnifications. (**c**) and (**d**) typical SEM image of the ZnO nanorods grown on Ag coated capillary using low temperature growth at different magnifications. Reproduced from ref. *(26)* with permission from the American Institute of Physics.

to interactions between the silver and organic molecules or an effect related to the miniaturization of the Ag/AgCl electrodes (*see* **Notes 1** and **2**). The Ag/AgCl reference microelectrode was calibrated externally versus an Ag/AgCl bulk reference electrode, which shows approximately constant potential difference using buffer solutions through the pH range 4–11. The fabrication of electrochemical potential ZnO nanorods was performed by the growing of a hexagonal single crystal of ZnO nanorods on silver-coated capillary glass using a low-temperature growth method described previously *(27–29)* (*see* **Notes 1** and **3**). The nanostructure is a rodlike shape with a hexagonal cross section and primarily aligned along the perpendicular direction of the capillary, a typical morphology of wurtzite ZnO structure. The nanorods are uniform in size, with a diameter of 60–80 nm and a length of 500–700 nm.

3.2. Results

3.2.1. Construction of the Electrodes (Potentiometric Measurement)

A two-electrode configuration was used for microliter volumes in electrochemical studies consisting of ZnO nanorods as the working electrode and Ag/AgCl as a reference microelectrode. All electrochemical experiments were conducted using a Metrohm pH meter, model 827 at room temperature (22 ± 2°C).

The response of the ZnO nanorod electrochemical potential difference (as a working electrode versus the Ag/AgCl reference microelectrode) to the changes in standard buffers at room temperature (potassium phthalate pH 4.0, sodium-potassium phosphate pH 6.0, sodium phosphate-potassium phosphate pH 7.0, hydrochloric acid-borate pH 8.0, boric acid-sodium-potassium borate pH 9.0, and boric acid-sodium-potassium borate pH 11) was measured and shows that this pH dependence is linear and has a sensitivity equal to 51.881 mV/pH at 22°C (*see* **Fig. 4**). Electrodes reading less than 50 mV per pH were discarded. The measurements were started immediately after placing the ZnO nanorods and the Ag/AgCl reference microelectrode in the electrolyte drop. To make certain that variations in the tip potential of the reference side caused by differences in the ionic strength of the standard buffers and the cytoplasm of the cell would not cause errors in pH measurement, we tested the electrodes in a potassium phosphate buffer simulating the internal environment of the cell. The measurement duration did not exceed 5 min to avoid significant changes in electrolyte concentration due to evaporation and to maintain the dissolving behavior and stability of the ZnO nanorods *(30)*.

Both the ZnO nanorod pH sensor and the Ag/AgCl reference microelectrode were immersed inside a 30 μL drop of distilled water as a test sample. During data acquisition, a 1-μL drop of 0.1 M HCl or NaOH was added to the distilled water on the glass slide. The signal change from one level to another was recorded, giving the response of the ZnO nanorod pH sensor without stirring (controlled by diffusion to the sensor). We noted that this response was determined both by diffusion to the sensor and within the sensor. The reversibility of the ZnO nanorod pH

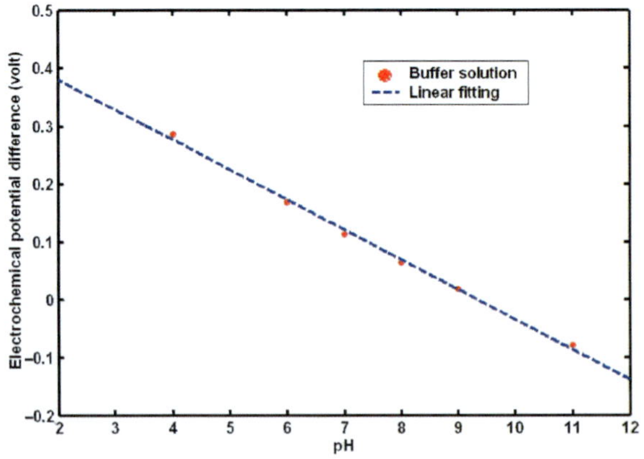

Fig. 4. Calibration curve showing the electrochemical potential difference for the ZnO nanorods as a working electrode with an Ag/AgCl reference microelectrode, versus pH changes for buffer solution. Reproduced from ref. *(26)* with permission from the American Institute of Physics.

sensor is good. The history, i.e., the order of how data are collected, does not affect the electrochemical potential for a specific pH buffer solution. This greatly enhances its ability to work with biological cells, where an abrupt change often occurs and its suitability for applications as a pH mapping sensor.

3.2.2. Intracellular pH in a Single Human Adipocyte or Fat Cell

Cells have multiple mechanisms for intracellular pH regulation that act to adjust pH to changing metabolic conditions. A redundancy of regulatory mechanisms reflects the crucial importance of pH control for cell function and survival, which, to a large extent, depends on enzymes exhibiting more or less distinct pH optima. The control of intracellular pH is thus influenced by such things as hormonal control of cellular function *(31,32)*. There is a particular potential for the local transient control of pH through local production and consumption of protons.

We used the ZnO nanosensor to measure intracellular pH in a single human adipocyte or fat cell *(33)* (*see* **Note 4**). Cells were incubated overnight before use as previously described *(34)*. Informed consent was obtained from all participating individuals and the procedures were approved by the local ethics committee. A glass slide substrate (5 cm in length, 4 cm in width, and 0.17 mm in thickness) with sparsely distributed fat cells was placed on the prewarmed microscope stage set at 37°C. The pH nanoelectrode, mounted on a micropipette holder of a micromanipulation system, was moved into position in the same plane as the cells. The ZnO nanoelectrode and reference electrodes were then gently micromanipulated into the cell using the hydraulic fine adjustments. They were inserted past the cell membrane and extended a short way into the cell. A signal reading was taken with the nanoelectrode inside the cell (pH = 6.81) (*see* **Fig. 5**). Once the ZnO nanorod working electrode and the Ag/AgCl reference microelectrode were inside the cell, the electrochemical potential difference signal detected and identified the proton activity (pH). The pH signal generated was detected with a Metrohm pH meter, model 827. Additionally, the entire experiment was imaged using a charge-coupled device (CCD) camera coupled to the side port of the Nikon inverted microscope, as verification that individual cells were probed. A typical experimental measurement required approximately 5 min. In this work, we found the measured pH value (6.81) to be close to reported values for intracellular pH, 6.95–7.57 in rat brown adipocytes *(31)* or 6.85–7.05 in rat hepatocytes *(35)* using an indirect determination of pH.

3.2.3. Cell Viability

Undoubtedly, placing pH microelectrodes inside cells causes some damage, but the membrane potential is measured and the damage can be assessed. The damage usually consists of a "leak" around the electrodes. This leak rarely causes a large change in intracellular pH because the intracellular buffering power is high

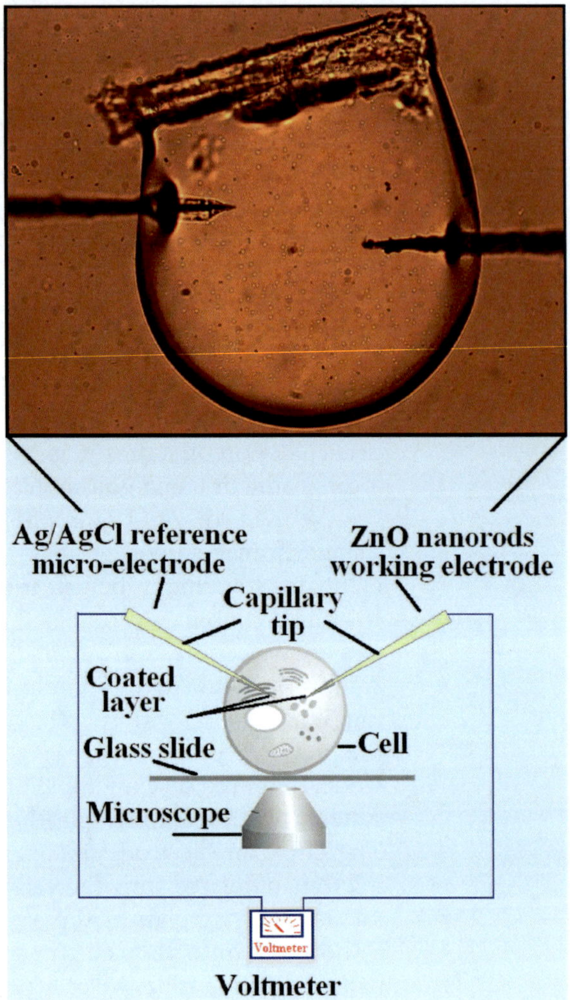

Fig. 5. Optical image and schematic diagram illustrating intracellular pH measurements performed in a single human fat cell using ZnO nanorods as a working electrode with an Ag/AgCl reference microelectrode. Reproduced from ref. *(26)* with permission from the American Institute of Physics.

and the pH gradient across the cell membrane is low. The extent to which the membrane potential reflects the amount of damage is, of course, dependent on the input resistance of the cell. The damage to large cells is less than the damage to small cells. However, in both small and large cells, the damage, if sufficient, will lead to a large influx of other ions, such as calcium and sodium. This physical damage to the integrity of the cell membrane has limited the used of pH microelectrodes to large cells and represents a constant source of anxiety for those using ion-sensitive microelectrodes.

The viability of the penetrated cells depends strongly on the size of the ZnO nanorods. We used ZnO nanorods (80 nm in diameter and 700 nm in length) grown on one side of the capillary glass with a 0.7-μm tip diameter, so the total diameter of the tip is 1.5 μm. By reducing the size of the ZnO nanorods, the total diameter of the tip was reduced, which, in turn, increased cell viability and the sensitivity of the device increases. Increasing the size of the ZnO nanorods to 500–600 nm in diameter and 3–5 μm in length caused the cells to die immediately (*see* **Fig. 6**).

The introduction of ZnO nanorod pH sensors into a single cell's cytoplasm to measure the intracellular pH does not visibly seem to affect cellular viability. This has been empirically established in several experiments in which, after ZnO nanorod penetration and equilibration for 5 min, the electrode was withdrawn and the cells were monitored by microscope. This study demonstrated that ZnO nanorods are minimally invasive tools appropriate for monitoring pH changes inside living cells.

3.2.4. Probe Usability

The ZnO nanorod electrode was used to obtain only one measurement at a time and was not reused. We made calibration measurements of the solution surrounding the cell after the measurement inside the cell to obtain a quantitative estimation of the detection signal. For these calibration measurements, the ZnO nanorod electrode was placed in the solution surrounding cells directly after the intracellular measurement, and a pH reading of 6.77 was obtained where the actual pH value of the surrounding solution was 7.4. We think that the difference between actual and measured pH values resulted from the strong association of cell materials with the electrode (*see* **Fig. 7**).

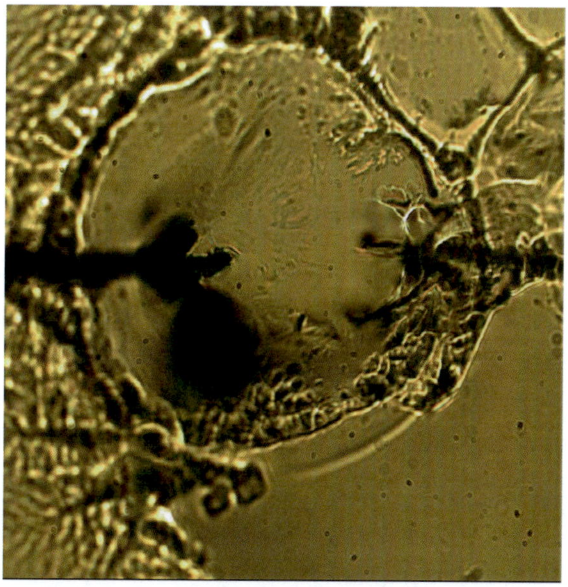

Fig. 6. Dead adipocyte during the intracellular pH measurement using a ZnO electrode.

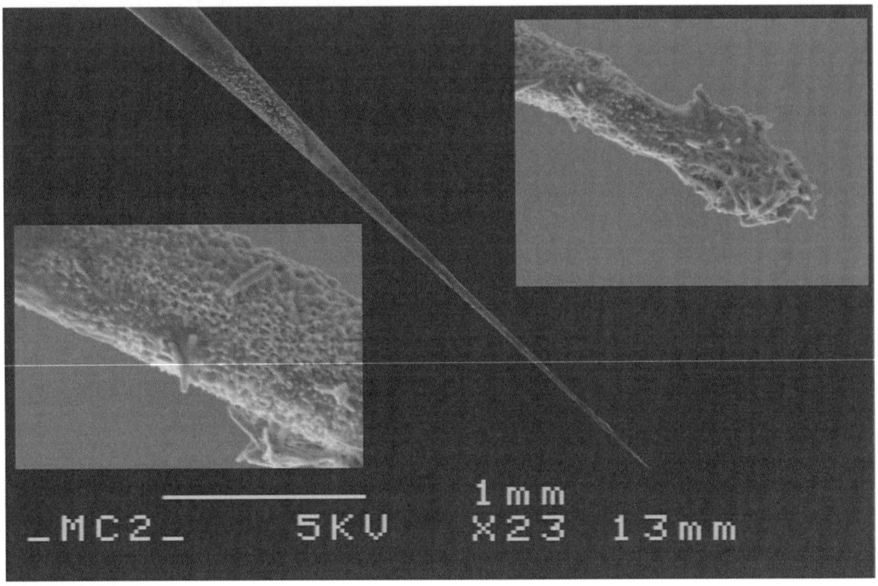

Fig. 7. The strong association of binding to cell materials with the ZnO nanorod electrode after the experiment. Reproduced from ref. *(26)* with permission from the American Institute of Physics.

4. Notes

1. Sterile borosilicate glass capillaries (Femtotip® II) were fixed on a flat support in the vacuum chamber of a sputtering system (AVAC HVC 600 e-beam evaporator), so that a thin silver film with a thickness of 100 nm was uniformly deposited onto their outer surface.
2. By introducing some optimization steps, the AgCl tip coating was prepared electrochemically by dipping the coated end of a capillary in 0.2 M HCl solution and then the silver film was electrolyzed to form AgCl by polarizing it at 1.0 V for 30 s. A 3-cm-long Ag/AgCl layer was coated on the tip of the capillary to serve as a reference electrode, leaving 3 mm of Ag/AgCl exposed at the very tip. The remainder of the Ag/AgCl layer was coated with insulation. The other end of the Ag/AgCl layer was connected with a copper wire (0.5 mm in diameter and 10 cm in length) and fixed by means of high-purity silver conductive paint.
3. The construction of ZnO nanorods working electrode was made by the growing of ZnO nanorods on silver-coated capillary glass using a low-temperature growth method described previously in the text. ZnO nanorods were grown on borosilicate glass Femtotip® II capillary coated with silver film. The ZnO

nanorods cover 3 mm of the silver-coated film that is 3-cm long. The electrical contacts are made by deposition of gold-coated film (100 nm thickness and 1 cm length) on the other end of the Ag film and then connected to the 0.5-mm-diameter insulated copper wire using high-purity silver conductive paint.

4. The cells were isolated by collagenase digestions of pieces of subcutaneous adipose tissue obtained during elective surgery at the University Hospital in Linköping.

Acknowledgments

The authors gratefully acknowledge the financial support from the Swedish Research Council and Molecular Skin Research Platform within the Faculty of Science at Göteborg University.

References

1. Zheng, G., Patolsky, F., Cui, Y., Wang, W. U., and Lieber, C. M. (2005). Multiplexed electrical detection of cancer markers with nanowire sensor arrays. *Nature Biotechnology* **23**, 1294–1301
2. Cui, Y., Wei, Q., Park, H., Lieber, C. M. (2001). Nanowire nanosensors for highly sensitive and selective detection of biological and chemical species. *Science* **293**, 1289–1292
3. Popovtzer, R., Neufeld, T., Ron, E. Z., Rishpon, J., and Shacham-Diamand, Y. (2006). Electrochemical detection of biological reactions using a novel nano-bio-chip array. *Sensors and Actuators B* **119**, 664–672
4. Cai, X., Klauke, N., Glidle, A., Cobbold, P., Smith G. L., and Cooper, J. M. (2002). Ultra-low-volume, real-time measurements of lactate from the single heart cell using microsystems technology. *Analytical Chemistry* **74**, 908–914
5. Zwicker, G. and Jacobi, K. (1983). Site-specific interaction of H2O with ZnO single-crystal surfaces studied by thermal desorption and UV photoelectron spectroscopy. *Surface Science* **131**, 179–194
6. Parker, T. M., Condon, N. G., Lindsay, R., Leibsle, F.M., and Thornton, G. (1998). Imaging the polar $(000\bar{1})$ and non-polar $(10\bar{1}0)$ surfaces of ZnO with STM. *Surface Science* **415**, L1046–L1050
7. Matsunaga, K., Oba, F., Tanaka, I., and Adachi, H. (1999). Valence band structure of ZnO (1010) surface by cluster calculation. *Journal of Electroceramics* **4**, 69–80
8. Wander, A. and Harrison, N. M. (2001). The stability of polar oxide surfaces: The interaction of H_2O with ZnO (0001) and ZnO$(000\bar{1})$. *Journal of Chemical Physics* **115**, 2312–2316
9. Wander, A., Schedin, F., Steadman, P., Norris, A., McGrath, R., Turner, T.S., Thornton, G., and Harrison, N. M. (2001). Stability of polar oxide surfaces. *Physical Review Letters* **86**, 3811–3814
10. Dulub, O., Boatner, L. A., and Diebold, U. (2002). STM study of the geometric and electronic structure of ZnO (0001)-Zn, $(000\bar{1})$-o,$(10\bar{1}0)$, and (11\2̄0) surfaces. *Surface Science* **519**, 201–207
11. Kresse, G., Dulub, O., and Diebold, U. (2003). Competing stabilization mechanism for the polar ZnO(0001)-Zn surface. *Physical Review B* **68**, 245409–(15 pages)
12. Dulub, O., Diebold, U., and Kresse, G. (2003). Novel stabilization mechanism on polar surfaces: Zn(0001)-Zn. *Physical Review Letters* **90**, 016102–(4 pages)

13. Meyer, B., Marx, D., and Dulub, O. (2004). Partial dissociation of water leads to stable superstructures on the surface of zinc oxide. *Angewandte Chemie (International ed. in English)* **43**, 6642–6645
14. Meyer, B. (2004). First principles study of the polar O-terminated ZnO surface in thermodynamic equilibrium with oxygen and hydrogen. *Physical Review B* **69**, 045416–(10 pages)
15. Dulub, O., Meyer, B., and Diebold, U. (2005). Observation of the dynamical in a water monolayer adsorbed on a ZnO surface. *Physical Review Letters* **95**, 136101–(4 pages)
16. Glab, S., Hulanicki, A., Edwall, G., and Ingman, F. (1989) Metal-metal oxide and metal oxide electrodes as pH sensors. *Analytical Chemistry* **21**, 29–46
17. Cammann, K., Ross, B., Katerkamp, A., Reinbold, J., Grundig, R., and Renneberg, R. (2002). Chemical and Biochemical Sensors. Wiley-VCH verlag GmbH & Co. KGaA, Ullmann's encyclopedia of industrial chemistry
18. Wang, H., Nakamura, H., Yao, K., Uehara, M., Nishimura, S., Maeda, H., and Abe, E. (2002). Effect of polyelectrolyte dispersants on the preparation of silica-coated zinc oxide particles in aqueous media. *Journal of the American Ceramic Society* **85**, 1937–1940
19. Caldwell, P. C. (1954). An investigation of the intracellular pH of crab muscle fibres by means of micro-glass and micro-tungsten electrode. *Journal of Physiology* **126**, 169–180
20. Thomas, R. C. (1974). Intracellular pH of snail neurons measured with a new pH-sensitive glass micro-electrode. *Journal of Physiology* **238**, 159–180
21. Yan, Y. and Al-Jassim, M. M. (2005). Structure and energetics of water adsorbed on the ZnO($10\bar{1}0$) surface. *Physical Review B* **72**, 235406–(6 pages)
22. Martins, J. B. L., Longo, E., Salmon, O. D. R., Espinoza, V. A. A., and Taft, C. A. (2004).w The interaction of H2, CO, CO2, H2O and NH3 on ZnO surfaces: an Oniom study. *Chemical Physics Letters* **400**, 481–486
23. Martins, J. B. L., Andres, J., Longo, E., and Taft, C. A. (1996). Properties, dynamics, and electronic structure of condensed systems H2O and H2 interaction with ZnO surfaces: A MNDO, AM1, and PM3 theoretical study with large cluster models. *International Journal of Quantum Chemistry* **57**, 861–870
24. Martins, J. B. L., Moliner, V., Andres, J., Longo, E., and Taft, C. A. (1995). A theoretical study of water adsorption on ($10\bar{1}0$) and (0001) ZnO surfaces: molecular cluster, basis set and effective core potential dependence. *Journal of Molecular Structure (Theochem)* **330**, 347–351
25. Stumm, W. and Morgan, J. J. (1981). Precipitation and dissolution. In: Aquatic Chemistry: An Introduction Emphasizing Chemical Equilibria in Natural Waters. (John Wiley&Sons), New York, pp. 230–322
26. Al-Hilli, S., Öst, A., and Strålfors, P., and Willander, M. (2007). ZnO nanorods as an intracellular sensor for pH measurements. *Journal of Applied Physics* **102**, 084304–(5 pages)
27. Greene, L. E., Law, M., Goldberger, J., Kim, F., Johnson, J. C., Zhang, Y., Saykally, R. J., and Yang, P. (2003). Low-temperature wafer-scale production of ZnO nanowire arrays. *Angewandte Chemie (International ed. in English)* **42**, 3031–3034
28. Vayssieres, L., Keis, K., Lindquist, S., and Hagfeldt, A. (2001). Purpose-built anisotropic metal oxide material: 3D highly oriented microrod array of ZnO. *The Journal of Physical Chemistry B* **105**, 3350–3352
29. Li, Q., Kumar, V., Li, Y., Zhang, H., Marks, T. J., and Chang, R. P. H. (2005). Fabrication of ZnO nanorods and nanotubes in aqueous solutions. *Chemistry of Materials* **17**, 1001–1006
30. Zhou, J., Xu, N., and Wang, Z. L. (2006). Dissolving behavior and stability of ZnO wires in biofluids: a study on biodegradability and biocompatibility of ZnO nanostructures. *Advanced Materials* **18**, 2432–2435
31. Lee, S. C., Hamilton, J. S., Trammell, T., Horwitz, B., and Pappone, P. A. (1994). Adrenergic modulation of intracellular pH in isolated brown fat cells from hamster and rat. *American Journal of Physiology Cell Physiology* **267**, C349–C356
32. Shrode, L. D., Tapper, H., and Grinstein, S. (1997). Role of intracellular pH in proliferation, transformation, and apoptosis. *Journal of Bioenergetics and Biomembranes* **29**, 393–399
33. Strålfors, P. and Honno, R. C. (1989). Insulin-induced dephosphorylation of hormone-sensitive lipase-correlation with lipolysis and cAMP-dependent protein kinase activity. *European Journal of Biochemistry* **182**, 379–385
34. Danielsson, A., Öst, A., Lystedt, E., Kjolhede, P., Gustavsson, J., Nystrom, F. H., and Strålfors, P. (2005). Insulin resistance in human adipocytes occurs downstream of IRS1 after surgical cell isolation but at the level of phosphorylation of IRS1 in type 2 diabetes. *FEBS Journal* **272**, 141–151
35. Pollock, A.S. (1984). Intracellular pH of hepatocytes in primary monolayer culture. *American Journal of Physiology. Renal Physiology* **246**, F738–F744

Chapter 14

Analysis of Biomolecules Using Surface Plasmons

M. Willander and Safaa Al-Hilli

Summary

Surface plasmon resonance (SPR) biosensors are optical sensors that use special electromagnetic waves (surface plasmon-polaritons) to probe interactions between an analyte in solution and a biomolecular recognition element immobilized on the SPR sensor surface. Major application areas include the detection of biological analytes and analysis of biomolecular interactions, where SPR biosensors provide benefits of label-free real-time analytical technology. The information obtained is both qualitative and quantitative and it is possible to obtain the kinetic parameters of the interaction. This new technology has been used to study a diverse set of interaction partners of biological interest, such as protein–protein, protein–lipids, protein–nucleic acids, or protein and low molecular weight molecules such as drugs, substrates, and cofactors. In addition to basic biomedical research, the SPR biosensor has recently been used in food analysis, proteomics, immunogenicity, and drug discovery. This chapter reviews the major developments in SPR technology. The main application areas are outlined and examples of applications of SPR sensor technology are presented. Future prospects of SPR sensor technology are discussed.

Key words: Surface plasmons resonance, Biosensors, Optical sensor, Biomolecular interaction

1. Introduction

Surface plasmons (SPs) are of interest to a wide spectrum of scientists ranging from physicists, chemists, and material scientists to biologists. Renewed interest in SPs comes from recent advances that allow metals to be structured and characterized on the nanometer scale. In turn, this has enabled us to control SP properties to reveal new aspects of their underlying science and tailor them for specific applications.

SPs were widely recognized in the field of surface science following the pioneering work of Ritchie in the 1950s *(1)*. SPs are waves that propagate along the surface of a conductor, usually a

James Weifu Lee and Robert S. Foote (eds.), *Micro and Nano Technologies in Bioanalysis*,
Methods in Molecular Biology, vol. 544
DOI 10.1007/978-1-59745-483-4_14,

metal. These are essentially light waves that are trapped on the surface because of their interaction with free electrons of the conductor (strictly speaking, they should be called surface plasmon-polaritons to reflect this hybrid nature *(2)*). In this interaction, the free electrons respond collectively by oscillating in resonance with the light wave. The resonant interaction between the surface charge oscillation and the electromagnetic field of the light constitutes the SP and gives rise to its unique properties.

SPs help to concentrate light in subwavelength structures by using the different (relative) permittivities, ε, of the metals and the surrounding nonconducting media (ε is the square of the complex index of refraction). In this manner, concentrating light promotes an electric field enhancement that can be used to manipulate light-matter interactions and boost nonlinear phenomena. For example, metallic structures that are much smaller than the wavelength of light are vital to the massive signal enhancement achieved in surface-enhanced Raman spectroscopy (SERS)—a technique that can now detect a single molecule *(3, 4)*. Furthermore, the enhanced field associated with SPs makes them suitable for use as sensors, and commercial systems have already been developed for sensing biomolecules.

Surface plasmon resonance (SPR) biosensors are optical sensors that use special electromagnetic waves—surface plasmon-polaritons—to probe interactions between an analyte in solution and a biomolecular recognition element immobilized on the SPR sensor surface. Major application areas include the detection of biological analytes and analysis of biomolecular interactions, where SPR biosensors provide benefits for label-free real-time analytical technology.

An affinity biosensor consists of a transducer (electrochemical *(5)*, piezoelectric *(6)*, or optical *(7)*) and a biological recognition element that interacts with a selected analyte. Various optical methods have been exploited in biosensors including fluorescence spectroscopy *(8)*, interferometry (reflectometric white light interferometry *(9)* and modal interferometry in optical waveguide structures *(10)*, spectroscopy of guided modes of optical waveguides (grating coupler *(11)* and resonant mirror *(12)*, and surface plasmon resonance (SPR) *(13, 14)*. Fluorescence-based biosensors offer high sensitivity but, because of the use of labels, they require either multistep detection protocols or delicately balanced affinities of interacting biomolecules for displacement assays, causing sensor cross-sensitivity to nontarget analytes *(15)*. Sensors such as optical interferometers, grating couplers, resonant mirrors, and SPR rely on the measurement of binding-induced refractive index changes, and thus, are label-free technologies.

This application arises from the unique nanoparticle optical properties of certain materials, such as silver and gold *(16)*.

It is now well established that optical excitation of the localized surface plasmon resonance (LSPR) of silver and gold nanoparticles results in absorption with extremely large molar extinction coefficients of ~3×10^{11} M^{-1} cm^{-1} *(17, 18)*, resonant Rayleigh scattering *(19, 20)* with an efficiency equivalent to that of 10^6 fluorophores *(21, 22)*, and strong enhancement of the local electromagnetic fields near the nanoparticle surface *(23, 24)*. Furthermore, the peak extinction or resonant Rayleigh scattering wavelength, λ_{max}, intensity, and line width of these LSPR spectra are strongly dependent on their size, shape, interparticle spacing, and local dielectric environment *(17, 24–28)*. Consequently, five different nanoparticle-based sensing mechanisms that enable the transduction of macromolecular or chemical binding events into optical signals have been introduced. These mechanisms are:

1. Resonant Rayleigh scattering from nanoparticle labels in a manner analogous to fluorescent dye labels *(16, 19, 21, 22)*.
2. Nanoparticle aggregation *(29, 30)*.
3. Charge-transfer interactions at nanoparticle surfaces *(25, 31, 32)*.
4. Local refractive index changes *(32–35)*.
5. Surface-enhanced Raman scattering *(36)*.

The optical-field enhancement from plasmon resonance at spheroids is studied by solving Maxwell's equations using spheroidal vector wave functions *(37)*. This treatment is an extension of the Mie theory for spheres. The optical-field enhancement at the end of a sharp tip has been used in surface-enhanced Raman spectroscopy (SERS). The optical-field enhancement will increase the Raman signals such that spectra of single molecules are possible. Single-molecule SERS experiments on Ag and Au colloids have indicated signal-to-noise enhancement factors of 10 to 15 orders of magnitude. Because the spectroscopic signal-to-noise enhancement goes as the fourth power of the field enhancement (light power two ways), the field enhancement would then be approximately three orders of magnitude at the detection spots. Theoretical studies of the optical-field enhancement at prolate spheroids show strong enhancements in the quasistatic limit. The field enhancement factor at the end of a spheroid is calculated by the finite-element time-domain (FETD) method. The enhancement factor strongly depends on the ratio of the wavelength to the structure size. The enhancement falls off rapidly as the structure size is enlarged; this is confirmed by our calculations. This is attributed to phase retardation or dephasing effects. The exactly solvable model of a spheroid shows that there are plasmon resonance regions with strong-field enhancements in the nonstatic regime and shows how their positions in parameter space and their magnitudes are related. In fact, there is a strong dependence on the parameters' wavelength, spheroidal length, axial ratio, polarization, angle of incidence, and material properties. These sometimes very sensitive spots in the space spanned by these parameters may be used for specific sensing of the near environment of the spheroids.

It was also suggested in **ref.** *(37)* that the strong enhanced optical field caused by surface plasmon resonance could be used for trapping single fluorescent molecules. Because both attractive as well as negative forces are possible depending on the wavelength of the optical field, a molecule could be taken into the trap and then, by changing the optical wavelength, taken out of the trap. At the same time, the molecule could be investigated through SERS.

2. Fundamentals of Surface Plasmon Resonance

2.1. Coupling to Surface Plasmons

The interaction between the surface charges and the electromagnetic field that constitutes SP has two consequences (*see* **Fig. 1**). First, the interaction between the surface charge density and the electromagnetic field results in the momentum of the SP mode, $\hbar k_{SP}$, which is greater than that of a free-space photon of the same frequency, $\hbar k_0$ ($k_0 = \omega/c$ is the free-space wave vector). The charge density wave is associated with an electromagnetic wave; the field vectors of which reach their maxima at the interface and decay evanescently into both media. This surface plasma wave (SPW) is a transverse-magnetic (TM) polarized wave (the magnetic vector is perpendicular to the direction of propagation of the SPW and parallel to the plane of interface) and is characterized by the propagation constant and electromagnetic field distribution. Solving Maxwell's equations under the appropriate boundary conditions yields the SP dispersion relation *(38)*, that is, the frequency-dependent SP wave vector, k_{SP},

$$k_{SP} = k_0 \sqrt{\frac{\varepsilon_d \varepsilon_m}{\varepsilon_d + \varepsilon_m}}. \tag{1}$$

The frequency-dependent permittivity of the metal, ε_m, and the dielectric material, ε_d, must have opposite signs if SPs are possible at such an interface. This condition is satisfied for metals (gold and silver are the most commonly used) because ε_m is both negative and complex (the latter corresponding to absorption in the metal). The real and imaginary parts of the propagation constant describe spatial periodicity and attenuation of an SPW in the direction of propagation, respectively *(39)*. As an example, using **Eq.1**, the SP wave vector for a silver-air interface in the red part of the visible spectrum is found to be $k_{SP} \cong 1.03k_0$. This increase in

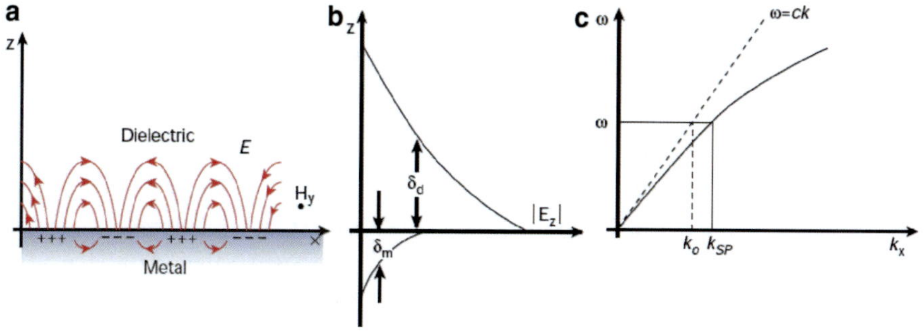

Fig. 1. SPs at the interface between a metal and a dielectric material have a combined electromagnetic wave and surface charge character, as shown in (**a**). They are transverse magnetic in character (H is in the *y* direction), and the generation of surface charge requires an electric field normal to the surface. This combined character also leads to the field component perpendicular to the surface being enhanced near the surface and decaying exponentially with distance away from it. (**b**) The field in this perpendicular direction is said to be evanescent, reflecting the bound, nonradiative nature of SPs, and prevents power from propagating away from the surface. In the dielectric medium above the metal, typically air or glass, the decay length of the field, δ_d, is on the order of half the wavelength of light involved, whereas the decay length into the metal, δ_m, is determined by the skin depth. (**c**) The dispersion curve for a SP mode shows the momentum mismatch problem that must be overcome to couple light and SP modes together with the SP mode lying beyond the light line, that is, it has greater momentum $\hbar k_{SP}$ than a free space photon $\hbar k_{SP}$ of similar frequency ω *(40)*. Reproduced from **ref.*(40)*** with permission from Nature Publishing Group.

momentum is associated with SP binding to the surface and the resulting momentum mismatch between light and SPs of similar frequency must be bridged if light is used to generate SPs.

The second consequence of the interaction between the surface charges and the electromagnetic field is that, in contrast to the propagating nature of SPs along the surface, the field perpendicular to the surface decays exponentially with distance from the surface. The field in this perpendicular direction is said to be evanescent or near field in nature and is a consequence of the bound, nonradiative nature of SPs, which prevents power from propagating away from the surface.

There are three main techniques by which the missing momentum can be provided. The first makes use of prism coupling to enhance the momentum of the incident light *(41, 42)*. The second involves scattering from a topological defect on the surface, such as a subwavelength protrusion or hole, which provides a convenient way to generate SPs locally *(43, 44)*. The third makes use of a periodic corrugation on the metal's surface *(45)*.

2.2. Surface Plasmon Propagation

Once light has been converted into an SP mode on a flat metal surface, it will propagate but will gradually attenuate because of losses arising from absorption in the metal. This attenuation depends on the dielectric function of the metal at an oscillation frequency of SP. The propagation length, δ_{SP}, can be found by seeking the imaginary part, k''_{SP}, of the complex surface plasmon wave vector, $k_{SP} = k'_{SP} + ik''_{SP}$, from the SP dispersion **Eq.1** *(46)*,

$$\delta_{\mathrm{SP}} = \frac{1}{2k''_{\mathrm{SP}}} = \frac{c}{\omega}\left(\frac{\varepsilon'_{\mathrm{m}} + \varepsilon_{\mathrm{d}}}{\varepsilon'_{\mathrm{m}}\varepsilon_{\mathrm{d}}}\right)^{\frac{3}{2}} \frac{(\varepsilon'_{\mathrm{m}})^2}{\varepsilon''_{\mathrm{m}}}, \tag{2}$$

where $\varepsilon'_{\mathrm{m}}$ and $\varepsilon''_{\mathrm{m}}$ are the real and imaginary parts of the dielectric function of the metal, $\varepsilon_{\mathrm{m}} = \varepsilon'_{\mathrm{m}} + i\varepsilon''_{\mathrm{m}}$. Silver is the metal with the lowest losses in the visible spectrum: propagation distances are typically in the range of 10–100 μm, increasing toward 1 mm as one moves into the 1.5-μm near-infrared telecom band (*see* **Fig. 2**). In the past, absorption by the metal was considered a significant problem because SPs were not thought to be viable for photonic elements; the SP propagation length was smaller than the size of components at that time. This view is now changing thanks to recent demonstrations of SP-based components that are significantly smaller than the propagation length *(47)*. Such developments pave the way for the integration of many SP-based devices into circuits before propagation losses become too significant. In addition to dealing with the problem of loss owing to absorption in the metal, another key loss mechanism must be considered, that is, unwanted coupling to radiation. To build SP-based circuits one will need components that convert one SP mode into another, for example, a switch to reroute SPs without

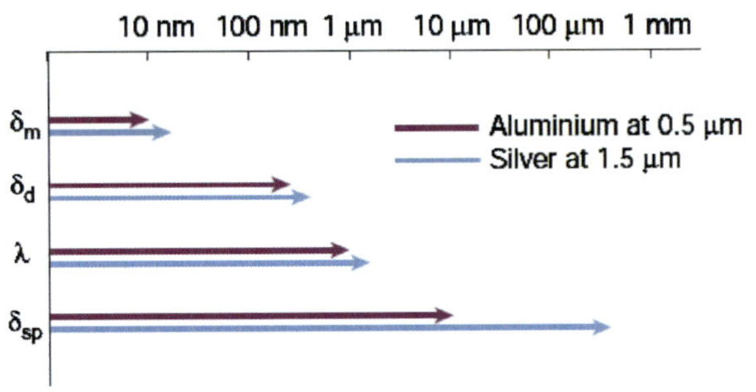

Fig. 2. Three characteristic length scales are important for SP-based photonics in addition to that of the associated light. The propagation length of the SP mode, δ_{SP}, is usually dictated by loss in the metal. For a relatively absorbing metal such as aluminum, the propagation length is 2 μm at a wavelength of 500 nm. For a low-loss metal, for example, silver, at the same wavelength, the propagation length is increased to 20 μm. By moving to a slightly longer wavelength, such as 1.55 μm, the propagation length is further increased toward 1 mm. The propagation length sets the upper size limit for any photonic circuit based on SPs. The decay length in the dielectric material, δ_{d}, is typically on the order of half the wavelength of light involved and dictates the maximum height of any individual features and components that might be used to control SPs. The ratio of δ_{SP} thus provides a measure of the number of SP-based components that may be integrated together. The decay length in the metal, δ_{m}, determines the minimum feature size that can be used; as shown in the diagram, this is between one and two orders of magnitude smaller than the wavelength involved, thus highlighting the need for good control of fabrication on the nanometer scale. These combinations provide an indication of range for poor (Al at 0.5 μm) to good (Ag at 1.5 μm) SP performance ***(40)***. Reproduced from **ref.*(40)*** with permission from Nature Publishing Group.

scattering the SP mode in a manner that results in the lose of energy to freely propagating light.

2.3. Surface Plasmon Band Structure and Periodic Surfaces

One of the key developments in photonics in the past 15 years has been photonic bandgap (PBG) materials. These synthetic materials use wavelength-scale periodic structures to manipulate the interaction between light and matter to build new photonic structures, such as photonic crystal fiber *(48)*. These developments have been predominantly made in periodically structured insulating and semiconducting materials. By making use of SPs, metals can also be used as PBG materials in the form of photonic surfaces *(49)*.

The nature of SPs changes when they propagate on metal surfaces that are periodically textured on the scale of the wavelength of light. When the period of the nanostructure is half that of the effective wavelength of the SP mode, scattering may lead to the formation of SP standing waves and the opening of an SP stop band *(49) see* **Fig. 3**). When the surface is modulated in both in-plane directions, for example by a periodic array of bumps, SP modes may be prevented from traveling in any in-plane direction, thus leading to a full PBG for SP modes *(49, 51)* (*see* **Fig. 4**). **Figure 4** provides a nice demonstration of how the problems associated with absorption by the metal can be overcome. It demonstrates that, although finite, the SP propagation length is more than enough to enable the SP to experience many periods of the textured surface, and thus, display the PBG phenomenon.

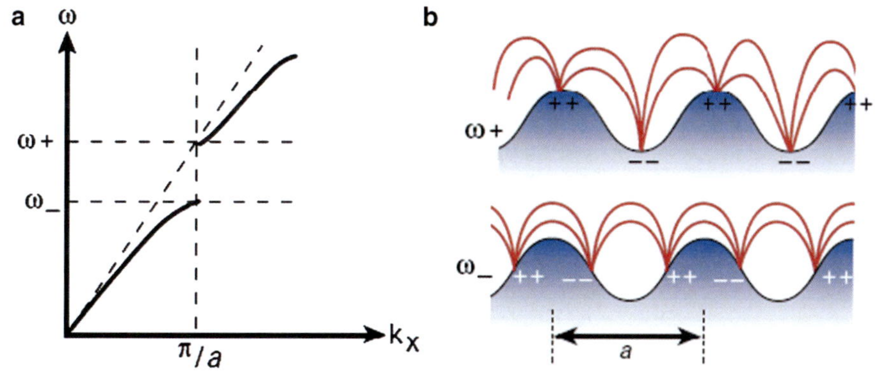

Fig. 3. Periodic texturing of the metal surface can lead to the formation of an SP photonic bandgap when the period, *a*, is equal to half the wavelength of the SP, as shown in the dispersion diagram (**a**). Similar to electron waves in crystalline solids, there are two SP standing wave solutions, each with the same wavelength but, owing to their different field and surface charge distributions, they are of different frequencies. The upper frequency solution, *v*&, is of higher energy because of the greater distance between the surface charges and the greater distortion of the field, as shown schematically in (**b**). SP modes with frequencies between the two band edges, *v*& and *v*–, cannot propagate, and, as a result, this frequency interval is known as a stop gap. By providing periodic texture in two dimensions, SP propagation in all in-plane directions can be blocked, leading to the full bandgap for SPs. At the band edges, the density of SP states is high, and there is a significant increase in the associated field enhancement ***(40)***. Reproduced from **ref.*(40)*** with permission from Nature Publishing Group.

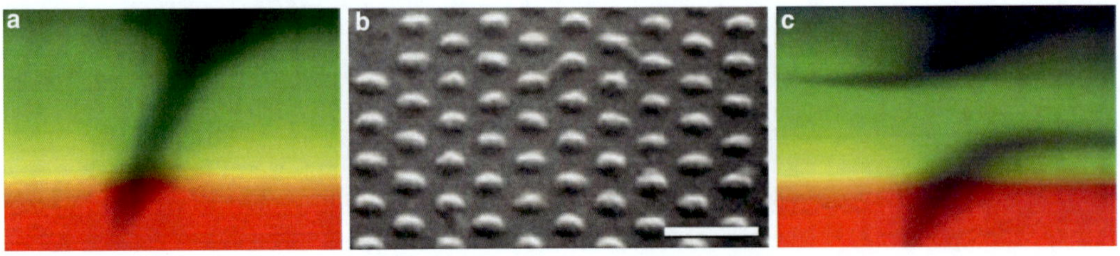

Fig. 4. SP photonic bandgap. The SP dispersion curve shown in **Fig. 1** was directly imaged using a modified prism coupling technique. (**a**) The dispersion curve (here shown as inverse wavelength versus angle) for a flat surface is shown in the upper picture; here, dark regions correspond to the coupling of incident light to the SP mode, and the colors are produced on a photographic film by the wavelength of the light used. (**b**) If the metal surface is textured with a two-dimensional pattern of bumps on an appropriate length scale (roughly half the wavelength of light) as shown in this scanning electron microscopy (SEM) image, a bandgap is introduced into the dispersion curve of the associated SP modes. Bar, 0.7 mm. (**c**) The bandgap is clearly seen in the lower pictures, where there is a spectral region in which no SP mode (as indicated by the dark regions) exists. Also, note the distortion of the SP mode and the edges of the bandgap ***(40)***. Reproduced from **ref.*(40)*** with permission from Nature Publishing Group.

We also note with interest that recent developments in the fabrication of periodic nanostructures via self-assembly offer the prospect of easily producing SP PBG substrates to act as photonic substrates on which to define SP photonic circuits.

At frequencies within a band gap, the density of SP modes is zero—no SP modes can be supported. However, at the band edges, the SP mode dispersion is flat and the associated density of SP modes is high, corresponding to a high field enhancement close to the metal surface. Additionally, the nature of this flat band signifies that such modes can be excited by incident light over a wide range of angles, making them good candidates for frequency-selective surfaces. Flat bands are also associated with the localized SP modes of metallic nanoparticles *(52)*. The frequency and width of these modes are determined by the particle's shape, material, size and environment *(52–54)*, and, for this reason, they are being pursued as tags for biosensing *(54, 55)*, and as substrates for SERS *(57)*, and potentially as aerials for fluorophores *(57, 58)*. The interaction between two or more nanoparticles can lead to additional levels of field enhancement *(60–62)*, with even more dramatic effects associated with hot spots in random structures *(63)*.

3. Materials

1. Gold substrates for SPR experiments were prepared using vapor deposition to deposit 48-nm-thick Au films onto 18 × 18-mm × 0.45-mm-thick SF-10 glass (η = 1.727) cover slips (Schott Glass Technologies, Durea, PA).

2. Water used in making solutions and for washing the samples was reverse-osmosis purified (18 MΩ cm).
3. A standard hybridization buffer and contains 0.3 M NaCl, 20 mM sodium phosphate, 2 mM EDTA, adjusted to pH 7.4 with NaOH (2-SSPE), plus 0.2% (w/v) (6.9 mM) sodium dodecyl sulfate (SDS).
4. 11-Mercaptoundecanoic acid (MUA) was from Aldrich (Milwaukee, WI), poly(L-lysine) (PL) was from Sigma (St. Louis, MO), and 1,4-phenylenediisothiocyanate (PDITC) was from Aldrich.
5. MUA monolayers were prepared on clean vapor-deposited gold substrates, and a layer of PL was electrostatically adsorbed onto the MUA surface (*see* **Note 1**).
6. Pyridine.
7. Dimethylformamide (DMF).
8. 0.1 M sodium bicarbonate buffer, pH 9.0, containing 0.5 M NaCl.
9. NH_4OH.
10. A Kretschmann-configuration imaging SPR instrument.
11. Resistive thermofoil heating elements (Minco Products Inc., Minneapolis, MN) attached to the cell, with miniature platinum RTD sensors (Minco) embedded in the cell for feedback temperature control with a Love 1600 series self-tuning PID temperature controller (Love Controls, Wheeling, IL).
12. Molecular Dynamics FluorImager 575 (Sunnyvale, CA).

4. The Principle Underlying Surface Plasmon Resonance Detection

Because the vast majority of the field of an SPW is concentrated in the dielectric, the propagation constant of the SPW is extremely sensitive to changes in the refractive index of the dielectric. This property of SPW underlies the physical principle of affinity SPR biosensors—biomolecular recognition elements on the surface of metal recognize and capture analytes present in a liquid sample producing a local increase in the refractive index at the metal surface. The refractive index increase promotes an increase in the propagation constant of SPW propagating along the metal surface (*see* **Fig. 5**), which can be accurately measured by optical means.

The magnitude of the change in the propagation constant of an SPW depends on the refractive index change and its distribution with respect to the profile of the SPW field. There are two limiting cases:

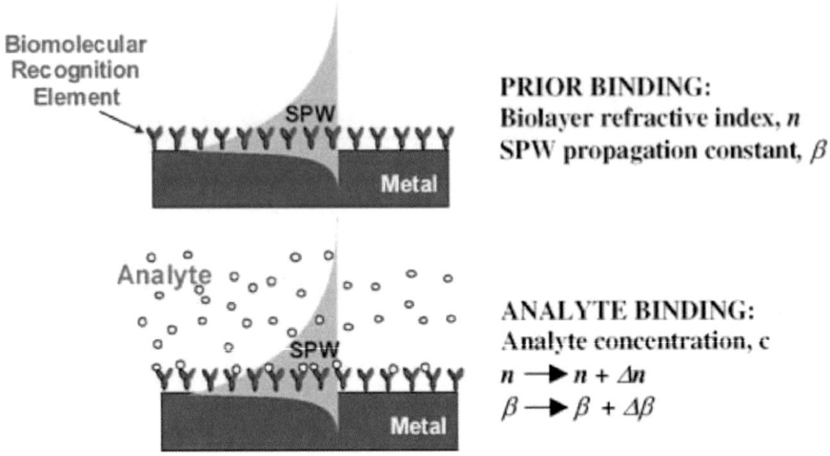

Fig. 5. Principle of SPR biosensing *(64)*. Reproduced from **ref.***(64)* with permission from Springer.

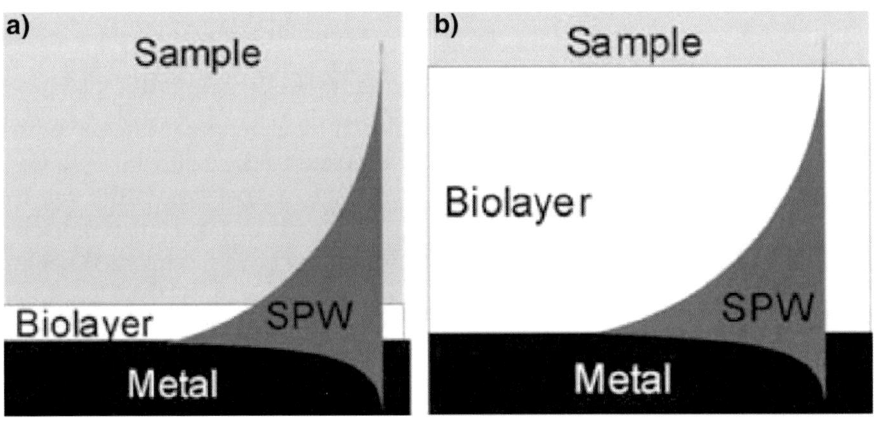

Fig. 6. Surface plasmon-polariton probing: (**a**) biomolecular interactions occurring within a short distance from the metal surface, and (**b**) biomolecular interactions occurring within the entire extent of the SPW field *(64)*. Reproduced from **ref.** *(64)* with permission from Springer.

1. Analyte capture occurs only within a short distance from the metal surface **(Fig. 6a)**.
2. Analyte capture occurs within the whole extent of the SPW field **(Fig. 6b)**.

Perturbation theory *(65)* suggests that if the binding occurs within the entire depth of the SPW field (**Fig. 6b**), the binding-induced refractive index change, Δn, produces a change in the real part of the propagation constant, $\Delta\beta$, which is directly proportional to the refractive index change:

$$\mathrm{Re}\{\Delta\beta\} \cong k\Delta n\,, \tag{3}$$

where k denotes the free-space wave number *(66)*. If the refractive index change is caused by a binding event occurring within a distance from the surface d that is significantly smaller than the penetration depth of the SPW, the corresponding change in the real part of the propagation constant can be expressed as follows:

$$\mathrm{Re}\{\Delta\beta\} \cong \frac{2n_s n_f k^2 d}{\sqrt{\left|\mathrm{Re}\{\varepsilon_m\}\right|}}\Delta n = Fk\Delta n\,, \tag{4}$$

where n_f and n_s denote the refractive index of the biolayer and the refractive index of the background dielectric medium (sample), respectively. The binding-induced change in the propagation constant of SPW is proportional to the refractive index change and the depth of the area within which the change occurs. The factor F ($F < 1$) accounts for the fact that the interaction occurring within a thin layer is probed only by a fraction of the SPW field.

5. Concept of Surface Plasmon Resonance Optical Chemical Sensors and Biosensors (Excitation and Interrogation of Surface Plasmon-Polaritons)

In SPR sensors, an SPW is excited by a light wave and the effect of this interaction on the characteristics of the light wave is measured. From these measurements, changes in the propagation constant of the SPW can be determined. SPW excitation by light can occur only if the component of the light's wave vector parallel to the metal surface matches that of the SPW. This can be achieved by means of prism coupling, waveguide coupling, and grating coupling.

Generally, an SPR optical sensor comprises an optical system; a transducing medium, which interrelates the optical and (bio) chemical domains; and an electronic system supporting the opto-electronic components of the sensor and allowing data processing. The transducing medium transforms changes in the quantity of interest into changes in the refractive index, which may be determined by optically interrogating the SPR. The optical part of the SPR sensor contains a source of optical radiation and an optical structure in which SPW is excited and interrogated. In the process of interrogating the SPR, an electronic signal is generated and processed by the electronic system. Major properties of an SPR sensor are determined by properties of the sensor's subsystems. The sensor sensitivity, stability, and resolution depend on properties of both the optical system and the transducing medium. The selectivity and response time of the sensor are primarily determined by the properties of the transducing medium.

The interaction of a light wave with an SPW can alter the light's characteristics, such as amplitude, phase, polarization, and spectral distribution. Changes in these characteristics can be correlated with changes in the propagation constant of the SPW. Therefore, binding-induced changes on the refractive index at the sensor surface and, consequently, the propagation constant of the SPW can be determined by measuring changes in one of these characteristics. Based on the measured characteristic, SPR biosensors can be classified as angle, wavelength, intensity, phase, or polarization modulation-based sensors. In SPR sensors with angular modulation, the component of the light wave's wave vector parallel to the metal surface matching that of the SPW is determined by measuring the coupling strength at a fixed wavelength and multiple angles of light wave incidence and determining the angle of incidence yielding the strongest coupling (**Fig. 7a**, upper plot). In SPR sensors with wavelength modulation, the component of the light wave's wave vector parallel to the metal surface matching that of SPW is determined by measuring the coupling strength at a fixed angle of incidence and multiple wavelengths and determining the wavelength yielding the strongest coupling (**Fig. 7b**). In SPR sensors with intensity modulation, the change in the intensity of the light wave interacting with SPW is measured at a fixed wavelength and angle of incidence (**Fig. 7b**). In SPR sensors with phase modulation, a shift in the phase of the light wave interacting with the SPW is measured at a fixed wavelength and angle of incidence (**Fig. 7a**, lower plot). In SPR sensors with polarization modulation, changes in the polarization are measured at a fixed wavelength and angle of incidence.

6. Methods

6.1. Surface Plasmon Resonance Sensors Using Attenuated Total Reflection Optical Prism Couplers

In configurations based on prism coupling, a light wave passes through a high-refractive index prism and is totally reflected at the interface between a prism coupler and a thin metal layer (of the thickness of ~50 nm) and excites an SPW at the outer boundary of the metal by evanescently tunneling through the thin metal layer (**Fig. 8a**). This evanescent wave propagates along the interface with a propagation constant, which can be adjusted to match that of SPW by controlling the angle of incidence. This method is

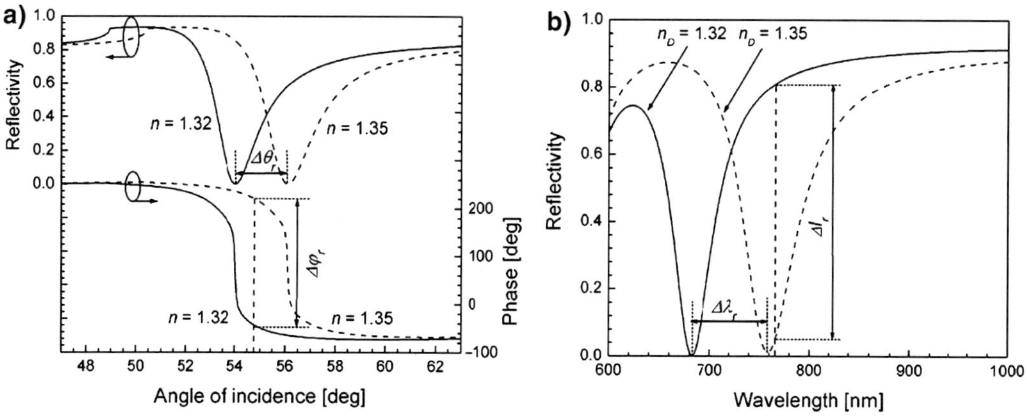

Fig. 7. Reflectivity and phase for a light wave exciting an SPW in the Kretschmann geometry (SF14 glass prism—50-nm-thick gold layer—dielectric) versus (**a**) the angle of incidence for two different refractive indices of the dielectric (wavelength 682 nm), and (**b**) wavelength for two different refractive indices of the dielectric (angle of incidence 54°) ***(64)***. Reproduced from **ref.*(64)*** with permission from Springer.

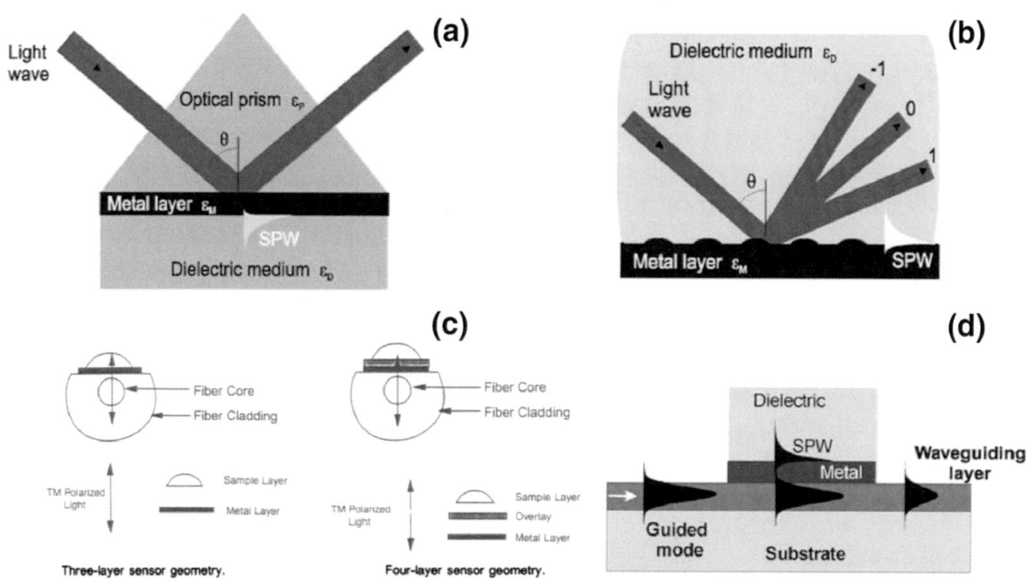

Fig. 8. Excitation of surface plasmon-polaritons: (**a**) by a light beam via prism coupling; (**b**) by light diffraction on a diffraction grating; (**c**) by optical fibers of a waveguide; and (**d**) by a guided mode of an optical waveguide ***(64,99)***. Reproduced from **ref.*(64)*** with permission from Springer and from **ref.*(99)*** with permission from American Chemical Society.

referred to as the attenuated total reflection (ATR) method *(67)*. Particularly, the Kretschmann geometry of the ATR method has been found to be suitable for sensing and has become the most widely used geometry in SPR sensors. In this configuration, a light wave is totally reflected at the interface between a prism coupler and a thin metal layer (of the thickness of ~50 nm) and excites an SPW at the outer boundary of the metal by evanescently tunneling through the thin metal layer. All of the main detection approaches have been demonstrated in SPR prism-based sensors, such as measurement of the intensity of the reflected light wave *(68, 69)*, measurement of the resonant angle of incidence of the light wave *(70, 71)*, and measurement of the resonant wavelength of the incident light wave *(72)*. Prism-based SPR sensors using angular interrogation have been extensively studied at Linköping University (Sweden) *(71)* and by Biacore (Sweden) *(73, 74)*.

The sample (*see* **Note 2**) was illuminated with p-polarized, collimated light from a 632.8-nm, 1.5-mW HeNe laser (Uniphase Model 1101P, Manteca, CA), coupled through a MgF_2 antireflection-coated SF-10 (η = 1.727) 30 × 30 × 30-mm equilateral glass prism (Ealing Electro-Optics, Holliston, MA) and η = 1.725 index-matching fluid (Cargille, Cedar Grove, NJ) to the sample. The reflected light from the surface was imaged onto a CCD camera (either a Panasonic Model WV-BL200 [Panasonic Broadcast & Television Systems Co., Secaucus, NJ] or an iSight iSC2050 [iSight, Inc., Cedar Knolls, NJ] was used, depending on when the experiment was performed), and the image was digitized by a frame grabber (Video Blaster RT300 [Creative Labs, Inc., Milpitas, CA] or Data Translation DT3155 [Data Translation, Marlboro, MA]) and AdobeCap Video Capture Utility v1.1 (Adobe Systems, San Jose, CA) or Data Translation Acquire To Host Application v1.0 software running on a Dell computer platform (*see* **Note 3**).

In traditional multichannel SPR sensors, SPWs were excited via a prism coupler in multiple areas that were arranged perpendicularly to the direction of propagation of SPWs; the angular or spectral distribution of reflected light was analyzed to yield information about the measurement in each channel. Although this spectroscopic approach led to the development of high-performance SPR sensing devices, the number of sensing channels that could be realized using this approach was rather limited (<10). To increase the number of sensing channels, various SPR sensor approaches have been proposed. One approach is based on SPR imaging in which a collimated light beam from a polychromatic light source passes through a prism and is made incident on an SPR-active metal layer; the reflected light is detected with a CCD camera after passing through a narrowband interference filter. This approach has been demonstrated for monitoring adsorption of a single-stranded DNA-binding protein onto single-stranded DNA patterned into an array of 500 × 500-μm squares (*see* **Note 4**). The operating wavelengths for imaging SPR sensors have been selected for different applications. An alternative approach is based on detection of spatial changes in the phase of light exciting an SPW and interferometry. Two interferometric schemes have been proposed. In the Mach-Zehnder interferometer-based scheme, monochromatic light was split into reference and signal beams; the signal beam passed through a prism and, after reflection from an SPR-active metal layer, was recombined with the reference beam, producing an interference pattern on a CCD camera. In the transverse electric (TE)–TM polarization interferometer, TE and TM polarized beams passed through a prism and, after reflection from an SPR-active surface, were shifted with respect to each other and recombined by means of a polarizer producing an interference pattern on a CCD camera. Another interesting approach is based on SPR microscopy and uses surface scanning and SPWs excited by means of an objective lens. Most recently, a new approach has been proposed that is based on spectroscopy of SPW on an array of diffraction gratings.

6.2. Surface Plasmon Resonance Sensors Using Grating Couplers

An SPW can be excited by diffraction on a grating, **Fig. 8b**. The component of the wave vector of the diffracted waves parallel to the interface is diffraction increased by an amount that is inversely proportional to the grating period and can be matched to that of an SPW *(75)*. If a metal–dielectric interface is periodically distorted, the incident optical wave is diffracted, forming a series of beams directed away from the surface at a variety of angles *(76)*. The component of momentum of these diffracted beams along the interface differs from that of the incident wave by multiples of the grating wave vector. If the total component of momentum along the interface of a diffracted order is equal to that of the SPW, the optical wave may couple the SPW *(77)*. The mathematics involved in modeling the grating SPR-sensing structures is more complex than that of planar prism-based systems *(78, 79)*; therefore, modeling the response of grating-based SPR structures and analysis of sensor data is more difficult. Grating-based optical SPR sensors that use the measurement of SPR light intensity variations *(80, 81)* and the wavelength interrogation *(82, 83)* have been demonstrated. Although the interrogation optical systems for SPR sensors using prisms and grating couplers are essentially the same, accurate control of the thickness of the plasmon active metal layer is not required in the grating-based SPR sensors. A drawback for certain applications is that in grating-based SPR systems, unlike prism-based systems, the light beam is incident through the sample solution, and, therefore, the analyte and flow-cell need to be optically transparent.

6.3. Surface Plasmon Resonance Sensors Using Optical Waveguides

The use of optical waveguides in SPR sensors provides numerous attractive features such as a simple method of controlling the optical path in the sensor system (efficient control of properties of the light, suppression of the effect of stray light, etc.), small size, and ruggedness. The process of exciting an SPW in optical waveguide-based SPR-sensing structures is, in principle, similar to that in the ATR coupler. A light wave is guided by the waveguide and, entering the region with a thin metal overlayer, it evanescently penetrates through the metal layer. If the SPW and the guided mode are phase matched, the light wave excites an SPW at the outer interface of the metal. Theoretically, the sensitivity of waveguide-based SPR devices is approximately the same as that of the corresponding ATR configurations. Despite increased design constraints compared with bulk prism-based SPR-sensing devices, all of the main SPR detection approaches have been implemented in waveguide SPR sensors.

6.3.1. Surface Plasmon Resonance Sensors Based on Optical Fibers

Currently, optical fiber SPR probes present the highest level of miniaturization of SPR devices, allowing for chemical and biological sensing in inaccessible locations where the mechanical flexibility and the ability to transmit optical signals over a long distance make the use of optical fibers very attractive (*see* **Fig. 8c**). The use of optical fibers for SPR sensing was first proposed by Jorgenson and Yee *(84)*. They used the wavelength interrogation technique and formed an SPR-sensing structure by using a conventional polymer-clad silica fiber with partly removed cladding and an SPR active metal layer deposited symmetrically around the exposed section of the fiber core. This approach allows construction of miniaturized optical fiber SPR probes with a limited interaction area at a length of approximately 10 mm. It has been demonstrated that the operating range of the sensor may be customized for sensor applications in the refractive index range 1–1.7 using a thin high-refractive index dielectric overlayer and high refractive index core fibers *(85)*. A similar structure, in which the SPR-sensing area is formed not at the tip but in the middle of an optical fiber, has been reported to be used as an intensity measurement-based sensor *(86, 87)*. In this configuration, a collimated monochromatic light beam is launched into a fiber in a fashion such that only modes with propagation constants within a narrow range are efficiently excited. Variations in the refractive index of the analyte are determined by measuring the transmitted optical power. The SPR sensor sensitivity is negatively influenced by exciting an SPW by fiber modes that are incident on the metal surface at slightly different angles. Generally, both types of multimode fiberoptic SPR-sensing devices may suffer from rather low stability. The modal distribution of light in the fiber is very sensitive to mechanical disturbance and disturbances occurring close to the sensing area of the fiber may cause intermodal coupling and modal noise. Because of the cylindrical shape of the sensing area, fabrication of homogeneous SPR coatings and functionalization of the sensor's surface pose technological challenges. To overcome these drawbacks and allow for further reduction of the sensing area, SPR sensors based on single-mode optical fibers have been proposed *(88–90)*. A major drawback of optical fiber SPR sensors of this type is that they require reliable control of polarization state of the optical wave propagating in the fiber (e.g., by using polarization-maintaining optical fibers). Recently, an alternative configuration of a wavelength interrogation-based SPR sensor using a single-mode optical fiber was described *(91)*. In this configuration, SPW is excited on a metal-coated tapered single-mode optical fiber. It has been demonstrated that if the sensor is operated at wavelengths below 600 nm, it may be used for monitoring variations in the refractive index of aqueous media.

6.3.2. Surface Plasmon Resonance Sensors Based on Integrated Optical Waveguides

The process of exciting an SPW in an optical waveguide-based SPR structure **(Fig. 8d)** is similar to that in the ATR coupler. The light wave is guided by an optical waveguide and, when entering the region with a thin metal layer, it evanescently penetrates through the metal layer, exciting an SPW at its outer boundary. Integrated optical waveguide SPR sensors appear promising for the development of multichannel sensing devices on one chip with the potential for efficient referencing and multicomponent sensor analysis of complex samples. Various groups have developed SPR-sensing devices using slab *(92, 93)* and channel *(94, 95)* single-mode integrated optical waveguides. Similar to single-mode optical fiber SPR sensors, the integrated optical SPR-sensing devices exhibit a rather limited operating range. Various possibilities for tuning the operating range of the sensor have been explored, such as using waveguides fabricated on low refractive index glass *(94)*, a buffer layer *(93)*, a high refractive index overlayer *(96)*, and multilayer structures that are more complex *(97, 98)*. However, all of the approaches that introduced additional layers were found to yield less-sensitive SPR-sensing devices because of a relatively lower concentration of electromagnetic field in the analyte.

6.4. Imaging Fluorescence Detection

For detection of hybridization with fluorescently labeled targets, hybridization was performed by pipetting a small volume (20 μL) of hybridization solution onto the gold surface and "sandwiching" the droplet between the gold surface and a plain glass cover slip, which was laid on top. The sandwich assembly was enclosed in a petri dish on a platform above several milliliters of water, to provide a humid atmosphere to prevent drying of the sample during the hybridization period. After hybridization, the cover slip was removed and the gold surface washed by immersion in a beaker of 2-SSPE/0.2% SDS (*see* **Notes 4–6**). Finally, after washing, the sample was placed face down on top of the glass scanner tray provided with a Molecular Dynamics FluorImager 575 (Sunnyvale, CA) with a droplet of 2-SSPE/0.2% SDS between the gold surface and the glass scanner tray to provide an aqueous environment during the scan. The sample was scanned in the FluorImager using the standard Scanner Control v.1.1 software provided with the instrument, using a 530 (15-nm) band-pass interference filter to select for the fluorescein label's fluorescence emission. The SPR images were imported into NIH Image v.1.61 software where they were filtered and cropped as necessary, and "Plot Profile" analysis was performed on selected regions. In a plot profile, the average pixel value for each column of pixels within a rectangular region is calculated and plotted against that column's position. Then, numerical values from each plot profile were exported as a text file, which was subsequently opened in Igor v.1.25 graphing software (WaveMetrics, Lake Oswego, OR), where plot profiles could be overlaid, offset, ratioed, etc. For analysis of fluorescence images, ImageQuaNT v.4.1 software (provided with the FluorImager) was used.

7. Performance Characteristics

The main performance characteristics relevant for SPR biosensors include sensitivity, accuracy, precision, repeatability, and the lowest detection limit. Sensor sensitivity, S, is the ratio of the change in sensor output, P (e.g., angle of incidence, wavelength, intensity, phase, and polarization of light wave interacting with an SPW), to the change in measurand (e.g., analyte concentration, c). SPR biosensor sensitivity can be broken down into two components—sensitivity to refractive index changes produced by binding of the analyte to biomolecular recognition elements on the sensor surface, S_{RI}, and the efficiency, E, with which the presence of analyte at a concentration c is converted into the change in the refractive index n:

$$S = \frac{\partial P}{\partial n}\frac{\partial n}{\partial c} = S_{RI}E. \tag{5}$$

The efficiency E depends on the properties of the biomolecular recognition elements and the target analyte. The refractive index sensitivity S_{RI} can be separated into two terms:

$$S_{RI} = \frac{\partial P}{\partial \operatorname{Re}\{\beta\}}\frac{\partial \operatorname{Re}\{\beta\}}{\partial n} = S_1 S_2. \tag{6}$$

The first term, S_1, depends on the modulation method and the method of SPW excitation *(100–103)*. The S_2 term is independent of the modulation method and the method of SPW excitation and describes the sensitivity of SPW's propagation constant to refractive index changes, **Eqs. 3** and **4**.

Accuracy describes the degree to which a sensor output represents the true value of the measurand (analyte concentration). Accuracy is often confused with *precision*, which refers to the manner in which repeated measurements conform without a reference to any true value. *Repeatability* refers to the capacity of a sensor to reproduce output reading under the same measurement conditions over a short interval of time. The *lowest detection limit* describes the lowest concentration of analyte that can be measured by the sensor.

8. SPR Biosensing Formats

An interaction between a biomolecular recognition element on an SPR sensor surface and an analyte in a liquid sample is governed by kinetic equations. To illustrate the fundamental properties of the interaction, we discuss the pseudo-first-order kinetic equation:

$$\frac{\mathrm{d}R}{\mathrm{d}t} = k_\mathrm{a} c(1 - R) - k_\mathrm{d} R, \tag{7}$$

where R is the relative amount of bound analyte, c is the analyte concentration, t is time, and k_a and k_d are the association and dissociation kinetic rate constants, respectively *(104)*. This interaction model assumes a 1:1 binding, rapid mixing of the analyte from the bulk phase to the sensor surface layer, and single-step binding. Observed binding, however, may deviate from this simple model because of more complex mechanisms of interaction and mass transport limitations *(105)*. **Equation** 7 yields for R:

$$R(t) = \left[\frac{k_\mathrm{a} c}{k_\mathrm{a} c + k_\mathrm{d}} - R_0\right]\left(1 - e^{-(k_\mathrm{a} c + k_\mathrm{d})t}\right) + R_0, \tag{8}$$

where R_0 denotes the initial amount of analyte bound at the time $t = 0$ *(104)*.

Various measurement formats have been adopted in SPR biosensing to ensure that the monitored binding event produces a measurable sensor response. The most frequently used measurement formats are direct detection, sandwich assay, and inhibition assay. In the direct detection format, the analyte in a sample interacts with a biomolecular recognition element (antibody) immobilized on the sensor surface, **Fig. 9**. The resulting refractive index change is directly proportional to the concentration of the analyte.

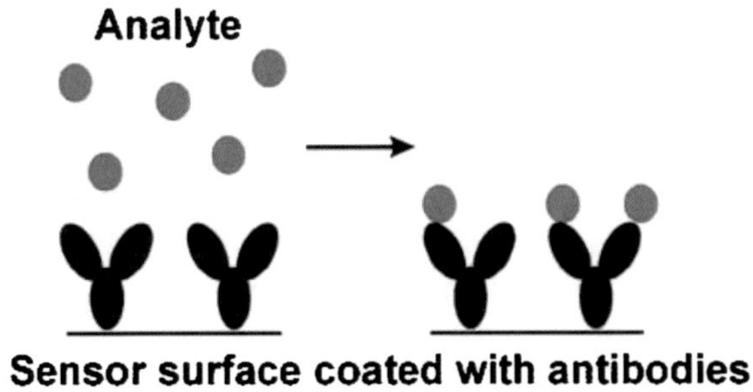

Fig. 9. Direct detection ***(64)***. Reproduced from **ref. *(64)*** with permission from Springer.

The kinetic model of the interaction between antibody and analyte suggests that the binding between the target analyte and the antibody is fast initially. As the interaction progresses, the binding rate gradually decreases and eventually reaches a state in which the association and dissociation processes are in equilibrium. The time required for the interaction to reach the equilibrium depends on the concentration of analyte and is longer for lower concentrations of analyte.

The dependence of the relative binding at equilibrium on the concentrations of analyte can be explained as follows: at low analyte concentrations, the equilibrium binding increases linearly with analyte concentration, whereas, at higher analyte concentrations, the binding sites provided by the biomolecular recognition elements are saturated and a further increase in the analyte concentration produces a smaller increase in the amount of bound analyte. The initial binding rate $\mathrm{d}R/\mathrm{d}t\ (t = 0)$ is directly proportional to the association rate constant and the analyte concentration. Both the amount of analyte bound at equilibrium and the initial binding rate can be used to determine analyte concentration. The measurement of the binding rate is faster and offers a larger dynamic range than the measurement of equilibrium binding. In the *sandwich assay format*, the measurement consists of two steps. In the first step, sample containing analyte is brought into contact with the sensor and the analyte molecules bind to the antibodies on the sensor surface. Next, the sensor surface is incubated with a solution containing secondary antibodies. The secondary antibodies bind to the previously captured analyte, further increasing the number of bound biomolecules **(Fig. 10)**, and, thus, the sensor response.

The *inhibition assay* is an example of a competitive assay. In this detection format, a sample is initially mixed with respective antibodies and then the mixture is brought into contact with the sensor surface coated with analyte molecules so that the unoccupied antibodies can bind the analyte molecules **(Fig. 11)**.

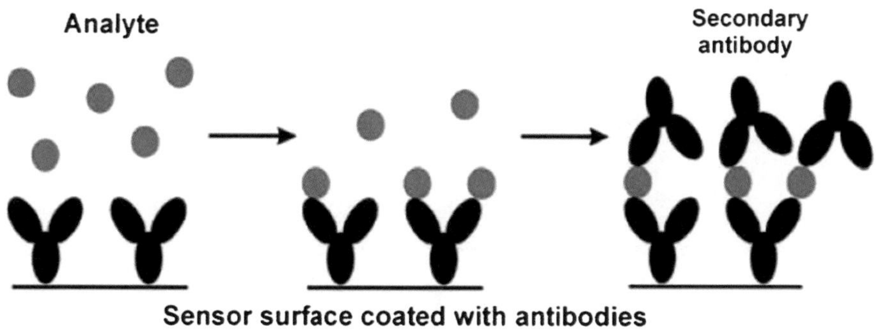

Fig. 10. Sandwich assay *(64)*. Reproduced from **ref.** ***(64)*** with permission from Springer.

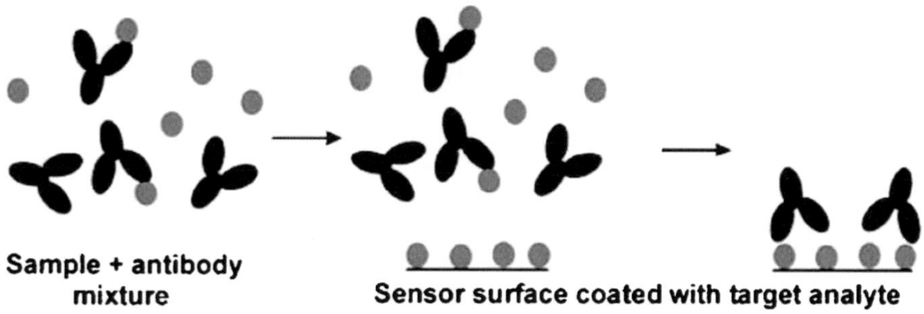

Fig. 11. Inhibition assay *(64)*. Reproduced from **ref.** ***(64)*** with permission from Springer.

The amount of bound analyte over time may be estimated by calculating the equilibrium concentration of antibody that did not bind the analyte in the sample and then simulating the interaction between the unbound antibody and the analyte-derivatized surface.

9. Features and Challenges

SPR biosensor technology exhibits various advantageous features. In particular, these include:

1. Versatility—generic SPR sensor platforms can be tailored for the detection of any analyte, thereby providing a biomolecular recognition element that recognizes that the analyte is available; the analyte does not have to exhibit any special properties such as fluorescence or characteristic absorption and scattering bands.

2. No labels required—binding between the biomolecular recognition element and analyte can be observed directly without the use of radioactive or fluorescent labels.
3. Speed of analysis—the binding event can be observed in real time, providing a potentially rapid response.
4. Flexibility—SPR sensors can perform continuous monitoring as well as one-time analyses.

SPR biosensors exhibit two inherent limitations:

1. Specificity of detection—specificity is solely based on the ability of biomolecular recognition elements to recognize and capture analyte. Biomolecular recognition elements may exhibit cross-sensitivity to structurally similar but nontarget molecules. If the nontarget molecules are present in a sample at high concentrations, the sensor response caused by the nontarget analyte molecules may conceal specific responses produced by low levels of target analyte.
2. Sensitivity to interfering effects—similar to other affinity biosensors relying on the measurement of refractive index changes, SPR biosensor measurements can be compromised by interfering effects, which produce refractive index variations. These include nonspecific interactions between the sensor surface and sample (adsorption of nontarget molecules by the sensor surface) and background refractive index variations (caused by sample temperature and composition fluctuations).

10. Applications of SPR Biosensors

Two major application areas for SPR biosensing are the detection and identification of biological analytes and the biophysical analysis of biomolecular interactions. This review focuses on applications for detection and identification of biological analytes; recent advances in SPR-based biomolecular interaction analysis can be found in **ref.** ***(106)***.

Numerous SPR biosensors have been developed for the detection and identification of specific analytes. These biosensors use a number of platform designs, biomolecular recognition elements, and detection formats. The choice of detection format for a particular application depends on the size of the target analyte molecules, the binding characteristics of the biomolecular recognition elements, and the range of measured analyte concentrations. Direct detection is usually preferred in applications where direct binding of known concentrations of analyte produce a sufficient response. If necessary, the lowest detection limits of the direct SPR biosensors can be improved by using a sandwich assay. The secondary antibodies may also be coupled to large particles such as latex particles *(107)* and gold beads *(108)* to further enhance the SPR sensor response. Smaller analytes (molecular weight <1,000) are usually measured using an inhibition assay.

11. Future Trends in Development of Surface Plasmon Resonance Sensors

Detection and analysis of chemical and biochemical substances are needed in many important areas, including medicine, environmental monitoring, biotechnology, and drug and food monitoring. Surface plasmon resonance sensor technology has potential for applications in these areas. Currently, SPR biosensor devices compete with other types of biosensors *(109)*, however, the major competitors of biosensors are immunoassays, which are commonly and widely used for the determination of numerous important substances and offer low-cost tests of high specificity and sensitivity. Today, commercially available biosensors cover a very limited area of the biochemical monitoring market, aiming primarily at research and analytical laboratories. To expand the market from specialized laboratories and centralized testing sites and gain a fair share of the biochemical monitoring market, SPR sensors have to compete with existing technologies on the basis of factors such as low cost, ease of use, robustness, sensitivity, and stability. It is envisaged that this will drive research and development of SPR-sensing devices in the following directions:

1. Improvement of detection limits. Current direct SPR biosensors are limited to the detection of approximately 1 pg mm^{-2} surface coverage of biomaterials, which is not sufficient for detecting low concentrations of low molecular weight analytes. Although further optimization of SPR optical instruments and development of efficient referencing concepts and sophisticated data-processing methods may improve the resolution of SPR-sensing devices and lower the detection limits, at present, no approach that can decrease this limit of detection by orders of magnitude is available.
2. Multichannel performance. Multichannel SPR sensors are required for direct detection in high-throughput screening systems in the search for new pharmaceuticals. The first steps in this direction include the system introduced by Biacore and the approach proposed by Berger *(109)*, which uses a 4-channel chip that can be rotated by 90° to provide 16 channels.
3. Development of advanced recognition elements. For applications of SPR sensors in complex realistic samples (e.g., blood),

advanced stable receptor matrices, which allow for resolving sensor responses from nonspecific background effects, will have to be developed.

Undoubtedly, in the future, SPR technology will benefit from the use of optical waveguide technology, which offers the potential for the development of miniaturized, compact, and rugged sensing elements with prospects of fabricating multiple sensors on one chip.

12. Notes

1. PDITC was then reacted with the surface by immersion for 2 h in a room temperature solution of 0.2% (w/v) PDITC in 10% pyridine/90% dimethylformamide (DMF), after which the sample was washed (first with DMF and then with ethanol) and then dried under a stream of nitrogen, leaving an isothiocyanate-derivatized surface.
2. Test droplets in 0.1 M sodium bicarbonate buffer, pH 9.0, containing 0.5 M NaCl were spotted onto the hydrophobic surface and left to react for 2 h at 37°C in a humid environment to prevent the spots from drying out. To stop the coupling reaction, the samples were immersed for 2 min in room temperature 1% NH_4OH, washed two times for 5 min each wash in room temperature H_2O, and finally blown dry under a stream of nitrogen, at which point they were ready to use for hybridization.
3. In all cases, the image data were recorded as an 8-bit grayscale image. The image acquisition systems were tested and shown to have a linear response to incident light intensity in the range used in the experiments.
4. For the DNA hybridization experiments, a special in situ cell was constructed, in which a Kel-F block (PCTFE, 3 M Industrial Chemical Products Division, St. Paul, MN) and a ½-in.-inner diameter Viton fluoroelastomer *O*-ring (Du Pont Dow Elastomers, L.L.C., Wilmington, DE) were sealed against the gold surface, to contain a small volume of liquid in contact with the central region of the gold surface. The cell was fitted with an inlet and outlet for replacing the cell contents by means of a syringe. The total system volume (inlet to outlet) was found to be 200 μL. The sample stage was rotatable through a large angular distance to find the optimal observation angle for a given sample but was locked at a fixed angle for the course of a hybridization experiment.

5. The cell could be flushed with 2-SSPE/0.2% SDS buffer to remove any unhybridized target DNA if desired.
6. To begin hybridization, a total volume of at least 300 μL of hybridization solution was injected into the in situ SPR cell using a syringe. Hybridization proceeded at room temperature while the SPR image was continuously monitored. After hybridization, in experiments where it was necessary to remove mismatched hybrids, the cell temperature was gradually raised using resistive thermofoil heating elements attached to the cell, with miniature platinum RTD sensors (Minco) embedded in the cell for feedback temperature control with a Love 1600 series self-tuning PID temperature controller. After heating was complete, the sample was passively cooled back to room temperature.

References

1. Ritchie, R. H. (1957). Plasma losses by fast electrons in thin films. *Phys. Rev.* **106**, 874–881
2. Burstein, E. (1974). Polaritons (Burstein, E. and De Martini, F., eds.). 1–4 Pergamon, New York
3. Kneipp, K., Wang, Y., Kneipp, H., Perelman, L. T., Itzkan, I., Dasari, R. R., and Feld, M. S. (1997). Single molecule detection using surface-enhanced Raman scattering (SERS). *Phys. Rev. Lett.* **78**, 1667–1670
4. Nie, S. M. and Emery, S. R. (1997). Probing single molecules and single nanoparticles by surface-enhanced Raman scattering. *Science.* **275**, 1102–1106
5. Ghindilis, A. L., Atanasov, P., Wilkins, M., and Wilkins, E. (1998). Immunosensors: Electrochemical sensing and other engineering approaches. *Biosens. Bioelectron.* **13**, 113–131
6. Chu, X., Lin, Z. H., Shen, G. L., and Yu, R. Q. M. (1995). Piezoelectric immunosensor for the detection of immunoglobulin M. *Analyst.* **120**, 2829–2832
7. Gauglitz, G. (1996). Opto-chemical and opto-immuno sensors, Sensor Update, Vol 1. VCH, Weinheim
8. Rowe-Taitt, C. A., Hazzard, J. W., Hoffman, K. E., Cras, J. J., Golden, J. P., and Ligler, F. S. (2000). Simultaneous detection of six biohaz ardous agents using a planar waveguide array biosensor. *Biosens. Bioelectron.* **15**, 579–589
9. Piehler, J., Brecht, A., and Gauglitz, G. (1996). Affinity detection of low molecular weight analytes. *Anal. Chem.* **68**, 139–143
10. Heideman, R. G., Kooyman, R. P. H., and Greve, J. (1993). Performance of a highly sensitive optical waveguide Mach-Zehnder interferometer immunosensor. *Sens. Actuators B* **10**, 209–217
11. Clerc, D. and Lukosz, W. (1994). Integrated optical output grating coupler as biochemical sensor. *Sens. Actuators B* **19**, 581–586
12. Cush, R., Cronin, J. M., Stewart, W. J., Maule, C. H., Molloy, J., and Goddard, N. J. (1993). The resonant mirror: a novel optical biosensor for direct sensing of biomolecular interactions part I: principle of operation and associated instrumentation. *Biosens. Bioelectron.* **8**, 347–353
13. Homola, J., Yee, S., and Gauglitz, G. (1999). Surface plasmon resonance sensors: review. *Sens. Actuators B* **54**, 3–15
14. Homola, J., Yee, S., and Myszka, D. (2002). Surface plasmon biosensors. In: Optical Biosensors: Present and Future (Ligler, F. S., and Taitt, C. R., eds.)., Elsevier, Amsterdam
15. Rabbany, S. Y., Lane, W. J., Marganski, W. A., Kusterbeck, A. W., and Ligler, F. S. (2000). Trace detection using a membrane-based displacement immunosassay. *J. Immunol. Methods* **246**, 69–77
16. Taton, T. A., Lu, G., Mirkin, C. A. (2001). Two-color labeling of oligonucleotide Arrays via size-selective scattering of nanoparticle probes. *J. Am. Chem. Soc.*, **123**, 5164–5165

17. El-Sayed, M. A. (2001). Some interesting properties of metals confined in time and nanometer space of different shapes. *Acc. Chem. Res.* **34**, 257–264
18. Jensen, T. R., Malinsky, M. D., Haynes, C. L., Van Duyne, R. P. (2000). Nanosphere lithography: tunable localized surface plasmon resonance spectra of silver nanoparticles. *J. Phys. Chem. B*, **104**, 10549
19. Schultz, S., Smith, D. R., Mock, J. J., Schultz, D. A. (2000). Single-target molecule detection with nonbleaching multicolor optical immunolabels. *Proc. Natl. Acad. Sci. U.S.A.*, **97**, 996
20. Michaels, A. M., Nirmal, M., Brus, L. E. (1999). Surface enhanced raman spectroscopy of individual rhodamine 6G molecules on large Ag nanocrystals. *J. Am. Chem. Soc.* **121**, 9932–9939
21. Yguerabide, J. and Yguerabide, E. E. (1998). Light-scattering submicroscopic particles as highly fluorescent analogs and their use as tracer labels in clinical and biological applications. *Anal. Biochem.* **262**, 157–176
22. Yguerabide, J. and Yguerabide, E. E. (1998). Light-scattering submicroscopic particles as highly fluorescent analogs and their use as tracer labels in clinical and biological applications. *Anal. Biochem.* **262**, 137–156
23. Schatz, G. C. and Van Duyne, R. P. (2002). Electromagnetic Mechanism of Surface-Enhanced Spectroscopy, Vol. 1. Wiley: New York
24. Kelly, K. L., Coronado, E., Zhao, L., and Schatz, G. C. (2003). The optical properties of metal nanoparticles: the influence of size, shape, and dielectric environment. *J. Phys. Chem. B*, **107**, 668–677
25. Kreibig, U., Gartz, M., and Hilger, A. (1997). Mie resonances. Sensors for physical and chemical cluster interface properties. *Ber. Bunsen-Ges.* **101**, 1593–1604
26. Link, S. and El-Sayed, M. A. (1999). Spectral properties and relaxation dynamics of surface plasmon electronic oscillations in gold and silver nanodots and vanorods. *J. Phys. Chem. B.* **103**, 8410–8426
27. Haynes, C. L. and Van Duyne, R. P.(2001). Nanosphere lithography: a versatile nanofabrication tool for studies of size-dependent nanoparticle optics. *J. Phys. Chem. B* **105**, 5599–5611
28. Mulvaney, P. (2001). Not all that's gold does glitter. *MRS Bull.* **26**, 1009–1014
29. Connolly, S., Cobbe, S., and Fitzmaurice, D. (2001). Effects of ligand-receptor geometry and stoichiometry on protein-induced aggregation of biotin-modified colloidal gold. *J. Phys. Chem. B.* **105**, 2222–2226
30. Storhoff, J. J., Lazarides, A. A., Mucic, R. C., Mirkin, C. A., Letsinger, R. L., and Schatz, G. C. (2000). What controls the optical properties of DNA-Linked gold nanoparticles assemblies. *J. Am. Chem. Soc.* **122**, 4640–4650
31. Henglein, A. and Meisel, D. (1998). Spectrophotometric observations of the adsorption of organosulfur compounds on colloidal silver nanoparticles. *J. Phys. Chem. B* **102**, 8364–8366
32. McFarland, A. D. and Van Duyne, R. P. (2003). Single silver nanoparticles as real-time optical sensors with zeptomole sensitivity. *Nano Lett.* **3**, 1057–1062
33. Malinsky, M. D., Kelly, K. L., Schatz, G. C., and Van Duyne, R. P. (2000). Chain length dependence and sensing capabilities of the localized surface plasmon resonance of silver nanoparticles chemically modified with alkanethiol self-assembled monolayers. *J. Am. Chem. Soc.* **123**, 1471–1482
34. Nath, N. and Chilkoti, A. (2002). A colorimetric gold nanoparticle sensor to interrogate biomolecular interactions in real time on a surface. *Anal. Chem.* **74**, 504–509
35. Riboh, J. C., Haes, A. J., McFarland, A. D., Yonzon, C. R., and Van Duyne, R. P. (2003). A nanoscale optical biosensor: real-time immunoassay in physiological buffer enabled by improved nanoparticle adhesion. *J. Phys. Chem. B* **107**, 1772–1780
36. Cao, Y. W., Jin, R. C., and Mirkin, C. A. (2002). Nanoparticles with Raman spectroscopic fingerprints for DNA and RNA Detection. *Science* **297**, 1536–1540
37. Calander, N. and Willander, M. (2002). Theory of surface-plasmon resonance optical-field enhancement at prolate spheroids. *J. Appl. Phys.* 92, 4878. Nils Calander, N. and Willander, M. (2002). Optical trapping of single fluorescent molecules at the detection spot of nanoprobes. *Phys. Rev. Lett.* **89**, 1436031–1436034
38. Sambles, J. R., Bradbery, G. W., and Yang, F. Z. (1991). Optical-excitation of surface-plasmons - an introduction. *Contemp. Phys.* **32**, 173–183
39. Boardman, A. D. (1982). Electromagnetic surface modes. Wiley, Chichester
40. Barnes, W. L., Dereux, A., and Ebbesen, T. W. (2003). Surface plasmon subwavelength optics. *Nature* **424**, 824–830
41. Kretschmann, E. and Raether, H. (1968). Radiative decay of nonradiative surface plasmons excited by light. *Z. Naturforsch. A* **23**, 2135–2136

42. Otto, A. (1968). Exitation of nonradiative surface plasma waves in silver by the method of frustrated total reflection. *Z. Phys.* **216**, 398
43. Hecht, B., Bielefeldt, H., Novotny, L., Inouye, Y., and Pohl, D. W. (1996). Local excitation, scattering, and interference of surface plasmons. *Phys. Rev. Lett.* 77, 1889–1892
44. Ditlbacher, H., Krenn, J. R., Felidj, N., Lamprecht, B., Schider, G., Salerno, M., Leitner, A., and Aussenegg, F. R. (2002). Fluorescence imaging of surface plasmon fields. *Appl. Phys. Lett.* **80**, 404–406
45. Ritchie, R. H., Arakawa, E. T., Cowan, J. J.&Hamm, R. N. (1968). Surface-plasmon resonance effect in grating diffraction. *Phys. Rev. Lett.* **21**, 1530–1533
46. Raether, H. (1988). Surface Plasmons (Hohler, G., ed.)., Springer, Berlin
47. Ditlbacher, H., Krenn, J. R., Schider, G., Leitner, A., and Aussenegg, F. R. (2002). Two-dimensional optics with surface plasmon polaritons. *Appl. Phys. Lett.* **81**, 1762–1764
48. Cregan, R. F., Mangan, B. J., Knight, J. C., Birks, T. A., Russell, Roberts, P. J., and Allan, D. C. (1999). Single-mode photonic band gap guidance of light in air. *Science* **285**, 1537–1539
49. Kitson, S. C., Barnes, W. L., and Sambles, J. R. (1996). A full photonic band gap for surface modes in the visible. *Phys. Rev. Lett.* 77, 2670–2673
50. Barnes, W. L., Preist, T. W., Kitson, S. C., and Sambles, J. R. (1996). Physical origin of photonic energy gaps in the propagation of surface plasmons on gratings. *Phys. Rev. B* **54**, 6227–6244
51. Barnes, W. L., Kitson, S. C., Preist, T. W., and Sambles, J. R. (1997). Photonic surfaces for surface plasmons polaritons. *J. Opt. Soc. Am. A* **14**, 1654–1661
52. Kreibig, U. and Vollmer, M. (1995). Optical properties of metal clusters. Springer, Berlin
53. Haynes, C. L. and Van Duyne, R. P. (2001). Nanosphere lithography: a versatile nanofabrication tool for studies of size-dependent nanoparticle optics. *J. Phys. Chem. B* **105**, 5599–5611
54. Sonnichsen, C., Geier, S., Hecker, N. E., von Plessen, G., Feldmann, J., Ditlbacher, H., Lamprecht, B., Krenn, J. R., Aussenegg, F. R., Chan, V. Z.-H., Spatz, J. P., and Moller, M. (2000). Spectroscopy of single metallic nanoparticles using total internal reflection microscopy. *Appl. Phys. Lett.* 77, 2949–2951
55. Schultz, D. A. (2003). Plasmon resonant particles for biological detection. *Curr. Opin. Biotechnol.* **14**, 13–22
56. Oldenburg, S. J., Genick, C. C., Clark, K. A., and Schultz, D. A. (2002). Base pair mismatch recognition using plasmon resonant particle labels. *Anal. Biochem.* **309**, 109–116
57. Félidj, N., Aubard, J., Levi, G., Krenn, J. R., Hohenau, A., Schider, G., Leitner, A., and Aussenegg, F. R. (2003). Optimized surface-enhanced Raman scattering on gold nanoparticle arrays. *Appl. Phys. Lett.* **82**, 3095–3097
58. Levi, S., Mourran, A., Spatz, J. P., Veggel van, F., Reinhoudt, D., and Moller, M. (2002). Fluorescence of dyes adsorbed on highly organized, nanostructured gold surfaces. *Chem. Eur. J.* **8**, 3808–3814
59. Vargas-Baca, I., Brown, A. P., Andrews, M. P., Galstian, T., Li, Y., Vali, H., and Kuzyk, M. G. (2002). Linear and nonlinear optical responses of a dye anchored to gold nanoparticles dispersed in liquid and polymeric matrixes. *Can. J. Chem.* **80**, 1625–1633
60. Rechberger, W., Hohenau, A., Leitner, A., Krenn, J. R., Lamprecht, B., and Aussenegg, F. R. (2003). Optical properties of two interacting gold nanoparticles. *Opt. Commun.* **220**, 137–141
61. Kottmann, J. P. and Martin, O. J. F. (2001). Retardation-induced plasmon resonances in coupled nanoparticles. *Opt. Lett.* **26**, 1096–1098
62. García-Vidal, F. J. and Pendry, J. B. (1996). Collective theory for surface enhanced Raman scattering. *Phys. Rev. Lett.* 77, 1163–1166
63. Gresillon, S., Aigouy, L., Boccara, A. C., Rivoal, J. C., Quelin, X., Desmarest, C., Gadenne, P., Shubin, V. A., Sarychev, A. K., and Shalaev, V. M. (1999). Experimental observation of localized optical excitations in random metal-dielectric films. *Phys. Rev. Lett.* **82**, 4520–4523
64. Homola, J. (2003). Present and future of surface plasmon resonance biosensors. *Anal. Bioanal. Chem.* **377**, 528–539
65. Snyder, A. W. and Love, J. D. (1983). Optical waveguide theory. Chapman and Hall, London
66. Parriaux, O. and Voirin, G. (1990). Plasmon wave versus dielectric waveguiding for surface wave sensing. *Sens. Actuators A* **23**, 1137–1141
67. Reather, H. (1983). Surface Plasmons on Smooth and Rough Surfaces and on Gratings, Springer Tracts in Modern Physics, Vol. 111, Springer, Berlin
68. Nylander, C., Liedberg, B., and Lind, T. (1982). Gas detection by means of surface plasmons resonance. *Sens. Actuators* **3**, 79–88

69. Liedberg, B., Nylander, C., and Lundstro¨m, I. (1983). Surface plasmons resonance for gas detection and biosensing. *Sens. Actuators* **4**, 299–304
70. Matsubara, K., Kawata, S., and Minami, S. (1988). Optical chemical sensor based on surface plasmon measurement. *Appl. Opt.* **27**, 1160–1163
71. Liedberg, B., Lundstrom, I., and Stenberg, E. (1993). Principles of biosensing with an extended coupling matrix and surface plasmon resonance. *Sens. Actuators B* **11**, 63–72
72. Zhang, L. M. and Uttamchandani, D. (1988). Optical chemical sensing employing surface plasmon resonance. *Electron. Lett.* **23**, 1469–1470
73. Jonsson, U., Fagerstam, L., Ivarsson, B., Johnsson, B., Karlsson, R., Lundh, K., Lofas, S., Persson, B., Roos, H., Rönnberg, I., et al. (1991). Real-time biospecific interaction analysis using surface plasmon resonance and a sensor chip technology. *Biotechniques* **11**, 620–627
74. Löfås, S., Malmqvist, M., Rönnberg, I., Stenberg, E., Liedberg, B., and Lundström, I. (1991). Bioanalysis with surface plasmon resonance. *Sens. Actuators B* **5**, 79–84
75. Hutley, M. C. (1982). Diffration gratings. Academic Press, London
76. Hutley, M. C. (1982). Diffration gratings, Academic Press, London
77. Raether, H. (1988). Surface Plasmons on Smooth and Rough Surfaces and on Gratings. Springer-Verlag, Berlin
78. Moharam, M. G. and Gaylord, T. K. (1986). Rigorous coupled-wave analysis of metallic surface-relief gratings. *J. Opt. Soc. Am A.* **3**, 1780–1787
79. Chandezon, J., Dupuis, M. T., Cornet, G., and Maystre, D. (1982). Multicoated gratings: a differential formalism applicable an the entire optical region. *J. Opt. Soc. Am.* **72**, 839–846
80. Cullen, D. C., Brown, R. G., and Lowe, C. R. (1987). Detection of immunocomplex formation via surface plasmon resonance on goldcoated diffraction gratings. *Biosensors* **3**, 211–225
81. Cullen, D. C. and Lowe, C. R. (1990). A direct surface plasmon-polariton immunosensor: preliminary investigation of the non-specific adsorption of serum components to the sensor interface. *Sens. Actuators B* **1**, 576–579
82. Vukusic, P. S., Bryan-Brown, G. P., and Sambles, J. R. (1992). Surface plasmons resonance on grating as novel means for gas sensing. *Sens. Actuators B* **8**, 155–160
83. Jory, M. J., Vukusic, P. S., and Sambles, J. R. (1994). Development of a prototype gas sensor using surface plasmon resonance on gratings. *Sens. Actuators B* **17**, 1203–1209
84. Jorgenson, R. C. and Yee, S. S. (1993). A fiber-optic chemical sensor based on surface plasmon resonance. *Sens. Actuators B* **12**, 213–220
85. Jorgenson, R. C. and Yee, S. S. (1994). Control of the dynamic range and sensitivity of a surface plasmon resonance based fiber optic sensor. *Sens. Actuators A* **43**, 44–48
86. Trouillet, A., Ronot-Trioli, C., Veillas, C., and Gagnaire, H. (1996). Chemical sensing by surface plasmon resonance in a multimode optical fibre. *Pure Appl. Opt.* **5**, 227–237
87. Ronot-Trioli, C., Trouillet, A., Veillas, C., and Gagnaire, H. (1996). Monochromatic excitation of surface plasmon resonance in an optical-fibre refractive-index sensor. *Sens. Actuators A* **54**, 589–593
88. Dessy, R. E. and Bender, W. J. (1994). Feasibility of a chemical microsensor based on surface plasmon resonance on fiber optics modified by multilayer vapor deposition. *Anal. Chem.* **66**, 963–970
89. Homola, J. (1995). Optical fiber sensor based on surface plasmon excitation, Second European Conference on Optical Chemical Sensors and Biosensors, Florence, Italy, April 1994. *Sens. Actuators B* **29**, 401–405
90. Homola, J. and Slavik, R. (1996). Fibre-optic sensor based on surface plasmon resonance. *Electron. Lett.* **32**, 480–482
91. Tubb, A. J. C., Payne, F. P., and Millington, R. B. (1997). C. R. Lowe, Singlemode optical fiber surface plasma wave chemical sensor. *Sens. Actuators B* **41**, 71–79
92. Homola, J., Tyroky, J. C., Skalsky, M., Hradilova, J., and Kola´r' ova´P. (1997). A surface plasmon resonance based integrated optical sensor. *Sens. Actuators B* 38–39, 286–290
93. Lavers, C. R. and Wilkinson, J. S. (1994). A waveguide-coupled surface plasmon sensor for an aqueous environment. *Sens. Actuators B* **22**, 75–81
94. Harris, R. D. and Wilkinson, J. S. (1995). Waveguide surface plasmon resonance sensors. *Sens. Actuators B* **29**, 261–267
95. Mouvet, C., Harris, R. D., Maciag, C., Luff, B. J., Wilkinson, J. S., Piehler, J., Brecht, A., Gauglitz, G., Abuknesha, R., and Ismail, G. (1997). Determination of simazine in water samples by waveguide surface plasmon resonance. *Anal. Chim. Acta* **338**, 109–117

96. Ctyroky, J., Homola, J., and Skalsky, M. (1997). Tuning of spectral operation range of a waveguide surface plasmon resonance sensor. *Electron. Lett.* **33**, 1246–1248
97. Weiss, M. N., Srivastava, R., Groger, H., Lo, P., and Luo, S. F. (1996). A theoretical investigation of environmental monitoring using surface plasmon resonance waveguide sensors. *Sens. Actuators A* **51**, 211–217
98. Weiss, M. N., Srivastava, R., and Groger, H. (1996). Experimental investigation of a surface plasmon-based integrated-optic humidity sensor. *Electron. Lett.* **32**, 842–843
99. Bender, W. J. H., Dessy, R. E., Miller, M. S., and Claus, R. O. (1994). Feasibility of a chemical microsensor based on surface plasmon resonance of fiber optics modified by multilayer vapor deposition. *Anal. Chem.* **66**, 963–970
100. Kooyman, R. P. H., Kolkman, H., van Gent, J., and Greve, J. (1988). Surface plasmon resonance immunosensors: sensitivity considerations. *Anal. Chim. Acta* **213,** 35–45
101. Yeatman, E. M. (1996). Resolution and sensitivity in surface plasmon microscopy and sensing. *Biosens. Bioelectron.* **11**, 635–649
102. Homola, J. (1997). On the sensitivity of surface-plasmon resonance sensors with spectral interrogation. *Sens. Actuators B* **41**, 207–211
103. Homola, J., Koudela, I., and Yee, S. (1999). Surface plasmon resonance sensors based on diffraction gratings and prism couplers: sensitivity comparison. *Sens. Actuators B* **54**, 16–24
104. Edwards, P. R. and Leatherbarrow, R. J. (1997). Determination of association rate constants by an optical biosensor using initial rate analysis. *Anal. Biochem.* **246,** 1–6
105. Vijayendran, R. A., Ligler, F. S., and Leckband, D. E. (1999). A computational reaction-diffusion model for the analysis of transport limited kinetics. *Anal. Chem.*.**71**, 5405–5412
106. Rich, R. L. and Myszka, D. G. (2002). Survey of the year 2001 commercial optical biosensor literature. *J. Mol. Recognit.* **15,** 352–376
107. Severs, A. H. and Schasfoort, R. B. M. (1993). Enhanced surface plasmon resonance inhibition test (ESPRIT). using latex particles. *Biosens. Bioelectron.* 8, 365–370
108. Leung, P. T., Pollard-Knight, D., Malan, G. P., and Finlan, M. F. (1994). Modelling of particle-enhanced sensitivity of the surface-plasmon-resonance biosensor. *Sens. Actuators B* **22,** 175–180
109. Owen, V. (1997). Real-time optical immunosensors – a commercial reality, *Biosens. Bioelectron.* **12,** i–ii
110. Berger, C. E. H., Baumer, T. A. M., Kooyman, R. P. H., and Greve, J. (1998). Surface plasmon resonance multisensing. *Anal. Chem.* **70,** 703–706

Chapter 15

Use of Residual Dipolar Couplings in Structural Analysis of Protein–Ligand Complexes by Solution NMR Spectroscopy

Nitin U. Jain

Summary

Investigation of structure–function relationships in protein complexes, specifically protein–ligand interactions, carry great significance in elucidating the structural and mechanistic bases of molecular recognition events and their role in regulating cell processes. Nuclear magnetic resonance (NMR) spectroscopy is one of the leading structural and analytical techniques in in-depth studies of protein–ligand interactions. Recent advances in NMR methodology such as transverse relaxation–optimized spectroscopy (TROSY) and residual dipolar couplings (RDCs) measured in liquid crystalline alignment medium, offer a viable alternative to traditional nuclear Overhauser enhancement (NOE)-based approaches for structure determination of large protein complexes. RDCs provide a way to constrain the relative orientation of two molecules in complex with each other by aligning their independently determined order tensors. The potential for utilization of RDCs can be extended to proteins with multiple ligands or even multimeric protein–ligand complexes, where symmetry properties of the protein can be taken advantage of. Availability of effective RDC data collection and analysis protocols can certainly aid this process by their incorporation into structure calculation protocols using intramolecular and intermolecular orientational restraints. This chapter discusses in detail some of these protocols including methods for sample preparation in liquid crystalline media, NMR experiments for RDC data collection, as well as software tools for RDC data analysis and protein–ligand complex structure determination.

Key words: NMR spectroscopy, Residual dipolar couplings, RDC, Protein–ligand complexes, Liquid crystalline media

1. Introduction

NMR spectroscopy is one of the two current established methods for high-resolution structure determination of proteins and protein complexes, X-ray crystallography being the other method.

James Weifu Lee and Robert S. Foote (eds.), *Micro and Nano Technologies in Bioanalysis*,
Methods in Molecular Biology, vol. 544
DOI 10.1007/978-1-59745-483-4_15, © Humana Press, a part of Springer Science+Business Media, LLC 2009

With the development of higher magnetic strengths, new pulse sequence technology and isotopic labeling methods, structure determination by NMR spectroscopy is no longer limited to molecules small in size, but is now routinely being extended to sizes >30 kDa *(1–4)*. In particular, NMR has increasingly been applied to the direct study of large proteins and protein complexes in the last few years *(5–9)*. Within this context, structural studies of protein–ligand complexes have assumed great significance. Protein–ligand interactions are critical to the specific function of many biological systems and their characterization is important to understand the process of molecular recognition and its role in regulating cell processes. Several molecular recognition events are connected to disease, validating the associated proteins as potential drug targets. Availability of individual protein structures for many of these targets has therefore prompted detailed analyses of structures of their ligand complexes, incorporated within the general strategy of structure-based drug design. It is possible to use this structural information to determine ligand conformation and alter its structural and binding properties, leading to development of effective drugs. NMR spectroscopy has enjoyed considerable success in this endeavor to probe the specific nature of protein–ligand interactions, especially those transient in nature, which are difficult to study by other structural methods.

There are numerous NMR methods to detect and study protein–ligand interactions, prominent among them being chemical shift mapping, changes in relaxation parameters (T_1, T_2 relaxation times), and saturation transfer in chemically exchanging systems *(10–14)*. In addition, detailed NMR structural investigations of protein–ligand interactions have traditionally relied on intramolecular and intermolecular nuclear Overhauser enhancement (NOE) data to characterize binding modes of the ligand in a structure-specific manner *(15)*. Considerable effort can be expended for de novo structural studies of large protein–ligand complexes (>50 kDa) by the traditional NOE approach because of the a priori requirement of extensive assignments for the protein and a large number of intermolecular restraints for the complex. Lately, residual dipolar coupling (RDC) measurements in aligned media have emerged as an alternative to NOE-based distance restraints for conformational and structural studies of biomolecules. Applications of RDCs in structural studies have been plenty, some of them being structural refinement of both small and large proteins, characterization of interdomain relationships in multidomain proteins, and structure elucidation of protein complexes *(16–21)*. RDCs report on the angular dependence of an internuclear vector relative to the external magnetic field and occur because of incomplete averaging of the dipolar interaction between two nuclei. These couplings are usually not observed in solution spectra because of isotropic averaging, however, if a small amount of molecular order is introduced in the system, they can be then measured as positive or negative contributions to scalar couplings. The degree of molecular order that gives rise to RDCs can be described by a Saupe's order tensor *S*, which describes both the magnitude and direction of ordering in terms of direction cosines:

$$D_{ij} = -\left(\frac{\mu_0}{4\pi}\right)\frac{\gamma_i\gamma_j h}{2\pi^2 r_{ij}^3}\sum_{kl} S_{kl}\cos(\alpha_k)\cos(\alpha_l). \quad (1)$$

One can infer orientational information from **Eq. 1** by diagonalization of an order tensor providing principal values of the order tensor in terms of the axes S_{zz}, S_{yy}, and S_{xx}, and an asymmetry parameter η defined by $(S_{xx}-S_{yy})/S_{zz}$ *(22)*. Here, S_{zz} is designated as the highest ordered axis and η describes the deviation from an axially symmetric ordering. The order tensor consists of five independent elements, three for magnitude and two for direction, that require a minimum of five mutually independent measured RDC values for tensor solution. The principal values of the order tensor are referenced with respect to the initial molecular frame and hence they correspond indirectly to the mean orientation of the molecule with reference to the external magnetic field related by three Euler angles.

Structure determination of protein–ligand complexes offers an excellent opportunity for use of RDC-based methodology. When two interacting molecules are oriented relative to the magnetic field, such as in a protein–ligand complex, both molecules experience the same degree of ordering as a result of the interaction and hence are influenced by a common order tensor. The ordering of each molecule within the complex can be individually determined by measuring RDCs for each molecule in the complex form and defining their individual tensors of ordering. This allows assembly of the complex by orienting the individual structures of the two molecules such that their directions of ordering match, as shown in **Fig. 1**. RDCs thus represent powerful long-range structural restraints that can help define the relative orientation of two structural elements that are spatially remote from each other. For weak-affinity complexes, the backbone structure of the protein does not usually change significantly upon complex formation. In such a case, if the high-resolution structure of the protein is available, then it is feasible to dock a ligand onto the protein on the basis of these mutually connected orientational restraints. With advances in modeling and data-driven docking procedures, this can be done in an automatic fashion by incorporating RDC-derived orientational and other experimental restraints in structure calculation protocols, helping the deduction of biologically relevant protein complex structures *(23–27)*.

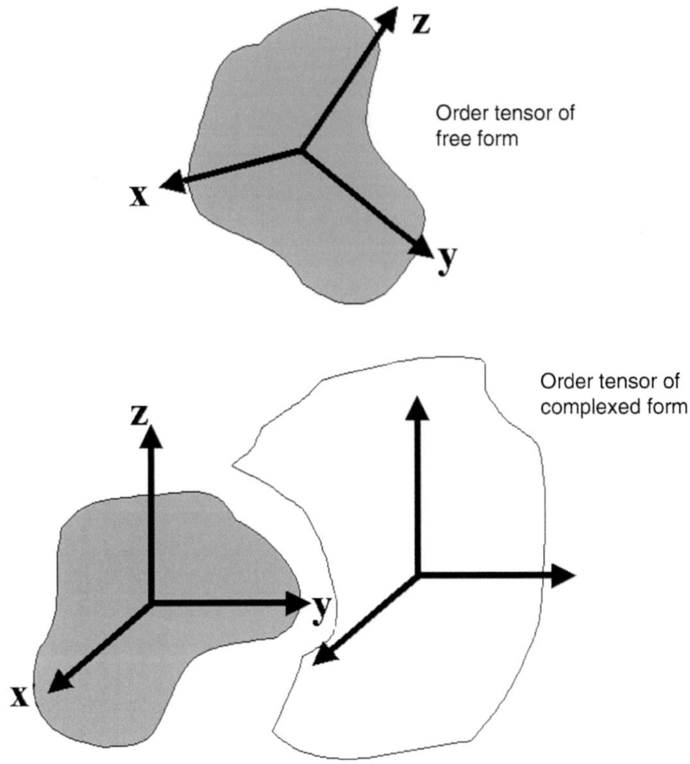

Fig. 1. Schematic for assembly of protein complexes using RDC-derived orientational constraints. Rotation of the order tensor in free form to match the alignment tensor in the complexed form gives the orientation of each protein in the complex.

Although RDCs can be measured even in isotropic solutions, since molecules tend to align to a small extent in the external magnetic field because of the inherent magnetic susceptibility anisotropy, RDCs are usually too small to be measured with great accuracy, this is especially true for multiple bond couplings. A preferred option is to orient molecules artificially to amplify RDCs to an extent that they can be measured with sufficient accuracy, but do not cause drastic line broadening in spectra from increased ordering that can affect resolution. To accomplish this, a variety of dilute liquid crystalline media have been introduced in the last few years that induce molecular orientation that can be used in high-resolution applications for efficient measurement of RDCs *(28–32)*. Use of these media permits measurement of a multitude of couplings corresponding to both homonuclear and heteronuclear, one-bond, two-bond, or even three-bond couplings *(33)*. It is realized now that successful alignment of protein–ligand samples and accurate measurement of the various RDCs is key to the high-resolution structural and functional characterization of protein–ligand complexes, and knowledge of the methods to do so is necessary.

This chapter reviews methods for solution structure determination of protein–ligand complexes using residual dipolar couplings from a practical point of view. Methods for preparation of commonly used alignment media are outlined, as are techniques for stable alignment of protein–ligand complexes in these media. NMR pulse schemes for measurement of one-bond RDCs in large proteins and ligands at natural isotopical abundance are summarized along with subsequent data analysis protocols. Finally, use of measured RDCs is illustrated with examples of structure determination of large protein–ligand complexes with special structural characteristics such as molecular symmetry and multiple ligand binding properties.

2. Materials

2.1. Alignment Media

1. To prepare 1 mL of 20%, w/v bicelle stock solution, the following recipe can be followed: weigh out the two lipids, 1,2-dihexanoyl-*sn*-glycero-3-phosphocholine (DHPC) (MW ~453.5) and 1,2-dimyristoyl-*sn*-glycero-3-phosphocholine (DMPC) (MW ~677.9) from Avanti Polar Lipids (Alabaster, AL) to prepare a DMPC:DHPC lipid mixture in a molar ratio of 3:1. For a 1-mL solution, you will need 203 mg of DMPC and 45 mg of DHPC.
2. The lipid mixture is then added to 1 mL of buffer solution in a glass vial. Typical buffer solutions that can be used are 25–100 mM Tris–Cl, pH 7.0–8.0 or 10–50 mM phosphate, pH 6.5–7.0. Low amounts of salt, NaCl/KCl (50 mM) can also be added to these buffer solutions. Higher amounts of salt tend to lower the temperature at which the bicelles spontaneously order (*see* **Note 1**).
3. For phage alignment media preparation: 1.05 L Luria-Bertini medium (LB) culture of *Pseudomonas aeruginosa* (ATCC); 50 μL of pure Pf1 phage (Asla Biotech Ltd, Riga, Latvia); 500 mL of 60 g/L NaCl; 100 mL of 20 g/L PEG8000; 100 mL each of 20%, 28%, 36%, 42%, and 50% potassium bromide (KBr) solutions; 10 mM Tris–Cl, pH 7.8 buffer.

2.2. Preparation of Samples of Protein–Ligand Complexes

1. Preparation of protein and ligand samples: In the case of mannose binding protein (MBP), 200 μL of 1 mM $^{15}N^{2}H$-labeled protein in 50 mM Tris–Cl, pH 7.4 buffer with 10% D_2O is mixed with 100 μL of 20% (w/v) bicelle solution (DMPC:DHPC in a molar ratio of 3:1) to yield an effective bicelle concentration of 6% (w/v). A 50-mM stock solution of trimannoside ligand is prepared in the same buffer.

2. For *N*-acetyl-glucosaminosyltransferase V (GnTV), 200 μL of 0.2 mM protein with stock solutions of 5 mM UDP-GlcNAc and 5 mM synthetic trisaccharide acceptor [βGlcNAc*(1,2)*αMan*(1,6)*βMan(OR)] is prepared in phosphate-buffered saline (PBS) buffer.

3. Methods

3.1. Preparation of Protein–Ligand Samples in Alignment Media

To carry out efficient measurement of RDCs, it is important to find a suitable alignment medium that can impart molecular order to the complex. Such a medium should possess certain properties. For one, it has to be preferably aqueous and stably align under physiological conditions where most proteins are studied. Additionally, the order introduced by the medium should be small for acquisition of high-resolution spectra. And most importantly, the medium should be compatible with the protein–ligand system under study in that it should not cause substantial changes in properties of the sample (protein–ligand) upon alignment, such as folding topology, interaction between binding partners, or any other change in function. A wide array of alignment media have been introduced over the last few years that satisfy these properties. These belong to the liquid crystalline solvent systems that exhibit spontaneous cooperative ordering in an external magnetic field typical of high-field NMR spectrometers. An extensive list of such media can be found in a review by Prestegard and coworkers *(33, 34)*. Our discussion here is focused on two of the most commonly used media for aligning proteins and associated complexes: lipid bilayer discs known as bicelles and filamentous bacteriophage Pf1.

Bicelles are disc-shaped assemblies with diameters of 200–300 Å and thicknesses of approximately 40 Å. In the liquid crystalline phase, bicelles align with their normal perpendicular to the magnetic field with spacing between discs close to few hundred angstroms, which is filled with liquid. Proteins and other small biomolecules experience a solution-like environment within this spacing and orient as a result of weak steric interactions with the bilayer discs. Filamentous bacteriophage Pf1, on the other hand, is the most commonly used phage medium for alignment of biomolecules. It consists of single-stranded circular DNA genome packaged in a coat protein that form long rods, 60 Å in diameter and 20,000 Å in length, which spontaneously align with their long axes parallel to the external magnetic field. The techniques for preparation of protein–ligand samples in these two media are described next.

3.1.1. Preparation of Protein–Ligand Sample in Bicelle Medium

1. Once the DMPC:DHPC lipid mixture is prepared in buffer, it can be left at room temperature with gentle shaking for 24 h to achieve complete hydration. Upon hydration, the solution transforms from a white cloudy mix to a clear opalescent solution, which signifies formation of the liquid crystalline gel phase. The solution is extremely viscous at this stage.
2. Alternatively, to achieve accelerated hydration (within an hour), the mixture can be taken through three or four heating and cooling cycles by heating up to 40°C for 10 min and then cooling to 4°C for 5 min followed by brief vortexing.
3. The alignment of bicelle solution prepared by the above method should be checked by observing the quadrupolar splitting on the deuterium resonance in D_2O to confirm proper alignment before using with the protein–ligand solution (*see* **Note 2**). For this, a 5%, w/v bicelle solution can be prepared from the 20% stock solution using the buffer intended to be used with the protein sample. The temperature range in which quadrupolar splitting is observed should be noted. The bicelles prepared in this fashion normally transition from the isotropic phase (no quadrupolar splitting) to the aligned phase at approximately 35°C (*see* **Note 1**).
4. The alignment of bicelles (5%, w/v) with a protein–ligand solution can be checked in a similar fashion by mixing in three volumes of protein–ligand solution with one volume of 20% bicelle solution. If the temperature of maximal alignment does not change substantially compared with the bicelle solution alone (i.e., by more than 2–3°C), then the addition of protein–ligand does not affect the ordering of the lipid bilayer discs and thus constitutes a stable system for RDC measurements (*see* **Note 3**).

3.1.2. Preparation of Protein–Ligand Sample in Phage Medium

The general steps in preparation of Pf1 phage from *Pseudomonas aeruginosa* are as follows (*see* **Note 4**):

1. Infect a 50-mL LB culture of *P. aeruginosa* with 50 μL of pure Pf1 phage, culture for 2 h, and then transfer to 1 L LB medium.
2. Culture for another 12–16 h until $OD_{600} > 1.0$ and then harvest by centrifugation at 6000g for 1 h. The phage is in the yellow supernatant solution.
3. Precipitate phage from the supernatant by slowly adding 60 g/L NaCl and 20 g/L PEG8000 with gentle stirring. Stir for another 3 h.
4. Centrifuge again at 6000g for 1 h. Phage pellets at the bottom. Discard the supernatant and resuspend the pellet in distilled water.

5. Centrifuge again at 30000g for 1 h and then resuspend the pellet in 40 mL of 20% KBr solution. Divide the resulting solution into six portions in 50-mL ultracentrifuge tubes.
6. To further purify phage, the KBr density gradient method is used. Create a KBr density gradient in each centrifuge tube by gently adding 7 mL of each KBr solution (20%, 28%, 36%, 42%, and 50%) in increasing order of concentration with a syringe.
7. Centrifuge the six tubes in an ultracentrifuge at 250,000g for 20 h at 25°C. A bluish-white band forms near top of the gradient. Extract the band carefully from each tube using a glass pipet. Pool the bands together in a small piece of dialysis tubing.
8. Dialyze the phage against 10 mM Tris–Cl, pH 7.8 buffer with three changes of buffer over a period of 24 h.
9. Pellet phage from this dialyzed solution again by ultracentrifugation at 250,000g for 8 h. Resuspend phage in dialysis buffer and store at 4°C at a concentration of 50 mg/mL. The concentration of phage can be estimated by observing the quadrupolar splitting of deuterium resonance in D_2O. The phage concentration follows a linear relationship with the observed splitting in Hertz up to a concentration of 50 mg/mL *(29)*. For example, an observed splitting of 10 Hz corresponds roughly to a 10 mg/mL phage concentration. We suggest making a dilute solution of phage from the stock to check the concentration.
10. Phage prepared in this fashion can be directly used for NMR studies by adding an appropriate amount of phage solution to a protein–ligand sample to give an effective phage concentration between 5 and 10 mg/mL. The alignment of phage with the sample can then be checked again by D_2O quadrupolar splitting. If splitting of the sample plus the phage is of same magnitude as with the phage alone, then the addition of protein–ligand does not affect the ordering of the Pf1 phage long rods and thus constitutes a stable system for RDC measurements (*see* **Note 5**).

3.2. NMR Experiments for Measurement of RDCs for Protein–Ligand Complexes

Extraction of couplings from frequency domain spectra is the most commonly used method for direct RDC measurements on protein–ligand complexes. Although a large number of NMR experiments have been developed to measure several single and multiple bond RDCs, such as D_{NH}, D_{CH}, D_{NC}, D_{HH}, and D_{CC} *(33, 35)*, routine measurements involve $^1D_{NH}$ couplings for proteins and $^1D_{CH}$ couplings for ligands and will be the focus of our discussion here. We illustrate the use of such measurements in the structure determination of a complex between the 53-kDa homotrimer, mannose-binding protein (MBP), and its trimannoside ligand, where a limited set of ^{1}H–^{15}N RDCs for MBP and ^{1}H–^{13}C couplings for the trimannoside ligand were measured and used as orientational constraints to develop a structural model for the MBP–trimannoside complex *(36)*. This application demonstrates the utility of RDCs to deduce a preferred orientation of the ligand from orientational constraints in an axially symmetric system such as MBP.

The 53-kDa MBP–trimmanoside complex is an example of a large protein–ligand complex. For large protein–ligand complexes, use of pulse sequences that use transverse relaxation optimized spectroscopy (TROSY) can be more advantageous than standard heteronuclear single quantum coherence (HSQC) experiments to improve resolution and accuracy of measurements. Because of the large molecular weight of MBP, a potential problem arises when using the standard scheme of measuring RDCs via ^{1}H–^{15}N splittings from the frequency domain of HSQC spectra. Severe line broadening of the up-field component of the splitting can occur, considerably decreasing the precision of the measured RDCs. To overcome this problem, a new experimental scheme known as coupling-enhanced TROSY (CE-TROSY) has been developed that retains the line-narrowing characteristic of TROSY in RDC measurements *(36)* (*see* **Note 6**). The CE-TROSY experiment is based on the well-known principles of accordion spectroscopy, wherein an extra J-coupling evolution period is incorporated in a regular TROSY experiment and controlled by a parameter κ that varies between 0 and 1. As a result, the J-coupling values are scaled depending on value of κ, resulting in displacement of the TROSY peak by an amount proportional to the J-coupling. The RDC value can then be extracted by measuring the difference in peak positions of the CE-TROSY peaks in a pair of spectra corresponding to the isotropic and aligned phase. Even though only the TROSY peak is observed in this experiment, mixing of TROSY and anti-TROSY peaks can lead to additional line broadening as the value of κ is increased. This can be minimized by choosing appropriate values of κ (*see* **Note 7**).

For RDC measurements on the ligand bound to MBP, the ^{1}H–^{13}C couplings can be measured from splittings in the ^{13}C or ^{1}H frequency dimensions of standard HSQC spectra. The CE-TROSY experiment is not required in this case, because the ligand is in excess of the protein (up to tenfold) and exhibits narrow line-width in the spectra. The excess is required to saturate the binding of ligand to the protein as a consequence of its weaker affinity (~1 mM). The narrow line-widths observed are therefore reflective mostly of the free ligand. The excess use of ligand provides another advantage in that ^{1}H–^{13}C spectra for the free and bound ligands can be acquired at ^{13}C natural abundance. This can open up new avenues for structure determination of several low affinity protein–ligand complexes without recourse to ^{13}C labeling of ligands, which can be fairly difficult as well as expensive.

The methods for RDC measurement on both MBP and its trimannoside ligand are described next:

1. The protein–ligand samples can be prepared with the alignment media as discussed above (*see* **Note 8**). A small volume of the stock solution of trimannoside ligand is added to the protein:bicelle mixture such that the ligand is in fivefold to tenfold molar excess of the protein concentration (*see* **Note 9**). The mixing is done at 4°C because it makes the bicelle solution less viscous and easier to pipet.
2. Two sets of measurements can be carried out for MBP in bicelles, in isotropic phase at 25°C and aligned phase at 35°C. CE-TROSY experiments can be performed in each phase with distinct values of κ, as shown in **Fig. 2**. For example, an RDC data set can be collected for a value of κ = 0 and another where κ = 1.0. The difference in peak positions between the two data sets can then be used to calculate the value of J_{NH}. This can be carried out in both isotropic and aligned phase (*see* **Note 10**). The difference in J_{NH} in the two phases gives the ^{1}H–^{15}N RDC value scaled by a factor of κ.

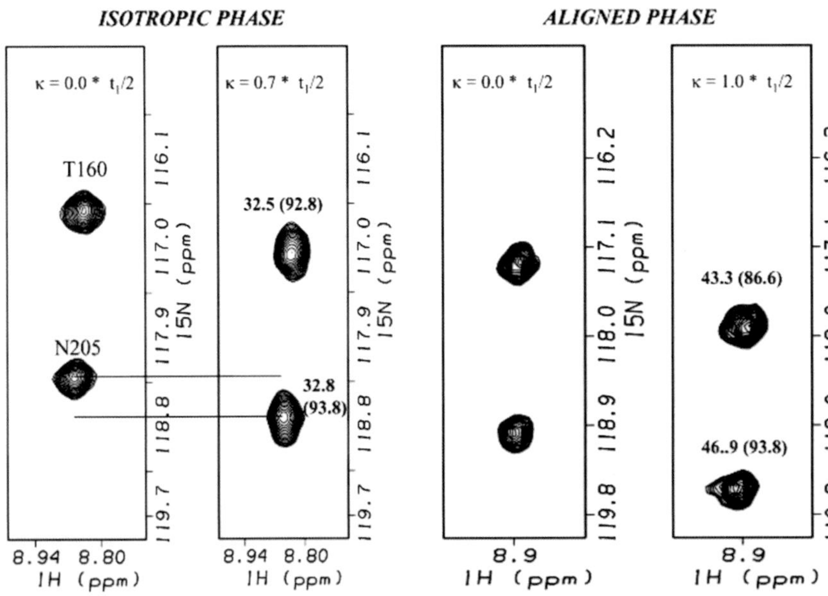

Fig. 2. RDC measurements from spectra recorded using the CE-TROSY sequence at 800 MHz on a 0.7 mM sample of ^{15}N,^{2}H (50% randomly deuterated)-labeled sample of the MBP trimer. A 6% (w/v) bicelle solution was used for collection of data in the aligned phase. J-coupling values are measured from the difference in peak positions in the pair of spectra for the isotropic and aligned phase, and are scaled depending on the value of κ. The scaled values are shown within parentheses for two representative peaks (reproduced from **ref. *(36)*** with permission from Elsevier Science).

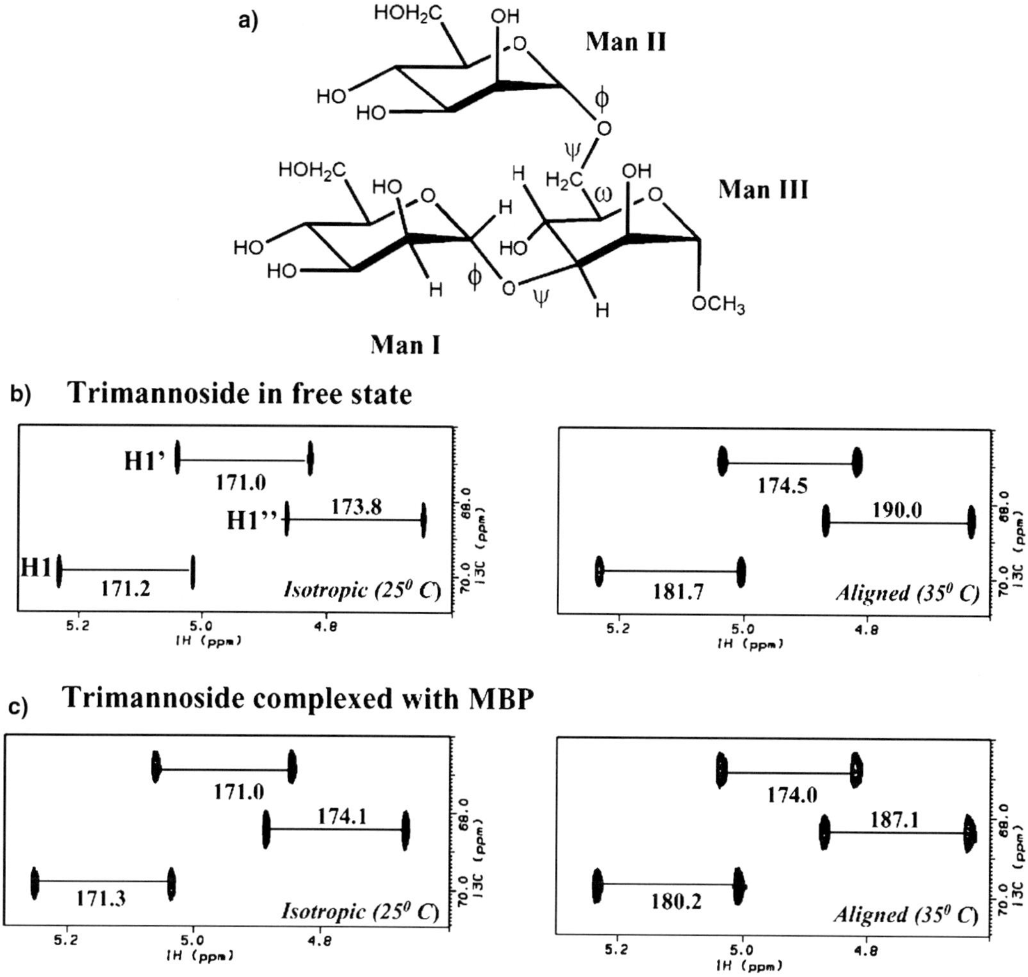

Fig. 3. (**a**) Trimannoside ligand used for binding studies with MBP. *Man I*, *Man II*, and *Man III* represent rings 1, 2, and 3 of trimannoside, respectively, whereas ϕ, ψ, and ω represent the glycosidic dihedral angles. (**b**) Section of a ^{1}H–^{13}C-coupled HSQC spectra of trimannoside showing ^{1}H–^{13}C couplings in the anomeric region of the three mannose units in the free state and (**c**) in complex with trimeric MBP. Spectra were recorded at ^{13}C natural abundance for the two samples in isotropic phase and in aligned phase (6% [w/v] bicelle solution). RDCs were calculated as the difference between the coupling values in isotropic and aligned phase shown on the spectra (reproduced from **ref. *(36)*** with permission from Elsevier Science).

3. ^{1}H–^{13}C couplings can be measured from splittings in a ^{1}H-detected ^{13}C HSQC spectrum with no decoupling in direct or indirect dimensions, as shown in **Fig. 3**. Measurements are carried out in isotropic and aligned phase with a similar complex sample used for ^{1}H–^{15}N RDC measurements of MBP (*see* **Note 11**). Because the ligand is in excess, the measured RDC values are an average between the free and bound states of the ligand. To separate the contribution from the free ligand, the measurements are repeated on a ligand sample without MBP.

4. The bound state RDCs for the ligand can be calculated from:

$$D_b = (D_{obs} - x_f D_f) / x_b, \tag{2}$$

where D_b is the RDC for the ligand in bound state, D_{obs} is the measured RDC, D_f is the RDC for the ligand in free state, x_f is the fraction of ligand in free state, and x_b is the fraction of ligand in bound state. The free and bound fractions of the ligand can be calculated from known binding constant of the protein–ligand complex (0.94 mM) (*see* **Note 12**). The bound state couplings are then used to determine the order tensor and hence the orientation of ligand relative to the protein.

3.3. Determination of Order Tensors for Protein and Ligand in Complex

To deduce the relative orientation of the protein and ligand in the complex, RDC measurements need to be analyzed to surmise the direction of ordering of the molecules in terms of their order tensors. This can be accomplished by calculating the magnitude and direction of the principal axes of the ordering tensor, namely S_{zz}, S_{yy}, and S_{xx}. There are several useful software packages available for RDC data analysis (most can be downloaded for free) that allow calculation of these order tensors *(37–40)*. A software package that we find very user friendly and versatile for analysis is a program called REDCAT, developed in James Prestegard's group *(40)*. In addition to basic utilities to solve for order tensors and back-calculate couplings from a given order tensor and proposed structure, it includes error analysis for identification of problematic measurements and simulation of effects of dynamic averaging process. Inherent to the software is a nice graphical user interface that allows the user to input RDC data for most biomolecules (proteins, nucleic acids, carbohydrates) along with their three-dimensional structures. Typical input into REDCAT consists of a table of the *x,y,z* PDB or Cartesian coordinates for individual N–H or C–H bond vectors (or any other combination of single or multiple bond vectors) for which the RDCs have been measured, along with corresponding RDC values and measurement errors. REDCAT uses singular value decomposition (SVD) and Monte Carlo sampling as core methods for solving a system of linear equations to determine the principal axes values of S_{zz}, S_{yy}, and S_{xx} as well as the Euler angles that allow transformation of a structure into its principle alignment frame. The order parameters along with their directions of ordering are calculated from this input data and output into a file. They are also displayed as a Sauson–Flamsteed projection map that basically relates the direction of S_{zz}, S_{yy}, and S_{xx} axes with respect to the *x,y,z* axes of the molecular coordinate frame (*see* **Note 13**). The Sauson–Flamsteed projection maps for MBP and trimannoside ligand are shown in **Fig. 4**. As can be seen from the figure, the order tensor of the ligand reflects the trimeric axial symmetry of MBP, indicating its binding to MBP in a symmetric fashion. The order tensor for MBP also displays a similar axial symmetry and the direction of

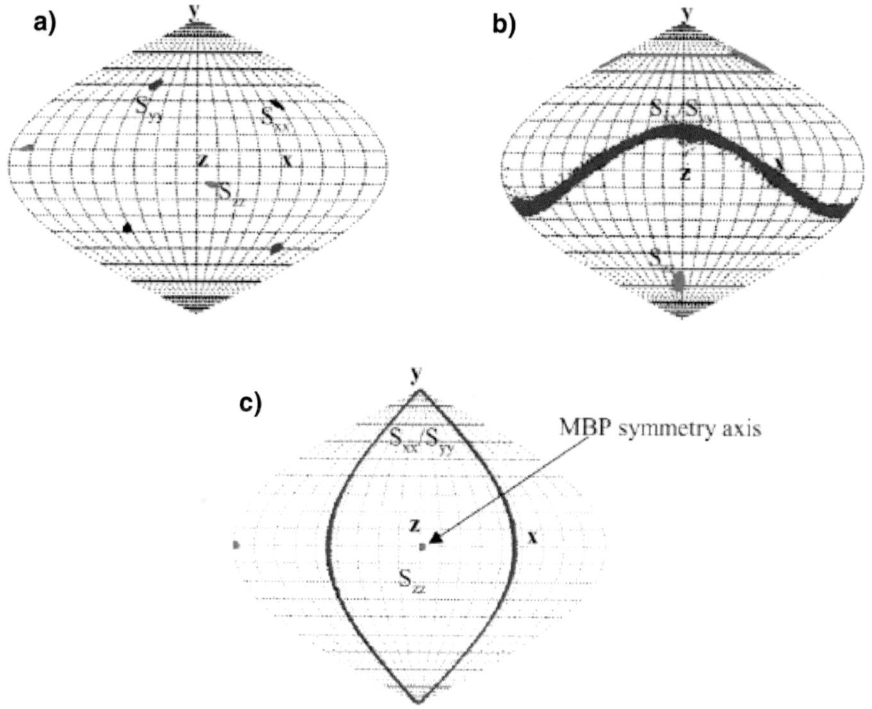

Fig. 4. Sauson–Flaumsteed projection of the directions of highest order for oriented trimannoside (**a**) in the free state and (**b**) bound to MBP. The continuous band in the latter plot reflects the axial symmetry of MBP; *x*, *y*, and *z* represent the molecular coordinate frame of the trimannoside model in each case. (**c**) Sauson–Flaumsteed projection of the directions of ordering (S_{zz}, S_{yy}, and S_{xx}) for oriented MBP. The direction of highest order S_{zz} is coincident with the rotational symmetry axis as viewed with respect to the molecular frame of trimeric MBP (shown as *x*, *y*, *z* in the Figure) (reproduced from ref. ***(36)*** with permission from Elsevier Science).

highest order is seen to point along the direction of central axis of symmetry. This implies that for homomultimeric and other symmetric systems with point group symmetry, the direction of highest order is likely to be along the highest symmetry axis and thus determines the direction of ordering for the protein as well as the bound ligand. If a three-dimensional structure of the protein is available, the direction of ordering of the protein along its axis of symmetry is known in advance (*see* **Note 14**). This feature can then be exploited to determine the bound geometry of the ligand and its relative orientation to the protein, by just RDC measurements on the small ligand and without any RDC measurements on the protein itself, as demonstrated in the structure determination of the MBP–trimannoside complex.

3.4. Structure Determination of Protein–Ligand Complex

A number of software applications are available that allow assembly of biomolecular complexes by constraining protein and ligand relative to each other with the help of orientational data *(25–27)*. For the assembly of MBP–trimannoside complex, the program

AUTODOCK was used, because it is very efficient in defining ligand conformation and guiding ligand docking onto the protein target while modeling in flexibility for both protein and ligand *(41, 42)*. Stepwise instructions for setting up the docking run are outlined in great detail in the AUTODOCK manual and will not be repeated here. The docking of trimannoside was performed by keeping the geometry of trimannoside constant within ±20° of its torsional angle values (*see* **Note 15**). A search was performed for the minimum energy configuration of the ligand without constraints on relative orientation of the protein and ligand. The AUTODOCK search produced a cluster of structures with the lowest energy that matched the orientation predicted by the RDC-derived axially symmetric order tensor for bound trimannoside, validating the usefulness of this approach in determining protein–ligand complex structures (*see* **Note 16**). The structure of MBP–trimannoside complex from the RDC approach compared very well (RMSD < 3 Å) with that from an intermolecular NOE and molecular modeling approach *(43)*. The utility of RDC approach is apparent when one considers that most large proteins are multimeric in nature, with quite a few that are homomultimeric and possess inherent symmetry. RDC data in combination with symmetry properties of the protein could then avoid need for extensive protein assignments or collection of NMR data on the protein entirely, providing an efficient and rapid way of determining structures of large protein complexes.

3.5. Structure Determination of Protein Complex with Multiple Ligands

In some cases, there may be multiple binding interactions of the protein molecule with ligand. Two or more ligands may be able to bind to the protein, at similar or different binding sites. In this situation, all ligands bound to the protein experience a common dominant ordering force caused by the orientation of the much larger protein. This is a useful property because the magnitude and direction of the ordering tensor for each ligand is going to be the same. The order tensors along with the associated molecular frames for the bound ligands can then be rotated to match each other to obtain a relative binding geometry for the ligands, without any measurements on the protein. If the ligand binding sites happen to be in close proximity on the protein surface, then a structural linker can be designed to bridge the two ligands such that a new molecule is generated with higher affinity to the protein than either ligand alone. This corresponds to structure-based drug design, in an approach similar to structure-activity relationship (SAR) by NMR *(44)*. An application of this design strategy is illustrated in the development of drugs against the cancer target, *N*-acetyl-glucosaminosyltransferase V (GnTV). GnTV is a 98-kDa glycoprotein involved in the biosynthesis of branched *N*-linked oligosaccharides *(45)*. Regulation of branching affects cell invasiveness and metastatic potential, making GnTV a target for rational drug design. Binding of GnTV to its two ligands, UDP-GlcNAc (K_d = 1 mM) and a synthetic trisaccharide acceptor [βGlcNAc*(1,2)*αMan*(1,6)*βMan(OR)] (K_d = 70 μM) is focus of NMR investigation *(46)*. It is thought that both of these ligands bind to the same site on GnTV in close proximity for transfer of GlcNAc group from UDP-GlcNAc to the acceptor. ^{1}H–^{13}C RDC measurements on both GnTV-bound ligands at ^{13}C natural abundance have been performed and can be used to obtain the relative orientation of the two ligands on the surface of GnTV from their RDC-derived order tensors, as shown in **Figs. 5 and 6** (*see* **Note 17**). The preparation of protein–ligand sample follow the same protocol as described for the MBP–trimannoside complex, except that now there are two ligands in the sample instead of one, and phage is used as an alignment medium instead of bicelles. The RDCs, however, do not constrain the ligands translationally relative to the protein. The exact placement of the ligands in the binding pocket can be discerned with the help of

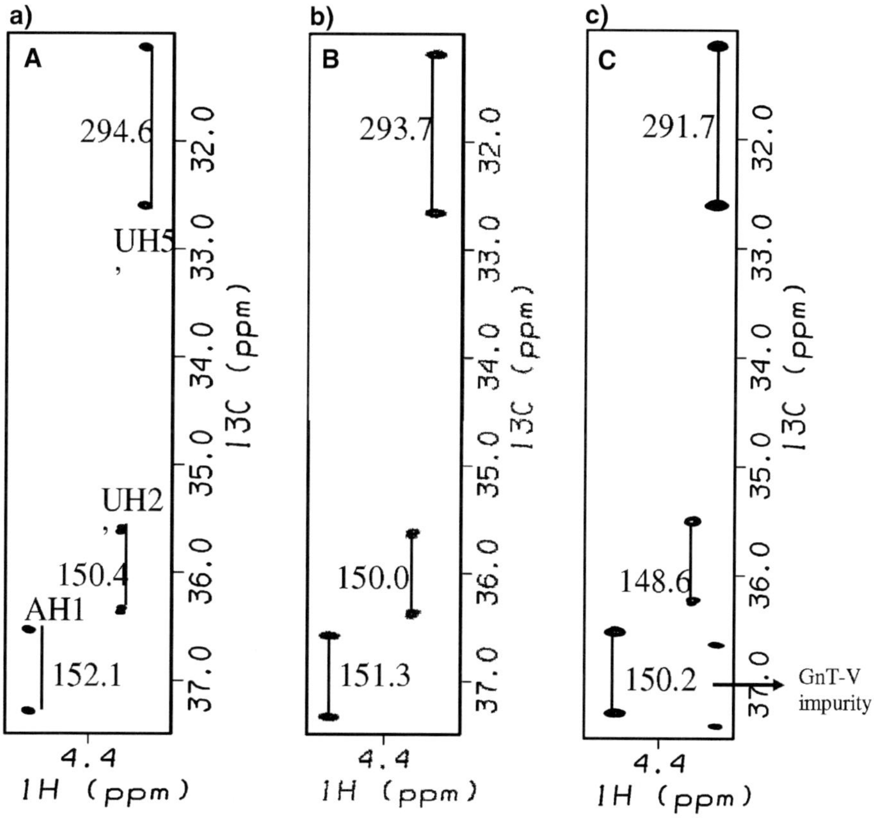

Fig. 5. Sections of ^{1}H–^{13}C-coupled HSQC spectra showing ^{1}H–^{13}C couplings of (**a**) UDP (6 mM) and acceptor (6 mM) in solution, (**b**) UDP (6 mM) and acceptor (4 mM) in 10 mg/mL phage, and (**c**) UDP (6 mM) and acceptor (4 mM) bound to GnTV (0.4 mM) in 10 mg/mL phage. All spectra were acquired at ^{13}C natural abundance. RDCs were calculated as the difference between the coupling values in isotropic and aligned phase shown on the spectra.

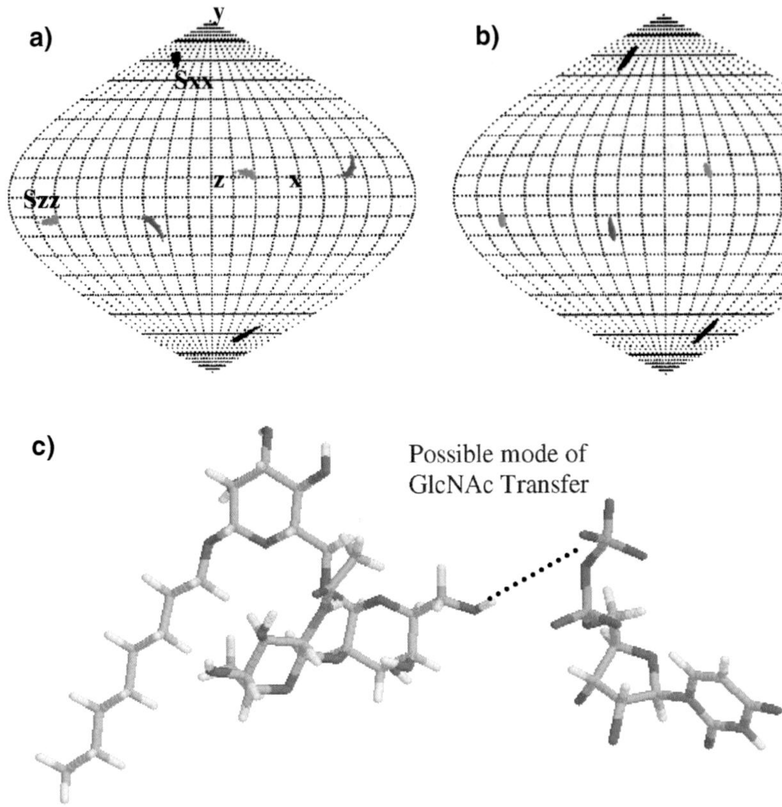

Fig. 6. Order tensor determination of bound state ligands of GnTV. (**a**) Order tensor for acceptor and (**b**) order tensor for UDP. (**c**) Relative orientation for the two ligands determined from the order tensors, showing the possible mode of GlcNAc transfer activity of GnTV.

a few translational constraints such as intermolecular NOEs or paramagnetic relaxation enhancement (PRE) constraints via spin-labeling one of the ligands *(47)* (*see* **Note 18**). Restraining the ligands translationally to side-chain or main-chain atoms on the protein then gives an idea of the binding surface topology of the protein (*see* **Note 19**). This complementary methodology incorporating RDC orientational data and NOE or PRE distance data is very useful for drug design and precludes the need for solving the structure for the entire protein–ligand complex.

4. Notes

1. We have noticed that the temperature of transition for bicelles from an isotropic to an aligned phase can be altered by changing the molar ratio of DMPC:DHPC. For example, if the ratio is

increased to 3.2:1, the temperature of transition drops down to 32°C. Also, the transition temperature can be altered by addition of salt. An additional 50 mM salt typically drops the transition temperature by approximately 2–3°C, depending on the buffer used. This may be very useful for study of samples that are more stable at lower temperatures.

2. The homogeneity of the bicelle preparation can be deduced from lineshape analysis of the quadrupolar splitting of D_2O. If both lines of the observed doublet are of equal intensity and similar line widths, then it indicates that the bicelles are homogenous. It is important to have homogeneity in the bicelle preparation for maintaining long-term stability of bicelle alignment during the NMR experiment and also to get the best resolution for resonances in the NMR spectra of protein–ligand samples.
3. Presence of charged solutes can sometimes be a source of instability for bicelles because of nonspecific interactions. In such cases, bicelle stability can be improved by doping with charged amphiphiles such as cetyltrimethylammonium bromide (CTAB) in a molar ratio of 3:0.1 of DMPC:CTAB, which prevents phase separation *(48)*. For measurement of RDCs at more acidic or basic pH values, a lipid mixture of didodecyl-phosphatidylcholine and dihexyl-phosphatidylcholine in a molar ratio of 3:1 can be prepared similar to the DMPC: DHPC bicelles *(49)*. Measurements can be carried out in a pH range of 2.5–10.5 for these bicelles.
4. Pf1 phage is also now available commercially from companies such as Asla Biotech.
5. Pf1 aligns over a wide range of temperatures (5–70°C) and under a variety of sample conditions. It is one of the most preferred aligning media because it is highly stable. However, because Pf1 is negatively charged, it is unsuitable for alignment of positively charged solutes, especially those with Ca^{+2}, Mg^{+2}-containing buffers and proteins.
6. A similar TROSY-based scheme for measurement of RDCs, incorporated as part of a three-dimensional (3D) HNCO experiment, has been used by Kay and co-workers, which differs slightly from our sequence *(35)*. Their experimental scheme is particularly advantageous for very large proteins, where considerable resonance overlap is observed and hence going to a 3D scheme versus a two-dimensional (2D) scheme alleviates some of the resolution problems. The flip side of using a 3D versus a 2D scheme is that triply labeled (^{15}N, ^{13}C, ^{2}H) samples are required, which are difficult and expensive to obtain. The 3D experiments also require longer experimental times for data acquisition, which may not be good for unstable samples.

7. The value of κ should be chosen carefully so as not to introduce too much line broadening in the CE-TROSY spectra. Usually, a value of $\kappa = 0.5$ works fine for most measurements without sacrificing precision because of loss in sensitivity or line broadening.
8. The volumes given here are in reference to sample use in a Shigemi tube, which requires a minimum of 300 μL for data acquisition.
9. The amount of molar excess is determined by the binding affinity of the protein–ligand complex and the magnitude of ordering for the complex. The contribution from bound-state orientation is dependent on the magnitude of ordering and should be maximized to measure accurately differences between free ligand RDCs and observed RDCs (averaged over free and bound). This is easily achieved in cases of complexes that orient strongly (bound state RDCs >20 Hz) and here molar excess of ligand can be used to the point of saturation. For weak-orienting complexes, the molar excess has to be reduced accordingly, so at least a difference of 1–2 Hz can be measured between free ligand RDCs and the averaged ligand RDCs. For tight-binding ligands ($K_d \sim 10^{-5}$ to 10^{-4} mM), a threefold to fivefold molar excess of ligand saturates the protein binding site and gives measurable RDCs. For weaker binding ligands ($K_d > 10^{-4}$ mM), a fivefold to tenfold molar excess is suitable for measurements.
10. The free and bound RDCs for MBP need not be determined like the ligand in this case, because it is presumed that if the protein–ligand complex aligns via weak steric interactions, the alignment is dominated by the orientation of the much larger protein and it should not change upon binding of the ligand. Thus, separate measurements for the free and bound protein are not necessary. Either of the measurements will suffice, because they would provide the same order tensor within experimental error.
11. If possible, it is a good idea to do both the protein and ligand RDC measurements on the same protein–ligand complex sample to minimize errors resulting from varying protein:ligand ratios, buffer conditions, quality of samples, etc.
12. In some cases, the binding affinity of the protein–ligand complex may not be known accurately and an estimate has to be used. The extraction of bound-state RDCs from **Eq. 2** in this case will not be as precise. However, if the magnitude of bound state RDCs measured is >10 Hz and a large number of measurements are available, an error of up to ±3 Hz can be tolerated without significantly affecting the results from the order tensor calculation in terms of overall magnitude and direction. Similarly, large errors in a few individual

measurements are also tolerated (possible with resonance overlap situations), because this process averages out errors over the various RDCs during calculation. The precision of the RDC measurements thus need not be very high for structure determination of the complex.

13. The *x,y,z* axes of Sauson–Flamsteed projection maps are not the same as *x,y,z* coordinate axes of the PDB structure file used for display. To make them coincident, the molecule can be rotated and new PDB coordinates written out iteratively, until the red, blue, and black spots coincide with the *x,y,z* axes direction in the projection map.
14. This is true for most homomultimeric systems, with point group symmetry. For other multimeric systems, this approach is unambiguous only if they also have point group symmetry. A more detailed analysis would be needed to separate out the different symmetry contributions in such cases.
15. Small leeway in torsion angles is given to make the ligand a bit flexible for docking. Use of torsional angle errors >20° creates too many solutions and increases the docking time dramatically, because the program samples configurations at five intervals.
16. The clustering feature is a very valuable feature of AUTODOCK, because it allows a look at different binding modes of the ligand, including analysis of population in the various modes. This can be very helpful in structure-based drug design, where other weak affinity modes and sites for the ligand may be detected on the protein. Alternatively, a program such as HADDOCK can now be used for incorporating orientational data directly into the structure calculations and clustering results.
17. When the difference in measured RDCs for free and averaged RDCs is small, such as in this case, the data has to be used with caution to determine corresponding order tensors. Although we were able to determine order tensors for the ligands bound to GnTV and obtain a relative orientation on the protein surface, the structural characterization is very preliminary. Obviously, better quality measurements are needed to have a more accurate representation of the binding geometry. Our main aim here is to demonstrate the quality of data that can be obtained for ligands bound to large proteins even at natural ^{13}C abundance because of high sensitivity and narrow line widths observed for various resonances in the spectra. With considerably stronger ordering of the protein, the differences between free and averaged RDCs can be increased significantly without affecting the quality of the spectra too much.

18. Recently, nickel chelate-carrying lipid tags have been developed that can be attached to His-tagged proteins to anchor them into lipid bilayers of bicelles *(50, 51)*. As a result, an increase in the weighting of bound-state contribution in averaged ligand RDCs is observed, allowing use of greater free-to-bound ligand ratios, leading to better sensitivity and precision of RDC measurements at ^{13}C natural abundance. This could potentially open up applications to GnTV and increasingly larger protein–ligand complexes.
19. A study incorporating intermolecular NOEs and PREs from spin-labeled ligand for structural characterization of substrates bound to GnTV has appeared recently *(52)*. In this study, transferred nuclear Overhauser effect (trNOE) and saturation transfer difference (STD) experiments, were used to characterize the ligand conformation and ligand–protein contact surfaces. In addition, a spin-labeled ligand analog, 5′-diphospho-4-O-2,2,6,6-tetramethylpiperidine 1-oxyl (UDP-TEMPO), was used to characterize the relative orientation of the two bound ligands. Results from this study are comparable to the orientation determined independently by RDC measurements.

Acknowledgements

The author thanks Dr. James Prestegard and Dr. Michael Pierce (University of Georgia) for access to samples and data on GnTV-substrate interactions. Portions of this work were supported by National Institutes of Health (NIH) grants GM03325 and RR005351.

References

1. Clore, G. M., and Gronenborn, A. M. (1998). NMR structure determination of proteins and protein complexes larger than 20 kDa. *Current Opinion in Chemical Biology* **2**, 564–70
2. Tugarinov, V., Hwang, P. M., and Kay, L. E. (2004). Nuclear magnetic resonance spectroscopy of high-molecular-weight proteins. *Annual Review of Biochemistry* **73**, 107–46
3. Tzakos, A. G., Grace, C. R. R., Lukavsky, P. J., and Riek, R. (2006). NMR techniques for very large proteins and RNAs in solution. *Annual Review of Biophysics and Biomolecular Structure* **35**, 319–42
4. Bonvin, A. M. J. J., Boelens, R., and Kaptein, R. (2005). NMR analysis of protein interactions. *Current Opinion in Chemical Biology* **9**, 501–08
5. Riek, R., Pervushin, K., and Wuthrich, K. (2000). TROSY and CRINEPT: NMR with large molecular and supramolecular structures in solution. *Trends in Biochemical Sciences* **25**, 462–68
6. van Dijk, A. D. J., de Vries, S. J., Dominguez, C., Chen, H., Zhou, H. X., and Bonvin, A. M. J. J. (2005). Data-driven docking: HADDOCK's adventures in CAPRI. *Proteins: Structure Function and Bioinformatics* **60**, 232–38
7. Mackereth, C. D., Simon, B., and Sattler, M. (2005). Extending the size of protein-RNA

complexes studied by nuclear magnetic resonance spectroscopy. *Chembiochem* **6**, 1578–84

8. Tang, C., Iwahara, J., and Clore, G. M. (2006). Visualization of transient encounter complexes in protein-protein association. *Nature* **444**, 383–86
9. Sprangers, R., Velyvis, A., and Kay, L. E. (2007). Solution NMR of supramolecular complexes: providing new insights into function. *Nature Methods* **4**, 697–703
10. Mayer, M., and Meyer, B. (1999). Characterization of ligand binding by saturation transfer difference NMR. Spectroscopy. *Angewandte Chemie-International Edition* **38**, 1784–88
11. Wyss, D. F., McCoy, M. A., and Senior, M. M. (2002). NMR-based approaches for lead discovery. *Current Opinion in Drug Discovery & Development* **5**, 630–47
12. Betz, M., Saxena, K., and Schwalbe, H. (2006). Biomolecular NMR: a chaperone to drug discovery. *Current Opinion in Chemical Biology* **10**, 219–25
13. Vajda, S., and Guarnieri, F. (2006). Characterization of protein–ligand interaction sites using experimental and computational methods. *Current Opinion in Drug Discovery & Development* **9**, 354–62
14. Pintacuda, G., John, M., Su, X. C., and Otting, G. (2007). NMR structure determination of protein–ligand complexes by lanthanide labeling. *Accounts of Chemical Research* **40**, 206–12
15. Zabell, A. P. R., and Post, C. B. (2002). Docking multiple conformations of a flexible ligand into a protein binding site using NMR restraints. *Proteins: Structure Function and Genetics* **46**, 295–307
16. Fischer, M. W. F., Losonczi, J. A., Weaver, J. L., and Prestegard, J. H. (1999). Domain orientation and dynamics in multidomain proteins from residual dipolar couplings. *Biochemistry* **38**, 9013–22
17. Hus, J. C., Marion, D., and Blackledge, M. (2000). De novo determination of protein structure by NMR using orientational and long-range order restraints. *Journal of Molecular Biology* **298**, 927–36
18. Jain, N. U., Wyckoff, T. J. O., Raetz, C. R. H., and Prestegard, J. H. (2004). Rapid analysis of large protein–protein complexes using NMR-derived orientational constraints: the 95 kDa complex of LpxA with acyl carrier protein. *Journal of Molecular Biology* **343**, 1379–89
19. Lipsitz, R. S., and Tjandra, N. (2004). Residual dipolar couplings in NMR structure analysis. *Annual Review of Biophysics and Biomolecular Structure* **33**, 387–413
20. Bax, A., and Grishaev, A. (2005). Weak alignment NMR: a hawk-eyed view of biomolecular structure. *Current Opinion in Structural Biology* **15**, 563–70
21. Getz, M., Sun, X. Y., Casiano-Negroni, A., Zhang, Q., and Al-Hashimi, H. M. (2007). NMR studies of RNA dynamics and structural plasticity using NMR residual dipolar couplings. *Biopolymers* **86**, 384–402
22. Prestegard, J. H., Al-Hashimi, H. M., and Tolman, J. R. (2000). NMR structures of biomolecules using field oriented media and residual dipolar couplings. *Quarterly Reviews of Biophysics* **33**, 371–424
23. Clore, G. M. (2000). Accurate and rapid docking of protein–protein complexes on the basis of intermolecular nuclear Overhauser enhancement data and dipolar couplings by rigid body minimization. *Proceedings of the National Academy of Sciences of the United States of America* **97**, 9021–25
24. McCoy, M. A., and Wyss, D. F. (2002). Structures of protein–protein complexes are docked using only NMR restraints from residual dipolar coupling and chemical shift perturbations. *Journal of the American Chemical Society* **124**, 2104–5
25. Dominguez, C., Boelens, R., and Bonvin, A. M. J. J. (2003). HADDOCK: a protein–protein docking approach based on biochemical or biophysical information. *Journal of the American Chemical Society* **125**, 1731–37
26. Schwieters, C. D., Kuszewski, J. J., Tjandra, N., and Clore, G. M. (2003). The Xplor-NIH NMR molecular structure determination package. *Journal of Magnetic Resonance* **160**, 65–73
27. Fahmy, A., and Wagner, G. (2002). TreeDock: a tool for protein docking based on minimizing van der Waals energies. *Journal of the American Chemical Society* **124**, 1241–50
28. Tjandra, N., and Bax, A. (1997). Direct measurement of distances and angles in biomolecules by NMR in a dilute liquid crystalline medium. *Science* **278**, 1111–14
29. Hansen, M. R., Hanson, P., and Pardi, A. (2000). Filamentous bacteriophage for aligning RNA, DNA, and proteins for measurement of nuclear magnetic resonance dipolar coupling interactions. *RNA–Ligand Interactions Pt A* **317**, 220–40
30. Fleming, K., Gray, D., Prasannan, S., and Matthews, S. (2000). Cellulose crystallites: a new and robust liquid crystalline medium for the measurement of residual dipolar couplings. *Journal of the American Chemical Society* **122**, 5224–25

31. Sass, H. J., Musco, G., Stahl, S. J., Wingfield, P. T., and Grzesiek, S. (2000). Solution NMR of proteins within polyacrylamide gels: diffusional properties and residual alignment by mechanical stress or embedding of oriented purple membranes. *Journal of Biomolecular NMR* **18**, 303–09

32. Ruckert, M., and Otting, G. (2000). Alignment of biological macromolecules in novel nonionic liquid crystalline media for NMR experiments. *Journal of the American Chemical Society* **122**, 7793–97

33. Prestegard, J. H., Bougault, C. M., and Kishore, A. I. (2004). Residual dipolar couplings in structure determination of biomolecules. *Chemical Reviews* 104, 3519–40

34. Prestegard, J. H., and Kishore, A. I. (2001). Partial alignment of biomolecules: an aid to NMR characterization. *Current Opinion in Chemical Biology* **5**, 584–90

35. Yang, D. W., Venters, R. A., Mueller, G. A., Choy, W. Y., and Kay, L. E. (1999). TROSY-based HNCO pulse sequences for the measurement of (HN)-H-1-N-15, N-15-(CO)-C-13,(HN)-H-1-(CO)-C-13,(CO)-C-13-C-13(alpha) and (HN)-H-1-C-13(alpha) dipolar couplings in N-15, C-13, H-2-labeled proteins. *Journal of Biomolecular NMR* **14**, 333–43

36. Jain, N. U., Noble, S., and Prestegard, J. H. (2003). Structural characterization of a mannose-binding protein-trimannoside complex using residual dipolar couplings. *Journal of Molecular Biology* 328, 451–62

37. Losonczi, J. A., Andrec, M., Fischer, M. W. F., and Prestegard, J. H. (1999). Order matrix analysis of residual dipolar couplings using singular value decomposition. *Journal of Magnetic Resonance* **138**, 334–42

38. Zweckstetter, M., and Bax, A. (2000). Prediction of sterically induced alignment in a dilute liquid crystalline phase: aid to protein structure determination by NMR. *Journal of the American Chemical Society* **122**, 3791–92

39. Dosset, P., Hus, J. C., Marion, D., and Blackledge, M. (2001). A novel interactive tool for rigid-body modeling of multi-domain macromolecules using residual dipolar couplings. *Journal of Biomolecular NMR* 20, 223–31

40. Valafar, H., and Prestegard, J. H. (2004). REDCAT: a residual dipolar coupling analysis tool. *Journal of Magnetic Resonance* **167**, 228–41

41. Goodsell, D. S., Morris, G. M., and Olson, A. J. (1996). Automated docking of flexible ligands: applications of AutoDock. *Journal of Molecular Recognition* **9**, 1–5

42. Morris, G. M., Goodsell, D. S., Halliday, R. S., Huey, R., Hart, W. E., Belew, R. K., and Olson, A. J. (1998). Automated docking using a Lamarckian genetic algorithm and an empirical binding free energy function. *Journal of Computational Chemistry* **19**, 1639–62

43. Sayers, E. W., and Prestegard, J. H. (2002). Conformation of a trimannoside bound to mannose-binding protein by nuclear magnetic resonance and molecular dynamics simulations. *Biophysical Journal* **82**, 2683–99

44. Shuker, S. B., Hajduk, P. J., Meadows, R. P., and Fesik, S. W. (1996). Discovering high-affinity ligands for proteins: SAR by NMR. *Science* **274**, 1531–34

45. Kaneko, M., Alvarez-Manilla, G., Kamar, M., Lee, I., Lee, J. K., Troupe, K., Zhang, W. J., Osawa, M., and Pierce, M. (2003). A novel beta(1,6)-N-acetylglucosaminyltransferase V (GnT-VB). *FEBS Letters* **554**, 515–19

46. Pierce, M., Arango, J., Tahir, S. H., and Hindsgaul, O. (1987). Activity of Udp-Glcnac – alpha-mannoside beta-(1,6)N-acetylglucosaminyltransferase (Gnt V) in cultured-cells using a synthetic trisaccharide acceptor. *Biochemical and Biophysical Research Communications* **146**, 679–84

47. Jain, N. U., Venot, A., Umemoto, K., Leffler, H., and Prestegard, J. H. (2001). Distance mapping of protein-binding sites using spin-labeled oligosaccharide ligands. *Protein Science* **10**, 2393–400

48. Losonczi, J. A., and Prestegard, J. H. (1998). Improved dilute bicelle solutions for high-resolution NMR of biological macromolecules. *Journal of Biomolecular NMR* **12**, 447–51

49. Ottiger, M., and Bax, A. (1999). Bicelle-based liquid crystals for NMR-measurement of dipolar couplings at acidic and basic pH values. *Journal of Biomolecular NMR* **13**, 187–91

50. Seidel, R. D., Zhuang, T. D., and Prestegard, J. H. (2007). Bound-state residual dipolar couplings for rapidly exchanging ligands of His-tagged proteins. *Journal of the American Chemical Society* **129**, 4834–39

51. Zhuang, T. D., Leffler, H., and Prestegard, J. H. (2006). Enhancement of bound-state residual dipolar couplings: conformational analysis of lactose bound to Galectin-3. *Protein Science* **15**, 1780–90

52. Macnaughtan, M. A., Kamar, M., Alvarez-Manilla, G., Venot, A., Glushka, J., Pierce, J. M., and Prestegard, J. H. (2007). NMR structural characterization of substrates bound to *N*-acetylglucosaminyltransferase V. *Journal of Molecular Biology.* **366**, 1266–81

Chapter 16

Raman-Assisted X-Ray Crystallography for the Analysis of Biomolecules

Dominique Bourgeois, Gergely Katona, Eve de Rosny, and Philippe Carpentier

Summary

In this chapter, we describe Raman microspectrophotometry applied to crystals of biomolecules. Raman spectra collected in crystallo provide structural information highly complementary to X-ray diffraction, relate the crystalline state to the solution state, and allow the identification of ligand-bound or intermediate states of macromolecules. Nonresonant Raman spectroscopy is particularly suitable to the study of macromolecular crystals, and therefore applies to a wide range of noncolored crystalline proteins. Practical issues related to the investigation of crystals by Raman microspectrophotometry are reviewed, and the current limitations are highlighted.

Key words: In crystallo Raman spectroscopy, Macromolecules, Microspectrophotometers, Crystallography, Complementary methods, Nonresonant Raman spectroscopy

1. Introduction

Applying complementary techniques that probe different properties of the same biomolecule helps understanding the relationship between structure, dynamics, and function. Optical spectroscopy, encompassing the ultraviolet (UV) to infrared (IR) range, constitutes a particularly powerful tool to be used jointly with X-ray crystallography, the central technique to decipher the three-dimensional (3D) structure of biological macromolecules. Microspectrophotometers have been developed to investigate protein crystals in experimental conditions identical to those typically

James Weifu Lee and Robert S. Foote (eds.), *Micro and Nano Technologies in Bioanalysis*,
Methods in Molecular Biology, vol. 544
DOI 10.1007/978-1-59745-483-4_16, © Humana Press, a part of Springer Science+Business Media, LLC 2009

used on synchrotron beamlines dedicated to macromolecular crystallography *(1–6)*. Many studies with functional or mechanistic perspectives have benefited from complementing X-ray diffraction data with in crystallo UV-visible absorption *(7–12)* or fluorescence spectroscopy *(5, 13, 14)*. Electronic spectroscopy, however, is usually restricted to colored samples, typically metalloproteins or photoreceptors. Vibrational spectroscopy, on the other hand, applies to a much broader range of biological molecules. Infrared absorption spectroscopy *(15)* and Raman spectroscopy *(14, 16–18)* have been applied to macromolecular crystals for a long time. However, the practical difficulty in implementing IR microspectrophotometers (in particular due to the extreme sensitivity to moisture around the sample), the poor sensitivity of Raman scattering, and the cost and lack of portability of spectrometers have hampered specific developments targeting crystalline biomolecules.

Nonetheless, crystals of macromolecules constitute excellent samples for Raman spectroscopy, because of their very high concentration in biological material associated with a relatively low solvent content *(17)* (this is as opposed to UV–visible absorption and fluorescence spectroscopy, in which the large concentration of chromophores in crystals often results in poor or distorted signals). In parallel, recent progress in lasers, optics, and detectors has opened the door to the rapid recording of Raman spectra of outstanding quality. Whereas studies with dilute solutions are generally based on resonantly enhanced Raman scattering (by exciting at a wavelength close to an electronic absorption maximum), these improvements, when applied to crystalline samples, offer the possibility to use the nonresonant Raman mode, which is a priori applicable to any macromolecule. In turn, this mode simplifies the experimental setup, because only one wavelength is needed for most samples (typically in the near infrared). It also minimizes potential photodamage, spurious actinic effects, and contamination by fluorescence. On the other hand, nonresonant Raman spectra may be very complex to analyze, and comparisons with resonant spectra of solution samples may prove difficult in some cases. Therefore, in crystallo resonance Raman spectroscopy remains beneficial for a number of projects, e.g., the investigation of crystalline heme proteins.

Applications of in crystallo Raman spectroscopy include:

(i) The comparison between the solution state and the crystalline state, as a way to assess the biological relevance of X-ray structures.

(ii) The identification of ligands or intermediate states in a protein crystal. In particular, difference spectroscopy can be applied to identify bands originating from a bound ligand or from induced conformational changes that result from binding. Such an experiment can be conducted before X-ray data collection to efficiently screen ligands or to establish a suitable protocol for the accumulation and trapping of transient states in kinetic crystallography. Alternatively, it can be conducted after X-ray data collection to identify unexpected ligands that sometimes happen to bind to the protein or to check for the presence of chemical bonds (e.g., coordination bonds) or radical states that are unclear from electron density maps. In this way, Raman data can be used to impose restraints during crystallographic model refinement.

(iii) The monitoring of chemical changes affecting specific bonds, which are induced by radiation or by chemical treatment; of special interest is the online monitoring by Raman spectroscopy of X-ray induced radiation damage such as the breaking of disulfide bridges or the reduction of metal centers.

(iv) The local enhancement of the accuracy of the atomic model, because changes in the frequency of Raman bands can be linked to changes in bond length with subatomic resolution (up to 0.001 Å).

(v) The monitoring of slow kinetics in crystals, e.g., during a ligand-soaking procedure.

(vi) The possibility to use the X-ray structure as a tool to assist assignments of Raman bands or quantify their wavenumbers. In this case, one may talk about "crystallography-assisted Raman spectroscopy."

At the European Synchrotron Radiation Facility (ESRF, Grenoble, France), we have set up a laboratory dedicated to in crystallo UV–visible spectroscopy called the "Cryobench" (http://www.esrf.eu/UsersAndScience/Experiments/MX/Cryobench/). The central device of the laboratory is a microspectrophotometer that allows analyzing nano-volumic samples (in the liquid or crystalline state), at room or cryo temperatures, in the same experimental conditions as those used on X-ray crystallography beamlines **(Fig. 1)**. This offline configuration (remote from X-ray instruments) can be complemented by online configurations (where the microspectrophotometer is directly inserted onto the X-ray instrument for quasi-simultaneous data collection by spectroscopy and crystallography (*see* **Note 1**)). It should be noted that progress in Raman instrumentation has recently allowed the development of a series of instruments that complement synchrotron-based X-ray techniques such as microdiffraction *(19)*, X-ray absorption spectroscopy *(20)*, or powder diffraction *(21)*.

At the Cryobench, the recent installation of a Raman spectrometer essentially working in nonresonant conditions with excitation at 785 nm has already proven to be extremely promising *(22)*, both offline *(23)* and online *(24)*. In this chapter, we describe the procedures that allow the collection of Raman spectra from

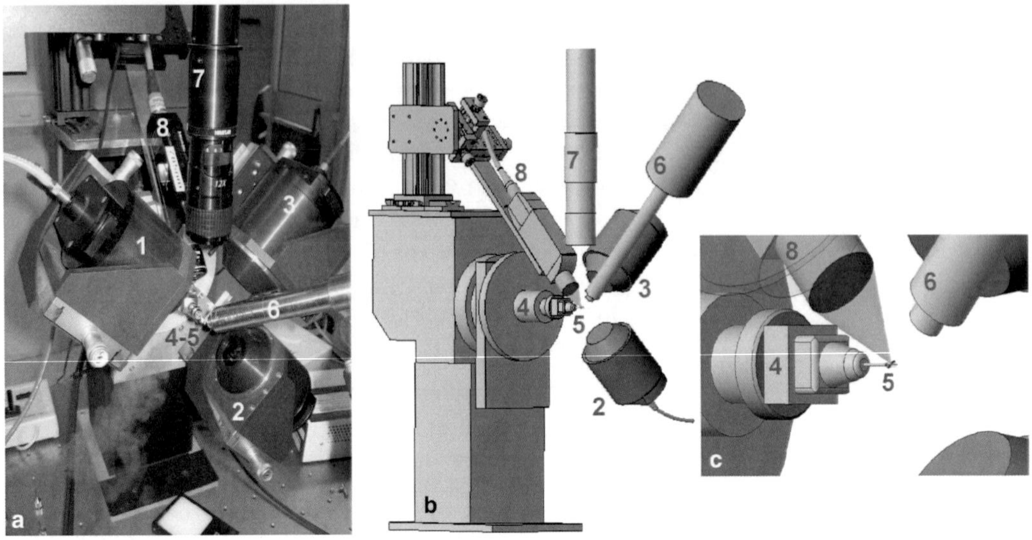

Fig. 1. Microspectrophotometer of the Cryobench laboratory (ESRF, Grenoble, France). (**a**) Picture of the instrument. (**b**) Schematic representation of the microspectrophotometer. Objective 1 has been removed for clarity. (**c**) Zoom at sample position showing details of the goniometer head, the sample, the collection optics, and the backscattering geometry of the Raman head. Parts *1*, *2*, *3*: objectives used for absorption/fluorescence spectroscopy; *4*: goniometer; *5*: biological sample; *6*: cryogenic cooling device; *7*: camera for sample alignment; and *8*: backscattering Raman probe.

protein crystals, emphasizing the critical steps and potential difficulties of the method. Our description is largely based on our experience with the non-heme iron enzyme superoxide reductase (SOR), for which the offline nonresonant Raman data were decisive to establish the build-up of iron peroxide species in the crystal (*23*). Although our experience with the resonant mode is very recent, some preliminary considerations on resonant in crystallo Raman spectroscopy are also outlined (*see* **Note 2**).

2. Materials

In this section, we list the properties of macromolecular crystals that strongly influence the quality of Raman spectra.

2.1. Crystal Size

In the nonresonant mode, large, bulky crystals are the best candidates, on the order of at least $100 \times 100 \times 100\ mm^3$ for a $20\times$ objective. Smaller, plate-like crystals are a priori more appropriate in the resonant mode.

2.2. Crystal Morphology

Crystal morphology will affect the quality of Raman data, for example because of Fresnel reflections and refraction effects. Plate-like or rectangular crystals are a priori more favorable than, e.g., bi-pyramidal crystals for which a relatively large flat face is missing.

2.3. Concentration of the Macromolecules in the Crystal

The very high concentration (typically several tens of millimolar) usually encountered in crystals represents a strong advantage for the nonresonant Raman mode, but it may be detrimental in the resonant mode because of a lack of penetration of the exciting light, reabsorption of Raman scattered photons, and potential photodamage.

2.4. Fraction of Solvent

Although the fraction of solvent in a crystal is usually moderate (20–80%) relative to what is found in solution, its contribution may dominate the spectrum, depending on its composition. More densely packed crystals with minimum solvent content tend to produce better nonresonance Raman spectra.

2.5. Solvent Composition and Cryoprotection

The composition of the mother liquor must be carefully checked, because some popular compounds may produce very intense Raman bands in the nonresonant mode (e.g., ammonium sulfate), or produce residual fluorescence. Likewise, Raman bands from cryoprotectants such as glycerol may severely interfere with signal from the macromolecule. A Raman spectrum from a flash-cooled film of the cryoprotected mother liquor solution should be collected for reference: dip a cryoloop into a microliter droplet of the mother liquor solution and mount on the Raman microspectrophotometer. If necessary, recording Raman data from noncryoprotected crystals should be considered, at least for offline experiments for which ice and crystal disorder is generally not a serious concern. *See* **Fig. 2** for the case of superoxide reductase.

2.6. Isotopic Replacement

Isotope-induced shifts of Raman bands can be measured in crystals in the same way as in solution samples, for example, by soaking the crystal in a solution containing an isotopically labeled ligand **(Fig. 2)**.

2.7. Crystal Holder

For low-temperature measurements, the use of loops provides a windowless sample environment, which is highly favorable to minimize background signals. Nylon loops are typically used, and care should be taken with loops made of strongly absorbing and possibly fluorescent material, e.g., loops made out of mylar (such as litholoops™), which may melt under the intense IR beam or may produce background signal under green or red beams. For ambient temperature measurements, crystals can be mounted between a pair of coverslips sealed with a grease gasket and mounted on a standard magnetic base. The coverslip facing the Raman probe should be in quartz, whereas the opposite plate can be a conventional glass coverslip typically used for crystallogenesis.

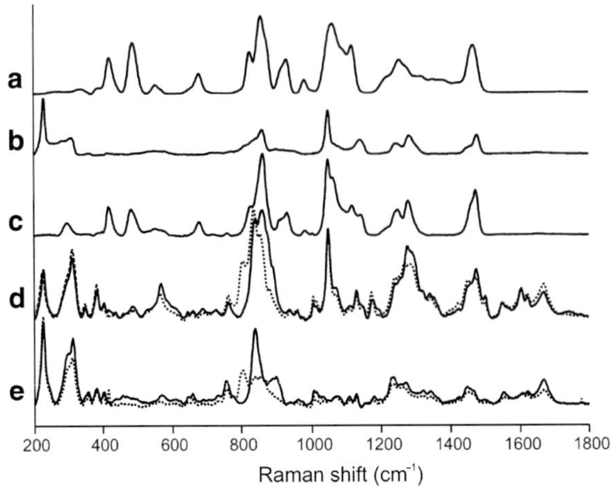

Fig. 2. Nonresonant Raman spectra of superoxide reductase crystals. Baseline-corrected spectra collected with the Synchroscan mode in the 200–1,800 cm^{-1} range using 785-nm excitation, with a 20× magnification objective. Samples (3–6 nL) were flash-frozen in nylon loops and the temperature was kept at 100 K throughout data collection. (**a**) Spectrum taken from SOR crystallization buffer (16% PEG 4000, 100 mM Tris/HNO_3 pH 9.0, 200 mM Ca[NO_3]$_2$), 300-s exposure; (**b**) spectrum from pure glycerol, 100-s exposure; (**c**) spectrum from cryoprotected crystallization buffer (30% glycerol added) with the addition 10 mM H_2O_2, to test the reaction between H_2O_2 and other components than SOR, 300-s exposure; (**d**) spectra from SOR crystals (~300 × 200 × 50 μm^3) treated in crystallization buffer with the addition of 10 mM H_2O_2 (*plain line*) and $H_2{}^{18}O_2$ (*dotted line*) for 3 min and subsequently flash-frozen in cryoprotected buffer. Spectra collected with 300-s exposure, with five and seven times averaging, respectively. (**e**) Spectra from SOR solutions treated with H_2O_2 (*plain line*) and $H_2{}^{18}O_2$ (*dotted line*). Reactions were triggered by rapid mixing of 12 mM (respectively 7 mM) SOR and 1 M H_2O_2 (respectively 100 mM $H_2{}^{18}O_2$) solutions in 2.7:0.2 (respectively 3:2) ratio, followed by freezing after a 1-min incubation.

Despite the high optical quality and weak scattering of quartz, this arrangement produces significantly more background than windowless loop-based mounting.

3. Methods

In this section, we describe our Raman microspectrophotometer and the methodology to successfully record Raman spectra from biological crystals.

3.1. Description of the Raman Microspectrophotometer

Our Raman microspectrophotometer is adapted for Raman studies with biological solutions or crystals, at temperatures ranging from cryogenic to ambient (**Fig. 1**). The device consists of an InVia Raman spectrometer (Renishaw, Gloucestershire, UK), and a series of excitation lasers optically coupled to dedicated Raman probes working in the backscattering mode. The Raman probes focus light onto the nano-volumic sample that is mounted on a motorized one-circle goniometer equipped with a nitrogen cryostream and a videocamera. Three wavelengths are currently available: 514, 633, and 785 nm, and the corresponding lasers deliver a maximum of ~60, ~18, and ~50 mW at the sample position, respectively. The 785-nm wavelength mainly targets nonresonant applications; the 633-nm wavelength can be used in nonresonant, preresonant, or resonant conditions; and the 514-nm wavelength is more appropriate for preresonant or resonant experiments, in particular for heme proteins. Higher power is generally needed to collect spectra with sufficient signal-to-noise ratio in the nonresonant mode relative to the resonant mode. Therefore, depending on the type of measurements, filters can be inserted into the optical path to attenuate the beam (from 5×10^{-6} % to 100% transmission, adjusted on a logarithmic scale). Beam attenuation should be tuned to achieve a compromise between quality of Raman spectra, exposure time, and sample damage. The excitation beam and Raman scattered light are transported between the laser/spectrometer and the sample through optical fibers, which can reach 100 m in length. Thus, experiments can be carried out on samples remote from the laser/spectrometer, which greatly facilitates the setup of online experiments.

The compactness of the Raman heads is important to allow their integration onto goniometers dedicated to biological crystallography. The backscattering excitation/collection geometry provides a self-alignment of the excited and the scattering volumes. The Rayleigh band is rejected by a dielectric filter so that accessible Raman shifts are comprised between 200 and 4,000 cm^{-1}, while anti-Stokes vibrations are not covered. Two objectives can be mounted on the Raman heads. A low-magnification objective (20× magnification; focal spot: $\sim 50 \times 50 \times 100$ μm^3; working distance: 21 mm; numerical aperture: 0.35) is adapted to large samples that exceed 1 nL in volume ($100 \times 100 \times 100$ μm^3). Rod-shaped crystals, plates, or small crystals of volumes down to 10 pL are better studied with a higher magnification objective (50× magnification; focal spot: $\sim 20 \times 20 \times 50$ μm^3; working distance: 8 mm; numerical aperture: 0.5), which nevertheless is more difficult to align.

The Raman scattered light enters the spectrometer through entrance slits, diffracts on a grating (1,200 lines/mm at 785 nm; 1,800 lines/mm at 633 and 514 nm), and reaches a Peltier-cooled CCD camera. Overall, the spectral resolution of the spectrometer is approximately 4 cm^{-1}, an acceptable value for studying biological samples. Switching from one wavelength to another is semiautomatic but requires care; whereas the exchange of grating is motorized, few lenses on kinematic mounts need to be manually installed, and the optical alignment and CCD setting need to be tuned. Such an adjustment is done by measuring the band position and intensity of a silicon calibration sample (see below).

3.2. Raman Data Acquisition Modes

Three different data-collection modes can be chosen:

1. Static acquisition: A fixed grating position is used so as to select a restricted spectral window centered on a defined Raman band. This mode is useful for adjustment purposes, or for rapid kinetic measurements.
2. "Step and stitch method": The grating is moved in discontinuous steps along the acquisition so as to cover a wide spectral range. This method allows collecting wide-range spectra, but induces potential problems at the junction between individual spectral windows.
3. "Synchroscan™": This method also allows collecting wide-range spectra but is based on a continuous rotation of the grating synchronized with CCD readout, avoiding jumps in the reconstituted spectra.

3.3. Successfully Collecting Raman Spectra from Macromolecular Crystals

3.3.1. Step 1: Wavelength Calibration of the Raman Spectrometer

Wavelength calibration needs to be done regularly (once a month) to check for drifts in the spectrometer optics. Drifts in wavelength calibration may appear upon temperature changes, so it is important to work in a temperature-controlled laboratory. For this reason, it is also much preferable to perform all measurements that need to be compared in terms of band shifts (e.g., isotopic shifts or difference spectroscopy) on the very same day. To calibrate the spectrometer, a small silicon crystal is centered on the goniometer and a spectrum is collected in the 200–800 cm^{-1} region (data acquisition mode 1). At 20°C, silicon displays a single band at 520 cm^{-1} (this band shifts with temperature, so be careful to remove the cryostream to do the measurement). A software-controlled adjustment of the grating allows setting the measured value to the theoretical one.

3.3.2. Step 2: Cryostream Adjustment

Turn on cryo-cooling if the experiment is planned at cryo temperature, and precisely adjust the cryo-nozzle position, at very close distance from the sample (~8 mm). This is important because only spectra recorded at the same temperature can be reliably compared.

3.3.3. Step 3: Raman Probe Adjustment

Depending on the crystal size, choose and install the proper objective (20× or 50×) on the Raman probe. Use a pinhole of a diameter approximately equal to the section of the excitation beam at the focal point (our pinholes are made of small aluminum foils drilled by lithography and glued onto standard pins that fit onto a goniometer head) and center the hole on the goniometer rotation axis, at the temperature planned for the experiment. Position the Raman probe so that as much laser light as possible passes through the pinhole. This is done by adjusting the three translation stages X, Y, and Z onto which the Raman probe is mounted and by looking at residual light reflected by the aluminium surface with the videocamera. Only 1% of the full laser power is typically used at this stage.

3.3.4. Step 4: Finding Suitable Orientations of the Sample

This step is critical. The position and orientation of the crystal relative to the Raman beam is of utmost importance. At certain orientations, Fresnel reflections and refraction effects may considerably degrade the spectral quality. Polarization effects also play a role, because biological crystals are highly anisotropic (*see* **Note 3**). While fishing the crystal from its crystallization drop, try to minimize the amount of (cryo)solvent. If the sample gets covered with a large amount of solvent, this will make the alignment more tricky and may seriously degrade the data quality. If the crystal is not intended to be used for X-ray data collection, it might be advantageous not to use any cryoprotectant (a moderate amount of ice is generally not a problem, however, the modified rate of cooling may affect the final equilibrium between conformational states *(25)*). Mount and center the crystal on the goniometer axis. If possible, the crystal should a priori be oriented so that the excitation light is perpendicular to its largest face. Try to avoid the sample holder loop ending up in the laser path.

Collect coarse Raman data in the window 1,600–1,700 cm^{-1} using 1–3 s acquisition time (acquisition mode 1). Signal from the strong amid band I (~1,660 cm^{-1}) is taken as a sign of crystalline biological material at the focal point, because most reservoir solutions do not produce vibration bands in this region. Alternatively, concentrate on a band known to be specific from the macromolecule under study, of high intensity and preferably of low depolarization ratio. Progressively rotate the crystal while monitoring the Raman signal until the strongest intensity is observed. Also look at the background slope: a strong slope is generally a bad sign, e.g., from residual fluorescence. In the resonant mode, pay attention to photodamage and attenuate the laser beam as much as possible during the alignment stage.

3.3.5. Step 5: Data Collection

Work in complete darkness (dim or even turn off computer screens). To collect spectra in the 200–2,000 cm^{-1} spectral range, the "Synchroscan" acquisition mode 3 is used with a typical overall acquisition time of approximately 15 min. Multiple acquisitions can be performed and averaged to improve the signal-to-noise ratio without saturating the CCD detector. For publication-quality data, multiple acquisitions help to identify bands originating from cosmic rays and stray light sources. Photodamage can also be monitored by comparing sequentially collected Raman spectra.

3.3.6. Step 6: Control Experiments

Control experiments are often necessary to optimize the setup, to interpret the data, and *in fine* to validate the results. Collect reference spectra of the sample holder, of the mother liquor and of the cryoprotected mother liquor before measuring crystals. This will help to identify non-specific bands and possibly to improve the experimental protocol.

3.3.7. Step 7: Special Experiments

1. Isotopic shifts: To provide firm evidence for the chemical origin of specific bands, measuring shifts induced by isotope labeling is a common and efficient (sometimes costly) procedure that also works in crystals **(Fig. 2)**. As opposed to band intensities that tend to vary to some degree from crystal to crystal, band shifts can be quantitatively measured, allowing isotopic effects to be precisely assessed, provided identical protocols are applied to the various isotopes and spectra are collected on the same day.
2. Assessment of X-ray radiation damage: If the Raman signatures of a crystal before and after X-ray data collection are to be compared to assess potential X-ray-induced radiation damage, the online setup is much preferable and allows quantitative measurements to be carried out (*see* **Note 1**). With the offline setup, crystals need to be transported twice (from the Cryobench laboratory to the beamline, and back) and it is crucial to keep track of the crystal orientation used during recording of the first Raman spectrum and to keep track of the volume of the crystal probed by the X-rays to investigate that same volume once back at the spectrophotometer (take snapshots of the crystal to visualize the X-ray fingerprint on the crystal surface). In any case, such a comparison made offline will remain only qualitative.
3. Nonresonance Raman studies on nano-volumic solution samples: Measurements can be attempted in the nonresonant mode using flash-cooled films or droplets of highly concentrated protein solutions. However, protein concentration as high as, and solvent content as low as in a crystal, respectively, can never be attained; hence a much reduced signal-to-noise ratio is to be expected. Glycerol is often prohibited for such experiments. This points to a potential flaw in using Raman spectroscopy to compare the crystalline and solution states of a macromolecule: in some instances, it is expected that only the nonresonance mode might be used with the crystal, whereas only the resonant mode might be used with the corresponding solution, making the comparison difficult. This is not always the case, fortunately, and for SOR, a clear nonresonant solution spectrum **(Fig. 2)** could be obtained by flash-freezing in a nylon loop ~3 nL of a noncryoprotected ~11 mM protein solution containing 70 mM H_2O_2. However,

5–10 spectra (15 min acquisition time each) were required to obtain a sufficiently detailed spectrum.

4. Kinetic measurements at room temperature: One interesting application of in crystallo Raman spectroscopy at room temperature is to follow kinetics of heavy atom or ligand binding in the protein crystal when reaction times are in the minutes time scale *(22, 26)*. At the zero time point, a small volume of a solution of the ligand is injected into the crystallization drop to be mounted on the Raman microspectrophotometer (e.g., using the coverslip sandwich method described above). The initial concentration of the ligand in the drop should be approximately 10 times the Michaelis binding constant K_m, but should remain much less than the concentration of the crystalline protein. Raman spectra are sequentially recorded during the estimated time of the binding event. As the ligand binds to the protein, its concentration increases in the crystal, and either new ligand-specific peaks or protein-specific band shifts appear in the Raman spectra. Those spectral changes are best revealed by difference Raman spectroscopy. Real-time binding kinetics may be assessed by following the evolution of the difference integrated intensity from a selected Raman band.

3.3.8. Step 8: Processing of Spectra

Raw Raman spectra from macromolecular crystals are processed in much the same way as solution spectra. For qualitative data evaluation (as in **Fig. 2**), standard baseline corrections using, e.g., cubic spline fits are generally performed (although they carry a certain level of user bias). Spurious bands from cosmic rays and stray light are removed manually. When spectra are averaged, cosmic ray removal is performed on the individual spectra but baseline correction is done on the averaged spectrum. For quantitative data evaluation of intensity changes in Raman bands (which requires working on a single crystal), a different processing strategy is applied. Local baseline correction around the band(s) of interest is done, followed by, e.g., Gaussian deconvolution. It should be noted that, even for a single crystal, the baseline correction may vary from spectrum to spectrum, for example because of the accumulation of radical species in the crystal exposed to X-rays at low temperature.

4. Notes

1. Online or offline in crystallo Raman spectroscopy? Whereas Raman measurements performed offline (at the ESRF Cryobench laboratory) are extremely useful in many cases, online

data (collected directly on macromolecular crystallography beamlines) provide the opportunity to combine X-ray crystallography and Raman spectroscopy on the same sample without manipulation. The compact Raman probes described above, in combination with long optical fibers that allow maintaining the laser sources and the spectrometer at the offline location (i.e., no recalibration is needed) can be easily integrated on the biocrystallography ESRF beamlines. Such an online setup provides two main advantages: (a) the absence of handling between X-ray and Raman measurements avoids any alteration of the sample, for example, caused by an unforeseen transient temperature rise; and (b) the crystal volumes investigated by the Raman and X-ray beam can be more accurately overlaid and many interleaved measurements can be performed, which is essential when X-ray-induced chemistry is investigated. Raman data recorded online before X-ray exposure provide a "zero dose" reference spectrum to be compared with spectra recorded after diffraction data collection. Such zero dose data may be very useful to correct for early X-ray-induced structural damage.

However, offline measurements, made in a more relaxed way, should always be performed beforehand to assess in detail the behavior of the sample under consideration. It should also be kept in mind that only properly cryoprotected samples can be studied online, which may complicate the spectra. In addition, the formation of radicals that results from X-ray exposure may produce significant background that may progressively alter the Raman spectral quality.

We list some practical considerations pertaining to online Raman data collection. (a) To properly superimpose measurement volumes, it is wise to keep the Raman volume much smaller than the X-ray volume, so that the X-ray flux density remains homogeneous within the Raman beam. At the same time, owing to constraints imposed by the sample environment on beamlines, only lenses with relatively long working distances and low magnification can be used (20× objective in our case). As a consequence, performing online Raman measurements on a microfocus beamline would necessitate special developments to tailor a strongly microfocused Raman beam. It should also be noted that the use of the resonant mode with crystals of high optical density may pose delicate issues of volume superimposition caused by the lack of penetration of the optical beam. (b) As mentioned above, good-quality Raman spectra are only obtained at some orientations of the crystal. This means that Raman and X-ray data cannot be collected strictly at the same time. Rather, all Raman spectra should be collected at the same favorable spindle rotation position, meaning that special data collection software need to be devised where the crystal is rotated back and forth between alternate Raman and X-ray measurements. (c) On modern synchrotron beamlines, the X-ray data collection time is often reduced to minutes. The experimental time dedicated to Raman spectroscopy may therefore easily exceed the one dedicated to X-ray crystallography, which may be problematic when beamtime is limited. To shorten the Raman data collection time, spectra should be collected over a reduced spectral range of interest.

2. Resonant or nonresonant in crystallo Raman spectroscopy? Nonresonant in crystallo Raman spectroscopy (at 785 nm or possibly 633 nm) is easy to perform, applicable to a wide variety of (noncolored) biological molecules, and provides spectra of excellent quality with minimal photodamage. However, bands of interest may be obscured by a dominant contribution from the solvent/cryoprotectant and a comparison with the solution state may prove impossible. Resonant in crystallo Raman spectroscopy, on the other hand, only targets those proteins containing chromophores absorbing close to the available wavelengths (514 or 633 nm). The strong enhancement of the vibration modes coupled with the excited electronic transitions provides sensitivity and selectivity, greatly simplifying interpretation of the Raman spectra, and allows easy comparison between crystals and diluted solutions. However, these advantages might be severely offset by penetration depth problems (leading to a strong reduction in signal intensity, only the crystal surface being probed), laser induced photodamage (or photochemistry), and by strong fluorescence background originating from the cryosolvent or from the sample holder.

3. Polarized or nonpolarized in crystallo Raman spectroscopy? Polarized Raman measurements are helpful for assigning some specific vibration bands and for providing quantitative structural parameters of oriented samples. However, in crystals, the measurement of depolarization ratios is challenging because it requires a cautious alignment of the crystallographic axis with respect to the polarized excitation beam. Furthermore, depending on the crystal space group and molecular content of the asymmetric unit, analysis of polarized Raman data may be intricate. In fact, polarized Raman measurements have rarely been reported in the literature *(27, 28)*. Our offline Raman microspectrophotometer is equipped with a single-axis goniometer, which does not allow precise crystal alignment. Therefore, polarized measurements are not attempted and our Raman probes are presently not equipped with polarizers/analyzers. It should be noted that although the excitation light supplied by our laser sources is to a great extent depolarized throughout the optical fibers, a low degree of polarization (~30% at 785 nm) persists at the focal point. Therefore, vibration modes with

high polarization ratios display intensity fluctuations that are slightly dependent on crystal orientation. Such effects, however, tend to be buried under the influence of other parameters such as solvent shell thickness or crystal morphology. The possibility of measuring polarized Raman spectra in crystallo could benefit in the future from the online setup, where crystals can be precisely oriented using a kappa goniometer while monitiring the diffraction pattern.

Acknowledgments

This work received financial support from the European Molecular Biology Organisation (EMBO), the European Synchrotron Radiation Facility (Grenoble, France), "Ministère de l'Enseignement et de la Recherche," and the "Région Rhônes-Alpes" (France, CPER and CIBLE contracts). Contributions by Antoine Royant, Vincent Nivière, Jeremy Ohana, David Annequin, and Michel Belleil are acknowledged.

References

1. Hadfield, A. and Hajdu, J. (1993). A fast and portable microspectrophotometer for protein crystallography, *J. Appl. Cryst.*. **26**, 839–842.
2. Chen, Y., Srajer, V., Ng, K., Legrand, A. and Moffat, K. (1994). Optical monitoring of protein crystals in time-resolved X-ray experiments: microspectrophotometer design and performance, *Rev. Sci. Instrum.* **65**, 1506–1511.
3. Bourgeois, D., Vernede, X., Adam, V., Fioravanti, E. and Ursby, T. (2002). A microspectrophotometer for absorption and fluorescence studies of protein crystals, *J. Appl. Cryst.* **35**, 319–326.
4. Sakai, K., Matsui, Y., Kouyama, T., Shiro, Y. and Adachi, S. (2002). Optical monitoring of freeze-trapped reaction intermediates in protein crystals: a microspectro-photometer for cryogenic protein crystallography, *J. Appl. Cryst.* **35**, 270–273.
5. Klink, B. U., Goody, R. S. and Scheidig, A. J. (2006). A newly designed microspectrofluorometer for kinetic studies on protein crystals in combination with X-ray diffraction, *Biophys. J.* **91**, 981–992.
6. Royant, A., Carpentier, P., Ohana, J., McGeehan, J., Paetzold, B., Noirclerc-Savoye, M., Vernede, X., Adam, V. and Bourgeois, D. (2007). Advances in spectroscopic methods for biological crystals. Part 1. Fluorescence lifetime measurements, *J. Appl. Crystallogr.* **40**, 1105–1112.
7. Berglund, G. I., Carlsson, G. H., Smith, A. T., Szoke, H., Henriksen, A. and Hajdu, J., (2002). The catalytic pathway of horseradish peroxidase at high resolution, *Nature.* **417**, 463–468.
8. Kuhnel, K., Derat, E., Terner, J., Shaik, S. and Schlichting, I. (2007). Structure and quantum chemical characterization of chloroperoxidase compound 0, a common reaction intermediate of diverse heme enzymes, *Proc. Natl Acad. Sci. U. S. A.* **104**, 99–104.
9. Wilmot, C. M., Sjogren, T., Carlsson, G. H., Berglund, G. I. and Hajdu, J. (2002). Defining redox state of X-ray crystal structures by single-crystal ultraviolet-visible microspectrophotometry, *Methods Enzymol.* **353**, 301–318.
10. Adam, V., Royant, A., Niviere, V., Molina-Heredia, F. P. and Bourgeois, D. (2004). Structure of superoxide reductase bound to ferrocyanide and active site expansion upon X-ray-induced photo-reduction, *Structure (Camb)* **12**, 1729–1740.
11. Beitlich, T., Kuhnel, K., Schulze-Briese, C., Shoeman, R. L. and Schlichting, I. (2007).

Cryoradiolytic reduction of crystalline heme proteins: analysis by UV–Vis spectroscopy and X-ray crystallography, *J. Synchrotron Radiat.* **14**, 11–23.

12. Pearson, A. R., Mozzarelli, A. and Rossi, G. L. (2004). Microspectrophotometry for structural enzymology, *Curr. Opin. Struct. Biol.* **14**, 656–662.
13. Weik, M., Vernede, X., Royant, A. and Bourgeois, D. (2004). Temperature derivative fluorescence spectroscopy as a tool to study dynamical changes in protein crystals, *Biophys. J.* **86**, 3176–3185.
14. Pascal, A. A., Liu, Z., Broess, K., van Oort, B., van Amerongen, H., Wang, C., Horton, P., Robert, B., Chang, W. and Ruban, A. (2005). Molecular basis of photoprotection and control of photosynthetic light-harvesting, *Nature* **436**, 134–137.
15. Sage, J. T. and Jee, W. (1997). Structural characterization of the myoglobin active site using infrared crystallography, *J. Mol. Biol.* **274**, 21–26.
16. Zhu, L., Sage, J. T. and Champion, P. M. (1993). Quantitative structural comparisons of heme protein crystals and solutions using resonance Raman spectroscopy, *Biochemistry* **32**, 11181–11185.
17. Carey, P. R. and Dong, J. (2004). Following ligand binding and ligand reactions in proteins via Raman crystallography, *Biochemistry* **43**, 8885–8893.
18. Smulevich, G., Wang, Y., Mauro, J. M., Wang, J. M., Fishel, L. A., Kraut, J. and Spiro, T. G. (1990). Single-crystal resonance Raman spectroscopy of site-directed mutants of cytochrome c peroxidase, *Biochemistry* **29**, 7174–7180.
19. Davies, R. J., Burghammer, M. and Riekel, C. (2005). Simultaneous microRaman and synchrotron radiation microdiffraction: tools for materials characterization, *Appl. Phys. Lett.* **82**, 264105.
20. Briois, V., Vantelon, D., Villain, F., Couzinet, B., Flank, A. M. and Lagarde, P. (2007). Combining two structural techniques on the micrometer scale: micro-XAS and micro-Raman spectroscopy, *J. Synchrotron Radiat.* **14**, 403–408.
21. Boccaleri, E., Carniato, F., Croce, G., Viterbo, D., van Beek, W., Emerich, H. and Milanesio, M. (2007). In situ simultaneous Raman/high-resolution X-ray powder diffraction study of transformations occurring in materials at non-ambient conditions, *J. Appl. Cryst.* **40**, 684–693.
22. Carpentier, P., Royant, A., Ohana, J. and Bourgeois, D. (2007). Advances in spectroscopic methods for biological crystals. Part 2. Raman spectroscopy, *J. Appl. Cryst.* **40**, 1113–1122.
23. Katona, G., Carpentier, P., Niviere, V., Amara, P., Adam, V., Ohana, J., Tsanov, N. and Bourgeois, D. (2007). Raman-assisted crystallography reveals end-on peroxide intermediates in a nonheme iron enzyme, *Science* **316**, 449–453.
24. McGeehan, J., Carpentier, P., Royant, A., Bourgeois, D. and Ravelli, R. B. (2007). X-ray radiation-induced damage in DNA monitored by online Raman, *J. Synchrotron Radiat.* **14**, 99–108.
25. Halle, B. (2004). Biomolecular cryocrystallography: structural changes during flash-cooling, *Proc. Natl Acad. Sci. U. S. A.* **101**, 4793–4798.
26. Helfand, M. S., Totir, M. A., Carey, M. P., Hujer, A. M., Bonomo, R. A. and Carey, P. R. (2003). Following the reactions of mechanism-based inhibitors with beta-lactamase by Raman crystallography, *Biochemistry* **42**, 13386–13392.
27. Smulevich, G., Wang, Y., Edwards, S. L., Poulos, T. L., English, A. M. and Spiro, T. G. (1990). Resonance Raman spectroscopy of cytochrome c peroxidase single crystals on a variable-temperature microscope stage, *Biochemistry* **29**, 2586–2592.
28. Kudryavtsev, A. B., Mirov, S. B., DeLucas, L. J., Nicolete, C., van der Woerd, M., Bray, T. L. and Basiev, T. T. (1998). Polarized Raman spectroscopic studies of tetragonal lysozyme single crystals, *Acta Crystallogr. D Biol. Crystallogr.* **54**, 1216–1229.

Chapter 17

Methods and Software for Diffuse X-Ray Scattering from Protein Crystals

Michael E. Wall

Summary

Proteins in thermal equilibrium are associated with conformational distributions rather than single, static structures. Although there are no experimental methods to measure the full protein conformational distribution, several methods exist to probe important aspects. Diffuse X-ray scattering is one such method. We have measured the first three-dimensional reciprocal-space maps of the intensity of diffuse X-ray reflections from protein crystals, and used them to characterize protein conformational distributions. With straightforward modifications, X-ray beamlines can be engineered to enable diffuse scattering measurements for protein crystals. To facilitate future studies, the Lunus software package, used to create the first three-dimensional maps of diffuse X-ray reflections from protein crystals, has been made publicly available (http://lunus.sourceforge.net).

Key words: Protein dynamics, Conformational distribution, Diffuse X-ray scattering, Protein crystallography, Synchrotron

1. Introduction

Disorder in protein crystals leads to diffuse X-ray reflections at scattering vectors $\mathbf{q}$ away from the Bragg reflections in diffraction images. For example, consider a crystal in which the structure factor $f_n(\mathbf{q})$ of each unit cell n may be different. Define $f_n(\mathbf{q})$ as the difference between $f_n(\mathbf{q})$ and its mean value,

$$\Delta f_n(\mathbf{q}) = f_n(\mathbf{q}) - \langle f_n(\mathbf{q}) \rangle_n . \qquad (1)$$

Assume that the differences are uncorrelated across unit cell boundaries. Guinier's analysis of this case *(1)* leads to a total reflected intensity $I(\mathbf{q})$,

James Weifu Lee and Robert S. Foote (eds.), *Micro and Nano Technologies in Bioanalysis*,
Methods in Molecular Biology, vol. 544
DOI 10.1007/978-1-59745-483-4_17, © Humana Press, a part of Springer Science+Business Media, LLC 2009

$$I(\mathbf{q}) = N\left\langle \left|\Delta f_n(\mathbf{q})\right|^2 \right\rangle_n + \left|\left\langle f_n(\mathbf{q})\right\rangle_n\right|^2 \sum_{n,n'} e^{i\mathbf{q}\cdot(\mathbf{R}_n - \mathbf{R}_{n'})}, \quad (2)$$

where $\mathbf{R}_n$ is the vector from the origin to the position of the nth unit cell, N is the total number of unit cells in the crystal, and the averages are calculated over all unit cells. The first term in **Eq. 2** corresponds to the intensity of the diffuse reflections, $I_\text{D}(\mathbf{q})$,

$$I_\text{D}(\mathbf{q}) = N\left\langle \left|\Delta f_n(\mathbf{q})\right|^2 \right\rangle. \quad (3)$$

The second term in **Eq. 2** corresponds to the sharply peaked interference function that describes the intensity of the Bragg reflections, $I_\text{B}(\mathbf{q})$,

$$I_B(\mathbf{q}) = \left|\left\langle f_n(\mathbf{q})\right\rangle\right|^2 \frac{\sin^2(N_1\mathbf{q}\cdot\mathbf{a}_1/2)}{\sin^2(\mathbf{q}\cdot\mathbf{a}_1/2)} \times \quad (4)$$
$$\frac{\sin^2(N_2\mathbf{q}\cdot\mathbf{a}_2/2)}{\sin^2(\mathbf{q}\cdot\mathbf{a}_2/2)} \frac{\sin^2(N_3\mathbf{q}\cdot\mathbf{a}_3/2)}{\sin^2(\mathbf{q}\cdot\mathbf{a}_3/2)}$$

where $\mathbf{a}_i$ is the ith lattice vector, and N_i is the number of unit cells in the crystal counted along $\mathbf{a}_i$ ($N = N_1 N_2 N_3$).

In the immediate neighborhood of a Bragg peak, the Bragg intensity in **Eq. 4** is proportional to N^2, whereas the diffuse intensity in **Eq. 3** is proportional to N. Raman used this observation to argue that the diffuse intensity is N-fold smaller than the Bragg intensity, and is therefore not observable in diffraction images *(2)*. However, Lonsdale successfully refuted this argument in her definitive review of early studies of diffuse scattering *(3)*, and in the second part of a two-part letter published with Born and Smith *(4)*. To see the essence of the refutation, note that **Eq. 4** indicates that the width of a Bragg peak along direction $\mathbf{a}_i$ is equal to:

$$\delta q_i = \frac{2\pi}{N_i a_i}, \quad (5)$$

where $N_i a_i$ is the size of the crystal along the lattice vector $\mathbf{a}_i$. This width is smaller than can be resolved in a typical experiment—a detector with an angular resolution of 10^{-3} rad would barely resolve a Bragg peak from a 1-Å beam scattered from a 100-nm crystal [$\delta q = 2\pi(10^{-3}\text{ rad})/(1\text{ Å}) = 2\pi/100\text{ nm}^{-1}$]. Measured intensities of Bragg reflections therefore scale like the integrated intensity $\int_{\text{peak}} d^3\mathbf{q} I_\text{B}(\mathbf{q})$, which, by **Eqs. 4** and **5**, is proportional to N, just like the diffuse intensity. The diffuse intensity is expected to be observable.

In the early 1990s, a new area X-ray detector for use at synchrotrons was developed at Princeton University and was tested for macromolecular crystallography applications at the Cornell high-energy synchrotron source (CHESS) *(5, 6)*. The detector was based on a charge-coupled device (CCD) and had a higher dynamic range than previous detectors. Images of X-ray diffraction from protein crystals obtained using this detector revealed striking diffuse features, motivating a study of diffuse scattering from protein crystals. Over the course of several years, beamline and crystallography methods were refined to minimize systematic sources of error in diffuse scattering measurements, and methods were developed to obtain three-dimensional reciprocal-space maps of the intensity of diffuse reflections from protein crystals *(7)*. The methods were applied to measure diffuse reflections for crystalline hen egg-white lysozyme (unpublished) and Staphylococcal (staph) nuclease *(8)*; similar methods were later used to obtain more detailed maps of diffuse scattering from calmodulin crystals in the neighborhood of Bragg peaks, and to relate them to crystalline dynamics *(9)*.

Early studies of diffuse scattering from protein crystals made use of single diffraction images to yield insight into protein conformational distributions, as reviewed in *ref. 10*. Importantly, using the diffuse reflections for staph nuclease, it was demonstrated that diffuse scattering data can be used to refine parameters that characterize the protein conformational distribution in a manner similar to the traditional use of three-dimensional maps of Bragg reflections for protein structure refinement *(8)*. This demonstration supports the argument for integrating diffuse reflections with Bragg reflections to improve refinement of structural models *(10)*.

Recently, the staph nuclease diffuse reflections obtained in **ref.** *8* were used to validate molecular dynamics simulations *(11)*. In subsequent studies, the staph nuclease diffuse reflections were further used to validate models of solvent and protein dynamics *(12, 13)*. **References** *8* and *11–13* emphasize the utility of three-dimensional diffuse reflection data for validating models of protein dynamics, and motivate the measurement of three-dimensional maps of diffuse X-ray reflections for other protein crystals.

The methods used to obtain three-dimensional maps of diffuse reflections from crystals of staph nuclease are documented here. The Lunus software package for processing diffuse scattering data (http://lunus.sourceforge.net) is released in parallel with the publication of the methods. These resources are made available to facilitate future experiments and to help equip beamlines for routine measurement of the intensity of diffuse X-ray reflections in protein crystallography studies.

2. Materials

2.1. Data Collection

1. The crystal of staph nuclease was grown over 23% 2-methyl-2,4-pentanediol (MPD) 10.5 μM potassium phosphate (*see* **Note 1**).
2. Data were collected on the F1 and A1 beamlines at CHESS.
3. Diffraction patterns were imaged using the second Princeton University CCD area X-ray detector designed for protein crystallography at CHESS. The detector was similar in design to the CCD detector described in *ref. 5*. It had a dynamic range of roughly 10^4 X-rays/pixel, and an active area of about 80 × 80 mm^2, subdivided into an array of 2,048 × 2,048 pixels. Pixel values mapped to X-ray counts by a ratio of roughly one-to-one.
4. The intensities of Bragg reflections were measured and were assigned Miller indices from oscillation exposures using the programs DENZO and SCALEPACK *(14)*.

2.2. Image Processing

1. Image processing steps were carried out using the Lunus software package, available at (http://lunus.sourceforge.net).
2. Processed diffraction images for staph nuclease are distributed with the Lunus software package.

2.3. Obtaining the Intensity of Diffuse Reflections

1. The diffuse reflections were obtained for staph nuclease and were stored in a diffuse lattice data structure using **genlat** software in the Lunus software package.
2. The diffuse lattice for staph nuclease is distributed with the Lunus software package (*see* **Note 2**).

3. Methods

3.1. Data Collection

1. The staph nuclease crystal was transferred from a hanging drop to a capillary shortly before experimentation at CHESS. Excess buffer was wicked away to minimize scattering from the surrounding solvent. The crystal was mounted with the *c*-axis nearly parallel to the capillary. Data were collected at room temperature.
2. The beam was tuned to 0.91 Å, had a polarization of 0.8–0.93 perpendicular to the beam in the plane of the synchrotron ring, and was collimated to a 100-μm diameter. Great care was taken to minimize the contribution of unwanted X-rays to diffuse scattering images (*see* **Note 3**).

3. The detector was positioned 57.4 mm downstream of the crystal, with the detector face approximately perpendicular to the beam, and the beam approximately centered on the detector. To save disk space and decrease image transfer times, CCD pixel values were binned both horizontally and vertically during readout, producing a 1,024 × 1,024 image.
4. An "anti-blooming" procedure (described in a personal communication by James Janesick of Pixel Vision, Huntington, CA), where overflowing electrons from saturated wells in the CCD are channeled off the device during integration, was implemented to improve the handling of strong Bragg reflections, which can otherwise distort intensity measurements in the neighborhood of the peak.
5. The data collection protocol involved interleaving the collection of two data sets, one being a set of 2° *oscillation exposures* to be used in calculating the orientation of the crystal and in refining a structural model, and the other being a set of *stills* spaced 1° apart in spindle rotation, from which measurements of the intensity of diffuse reflections were obtained (*see* **Note 4**). DENZO and SCALEPACK were used to obtain parameters from oscillations, and DENZO was used to output the crystal orientation for each of the stills. The data set spanned 90° of spindle rotation. All exposures were 5-s long.
6. Bragg peaks were indexed from the oscillation exposures using standard methods, yielding best-fit values for unit cell parameters, mosaicity, and crystal orientation.

3.2. Image Processing

1. Overflow pixels were marked in diffuse scattering images (*see* **Note 5**).
2. To define which pixels contain data and which pixels define the edges of the image, border pixels were marked in diffuse scattering images (*see* **Note 6**).
3. Bragg peaks were eliminated from diffuse scattering images using a mode-filtering image-processing technique, where pixels in a new image are given the value of the mode (most common value) of the distribution of pixel values in a 15 × 15 patch about the same pixel in the original image (*see* **Notes 7** and **8**).
4. The polarization of the beam was estimated by analyzing a typical mode-filtered diffraction image. An azimuthal intensity distribution was calculated in a thin annulus at a scattering angle θ about the beam spot, and the polarization was obtained by fitting the distribution to a standard equation that characterizes the polarization effect (*see* **Note 9**). Diffuse scattering images were then corrected for the beam polarization effect (*see* **Note 10**).

5. Diffuse scattering images were corrected for the dependence of measured pixel intensity on the solid angle subtended by the pixel (*see* **Note 11**).

3.3. Obtaining the Intensity of Diffuse Reflections

1. To correct for beam intensity variations over time and for variations in scattering as the crystal is rotated, each diffuse scattering image was scaled to an arbitrary reference image by multiplying each pixel value in the image by an image-dependent scale factor. To calculate scale factors, one-dimensional distributions of circularly averaged intensity versus distance from the beam spot were calculated from diffuse scattering images (*see* **Note 12**). Scale factors were calculated by minimizing the root mean square deviation (RMSD) between each circularly averaged intensity profile and the reference profile (*see* **Notes 13** and **14**).
2. Each pixel in each image was mapped to a scattering vector **q**, which was oriented relative to the crystal lattice using both the detector face rotation angles (*see* **Note15**) and the crystal orientation matrix elements calculated using DENZO (*see* **Note 16**). The scattering vector was converted to fractional Miller indices $\mathbf{h}' = (h', k', l')$ using the unit cell parameters from DENZO. The Bragg reflection $\mathbf{h} = (h, k, l)$ nearest to (h', k', l') was identified, and the scattering vector was thus associated with a Bragg reflection. If a ½ × ½ × ½ cube centered on reflection **h** enclosed the point **h**′, the pixel was rejected as being too close to a Bragg reflection. All pixels not rejected were multiplied by an image-dependent scale factor (**step 1**) and were averaged together to determine the measured diffuse intensity in the neighborhood of reflection **h** in reciprocal space. The collection of values of diffuse intensity for all reflections **h** that span the data set was stored as a three-dimensional diffuse lattice, representing a three-dimensional map of the intensity of diffuse reflections from crystalline staph nuclease (*see* **Note 17**).

4. Notes

1. The crystal was selected from a batch of average size approximately 0.4 mm × 0.2 mm × 0.2 mm, and had unit cell parameters $a = b = 48.2$ Å, $c = 63.9$ Å, $\alpha = \beta = \gamma = 90°$. The crystal had a tetragonal unit cell with space group $P4_1$.
2. Lunus library routines **lreadlt** and **lwritelt** show examples of how to read and write a three-dimensional diffuse lattice created using Lunus.

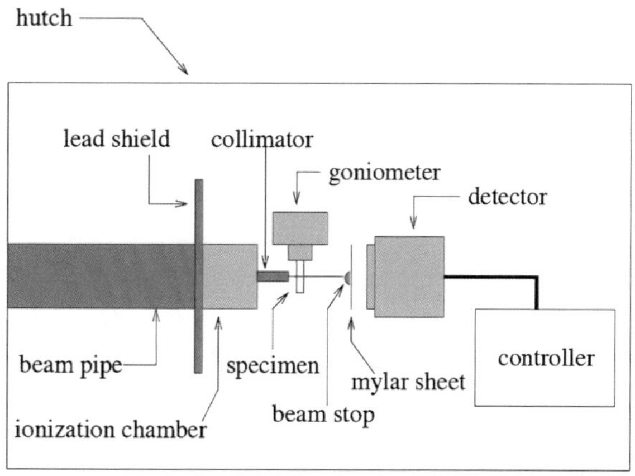

Fig. 1. Schematic of beamline instrumentation. The incident beam comes from the left through the beam pipe. It travels through an ionization chamber and is collimated before striking the specimen. The resulting scattering pattern is imaged on a CCD detector with electronic readout. A beam stop shields the detector from the main beam. Special steps were taken to minimize the contribution of unwanted X-rays to images of diffuse X-ray scattering (*see* **Note 3**).

3. *See* **Fig. 1**. The beam stop was adjusted to completely block the main beam. A lead sheath with a 1-mm hole at the end was slipped over the collimator, which was a known source of parasitic scattering. A large lead shield was placed before the collimator to eliminate contamination of background features by static, hard X-ray patterns. In addition, we found that absorption by the narrow strip of kapton tape holding the beam stop cast a shadow, causing static background contamination; to eliminate the shadow, the strip was replaced by a wide mylar sheet stretched across the face of an adjustable aluminum goal post, and the beam stop was fixed to the sheet with superglue.
4. Because, in this experiment, we were not interested in studying variations in diffuse intensity on a smaller scale than the separation between Bragg peaks, the set of stills adequately sampled reciprocal space.
5. The **thrshim** software marks overflow pixels.
6. The **windim** software marks the image borders.
7. The **modeim** software eliminates Bragg peaks from diffraction images. The mode filter was borrowed from astrophysics methods, where it is used to "de-star" night sky images *(15)*, yielding background intensities in an image "contaminated" by stars.
8. Saturated peaks can leave residual intensity in mode-filtered images *(7)*. In the case of the CCD detector used in the experiments described here, a non-exponential tail in the point-spread function of the detector might lead to such a

problem. As is described in **ref. 5**, a similar detector showed that, at a distance of 450 μm from a peak, the pixel value was still 0.1% of the maximum. Assuming that saturated peaks have 10^5 analog-to-digital unit (ADU) maximum values, this amounts to a 100~ADU effect, which is on the order of the measured intensity of diffuse features. Fortunately, 450 μm corresponds to less than 6 pixels on the detector, which had a pixel size of 80 μm, whereas the 15 × 15 mode-filter mask extends between 7.5 pixels (on a side) and 10.6 pixels (on a diagonal) from the center of the square. Even with the mask centered on a saturated peak, therefore, there is a good chance that the mode will not contain significant contributions from the tails of the peak. If there is any observable effect, it would be expected only in the immediate neighborhood of the saturated pixel, and would have a high likelihood of being eliminated in the rejection of measurements too close to a Bragg peak. Contamination of diffuse maps caused by saturated Bragg peaks, therefore, was expected to be an insignificant source of systematic errors for this staph nuclease, although it might present problems in measuring diffuse scattering from other crystals.

9. The beam polarization was determined by fitting the following equation from **ref. 16** to the azimuthal intensity distribution $I(\varphi,\theta)$ measured at scattering angle θ,

$$I(\varphi,\theta) = \frac{a}{2}\left[1 + \cos^2 2\theta - \varepsilon \cos 2(\varphi - \varphi_0)\sin^2 2\theta\right] \qquad (6)$$

In **Eq. 6**, φ is the azimuth as defined in **Fig. 2**, φ_0 is azimuth of a vector parallel to the plane of the synchrotron ring, a is a scale factor, and ε is the beam polarization.

10. The **polarim** software uses **Eq. 6** to correct for the polarization effect.

11. The **normim** software corrects for solid angle effects. The intensity measured at a pixel of area A at distance l from the crystal is proportional to the solid angle $\delta\Omega$ subtended by the pixel,

$$\delta\Omega = \frac{A}{l^2}\cos\psi, \qquad (7)$$

where ψ is the angle between a normal to the face of the pixel and the scattered X-ray. When the detector face is perpendicular to the incident beam, ψ is equal to the angle between the scattered and incident beam (i.e., $\psi = 2\theta$). In this case, **Eq. 7** becomes

$$\delta\Omega = \frac{A}{d^2}\cos^3\psi, \qquad (8)$$

where d is the shortest distance between the crystal and the detector face.

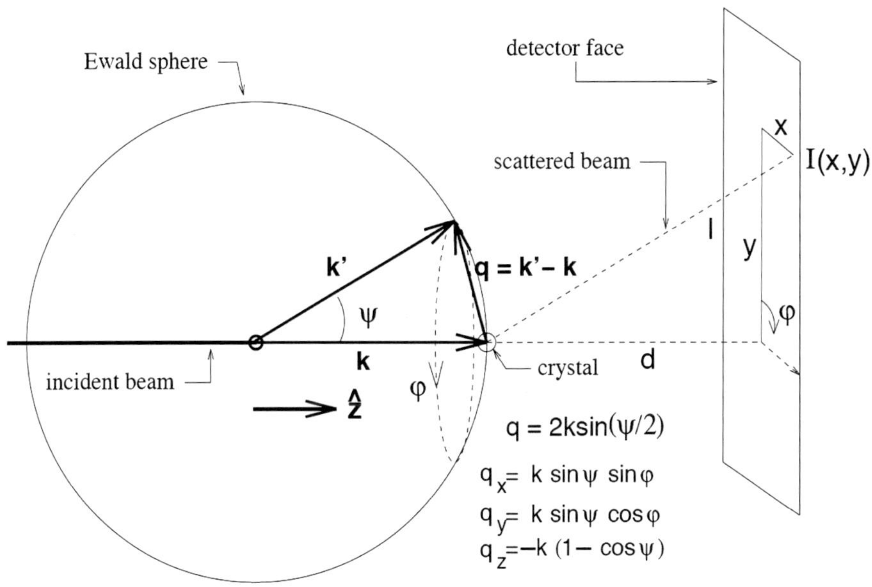

Fig. 2. Illustration of scattering geometry. The incident beam comes from the left and defines the *z*-direction. In this illustration, the detector face is perpendicular to the incident beam, the *x*-direction is along the detector horizontal, and the *y*-direction is along the detector vertical. The azimuthal angle φ is defined about the *z*-axis with respect to the vertical in a right-handed manner. X-rays detected at position (*x*, *y*) on the detector are related to the scattering vector **q** by **Eqs. 9** and **10**.

12. The **avgrim** software calculates the circularly averaged intensity versus distance from the beam spot. For each pixel in an image, the distance in pixels to the beam spot is calculated and rounded to the nearest integer. All pixels with the same distance are binned together, and their values are averaged to create an intensity profile.
13. One image is arbitrarily chosen as a reference image, and its profile is chosen as a reference profile. For each image, a linear least-squares fit was used to find a multiplicative constant that best scales the intensity profile to the reference profile.
14. This procedure may be used as an alternative method for calculating scale factors for Bragg reflections. Because of increased statistics, the method should provide a more precise measurement of the scale factors than is obtained by sole analysis of the intensity of Bragg reflections, as is routinely done in crystallographic analysis.
15. The correction for the detector face rotation angles as currently implemented in the **lgensv** library routine of Lunus is not exact and is only valid for small angles.
16. **Figure 2** shows the scattering geometry for our diffraction experiments. The elements of the scattering vector **q** can be expressed in terms of the experimental parameters as:

$$q_x = k \sin \psi \sin \varphi,$$
$$q_y = k \sin \psi \cos \varphi,$$
$$q_z = -k\ (1 - \cos \psi), \tag{9}$$

where k is the spatial frequency of the X-rays, $\psi = 2\theta$ is the angle between the incident and scattered beam, and φ is the azimuthal angle about the z-axis, which is aligned with the incident beam. If the detector is perpendicular to the incident beam, **Eq. 9** can be expressed simply in terms of the pixel coordinates (x, y) and the sample-detector distance d as

$$\begin{aligned} q_x &= \frac{kx}{\sqrt{x^2 + y^2 + d^2}}, \\ q_y &= \frac{ky}{\sqrt{x^2 + y^2 + d^2}}, \\ q_z &= -k\left(1 - \frac{d}{\sqrt{x^2 + y^2 + d^2}}\right). \end{aligned} \tag{10}$$

To properly orient each scattering vector **q**, it was multiplied by the crystal orientation matrix **U** as determined using DENZO.

17. The **genlat** software carries out the tasks in this step to generate the three-dimensional diffuse lattice.

Acknowledgments

I am grateful to Clarence E. Schutt for introducing me to the phenomenon of diffuse X-ray scattering, and to Sol M. Gruner, George N. Phillips, Jr., and Donald L. D. Caspar for support and advice in the development of diffuse scattering methods. The writing of this chapter was supported by the US Department of Energy through the LANL/LDRD program.

References

1. Guinier, A. (1963). *X-Ray Diffraction*, W. H. Freeman and Company, San Francisco.
2. Raman, C. V. (1942). Reflexion and scattering of X-rays with change in frequency. II. Experimental. *Proc. R. Soc. Lond. A* **179**, 302–14.
3. Lonsdale, K. (1942). X-ray study of crystal dynamics: an historical and critical survey of experiment and theory. *Proc. Phys. Soc.* **54**, 314–53.
4. Born, M., Lonsdale, K., and Smith, H. (1942). Quantum theory and diffuse X-ray reflections. *Nature* **149**, 402–05.
5. Tate, M. W., Eikenberry, E. F., Barna, S. L., Wall, M. E., Lowrance, J. L., and Gruner, S. M. (1995). A large-format high-resolution area X-ray detector based on a fiber-optically bonded charge-coupled device (CCD). *J. Appl. Cryst.* **28**, 196–205.
6. Walter, R. L., Thiel, D. J., Barna, S. L., Tate, M. W., Wall, M. E., Eikenberry, E. F., Gruner, S. M., and Ealick, S. E. (1995). High-resolution macromolecular structure determination using CCD detectors and synchrotron radiation. *Structure* **8**, 835–44.

7. Wall, M. E. (1996). Diffuse Features in X-Ray Diffraction from Protein Crystals, Ph.D. Thesis, Physics Department, Princeton University, Princeton, NJ.
8. Wall, M. E., Ealick, S. E., and Gruner, S. M. (1997). Three-dimensional diffuse X-ray scattering from crystals of Staphylococcal nuclease. *Proc. Natl Acad. Sci. U. S. A.* **94**, 6180–4.
9. Wall, M. E., Clarage, J. B., and Phillips, G. N. (1997). Motions of calmodulin characterized using both Bragg and diffuse X-ray scattering. *Structure* **5**, 1599–612.
10. Clarage, J. B., and Phillips, G. N., Jr. (1997). Analysis of diffuse scattering and relation to molecular motion. *Methods Enzymol.* **277**, 407–32.
11. Meinhold, L., and Smith, J. C. (2005). Fluctuations and correlations in crystalline protein dynamics: a simulation analysis of Staphylococcal nuclease. *Biophys. J.* **88**, 2554–63.
12. Meinhold, L., and Smith, J. C. (2005). Correlated dynamics determining X-ray diffuse scattering from a crystalline protein revealed by molecular dynamics simulation. *Phys. Rev. Lett.* **95**, 218103.
13. Meinhold, L., and Smith, J. C. (2007). Protein dynamics from X-ray crystallography: anisotropic, global motion in diffuse scattering patterns. *Proteins* **66**, 941–53.
14. Otwinowski, Z., and Minor, W. (1997). Processing of X-ray diffraction data collected in oscillation mode. *Methods Enzymol.* **276**, 307–26.
15. Stetson, P. B. (1995). User's Manual for DAOPHOT II: The Next Generation, *Dominion Astrophysical Observatory*, Herzberg Institute of Astrophysics, Victoria, BC.
16. Giacovazzo, C., Monaco, H. L., Artoli, G., Viterbo, D., Ferraris, G., Gilli, G., Zanotti, G., and Catti, M. (2002). *Fundamentals of Crystallography*, Oxford University Press, Oxford.

Chapter 18

Deuterium Labeling for Neutron Structure–Function–Dynamics Analysis

Flora Meilleur, Kevin L. Weiss, and Dean A.A. Myles

Summary

Neutron scattering and diffraction provide detailed information on the structure and dynamics of biological materials across time and length scales that range from picoseconds to nanoseconds and from 1 to 10,000 Å, respectively. The particular sensitivity of neutrons to the isotopes of hydrogen makes selective deuterium labeling of biological systems an essential tool for maximizing the return from neutron scattering experiments. In neutron protein crystallography, the use of fully deuterated protein crystals improves the signal-to-noise ratio of the data by an order of magnitude and enhances the visibility of the molecular structure (Proc Natl Acad Sci U S A 97:3872–3877, 2000; Acta Crystallogr D Biol Crystallogr 61:1413–1417, 2005; Acta Crystallogr D Biol Crystallogr 61:539–544, 2005). In solution and surface scattering experiments, the incorporation of deuterium-labeled subunits or components into complex assemblies or structures makes it possible to deconvolute the scattering of the labeled and unlabeled subunits and to determine their relative dispositions within the complex (J Mol Biol 93:255–265, 1975). With multiple labeling patterns, it is also possible to reconstruct the locations of multiple subunits in ternary and higher-order complexes (Science 238:1403–1406, 1987; J Mol Biol 271:588–601, 1997; J Biol Chem 275:14432–14439, 2000; Biochemistry 42:7790–7800, 2003). In inelastic neutron scattering experiments, which probe hydrogen dynamics in biological materials, the application of site, residue, or region-specific hydrogen–deuterium-labeling patterns can be used to distinguish and highlight the specific dynamics within a system (Proc Natl Acad Sci U S A 95:4970–4975, 1998).

Partial, selective, or fully deuterated proteins can be readily produced by endogenous expression of recombinant proteins in bacterial systems that are adapted to growth in D_2O solution and using selectively deuterated carbon sources. Adaptation can be achieved either by gradual step-wise increase in D_2O concentration or, more directly, by plating cells on media of choice and selecting colonies that perform best for subsequent culture and inoculation. Scale-up growth and expression is typically performed in standard shaker flasks using either commercial or "home-grown" rich media (derived, for example, from cell lysates produced from algae grown in D_2O) or under more controlled conditions in defined minimal media. Cell growth is typically slower in deuterated media (>5 times slower) and yields are correspondingly lower. Once the target protein has been expressed, purification proceeds by the protocols developed for the hydrogenated protein. The deuteration levels of the final product are determined by mass spectrometry.

Key words: Protein, Membrane, Nucleic acid, Deuterium labeling, Cell culture, Over-expression, Deuterium exchange, Crystallography, Reflectometry, Small angle scattering, Contrast variation

James Weifu Lee and Robert S. Foote (eds.), *Micro and Nano Technologies in Bioanalysis*,
Methods in Molecular Biology, vol. 544
DOI 10.1007/978-1-59745-483-4_18, © Humana Press, a part of Springer Science+Business Media, LLC 2009

1. Introduction

Neutron scattering is exquisitely sensitive to the position, content, and dynamics of hydrogen atoms in materials and thus is a powerful tool for the characterization of structure-function and interfacial relationships in biological systems. Applications in biology range from the atomic resolution analysis of individual hydrogen atoms in enzymes through to meso- and macro-scale analysis of complex biological structures, membranes, and assemblies. Because neutrons interact with and scatter from nuclei, rather than with electrons, neutron scattering lengths (*b*) show little variation across the periodic table. Most importantly for biology, neutrons are extremely sensitive to hydrogen atoms and to the deuterium isotope, whose scattering length differs in both magnitude and phase, and selective substitution of hydrogen with deuterium can therefore be used to distinguish and highlight the position, structure, or dynamics of individual components within complex macromolecular systems or assemblies. Neutron studies are thus greatly enhanced by the design and production of specific, random, and uniform hydrogen/deuterium (H/D)-labeled biological macromolecules.

The degree and extent of deuterium labeling required for neutron scattering depends on the specific application. Neutron protein crystallography has the most stringent demands, requiring complete isotopic substitution of deuterium for hydrogen, which greatly reduces the hydrogen incoherent scattering background and significantly increase the signal-to-noise ratio of the diffraction data ***(1–3)***. In small-angle neutron scattering (SANS) applications, which rely on neutron contrast variation techniques, partial (~70–80%) deuterium labeling is generally sufficient to label, highlight, and map chemically distinct or D-labeled components of larger protein/protein or protein/lipid/nucleic acid complexes and assemblies ***(4–8)***. Similarly, neutron reflectometry using specific labeling and contrast variation allows the structure, composition, and organizational changes of membranes and of integral or membrane associated proteins to be dissected and examined in situ. In neutron spectroscopy, which accesses molecular dynamics in the nanosecond to picosecond range, more elegant amino acid residue, site, or regio-specific H-labeling of otherwise fully deuterated complexes can allow the internal dynamics of functional components to be analyzed in situ ***(9)***. Total, partial, or selective isotopic labeling is thus a powerful tool in neutron scattering analysis, which promises to provide new and more sophisticated ways to tackle complex problems in biology.

Here we describe approaches developed for the production of fully, partially, and selectively deuterated protein by endogenous expression of recombinant proteins in bacterial systems grown in D_2O solution using deuterated carbon sources.

2. Materials

2.1. Deuterated Media Preparation

1. D_2O (Cambridge Isotope Laboratories).
2. Minimal medium salts as shown in **Table 1**.
3. Trace metal solution as shown in **Table 1**.
4. Rotary Evaporator (Heidolph).
5. D_8-glycerol (Cambridge Isotope Laboratories).

2.2. Cell Adaptation

1. LB medium plate: 20 g/L LB powder (DIFCO, Lennox or Miller), 15 g/L Bacto-agar (DIFCO) autoclaved at 121°C for 15 min; antibiotic (1,000×) sterilized by filtration (*see* **Note 1**).
2. Minimal medium (2×): salt solution (2×), trace element solution (1,000×) as shown in **Table 1**. Trace element solution should be prepared fresh (*see* **Note 2**). The minimal medium is sterilized by filtration.
3. Hydrogenated minimal medium plate: 2× minimal medium sterilized by filtration; 2× warm (52°C) agar solution sterilized at 121°C for 15 min; 1,000× antibiotic solution sterilized by filtration (*see* **Note 3**).
4. Deuterated minimal medium (2×): 2× salt solution prepared in D_2O, 1,000× trace element solution prepared in D_2O as shown in **Table 1** (*see* **Note 4**).

Table 1
Minimal medium *(10)*

Component	Initial concentration
$(NH_4)_2SO_4$	6.86 g/L
KH_2PO_4	1.56 g/L
$Na_2HPO_4 \cdot 2H_2O$	6.48 g/L
$(NH_4)_2$-H-citrate	0.49 g/L
$MgSO_4 \cdot 7H_2O$	0.25 g/L
Trace metal solution	1.0 mL/L
Glycerol	5.0 g/L
Trace metal solution 0.5 g/L $CaCl_2 \cdot 2H_2O$, 16.7 g/L $FeCl_3 \cdot 6H_2O$, 0.18 g/L $ZnSO_4 \cdot 7H_2O$, 0.16 g/L $CuSO_4 \cdot 5H_2O$, 0.15 g/L $MnSO_4 \cdot 4H_2O$, 0.18 g/L $CoCl_2 \cdot 6H_2O$, 20.1 g/L EDTA	

5. Deuterated minimal medium plate: 2× deuterated minimal medium sterilized by filtration; 2× warm (52°C) agar solution prepared in D_2O sterilized at 121°C for 15 min; 1,000× antibiotic solution prepared in D_2O sterilized by filtration (*see* **Note 3**).
6. 15-mL BD Falcon™ conical-bottom polypropylene tubes.
7. Vacuum-driven filtration and storage devices (Stericup Filter Units; Millipore).

2.3. Cell Culture

2.3.1. Flask

1. 1× deuterated minimal medium supplemented with antibiotic.
2. Fernbach flask.
3. Induction solution: 1,000× IPTG prepared in D_2O.

2.3.2. Bioreactor

1. 1× deuterated minimal medium supplemented with antibiotic.
2. Feeding solution: 10% deuterated glycerol, 0.2% $MgSO_4$ in D_2O supplemented with antibiotic.
3. Base solution: 10% NaOD in D_2O.
4. Induction solution: 1,000× IPTG prepared in D_2O.
5. 1.25-L Bioflo 3000 Bioreactor (New Brunswick Scientific).
6. Polypropylene glycol (PPG) (Sigma-Aldrich).
7. Air, nitrogen (Air Liquide).
8. Storage bottle headpiece (Sartorius BBI Systems).
9. Dissolved oxygen (DO) probe (Broadley James).
10. pH probe (Broadley James).

2.4. Cell Lysis and Purification

1. Hydrogenated purification buffers (*see* **Note 5**).

2.5. Deuterium Back-Exchange

1. Final protein buffer prepared in D_2O (*see* **Note 6**).
2. Centrifugal filter units (Amicon, Millipore).

3. Methods

This section describes a protocol for production of fully (per) deuterated protein. The levels of deuteration reached are greater than 95%. Substitution of the deuterated carbon source with a hydrogenated carbon source, and/or using D_2O/H_2O mixed solutions to prepare the medium will produce lower levels of deuteration, which are sufficient for neutron contrast variation experiments such as SANS and reflectometry. Alternatively, using protocols originally developed for nuclear magnetic resonance (NMR) applications, (per)deuterated medium can be supplemented with hydrogenated amino acids or their precursors for the preparation of selectively hydrogenated, deuterated proteins ***(11)***.

Culture growth in fully deuterated medium is typically slower than in hydrogenated media, and growth time can be can be long, especially when using a bioreactor. Therefore, plasmid loss over time can be a major problem in deuterated culture and can explain low or even null overexpression levels. The stability of the plasmid used requires careful consideration ***(12,13)***. An important parameter of plasmid stability is the selection marker. Under ampicillin selection, the β-lactamase protein that confers resistance is stored in the periplasmic space. An inner and an outer cell membrane limit the periplasmic space. "Leakiness" of the outer membrane leads to β-lactamase secretion into the culture medium. Selection then becomes rapidly ineffective because ampicillin can then be degraded by secreted β-lactamase. This can result in growth of bacteria that have lost their plasmids or contaminant cell growth. In contrast, the protein that confers kanamycin resistance is cytosolic and therefore less likely to leak into the culture medium. Expression vectors that confer ampicillin resistance to the host cell should therefore be avoided (*see* **Note 7**). The expression vector should also allow for overexpression induction in the last phase of the deuterated culture to avoid possible degradation of protein over time.

3.1. Preparation of Fully Deuterated Minimal Medium

Production of fully (per)deuterated protein requires media prepared from 100% D_2O and a perdeuterated carbon source. To prepare a "hydrogen-free" medium, special precautions need to be taken.

1. Dissolve hydrogenated and hydrated mineral salts in D_2O so that labile hydrogen atoms are exchanged for deuterium and dry using a rotary evaporator. Repeat twice for a more complete exchange. The deuterated salts are then dissolved in D_2O to make up the medium salt solution (*see* **Note 8**). A 2× salt solution can be prepared.
2. Similarly, hydrogenated and hydrated trace element salts should be dissolved in D_2O to exchange hydrogen for deuterium and then dried using a rotary evaporator. Prepare a 1,000× solution and add to the salt solution.
3. Prepare a 1,000× antibiotic solution in D_2O and add to the medium.
4. Any chemicals required for protein overexpression (substrate, cofactor, metal ion) can be dissolved in D_2O and added to the medium. Again, these should be deuterated if possible.
5. Add D_8-glycerol and sterilize the medium by filtration using a vacuum-driven filtration and storage device (Stericup Filter Units; Millipore).

3.2. Cell Adaptation on Solid Medium

Expression of (per)deuterated protein requires first an adaptation of the cells to growth in fully deuterated media. This adaptation can be made in a three-step process using solid media in standard plates at 37°C. Adaptation can also be made in liquid media (*see* **Note 9**).

1. Plate freshly transformed cells on a hydrogenated solid LB medium plate.
2. Select a colony and plate on hydrogenated solid minimal medium.
3. After overnight growth, plate cells on fully deuterated (heavy water and deuterated carbon source) solid minimal medium. To prepare solid deuterated minimal medium plates, autoclave a 2× mixture of agar in D_2O. In parallel, prepare a 2× liquid deuterated minimal medium that has been supplemented with antibiotic and filter sterilized. Combine equal volumes of the warmed solutions and pour the plates. Once plated, cell growth on the deuterated plates is observed after 2–4 days of incubation.
4. Select a colony and transfer adapted cells to fully deuterated liquid medium. Once growth is established, fresh deuterated minimal medium can be inoculated in a 1:20 ratio. Cycling this step increases the initial growth rate.
5. At this point, large volume cultures can be inoculated. If required, adapted cells can be stored in 10-mL aliquots at −80°C after flash freezing in liquid nitrogen.
6. Thaw adapted cells stored at −80°C slowly on ice (*see* **Note 10**).
7. Before growing the actual inoculum, perform up to four transfers in freshly prepared deuterated media to refresh the cells, complete the adaptation, and improve the growth rates.
8. Prepare inoculum in standard sterilized and dried flasks in shaking incubators. The deuterated culture inoculum volume can be up to 1/10th of the starting culture volume, presuming that the inoculum is in the exponential growth phase and free of toxic byproduct (*see* **Note 11**). An OD_{600} of 4 has been used successfully.

3.3. Culture

Deuterated cultures can be grown in flasks or, to reach higher cell density, in bioreactors. When using a bioreactor, the yield can be improved by first running a batch phase, followed by a fed-batch phase. Care should be taken to avoid any source of hydrogen (water drops, vapor) contamination.

3.3.1. Flask

1. Sterilize and dry flasks.
2. Prepare deuterated minimal medium supplemented with antibiotic.
3. Inoculate the medium with a D_2O adapted culture.
4. Shake at 180 rpm until the OD reaches 2.0.
5. Induce overexpression.
6. Stop the culture and harvest the cells when the expression level is satisfactory.

3.3.2. Bioreactor

1. Place and properly align the head plate on the 1.25-L vessel and tighten the screws.
2. Place the protective stainless steel cap over the bearing housing. Do not sterilize the rubber motor coupling.
3. Connect a 37-mm inlet filter (0.2 μm) to the sparger and a 50-mm exhaust filter (0.2 μm) to the condenser via a short length of tubing.
4. Attach tubing for the feed, base, and acid solutions to the appropriate ports. Wrap or clamp any open ends to maintain a sterile reactor (*see* **Note 12**). No solutions or probes should be added to the vessel before sterilization.
5. Prepare bottles with a storage bottle headpiece for feed, base, and acid solutions.
6. Autoclave the bioreactor and bottles at 121°C for 15 min.
7. Connect the water lines to the exhaust condenser and vessel jacket.
8. Turn on the water supply and then the bioreactor control unit.
9. Attach the inlet tubing to the inlet filter, remove the inlet clamp, and dry the vessel thoroughly with sterile-filtered, compressed air. Dry the feed, base, and acid bottles in a drying oven.
10. Check the tip of the DO probe for punctures or tears. Refill the tip with electrolyte solution if needed. Polarize the probe according to manufacturer's specifications (~6 h).
11. Calibrate the pH probe with standard solutions of known (pH 4 and 7).
12. Carefully sterilize the DO and pH probes with a 70% ethanol solution and insert them into the vessel.
13. Fill the vessel with deuterated medium through the inoculation port. Fill the feed, base, and acid bottles with solutions sterilized by filtration. Remove clamps and attach the bottles to the vessel using the corresponding tubing and pumps.
14. Calibrate the DO probe by sparging N_2 and air into the vessel to set the 0 and 100% calibrations, respectively.
15. Insert the temperature probe into the thermowell and set the bioreactor to the desired temperature.
16. Remove the protective stainless steel cover from the bearing housing and attach the rubber motor coupling. Attach the motor and connect it to the control unit.
17. Set the airflow to 0.5 L/min and the agitation rate to 200 rpm.
18. Inoculate the medium with a D_2O adapted culture.
19. Begin controlling the pH and DO% with the bioreactor's control unit or PC-based control software. The initial DO% in the bioreactor vessel is 100% (no oxygen consumption) and the culture medium defines the pH. During the batch

phase, cell growth is not controlled and is close to the maximum growth rate. After inoculation, the DO% is automatically adjusted to 30% *(14)* by controlling the stirring rate (*see* **Note 13**). The pH, which generally changes as the carbon source is metabolized, is kept within 0.1 U of initial pH of the medium. This is controlled by automatic addition of base or acid solution (*see* **Note 14**).

20. Add 200 μL of PPG to prevent foam formation.
21. Occasionally sample and check the OD_{600} of the culture during the growth (*see* **Note 15**).
22. During the fed-batch phase, provide the cells with fresh carbon source solution. The growth rate should be maintained constant but slightly lower than the maximum growth rate. This is performed by providing a limited amount of carbon source. A deuterated feeding solution is prepared with 10% deuterated glycerol, 0.2% $MgSO_4$, and antibiotic (*see* **Note 16**). The feeding rate can be determined manually by taking into consideration the regulation of the pH and of the DO%. The feeding rate is increased when a decrease in the base solution addition frequency and in the aeration is observed, both sign of depletion of the carbon source. The fed-batch phase can last up to 5 days. Alternatively, a control sequence can be used to automatically estimate appropriate feeding rates as the culture progresses. Expected yields are 1 g of cell paste per gram of carbon source used.
23. Protein overexpression can be induced at any time during the fed-batch phase because the cell growth rate is constant.
24. Stop the culture at the end of the induction period and harvest.

3.4. Cell Lysis and Purification

Standard protocols can be used for cell lysis using buffers and solutions prepared in H_2O and hydrogenated media. Subsequent purification steps for the target protein can also be done in hydrogenated buffers and solutions following the protocols established for the "native" hydrogenated form of the protein. Although this allows labile deuterium atoms on amide or hydroxyl groups to exchange for hydrogen during the protein purification steps (accounting for ~20% of the deuterium content of typical proteins), these can be readily back-exchanged to deuterium by equilibration with a final deuteration buffer.

3.5. Deuteration Level

Mass spectrometry is used to calculate the deuteration level of the (per)deuterated protein. The theoretical molecular weight of (per) deuterated protein purified in hydrogenated buffer is given by $MW_{\text{partially deuterated}} = MW_{\text{hydrogenated}}$ + (number of non-exchangeable deuterium) × 1.006, where with 1.006 is the mass difference between deuterium and hydrogen. An example is given in **Table 2**. The deuteration level is then given by $(MW_{\text{partially deuterated}} - MW_{\text{hydrogenated}})_{\text{determined by mass spectroscopy}} / (MW_{\text{partially deuterated}} - MW_{\text{hydrogenated}})_{\text{theoretical}}$.

Table 2
Theoretical molecular weight calculation of hydrogenated and (per)deuterated protein purified in hydrogenated buffer (all exchangeable deuterium atoms are considered to have exchanged to hydrogen during the purification). An example is given for rubredoxin, a 53-amino acid protein

AA	AA(H) MW	Number of AA	MW(H)	Non-exch. H/D	AA(H/D) MW	MW(H/D)
Ala	71.0788	3	213.2364	4	75.1028	225.3084
Arg	156.1876	0	0	7	163.2296	0
Asn	114.1039	1	114.1039	3	117.1219	117.1219
Asp	115.0886	7	805.6202	3	118.1066	826.7462
Cys	103.1448	4	412.5792	3	106.1628	424.6512
Glu	129.1155	6	774.693	5	134.1455	804.873
Gln	128.1308	0	0	5	133.1608	0
Gly	57.052	5	285.26	2	59.064	295.32
His	137.1412	0	0	5	142.1712	0
Ile	113.1595	4	452.638	10	123.2195	492.878
Leu	113.1595	2	226.319	10	123.2195	246.439
Lys	128.1742	5	640.871	9	137.2282	686.141
Met	131.1986	0	0	8	139.2466	0
Phe	147.1766	2	294.3532	8	155.2246	310.4492
Pro	97.1167	5	485.5835	7	104.1587	520.7935
Ser	87.0782	2	174.1564	3	90.0962	180.1924
Thr	101.1051	1	101.1051	5	106.1351	106.1351
Trp	186.2133	2	372.4266	8	194.2613	388.5226
Tyr	163.176	2	326.352	7	170.218	340.436
Val	99	2	198	8	107.048	214.096
End effect	18	1	18		18	18
Rubredoxin		53	5,895.2975	120		6,198.1035

AA amino acid; *AA(H) MW* hydrogenated amino acid molecular weight; *Number of AA* number of amino acid residues in the protein; *MW(H)* total contribution of amino acid to the protein molecular weight; *Non-exch. H/D* number of non-exchangeable hydrogen/deuterium in amino acid; *AA(H/D) MW* partially deuterated amino acid molecular weight; *MW(H/D)* total contribution of amino acid to the partially deuterated protein molecular weight

3.6. Deuterium Back-Exchange

Labile deuterium atoms exchange to hydrogen during the purification process and need to be exchanged back to deuterium. Back exchange to deuterium is completed by three dilution–concentration cycles of the protein in deuterated buffer using a centrifugal filter unit with the appropriate molecular weight cutoff (Amicon, Millipore). An overnight break before the last cycle may favor back-exchange of buried and protected hydrogen atoms (*see* **Note 17**).

3.7. Site-Specific Hydrogenation of Deuterated Proteins

Selectively methylated, triple-labeled proteins have been previously prepared for NMR applications. In those methods, the ketoacid precursor, [3-^{2}H] ^{13}C α-ketoisovalerate, was used ***(11)***. By simply changing the isotopic composition of the α-ketoisovalerate precursor and the minimal medium, (^{1}H-δ methyl)-leucine and (^{1}H-γ methyl)-valine can be selectively incorporated into an otherwise deuterated protein for neutron protein crystallography or spectroscopy applications using the method below.

1. Sterilize and dry flasks.
2. Prepare [3-^{2}H] α-ketoisovalerate from unlabeled α-ketoisovalerate by warming a 25 mM solution of α-ketoisovalerate in D_2O at pH 12.5, 45°C for 3 h ***(11)***.
3. Prepare fully deuterated minimal medium using 0.3%, w/v D-glucose (1,2,3,4,5,6,6,-D7, 98%) from Cambridge Isotope Laboratories supplemented with antibiotic.
4. Inoculate the medium with a D_2O-adapted culture and incubate at 37°C with 250-rpm shaking.
5. Approximately 1 h before induction, add 100 mg of [3-^{2}H] α-ketoisovalerate/L culture.
6. Induce overexpression with 1 mM IPTG at an OD_{600} of 0.9.
7. Harvest cells after 4 h of induction.

4. Notes

1. Antibiotic is added when the medium cools below 52°C.
2. Metalloprotein overexpression experiments may require the addition, subtraction, or substitution of certain salts in the trace element solution.
3. Add sterile-filtered minimal medium to warm autoclaved agar solution while stirring to avoid agar lumps.
4. Ensure that all of the glassware used is free of any trace of water (drops, vapor).
5. Although no significant modifications are expected compared with purifying hydrogenated material, a small-scale preparation

is recommended to check that the affinity of the deuterated material for the resins is not altered by isotopic effects. pH is the parameter most likely affected if differences in affinity are observed.

6. pH and pD differ by ~0.4 U in 100% D_2O. The relation between the pH read on a pH meter ($pH_{measured}$) and the pD of a solution is given by pD = $pH_{measured}$ + 0.4.
7. Antibiotic resistance can be switched using the Ez-Tn5 Kan-2e insertion kit (Epicentre Biotechnologies).
8. $MgSO_4$ must be dissolved last or precipitation will be observed in the medium.
9. Cells can be adapted using liquid media exclusively, starting from liquid LB, to liquid hydrogenated minimal medium and to liquid deuterated minimal medium. This may require increasing stepwise the D_2O concentration of the liquid deuterated medium (e.g., 10, 50, 80, and 100%). This alternate protocol requires the cells to be transferred when in exponential growth phase—or adaptation may fail.
10. The inoculum can be started from freshly adapted cells.
11. The pH of the inoculum should be around the pH of the medium. A large pH shift indicates the presence of growth by-products that may eventually be toxic as they accumulate.
12. The exhaust tubing should be covered with aluminum foil and not clamped.
13. Depending on the starting optical density, the consumption of the O_2 initially present can range from several minutes to several hours.
14. The pH decreases when glycerol is used as carbon source. An increase of pH can be used as an indicator for increasing the feeding rate.
15. The OD_{600} can be continuously monitored if a probe is available.
16. Cost of the deuterated carbon source should be considered when preparing the feeding solution. A higher concentration can be used if cost is not an issue.
17. Dialysis can also be used, but the required volume of deuterated buffer is larger.

Acknowledgments

This work was supported by the Office of Biological and Environmental Research of the U.S. Department of Energy project KP1102010 and the Laboratory Directed Research and Development program of Oak Ridge National Laboratory, managed by UT-Battelle, LLC under contract No. DE-AC05-00OR22725 with Oak Ridge National Laboratory. The submitted manuscript has been authored by a contractor of the U.S. Government under Contract DE-AC05-00OR22725. Accordingly, the U.S. Government retains a nonexclusive royalty-free license to publish or reproduce the published form of this contribution, or allow others to do so, for U.S. Government purposes.

References

1. Shu, F., Ramakrishnan, V., and Schoenborn, B.P. (2000). Enhanced visibility of hydrogen atoms by neutron crystallography on fully deuterated myoglobin. *Proc Natl Acad Sci U S A* **97**, 3872–3877.
2. Hazemann, I., Dauvergne, M.T., Blakeley, M.P., Meilleur, F., Haertlein, M., Van Dorsselaer, A., Mitschler, A., Myles, D.A., and Podjarny, A. (2005). High-resolution neutron protein crystallography with radically small crystal volumes: application of perdeuteration to human aldose reductase. *Acta Crystallogr D Biol Crystallogr* **61**, 1413–1417.
3. Meilleur, F., Dauvergne, M.T., Schlichting, I., and Myles, D.A. (2005). Production and X-ray crystallographic analysis of fully deuterated cytochrome P450cam. *Acta Crystallogr D Biol Crystallogr* **61**, 539–544.
4. Ibel, K., and Stuhrmann, H.B. (1975). Comparison of neutron and X-ray scattering of dilute myoglobin solutions. *J Mol Biol* 93, 255–265.
5. Capel, M.S., Engelman, D.M, Freeborn, B.R., Kjeldgaard, M., Langer, J.A., Ramakrishnan, V., Schindler, D.G, Schneider, D.K., Schoenborn, B.P., Sillers, I.Y., Yabuki, S., and Moore, P.B. (1987). A complete mapping of the proteins in the small ribosomal subunit of *Escherichia coli*. *Science* **238**, 1403–1406.
6. Svergun, D.I., Burkhardt, N., Pedersen, J.S., Koch, M.H., Volkov, V.V., Kozin, M.B., Meerwink, W., Stuhrmann, H.B., Diedrich, G., and Nierhaus K.H. (1997). Solution scattering structural analysis of the 70S *Escherichia coli* ribosome by contrast variation. I. Invariants and validation of electron microscopy models. *J Mol Biol* **271**, 588–601.
7. Svergun, D.I., and Nierhaus, K.H. (2000). A map of protein-rRNA distribution in the 70S *Escherichia coli* ribosome. *J Biol Chem* **275**, 14432–14439.
8. Heller, W.T., Finley, N.L., Dong, W.J., Timmins, P., Cheung, H.C., Rosevear, P.R., and Trewhella, J. (2003). Small-angle neutron scattering with contrast variation reveals spatial relationships between the three subunits in the ternary cardiac troponin complex and the effects of troponin I phosphorylation. *Biochemistry* **42**, 7790–7800.
9. Réat, V., Patzelt, H., Ferrand, M., Pfister, C., Oesterhelt, D., and Zaccai, G. (1998) Dynamics of different functional parts of bacteriorhodopsin: H-^{2}H labeling and neutron scattering. *Proc Natl Acad Sci U S A* **95**, 4970–4975.
10. Enfors, S.O., and Häggström, L. (2000). *Bioprocess technology: fundamentals and applications*, Högskoletryckeriet, Royal Institute of Technology, Stockholm.
11. Goto, N.K., Gardner, K.H., Mueller, G.A., Willis, R.C., and Kay, L.E. (1999). A robust and cost-effective method for the production of Val, Leu, Ile (δ1) methyl-protonated ^{15}N-, ^{13}C-, ^{2}H-labeled proteins. *J Biomol NMR* **13**, 369–374.
12. Tierny, Y., Hounsa, C.G., and Hornez, J.P. (1999). Effects of a recombinant gene product and growth conditions on plasmid stability in pectinolytic *Escherichia coli* cells. *Microbios* **97**, 39–53.
13. Park, S.H., Ryu, D.D.Y., and Lee, S.B. (1991). Determination of kinetic parameters related to plasmid instability: for the recombinant fermentation under repressed condition. *Biotech Bioeng* **37**, 404–414.
14. Riesenberg, D., Schulz, V., Knorre, W.A., Pohl, H.D., Korz, D., Sanders, E.A., Ross, A., and Deckwer, W.D. (1991). High cell density cultivation of *Escherichia coli* at controlled specific growth rate. *J Biotechnol* **20**, 17–27.

Chapter 19

Small-Angle Neutron Scattering for Molecular Biology: Basics and Instrumentation

William T. Heller and Kenneth C. Littrell

Summary

As researchers strive to understand the interplay between the complex molecular systems that make up living cells, tools for characterizing the interactions between the various players involved have developed. Small-angle neutron scattering (SANS) plays an important role in building a molecular-level understanding of the structures of macromolecular systems that make up cells. SANS is widely applicable to the study of biological structures including, but by no means limited to, protein–protein or protein–nucleic acid complexes, lipid membranes, cellular scaffolding, and amyloid plaques. Here, we present a brief description of the technique as it is commonly applied to the study of biological systems and an overview instrumentation that is available at the various facilities around the world.

Key words: Small-angle scattering, Neutrons, Contrast variation, User facilities

1. Introduction

Neutron scattering has long been applied to the study of the structure and dynamics of a vast array of materials, including polymers, magnetic materials, and biological systems. As a result, neutron scattering has a broad impact on a wide variety of scientific disciplines. This versatility stems from both the array of experimental methods that neutron scattering encompasses and the wide variety of materials to which these techniques can be applied. The specific properties of the neutron, being charge neutral, highly penetrating, and having a magnetic moment, make it a very powerful probe of matter that is capable of providing information not available to other techniques. Small-angle neutron scattering (SANS), which is one of the broadest applications

James Weifu Lee and Robert S. Foote (eds.), *Micro and Nano Technologies in Bioanalysis*,
Methods in Molecular Biology, vol. 544
DOI 10.1007/978-1-59745-483-4_19, © Humana Press, a part of Springer Science+Business Media, LLC 2009

of neutron scattering, is a structural probe applicable to length scales ranging from one to hundreds of nanometers.

Biological systems, in particular, are excellently suited for study by SANS. One common application of SANS to biological systems probes homogeneous, dilute solutions where the non-interacting particles are free to diffuse in their native structural state. SANS is a particularly powerful tool for investigating the interactions between binding partners, such as ions, small molecules, or other biological macromolecules. SANS can also probe the interplay between membranes and proteins. It is also applicable to larger molecular assemblies including cellular scaffolding and amyloid plaques. For this reason, SANS is used to obtain information complementary to that obtained by other structural biology methods.

Crystallography and nuclear magnetic resonance (NMR) can provide atomic-resolution information for biological macromolecules. Aside from the requirement that a sample crystallize, there are no practical limitations on the size of systems that can be addressed using crystallographic methods. Unfortunately, obtaining crystals is not always trivial, and, frequently, truncated variants of biological macromolecules must be studied to obtain crystals that diffract sufficiently. Crystallography is also a poor choice for studying dynamic biological macromolecular complexes that do not exhibit tight binding. NMR is the one competing technique in structural biology that can provide atomic resolution information on biological macromolecules in solution, but it also has limitations. The maximum size of systems that can be studied by NMR is much smaller than the multisubunit complexes that are the molecular machines of living cells. Although it does not provide atomic-resolution structures, SANS excellently complements both methods by expanding the size range of structures that can be studied and by being applicable to dynamic structures and processes.

Electron microscopy (EM) is another structural technique commonly applied to proteins and other biological macromolecular systems. EM is a static technique in which the samples are adsorbed to a mounting grid and frozen. Unlike crystallography, NMR, and SANS, EM is a direct imaging technique. The charge of the electron makes it possible to use electromagnetic lenses to focus the beam, as well as resolve and magnify the image of the structure being probed. The difference in electron density between the protein and the surrounding solution is often low, so samples require staining with a heavy atom salt that can lead to distortions in the structure. However, EM's ability to provide structures at up to 5 Å resolution makes it a very powerful technique for structural molecular biology. SANS complements EM by enabling the study of very large complexes in a native-like solution state while providing structural information of comparable resolution.

2. SANS Basics

SANS probes density in inhomogeneities in materials, whether they be between two components in a binary polymer mixture or proteins in aqueous solution. Unlike X-rays or electrons, which interact with the electrons of atoms, neutrons interact with atomic nuclei. The strength of the interaction between the neutron and an atomic nucleus that is important to small-angle scattering is characterized by the coherent scattering length of the nucleus. The scattering length depends on both the atomic species and the isotope of the atom. Unlike the X-ray-scattering length of an atom, which scales linearly with the number of electrons, the neutron scattering length does not vary in a predictable manner with atomic number or isotope. More importantly, the magnitudes of the scattering lengths of all of the atoms and their isotopes are comparable, making visualization of the light atoms easier with neutrons than with X-rays. The coherent scattering lengths of atoms and isotopes relevant to biology are shown in **Table 1**. Special attention should be given to the values for hydrogen and its isotope deuterium, which differ in sign. This large difference and the ability to substitute deuterium for hydrogen in biological structures with limited impact on structure and function makes SANS a very powerful method for investigating complex biological systems.

In SANS, the neutrons are incident on the sample as plane waves. From the de Broglie relation, the neutron has a wavelength

Table 1
Neutron-scattering lengths for atoms and isotopes of interest to biology *(1)*. If no isotope is specified, the average resulting from the natural isotopic abundance is assumed

Isotope	Scattering length (10^{-15} m)
^{1}H	−3.74
^{2}H (D)	6.67
C	6.65
N	9.36
O	5.80
S	2.84
P	5.13

$\lambda = h/mv$, where h is Plank's constant, m is the mass of the neutron, and v is the neutron velocity. The neutrons commonly used for scattering experiments have wavelengths ranging from 2 to 20 Å, which corresponds to velocities ranging from a few hundred to a few thousand meters per second. SANS results from the interference of the secondary (scattered) waves from the atomic nuclei in the scattering object. The strength of the scattering is a function of the difference in the scattering length density of the particle relative to that of the solvent and the direction relative to the incident beam. The scattered signal is measured as a function of $\vec{q}$, the momentum transfer of the scattered neutron, which has a magnitude $q = 4\pi\sin(\theta)/\lambda$. 2θ is the angle between the scattered neutron and the incident beam. The most general form for describing the scattering resulting from particles in solution, assuming no distance correlations among them, is given by **Eq. 1** *(2, 3)*:

$$I(q) = n\left\langle \left| \int_V (\rho(\vec{r}) - \rho_s) e^{-i\vec{q}\cdot\vec{r}} \, d^3r \right|^2 \right\rangle, \qquad (1)$$

where $I(q)$ is the scattered intensity, n is the number of particles per unit volume, $(\rho\vec{r})$ is the scattering length density of the particle at position $\vec{r}$, ρ_s is the scattering length density of the solvent, and the integral is taken over the particle volume *V(2, 3)*. The integral is averaged over time, all orientations, and the ensemble of structures present in the solution during the measurement.

Equation 1 describes the scattering relative to the background solvent, which is generally water or buffer solution for biological materials. As a result, contrast variation methods *(4)* can leverage the substitution of deuterium for hydrogen in the solvent with minimal structural or functional impact. It is also possible to vary the scattering length density of a biological macromolecule by substituting deuterium for hydrogen. Contrast variation is a particularly powerful technique for studying multisubunit complexes composed of components having different scattering lengths because it enables the separation of the scattering signals of the components within the complex. In a contrast variation experiment, a complex, such as a selectively deuterated protein-protein complex or a protein–DNA complex, is measured in a series of solutions consisting of mixtures of H_2O and D_2O. The intensity profiles in the contrast variation series can be written in the following manner:

$$I(q) = \Delta\rho_1^2 I_1(q) + \Delta\rho_2^2 I_2(q) + \Delta\rho_1 \Delta\rho_2 I_{12}(q). \qquad (2)$$

Here, $\Delta\rho_1$ and $\Delta\rho_2$ are the scattering length density differences of the two components of the complex having different average scattering length densities relative to the scattering length density of solvent. $I_1(q)$, $I_2(q)$, and $I_{12}(q)$ are the basic scattering functions. $I_1(q)$ and $I_2(q)$ are the scattering from the components having different scattering length densities, and the cross-term $I_{12}(q)$ contains information about the relative disposition of the components. The measured contrast variation series data defines a set of linear equations that can be solved for the basic scattering functions for further analysis.

2.1. Other Small-Angle Scattering Methods

Although SANS can provide unique information, other small-angle scattering techniques exist that provide complementary information. Small-angle X-ray scattering (SAXS) is the most directly comparable technique, providing information similar to a SANS experiment. The options for contrast variation in a SAXS experiment are much more limited than for SANS experiments, however. Although proteins and nucleic acids have inherently different electron densities, it is not possible to vary the electron densities of proteins to enable contrast variation studies of protein-protein complexes. Additionally, changing the electron density of the background solution requires addition of high concentrations of a small molecule to the solution, such as salt, glycerol, or a sugar (i.e., sucrose). These additives have a much stronger effect on the chemical behavior of the background solution than D_2O and can produce undesirable behavior, such as aggregation. Light scattering is another complementary experimental method, but it is generally more applicable to larger systems because of the much longer wavelength of the laser light relative to X-rays and neutrons.

3. SANS Instrumentation

SANS instruments are conceptually simple. A schematic is shown in **Fig. 1**. After the source, monochromator/chopper systems select the desired wavelengths for the experiment. Neutron guides (not shown), which are made of nickel-coated borosilicate glass, can be used to effectively bring the source closer to the sample. Neutron guides function by total internal reflection, which is possible because of the wave-like behavior of the neutron. In a manner similar to light transmission down a fiber optic cable, neutrons reflect off the nickel coating with no loss in flux when they are incident onto the nickel surface at shallow angles. The nickel, specifically the isotope ^{58}Ni, provides the change in the neutron's index of refraction relative to the vacuum inside the guide that allows the total internal reflection to take place. Optical elements, such as pinholes, produce a tightly collimated beam that is incident on the sample. The sample-scattered neutrons then spread out from the direction of the incident beam, where they are collected by the detector as an intensity pattern. Brief descriptions of these

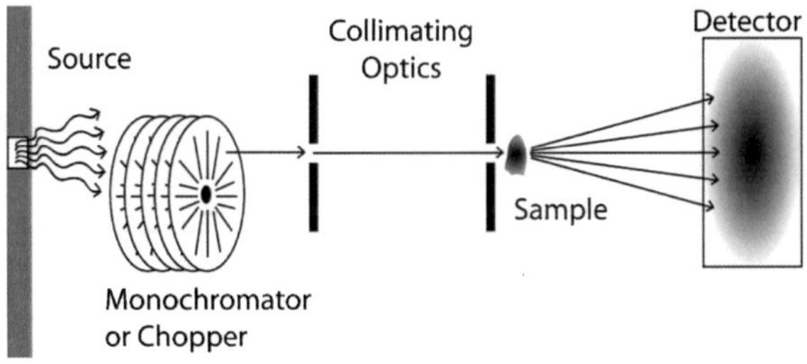

Fig. 1. Schematic of a SANS instrument.

elements and a short discussion of SANS instrument resolution considerations follow.

3.1. Neutron Sources

Unlike SAXS and light scattering, which can be accomplished using laboratory-based instrumentation, the production of neutrons requires large, specialized facilities. Nuclear reactors, in this case being research reactors rather than power-generating reactors, can be used to generate neutrons. There are several research reactors with neutron-scattering programs that include SANS instruments (*see* **Table 2**), including the National Institute of Standards and Technology's Center for Neutron Research, the High-Flux Reactor at the Institute Laue-Langevin, the High Flux Isotope Reactor at Oak Ridge National Laboratory, FRM-II at the Heinz Maier-Leibnitz Institute, and the OPAL reactor of the Australian Nuclear Science and Technology Organisation. Alternatively, neutrons can be generated by colliding particles with a target by means of a particle accelerator. The collision boils off neutrons through a process termed spallation that are subsequently delivered to neutron-scattering instruments (*see* **Table 2**). Examples of spallation sources include the Spallation Neutron Source at Oak Ridge National Laboratory, ISIS at Rutherford Appleton Laboratory, LANSCE at Los Alamos National Laboratory, SINQ at the Paul Scherer Institut, and the Intense Pulsed Neutron Source at Argonne National Laboratory.

Although the two above methods for producing neutrons would seem to provide a clear division between classes of neutron-scattering facilities, a more logical means of classifying neutron sources is to group them into continuous and pulsed sources because this has the greatest impact on the instrumentation that is best served by the facility. As the name implies, a continuous source provides a constant flow of neutrons with a spectrum of wavelengths to the instruments. Choosing a specific wavelength for experiments involves the use of a monochromator such as a silicon or pyrolytic graphite crystal for diffracting a specific wavelength by varying the diffraction angle, or a velocity selector.

Table 2
Neutron-scattering facilities with SANS instrumentation. The information here was adapted from the "World Directory of SANS Instruments" at http://www.ill.eu/html/lss/more/world-directory-of-sans-instruments/ which contains links to the facilities

Facility and country	Instruments	Source type	User program
ILL, France	D11 *(5, 6)*, D22	Reactor	Yes
LLB, France	PACE, PAXE, PAXY, Papyrus	Reactor	Yes
ISIS, UK	LOQ *(7)*	Spallation, TOF	Yes
IPNS, USA	SASI *(8)*, SAND	Spallation, TOF	Yes
LANSCE, USA	LQD *(9)*	Spallation, TOF	Yes
NCNR, USA	NG-3 SANS *(10)*, NG-7 SANS *(10)*	Reactor	Yes
HFIR, USA	CG-2SANS *(11)*, BioSANS *(12)*	Reactor	Yes
FRJ-2, Germany	KWS-1 *(13)*, KWS-2 *(13)*	Reactor	Yes
FRM-II, Germany	SANS-1 *(14)*, REFSANS *(15)*	Reactor	Yes
BENSC, Germany	V4-SANS *(16, 17)*	Reactor	Yes
GKSS, Germany	SANS-1 *(18)*, SANS-2	Reactor	Yes
SINQ, Switzerland	SANS-I *(19)*, SANS-II	Spallation, continuous	Yes
BATAN, Indonesia	SMARTer *(20)*	Reactor	Yes
OPAL, Australia	Quokka *(21)*	Reactor	Yes
BNC, Hungary	SANS *(22)*	Reactor	Yes
JAERI, Japan	SANS-U *(23)*, SANS-J *(24)*	Reactor	Yes
JINR, Russia	YuMO *(25, 26)*	Reactor, pulsed	Yes
HANARO, Korea	SANS *(27, 28)*	Reactor	Yes
JEEP-II, Norway	SANS	Reactor	Collaborative

Neutron velocity selectors use stacks of rotating neutron-absorbing material with regular holes. The offset of the holes between successive rotating elements, the speed of rotation, and the tilt of the rotation axis with respect to the beam direction defines the wavelength and wavelength distribution that is transmitted through the device.

Pulsed sources produce bursts of neutrons at regular intervals. The burst of neutrons contains a spectrum of wavelengths (velocities) that then travel to the instrument. The different neutron velocities result in different arrival times at the instrument and detector. By noting the location of a detected neutron relative to the incident beam direction and the time of arrival relative to the originating pulse, it is possible to use a very broad spectrum of neutrons, rather than a relatively monochromatic beam that represents a much smaller fraction of the total spectrum produced by the neutron source. This approach is termed time-of-flight (TOF) and can be applied to a wide variety of neutron-scattering applications, including SANS.

Each type of source has benefits for SANS applications. Continuous sources are in some respects simpler on a conceptual level because of the use of a single wavelength. Additionally, continuous sources provide higher flux on sample than a pulsed source using TOF at the longest wavelengths provided by the source. As a result, continuous sources are more effective at measuring very low q-values. Pulsed sources must always balance the total flux produced by the source with detecting the entire spectrum. The longest wavelength neutrons produced often cannot transverse the entire length of the instrument without being overtaken by the shortest wavelength neutrons from the following pulse, resulting in a condition known as frame overlap. As a result, the range of wavelengths is restricted through the use of choppers. The loss of the long wavelength neutrons has the strongest impact at small q-values, limiting signal to noise and the effective minimum q-value. Still, pulsed TOF SANS instruments collect a very broad range of q-values at a single instrument setting and the inherent time structure of the instrument makes dynamic SANS measurements possible without the addition of choppers into the instrument to create a timing mechanism. A continuous-source SANS often requires multiple instrument configurations to collect data over the desired q-range, which makes them somewhat less efficient. A list of SANS instruments with the type of neutron source is shown in **Table 2**.

3.2. Collimation

One of the most important elements of a SANS instrument is the collimation system. The collimation defines the beam divergence, which in turn determines the minimum scattering angle, and thus q-value, that can be probed. The most commonly used method of beam collimation involves the use of two apertures, such as round holes or rectangular slits, to define the size of the beam at the source and at the sample. For this reason, SANS instruments are often called pinhole cameras. However, the camera (the sample aperture) images the source aperture, not the sample being studied. By assuming round source and sample apertures with diameters A_1 and A_2, respectively, that are separated by a distance L_1 with the detector a distance L_2 away, the size of the beam B at the detector is given by:

$$B = \frac{L_2}{L_1}(A_1 + A_2) + A_2. \tag{3}$$

Many SANS instruments have variable collimation. Different beam divergences are produced by changing the distance between the source and sample apertures, the sizes of the apertures, or both. Most commonly, the available distances between the source and sample apertures are discreet values. This step-wise variability results from the use of additional neutron guide sections to ensure the maximum flux at the sample position. Such systems retain enough versatility to provide an enormous number of possible instrument configurations.

Collimation can also be achieved through the use of lenses *(29, 30)*. Neutron lenses are made of materials such as MgF_2 *(29)*. Most materials have indices of refraction for neutrons very near unity because of the weak interaction of the neutron with matter. As a result, the focal length of a single lens can be very long (~200 m), necessitating the use of compound lens systems to reduce the effective focal length to more manageable distances. Alternatively, magnetic fields can be used to focus the neutron beam through interaction with the neutron magnetic moment with the applied field *(30)*. In either case, the focal plane of the lens system is the detector, rather than the sample, making it possible to significantly reduce the minimum q-value that can be measured. Because lenses either absorb and scatter neutrons or require very clean polarization that reduces the available flux by more than 50%, this gain comes at the expense of increased background and decreased signal to noise. In the case of polarized neutrons, incomplete polarization can also lead to a structured background. Furthermore, the use of lenses necessitates advanced instrument resolution corrections during data analysis and modeling, corrections for which there is not yet full agreement in the literature. On the other hand, all treatments to date agree that the resolution at a given value of q is substantially improved through the use of lenses, a conclusion that is experimentally verified. Lenses have typically been reserved for use on relatively strongly scattering systems with very large scattering particles that need a small minimum q-value.

3.3. SANS Detectors

Neutron detectors for SANS are either linear positions sensitive detectors, or two-dimensional (2D) (area) position sensitive detectors. Such detectors are wire detectors that collect charge that results from ionization events caused by neutrons impinging on a gas. The most efficient SANS detectors are ^{3}He detectors. The pixel size of most SANS detectors is a few millimeters, which provides sufficient angular resolution when compared with the relatively large distances involved in most instrument configurations. However, the large size of the detector pixels and the source and sample apertures means that the features measured on a sample by SANS are typically less sharp than would be observed with X-rays in a system with comparable contrast between the components. Scintillator detectors comparable to CCDs for detecting X-rays are available, but these have a much lower neutron detection efficiency and, hence, are not typically chosen for SANS.

At all of the major user facilities, area detectors are preferred. Area detectors maximize the number of neutrons collected and can be used to collect full scattering patterns from nonisotropic samples, such as those that result from an applied magnetic, electric, or shear field. In the case of isotropic samples, such as dilute protein solutions, 2D data is azimuthally averaged around the beam center, which serves to improve the statistical quality of the reduced data. This statistical improvement is often critical for SANS of dilute protein solutions, which do not scatter strongly.

3.4. Comments About Instrument Resolution

Researchers new to SANS are occasionally confused about the term "resolution," which is used in the community with at least two distinct but related meanings. Most commonly within the community the term resolution is used—somewhat erroneously —to describe the minimum q-value measured in a SANS measurement and thus to characterize the longest length scale or distance probed. As with diffraction measurements, $2\pi/q$ is the real space distance being probed. It is important to have a high enough resolution in the sense of a low enough minimum q to ensure that the instrument is capable of measuring the full extent of the scattering particle and of detecting whether aggregation or unexpected correlations or other interactions are occurring, unfortunately a not-uncommon situation with biological samples. This use of the term leads to the correct but rather counterintuitive conclusion that high-resolution instrument settings are used to measure large-scale structures whereas lower-resolution settings are used to measure smaller objects.

The second, more proper but less common, use of the term resolution refers to the impact of the physical parameters of the instrument and distribution of wavelengths provided by the source and monochromator system on the sharpness of features measured in the scattered intensity curve. The resolution in this sense is important for distinguishing between competing models and resolving features in a priori reconstructions of the scattering particle. This is illustrated by the SANS of Cow Pea Mosaic Virus shown in **Fig. 2** (data taken from **ref.** ***(31)***). The proper knowledge of the instrument resolution also allows the effects of polydispersity and conformational instability to be separated and quantified.

In practice, the very similar construction of nearly all SANS instruments means that, in the qualitative sense, both definitions of resolution are interchangeable—an instrument configuration that allows one to access lower minimum q also allow features at a given value of q to be distinguished more sharply.

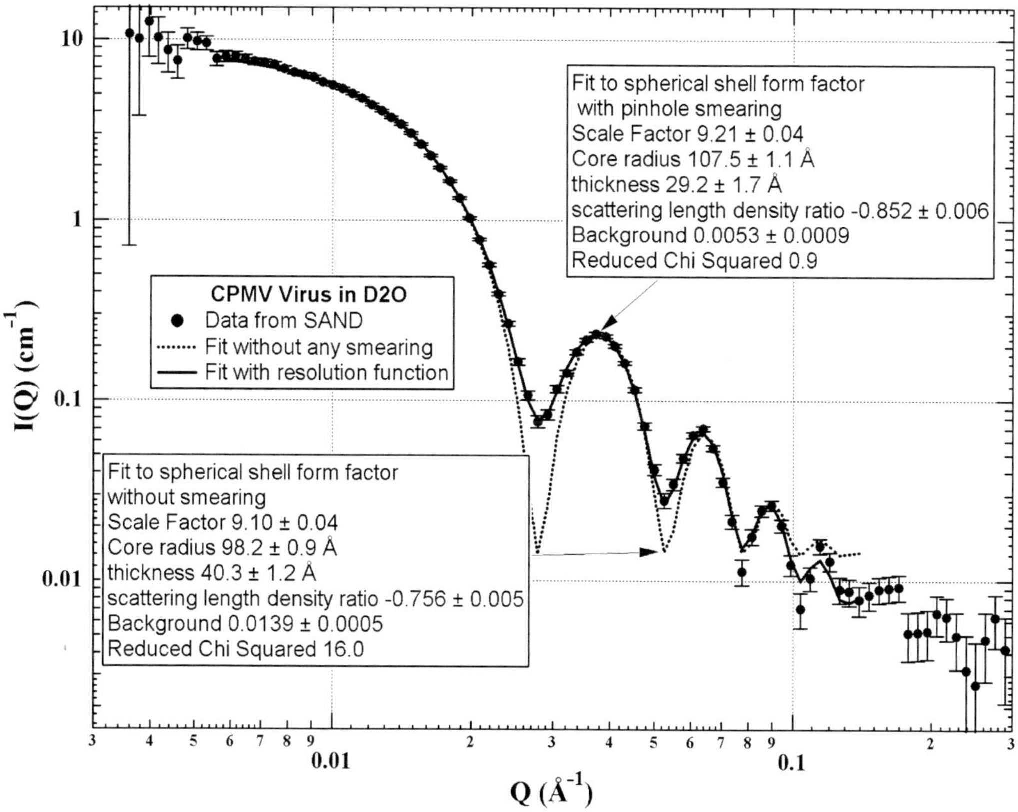

Fig. 2. Scattering from the CowPea Mosaic Virus in D_2O (data from **ref.** *(31)*). The proper inclusion of instrument resolution broadening dramatically improves the quality of the fit and shifts the parameters by several percent, far more than the estimated uncertainty in those parameters in most cases. The values obtained with instrument resolution taken into consideration are consistent with values estimated from knowledge of the structure.

Acknowledgments

This work was supported by Project KP1102010 of the Office of Biological and Environmental Research of the U.S. Department of Energy, under contract No. DE-AC05-00OR22725 with Oak Ridge National Laboratory, managed and operated by UT-Batelle, LLC. The submitted manuscript has been authored by a contractor of the U.S. Government under Contract DE-AC05-00OR22725. Accordingly, the U.S. Government retains a nonexclusive royalty-free license to publish or reproduce the published form of this contribution, or allow others to do so, for U.S. Government purposes.

References

1. Koester, L., Rauch, H. and Seymann, E. (1991). Neutron-scattering lengths - a survey of experimental-data and methods. *Atomic Data Nuclear Data Tables* **49**, 65–120
2. Debye, P. and Bueche, A. (1949). Scattering by an inhomogeneous solid. *J. Appl. Phys.* **20**, 518–525
3. Guinier, A. and Fournet, G. (1955). *Small-angle scattering of X-rays*, Wiley, New York
4. Ibel, K. and Stuhrmann, H. B. (1975). Comparison of neutron and X-ray-scattering of dilute myoglobin solutions. *J. Mol. Biol.* **93**, 255–265
5. Ibel, K. (1976). Neutron small-angle camera D11 at high-flux reactor, Grenoble. *J. Appl. Crystallogr.* **9**, 296–309
6. Lindner, P., May, R. P. and Timmins, P. A. (1992). Upgrading of the SANS instrument-D11 at the ILL. *Physica B* **180**, 967–972
7. Heenan, R. K., Penfold, J. and King, S. M. (1997). SANS at pulsed neutron sources: present and future prospects. *J. Appl. Crystallogr.* **30**, 1140–1147
8. Thiyagarajan, P., Epperson, J. E., Crawford, R. K., Carpenter, J. M., Klippert, T. E. and Wozniak, D. G. (1997). The time-of-flight small-angle neutron diffractometer (SAD) at IPNS, Argonne National Laboratory. *J. Appl. Crystallogr.* **30**, 280–293
9. Seeger, P. A., Hjelm, R. P. and Nutter, M. J. (1990). The low-Q diffractometer at the Los-Alamos-Neutron-Scattering-Center. *Mol. Cryst. Liq. Cryst.* **180**, 101–117
10. Glinka, C. J., Barker, J. G., Hammouda, B., Krueger, S., Moyer, J. J. and Orts, W. J. (1998). The 30 m small-angle neutron scattering instruments at the National Institute of Standards and Technology. *J. Appl. Crystallogr.* **31**, 430–445
11. Lynn, G. W., Buchanan, M. V., Butler, P. D., Magid, L. J. and Wignall, G. D. (2003). New high-flux small-angle neutron scattering instrumentation and the center for structural and molecular biology at Oak Ridge National Laboratory. *J. Appl. Crystallogr.* **36**, 829–831
12. Lynn, G. W., Heller, W., Urban, V., Wignall, G. D., Weiss, K. and Myles, D. A. A. (2006). Bio-SANS – a dedicated facility for neutron structural biology at oak ridge national laboratory. *Phys. B Condens. Matter* 385–386, 880–882
13. Schwahn, D., Meier, G. and Springer, T. (1991). SANS Instruments at the Julich Research Reactor Frj-2. *J. Appl. Crystallogr.* **24**, 568–570
14. Gilles, R., Ostermann, A., Schanzer, C., Krimmer, B. and Petry, W. (2006). The concept of the new small-angle scattering instrument SANS-1 at the FRM-II. *Phys. B Condens. Matter* 385–386, 1174–1176
15. Kampmann, R., Haese-Seiller, M., Kudryashov, V., Nickel, B., Daniel, C., Fenzl, W., Schreyer, A., Sackmann, E. and Radler, J. (2006). Horizontal ToF-neutron reflectometer REFSANS at FRM-II Munich/Germany: first tests and status. *Physica B: Condens. Matter* 385–386, 1161–1163
16. Keiderling, U. and Wiedenmann, A. (1995). New SANS instrument at the Ber-Ii Reactor in Berlin, Germany. *Physica B* **213**, 895–897
17. Keller, T., Krist, T., Danzig, A., Keiderling, U., Mezei, F. and Wiedenmann, A. (2000). The polarized neutron small-angle scattering instrument at BENSC Berlin. *Nucl. Instrum. Methods Phys. Res. Sect. A Accel. Spectrom. Detect. Assoc. Equip.* **451**, 474–479
18. Zhao, J., Meerwinck, W., Niinikoski, T., Rijllart, A., Schmitt, M., Willumeit, R. and Stuhrmann, H. (1995). The polarized target station at GKSS. *Nuclear Instrum. Methods Phys. Res. Sect. A Accel. Spectrom. Detect. Assoc. Equip.* **356**, 133–137
19. Kohlbrecher, J. and Wagner, W. (2000). The new SANS instrument at the Swiss spallation source SINQ. *J. Appl. Crystallogr.* **33**, 804–806
20. Putra, E. G. R., Ikram, A., Santoso, E. and Bharoto, B. (2007). Performance of the 36 m small–angle neutron scattering spectrometer at BATAN, Serpong, Indonesia. *J. Appl. Crystallogr.* **40**, S447–S452
21. Gilbert, E. P., Schulz, J. C. and Noakes, T. J. (2006). 'Quokka' - the small-angle neutron scattering instrument at OPAL. *Phys. B Condens. Matter* 385–386, 1180–1182
22. Rosta, L. (1995). Neutron-scattering for condensed-matter research and materials science at the Budapest-Research-Reactor. *Physica B* **213**, 848–850
23. Ito, Y., Imai, M. and Takahashi, S. (1995). Small-Angle Neutron-Scattering Instrument of the Institute for Solid-State Physics, University-of-Tokyo (SANS-U). *Physica B* **213**, 889–891
24. Koizumi, S., Iwase, H., Suzuki, J., Oku, T., Motokawa, R., Sasao, H., Tanaka, H., Yamaguchi, D., Shimizu, H. M. and Hashimoto, T. (2006). Focusing and

polarized neutron ultra-small-angle scattering spectrometer (SANS-J-II) at Research Reactor JRR3, Japan. *Phys. B Condens. Matter* 385–386, 1000–1006

25. Ostanevich, Y. M. (1988). Time-of-flight small-angle scattering spectrometers on pulsed neutron sources. *Makromol. Chem., Macromol. Symp.* **15**, 91–103
26. Serdyuk, I. N. (1995). Small-angle neutron instrument Yumo (Jinr, Dubna) – some new results and perspectives. *Physica B* **213**, 892–894
27. Seong, B. S., Han, Y. S., Lee, C. H., Lee, J. S., Hong, K. P., Park, K. N. and Kim, H. J. (2002). The small angle neutron spectrometer at the HANARO reactor, Korea. *Appl. Phys. A-Mater. Sci. Process.* **74**, S201–S203
28. Han, Y.-S., Choi, S.-M., Kim, T.-H., Lee, C.-H. and Kim, H.-R. (2006). Design of 40M SANS instrument at HANARO, Korea. *Phys. B Condens. Matter* 385–386, 1177–1179
29. Choi, S. M., Barker, J. G., Glinka, C. J., Cheng, Y. T. and Gammel, P. L. (2000). Focusing cold neutrons with multiple biconcave lenses for small-angle neutron scattering. *J. Appl. Crystallogr.* **33**, 793–796
30. Suzuki, J., Oku, T., Adachi, T., Shimizu, H. M., Hirumachi, T., Tsuchihashi, T. and Watanabe, I. (2003). Cold neutron beam focusing by a superconducting sextupole magnet. *J. Appl. Crystallogr.* **36**, 795–799
31. Russell, J. T., Lin, Y., Boker, A., Su, L., Carl, P., Zettl, H., He, J. B., Sill, K., Tangirala, R., Emrick, T., Littrell, K., Thiyagarajan, P., Cookson, D., Fery, A., Wang, Q. and Russell, T. P. (2005). Self-assembly and cross-linking of bionanoparticles at liquid-liquid interfaces. *Angew. Chem. Int. Ed. Engl.* **44**, 2420–2426

Chapter 20

Small-Angle Scattering and Neutron Contrast Variation for Studying Bio-Molecular Complexes

Andrew E. Whitten and Jill Trewhella

Summary

Structural molecular biology over the past several decades has progressed from studies of the individual proteins, subunits, and domains that accomplish specific biochemistry to seeking to understand the dynamic bio-molecular complexes and assemblies that are responsible for biological function. This progress has led to an expansion of the structural analysis "tool box" to include methods that complement the mainstay techniques of the field: X-ray crystallography, nuclear magnetic resonance (NMR), and cryo-electron microscopy. Small-angle scattering of X-rays or neutrons is one such complementary technique that provides information on the size and shape of scattering particles in solution. This low-resolution structural information can be a powerful complement to high-resolution structural data, especially for the study of bio-molecular interactions with ligands or each other. Further, exploitation of the different neutron-scattering properties of the stable isotopes of hydrogen (^{1}H and ^{2}H) can be used to enrich the information available from the small-angle scattering data, especially for bio-molecular complexes.

Key words: Small-angle scattering, X-ray scattering, Neutron contrast variation, Deuterium labelling, Protein complexes

1. Introduction

Despite the apparent simplicity of the small-angle scattering experiment, the demands placed on sample quality and instrumental precision make accurate data collection a challenging task. Developments in molecular biology and X-ray- and neutron-scattering instrumentation have alleviated these challenges, and the emergence of easy-to-use structural modelling programs has generated a surge in interest in small-angle solution scattering as

James Weifu Lee and Robert S. Foote (eds.), *Micro and Nano Technologies in Bioanalysis*,
Methods in Molecular Biology, vol. 544
DOI 10.1007/978-1-59745-483-4_20, © Humana Press, a part of Springer Science+Business Media, LLC 2009

a tool for studying bio-molecular structures *(1, 2)*. This chapter briefly outlines the basic theory of small-angle scattering essential for understanding the experiment, sample requirements, including those for a neutron contrast variation experiments, and the steps involved in data analysis and structural modelling. Descriptions of neutron small-angle instrumentation, as well as data acquisition and reduction procedures are covered in chapter 19, Small-Angle Neutron Scattering for Molecular Biology.

1.1. Small-Angle Scattering

Small-angle solution scattering involves measuring the intensity of radiation with wavelength λ, scattered by a sample through an angle of 2θ (**Fig. 1**), and yields information related to the time and ensemble average structure of the particles in solution. For an ensemble of structurally homogenous, randomly oriented particles, the intensity of scattered radiation can be expressed as:

$$I(q) = N\left(\Delta\bar{\rho}V\right)^2 P(q)S(q). \tag{1}$$

In **Eq.1**, N is the number of particles per unit volume; $\Delta\bar{\rho}$ is the contrast (discussed in detail in later sections), and V is the volume of each particle; $P(q)$, the "form factor", encodes the ensemble average structure of the particles in reciprocal space; and $S(q)$, the "structure factor", encodes correlation distances between particles in reciprocal space. Accurate interpretation of $I(q)$ in terms of structure requires that all particles in solution are identical (to within the resolution of the experiment), because significant structural and conformational heterogeneity preclude the interpretation of the scattering data in terms of a single structure. It is also important that the solutions are sufficiently dilute, such that the motions of the particles are essentially uncorrelated so that the structure factor can be safely neglected (S=1), and $I(q)$ can be directly related to $P(q)$.

1.2. Contrast Variation

Equation 1 shows that the scattered intensity is dependent on contrast ($\Delta\bar{\rho}$). The contrast of a particle is in essence its "scattering

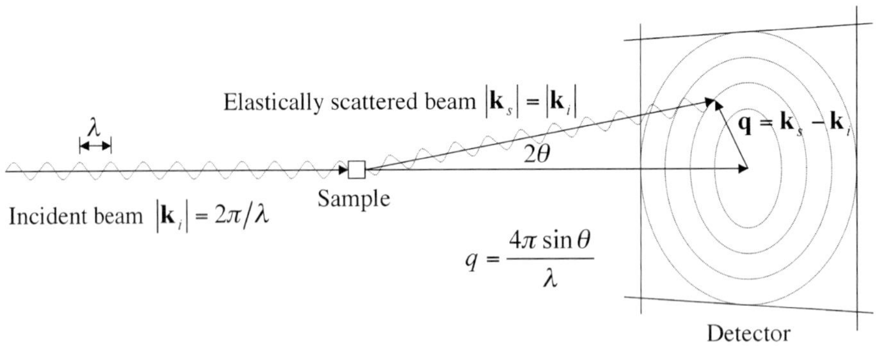

Fig. 1. Conceptual diagram of the small-angle scattering experiment.

power" and is defined as the difference between the average scattering density of the particle ($\langle\rho(\mathbf{r})\rangle$) and that of the medium surrounding it ($\bar{\rho}_s$):

$$\Delta\bar{\rho} = \langle\rho(\mathbf{r})\rangle - \bar{\rho}_s. \quad (2)$$

Because X-rays are scattered primarily by electrons, X-ray contrast is dependent on the difference between the electron density of the particle and the solvent, which is related to the elemental composition of each. Neutrons, on the other hand, are scattered primarily by nuclei, which means that neutron contrast is dependent on the isotopic composition of the particle and the solvent. Conveniently, hydrogen (^{1}H) is ubiquitous in bio-molecules and deuterium (^{2}H) is a stable and relatively abundant isotope that possesses a markedly different scattering density to ^{1}H. This difference creates an opportunity for easily varying contrast without altering the elemental composition.

Contrast variation provides additional information, inaccessible from a conventional scattering experiment, relating to the structure and arrangement of components within the scattering particle that possess differing scattering densities. These differences may be caused by natural compositional differences, such as in DNA–protein or RNA–protein complexes, or caused by selective isotopic labelling, such as a protein-deuterated protein complex. Neutron contrast variation data is measured on complexes with two components of differing scattering density, in solvents where the ^{1}H:^{2}H ratio is systematically varied (a "contrast series"). These experiments deliver additional structural information on each of the components and their arrangement in the particle by enhancing or diminishing the relative contribution each makes to the measured scattering data. A special application of contrast variation is when one component of a complex is made practically "invisible" by adjusting the solvent scattering density to match that component so that the measured scattering signal is from the unmatched ("visible") component alone. This special application of contrast variation is termed "contrast matching" (or solvent matching) and it yields structural information for the visible component within the complex.

1.3. Data Analysis and Modelling

One of the simplest analysis techniques that can be applied to small-angle scattering data is Guinier analysis *(3)*. Guinier analysis involves a linear fit to a plot of In $I(q)$ versus q^2 for small values of q (*see* **Note 1**):

$$\ln I(q) \approx \ln I(0) + q^2\left(\frac{R_g^2}{3}\right). \quad (3)$$

The intercept of the Guinier plot with the vertical axis is related to the zero-angle scattering intensity, $I(0)$. Given that the form factor is normalised such that $P(0)=1$, from **Eq.1**:

$$I(0) = N(\Delta\bar{\rho}V)^2 = C\Delta\rho^2\bar{v}^2 M, \quad (4)$$

where C is the concentration in units of mass per unit volume, $\bar{v}$ is the partial specific volume of a particle, and M is the mass of a particle — multiplication of M by Avagodro's number will yield the molecular mass of the particle. **Equation 4** shows that $I(0)$ is sensitive to the concentration and composition of the particles, and not to their shape. Hence, given concentrations and contrasts of particles in solution, calibrated $I(0)$ values (*see* **Subheading 2.1.4**) can be used to estimate the volume or mass of the scattering particles in solution. **Equation 4** also shows that doubling the concentration of particles doubles $I(0)$, whereas doubling the volume of the particles quadruples $I(0)$. This dependence is important in relation to sample purity because it demonstrates that small-angle scattering (of X-rays, neutrons or light) is extremely sensitive to large molecular mass contaminants.

The slope of the Guinier plot is related to the square of the radius of gyration (R_g^2), and provides information regarding the distribution of the contents of the particle (*see* **Note 2**). With respect to contrast variation, analysis of the dependence of R_g^2 on contrast can provide information regarding the distribution of scattering density within a particle. The Stuhrmann plot *(4)* is a quadratic plot of R_g^2 versus $\Delta\bar{\rho}^{-1}$:

$$R_{obs}^2 = R_m^2 + \frac{\alpha}{\Delta\bar{\rho}} - \frac{\beta}{\Delta\bar{\rho}^2}. \quad (5)$$

The coefficients of the quadratic expression are related to the radius of gyration of that object in which there are no density fluctuations (R_m^2), the second moment of the density fluctuations (α), and the first moment of the density fluctuations (β). The Stuhrmann plot provides a good test for the quality of the collected data and provides some insight regarding the arrangement of the different components. Another useful method for analysing the dependence of R_g^2 on contrast is based on a generalisation of the parallel-axis theorem *(5–7)*, which directly casts the radius of gyration at each contrast point in terms of the radius of gyration of each of the components and the distance between the two:

$$R_{obs}^2 = \frac{\Delta\bar{\rho}_1 V_1}{\Delta\bar{\rho}V} R_1^2 + \frac{\Delta\bar{\rho}_2 V_2}{\Delta\bar{\rho}V} R_2^2 + \frac{\Delta\bar{\rho}_1 V_1}{\Delta\bar{\rho}V}\frac{\Delta\bar{\rho}_2 V_2}{\Delta\bar{\rho}V} D^2. \quad (6)$$

Given R_{obs}^2 at a sufficient number of contrast points, it is possible to use **Eq. 6** to determine R_1^2, R_2^2 and D^2.

In addition to numerical parameters describing the arrangement and size of the components in a particle, some insight can be gained into the shape and disposition of the components before attempting to interpret the data in terms of three-dimensional models. For particles composed of two components (components 1 and 2), with substantially different average scattering densities, the total scattered intensity can be approximated by:

$$I(q) = N\left[\Delta\bar{\rho}_1^{\,2}V_1^2P_{11}(q) + \Delta\bar{\rho}_2^{\,2}V_2^2P_{22}(q) + 2\Delta\bar{\rho}_1V_1\Delta\bar{\rho}_2V_2P_{12}(q)\right].$$
$$= \Delta\bar{\rho}_1^{\,2}I_{11}(q) + \Delta\bar{\rho}_2^{\,2}\ I_{22}(q) + \Delta\bar{\rho}_1\Delta\bar{\rho}_2I_{12}(q). \qquad (7)$$

Equation 7 shows that the scattering from a two-component system is dependent on the contrast of each component, $\Delta\bar{\rho}_1$ and $\Delta\bar{\rho}_2$; their scattering profiles, $I_{11}(q)$ and $I_{22}(q)$; and an additional term arising from interference between scattering elements in each component, $I_{12}(q)$. **Equation** 7 can be used to decompose the entire contrast variation series into the composite scattering functions $I_{11}(q)$, $I_{22}(q)$ and $I_{12}(q)$.

Although particle shape information is encoded in $I(q)$, a more intuitive measure of shape is in the $p(r)$ function, which is the inverse Fourier transform of $I(q)$. The $p(r)$ is simply the distribution of inter-atomic distances within the particle, weighted by the product of the contrast values at each atom centre:

$$p(r) = \sum_{i,j} \Delta\rho(\mathbf{r}_i)\Delta\rho(\mathbf{r}_j), \text{for } r = \left|\mathbf{r}_i - \mathbf{r}_j\right|. \qquad (8)$$

The distribution of inter-atomic distances is characteristic of the shape of the particle and is only non-zero for $0 < r < D_{max}$, where D_{max} is the maximum dimension of the particle. Because the size of the particle is finite, calculation of $I(q)$ as the Fourier transform of $p(r)$ is well defined, but because q-space is infinite, and only a small region is measured in a scattering experiment, the more useful calculation to obtain $p(r)$ from $I(q)$ is not well defined. For this reason, $p(r)$ is obtained from $I(q)$ via well-established indirect transformation procedures *(8, 9)*. Once the $p(r)$ has been determined, $I(0)$ and R_g^2 also can be calculated as the zeroth and second moments of $p(r)$.

The Guinier, $p(r)$, R_g^2 analyses, and composite scattering function extraction provide information on the quality of the data and its inherent information content. The importance of these analyses cannot be overstated, but, as mentioned in the introduction, much of the increased interest in small-angle solution scattering can be attributed to the development of software capable of yielding three-dimensional models of proteins and other macromolecules in solution. Such three-dimensional structures are much more appealing than numerical parameters or profiles that describe characteristics of a particle in solution, largely because the three-dimensional solution structure can be interpreted intuitively in terms of its biological relevance. Modelling of solution scattering data can be broken down into two basic categories: shape restoration and rigid body modelling.

Shape restoration *(10, 11)*, involves optimising a distribution of "dummy residues", with homogenous scattering density, against the scattering data. In practice, there is no unique solution to this problem, but various restraints relating to particle connectivity and compactness can be enforced to ensure a realistic result. This method has been demonstrated to work extremely well in many cases, but in some circumstances many different classes of models can be obtained that fit the data equally well. In cases such as this, additional biochemical or biophysical data are required to eliminate models, or validate one class of models. More recently, this method has been extended to be used with contrast variation data (implemented in the program MONSA *(12)*), which decreases the likelihood of obtaining many different classes of models.

Rigid-body modelling takes advantage of the fact that proteins are usually composed of domains that have a well-defined fold and average conformation. Where high-resolution structures exist for each domain of the structure, the relative position and orientation of each can be optimised against the scattering data. Rigid body modelling, such as that used in the program SASREF *(2, 12)*, has the advantage that it better represents the distribution of scattering density within the particle and can model a structure using fewer degrees of freedom than *ab initio* techniques.

2. Materials

2.1. Sample Requirements

1. Extraction of reliable structural information from scattering data requires highly purified samples that contain monodisperse, structurally homogeneous particles of known concentration (*see* **Note 3**). Because $I(q)$ depends on the square of the particle volume, V^2 (*see* **Eq. 1**), a few percent of low molecular weight contaminants might be tolerated, but large molecular weight contaminants or aggregates, even at the level of a few percent, will bias the derived structural parameters.
2. Because of the relatively low flux of neutron sources, the sample volume must be large (200–300μL), and the particle concentration must be relatively high. The exact concentration requirement depends on the intensity of the neutron source, the molecular volume, and contrast of the scattering particle. For the NG3 instrument at the NIST Centre for Neutron Research *(13)*, using a 5-m detector position, some minimum recommendations for data acquisition times are:

(a) Proteated 10 kDa protein in $^{1}H_2O$, at 10 mg/mL (2 h collection).

(b) Perdeuterated 10 kDa protein in $^{1}H_2O$, at 10 mg/mL (1 h collection).

(c) Proteated 10 kDa protein in $^{2}H_2O$, at 10 mg/mL (<1 h collection).

Larger particles require shorter collection times and/or lower concentrations. Higher-flux instruments, such as those located at the Institute Laue Langevin, also facilitate shorter collection times by approximately a factor of four (based on our experience).

3. X-ray-scattering experiments using current state-of-the-art laboratory-based systems typically require 5- to 60-min measurement times, similar protein concentrations to those shown above for neutron experiments, and sample volumes of the order of 10–50 μL. Experiments using high-intensity synchrotron instrumentation take a few seconds or minutes and particle concentrations can be more dilute (by at least an order of magnitude).

4. Accurate interpretation of scattering data requires that the measured intensities are on an absolute scale, or compared with standards. Typically, neutron-scattering data are placed on an absolute scale via measurement of the primary beam flux. The primary beam flux can be collectively related to various instrumental parameters that are used to place data on an absolute scale. Secondary standards such as water can also be used to place X-ray-scattering data on an absolute scale *(14)* (*see* **Note 4**). Proteins that are known to exist as stable, monodisperse particles, such as lysozyme *(15)* and glucose isomerase *(16)*, can also be used as secondary standards (*see* **Note 4** and **5**).

5. In the case of protein complexes, accurate knowledge of the stoichiometry is essential, as is knowledge of the precise deuteration level of the protein components and their solvents (*see* **Note 6**).

6. A typical neutron-scattering sample cell is made of quartz, with a path length of 1 mm (*see* **Note 7**), and volumes are ~200–300 μL (*see* **Note 8**). For a neutron contrast variation experiment, a minimum of five samples is recommended. Neutron radiation is relatively benign for biological samples and, in cases of severe limitations to sample production, samples can be reused.

7. X-ray-scattering sample cells are typically made of quartz with path lengths of 1–2 mm and volumes of ~10–50 μL (*see* **Note 9**). Because X-rays are ionising radiation, free-radical formation during the exposure can result in bond breaking, protein unfolding, and aggregation (*see* **Note 10**). For this reason, samples are generally not used for further structural analysis after X-ray irradiation.

8. Solvents can contain most of the typical additives (up to a few hundred millimolar concentrations) used to store proteins; salts, buffering agents, and reducing agents. However, additives in solution may contribute to the background scattering. Because the sample scattering is always corrected for solvent scatting, any increase in background will increase the noise in the background-corrected data.

3. Methods

3.1. Initial Sample Characterisation

There are relatively few neutron-scattering facilities around the world, so it is essential to approach scheduled neutron beam time with fully characterised samples, as outlined in the following:

1. A final gel filtration step in sample preparation is recommended immediately before scattering experiments to minimise potential aggregation (*see* **Note 11**).
2. Establishing optimal solvent conditions for stable, monodisperse protein components and complexes is initially best done using dynamic light scattering (DLS). The measurements are fast and require only small sample volumes (as small as 10 μL) at concentrations of 2–5 mg/mL (smaller for larger particles), and are highly sensitive to the presence of aggregates (*see* **Note 12**). Evaluations must also be done for the deuterated and proteated components in deuterated and proteated solvents (*see* **Note 13**).
3. Small-angle X-ray scattering is used to evaluate samples for potential inter-particle interactions, to estimate the molecular mass of the components and complexes, and to provide information on the overall shapes of the components and the complexes to be studied. X-ray-scattering data also provide an additional contrast point that can be used in conjunction with neutron contrast variation data for model building and refinement.
4. The nature of the interactions between particles in solution can be determined using small-angle X-ray-scattering data from a dilution series. If the zero-angle scattering intensity divided by sample concentration (wt/vol) (*see* **Note 14**) increases with decreasing concentration, then the interactions between particles are repulsive, and $S(q)$ deviates from unity due to correlation distances between particles. If the zero-angle scattering divided by sample concentration increases with increasing concentration, then inter-particle forces are attractive and aggregates are being formed. In the neutron contrast

variation experiment, it is generally impractical to take data over a range of concentrations, hence it is essential to work with solutions where the protein concentration is high (to maximise signal) and intermolecular interactions are negligible. Trying various solution conditions and protein concentrations will help to establish the most appropriate solution conditions (*see* **Note 15**).

3.2. Planning the Neutron-Scattering Experiment

The information being sought from the neutron-scattering experiment will influence:

1. Data collection strategies (contrast matching versus contrast variation).
2. The amount of material required.
3. The subunit(s) that are to be labelled.
4. Deuteration levels of the labelled subunit(s).

3.2.1. Contrast Matching

The contrast matching (or solvent matching) experiment is the most appropriate collection strategy when information is sought regarding the shape of one component in a complex, or shape changes in response to complex formation or ligand binding *(17)*.

1. If the complex has components with a natural contrast difference, such as a DNA–protein or RNA–protein complex, then deuterium labelling is not required. However, for protein complexes, one of the protein components must be deuterated to introduce the necessary contrast difference (*see* **Note 16**).
2. The contrast match point *(18)* for:

   (a) An unlabelled protein is ~40% ^2H_2O.

   (b) A perdeuterated protein is not reachable (~120% ^2H_2O).

   (c) DNA or RNA is between 60 and 70% ^2H_2O (*see* **Note 17**).

   It is important that the scattering density of the components be separated as widely as is practical, so that the signal from the visible component is large at the match point of the other component.
3. Conventionally, for protein complexes, the unlabelled portion of the complex is contrast matched, allowing measurement of the signal from the deuterated portion of the complex; however, in some circumstances it may be necessary to contrast match the deuterated component. To contrast match a deuterated protein, the deuteration level must be carefully tuned so the contrast matching occurs below 100% ^2H_2O. Such tuning of the deuteration levels (*see* **Note 6**) can be accomplished by the correct balance of deuterated constituents in the growth media *(19)*, and can be quantified by mass spectrometry. This experiment also has the advantage that the incoherent background is reduced because the proportion of ^1H in the sample

is greatly reduced relative to contrast matching the unlabelled component (*see* **Note 7**).

3.2.2. Contrast Variation

A contrast variation series requires large quantities of protein and is best used when information on the shapes and dispositions of both components in the complex is sought *(20)*.

1. The choice of which subunit to deuterate and the deuteration level is a trade off between practicality (protein production and yield), cost, and the requirements for a successful measurement. Depending on the information sort, a contrast variation experiment typically consists of five or more well-spaced contrast points (0, 20, 40, 80, and 100% ^2H_2O). It is important to measure at: low ^2H_2O concentrations where the contrast of the deuterated component is maximised; at high ^2H_2O concentrations where incoherent scattering is low and scattering from the proteated component is maximised; and at, or close to, the contrast match point of the unlabelled component, where the relative signal from the deuterated component is maximised.
2. If the two components are similar in size and one component is perdeuterated, then the signal in 80% ^2H_2O will be very low, because this is close to the match point of the entire particle. It is more profitable to deuterate to a level such that the match point of the entire particle is between 55 and 65%. This strategy allows comfortable measurement at a variety of contrast points on either side of the total particle match point.
3. To aid in the planning of contrast variation experiments, it is extremely important to have a good idea of the contrast of the various components of the complex as a function of ^2H_2O content of the solvent. Knowledge of the contrast as a function of ^2H_2O content of the solvent allows decisions to be made regarding required deuteration levels, appropriate data collection strategies, and the required protein concentrations. Contrasts can be calculated from isotopic composition and volume of the particle; a tool for calculating these values is *Contrast*, a module of the program suite MULCh (ModuLes for the analysis of contrast variation data from biomolecular assemblies) *(21)* (http://www.mmb.usyd.edu.au/NCVWeb/).

3.3. Data Collection and Reduction

1. To accurately define the dimensions of a particle, the low-q scattering must be sufficiently sampled. Because of instrument factors, a typical bio-molecular complex (<200 kDa) would require a sample-to-detector distance of 5–7 m. However, these long detector positions result in limited sampling of the high-q scattering, which provide a crucial check for accurate solvent subtraction (see below). Hence, data are generally collected at two different detector positions, with collection times

at the shorter position normally being half that at the longer position. Another method of extending the q-range is to offset the position of the detector at a given sample to detector distance. The data collected at different detector positions can be merged during data reduction before solvent subtraction. For very large complexes, or small virus particles, even longer detector positions are required (as long as 13 m or more) and, in these cases, data may require three detector positions to cover the required q-range.

2. To obtain the scattering from particles or complexes in solution, it is necessary to accurately subtract the solvent scattering from the sample data. Any degassing of samples during an experiment (*see* **Note 18**), small variations in sample thickness (*see* **Note 8**), and incoherent scattering (*see* **Note 7**) can affect the solvent-corrected scattering data. The intensity level in the high-q region may be unusually high for some solvent subtracted data, whereas, in other cases, the data may be negative in value. In such cases, the subtracted data will need to be carefully evaluated and corrected.

3.4. Data Analysis and Modelling

1. Before scattering data analysis can proceed, an isotropic correction for incoherent scattering must be evaluated and may require addition or subtraction of a constant. This constant can be approximated by numerous methods. One recommended method is to study the behaviour of the $p(r)$ profile without constraining $p(0)=0$. A non-zero value at this point indicates that a background adjustment is required. If the background is adjusted such that $p(0)=0$ naturally comes to 0, the background level is approximately correct. Once this correction has been made, the Guinier plots *(22)* and $p(r)$ profiles can be interpreted appropriately.
2. Once the $p(r)$ and Guinier analyses have been performed, the dependence of $I(0)$ on the fraction of ^2H_2O in the solvent can be determined. A plot of $\sqrt{I(0)}\,/\,\text{concentration}$ versus the fraction of ^2H_2O will indicate the match point of the entire particle. This information can be used as a guide to adjust the parameters involved in the calculation of the contrast of the particle, because the calculated match point should agree with the experimental match point. The calculated contrasts can then be used to analyse the contrast variation data. This analysis is also incorporated into the program *Contrast* (*see* **Subheading 3.2.2.3**, *(21)*).
3. Given the radii of gyration determined from the Guinier or $p(r)$ analyses, and contrast values determined in the previous step, the Stuhrmann plot (**Eq.5**) and the parallel-axis theorem (**Eq.6**) can be used to obtain estimates of the radii of gyration for each component in the structure, and their separation distance. The *Rg* module of the program suite MULCh *(21)* can perform these analyses.

4. Application of **Eq. 7** to the contrast variation series yields the scattering profile for each component in the complex and an interference term related to the disposition of the two. This information should be sufficient to gain some insight into the structure and to build model structures that are qualitatively consistent with the scattering data. This analysis can be performed using the *Compost* module of the program suite MULCh *(21)*.
5. The ultimate goal of structural analysis is a three-dimensional model. There are now various programs available for three-dimensional model refinement against scattering data (*see* **Note 19**). The type of model refinement procedure will depend on whether there are available atomic structures for the components. If so, the positions and orientations of the available component structures can be optimised with respect to the contrast variation data using a program such as SASREF7 *(12)*. If there are no atomic structures available, MONSA *(12)* can be used to derive a model that represents possible shapes and dispositions of the components with different contrasts.

4. Notes

1. It is generally accepted that for globular particles the Guinier approximation is valid for $qR_g<1.3$. A Guinier plot of data collected from samples for which small amounts of aggregation or inter-particle interference are present may be linear, but the R_g and $I(0)$ values determined by Guinier analysis will be inaccurate. If these effects are sufficiently large, the Guinier approximation will break down and the Guinier plot will show clear curvature.
2. The radius of gyration of a homogenous sphere is $\frac{3}{5}R^2$, hence, $D_{\max}/\sqrt{\left(R_g^2\right)} \sim 2.6$; a ratio smaller than this signifies a particle with mass concentrated at the periphery, such as a hollow sphere; a ratio larger than this signifies an elongated particle.
3. Small-angle scattering requires accurate concentration measurements (to within 5%). In ideal circumstances, UV absorption can be used to determine concentrations to an accuracy of 5%, but small UV extinction coefficients (for example, caused by a protein containing no tryptophan residues) or the presence of solutes that absorb UV light (such as the reducing agents dithiothreitol [DTT] and β-mercaptoethanol [BME]) will impact negatively on the accuracy. Tris(2-carboxyethyl)

phosphine hydrochloride (TCEP) is an alternative reducing agent that is stable and does not absorb UV light. Colorimetric methods for concentration determination (such as the Bradford protein assay) are not generally sufficiently accurate. Quantitative amino acid analysis can be very accurate if the samples are handled without loss and the equipment is well serviced and operated, but it is advisable to make multiple determinations (possibly even at numerous concentrations) so that reliability can be evaluated.

4. The scattering from water is isotropic (independent of q), and the ratio between the experimentally and theoretical intensity levels *(23)* can be used as a scale factor to place data on an absolute scale.

5. Secondary protein standards are used to calibrate $I(0)$ and evaluate samples for potential aggregation. Although there are numerous ways to calibrate $I(0)$, typically the zero-angle scattering (with arbitrary units) for all samples is normalised by concentration (in units of weight per unit volume) and expected molecular weight (*see* **Eq. 4**). Comparison of these calibrated $I(0)$ values for particles of similar composition should be approximately constant. Normalised $I(0)$ values that differ significantly from the standard indicate the assumed average association state of the particles in solution is incorrect. If the composition of the sample and the standard are different (e.g. protein and DNA or RNA), corrections for differences in contrast and partial specific volume must be applied *(24)*.

6. Mass spectrometry is the best method for determining accurate deuteration levels in proteins. Quantitative amino acid analysis can provide accurate information on stoichiometry of protein complexes.

7. ^{1}H possesses a large incoherent scattering cross-section that gives rise to high isotropic background intensity levels for ^{1}H-rich samples, which increases the noise in the background-adjusted data relative to measurements made in pure ^2H_2O. Multiple scattering can also be a significant effect for ^{1}H-rich samples and may bias the scattering data. For all measurements of samples in ≥50% ^1H_2O, the total path length for the sample should not exceed 1 mm.

8. It is common practice in neutron-scattering experiments to collect sample and solvent scattering in different cells to help with the logistics of very long experiments (1–4 days). High-precision quartz cells are manufactured to a path length tolerance of ±10 μm (±1% for a 1-mm cell). For samples containing large amounts of hydrogen, this difference can cause inaccuracies in solvent scattering subtraction.

9. Solvent and sample scattering for X-ray experiments must always be done in the same sample cell that can be re-positioned with high precision in the X-ray beam. Some synchrotron beam lines have flow cells as an alternative and these generally require larger samples (~100 μL).
10. Because X-rays are ionising radiation, their interactions with samples will generate free radicals that can cause bond breakage. Bio-molecules have different levels of sensitivity to these events, and in some proteins they will cause local unfolding and aggregation. This type of radiation-induced aggregation will evolve with time during sample irradiation and can be evaluated by monitoring the zero-angle scattering intensity, which will increase as the square of the average molecular weight of the particles in the sample. Reducing agents and free-radical soaking agents (e.g. dithiothreitol, ascorbate) can reduce such radiation damage. Radiation damage also can be assessed after the experiment by gel electrophoresis on the irradiated samples.
11. If there is a significant time lapse between protein purification and data collection, it is generally advisable to keep the samples cold (4°C), but not frozen, unless previous experience has shown that freezing does not induce even small amounts of aggregation. If samples must be frozen for shipment, then 5–10% glycerol or sucrose may protect them from damage during the freeze–thaw cycle. Solution samples should also be transported in containers that are completely filled to prevent frothing.
12. The calculated mass fraction of aggregates in DLS measurements must be less than 0.01%.
13. It is important to test the solubility of proteins and protein complexes in ^2H_2O, because high deuterium levels can induce aggregation. The reason for this is not well understood, but the widely held belief is that stronger hydrogen bonding in ^2H_2O contributes to stabilising hydrophobic interactions between solute molecules. It is also important to check that the deuteration of a protein does not affect its solubility.
14. Concentration measurements in small-angle scattering are conventionally quoted as wt/vol (e.g. milligrams per milliliter), instead of in molar units as a convenience. Plots of $I(0)/c$ versus c (used to evaluate the extent of inter-particle interactions) for different sized molecules are easier to compare because they span similar ranges on the horizontal axis. Also, for compounds with similar contrasts, $I(0)$ normalised by concentration in wt/vol is directly proportional to molecular weight, whereas $I(0)$ normalised by concentration in molar units is proportional to the square of the molecular weight (*see* **Eq. 4**).

15. The structure factor caused by Coulombic repulsions is known as inter-particle interference, and is dependent on the charge distribution on the surface of the molecule. Adjusting the pH of the solution alters the ionisation state of specific groups, which can alter Coulombic repulsions between particles. Such repulsive effects can also be reduced by increasing the salt concentration of the solvent, thus increasing charge screening effects. Conversely, weak attractive forces that drive concentration-dependent aggregation can also be disrupted by adjusting the solvent conditions.

16. Deuterium labelling of proteins can be accomplished by bacterial expression in deuterium-enriched minimal media. High deuteration levels (85–90%) can be obtained using a pure ^2H_2O and a proteated carbon source. In addition to the use of pure ^2H_2O, perdeuteration requires the use of a deuterated carbon source, which may include deuterated glucose, glycerol, or deuterated algal hydrolysate (see Chap. 18, Deuterium Labeling for Neutron Structure – Funcition – Dynamics Analysis). Protein yields are affected by deuterium, but reasonable protein yields can be obtained by growing bacteria on minimal media made up with a proteated carbon source in ~90% ^2H_2O. The deuteration level of proteins in these circumstances will be ~75–80% *(19)*.

17. DNA and RNA have similar, but not identical, neutron and X-ray contrasts because the elemental composition of each is slightly different.

18. Often samples are stored at low temperatures before measurement, and there can be substantial amounts of dissolved gas in the sample. The faces of the quartz cells provide a large nucleation area for the formation of bubbles. If gas bubbles form during an experiment they can contribute to the small-angle scattering and also affect the transmission of the samples. Samples therefore should be loaded into the cells sometime before the experiment after they have warmed to the temperature of the experiment. If there is evidence of bubble formation inside the cells, short periods of sonication (1–5 s) should remove the bubble from the face of the cell without harming the sample. Even if bubbles are not visible, sonication is a good precaution because it will aid in degassing the sample and help prevent potential bubble formation during the experiment.

19. To be comparable with experimental data, numerous corrections may need to be applied to a calculated scattering profile, including wavelength distribution and beam shape. For a neutron experiment, the scattering instrument approximates point geometry and the smearing corrections are generally quite small. Nonetheless they can be incorporated easily into all data

analyses and model refinements and it is desirable to include them where practical *(25)*. Data collected in a slit geometry, sometimes used in laboratory-based X-ray instrumentation to maximise the flux seen by the sample, however, gives rise to substantial data smearing and must be corrected for.

References

1. Svergun, D. I. and Koch, M. H. J. (2003). Small-Angle Scattering Studies of Biological Macromolecules in Solution. *Rep. Prog. Phys.* **66**, 1735–1782
2. Petoukhov, M. V. and Svergun, D. I. (2005). Global Rigid Body Modeling of Macromolecular Complexes against Small-Angle Scattering Data. *Biophys. J.* **89**, 1237–1250
3. Guinier, A. and Fournet, G. (eds.) (1955). Small-Angle Scattering of X-rays. Wiley, New York
4. Ibel, K. and Stuhrmann, H. B. (1975). Comparison of Neutron and X-Ray-Scattering of Dilute Myoglobin Solutions. *J. Mol. Biol.* **93**, 255–265
5. Moore, P. B., Engelman, D. M., and Schoenborn, B. P. (1974). Asymmetry in 50s Ribosomal-Subunit of Escherichia coli. *Proc. Natl. Acad. Sci. U.S.A.* **71**, 172–176
6. Damaschun, G., Fichtner, P., Purschel, H. V., and Reich, J. G. (1968). Studies on Quaternary Structure of Proteins by Small Angle Scattering of X-Rays.2. On Structure of Complex of Trypsin with Soyabean Inhibitor. *Acta Biol. Med. Ger.* **21**, 309–316
7. Serdyuk, I. N. and Fedorov, B. A. (1973). New Method of Studying Structure of Block Copolymers in Solution. *J. Polym. Sci. Pol. Lett.* **11**, 645–649
8. Semenyuk, A. V. and Svergun, D. I. (1991). Gnom – a Program Package for Small-Angle Scattering Data-Processing. *J. Appl. Crystallogr.* **24**, 537–540
9. Bergmann, A., Fritz, G., and Glatter, O. (2000). Solving the Generalized Indirect Fourier Transformation (GIFT) by Boltzmann Simplex Simulated Annealing (BSSA). *J. Appl. Crystallogr.* **33**, 1212–1216
10. Svergun, D. I. (1999). Restoring Low Resolution Structure of Biological Macromolecules from Solution Scattering Using Simulated Annealing. *Biophys. J.* **76**, 2879–2886
11. Chacon, P., Moran, F., Diaz, J. F., Pantos, E., and Andreu, J. M. (1998). Low-Resolution Structures of Proteins in Solution Retrieved From X-Ray Scattering With a Genetic Algorithm. *Biophys. J.* **74**, 2760–2775
12. Petoukhov, M. V. and Svergun, D. I. (2006). Joint Use of Small-Angle X-Ray and Neutron Scattering to Study Biological Macromolecules in Solution. *Eur. Biophys. J. Biophys. Lett.* **35**, 567–576
13. Glinka, C. J., Barker, J. G., Hammouda, B., Krueger, S., Moyer, J. J., and Orts, W. J. (1998) The 30 M Small-Angle Neutron Scattering Instruments at the National Institute of Standards and Technology. *J. Appl. Crystallogr.* **31**, 430–445
14. Orthaber, D., Bergmann, A., and Glatter, O. (2000). SAXS Experiments on Absolute Scale With Kratky Systems Using Water as a Secondary Standard. *J. Appl. Crystallogr.* **33**, 218–225
15. Mylonas, E. and Svergun, D. I. (2007). Accuracy of Molecular Mass Determination of Proteins in Solution by Small-Angle X-Ray Scattering. *J. Appl. Crystallogr.* **40**, S245–S249
16. Kozak, M. (2005). Glucose Isomerase From *Streptomyces rubiginosus* – Potential Molecular Weight Standard for Small-Angle X-Ray Scattering. *J. Appl. Crystallogr.* **38**, 555–558
17. King, W. A., Stone, D. B., Timmins, P. A., Narayanan, T., Von Brasch, A. A. M., Mendelson, R.A., and Curmi, P.M.G. (2005). Solution Structure of the Chicken Skeletal Muscle Troponin Complex Via Small-Angle Neutron and X-Ray Scattering. *J. Mol. Biol.* **345**, 797–815
19. Leiting, B., Marsilio, F., and O'Connell, J. F. (1998). Predictable Deuteration of Recombinant Proteins Expressed in Escherichia coli. *Anal. Biochem.* **265**, 351–355
18. Jacrot, B. (1976). Study of Biological Structures by Neutron-Scattering From Solution. *Rep. Prog. Phys.* **39**, 911–953
20. Whitten, A. E., Jacques, D. A., Hammouda, B., Hanley, T., King, G. F., Guss, J. M., Trewhella, J., and Langley, D. B. (2007). The

Structure of the KinA-Sda Complex Suggests an Allosteric Mechanism of Histidine Kinase Inhibition. *J. Mol. Biol.* **368**, 407–420

21. Whitten, A. E., Cai, S., and Trewhella, J. (2008). MULCh: ModULes for the Analysis of Small-Angle Neutron Contrast Variation Data from Bio-molecular Assemblies. *J. Appl. Crystallogr.* **41**, 222–226
22. Konarev, P. V., Volkov, V. V., Sokolova, A. V., Koch, M. H. J., and Svergun, D. I. (2003). Primus: a Windows Pc-Based System for Small-Angle Scattering Data Analysis. *J. Appl. Crystallogr.* **36**, 1277–1282
23. http://physchem.kfunigraz.ac.at/sm/Service/Water/H2OI0.htm
24. Taraban, M., Zhan, H., Whitten, A. E., Langley, D. B., Matthews, K. S., Swint-Kruse, L., and Trewhella, J. (2008). Ligand-Induced Conformational Changes and Conformational Dynamics in the Solution Structure of the Lactose Repressor Protein. *J. Mol. Biol.* **376**, 466–481
25. Pedersen, J. S., Posselt, D., and Mortensen, K. (1990). Analytical Treatment of the Resolution Function for Small-Angle Scattering. *J. Appl. Crystallogr.* **23**, 321–333

Chapter 21

Protein Sequencing with Tandem Mass Spectrometry

Assem G. Ziady and Michael Kinter

Summary

The recent introduction of electrospray ionization techniques that are suitable for peptides and whole proteins has allowed for the design of mass spectrometric protocols that provide accurate sequence information for proteins. The advantages gained by these approaches over traditional Edman Degradation sequencing include faster analysis and femtomole, sometimes attomole, sensitivity. The ability to efficiently identify proteins has allowed investigators to conduct studies on their differential expression or modification in response to various treatments or disease states. In this chapter, we discuss the use of electrospray tandem mass spectrometry, a technique whereby protein-derived peptides are subjected to fragmentation in the gas phase, revealing sequence information for the protein. This powerful technique has been instrumental for the study of proteins and markers associated with various disorders, including heart disease, cancer, and cystic fibrosis. We use the study of protein expression in cystic fibrosis as an example.

Key words: Protein sequencing, Tandem mass spectrometry, Liquid chromatography, Electrospray ionization

1. Introduction

A comprehensive proteomic approach often makes use of protein gel electrophoresis and tandem mass spectrometry to profile all of the proteins expressed by a given cell type or tissue (the proteome), classifying them as overexpressed, underexpressed, or unchanged compared with a reference cell type under specific experimental conditions. The proteome of a cell type maps out the proteins specific to that cell type, and, once established, can be examined for changes in response to any condition over any period of time, and thus may be an excellent outcome measure for corrective therapy. Two experimental tools are commonly used to study the proteome: *(1)* two-dimensional (2D) gel electrophoresis to

James Weifu Lee and Robert S. Foote (eds.), *Micro and Nano Technologies in Bioanalysis*,
Methods in Molecular Biology, vol. 544
DOI 10.1007/978-1-59745-483-4_21, © Humana Press, a part of Springer Science+Business Media, LLC 2009

separate whole cell proteins based on their molecular weight and isoelectric point, and *(2)* mass spectrometric analysis of digests of any Coomassie or silver-stained band on the gel (femtomole quantities of proteins can be analyzed).

Gel electrophoresis allows for the separation and quantitation of proteins in complex mixtures, such as whole cell homogenates. A common approach is to characterize a biological system of interest and to select specific proteins for sequencing and identification. Changes in protein band density in response to experimental conditions, versus established baseline levels, indicate changes in protein expression. Furthermore, changes in the isoelectric point of a protein can indicate modifications, such as phosphorylation *(1)* or isoprenylation *(2)*. Proteins selected based on these criteria are excised from the gel and subjected to proteolytic digestion (typically with trypsin), and analyzed by mass spectrometry *(3)*.

For electrospray tandem mass spectrometric analysis, an ion trap mass spectrometer, such as Thermo Fisher's *LCQ* Deca XP, is most commonly used. These instruments, when calibrated, determine the mass to charge ratio (m/z) of ions derived from an analyte to values within ±0.1–0.2 Da. The *LCQ* Deca XP is typically connected to a liquid chromatography (LC) column that separates peptide digests before they enter the mass spectrometer. The ion trap is used to select ions based on their m/z and subjects them to collision-induced dissociation (CID), which results in a charged-site initiated fragmentation of the selected peptide at the amide bond positions *(3)*. Analysis provides information about the primary structure of analyzed ions because the m/z for both an unfragmented parent ion (MS^1) and daughter ions (MS^2) resulting from fragmentation is accurately measured by the mass analyzer in tandem. This approach is known as capillary column liquid chromatography tandem mass spectrometry (LC MS-MS). For peptide ions that exhibit poor initial fragmentation, further fragmentation of each generation of daughter ions may be carried out (MS^3, MS^4, and so on), but the diminishment of available ions and poorer fragmentation characteristic eventually limit the utility of these experiments.

Improvements in the field of proteomics provide better methods for the detection of post-translationally modified proteins. Investigators have used affinity purification with immobilized metal ion affinity chromatography (IMAC) columns *(4, 5)* or selection based on specific fragmentation reactions *(6)* to analyze selectively phos-phorylated peptides in a complex mixture. Similar techniques have been used to study proteins that are nitrated *(7)* or phosphorylated *(8–10)*, with good success. However, analysis of post-translational modification remains more difficult than protein identification *(3)*, because coverage of the peptide containing the modification is absolutely necessary and may not occur. Purification techniques can enhance the abundance of these peptides, thereby increasing the likelihood of detection by the instruments. If, however, the peptide containing the modified residue cannot be detected, the modification cannot be determined by tandem mass spectrometry.

2. Materials

Materials and reagents amounts reported in this section are based on the processing of one confluent 100-mm plate of cultured cells or ~50 mg of tissue.

2.1. Cell Culture and Homogenate Preparation

1. 10% fetal calf serum (FCS) in standard cell culture media as needed (*see* **Note 1**).
2. PBS: Dulbecco's phosphate-buffered saline.
3. For quantitative experiments, protein labels (i.e., nonabundant isotopes of carbon or nitrogen) may be added to cell culture media as a supplement.
4. Cell pellet or section of tissue (*see* **Note 2**).
5. Standard cell culture trypsin solution (Invitrogen, Corp., Carlsbad, CA).
6. 15-mL polypropylene centrifuge tubes. Polypropylene is required for acetone precipitation.
7. Protease inhibitor cocktail: one Complete Mini protease inhibitor tablet (Roche, Nutley, NJ) + 10 mL PBS.
8. Tris stock: 0.5 M Tris-HCl, 50 mM $MgCl_2$. Mix 3 g Tris base, 0.3 g $MgCl_2$ in 40 mL water. Adjust pH to 7.8 with 6 M HCl. Adjust total volume to 50 mL with water. Check the pH again.
9. Lysis buffer: 0.5% sodium dodecyl sulfate (SDS), 25 mM Tris–HCl (pH 7.8), 2.5 mM $MgCl_2$. Mix 300 μL of 10% SDS, 300 μL Tris stock, 0.5 mL protease inhibitor. Bring volume up to 5.0 mL and mix well. Approximately 1 mL per cell pellet sample is needed.
10. DNAse stock: 10 mg/mL bovine pancreatic DNAse I (Sigma-Aldrich Co., St. Louis, MO; ≥85% purity) in Tris stock.
11. RNAse stock: 10 mg/mL bovine pancreatic RNAse A (Sigma-Aldrich Co.; ≥85% purity) in Tris stock.
12. Nuclease reagent: 1 mg/mL DNAse, 1 mg/mL RNAse. Mix 20 μL DNAse stock, 20 μL RNAse stock, 100 μL Tris stock, 60 μL protease inhibitor; 60 μL per 0.5-mL sample is needed.
13. Acetone: high-performance liquid chromatography (HPLC)-grade acetone.

14. Solubilization buffer: 7 M urea, 2 M thiourea, 1% dithiothreitol (DTT), 1% CHAPS, 1% ampholytes, 1% Triton X-100. Mix 2.1 g urea, 0.8 g thiourea, 50 mg CHAPS, 50 mg DTT, 50 μL BioLytes, 50 μL Triton X-100. Bring volume up to 5 mL with water and dissolve with shaking (*see* **Note 3**).

2.2. One- and Two-Dimensional Gel Electrophoresis

1. Rehydration buffer (same as solubilization buffer above).
2. 1% BPB: 1% bromophenol blue in water.
3. Electrode wicks; wetted with water.
4. Equilibration buffer: Mix 5.4 g urea, 0.3 g SDS, 3.8 mL of 1.5M Tris–HCl (pH 8.8), 3 mL glycerol in a 50-mL centrifuge tube. Adjust the total volume to 15 mL with water. Dissolve with shaking, but do not use any heat.
5. Reducing reagent: 120 mg DTT in 7.5 mL equilibration buffer.
6. Alkylation reagent: 150 mg iodoacetamide in 7.5 mL equilibration buffer with 100 μL 1% BPB.
7. Agarose: 0.5% in modified running buffer. Dissolve 0.5 g agarose in 10 mL of 10× Biorad (Hercules, CA) Tris–glycine running buffer and 30 mL glycerol. Adjust the total volume to 100 mL with water. Add 1 mL of 1% BPB. This reagent can be stored at room temperature and used repeatedly over several months.
8. Running buffer: 1× Biorad Tris–glycine buffer. Dilute 100 mL 10× Biorad Tris-glycine buffer with 900 mL water. Cool on ice before use.
9. Criterion precast gels from Biorad: The gels are stored at 4°C (*see* **Note 4**).
10. Fixing solution: 50% ethanol, 10% acetic acid in water. Mix 240 mL water, 60 mL acetic acid, and 300 mL ethanol.
11. Protein label: Gel code blue (Pierce Biotechnology, Inc., Rockford, IL).

2.3. In Gel Digestion of Proteins (see Note 5)

1. Ethanol, 95% USP grade.
2. Water, MilliQ grade.
3. Acetic acid (99.9%).
4. Acetonitrile, HPLC grade (Burdick and Jackson, McGraw Park, IL).
5. Formic acid, 88% ACS reagent grade.
6. Siliconized Eppendorf tubes; rinsed before use with ethanol.
7. Sequencing grade modified trypsin (Promega Corp., Madison, WI); 20 μg as a lyophilized powder.
8. Wash reagent: 50% ethanol/5% acetic acid. Mix 9 mL water with 10 mL ethanol and 1 mL acetic acid.
9. 100 mM bicarbonate solution: Dissolve 0.2 g ammonium bicarbonate in 20 mL water.

10. 50 mM bicarbonate solution: Mix 3 mL 100 mM ammonium bicarbonate with 3 mL water. Typically prepared on ice for trypsin preparation.
11. Dithiothreitol (DTT) reagent: 5–10 mg DTT/mL in 100 mM ammonium bicarbonate.
12. Iodoacetamide reagent: 25–30 mg iodoacetamide/mL in 100 mM ammonium bicarbonate.
13. Trypsin reagent: 20 μg trypsin in 1,000 μL of 50 mM ammonium bicarbonate (20 ng trypsin/μL). Keep ice cold to prevent autolysis.
14. Extraction reagent: 50% acetonitrile/5% formic acid. Mix 9 mL water with 10 mL acetonitrile and 1 mL formic acid.
15. 1% acetic acid: Mix 20 mL water with 200 μL acetic acid.

2.4. Liquid Chromatography Tandem Mass Spectrometry

1. Ion trap mass spectrometer equipped with a nanoelectrospray source and HPLC pump.
2. Automatic or manual “loading bomb” *(3)* sample injector. The loading bomb is an aluminum cast chamber that contains two portions. The bottom portion consists of a chamber large enough to easily hold a 1.5- or 2-mL Eppendorf tube containing sample, which is connected to a nitrogen gas tank. The top portion consists of a rubber sealed cap that contains a hole at its center large enough for fitting a Teflon ferrule that will hold the packed LC column. When pressure is applied to the sealed sample-containing chamber, sample is forced onto the column.
3. Fused silica tubing (internal diameter 50–70 μm).
4. C18 beads (10 μm packing material). Pack columns that are 7–12 cm in length. Shorter columns will result in substandard separation, and longer columns will give unnecessarily longer experiment times.
5. Acetonitrile and 50 mM acetic acid.

2.5. Data Collection and Analysis

1. Data recording software such as Xcalibur ™ (Thermo Scientific, Waltham, MA).
2. Protein database search software such as Bioworks ™ (Thermo Scientific).

3. Methods

For demonstration purposes, we discuss the typical proteomic experiment for the identification of proteins that exhibit differential expression in cystic fibrosis (CF) versus genetically matched-normal cell pairs *(11)*. In CF, identification of the precise series of events that are differentially activated by airway infection in normal versus CF cells is incomplete, but identifying this series of events is a pressing need given the role that inflammation plays in the pathology of disease *(12)*. Novel means of identifying pathways uniquely activated or repressed in CF epithelial cells in response to inflammatory stimulation are of great potential interest. In this context, sampling the proteome of airway epithelial cells, non-CF and CF, and well matched except for the CF lesion *(11, 13–15)* is a logical approach (**Table 1**). Inflammatory pathways are often activated by changes in protein expression to a greater extent than gene expression, or by changes in the post-translational modification of key proteins in the cell *(13–15)*. Both of these changes can be observed by 2D gel electrophoresis *(1–3)* and tandem mass spectrometry (*see* **Figures 1 and 2**). Proteins that show altered expression levels in CF, or that shift on the gel in response to inflammatory stimulation, can be identified by mass spectrometry. Once the proteome of the epithelial cell is mapped, it is digitally recorded and available for future studies. For example, identifying changes in this established proteome in response to potential therapeutics is an excellent outcome measure.

3.1. Cell Culture and Collection

1. We use validated *(11)* airway epithelial cell pairs that model the CF or non-CF condition. One of these is the 9HTEo– pCEP and pCEP-R (CF phenotype) cell pair that share the same genetic background with the exception of the stable expression of the regulatory (R) domain of cystic fibrosis transmembrane

Table 1
Some redox and mitochondrial proteins that differ by more than twofold between CF and non-CF epithelia

Protein	CF vs. non-CF (≥2-fold changes)
Stress-70 protein	Elevated
TRX reductase	Decreased
ATP synthase coupling factor	Decreased
Glutamate dehydrogenase	Decreased
ATP synthase β chain	Elevated
Glutathione-*S*-transferase	Decreased
Acetyl CoA transferase	Elevated
Thioredoxin	Decreased
Superoxide dismutase-2	Elevated

conductance regulator (CFTR) *(11)*. Cells are maintained in 75-cm^2 culture flasks and passaged when near confluence to experimental cultures in 100-mm plates (*see* **Note 2**). Normal and CF cells are grown under identical conditions in Dulbecco's minimal Eagle's medium (DMEM) supplemented with 10% fetal bovine serum (FBS) and 2.5 mM L-glutamine and maintained under selection with 40 μg/mL hygromycin at 37°C in 95% air/5% CO_2.

2. When cells are 80% confluent, pour the growth media off of the monolayer and add 10 mL of PBS to each dish to wash the monolayer. After the wash, add 2 mL of 0.25% cell culture trypsin to each dish, let sit for 5–10 min, and harvest the cells. Monitor the reaction as needed to assure complete harvest. Add 2 mL ice-cold 10% FCS to stop the reaction.
3. Collect the cells/buffer in a 15-mL centrifuge tube; combining up to three dishes per centrifuge tube (*see* **Note 2**). Keep the collected cells/buffer on ice.
4. Pellet the cells by centrifugation at ~500 × *g* for 5 min (longer times or higher speeds can lead to cell lysis). Resuspend the cell pellets in cold PBS and then pellet the cells by centrifugation. Repeat this wash three times (*see* **Note 8**).
5. Add 1 mL of the protease inhibitor and resuspend the cells and incubate on ice for 10 min. Pellet the cells by centrifugation at ~500 × *g* for 5 min, pour off the supernatant, and blot any excess liquid out of the tube with a Kimwipe (do not disturb the pellet). The cell pellet may be processed immediately or frozen at –80°C for later analysis.

3.2. Sample Preparation for Gel Electrophoresis Analysis

1. Thaw frozen cell pellets to room temperature. Add 1–2 mL lysis buffer to each cell pellet (*see* **Note 9**). Mix well until sample becomes viscous because of the interaction of the DNA/RNA and SDS.
2. Heat the samples at 100°C for 5 min with intermittent mixing by gentle vortex. Then remove tubes from heating, mix well, and cool the samples to room temperature.
3. Add 100 μL nuclease reagent/mL of lysis solution while mixing the sample. The viscosity of the sample will begin to clear relatively fast, depending on the amounts of DNA/RNA. Allow the reaction to continue for 10 min at room temperature. Several samples may be combined to accumulate a desired amount of protein at this stage.
4. Remove an aliquot for the protein assay. Because of the high level of detergent in the lysis buffer, a detergent-compatible protein assay must be used.
5. Cool the homogenate on ice for 10 min, then add 6 mL ice-cold acetone to each milliliter of homogenate (to make acetone

>80%). The precipitation of the protein in the sample should be white and fluffy, and should not appear stringy if the DNA/RNA has been properly digested. Cool the sample at –20°C for 1 h to overnight.

6. Pellet the precipitated proteins by centrifugation at 1,000 rpm ~900 × *g* for 5 min. Discard the acetone without disturbing the protein pellet. Blot any residual acetone out of the tube with a Kimwipe, and allow the acetone to dry thoroughly at room temperature.
7. Once dried, solubilize the proteins in solubilization buffer at 5 mg/mL based on the protein measurement carried out in **step 4**. More time and agitation are necessary to solubilize drier protein pellets that contain higher levels of protein. Once solubilized, the sample can be stored at –20°C until use.

3.3. Isoelectric Focusing of Proteins on IPG Pi Strips

1. Thaw protein sample in solubilization buffer, as needed. For 2D SDS–polyacrylamide gel electrophoresis (PAGE), immobilized pH gradient (IPG) p*I* strips must be loaded with protein and focused using the following protocol:
2. Prepare the samples in 1.5-mL Eppendorf tubes:
   (a) For Coomassie blue-stained gels: Add 150 μL sample (5 mg/mL) to 100 μL rehydration buffer and 10 μL 1% BPB, and mix well.
   (b) For silver-stained gels: Add 30 μL sample (5 mg/mL) to 220 μL rehydration buffer and 10 μL of 1% BPB, and mix well.
3. Transfer the samples to a well in the isoelectric focusing (IEF) tray. Place the entire volume of the sample at one end of the well and coat the entire well by tipping the tray back and forth several times. Remove any bubbles with a pipette.
4. Place the IPG gel pH strip face down in the sample. Gently rock the tray to ensure good wetting of the gel surface, and force any bubbles out from under the gel with forceps or a pipette. Cover the strip with oil to protect the gel from drying during rehydration or focusing.
5. Rehydrate the strip overnight at room temperature, applying 50 V. A typical rehydration time is 12–16 h.
6. Once rehydration is complete, remove the strips from the tray, and place them on a wet paper towel. Clean the tray with soap and water, then ethanol, and dry with a Kimwipe. Place electrode strips, which are well wetted with water (to facilitate conduction), at each electrode. Reposition the rehydrated strips in the tray, face down and properly oriented (the strips are labeled with either a "+" or "–" end that should be aligned to the positive and negative electrodes of the focuser). Cover the strips with mineral oil and run the isoelectric focusing program.

7. A typical program for a 11-cm, 5–8 p*I* range strip loaded with 500–750 μg protein follows:

   Step 1: to 250 V in 0:15 h

   Step 2: to 8,000 V in 3:00 h

   Step 3: at 8,000 V for 2:30 h

   Total volt-hours = 44 kVh

   Higher volt-hour totals may be needed for heavier protein loads, whereas lower volt-hour totals may be needed for lighter protein loads. Other p*I* range strips may have different volt-hour optimums. In general, "rapid" focusing recommendations by the IEF manufacturer for varying strips and protein loads can be used to achieve good focusing.

   ◦ At the end of the focusing, remove the strips from the tray and blot the oil off the strips. At this stage, the strips must either be immediately processed for the SDS–PAGE as described below, or be wrapped in plastic, labeled, and stored at −20°C until processing for SDS–PAGE as described below.

3.4. SDS–PAGE

1. For one-dimensional (1D) SDS–PAGE, sample (from **Subheading 3.2, step 7**) can be loaded onto gel and routine methodology may be used as long as fresh reagents and precast gels are used to decrease contamination.
2. For 2D SDS–PAGE, equilibrate the strips by placing them face up in the equilibration tray. Cover each strip with approximately 3.5 mL of the reducing reagent. Incubate the strip at room temperature while shaking for at least 15 min. Remove the reducing agent and replace with 3.5 mL of the alkylating reagent. Incubate the strip at room temperature while shaking for at least 15 min. Reduction and alkylation are necessary to prevent protein refolding.
3. While strips are equilibrating, the gel apparatus may be assembled. Prepare the running buffer (if needed) and cool on ice. Melt soft agarose, and cool to approximately 45°C before use (*see* **Note 3**). Good timing is needed here, because the agarose begins to solidify at 40°C.
4. While the agarose is cooling, remove the Criterion gel from its package and wash briefly with deionized water. Place the gel in a stand. Remove the green comb and rinse three times with running buffer. Leave the well covered with running buffer.
5. At the end of equilibration, remove the strip from the alkylating reagent and blot away any excess reagent. Pour the running buffer out of the gel well, and place the strip in the well in the proper (+) and (−) orientation. Check that the agarose is at approximately 45°C, and cap the well with agarose. Using a glass pipette, force the agarose over the strip, filling the well. Then allow the agarose to solidify.

6. Assemble the gel system. Remove the tape covering the bottom of the gel cassette, place the gel in the Criterion cell, and fill the bottom and top chambers with cold running buffer to the appropriate levels (usually marked by fill lines). Cap the gel apparatus with the top cover cell, and run the gel at a constant voltage of 200 V. The total run time should be approximately 70 min. Stop the run when the tracking dye just leaves the bottom of the gel. If multiple gels are being run concurrently, it is possible to pause the run for gels that run faster, while slower running gels are run to the same extent. At the end of the run, immediately process the gels for the fixing and staining procedure described below.

3.5. Gel Processing

1. Place 300 mL of the fixing solution in the clean Pyrex dish, break the Criterion gel cassette using the green plastic comb taken out of the IPG well, and immerse the broken cassette in the fixing solution and begin removing one plate of the cassette. Taking care not to tear the gel, remove one plate from the cassette, remove and discard the IPG strip, and immerse the second plate, with the gel on it, into the fixing solution (if the gel is to be used for Western analysis, do not fix the gel). The gel will move off the second plate with a little agitation.
2. Fix the gel for 30 min at room temperature with gentle shaking, aspirate the fixing solution off the gel, and rinse the gel with water for 5–10 min. Aspirate the water and repeat the wash two times. Add 125 mL of Gelcode Blue to the gel and cover with plastic wrap. Stain overnight at room temperature with gentle shaking. Abundant bands may become visible within 15 min.
3. When staining is complete, aspirate the stain off the gel, and destain with several changes of water until background staining is removed. Each 2D gel is scanned using a Bio-Rad QS-800 Calibrated Densitometer. Acquired images are imported into the PDQuest™ 2D gel analysis software and compared. This software uses state-of-the-art iterations to background subtract, normalize, and compare imaged bands on up to 20 different gels, and is considered to be the software of choice for such analysis in the field of proteomics.
4. The software is used to compare sets of gels, from the non-CF (pCEP) and CF (pCEP-R) cell line pair, for example. Protein band volumes are calculated (with band intensity as the height, and band size as the circular base of a cone). The software uses this volume to assign each band a value that is a fraction of the overall volume of the entire gel. Comparing this value for matching bands on different gels is the basis for assessing differential protein expression in different samples. When comparing samples with common genetic backgrounds, such

as the pCEP and pCEP-R cells, many of the 2D gel bands should align easily (*see* **Fig. 1**). Some manual orientation may be necessary, but running equally loaded gels to identical extents should decrease the need for this. The software can be used to determine significant quantitative (band volume), and/or qualitative (changes in band p*I* or M_r shift).

5. Select bands of interest, for example, bands that change significantly in CF versus non-CF cells (**Fig. 2**). Cut the band from the gel as closely as possible with a clean band picker (The Gel Company, Inc.) or scalpel (especially for 1D gels) and divide the band into approximately 1-mm^3 pieces. Place the gel pieces in a rinsed, 1-mL Eppendorf tube. For Coomassie-stained gels, go to **step 7**.
6. For silver-stained gels, wash in 200 μL water two times, destain with 50–100 μL Farmer's reducer (Invitrogen Corp.) for 15 min at room temperature, then wash out destain reagent three times with 200 μL water. Continue to **step 7**.
7. Wash and destain by adding 200 μL wash reagent at room temperature for at least 30 min, then remove wash. Repeat this three times, then rinse bands in 200 μL of 100 mM bicarbonate. For 2D gels, go to **step 9**.

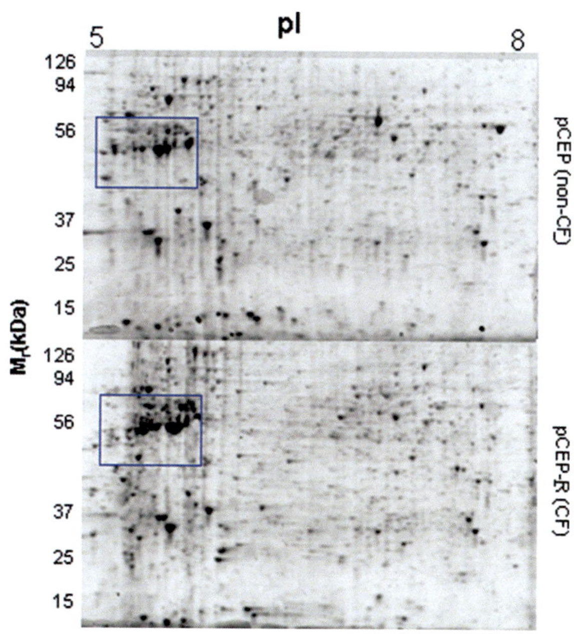

Fig. 1. Coomassie blue-stained 2D SDS–PAGE of whole cell homogenates from unstimulated 9HTEo⁻ cell line pairs. Cells were grown to 80% confluence, homogenized, and whole cell protein was prepared. Precast gels were equally loaded, run, fixed, and stained with GelCode Blue? Coomassie stain. Although protein abundance varies, band correspondence is evident. Boxed region is magnified in **Fig. 2**.

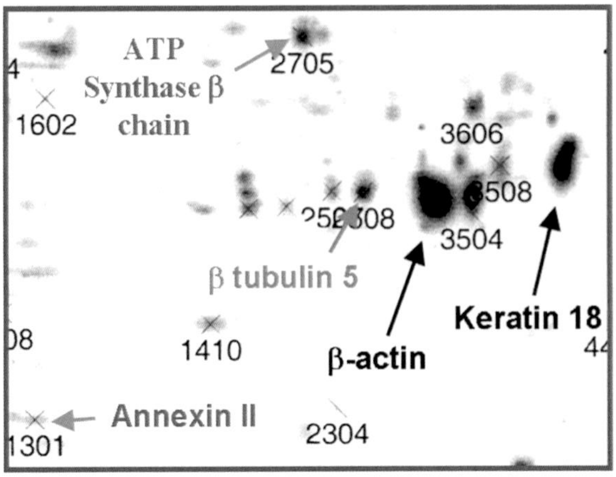

Fig. 2. Comparative overlay of 2D gels of CF and non-CF ($n = 5$ each) whole cell protein. Region magnified from **Fig. 1**. Band numbers connote proteins that are determined to be differentially expressed (we use these numbers for cataloging purposes). Some identifications are shown. *Green* identifications are downregulated, whereas red identifications are upregulated in CF (twofold or greater difference). *Black* identifications are unchanged.

8. For 1D gel-excised bands, proteins must be reduced and alkylated to prevent refolding. So, dehydrate each band by adding 200 μL acetonitrile for 5 min, remove acetonitrile, then dry bands in a SpeedVac for approximately 3 min. Add 50 μL DTT for 30 min at room temperature to reduce proteins. Remove DTT and add 50 μL iodoacetamide reagent for 30 min at room temperature to alkylate the proteins. Remove iodoacetamide before proceeding to **step 9**.
9. Dehydrate by adding 200 μL acetonitrile for 5 min. Remove acetonitrile and rehydrate by adding 200 μL bicarbonate for 5 min. Remove bicarbonate, and repeat this dehydration/hydration process two more times. Dry gel pieces in a SpeedVac for approximately 3 min, during which time, prepare sequencing grade trypsin solution by adding 1 mL of 50 mM bicarbonate to 20 μg modified trypsin (Promega Corp., Madison, WI) (concentration, 20 ng/μL trypsin) and keep on ice. Rehydrate the gel pieces in approximately 50 μL of trypsin reagent for 10 min on ice, then microfuge briefly and remove excess trypsin. Add 10 μL of 50 mM bicarbonate, vortex to mix and microfuge briefly, and let digest overnight at room temperature.
10. Microfuge gel pieces briefly, then add 30 μL extraction reagent, vortex to mix, and incubate for 10 min. Transfer supernatant to a 0.5-mL Eppendorf tube. Add another 30 μL extraction reagent, and vortex to mix, then combine the supernatants in the 0.5-mL Eppendorf tube. Microfuge

the extract briefly, and reduce the volume of the extract to less than 10 μL in a SpeedVac. Avoid drying completely because dissolving dried peptides can be difficult. Bring up the sample volume to approximately 25 μL with 1% acetic acid (*see* **Note 10**). At this stage, the sample is ready for analysis by electrospray LC MS-MS. To analyze samples by matrix-assisted laser desorption ionization (MALDI) mass spectrometry, simply desalt the samples on a desalting column or ZipTip™ (Millipore Corp., Billerica, MA).

3.6. Electrospray Ionization and LC MS-MS

The following methods assume the use of a Thermo Electron *LCQ* Classic, Fleet, Deca, Deca XP, or Deca XP plus mass spectrometer using a nanospray ion source. However, these methods can be applied to most ion trap mass spectrometers with minor changes.

1. Interfaced a pre- or self-packed 10 cm × 75 μm id Phenomenex Jupiter C18 reversed-phase capillary chromatography column to the Nanospray source (*see* **Note 11**). Load an aliquot (2 μL) from the extracted tryptic digest onto the column using the loading bomb.
2. Peptides are eluted off the C18 column by an acetonitrile/0.05 M acetic acid gradient at a flow rate of 0.2 μL/min with the nanospray ion source operated at 1.8–2.5 kV. Set the mass spectrometer to trap and fragment peptides using the data-dependent multitask capability of the instrument. Do not set extensive exclusion lists (i.e., background ions), because this will also limit the analysis of nonbackground ions of the same m/z.
3. The mass spectrometer will acquire full scan mass spectra containing parent peptide ion m/z, and product ion spectra that contain amino acid sequence information in successive instrument scans. This mode of analysis should produce approximately 1,000–3,000 collisionally induced dissociation (CID) spectra of ions ranging in abundance over several orders of magnitude (not all CID spectra are derived from peptides).
4. Once collected, these data files that contain each CID spectrum from a given sample are digitally stored and can be revisited at any time using the Xcalibur™ software. Data analysis is performed using all CID spectra collected in an experiment to search the relevant protein database with the Bioworks or other comparable search software (*see* **Note 12**).

3.7. Affinity Purification and Modified Peptide Identification by Observation of Specific Fragmentation Reactions

Affinity chromatography may be used to purify and concentrate peptides or proteins that contain modifications (i.e., phosphorylation, FLAG tag, etc.). This biochemical method of isolation gives higher yields than immunoprecipitation, and should contain less protein contaminants, allowing for good mass spectrometric analysis. A basic approach is to replace the reverse-phase column in **Subheading 3.6** with an affinity column for the modification of interest and elute bound peptides with the appropriate gradient (*see* **Note 13**). A method for purifying phosphopeptides is as follows:

1. Isolation of phosphopeptides with good success can be achieved *(4)*. Pack a fused silica column with a 20% ethanol slurry of chelating Sepharose and activate it with Fe(III) ions using 500 μL of a 30 mM $FeCl_3$ solution.
2. Wash excess ions and equilibrate the column with 0.1 M acetic acid (pH 3.3). Whole cell homogenate or a purified protein (20 μL) that is prepared and digested as described above, but always in the presence of phosphates inhibitor, is loaded onto the column in a loading buffer of 0.1 M acetic acid.
3. After a brief wash, attach the column to the *LCQ* mass spectrometer and elute the retained phosphopeptides into the instrument with a 0.1% ammonium acetate solution over a gradient.
4. Fragmentation of modified peptides can be complex *(8)*, and, in some cases, the position of the modification influences the fragmentation pattern, resulting in CID spectra that are not interpretable. For phosphoserine or phosphothreonine, loss of the phosphate group can occur in internal reactions within the spectrometer. The mass spectrometer can be set to monitor these losses and select the peptide ions that are losing the modification for extended fragmentation (i.e., MS-MS-MS) *(6)*. In this way, signals derived from unmodified peptides can be eliminated, resulting in a total-ion current chromatographic trace of only modified peptides present in the digest. In addition, selecting extended fragmentation is a good strategy for the analysis of modified peptides that will help obtain sufficient sequence information for identification.

4. Notes

1. For tissue preparations, blood should be flushed with PBS before processing to decrease levels of proteins from this extracellular fluid in the experimental sample.
2. Because of the acetone precipitation step, it is generally desirable to have at least 100 μg of protein present in a cell pellet. It is common to combine the contents of two to three plates to get the desired amount of protein in a pellet.
3. Do not heat any urea-containing solutions or samples.
4. The use of precast gels is recommended because it decreases sample contamination.

5. All reagents should be made fresh and all Eppendorf tubes should be rinsed with 95% ethanol before use.
6. Cell culture techniques that avoid conditions that impact cell proliferation and stress are crucial in proteomic analyses to avoid artifactual changes in protein expression.
7. When conducting proteomic experiments on the differential expression and/or modification of proteins, it is important to compare samples with common genetic backgrounds, because genetic variation alone can contribute heavily to protein expression. Stringent control of cell culture conditions, plating surface and density, monolayer confluence, and media components is necessary for the most meaningful comparisons. For in vivo analyses, tissues from subjects that are sex and age matched should be used.
8. It is important to completely remove extracellular contaminants, such as media or trypsin before analysis, to limit contamination by the proteins present in those fluids. These proteins may complicate data interpretation and introduce artifacts. For in vivo analyses, tissues should be perfused to remove as much extracellular biological fluid as possible.
9. The volume of lysis buffer used depends on the number of cells harvested and the amounts of protein, RNA, and DNA contain therein. Because these factors vary from cell type to cell type, the volume must be experimentally determined. High concentrations of protein and nucleic acids can decrease nuclease activity and result in incomplete digestion of RNA and/or DNA, which can affect protein migration on 1D and 2D gels. Therefore, pellets should be homogenized in volumes sufficient to dilute proteins and nucleic acids to suitable levels before the addition of nucleases.
10. Samples that are analyzed by electrospray ionization have to be in a charged state (either positive or negative). For most protein analyses, acetic or formic acids may be used to protonate peptides. Organic bases may be used for the analysis of species that are negatively charged (i.e., DNA or phospholipids).
11. When using the nanospray ion source, it is crucial to control the flow rate into the mass spectrometer to 100–300 nL/min. Higher flow rates reduce the sensitivity of the experiment *(3)*.
12. Once the search software has identified a protein and matched a CID spectra to a specific peptide sequence, it is important to verify selected results by manual interpretation (described extensively in **ref.** *(3)*). Our laboratory typically inspects at least two spectra for any protein identified.
13. Eluting buffers should not contain high concentrations of salt (i.e., ion exchange columns), which interfere with the ionization of eluting peptide ions and can damage the

mass spectrometer. For applications where such buffers are required, eluted peptides or proteins should be collected off of a first column and then loaded onto a reverse-phase column attached to the mass spectrometer. A typical gradient can then be used to elute the peptides into the instrument.

Acknowledgements

The authors thank Junnan Chen and Samuel Shank for technical assistance. This work was funded by the Cystic Fibrosis Foundation. Dr. Ziady is an inventor on patents that cover the construction of a nonviral gene delivery system and holds equity in Copernicus Therapeutics Inc., a company that has licensed this technology. This presents no conflict for the work discussed in this chapter.

References

1. He, L., and Lemasters, J. J. (2005). Dephosphorylation of the Rieske iron-sulfur protein after induction of the mitochondrial permeability transition. *Biochem Biophys Res Commun.* **334**(3), 829–837
2. Cicha, I., Schneiderhan-Marra, N., Yilmaz, A., Garlichs, C. D., and Goppelt-Struebe, M. (2004). Monitoring the cellular effects of HMG-CoA reductase inhibitors in vitro and ex vivo. *Arterioscler Thromb Vasc Biol.* **24**(11), 2046–2050
3. Kinter, M., and Sheman N. (2000). *Protein Sequencing and Identification Using Tandem Mass Spectrometry.* Wiley & Sons, New York, NY
4. Li, S., and Dass, C. (1999). Iron(III)-immobilized metal ion affinity chromatography and mass spectrometry for the purification and characterization of synthetic phosphopeptides. *Anal Biochem.* **270**(1), 9–14
5. Liu, H., Lin, D., and Yates, J. R.III. (2002). Multidimensional separations for protein/peptide analysis in the post-genomic era. *Biotechniques.* **32**(4), 898, 900, 902, passim
6. Steen, H., and Mann, M. (2003). A new derivatization strategy for the analysis of phosphopeptides by precursor ion scanning in positive ion mode. *J Am Soc Mass Spectrom.* 13(8), 996–1003, Erratum in: J Am Soc Mass Spectrom **14**(1), 83
7. Willard, B. B., Ruse, C. I., Keightley, J. A., Bond, M., and Kinter, M. (2003). Site-specific quantitation of protein nitration using liquid chromatography/tandem mass spectrometry. *Anal Chem.* **75**(10), 2370–2376
8. Ruse, C. I., Willard, B., Jin, J. P., Haas, T., Kinter, M., and Bond, M. (2002). Quantitative dynamics of site-specific protein phosphorylation determined using liquid chromatography electrospray ionization mass spectrometry. *Anal Chem.* **74**(7), 1658–1664
9. Wilkins, M. R., Gasteiger, E., Gooley, A. A., Herbert, B. R., Molloy, M. P., et al. (1999). High-throughput mass spectrometric discovery of protein post-translational modifications. *J Mol Biol.* **289**(3), 645–657
10. Aulak, K. S., Miyagi, M., Yan, L., West, K. A., Massillon, D., Crabb, J. W., et al. (2001). Proteomic method identifies proteins nitrated in vivo during inflammatory challenge. *Proc Natl Acad Sci U S A.* **98**(21), 12056–12061
11. Kube, D., Sontich, U., Fletcher, D., and Davis, P. B. (2001). *Proinflammatory* cytokune responses to P. aeruginosa infection in human airway epithelial cell lines. *Am J Physiol.* **280**(3), L493–L502
12. Chmiel, J. F., Berger, M., and Konstan, M. W. (2002). The role of inflammation in the pathophysiology of CF lung disease. *Clin Rev Allergy Immunol.* **23**(1), 5–27

13. Blackwell, T. S., Stecenko, A. A., and Christman, J. W. (2001). Dysregulated NF-kappaB activation in cystic fibrosis: evidence for a primary inflammatory disorder. *Am J Physiol Lung Cell Mol Physiol.* **281**(1), L69–L70
14. DiMango, B., Ratner, A. J., Bryan, R., Tabibi, S., and Prince, A. (1998). Activation of NF-kappaB by adherent *Pseudomonas aeruginosa* in normal and cystic fibrosis respiratory epithelial cells. *J Clin Invest.* **101**(11):2598–2605
15. Eidelman, O., Srivastava, M., Zhang, J., Leighton, X., Murtie, J., Jozwik, C., et al. (2001). Control of the proinflammatory state in cystic fibrosis lung epithelial cells by genes from the TNF-α/NFκB pathway. *Mol Med.* 7(8):523–534

Chapter 22

Metabolic Analysis

Vladimir V. Tolstikov

Summary

Analysis of the metabolome with coverage of all of the possibly detectable components in the sample, rather than analysis of each individual metabolite at a given time, can be accomplished by metabolic analysis. Targeted and/or nontargeted approaches are applied as needed for particular experiments. Monitoring hundreds or more metabolites at a given time requires high-throughput and high-end techniques that enable screening for relative changes in, rather than absolute concentrations of, compounds within a wide dynamic range. Most of the analytical techniques useful for these purposes use GC or HPLC/UPLC separation modules coupled to a fast and accurate mass spectrometer. GC separations require chemical modification (derivatization) before analysis, and work efficiently for the small molecules. HPLC separations are better suited for the analysis of labile and nonvolatile polar and nonpolar compounds in their native form. Direct infusion and NMR-based techniques are mostly used for fingerprinting and snap phenotyping, where applicable. Discovery and validation of metabolic biomarkers are exciting and promising opportunities offered by metabolic analysis applied to biological and biomedical experiments. We have demonstrated that GC–TOF–MS, HPLC/UPLC–RP-MS and HILIC–LC–MS techniques used for metabolic analysis offer sufficient metabolome mapping providing researchers with confident data for subsequent multivariate analysis and data mining.

Key words: Metabolic analysis, Metabolome, Biomarker, GC, HPLC, UPLC, RP, HILIC, MS, Mass spectrometry

1. Introduction

To understand the function of genes, biological systems should be analyzed at multiple levels of control for external parameters (environment, developmental stage, molecular signals, etc.) and for internal parameters (transcription and messenger RNA [mRNA] degradation, posttranslational modification, protein

James Weifu Lee and Robert S. Foote (eds.), *Micro and Nano Technologies in Bioanalysis*,
Methods in Molecular Biology, vol. 544
DOI 10.1007/978-1-59745-483-4_22,

dynamics, and metabolite concentrations and fluxes). Although there may be more than tens of thousands of genes, several hundred thousand transcripts, and up to one million proteins, it is estimated that there may be as few as several thousand small molecules in the metabolome of higher organisms. Analysis of the metabolome *(1)* looks very attractive from the point of view of the lower numbers of compounds to be identified and quantified. High chemical complexity, analytical and biological variance, and large dynamic range are very challenging, even for the latest analytical methods. The technology provides high analytical precision, comprehensiveness, and sample throughput. However, the diversity of metabolites necessitates the application of different complementary analytical methods. In most cases, analytical methods are based on chromatographic separation techniques such as gas chromatography (GC) and high-performance liquid chromatography (HPLC), and, in many cases, comprise Fourier-transform infrared spectroscopy, electron impact ionization-mass spectrometry (EI–MS), electrospray ionization-mass spectrometry (ESI-MS), and nuclear magnetic resonance (NMR) spectroscopy. Mass spectrometers are generally more sensitive and more selective than any other type of detector *(2)*. Before MS detection, the metabolites have to be separated and the separated compounds must be ionized. Ionization techniques may vary, especially for gas chromatography-mass spectrometry (GC–MS) and liquid chromatography–mass spectrometry (LC–MS) couplings. High throughput, robustness, and unmatched comprehensiveness for different small molecule classes make GC–MS coupling a superior technique for metabolic analysis *(3)*. It is, however, limited to molecules below 400–500 Da in size and cannot be used for labile compounds that can be destroyed during derivatization. The LC–MS technique does not require chemical modifications for metabolic analysis *(4)*. Applying high-throughput screening with GC–MS and LC–MS techniques generates large volumes of analytical data that require advanced software and hardware for successful data mining *(5)*. Metabolomics studies and analysis of the tissue crude extracts cannot be accomplished with the use of a single separation/detection method because of the high chemical diversity of the analyzed mixture. Hydrophobic components are typically separated by reversed-phase chromatography. Ultra performance liquid chromatography (UPLC) provides an extremely powerful separation tool *(6)* and allows short analysis times relative to GC–MS techniques. Hydrophilic and neutral compounds are best separated using hydrophilic interaction liquid chromatography (HILIC). We have recently demonstrated the feasibility of HILIC–ESI-MS analysis for different types of samples *(7–9)*.

Emerging trends in metabolic analysis with LC–MS are faster chromatography, elemental formula determination with accurate mass instruments, improved biomarker detection, and identification through advanced clustering analysis of data sets, and enhanced combinations of LC–MS with NMR/LC–NMR. Regarding one particularly strong emerging trend, LC–MS–MS-based analytical strategies are moving rapidly into focus for elucidating complex samples.

In the present work, we have focused on comprehensive metabolic profiling that includes three different techniques for the same set of the samples. Extracts from needle tissues from six different clones of the 1-year-old loblolly pine (*Pinus taeda* L.) seedling, grown under standard nursery conditions including irrigation and fertilizer, served as samples for metabolic analysis. The GC–MS technique requires sample derivatization. LC–MS techniques do not require any preprocessing for analysis. Applying reversed-phase (RP) and HILIC chromatography before MS detection enhanced coverage for polar and non-polar metabolites. Use of monolithic silica-based C-18 RP-columns (Onyx) provided stability, high performance, low bleeding/carryover, and good recovery and reproducibility for crude extract profiling. Using the combined GC–MS and LC–MS approach, we were able to detect and analyze several hundred compounds with acceptable recovery and variability. Preliminary results using data mining allowed the discrimination of phenotypes (clones) and identification of particular components contributing to discrimination.

2. Materials

2.1. Standards and Chemicals

1. Oligosaccharides kit, LC–MS grade solvents, methoxylamine hydrochloride, pyridine, *N*-methyl-*N*-trimethylsilyltrifluoroacetamide (MSTFA) and reserpine are all purchased from Sigma-Aldrich (St. Louis, MO, USA).
2. Ammonium acetate, ammonium hydroxide, and acetic acid are the highest purity grade available from Sigma-Aldrich.
3. Reserpine stock solution: 0.2 mg/mL in methanol.

2.2. Instrumentation

2.2.1. GC–MS Analysis

GC–MS analysis is performed using an Agilent 6890 N gas chromatograph (Atlanta, GA, USA) interfaced to a time-of-flight (TOF) Pegasus III mass spectrometer (Leco, St. Joseph, MI, USA). Automated injections are performed with a programmable robotic Gerstel MPS2 multipurpose sampler (Mülheim an der Ruhr, Germany). The GC is fitted with both an Agilent injector and a Gerstel temperature-programmed injector, cooled injection system (model CIS 4), with a Peltier cooling source. An automated liner exchange (ALEX) designed by Gerstel is used to eliminate cross-contamination from sample matrix occurring between sample runs. Multiple baffled liners for the GC inlet are deactivated with 1-μL injections of MSTFA. Initial peak detection and mass spectrum deconvolution are performed with ChromaTOF software (version 2.25, Leco), and later files are exported to the netCDF format for further data evaluation.

2.2.2. LC–MS Analysis

The LC–MS system consists of a Surveyor HPLC module (Thermo Electron, Boston, MA, USA) coupled to a Finnigan LTQ (Thermo Electron, San Jose, CA, USA) linear ion trap mass spectrometer without splitting. Normal flow HPLC operations with the use of conventional or microbore columns require pneumatically assisted electrospray ionization with the regular ESI source. The ESI ion source is equipped with a metal needle. The nitrogen sheath gas pressure is set to 7 bar at a flow rate of 2–3 L/min. The spray voltage is set to 5 kV. The temperature of the heated transfer capillary is maintained at 350°C. RP-LC–MS analysis is performed with the use of two identical Onyx Monolithic C-18 reversed-phase silica-based columns (100 × 3 mm, Phenomenex, Torrance, CA, USA) coupled in series with the column coupler. HILIC-LC–MS analysis is performed with the use of a normal phase amino-propyl modified silica-based column (Luna NH_2, 150 × 3 mm, 3-μm particle size, Phenomenex, Torrance, CA, USA).

3. Methods

3.1. Sample Preparation for Metabolic Analysis

Sample preparation is very important to the success of metabolic analysis. Live tissues, organs, or fluid must be deep flash frozen before extraction to prevent enzymatic alteration of components. The following extraction should be almost complete because an absence or insufficient amount of the component in the sample gives zero detector response. Proteins should be precipitated and removed unless they are of particular interest. Protein precipitation should not be harsh and abrupt because some of the small molecules associated with proteins will co-precipitate and will not be recovered. The final sample concentration should be high enough to allow sufficient column loading with a low injection volume.

3.1.1. Extraction

1. Place three metal balls (one 3-mm inner diameter [ID] and two 2-mm ID) in each of the 2-mL Eppendorf sample tubes before sampling.
2. Submerge the tubes into liquid nitrogen. Alternatively, keep the tubes in an 80°C freezer for an hour.

3. Fill tubes with 60 mg of fresh-weight frozen pine needles. Add 0.6 mL of the extraction mixture: acetonitrile:isopropanol:water (1:1:1, v:v:v).
4. Prechill Retsch ball-mill tube holders in liquid nitrogen. Alternatively, keep tube holders with the tubes in a –80°C freezer for an hour. Put tubes into each of the ball-mill tube holders and homogenize the tissue for 1 min at 60% speed. Add 50 μL of a reserpine stock solution (0.2 mg/mL methanol) as internal reference. Sonicate for 1 min at ambient temperature. Repeat the procedure unless the suspension is homogeneous.
5. Vortex vigorously. Shake the resulting suspension for 15 min at ambient temperature.
6. Centrifuge at 13,000 rpm for 5 min with Eppendorf microcentrifuge 5415D. Carefully transfer the green supernatant into a sample glass vial that is equipped with a screw cap.

3.2. GC–MS Metabolic Analysis

After the extracts are completely dried by speed vacuum concentrator or by freeze-drying, 20 μL of 40 mg/mL methoxylamine hydrochloride in pyridine is added, and samples are agitated at 30°C for 30 min. Subsequently, 180 μL of trimethylsilylating agent, MSTFA, is added, and samples are agitated at 37°C for 30 min (*see* **Note 1**). Analytical GC chromatography is performed with injections of 1 μL made in split (1:10) mode (purge time 120 s; purge flow 40 mL/min). The Agilent injector temperature is held constant at 250°C while the Gerstel injector is programmed (initial temperature 50°C, hold 0.1 min, and increase at a rate of 10°C/s to a final temperature of 330°C, hold time 10 min). Chromatography is performed on an Rtx-5Sil MS column (30 m × 0.25 mm ID, 0.25-μm film thickness) with an Integra-Guard column (Restek, Bellefonte, PA, USA). A helium carrier gas is used at a constant flow of 1 mL/min. The GC oven temperature program is 50°C initial temperature with 1-min hold time and ramping at 20°C/min to a final temperature of 330°C, with 5-min hold time. Both the transfer line and source temperatures are 250°C. After a solvent delay of 350 s, mass spectra are acquired at 20 scans per second with a mass range of 50–500 *m/z*. The ion source filament energy is set to 70 eV.

3.3. RP ESI–LC–MS Metabolic Analysis

Analytical LC is performed using 6.5 mM ammonium acetate (pH 5.5, adjusted by acetic acid) (A) and acetonitrile/propanol-2 mixture (3:1, v:v) (B) as the mobile phase at flow rates of 0.6–0.7 mL/min at 30°C (*see* **Notes 2** and **3**). After a 0.5-min isocratic run at 5% B, a linear gradient to 100% B is concluded at 18 min, followed by 100% B up to 25 min. After the run, the column is washed with 100% B for another 5 min. Equilibration with starting buffer takes 10 min. Full scan mass spectra are acquired from 150 to 2,000 amu at unit mass resolution. For MS^n experiments, data-dependent scans are chosen with the wideband activation turned off. The normalized collision energy is set to 35%, and the activation Q is set to 0.250 with the source fragmentation turned off. The mass spectrometer is tuned on sucrose solution (0.1 mg/mL) in a mixture of acetonitrile/ammonium acetate buffer pH 5.5 (1:1, v:v) before measurements.

1. Prepare serial dilutions of reserpine stock solution in methanol. Make fivefold dilutions starting from 0.1 mg/mL of reserpine in methanol. Six or more points are acceptable for creating the calibration curve.
2. Equilibrate the column with the starting buffer for at least 5 min. Inject 5 μL and acquire the data in the full scan mode for the positive and negative ions in the range of 100–1,500 amu.
3. Use reserpine as internal or external standard for instrument calibration by serial dilution. This procedure is essential for further semiquantitative analysis.
4. Prepare samples in accordance with **Subheading 3.1.** Inject 5 μL and acquire the data in the full scan mode for the positive and negative ions.
5. Introduce selected internal standards and repeat the analysis (*see* **Notes 2** and **3**).
6. Refer to **refs.** *2–4* for peaks annotation.
7. Include MS-MS or MS^n experiments for both positive and negative modes in the analytical run to annotate peaks through the MS-MS library search and/or collect fragmentation information for de novo identification.
8. Export data into netCDF format for further data evaluation.

3.4. HILIC ESI–LC–MS Metabolic Analysis

Analytical liquid chromatography is performed using acetonitrile (A), 75 mM ammonium acetate (pH 5.5, adjusted by acetic acid) (B), and 100 mM ammonium bicarbonate (pH 9.4 adjusted with ammonium hydroxide) (C) as the mobile phase at flow rates of 0.5 mL/min at 30°C (*see* **Notes 2** and **3**). After a 1-min isocratic run at 5% B, a linear gradient to 35% B is concluded at 11 min, and then B is replaced with C starting at 50%, followed by a linear gradient to 75% C in 20 min. After the run, the column is washed with 100% C for 2 min. Equilibration with starting buffer takes 10 min. Full scan mass spectra are acquired from 150 to 2,000 amu at unit mass resolution. For MS^n experiments, data-dependent scans are chosen with the wideband activation turned off. The normalized collision energy is set to 35%, and the activation Q is set to 0.250 with the source fragmentation turned off. The mass spectrometer is tuned on sucrose solution (0.1 mg/mL) in a mixture of acetonitrile/ammonium acetate buffer pH 5.5 (1:1, v:v) before measurements.

1. Prepare the standards mix using an oligosaccharides kit in a mixture of acetonitrile:water (1:1, v:v). All of the standards

present in this kit can be used in a single mix. For a simple chromatogram, one can use a smaller number of standards in a mix. The total concentration should not exceed 0.5 mg/mL.

2. Equilibrate the column with the starting buffer for at least 10 min. Inject 5 μL and acquire the data in the full scan mode for the positive and negative ions in the range of 100–1,500 amu.
3. Monosaccharides and oligosaccharides are detected as ammonia adducts in the positive mode and as [M-H]$^-$ ions in the negative mode. Oligosaccharides are eluted in the order of increasing monomer units; the largest oligomer is eluted last.
4. Use selected oligomers as internal or external standards for instrument calibration by serial dilution. This procedure is essential for further semiquantitative analysis.
5. Prepare samples in accordance with **Subheading 3.1.** Inject 10 μL and acquire the data in the full scan mode for the positive and negative ions.

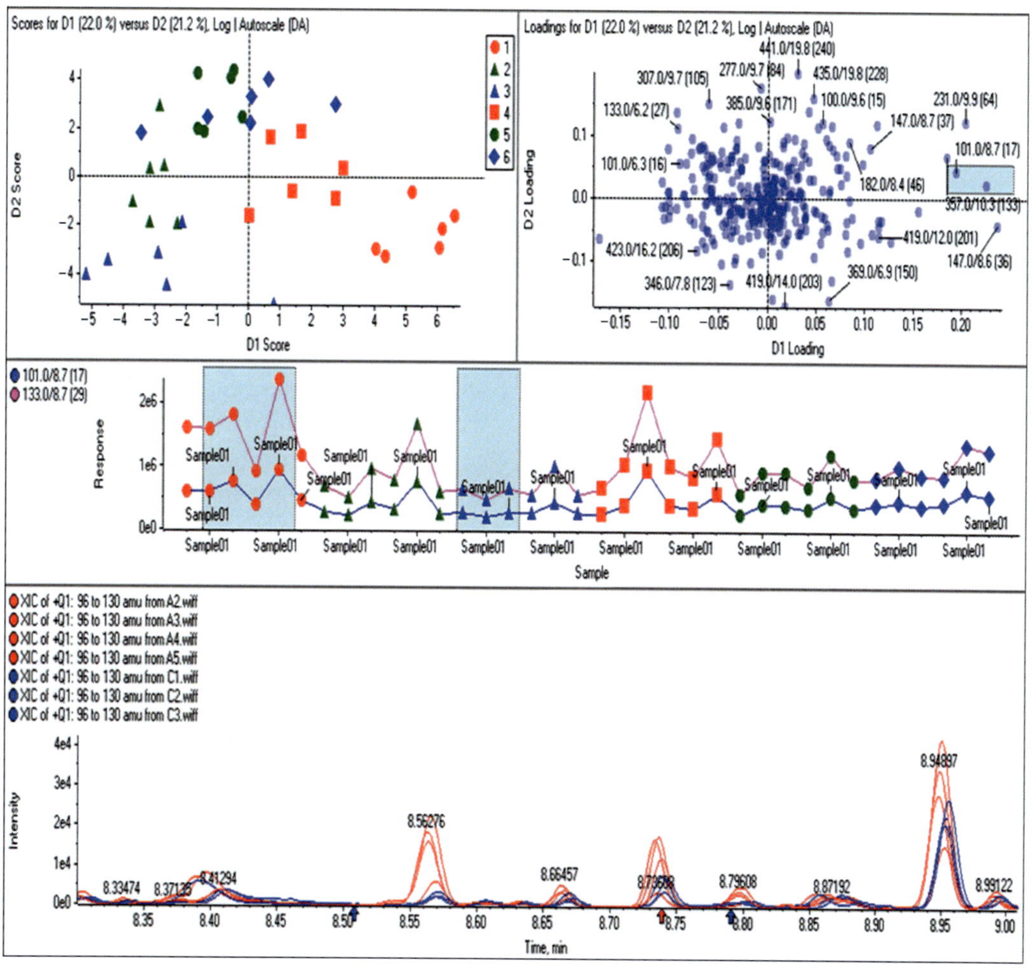

Fig. 1. Tolstikov, Metabolic Analysis.

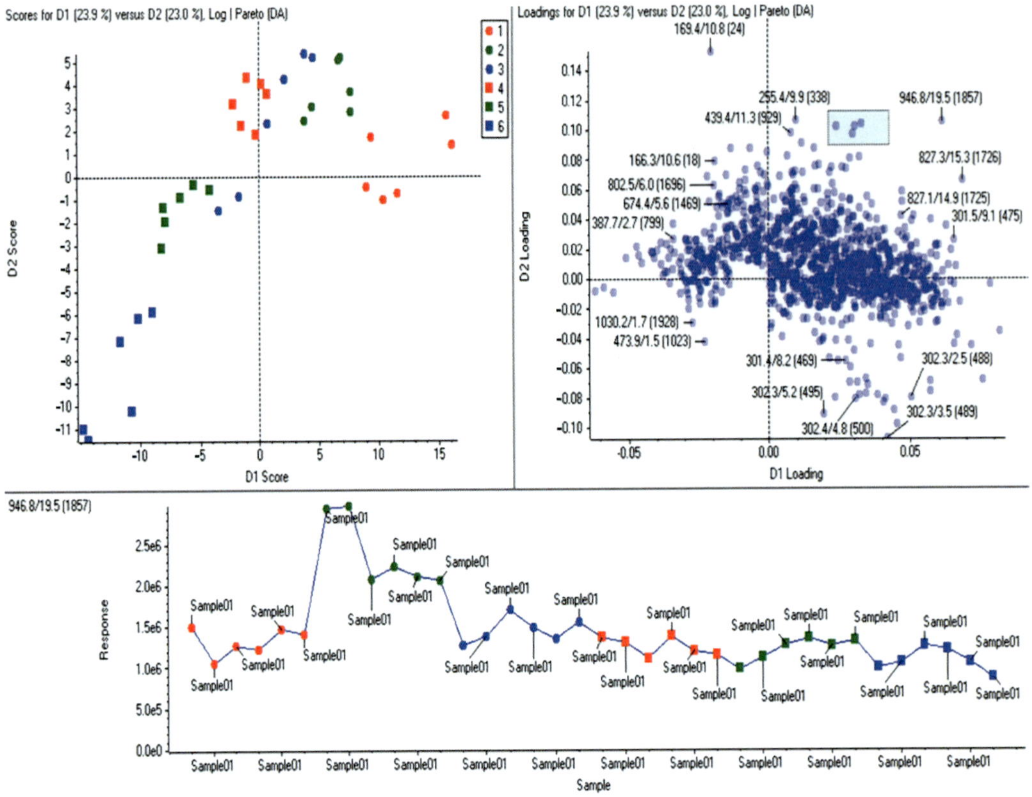

Fig. 2. Tolstikov, Metabolic Analysis.

6. Introduce selected internal standards and repeat the analysis (*see* **Notes 2** and **3**).
7. Refer to **refs.** ***2–4*** peak annotation.
8. Include MS–MS or MS^n experiments for both positive and negative modes in the analytical run to annotate peaks through the MS–MS library search and/or collect fragmentation information for de novo identification.
9. Export data into netCDF format for further data evaluation.

3.5. Basic Data Analysis

Exported data sets in netCDF format are further converted and loaded into the MarkerView 1.1 software (Applied Biosystems, Foster City, CA, USA). Peak alignment and peak picking parameters are adjusted in accordance with the chromatography quality (peak width, baseline, background noise, etc.). Principle component analysis (PCA) is performed. Data point reduction is applied to remove components possessing low relative variances of metabolite abundance. Modified peak lists are exported into Statistica Dataminer (StatSoft) for further analysis. **Figures 1–3** illustrate clones (populations) discriminated by means of PCA-DA

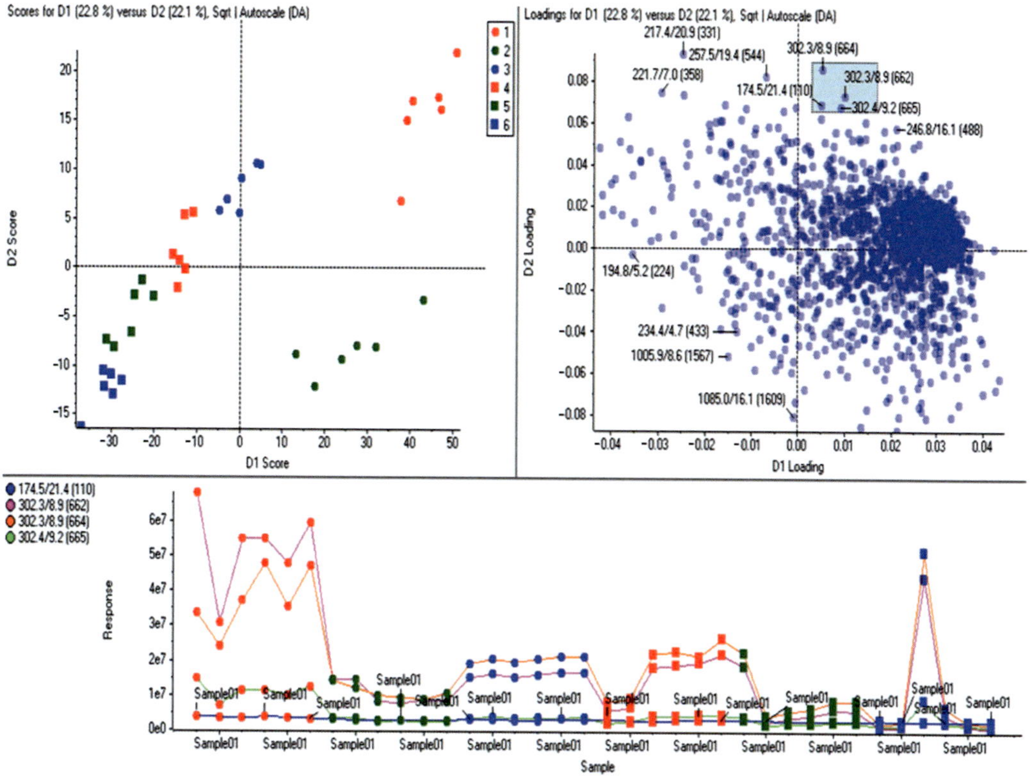

Fig. 3. Tolstikov, Metabolic Analysis.

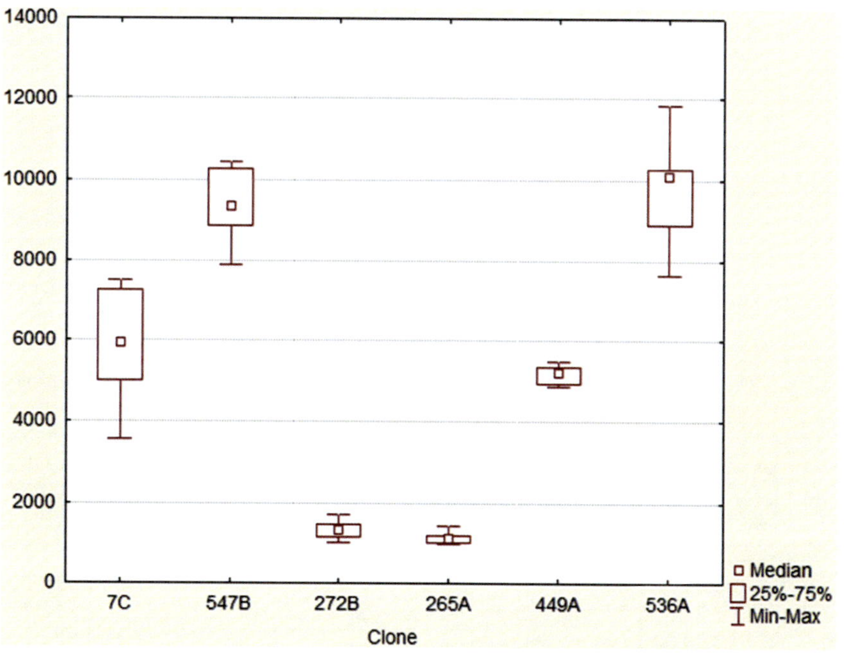

Fig. 4. Tolstikov, Metabolic Analysis.

of data obtained with the use of different analytical platforms for the same set of samples. **Figure 4** demonstrates variations found for the levels of 4-*O*-galactopyranosyl-d-mannopyranose in different clones of pine trees. Biochemical pathways affected by the plant genetic manipulations can be hypothesized.

4. Notes

1. MSTFA, pyridine, and/or mixed reagents should be stored under dry nitrogen after opening the bottle/ampoule or preparing the mix. The best way is to aliquot and seal small amounts for daily use.
2. Unless stated otherwise, aqueous buffers for HPLC should be refreshed daily.
3. Each lot of organic solvents should be investigated by LC/MS analysis. Purity control assessed by manufacturer is based on GC–MS analysis only.

Acknowledgments

Pine needle samples were provided by David Neale, Department of Plant Sciences, UC Davis. This work was supported by the UC Davis Genome Center.

References

1. Dunn, W.B., Bailey, N.J.C., Johnson, H.E. (2005). Measuring the metabolome: current analytical technologies. *Analyst*, **130**, 606–625.
2. Dettmer, K., Aronov, P.A., Hammock, B.D. (2007). Mass spectrometry-based metabolomics. *Mass Spectrometry Reviews*, **26**(1), 51–78.
3. Fiehn, O., Kind, T. (2007). Metabolite profiling in blood plasma. In: Metabolomics: Methods and Protocols. Weckwerth, W. (ed.), Humana, Totowa, NJ, ISBN13: 978-1-59745-244-1. Methods in Molecular Biology, **358**, 3–18.
4. Tolstikov, V.V., Fiehn, O., Tanaka, N. (2007). Reversed-phase monolithic capillary chromatography and hydrophilic chromatography coupled to electrospray ionization-mass spectrometry. In: Metabolomics: Methods and Protocols. Weckwerth, W. (ed.), Humana, Totowa, NJ, ISBN13: 978-1-59745-244-1. Methods in Molecular Biology, **358**, 141–156.
5. Last, L.L., Jones, A.D., Shahar-Hill, Y. (2007). Towards the plant metabolome and beyond. *Nature Reviews Molecular Cell Biology*, **8**, 167–174.
6. Plumb, R.S., Granger, J.H., Stumpf, C.L., Johnson, K.A., Smith, B.W., Gaulitz, S., Wilson, I.D., Castro-Perez, J. (2005). A rapid screening approach to metabonomics using UPLC and oa-TOF mass spectrometry: applica-

tion to age, gender and diurnal variation in normal/Zucker obese rats and black, white and nude mice. *Analyst*, **130**, 844–849.

7. Tolstikov, V.V., Fiehn, O. (2002). Analysis of highly polar compounds of plant origin: combination of hydrophilic interaction chromatography and electrospray ion trap mass spectrometry. *Analytical Biochemistry*, **301**(2), 298–307.
8. Kind, T., Tolstikov, V.V., Fiehn, O., Weiss, R.H. (2007). A comprehensive urinary metabolomics approach for identifying kidney cancer. *Analytical Biochemistry*, **363**, 185–195.
9. Tolstikov, V.V. (2007). Comprehensive metabolomics approach towards diagnostic tests. In: Biomarker Discovery Summit, Philadelphia, PA, September 17–19.

Part III

Applications of Quantum Dots and Molecular Fluorescence in Detection, Tracking and Imaging of Biomolecules

Chapter 23

Multicolor Detection of Combed DNA Molecules Using Quantum Dots

Christophe Escudé, Bénédicte Géron-Landre, Aurélien Crut, and Pierre Desbiolles

Summary

DNA combing is a useful strategy for manipulating single DNA molecules and has a wide range of applications in genetics, single molecule studies, and nanobiotechnology. Visualization of combed DNA molecules is usually performed by using DNA binding organic dyes. Such dyes are not suitable in all circumstances, especially because of their photoreactivity. We have developed a method for the detection of combed DNA molecules by fluorescence microscopy that avoids the use of DNA-staining agents and does not perturb the structure of the DNA molecule. Biotin- and/or digoxigenin-modified DNA fragments are covalently linked at both ends of a DNA molecule via sequence-specific hybridization and subsequent ligation. After the modified DNA molecules have been combed on a polystyrene-coated surface, their ends are visualized by multicolor fluorescence microscopy using conjugated quantum dots.

Keywords: Quantum dots, DNA combing, Single DNA molecule, Photobleaching, DNA detection

1. Introduction

During the past several years, manipulation and observation of single DNA molecules have opened new perspectives for studying the dynamics and function of DNA and for analyzing the organization of DNA sequences. DNA has also provided a new structural framework for nanobiotechnologies. Among the methods used to attach DNA to a surface, DNA combing occupies a privileged position and has been used for genetic *(1, 2)* and single DNA molecule studies *(3)* as well as for applications in nanobiotechnology *(4)*.

James Weifu Lee and Robert S. Foote (eds.), *Micro and Nano Technologies in Bioanalysis*,
Methods in Molecular Biology, vol. 544
DOI 10.1007/978-1-59745-483-4_23, © Humana Press, a part of Springer Science+Business Media, LLC 2009

Native combed DNA molecules can be visualized upon staining with DNA staining agents. It is also possible to locate the position of specific DNA sequences on combed DNA molecules after denaturation by fluorescence in situ hybridization.

DNA-binding organic dyes have several drawbacks: their photodestruction limits the observation of the DNA molecules to short time frames; this process results in the release of free radicals, which induce cleavage of the double-stranded DNA molecules; the presence of dyes results in changes in the electrostatic, structural, and mechanical properties of DNA, which are likely to modify its interaction with proteins; by the end, observation under physiological conditions (i.e., millimolar concentrations of divalent cations) is not possible because the dyes do not remain bound to DNA under these conditions.

We have developed a method for visualization of combed DNA molecules using quantum dots (QD) *instead* of organic DNA-staining agents *(5)*. QD display very interesting optical properties for use in various biomedical and biotechnological applications *(6, 7)*. In our approach, QD permit not only long-standing observation by fluorescence microscopy, but also a two-color determination of the orientation of single DNA molecules. Our method has a potentially wide application range, because the opportunities offered by DNA combing in single-molecule studies has still not been fully explored. Moreover, our method could also apply to the detection of not overstretched DNA molecules, which are well suited to the study of DNA dynamics and DNA-protein interactions *(3, 8, 9)*.

The procedure described in this chapter allows an investigator to prepare an end-modified DNA molecule (with biotin on one side and digoxigenin on the other) and to proceed to combing and detection of the DNA molecule under the fluorescence microscope using QD.

2. Materials

2.1. Preparation of Modified DNAa

1. DNA to be modified and primer oligonucleotides (*see* **Note 1**).
2. Restriction enzyme(s) and appropriate restriction buffer(s), from commercial suppliers (e.g., New England Biolabs, Ipswich, MA, USA or Fermentas, Glen Burnie, MD, USA) (*see* **Note 1**).
3. T4 DNA ligase and appropriate buffer from commercial supplier (e.g., from New England Biolabs or any other provider).

4. Taq polymerase, polymerase chain reaction (PCR) buffer, dNTPs from commercial supplier (e.g., New England Biolabs or Promega, Madison, WI, USA).
5. Biotin-16-dUTP and digoxigenin-11-dUTP are obtained from Roche (Nutley, NJ, USA).
6. PCR purification kit (Qiagen, Valencia, CA, USA).
7. 5× Precipitation buffer: 200 mM Tris-HCl pH 8.0, 400 mM NaCl, 80 mM $MgCl_2$.
8. Rinsing solution: 1:1 (v:v) mix of isopropanol and a solution containing 600 mM NaCl, 20 mM $MgCl_2$, and 50 mM EDTA.
9. Solutions for DNA staining are obtained from Invitrogen (Carlsbad, CA, USA): SybrGeen I, PicoGreen, and YOYO-1.
10. TE buffer: 10 mM Tris-HCl pH 8.0, 0.1 mM EDTA.

2.2. DNA Combing

1. Microscope coverslips (22 × 22 × 0.15 mm) from Fisher Scientific (Pittsburgh, PA, USA).
2. Polystyrene (from any source), dissolved in toluene (Sigma, St. Louis, MO, USA) to a final 5% weight:volume concentration.
3. 2-(*N*-morpholino)ethanesulfonic acid (MES) (Sigma). Prepare a 0.5 M solution by diluting 4.88 g in 50 mL deionized water, then prepare different batches (50 mL in Falcon tubes are convenient) of 50 mM solutions with pH adjusted to 5.4, 5.5, or 5.6 with NaOH. Solutions can be stored at 4°C for up to 2 months.
4. Microscope slides (any type).
5. Rubber cement.

2.3. DNA Labeling

1. Blocking reagent (Roche). Prepare a 1.5-mg/mL solution in 5 mM MES (pH 5.5), 1 mM EDTA by dissolving 15 mg in 10 mL buffer, heating to 37°C, and vortexing.
2. Revelation solution: 40 mM sodium borate (pH 8.3), 100 mM NaCl, 300 μg/mL blocking reagent.
3. Washing solution: 40 mM sodium borate (pH 8.3).
4. Anti-digoxigenin (Roche), 200 ng/mL final concentration in revelation solution.
5. QDs, streptavidin coated or antibody coated (from Invitrogen, *see* **Note 2**). We use QD 565 streptavidin for biotin detection and QD 655 goat $F(ab')_2$ anti-mouse IgG conjugate for indirect detection of digoxigenin. Dilute QD to a final concentration of 2 nM QD in revelation solution.

2.4. Fluorescence Microscopy

1. The coverslips are scanned using an inverted microscope (Olympus IX70 in our case) equipped with a 60× water-immersion objective (NA = 1.2). The light source is a mercury lamp.
2. Appropriate filters (Omegafilters) are required: we use a single excitation filter (475DF40) and dichroic mirror (505DRLP) and different emission filters depending on what has to be seen: 535DF45 for YOYO-1, 540DF27 for QD 565, and 645DF75 for QD 655 (*see* **Note 3**).
3. Images are captured by a Coolsnap camera (Roper Scientific, Tucson, AZ, USA) and processed with the Metamorph software package (Molecular Devices, Sunnyvale, CA, USA). Each pixel of the CCD camera chip corresponds to a 215 × 215-nm square on the sample.

3. Methods

All of the following steps can be carried out in a lab that is equipped for basic molecular biology with access to a PCR cycler and a fixed-temperature incubator.

3.1. Preparation of Modified DNA

1. In two different tubes, mix the primers (1.6 μM each) in 50 μL of Taq buffer with 2 mM $MgCl_2$; 50 μM of dATP, dCTP, and dGTP each; 33 μM of dTTP; 17 μM of biotin-dUTP in one tube and 17 μM of digoxigenin-dUTP in the other tube; 10 pg/μL of pBluescript SK+ as a template; and 0.1 U/μL of Taq polymerase (*see* **Note 4**).
2. Prepare the two modified fragments by PCR, using 30 cycles of amplification in three stages (30 s at 94°C, 30 s at 61°C, and 1 min at 72°C, increasing the last stage by 10 s per cycle, *see* **Note 5**) and a concluding extension of 10 min at 72°C.
3. Remove primers and unincorporated dNTP using PCR purification kits, following the protocols of the manufacturer.
4. Digest the PCR products in a total volume of 50 μL. In our case, digestion is conducted overnight at 50°C with 50 U Bsa I.
5. Remove the cleaved extremities by ethanol precipitation or by using a PCR purification kit.
6. Quantify the labeled fragments (*see* **Note 6**). The number of biotins or digoxigenins incorporated into the DNA fragment can also be assessed using a dot-blot assay if necessary *(10)*.
7. Digest your DNA of interest, leaving two different nonpalindromic single-stranded ends (*see* **Note 1**).

8. Ligate the digested DNA (2.5 nM, 100 ng for a 6,500-bp DNA) with a threefold excess of each labeled fragments (7.5 nM final concentration, which corresponds to 25 ng for 500-bp fragments) using 100 U of T4 DNA ligase in 10 μL of the recommended buffer. Ligation is carried out overnight at 20°C.
9. Remove the labeled fragments in excess using the spermidine precipitation procedure: add 20 μL of 5× precipitation buffer and 35 μL water to the ligation mix, then 50 μL of ice-cold spermidine, vortex, and incubate for 30 min at room temperature.
10. Centrifuge for 15 min at 10,000 × *g* and 4°C, eliminate the supernatant, rinse the resulting pellet with the rinsing solution, and centrifuge again for 2 min at 10,000 × *g* and 4°C (*see* **Note 7**).
11. Resuspend in 10 μL of TE supplemented with 1.5 μM YOYO-1.

3.2. DNA Combing

1. Clean the coverslips, preferentially using a Plasma Cleaner.
2. Spin-coat a drop of the polystyrene solution on each coverslip.
3. Bake all coverslips at 80°C for at least 1 h in an oven.
4. Add 10 ng of the YOYO-1-stained DNA in a reservoir containing 3 mL of 50 mM MES (pH 5.5), EDTA 1 mM (*see* **Note 8**).
5. Dip a polystyrene-coated coverslip into the reservoir for 2 min, and then slowly pull up the slide (*see also* **Note 7**). At appropriate pH, DNA molecules in the reservoir bind to the coverslip's surface by one of their extremities. As the coverslip is pulled up out of the reservoir, the DNA molecules are aligned and uniformly overstretched by the receding meniscus *(11, 12)*.
6. Stick the coverslip to a microscope slide, with the hydrophobic surface facing down, using two small bands of Parafilm on two opposite sides of the coverslip, which will melt when the slide is put at 50°C (*see* **Fig. 1**, right).
7. Check for the presence of combed DNA molecules under the microscope, using an FITC filter set. Combed DNA molecules should appear linear and aligned, with a length corresponding approximately to 0.5 μm per 1,000 bp.
8. If numerous DNA molecules are present in a coil conformation, then restart from the beginning and proceed with a MES solution at higher pH, and check again. If the combing efficiency is very low (i.e., very few molecules can be seen), then proceed again with a MES solution at more acidic pH.

If the combing works well, then the DNA solution can be used for processing several coverslips (usually between 10 and 20 coverslips). Check for the combing efficiency on your last coverslip. Prepare a new DNA solution when the number of combed DNA molecules becomes too low. The coverslips that were mounted for checking purposes will be discarded.

9. Stick the coverslips to a microscope slide with rubber cement (one point at each angle), with the DNA facing up, to facilitate further manipulation of the coverslips (*see* **Fig. 1**, left).

3.3. Detection of DNA Molecules

1. Add 100 μL blocking reagent and incubate for 10 min (*see* **Note 9**). There is no need for using coverslips to cover the solution. This is also true for all of the following steps.
2. Remove the blocking reagent with a pipet. Wash the coverslips twice with 200 μL washing solution. All washing steps are carried out by pipetting the solution down onto the slide and removing it by aspiration after the incubation time. Leave the sample to dry.
3. Add mouse anti-digoxigenin antibody, incubate for 10 min, and then wash twice with 200 μL washing solution.
4. Add the QD solution (100 μL) and incubate for 10 min.
5. Wash five times carefully with 200 μL washing solution.
6. Using a scalpel, remove the coverslips from the slide and mount upside down on another slide, using Parafilm as indicated in **Fig. 1**.
7. Examine the slides under a microscope. The YOYO-1 signal can be observed with a FITC filter set. QD are detected separately with different filter sets (see above). Because of the intermittent fluorescence of QD ("blinking"), a single image of the slide is not sufficient to localize all of the QD because some of them could be in a dark state during the shot. Therefore, it is necessary to record a 60-frame movie (exposure time for each

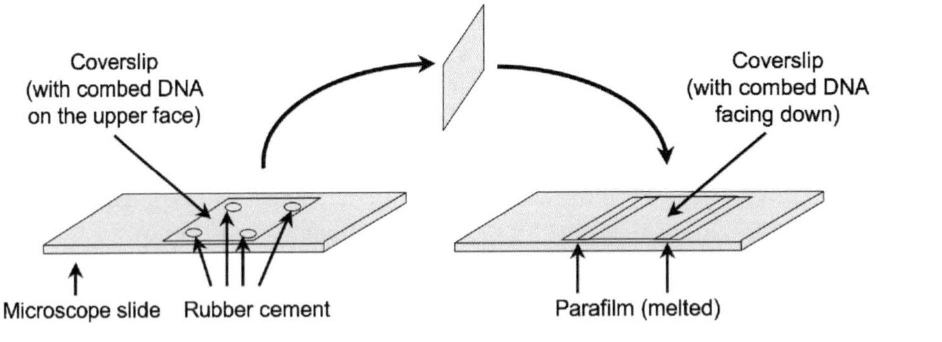

Fig. 1. Scheme for sticking a coverslip to a microscope slide. *Left* (DNA facing up): this scheme is used for all of the detection steps. *Right* (DNA facing down): this scheme is used for visualizing DNA under the microscope.

frame: 1 s). This movie is then processed to generate a QD maximum image where each pixel intensity is shown at the maximum intensity displayed in the 60-frame image stack. All of these tasks can be automated using microscopy software such as Metamorph, for example. Images can be further analyzed to localize QD. It is possible to use a method based on cross-correlation between the fluorescence image and a specific correlation template. For QD detection, this template is a two-dimensional isotropic Gaussian function that corresponds to the point spread function (PSF) of the optical system. Under carefully selected conditions (*see* **Note 10**), the position and orientation of individual DNA molecules can be inferred with good efficiency from the QD fluorescence signals alone. This is achieved by selecting QD pairs that have the distance and direction expected for the combed DNA molecules. Programs written in MATLAB (MathWorks) can easily perform all these tasks.

4. Notes

1. The DNA of interest can be any plasmid or other long DNA. Plasmid and phage DNA are commercially available or can be made in any lab equipped for microbiology. The DNA must contain target sites for restriction enzymes that leave nonpalindromic sticky ends, to obtain high ligation yields and avoid self-ligation. Class IIs restriction enzymes are interesting because they cleave DNA outside of their recognition sequence. An example of choice is provided in **Fig. 2**. In our example, BsmB I is used for cleaving the pTYB1 plasmid, and BsaI is used for cleaving the modified DNA fragments. The design of primers that provide complementary sticky ends is also highlighted. Custom primer sequences can be designed and synthesized to conform to any other restriction site.
2. Other sources of QD are available today, such those provided by Evident Technologies (Troy, NY, USA). Providers often change QD conditioning, which may result in unexpected experimental results.
3. More generally, YOYO-1 can be detected with any fluorescein filter set. Filter sets that are specially designed for QDs are now available. Alternatively, QD 565 and QD 655 can be detected with Cy3 and Cy5 filter sets, but the signal will not be as intense as by using a specially designed filter set, because of poor absorption of the QD at high wavelength.

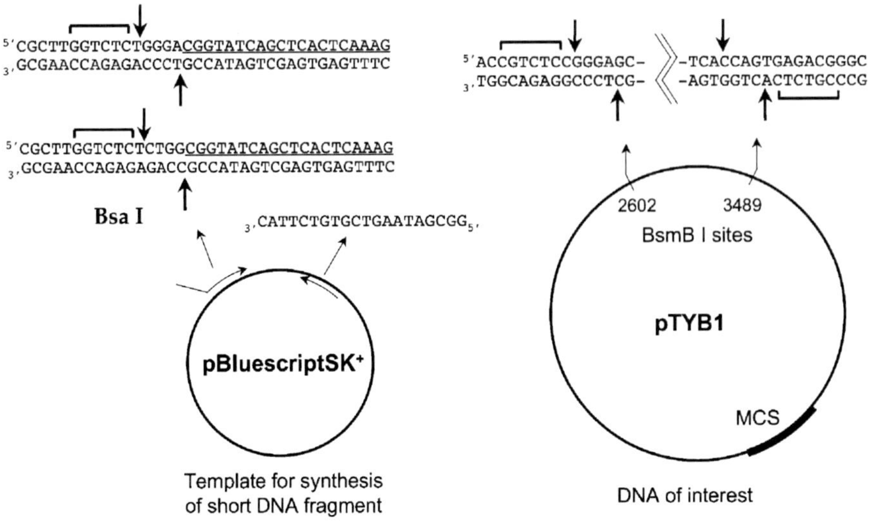

Fig. 2. System used for synthesis of modified DNA fragments (*left*) and an example of long DNA molecule to be modified (*right*). pBluescriptSK+ is commercially available from Stratagene (La Jolla, CA, USA). The sequence of two "forward" primers (*underlined*) as well as one "reverse" primer are indicated. A restriction site is introduced in the final sequence of the PCR product by the use of the flanking sequence in the primer. pTYB1 is also commercially available (New England Biolabs). *MCS* represents the position of the multicloning site where any DNA sequence of interest may be easily inserted.

4. It is possible to synthesize short DNA fragments that contain fewer modified nucleotides. For example, we obtained approximately ten biotins per fragment by using 50 μM dTTP and 2 μM biotin-dUTP in the PCR.

5. We have checked that using such an increase in elongation time resulted in a much higher quantity of DNA fragment. However, not all thermocyclers afford this possibility.

6. Quantification of nucleic acids can be done by UV spectroscopy. Because only a small quantity is available and to avoid further ethanol precipitation, it is convenient to use a nanodrop apparatus (Thermo Fisher Scientific), which measures absorption in a 1-μL volume. Alternatively, comparing the fluorescence intensity of the sample with that of a standard can be performed, in an agarose gel or using a microtiter plate reader. Ethidium bromide and SybrGreen are convenient as DNA stains for an agarose gel, whereas picogreen is suitable for a solution assay.

7. Spermidine precipitation can be tricky for non-experienced hands. for further details on how to optimize this process, *see* **ref.** *13*. It is probably possible to purify the modified DNA from the excess of short fragments using high-pressure liquid chromatography (HPLC), for example, with an AKTA purifier (General Electric, Piscataway, NJ, USA).

8. DNA combing is obtained as soon as a receding meniscus encounters a DNA molecule attached by one end. First combing experiments were performed by allowing a droplet to dry on the surface. A mechanical movement with a uniform speed is preferable because it leads to more uniformly extended DNA molecules. Reservoirs made of Teflon are convenient and easy to clean. The reservoir should closely fit the size of the coverslip to minimize the volume of the solution. Dipping and removing the coverslip can be done manually, but also with a homemade mechanical or electrical device.
9. This step is expected to reduce nonspecific interactions between the QD and the surface.
10. In fact, this probability depends on DNA length, DNA density, and the density of QD signals on the coverslip. For more details, *see* **ref.** *5.*

Acknowledgments

We thank Jean-François Allemand for introduction to the combing method and members of our groups for helpful discussions.

References

1. Michalet, X., Ekong, R., Fougerousse, F., Rousseaux, S., Schurra, C., Hornigold, N., van Slegtenhorst, M., Wolfe, J., Povey, S., Beckmann, J.S. and Bensimon, A. (1997). Dynamic molecular combing: stretching the whole human genome for high-resolution studies. *Science*, **277**, 1518–1523.
2. Xiao, M., Gordon, M.P., Phong, A., Ha, C., Chan, T.F., Cai, D., Selvin, P.R. and Kwok, P.Y. (2007). Determination of haplotypes from single DNA molecules: a method for single-molecule barcoding. *Hum Mutat*, **28**, 913–921.
3. Guéroui, Z., Place, C., Freyssingeas, E. and Berge, B. (2002). Observation by fluorescence microscopy of transcription on single combed DNA. *Proc Natl Acad Sci U S A*, **99**, 6005–6010.
4. Keren, K., Krueger, M., Gilad, R., Ben-Yoseph, G., Sivan, U. and Braun, E. (2002). Sequence-specific molecular lithography on single DNA molecules. *Science*, **297**, 72–75.
5. Crut, A., Geron-Landre, B., Bonnet, I., Bonneau, S., Desbiolles, P. and Escude, C. (2005). Detection of single DNA molecules by multicolor quantum-dot end-labeling. *Nucleic Acids Res*, **33**, e98.
6. Michalet, X., Pinaud, F.F., Bentolila, L.A., Tsay, J.M., Doose, S., Li, J.J., Sundaresan, G., Wu, A.M., Gambhir, S.S. and Weiss, S. (2005). Quantum dots for live cells, in vivo imaging, and diagnostics. *Science*, **307**, 538–544.
7. Weng, J. and Ren, J. (2006). Luminescent quantum dots: a very attractive and promising tool in biomedicine. *Curr Med Chem*, **13**, 897–909.
8. Crut, A., Lasne, D., Allemand, J.F., Dahan, M. and Desbiolles, P. (2003). Transverse fluctuations of single DNA molecules attached at both extremities to a surface. *Phys Rev E*, **67**, 051910.
9. Graneli, A., Yeykal, C.C., Prasad, T.K. and Greene, E.C. (2006). Organized arrays of individual DNA molecules tethered to supported lipid bilayers. *Langmuir*, **22**, 292–299.
10. Conti, C., Caburet, S., Schurra, C. and Bensimon, A. (2001). Current protocols in Cytometry. Wiley, New York, pp. 8.10.1–8.10.23.

11. Bensimon, A., Simon, A., Chiffaudel, A., Croquette, V., Heslot, F. and Bensimon, D. (1994). Alignment and sensitive detection of DNA by a moving interface. *Science*, **265**, 2096–2098.
12. Allemand, J.F., Bensimon, D., Jullien, L., Bensimon, A. and Croquette, V. (1997). pH-dependent specific binding and combing of DNA. *Biophys J*, **73**, 2064–2070.
13. Murphy, J.C., Wibbenmeyer, J.A., Fox, G.E. and Willson, R.C. (1999). Purification of plasmid DNA using selective precipitation by compaction agents. *Nat Biotechnol*, **17**, 822–823.

Chapter 24

Quantum Dot Molecular Beacons for DNA Detection

Nathaniel C. Cady

Summary

Molecular beacons have become an important fluorescent probe for sequence-specific DNA detection. To improve the sensitivity and robustness of molecular beacon assays, fluorescent semiconductor quantum dots (QDs) are now being used as the fluorescent moiety for molecular beacon synthesis. Multiple linkage strategies can be used for attaching molecular beacon DNA to QDs, and multiple quenchers, including gold particles, can be used for fluorescence quenching. Covalent attachment of QDs to DNA can be achieved through amide linkage, and affinity-based attachment can be achieved with streptavidin-biotin linkage. We have shown that these linkage strategies can be used to successfully create quantum dot molecular beacons that can be used in DNA detection assays with high specificity.

Key words: Quantum dot, Biosensor, Molecular beacon, DNA, Detection

1. Introduction

Semiconductor quantum dots (QDs) have become attractive fluorophores for biosensing and high-throughput bioanalytical systems. Replacing organic fluorophores with QDs can greatly improve detection sensitivity and reduce negative effects such as photobleaching and spectral overlap during excitation and detection. QDs commonly have broad absorption spectra, narrow emission spectra (25–30 nm full width at half maximum), and large Stokes shift, making them highly amenable to multiplex detection strategies *(1–3)*. Because of the broad absorption spectra and large emission shift, QDs can be excited at wavelengths far removed from their emission peak. Additionally, QDs that emit at multiple wavelengths in the visible spectrum can typically be excited using a single, short-wavelength ultraviolet (UV)

James Weifu Lee and Robert S. Foote (eds.), *Micro and Nano Technologies in Bioanalysis*,
Methods in Molecular Biology, vol. 544
DOI 10.1007/978-1-59745-483-4_24, © Humana Press, a part of Springer Science+Business Media, LLC 2009

excitation source. This property can result in less-complicated optical detection systems that use only one excitation wavelength. The narrow emission spectra of QDs can also reduce emission spectrum overlap, making it possible to differentiate more fluorophores within a small spectral range. Recently, there have been multiple reports of using QDs for biosensing applications *(1, 2, 4, 5)*, especially for multiplex detection *(3, 5–7)*.

To improve signal-to-noise ratio and to avoid complicated separation techniques, fluorescence resonant energy transfer (FRET)-based quenching of fluorescence has been used in multiple detection strategies. FRET is a distance-dependent phenomenon that occurs when a donor fluorophore and an acceptor chromophore (quencher) are in close proximity *(8)*. When the fluorophore and quencher are within a certain distance from each other, excitation energy can be transferred from the fluorophore to the quencher, preventing fluorescence emission. The distance at which there is a 50% transfer in energy is known as the Förster Radius (R_o):

$$R_o = 9.78 \times 10^3\ [k^2\ n^2\ Q_o\ J(\lambda)]^{(1/6)}.$$

R_o is dependent upon k, the orientation factor for dipole-dipole interactions; n, the refractive index of the medium; Q_o, the quantum efficiency of the donor; and $J(\lambda)$, the spectral overlap integral between the donor and acceptor *(4, 9)*. Once R_o is known, the FRET efficiency (E) can be calculated for various distances (R) between the donor and the acceptor *(9)*:

$$E = \frac{R_o^6}{R_o^6 + R^6}$$

Multiple detection techniques have been devised to harness the FRET effect, including molecular beacons *(10–13)*. Molecular beacons are comprised of a fluorophore and a quencher moiety attached to opposite ends of a single-stranded DNA oligonucleotide. The sequence of the oligonucleotide is designed such that it preferentially base pairs with itself, forming a stem-loop structure. Included inside the loop region is a probe sequence that can bind to a complementary target DNA sequence. This hybridization event causes the stem-loop structure to open and spatially distance the fluorophore from the quencher. This decreases the FRET efficiency and results in increased emission from the fluorophore (donor). Molecular beacons are extremely target specific, primarily because of the competition between internal hybridization within the stem structure and hybridization between the target and the loop structure *(11)*. These dynamic interactions require specific hybridization between the target and loop structure to stabilize the molecular beacon in the open position. Additionally, the stem and loop length of the molecular beacon can be altered to achieve either higher specificity and/or increased kinetic rate constants *(13)*. These features make molecular beacons attractive for biosensor detection applications.

For molecular beacon-based assays, the fluorescent labels or tags need to have suitable attachment chemistry to covalently link them to the nucleic acid probes. For multiplex applications, fluorophores with different emission maxima have been linked to different molecular beacons, allowing for simultaneous detection within one solution, provided the appropriate optical detectors are used. The coupling strategies that are commonly used for DNA-fluorophore linkages typically involve reactions between functional groups (such as carboxyl and amine groups) or the use of high-affinity interactions such as that between the small molecule biotin and the protein streptavidin *(14)*. Successful linkage using these coupling strategies requires careful choice of the proper linkage method and significant optimization of the coupling reaction. Given the unique fluorescence characteristics of QDs, as described above, our group and others have developed several QD-based molecular beacons (QDMBs) to be used for DNA-based detection strategies *(9, 14)*. Here we present several methods to use covalent and affinity-based methods to attach QDs to molecular beacon DNA hairpins and assay for their functionality in DNA detection assays.

2. Materials

2.1. Amide-Linked Quantum Dot Molecular Beacons

1. Phosphate buffered saline (PBS) pH 7.4 (Sigma-Aldrich, St. Louis, MO) autoclaved and filter sterilized with a 0.2-μm filter.
2. 1-Ethyl-3-(3-dimethylaminopropyl) carbodiimide hydrochloride (EDC) (Pierce Biotechnology Inc., Rockford, IL) stored with desiccant at –20°C. Prepare a daily working solution of 10 mg/mL in PBS (52.2 mM final concentration) (*see* **Note 1**).
3. Carboxyl-modified QDs (Quantum Dot Corporation, Hayward, CA and Evident Technologies, Troy, NY) stored at 4°C until use.
4. Lyophilized molecular beacon DNA (Integrated DNA Technologies, Coralville, IA) diluted to 500 μM with sterile, 0.2-μm filter-sterilized distilled water. DNA should have an organic quencher on the 3′ terminus and an amino linker on the 5′ terminus.
5. Microcon 50,000 molecular weight cut-off (MWCO) spin filters (Millipore Corporation, Bedford, MA). Spin filters should be prewet with PBS before filtration of QDs or QDMBs (*see* **Note 2**).

2.2. Streptavidin-Biotin-Linked Quantum Dot Molecular Beacons

1. PBS pH 7.4 (Sigma-Aldrich) autoclaved and filter sterilized in a 0.2-μm filter.
2. Bovine serum albumin (BSA) (New England Biolabs, Beverly, MA).
3. Streptavidin-modified QDs (Quantum Dot Corporation and Evident Technologies) stored at 4°C until use.
4. Lyophilized molecular beacon oligonucleotide DNA (Integrated DNA Technologies) diluted to 500 μM with sterile, 0.2-μm filter-sterilized distilled water. DNA should be modified with an organic quencher on the 3′ terminus and a biotin linker on the 5′ terminus.
5. Microcon 50,000 MWCO spin filters (Millipore Corporation) (*see* **Note 2**).

2.3. Strepatvidin-Biotin-Linked Quantum Dot Molecular Beacons with Gold Quencher

1. PBS pH 7.4 (Sigma-Aldrich) autoclaved and filter sterilized in a 0.2-μm filter.
2. BSA (New England Biolabs).
3. EDC (Pierce Biotechnology Inc.) stored with desiccant at –20°C. Prepare a daily working solution of 10 mg/mL in PBS (52.2 mM final concentration) (*see* **Note 1**).
4. Streptavidin-modified QDs (Quantum Dot Corporation and Evident Technologies) stored at 4°C until use.
5. Lyophilized molecular beacon and target oligonucleotide DNA (Integrated DNA Technologies) diluted to 500 μM with sterile, 0.2-μm filter-sterilized distilled water. DNA should have been modified with a 3′ amino linker and a 5′ biotin linker.
6. Mono-sulfo-N-hydroxy-succinimide (NHS) Nanogold conjugate (1.4-nm diameter) (Nanoprobes Inc., Yaphank, NY) stored at –20°C until use. Prepare a working solution of Nanogold with 100 μM concentration by adding Nanogold conjugate to PBS. Nanogold conjugate can be purchased in premeasured quantities, such that the working solution can be prepared by simply adding a known volume of PBS.
7. Glycine quenching reagent. Prepare 10 mM glycine solution (Sigma-Aldrich) in sterile water. Store at room temperature.
8. Microcon 100,000 MWCO spin filters (Millipore Corporation). Spin filters should be prewet with PBS before filtration of QDs or QDMBs.

2.4. Electrophoretic Gel Mobility Shift Assay

1. Horizontal gel electrophoresis system with DC power supply capable of 5 V/cm electric field generation.
2. Agarose gels for electrophoresis: 2%, w/v agarose (Sigma-Aldrich) in Tris-acetate EDTA (TAE) buffer (Sigma-Aldrich).

Prepare gel by adding 2 g of agarose to 100 mL of TAE buffer and heat to boiling to melt agarose. Pour into horizontal gel electrophoresis casting tray using comb with at least a 5-mm tooth width to create 5-mm-wide wells.

3. 60% Glycerol (Sigma-Aldrich) loading reagent for enabling QDMB sedimentation into wells of the agarose gel.
4. TAE gel running buffer (Sigma-Aldrich).
5. UV transilluminator and gel imaging system. The transilluminator should be capable of producing UV light at approximately 280–360 nm for excitation of QD fluorescence. The gel imaging system can include a simple digital camera with greater than 3-megapixel resolution or a commercial gel imaging system.

2.5. Fluorescence Detection Assay

1. Microplate fluorometer for fluorescence detection in a 96-well plate format. Suitable microplate fluorometers can be manufactured by Tecan (Durham, NC), Biotek Instruments (Winooski, VT), and Molecular Dynamics (Sunnyvale, CA), as well as other manufacturers (*see* **Note 3**).
2. Lyophilized molecular beacon DNA (Integrated DNA Technologies) diluted to 500 μM with sterile, 0.2-μm filter-sterilized distilled water. A positive-control sequence should be 100% complementary to the molecular beacon DNA sequence, whereas a negative-control sequence should have at least one mismatch to the molecular beacon sequence, or be completely noncomplementary.
3. Hybridization buffer: PBS pH 7.4 (Sigma-Aldrich) autoclaved and filter sterilized in a 0.2-μm filter (*see* **Note 4**).
4. 96-Well plastic microplate. Plates should be black to prevent cross-talk between wells and to reduce any autofluorescence from the plastic. Half-volume or full-volume 96-well plates from multiple manufacturers are acceptable.

2.6. Molecular Beacon Design Guidelines

Molecular beacons have been designed for a wide variety of DNA targets and their sequences can be obtained from the literature *(10, 11, 13)*. Alternatively, molecular beacons can be designed using nucleic acid manipulation software including Vector NTI (Invitrogen, Carlsbad, CA). The "loop" portion of the molecular beacon DNA hairpin is typically designed to contain between 15 and 25 bases and the "stem" portions of the hairpin can contain between 4 and 10 bases. The stem regions are added to the 5′ and 3′ ends of the loop sequence, enabling stem-loop formation through internal hybridization to create the molecular beacon. For QDMBs, a QD is used as the fluorophore and a variety of quenching moieties can provide fluorescence quenching. These include the Iowa Black FQ quencher (Integrated DNA

Technologies), dabcyl, 1.4-nm gold particles (Nanoprobes), and a variety of other quenchers. Dark quenchers, such as Iowa Black FQ, have a broad absorbance spectrum and can be used with a wide range of different color QDs. Similarly, gold quenchers, such as 1.4-nm Nanogold, can quench most QDs emitting in the visible spectrum. Other quenchers, such as dabcyl, must have their absorbance spectrum tightly overlapped with the fluorescence emission spectrum of the QD that is used. For attachment, the QDMB oligonucleotide should be modified at each end with the appropriate linkage chemistry. For carboxyl-modified QDs, the oligonucleotide should have an amino linker for amide linkage, and for streptavidin-modified QDs, the oligonucleotide should have a biotin linker for attachment. At the opposite end, the oligonucleotide can be modified with an appropriate organic quencher by the DNA supplier, or can be modified with an amine for amide linkage with sulfo-NHS-modified Nanogold particles.

3. Methods

These methods describe the linkage of QDs and Nanogold particles to presynthesized oligonucleotide DNA to create QDMBs. As described in **Subheading 2**, oligonucleotides can be obtained with the appropriate linkage chemistry and can be presynthesized with organic quenchers such as Iowa Black FQ, dabcyl, and others. Methods for three different attachment methods are described, including *(1)* amide linkage, (2) streptavidin-biotin linkage, and (3) streptavidin-biotin linkage to QDs with amide linkage to 1.4-nm Nanogold particles. **Figure 1** shows an example of each type of QDMB that is described. After preparation of the QDMBs following these protocols, the resulting QDMBs can be analyzed using an electrophoretic mobility shift assay. As performed in our laboratory, quantification of DNA attachment to the QDs could not be assayed using traditional absorbance or fluorometric DNA quantification assays. Several reagents, including the EDC crosslinker, interfered with both absorbance (at 260 nm) measurements and the Pico Green (Invitrogen) fluorescent quantification assay. Therefore, the gel mobility assay was used to qualitatively determine whether QDs had been successfully modified with the molecular beacon DNA hairpins. To observe the hairpin opening and subsequent increase in fluorescence of the QDMBs, we describe a simple assay using a microplate fluorometer. In this assay, single-stranded target DNA can be added to the QDMBs, resulting in a quantifiable increase in fluorescence.

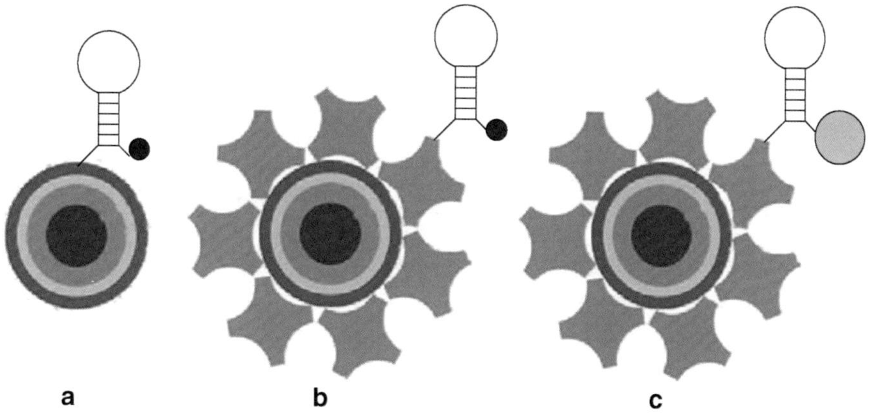

Fig. 1. Quantum dot molecular beacons (QDMBs) synthesized with various attachment chemistries. QDs are represented by the large circular structure, and the molecular beacon hairpins are shown, attached to various organic and gold-based quenchers. (**a**) Amino-modified DNA hairpin attached to a carboxyl-conjugated QD and organic quencher, (**b**) biotinylated DNA hairpin attached to a streptavidin-modified QD and organic quencher, (**c**) biotinylated DNA hairpin attached to a streptavidin-modified QD and a 1.4-nm Nanogold particle.

3.1. Amide Linkage of QD to Molecular Beacon Oligonucleotides

1. These instructions assume the use of carboxyl-modified QDs (Quantum Dot Corporation and/or Evident Technologies) for covalent linkage to a 5′ amino linker of molecular beacon DNA modified with the appropriate organic quencher/dye. The concentration of the QDs and the synthesized DNA is variable, so, whenever possible, use a high concentration of each of these reagents. This will allow for small volume additions (1–50 μL) and a small working volume of 4–500 μL. These small volumes will reduce the amount of excess, unused reagents and will allow for filtration-based washing in small volume MWCO filters.
2. Activate the carboxyl surface groups of the QDs with EDC by adding 100 pmol of QDs and 300 nmol (5.7 μL) of EDC working solution in 400 μL of PBS in an autoclave-sterilized 1-mL glass vial. Bring the total volume to 200 μL with PBS. Incubate for 15 min at room temperature.
3. Add 2 nmol (4 μL of 500 μM solution) of amino-labeled molecular beacon DNA to the mixture and mix thoroughly by inversion or vortexing. Allow this mixture to react for 2 h at room temperature.
4. Remove unlinked DNA and any remaining EDC from the reaction by spin filtration in Millipore Microcon 50,000 MWCO spin filters. Prewet the filter membrane with PBS, and then add the reaction mixture. Spin at 7,000 × *g* for 5 min, discard the flow-through, and resuspend the retentate (QD-modified DNA) in 400 μL PBS.

5. Repeat the washing step two times and resuspend the final retentate in 200 μL of PBS.
6. Visibly inspect the resulting QDMB conjugates and note any decrease in fluorescence intensity as compared with the unmodified QDs. Tubes containing the QDMB conjugates can be placed on a UV transilluminator for fluorescence excitation and visual observation of emission intensity.
7. Store QDMBs at 4°C. Stability was observed for 1 month or more after synthesis.

3.2. Streptavidin Linkage of QD to Biotinylated Molecular Beacon Oligonucleotides

1. These instructions assume the use of streptavidin-modified QDs (Quantum Dot Corporation and/or Evident Technologies) for affinity-based linkage to a 5′ biotin linker on molecular beacon DNA modified with the appropriate organic quencher/dye.
2. Mix 100 pmol of streptavidin-modified QDs with 1 nmol of biotinylated molecular beacon DNA (2 μL of 500 μM solution) and 40 μL of 10 mg/mL BSA. Bring the entire reaction volume to 400 μL with PBS. Incubate for 1 h at room temperature.
3. Remove unlinked DNA from the reaction by spin filtration in Millipore Microcon 50,000 MWCO spin filters. Prewet the filter membrane with PBS, and then add the reaction mixture. Spin at 7,000 × *g* for 5 min, discard the flow-through, and resuspend the retentate (QD-modified DNA) in 400 μL PBS.
4. Repeat the washing step two times and resuspend the final retentate in 200 μL of PBS.
5. Visibly inspect the resulting QDMB conjugates and note any decrease in fluorescence intensity as compared with the unmodified QDs. Tubes containing the QDMB conjugates can be placed on a UV transilluminator for fluorescence excitation and visual observation of emission intensity.
6. Store QDMBs at 4°C. Stability has been observed for 1 month or more after synthesis.

3.3. Streptavidin Linkage of QDMBs Using 1.4-nm Nanogold for Quenching

1. These instructions assume the use of streptavidin-modified QDs (Quantum Dot Corporation and/or Evident Technologies) for affinity-based linkage to a 5′ biotin linker on molecular beacon DNA. The opposite (3′) end of the DNA should be modified with an amine group for amide linkage to sulfo-NHS 1.4-nm Nanogold particles.
2. Mix 1 nmol of mono-sulfo-NHS Nanogold (10 μL of 100 μM working solution) with 10 nmol of amino-labeled, biotinylated molecular beacon DNA in 300 μL of PBS. Incubate for 2 h at room temperature.

3. Quench the reaction with 10 μL of 10 mM glycine to deactivate any remaining NHS on the surface of the Nanogold (*see* **Note 5**).
4. Apply the mixture to a prewet Millipore Microcon 10,000 MWCO spin filter, centrifuge at 7,000 × *g*, and resuspend the retentate in 200 μL PBS (*see* **Note 2**).
5. Repeat this washing procedure two times and resuspend in 100 μL of PBS.
6. Mix the resuspended Nanogold-DNA conjugate with 10 pmol streptavidin-modified QDs, 40 μL of 10 mg/mL BSA, and bring the reaction to a total volume of 400 μL with PBS. Incubate for 1 h at room temperature.
7. Add the mixture to a prewet Millipore Microcon 100,000 MWCO spin filter and centrifuge at 7,000 × *g* (*see* **Note 6**).
8. Repeat the wash procedure two times and resuspend the final retentate in 200 μL PBS.
9. Visibly inspect the resulting QDMB conjugates and note any decrease in fluorescence intensity as compared with the unmodified QDs. Tubes containing the QDMB conjugates can be placed on a UV transilluminator for fluorescence excitation and visual observation of emission intensity.
10. Store QDMBs at 4°C. Stability has been observed for 1 month or more after synthesis.

3.4. Gel Mobility Assay for Assessment of DNA Attachment to QDs

1. Electrophoretic gel mobility can be used to assess DNA attachment to QDs. This assay monitors the electrophoretic mobility of both modified and unmodified QDs, in which modified QDs exhibit decreased mobility in the agarose gel. This mobility shift is a qualitative method for determining whether QDs have been successfully modified with molecular beacon DNA hairpins or other DNA molecules. Unmodified QDs must be used as size standards to compare with the mobility of the modified QDs.
2. Mix QDMBs and unmodified QDs in a 1:1 ratio with 60% glycerol. This will allow the solutions to sink into individual wells in the cast gel.
3. Apply QD and QDMB glycerol mixtures to individual wells in the gel while submerged in TAE buffer, within the gel electrophoresis apparatus.
4. Assemble the gel electrophoresis device, add TAE running buffer, and apply 5 V/cm with the power supply. The wells of the gel should be oriented toward the positive pole of the system, similar to the orientation for DNA electrophoresis.
5. Apply voltage for approximately 30 min and then check QD mobility by observing gel on a standard UV transilluminator.

UV light in the range of 280–360 nm will be acceptable for excitation of QD fluorescence. The gel can be subjected to further electrophoresis if separation between modified and unmodified QD bands cannot be observed.

6. Document gel with standard gel electrophoresis imaging system, or by use of a digital camera while exciting fluorescence on a transilluminator. *See* **Fig. 2** for an example of an electrophoretic gel mobility shift assay for QDMBs. QDMBs with amide linkage show some mobility shift as compared with carboxyl-conjugated QDs, whereas streptavidin-biotin-linked QDMBs have a more pronounced mobility shift as compared with unmodified streptavidin QDs. As can be seen from **Fig. 2**, significant smearing is observed for the DNA-modified QDs, which is possibly caused by varying degrees of modification.

3.5. Fluorescence Measurement

1. Fluorescence intensity measurements can be performed to evaluate the ability of QDMBs to exhibit hybridization-based quenching and unquenching. In this assay, single-stranded DNA target is added to the QDMBs to initiate molecular beacon hairpin opening and hybridization to the target. Target sequences should be complementary to the molecular beacon sequence, and noncomplementary sequences can be used to determine the relative specificity of the QDMB.
2. Mix QDMB conjugates from the above synthesis (varying concentrations) with single-stranded DNA targets (varying concentrations) in PBS or another DNA hybridization buffer (*see* **Note 4**).
3. Allow QDMBs and target DNA to interact for 10 min or longer to facilitate hairpin opening and hairpin-target hybridization.

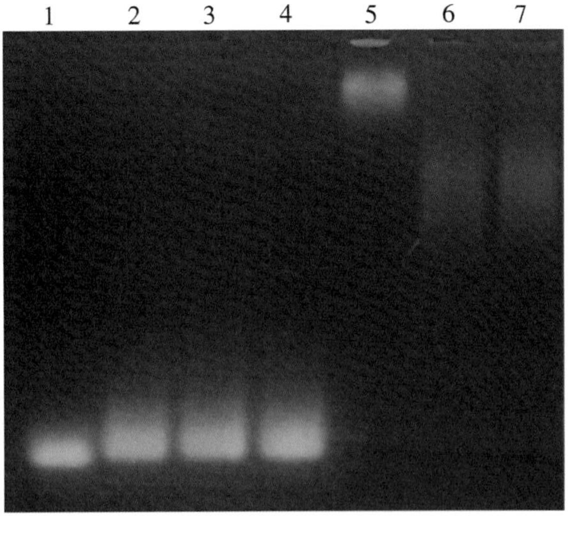

Fig. 2. Electrophoretic gel mobility assay for assessment of DNA attachment to QDs. Unmodified QDs and DNA-modified QDs were loaded into a 2% agarose gel and subjected to electrophoresis at 5 V/cm in TAE buffer. *Lane 1* unmodified, carboxyl-conjugated QDs, *lanes 2–4* DNA-modified QDs with amide linkage, *lane 5* streptavidin-conjugated QDs without DNA, *lane 6* streptavidin-conjugated QDs modified with biotinylated DNA, *lane 7* streptavidin-conjugated QDs modified with biotinylated, 1.4-nm Nanogold-conjugated DNA.

4. Apply 50–100 μL of each mixture to black 96-well fluorescence microplates. Use standards including unmodified QDs, QDMBs without target, and blank solution (PBS or other hybridization buffer only) (*see* **Note 7**).
5. Insert the microplate into the microplate fluorometer instrument.
6. Measure the fluorescence intensity of each well with excitation at the appropriate wavelength and detection at the appropriate emission wavelength for the QDs that are used (*see* **Note 3**).
7. Compare the fluorescence intensity of each sample to the blank solution to normalize the intensity data. After normalization, the fluorescence intensity of each sample can be compared with the other samples and standards. If the QDMBs are functioning properly, an increase in fluorescence intensity should be observed if a complementary target was added and allowed to hybridize with the QDMB.
8. If fluorescence intensity changes are not observed upon mixing with target DNA, hairpin opening can be achieved by slowly heating the solution to the melting temperature (T_m) of the hairpin or by adding denaturants such as urea or formamide. Hairpin opening through denaturants or temperature ramping should result in fluorescence intensity increases if the QDMBs were synthesized successfully and thev molecular beacon DNA sequence was designed properly. **Figure 3** shows an example of fluorescence intensity increases upon addition of chemical denaturants.

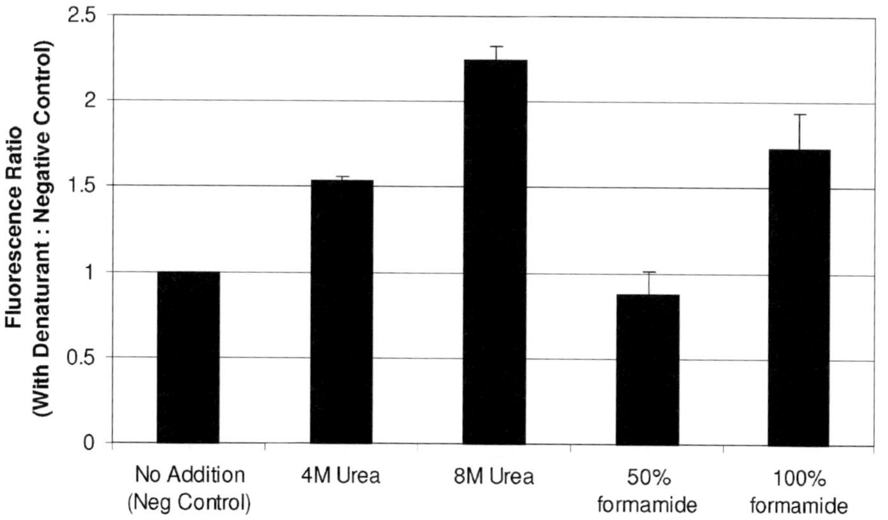

Fig. 3. Fluorescence intensity measurements of QDMBs and their response to chemical denaturants. QDMBs (amide linkage, 525-nm carboxyl-conjugated QD with Iowa Black FQ quencher molecular beacon DNA) were added to DNA denaturants of varying concentration. Fluorescence intensity was measured in a Tecan Genios FL microplate fluorometer with excitation at 360 nm and emission detection at 520 nm. All samples were compared with a negative control, which consisted of QDMBs suspended in PBS.

4. Notes

1. Fresh working solutions of 10 mg/mL (52.2 mM) are made daily in PBS and should not be used after 12 h of storage. If left longer than 12 h, the EDC will hydrolyze and will not be capable of activating carboxylic acid groups for amide linkage with amine groups.
2. Spin filters should be prewet with PBS before filtration of QDs or QDMBs. This can be achieved by filling the spin filter with 100 μL of PBS and centrifuging at 7,000 × g for 1 min. Discard the flow-through and then add the reaction mixture to this prewet filter for separation.
3. The fluorometer must have the appropriate filters for excitation of QDs in the UV range (approximately 280–360 nm) and fluorescence emission detection at the appropriate wavelength for the QDs that are used (e.g., 520- to 530-nm emission filter for Quantum Dot Corporation 525-nm QDs). Alternatively, monochromator-based instruments can be used for maximum flexibility in excitation and emission wavelength selection.
4. Other buffers, such as PCR buffer or buffers containing DNA stabilization salts, such as $MgCl_2$, can be used in place of PBS. Use of distilled water is not recommended because it is not buffered to control pH. Buffer choice can depend on the stringency of DNA hybridization that is desired. Therefore, factors such as Mg^{2+} concentration can be adjusted to reduce or increase nonspecific DNA hybridization. Typically, increasing the Mg^{2+} concentration will stabilize nonspecific interactions, lowering the stringency required for hybridization and successful hairpin opening. Additionally, the concentration of the QDMBs and DNA targets can be adjusted for the particular assay that is being developed. For previous studies in our group, 2 pmol of QDMBs have been mixed with 1–500 pmol of target DNA.
5. Glycine is used to inhibit further amide bond formation at the end of the reaction. Any remaining sulfo-NHS-modified carboxylic acids on the Nanogold particles will form amide linkages with the free amines on glycine. This prevents further reactions from taking place during subsequent streptavidin-biotin linkage of the QDs to the biotinylated, Nanogold-modified DNA.
6. It was noted that 100,000 MWCO filters could be used for the second wash step of this procedure. Although 50,000 MWCO filters could also be used, the 100,000 MWCO filters provided better flow-through and still retained the Nanogold-quenched QDMBs.

7. Make sure to fill each well of the 96-well plate with the same volume of fluid. Volumes ranging from 25 to 200 μL can be used in most 96-well plates, but care must be taken to ensure that each well has the same volume. Differences in volume can affect the fluorescence intensity measurement and skew the results of the assay.

Acknowledgments

The author thanks Prof. Carl Batt (Cornell University) as an advisor during development of these protocols and Dr. Aaron Strickland who helped to optimize the linkage strategies. This work was supported by USDA Grant #03-35201-13691, FDA Grant #06000002499A, and NIJ Grant #2004-DN-BX-K001.

References

1. Hohng, S., Ha, T. (2005). Single-molecule quantum-dot fluorescence resonance energy transfer. *Chemphyschem* **6**, 956–60.
2. Willard, D., M., Van Orden, A. (2003). Quantum dots: resonant energy-transfer sensor. *Nat Mater* **2**, 575–6.
3. Han, M., Gao, X., Su, J., Z., Nie, S. (2001). Quantum-dot-tagged microbeads for multiplexed optical coding of biomolecules. *Nat Biotechnol* **19**, 631–5.
4. Medintz, I.L., Clapp, A.R., Mattoussi, H., Goldman, E.R., Fisher, B., Mauro, J.M. (2003). Self-assembled nanoscale biosensors based on quantum dot FRET donors. *Nat Mater* **2**, 630–8.
5. Medintz, I.L., Uyeda, H.T., Goldman, E.R., Mattoussi, H. (2005). Quantum dot bioconjugates for imaging, labelling and sensing. *Nat Mater* **4**, 435–46.
6. Liu, X., Farmerie, W., Schuster, S., Tan, W. (2000). Molecular beacons for DNA biosensors with micrometer to submicrometer dimensions. *Anal Biochem* **283**, 56–63.
7. Robelek, R., Niu, L., Schmid, E.L., Knoll, W. (2004). Multiplexed hybridization detection of quantum dot-conjugated DNA sequences using surface plasmon enhanced fluorescence microscopy and spectrometry. *Anal Chem* **76**, 6160–5.
8. Epstein, J., Biran, I., Walt, D.R. (2002). Fluorescence-based nucleic acid detection and microarrays. *Anal Chim Acta* **469**, 3–36.
9. Kim, J.H., Morikis, D., Ozkan, M. (2004). Adaptation of inorganic quantum dots for stable molecular beacons. *Sens Actuators B Chem* **102**, 315–9.
10. Tyagi, S., Kramer, F.R. (1996). Molecular beacons: probes that fluoresce upon hybridization. *Nat Biotechnol* **14**, 303–8.
11. Tan, L., Li, Y., Drake, T.J., Moroz, L., Wang, K., Li, J., Munteanu, A., Chaoyong, J.Y., Martinez, K., Tan, W. (2005). Molecular beacons for bioanalytical applications. *Analyst* **130**, 1002–5.
12. Vet, J.A., Marras, S.A. (2005). Design and optimization of molecular beacon real-time polymerase chain reaction assays. *Methods Mol Biol* **288**, 273–90.
13. Tsourkas, A., Behlke, M.A., Rose, S.D., Bao, G. (2003). Hybridization kinetics and thermodynamics of molecular beacons. *Nucleic Acids Res* **31**, 1319–30.
14. Cady, N.C., Strickland, A.D., Batt, C.A. (2007). Optimized linkage and quenching strategies for quantum dot molecular beacons. *Mol Cell Probes* **21**, 116–24.

Chapter 25

Quantum Dot Hybrid Gel Blotting: A Technique for Identifying Quantum Dot-Protein/Protein-Protein Interactions

Tania Q. Vu and Hong Yan Liu

Summary

We describe an alternative to the molecular biology technique of polyacrylamide gel electrophoresis-based Western blotting and immunoprecipitation, which is an extensively used method for separating target proteins from complex cellular mixtures and for identification of protein expression and protein-protein interactions. This novel method, called quantum dot (QD) hybrid gel blotting, allows the purification and analysis of the action of QD bioconjugate-protein complexes in live cells. Moreover, these identified interactions can be correlated with spatial location in cells. QD hybrid gel blotting will be useful in the growing fields of molecular biology/proteomics and nanobiotechnology development in several respects: (1) as a method for identifying specific QD-protein interactions in cells, (2) as a method for correlating QD-protein interactions with their spatial location in live cells, (3) as a means to study the size and composition of QD bioconjugate probes/complexes; and, finally, (4) as an improvement over traditional bead-based immunoprecipitation methods for directly isolating and visualizing proteins from complex mixtures.

Key words: Quantum dot, Hybrid gel, Western blot, Gel electrophoresis, PAGE, Agarose, Immunoblotting

1. Introduction

The developing field of nanobiotechnology is in need of molecular techniques to study the interactions of nanostructures with cellular proteins and to confirm that nanoparticle bioconjugates bind to their intended cellular targets. Nanobiotechnology can offer unique capabilities to enhance traditional molecular biological

James Weifu Lee and Robert S. Foote (eds.), *Micro and Nano Technologies in Bioanalysis*,
Methods in Molecular Biology, vol. 544
DOI 10.1007/978-1-59745-483-4_25, © Humana Press, a part of Springer Science+Business Media, LLC 2009

techniques. Quantum dots (QDs) are fluorescent semiconductor nanoparticles whose size is on the scale of protein molecules and whose fluorescence emission significantly exceeds the brightness and photostability of conventional fluorophores typically used in molecular biological and biochemical applications ***(1–4)***. This combination of properties makes QDs ideal as imaging probes that can be tailored for specific interactions with cellular proteins ***(5–10)***. Methods that merge the information available from QD imaging techniques (such as cellular location of biomolecules) with the information available from gel separation techniques (such as protein-protein interactions) are needed.

Below, we describe the new technique of QD hybrid gel blotting. This technique is an alternative to the extensively used molecular technique of polyacrylamide gel electrophoresis (PAGE)-based Western blotting and immunoprecipitation, which is a traditional technique used to separate target proteins from a complex cellular mixture, to detect protein expression, and to identify in situ protein-protein interactions. QD hybrid gel blotting involves formulating a hybrid composition of polyacrylamide-agarose (PA-AGE) gels for high-resolution separation of QD bioconjugate complexes. Subsequently, the protein composition of QD bioconjugate complexes can be immediately imaged and analyzed in subsequent electroblotting and immuno-hybridization steps. This technique will be useful for nanobiotechnology development because QD bioconjugates that are complexed to other cellular proteins can be fractionated and their protein-QD composition rapidly identified. Moreover, this technique will be useful for molecular biology/proteomic analysis because QD hybrid gel blotting satisfies current traditional PAGE-based Western blotting and immunoprecipitation methods but further offers the new ability to correlate identified protein interactions with spatial location in live cells (e.g., extracellular membrane surface versus perinuclear sites). Furthermore, this technique is useful in practice at the lab bench because, by virtue of the "built-in" QD fluorescence, multiple hybridization and imaging steps are performed directly and immediately on the same membrane blot; this affords a significant savings in cost and times because time-consuming and extensive steps are eliminated (protein A/Sepharose bead separation, HRP chemiluminescent tags, X-ray developing).

Cell-internalized QD-bioconjugates are first visualized in situ and then extracted from cells and fractionated as cellular lysates containing discrete QD-protein complexes. A significant aspect of this new QD hybrid gel blotting method involves in-gel fractionation of QD bioconjugate probes using a hybrid gel mixture. This mixture consists of PA-AGE and allows QD separation that is otherwise difficult with PA gels, while offering a significant improvement in separation resolution that is characteristic of AGE gels. In additional to high separation resolution, these PA-AGE hybrid gels offer a second significant advantage in that gel-fractionated QD-bioconjugate complexes can be efficiently electroblotted onto membranes, making subsequent protein identification using immuno-hybridization possible. Thus, hybrid gel-fractionated QD-bound protein complexes can be isolated by QD-based PA–AGE electrophoresis, and blotted onto membranes for post-analysis of protein-protein interactions. We expect that QD hybrid gel blotting, combined with the exceedingly bright and photostable luminescence properties of QDs, will provide investigators with a rapid, sensitive new method for studying the interactions between QDs and proteins both in vitro and in situ.

The goal of the following protocol is to identify QD–ligand-receptor associations. This is illustrated using the specific example of QDs conjugated with the peptide growth hormone nerve growth factor (QD–NGFs) and identification of QD–NGF binding to in situ cognate TrkA receptors in PC12 cells. **Figure 1**

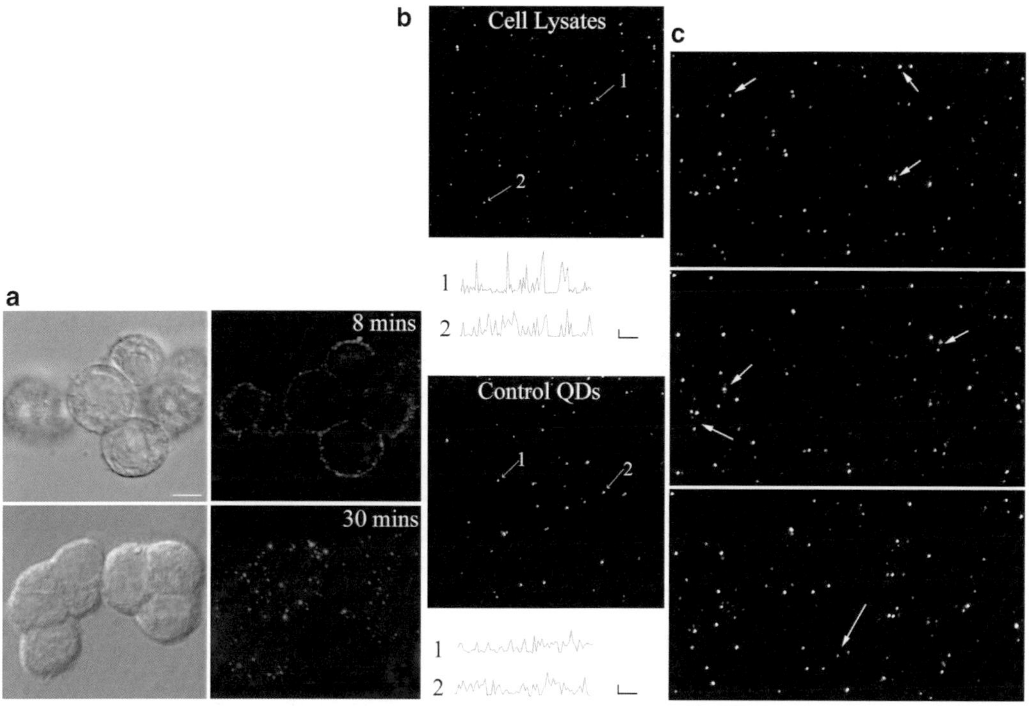

Fig. 1. QD–NGF bioconjugates can be retrieved as discrete complexes from cellular lysates. (**a**) Live PC12 cells treated with QD–NGF for 8 min exhibit discrete QD–NGFs that have bound to trkA receptor puncta on the cell membrane surface. After 30 min, discrete QD–NGFs are internalized into the cellular cytosol. *Dark regions* inside cells are nuclei. Scale bar: 5 μm. (**b**) Fluorescence microscope image of QD–NGFs from cellular lysates that have been cleared from PC12 cells treated with QD–NGFs for 30 min. Image shows that QD–NGF exhibit fluorescence blinking behavior similar to that for the positive control (COOH–QDs alone), suggesting that QD–NGFs remain discrete and do not aggregate after cellular lysis. Scale bar: *x*-axis is 2.5 s, *y*-axis is 10 pixel intensity value. (**c**) A sequence of video stills showing that QD–NGFs retrieved from cell lysates of PC12 cells treated with QD–NGFs (30 min) are discrete and blinking. *White arrows* point to examples of discrete QD–NGFs that blink and are not present in all three successive stills (Reprinted from **ref.** ***11*** with permission from ACS).

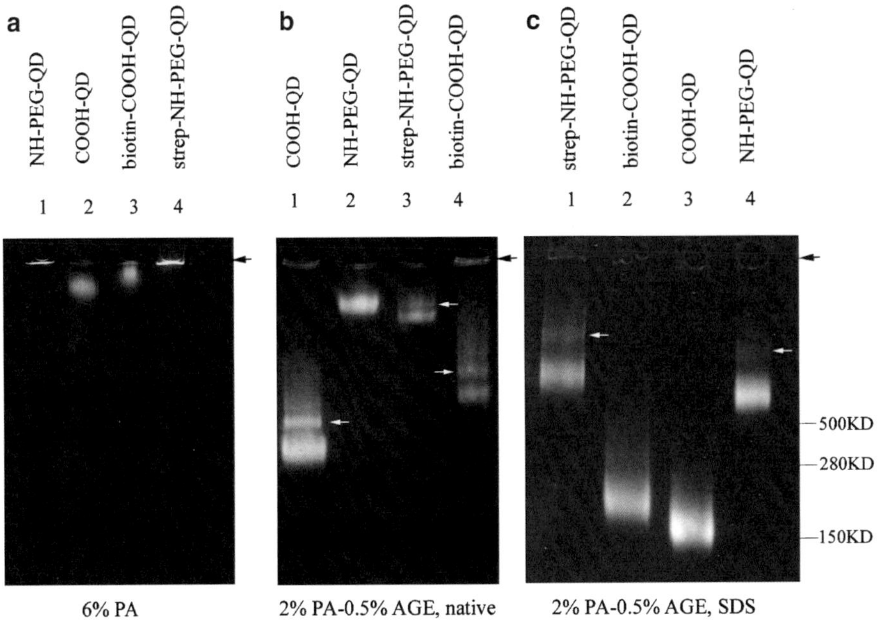

Fig. 2. Well-resolved separation of various quantum dot (QD) conjugates using hybrid PA-AGE gels. (**a**) Conventional native 6% PA gels severely retard QD entry and migration into gels for four types of QDs of different size and charge. (**b**) Hybrid 2% PA–0.5% AGE gels under native conditions (TBE, pH 8.3) separate QDs into tight bands. COOH–QDs and biotin–COOH–QDs are smaller and carry a greater negative charge density compared with streptavidin–NH-PEG-QDs and NH-PEG–QDs and migrate more quickly through the gel matrix. Further separation of QDs into subspecies (*white arrows*) highlights the separation capability of PA–AGE gels. (**c**) In the presence of SDS, larger and less negatively charged NH-PEG-QDs and streptavidin–NH-PEG-QDs migrate more quickly through gels than under native conditions. Good separation resolution is evident because subspecies can be resolved (*white arrows*). Biotin-QDs and COOH-QDs are also well separated, but migrate too quickly under these same conditions (100 V, 60 min) to resolve subspecies. Molecular weight markers show that PA-AGE gels can separate QD-conjugates with a wide range of migration rates. *Black arrows* show loading position (Reprinted from **ref. *11*** with permission from ACS).

shows the separation resolution of QD–NGF bioconjugates in these hybrid PA-AGE gels. **Figure 2** shows QD–NGF bioconjugates present in cellular lysates containing single or small groups of QD–NGF bioconjugates as obtained from cells treated with the QD–NGF bioconjugate probe. **Figure 3** shows an example of electrophoretic separation and electroblotting of cellular lysates of cells treated with QD–NGF bioconjugates and subsequent identification of QD–NGF–TrkA receptor association via immunoblotting. These examples illustrate the application of this technique and the methodology below is offered as a starting point for specific user applications.

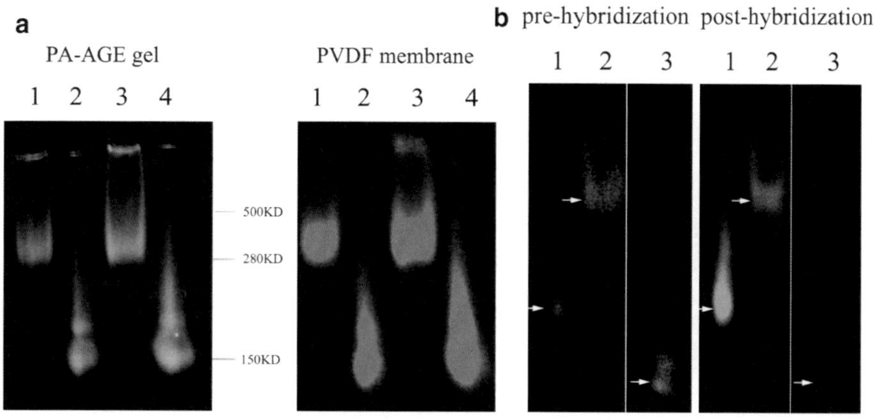

Fig. 3. QD-based PA–AGE Western blots for probing protein–protein interactions. (**a**) Left image shows PA–AGE separation of cellular lysates containing QD–NGFs from QD–NGF–treated cells (30 min) for two concentrations of cellular lysates (*lane 1* 40 μg, *lane 3* 80 μg). Cellular lysates containing QD–NGFs exhibit slower mobility compared with free QD–NGFs (*lane 2* 1 μL; *lane 4* 2 μL of 10 nM QD–NGFs) suggesting that cell–exposed QD–NGFs have complexes with trkAs. *Right hand image* is a fluorescence image of electroblotted PVDF membranes indicating faithful transfer of QD–NGFs from gels to membranes. (**b**) Western blots show that QD–NGFs obtained from cellular lysates are complexed to trkA receptors. *Left hand image, lane 2* is the red channel of a fluorescence image taken from PA–AGE fractionated cellular lysates from cells exposed to QD–NGFs (30 min). *Right hand image, lane 2* is the green channel of a fluorescence image of the same PVDF membrane after hybridization with biotinylated anti-trkA, followed by streptavidin-525 QDs. QD–NGF cellular lysates that hybridize in the same location and pattern with anti-trkA-biotin–streptavidin-525 QDs, indicating that QD-NGFs have bound to trkAs. *Lane 1* contains free biotin–QDs (2 μL, 5 nM) and serves as a positive control, demonstrating that biotin–QD hybridizes with streptavidin-525 QDs and, as expected, migrates more quickly than QD–NGF cellular lysates (*lane 2*). *Lane 3* contains free QD–NGFs (2 μL, 20 nM) and is a negative control showing lack of hybridization with either biotinylated anti-trkA or streptavidin-525 QDs (Reprinted from **ref. *11*** with permission from ACS).

2. Materials

2.1. Quantum Dots

1. Streptavidin-525 QDs and COOH-655 QDs (Invitrogen, Carlsbad, CA).
2. QD bioconjugates: QDs conjugated with NGF peptide ligand (QD–NGFs). Conjugate COOH–QDs with β-NGF (R&D Systems, Minneapolis, MN) by reacting NGF and COOH–QDs (2:1 molar ratio) in 1-ethyl-3-(3-dimethylaminopropoyl)-carbodiimide (EDAC, Sigma-Aldrich, St. Louis, MO) in borate buffer (10 mM, pH 7.4) at room temperature for 2 h. Free NGF and EDAC are removed by ultrafiltration with five exchange volumes of borate buffer (10 mM) (Microcon, MWCO 100kDA, Millipore, Billerica, MA) (*see* **Note 1**).

2.2. Cultured Cells

1. PC-12 cells (ATCC, Manassas, VA) grown in collagen-coated T-25 flasks in RPMI-1640 supplemented with 10% horse serum and 5% fetal bovine serum at 37°C.
2. Dulbecco's phosphate-buffered saline (D-PBS).

2.3. Cell Lysis Buffer

1. Phosphate-buffered saline (PBS) (0.1 M sodium phosphate, 0.15 M sodium chloride, pH 7.2).
2. Cell lysis buffer: PBS, 10% glycerol, 0.25% NP-40, 2 mM sodium orthovandate, 10 mM sodium fluoride, 10% Protease Inhibitor Cocktail (Sigma). Note that this mild cell lysis buffer does not contain sodium dodecyl sulfate (SDS) and is selected to retain protein-protein associations.

2.4. Antibodies

1. Biotinylated-polyclonal anti-trkA antibody.
2. 100 μL of 200 μg/mL anti-trkA (C-14, sc-11, Santa Cruz Biotechnology, Santa Cruz, CA) was biotinylated by incubation with 30-fold excess of NHS-PEO_4-biotin (Pierce, Rockford, IL), then dialyzed (Slide-A-Lyzer, 7 kDa MWCO, Pierce) against 500 mL PBS (pH 7.2) for 3 h to remove unbound biotin.

2.5. PA-AGE Hybrid Gels

1. 30% Acrylamide/bis gel stock: Prepare 30% acrylamide/bis by mixing 29 g of acrylamide and 1 g bis-acrylamide into dH_2O to make a final volume of 100 mL. Unpolymerized acrylamide is a neurotoxin so care should be taken to handle it with gloves.
2. *N,N,N,N′*-tetramethyl-ethylenediamine (TEMED, Bio-Rad, Hercules, CA).
3. 10% Ammonium persulfate: Prepare 10% solution by adding 1 g of ammonium persulfate into 10 mL ddH_2O and store at 4°C for up to 2 weeks.
4. 1% Agarose gel: Prepare by dissolving 0.1 g agarose (Bio-Rad) in 10 mL of distilled water, boil to melt, and cool to 55°C.
5. 4% Acrylamide gel: Mix 1.35 mL of 30% acrylamide/bis gel stock and 2 mL 10× Tris/borate/EDTA (TBE) buffer and 6.65 mL ddH_2O. Warm in a 55°C water bath for 10 min. Take care not to overheat the acrylamide because this will cause evaporation of TEMED in later steps.
6. Hybrid 2% acrylamide–0.5% agarose (2% PA–0.5% AGE) gel: Mix together 10 mL of 1% agarose and 4% acrylamide gel and add 100 μL of 10% ammonium persulfate and 7 μL of TEMED.

2.6. QD Hybrid Gel Blotting Equipment and Supplies

1. Loading buffer: Prepare by mixing 40% (w/v) sucrose and 0.25% (w/v) bromophenol blue in ddH_2O.
2. Running buffer: 1× TBE. Make 10× TBE by mixing 108 g Tris, 55 g boric acid, 9.3 g EDTA (disodium salt) in ddH_2O to a final volume of 1 L. Dilute 100 mL of 10× TBE with 900 mL ddH_2O to make 1× TBE.
3. Transfer buffer: 20% methanol in 0.5× TBE (pH 8.3). Make 1 L of transfer buffer by mixing 50 mL of 10× TBE, 750 mL ddH_2O, and 200 mL methanol. Cool the buffer to 4°C for 1 h before use (*see* **Note 2**).

4. Washing buffer: TBS-T (1× TBS/0.1% Tween-20). Prepare 10× TBS stock by mixing 87.6 g NaCl, 12.1 g Tris, and ddH_2O to a final volume of 1 L; adjust to pH 8.0 with HCL. Make 1× TBS by diluting 100 mL of 10× TBS with 900 mL of ddH_2O. Make washing buffer by mixing 1 mL of 100% Tween-20 with 1,000 mL of 1× TBS.
5. Blocking buffer: 3% (w/v) BSA in TBS-T.
6. PVDF membranes (Immobilon-FL, Millipore) (*see* **Note 3**).
7. Mini Trans-Blot filter paper (Bio-Rad).
8. Bio-Rad Protein Assay (Bio-Rad).
9. Mini Trans-Blot Cell and System (includes Trans-Blot Cell and Trans-Blot Module, Bio-Rad).
10. Power Pac HC High-current power supply (Bio-Rad).
11. UV trans-illuminator (Multi Doc-It Digital Imaging System, UVP, Upland, CA).

3. Methods

3.1. Obtaining Cellular Lysates of QD Bioconjugate-Treated Cells

1. Grow PC12 cells in collagen-coated T-25 flasks in RPMI-1640 supplemented with 10% horse serum and 5% fetal bovine serum at 37°C.
2. Aspirate culture media from PC12 cells and treat cells with 20 nM QD–NGFs in serum-free DMEM at 37°C for 30 min. Wash cells with D-PBS to remove the media containing unbound QD–NGFs.
3. If imaging of cells is desired, place QD-treated cells under a fluorescence microscope to capture images (*see* **Note 4**).
4. Harvest cells: Treat cells with 0.25% trypsin–EDTA for 5 min and then transfer to a 15-mL Falcon tube (*see* **Note 5**).
5. Pellet cells: Spin cells down in a centrifuge for 5 min at 1,000 rpm.
6. Obtain cellular lysates: Lyse cell pellet by adding lysis buffer and waiting for 2 h.
7. Clear cellular lysates: Remove insoluble materials by centrifugation (15 min, 13,000 rpm) and retain supernatant by transferring to a new Eppendorf tube.
8. If desired, QD–NGF bioconjugate complexes can be viewed under the microscope to check for QD fluorescence blinking, an indication that the retrieval of QD–NGF bioconjugate complexes from cellular lysates has resulted in numbers of single QD complexes: Place a 1-μL drop of cellular lysate

onto a coverslip and image under a fluorescence microscope to observe or record fluorescence blinking. Dilute cellular lysates concentrations if necessary.

9. Measure protein concentration of cellular lysates using the Bio-Rad Protein Assay: Prepare samples for spectrophotometry by adding 5 μL of cell lysate supernatant to 45 μL of distilled water in a 1.5-mL Eppendorf tube. Further dilute this sample into cuvettes by transferring 20 μL of diluted cell lysate to a cuvette and adding 780 μL of ddH_2O. Include one control blank (cuvette filled with 800 μL of ddH_2O). Add 200 μL protein assay reagent per cuvette and mix reagent with a transfer pipette (*see* **Note 6**). Perform spectrophotometry readings of the sample in the cuvette (*see* **Note 7**). Compare the spectrophotometry measurements of the cell lysate with standard curves made with serial dilutions of a protein of known concentrations (e.g., BSA) to calculate the concentration of protein in the cell lysate measurements.

3.2. Hybrid PA–AGE Gel Fractionation of Cellular Lysates Containing QD Bioconjugate Complexes

1. Mount clean glass plates into the gel caster: glass plates can be cleaned with detergent (Alconox), rinsed with water, and air-dried on a rack.
2. Cast hybrid PA–AGE gel: Pour 20 mL of the 2% PA–0.5% AGE gel into a gel caster, pouring slowly to avoid air bubbles. Insert a 10-well comb immediately and allow gel to polymerize for 30 min. After gels have hardened, remove comb with care to avoid breaking gels.
3. Wash gel wells several times using a syringe (3 mL, 25-gauge) filled with running buffer to remove gel debris, which can decrease fractionation resolution. Test wells by adding loading buffer (2 μL/well) and determine whether the loading buffer can run smoothly to the bottom of each well. If loading buffer does not pool at the bottom of the wells, then rewash the wells with running buffer.
4. Load samples into the gel wells: Add running buffer in the upper and lower chamber of gel apparatus. Dilute 3 μL of cell lysate samples with 1 μL loading buffer and gently load into the bottom of the wells.
5. Perform electrophoresis of cellular lysates containing QD bioconjugate complexes: cover the tank lid, connect the power supply to the electrophoresis electrodes, and run electrophoresis at 90–150 V for 1–2 h. Note that power and times can vary depending on the size of the QD complex.
6. View results of QD in-gel fractionation using a UV trans-illuminator.

3.3. Electroblotting QD Bioconjugate Complexes to PVDF Membranes

1. Prepare PVDF membrane for electroblotting: Cut a piece of PVDF membrane to a size that is a little bigger than that of the gel. PVDF membranes can be labeled with an ethanol-proof marker for later reference. Submerge the membrane in 100% methanol for 2 min with two exchanges to remove marker that readily bleeds from the membrane. Then pretreat the PVDF membrane by soaking in the transfer buffer for 10 min.
2. Assemble electroblotting apparatus: Disassemble electrophoresis apparatus and transfer the gel into a container filled with transfer buffer. Assemble electroblotting transfer cassette by sandwiching a fiber pad, a piece of mini trans-blot filter paper, the gel, a PVDF membrane, and a second piece of mini trans-blot filter paper, and a second fiber pad. Do this with the cassette submerged in a container filled with transfer buffer to avoid bubbles. Place the cassette into the electroblotting transfer module. Add a small stir bar inside the transfer module tank, fill the tank with transfer buffer, and then place the entire assembly into the large electrophoresis tank.
3. Electroblot: Place the electrophoresis tank containing the transfer module onto a stir plate and set to gently stir. Connect electrodes to the power supply and electroblot at 100 V for 2.5–3 h to transfer samples completely to the PVDF membrane. Note that duration of electroblotting will vary depending on the size of the QD complexes.

3.4. Western Blot Identification of QD Bioconjugate-Protein Interactions

1. Wash PVDF electroblots for 5 min and equilibrate with TBS at room temperature.
2. Block PVDF membrane: incubate PVDF membrane in blocking buffer for 2 h at room temperature or overnight at 4°C. Replace blocking buffer with 50 nM streptavidin in blocking buffer and block exposed biotin sites on the membrane for an 30 mins. Wash the membrane three times for 10 mins each with 15 mL washing buffer. Block streptavidin sites by incubating the membrane with 50 nM biotin in blocking buffer for a final 30 mins.
3. Wash three times for 10 min each with 15 mL washing buffer.
4. Hybridize the PVDF membrane: add 4 μg/mL biotinylated-polyclonal anti-trkA antibody in washing buffer to the membrane and incubate for 2 h with gentle agitation at RT.
5. Wash three times for 10 min each with 20 mL washing buffer.
6. Tagging antibody with QD probe: Add 1 nM streptavidin-525 QDs (green) in TBS buffer and incubate for 1 h at RT.
7. Wash the PVDF membrane four times for 10 min each with washing buffer.
8. Capture images of PVDF electroblots with a digital camera under UV trans-illumination.

4. Notes

1. QD-NGF bioconjugates serve as an example but a variety of other QD bioconjugates (e.g., QDs conjugated with antibodies/proteins/enzymes/amino acids) and bioconjugation schemes can be used.
2. We use 20% methanol in 0.5× TBE because conventional transfer buffers (1× Tris/glycine–20% methanol buffer) shift the QD emission spectrum. It is also essential to keep the transfer buffer cool (monitor current and adjust to <30 mA) because QD fluorescence emission will quench with heat. To aid this, the transfer buffer (0.5× TBE–20% methanol) can be cooled to 4°C before use.
3. This PVDF membrane has low autofluorescence compared with nitrocellulose membranes and is ideal for imaging QD fluorescence.
4. A number of different fluorescence microscope setups can be used to image QD fluorescence in live cells. A critical component of imaging single QD is ensuring that a fluorescence microscope equipped with appropriate excitation and emission filters, and a sensitive camera is used. Refer to http://probes.invitrogen.com/products/qdot/hints.html for suitable QD filter sets.
5. As an alternative, if live cell imaging is not desired, extracts of cellular lysates can be prepared and then treated with QD probes. As an additional variation, cellular lysates can be treated in sequence: first with a ligand/antibody, followed by a second QD probe that will specifically bind the first ligand/antibody complex.
6. Note also that the appropriate dilution of cellular lysates required for spectrophotometry readings will vary for different applications and depends on the starting concentration of cell lysates. These dilutions are offered as starting points.
7. Set the spectrophotometer excitation and emission wavelengths appropriately to detect protein assay absorption (the Bio-Rad Protein Assay uses a Coomassie dye, which shifts the absorption maximum from 465 to 595 nm when protein binding occurs and thus 595 nm is selected for reading the optical density). Controls performed with unconjugated QD samples indicate that QD fluorescence emission does not interfere with protein assay absorption readings.

References

1. Pons, T., Uyeda, H.T., Medintz, I.L.&Mattoussi, H. (2006). Hydrodynamic dimensions, electrophoretic mobility, and stability of hydrophilic quantum dots. *J Phys Chem B* **110**, 20308–20316.
2. Bruchez, M., Jr., Moronne, M., Gin, P., Weiss, S.&Alivisatos, A.P. (1998). Semiconductor nanocrystals as fluorescent biological labels. *Science* **281**, 2013–2016.
3. Chan, W.C.&Nie, S. (1998). Quantum dot bioconjugates for ultrasensitive nonisotopic detection. *Science* **281**, 2016–2018.
4. Michalet, X. et al. (2005). Quantum dots for live cells, in vivo imaging, and diagnostics. *Science* **307**, 538–544.
5. Vu, T.Q., Maddipati, R., Blute, T.A., Nehilla, B.J., Nusblat, L.&Desai, T. A. (2005). Peptide-conjugated quantum dots activate neuronal receptors and initiate downstream signaling of neurite growth. *Nano Lett* **5**, 603–607.
6. Zhou, M.&Ghosh, I. (2007). Quantum dots and peptides: a bright future together. *Biopolymers* **88**, 325–339.
7. Rosenthal, S.J. et al. (2002). Targeting cell surface receptors with ligand-conjugated nanocrystals. *J Am Chem Soc* **124**, 4586–4594.
8. Lidke, D.S., Nagy, P., Heintzmann, R., Arndt-Jovin, D.J., Post, J.N., Grecco, H.E., Jares-Erijman, E.A.&Jovin, T.M. (2004). Quantum dot ligands provide new insights into erbB/HER receptor-mediated signal transduction. *Nat Biotechnol* **22**, 198–203.
9. Dahan, M., Levi, S., Luccardini, C., Rostaing, P., Riveau, B.&Triller, A. (2003). Diffusion dynamics of glycine receptors revealed by single-quantum dot tracking. *Science* **302**, 442–445.
10. Yu, G., Liang, J., He, Z.&Sun, M. (2006). Quantum dot-mediated detection of gamma-aminobutyric acid binding sites on the surface of living pollen protoplasts in tobacco. *Chem Biol* **13**, 723–731.
11. Liu, H.Y.&Vu, T.Q. (2007). Identification of quantum dot bioconjugates and cellular protein co-localization by hybrid gel blotting. *Nano Lett* 7, 1044–1049.

Chapter 26

In Vivo Imaging of Quantum Dots

Isabelle Texier and Véronique Josserand

Summary

Noninvasive whole-body near-infrared fluorescence imaging is now acknowledged as a powerful method for the molecular mapping of biological events in live small animals such as mouse models. With outstanding optical properties such as high fluorescence quantum yields and low photobleaching rates, quantum dots (QDs) are labels of choice in the near-infrared domain. The main applications described in the literature for in vivo imaging of mice after injection of QDs encompass imaging of lymph nodes and tumors and cell tracking. Standard methods for the preparation, the purification, and the in vivo fluorescence whole-body imaging of QDs in the live mouse are described. Nanoparticles coated by PEG chains of different sizes and terminal groups are prepared using 705-nm-emitting commercial QDs. Their biodistribution after intravenous or intradermal injections in tumor-bearing mice is reported here.

Key words: Quantum dot functionalization, Nude mice, Molecular imaging, In vivo whole-body fluorescence imaging, Ts/Apc xenograft, Lymph nodes

1. Introduction

In vivo whole-body near-infrared fluorescence imaging is a very attractive tool for noninvasively assessing cells or drug biodistributions and molecular events such as target/probe recognition, enzymatic activity, and gene expression in small animal models *(1–3)*. The advantages of the optical methods in comparison with other molecular imaging modalities, such as nuclear techniques (single-photon emission computed tomography [SPECT], positron emission tomography [PET]), magnetic resonance imaging (MRI), or X-ray computed tomography (CT), lie in their low cost and easy handling, in the absence of ionizing radiations, in the possibility of using short acquisition times, and in their high

James Weifu Lee and Robert S. Foote (eds.), *Micro and Nano Technologies in Bioanalysis*,
Methods in Molecular Biology, vol. 544
DOI 10.1007/978-1-59745-483-4_26, © Humana Press, a part of Springer Science+Business Media, LLC 2009

sensitivities (nanomolar probe concentrations) *(4–6)*. Moreover, whole-body near-infrared fluorescence imaging can allow longitudinal studies in small rodents with a reduced number of animals and an improved reliability of the results in comparison with optical techniques assessing biomolecule biodistribution post mortem. Therefore, the development of this technique is tied to a faster identification of new biomarkers, to the understanding of biomolecular processes noninvasively in vivo, and to the development of new therapeutics.

Two classes of labels exist for the near-infrared domain, for which light absorption and diffusion by the tissues are minimal: organic dyes and quantum dots (QDs). QDs are inorganic luminescent semiconductor nanocrystals that display very attractive optical features *(7–10)*. They are now commercially available for in vivo tests in mice from companies such as Evident Technologies (http://www.evidenttech.com/) or Invitrogen (http://probes.invitrogen.com), and new QD compositions, with less toxic elements, are currently being developed.

The main applications described in the literature for in vivo imaging with QDs encompass imaging of lymph nodes *(11–14)* and tumors *(15–17)*, and in vivo cell tracking *(18,19)*.

In this chapter, experimental procedures for the preparation and imaging of several QDs in nude mice bearing subcutaneous tumors are described. Protocols for the covalent grafting of polyethylene glycol (PEG) chains onto the QDs are detailed. It should be noted that these functionalization procedures, hereafter described for specific PEG chains, can be extended to any other thiol-bearing compounds, especially peptides containing cysteine residues. The nanoparticles are imaged after intradermal (ID) or intravenous (IV) injection at low doses (20–40 pmol/mouse). The preparation and purification of the nanoparticles, the preparation of animals, and their injection and imaging using a fluorescence reflectance imaging (FRI) set up are detailed.

2. Materials

2.1. Quantum Dots (See Notes 1–3)

Aqueous soluble QDs are commercially available from companies such as Evident Technologies (http://www.evidenttech.com/) or Invitrogen (http://probes.invitrogen.com). We use commercial Qtracker ™ 705 nontargeted QDs and ITK705-amino (PEG) particles (Invitrogen). Both nanoparticles are 705-nm emitting CdTe/ZnS QDs, with a core diameter of ~10–15 nm, coated by a polymer.

Qtracker™705 nontargeted QDs are described by the manufacturer as having a PEG surface coating specially developed to minimize nonspecific interactions and to reduce immune response by the tissues *(20)*. They are supplied at 2 μM in borate buffer (50 mM borate, pH 8.3).

ITK705-amino (PEG) particles are coated by a polymer and 2,000-Da weight PEG chains bearing at their extremity primary amino groups, and are supplied at 8 μM in borate buffer (50 mM borate, pH 8.3). These nanoparticles can be functionalized to obtain nanoparticles coated by PEG chains of different sizes and terminal groups. Using cross-linkers and PEG polymers detailed in **Subheading 2.2**, the following nanoparticles are obtained: QD–PEG_{2750}–OCH_3, QD–PEG_{2000}–OH, QD–PEG_{5400}–OH, and QD–PEG_{7000}–OH (*see* **Note 4**), according to the global synthetic scheme of **Fig. 1**.

All QDs are preferentially manipulated and stored using low-binding siliconized microtubes (e.g., from VWR). The nanoparticles should never be frozen; otherwise they form aggregates, which cannot be further re-dispersed. The nanoparticles are stored at 4°C. Personal protection (lab coat, gloves, glasses) should be worn during manipulation, because the toxicity of the nanoparticles is not yet well documented *(21)*.

2.2. Cross-linkers and PEG Polymers Used for QD Modification

1. The cross-linkers and functional PEG polymers used in this study are the following:
   (a) NHS–maleimide (4-maleimidobutyric acid *N*-hydrosuccinimide ester, Sigma) (*see* **Note 5**).
   (b) NHS–PEG_{750}-OCH_3 (Iris Biotech) (*see* **Note 4–5**).
   (c) NHS–PEG_{3400}-maleimide (Nektar) (*see* **Notes 4–5**).
   (d) NHS–PEG_{5000}-maleimide (Nektar) (*see* **Notes 4–5**). 1 μmol of the product is dissolved in the minimum volume of anhydrous DMSO (typically 10–30 μL) (*see* **Note 6**).
2. 2-Mercaptoethanol (2-MCE) is used for obtaining a hydroxyl coating onto the nanoparticles. A 100-mM solution is prepared in a coupling buffer composed of 1× phosphate-buffered saline (PBS) and 1 mM EDTA (*see* **Note 7**).

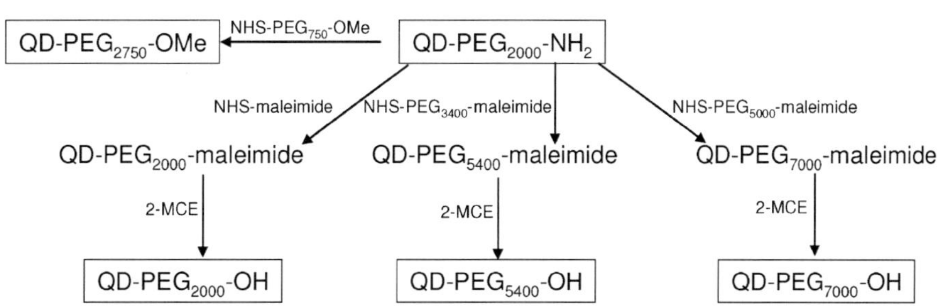

Fig. 1. Schematic representation of the functionalization process used for the preparation of the different QDs imaged in nude mice, and referred to in the text as QD–PEG_{2750}–OCH_3, QD–PEG_{2000}–OH, QD–PEG_{5400}–OH, and QD–PEG_{7000}–OH. The PEG cross-linkers used are described in **Subheading 2.2** (*see* **Notes 4** and **5**).

2.3. Concentration, Buffer Exchange, and Purification of the QD Particles after Functionalization

1. Concentration and buffer exchange of the QDs are achieved using Microcon YM100 spin columns from Millipore.
2. Purification of the QDs after cross-linker or PEG polymer grafting onto their surface is achieved using NAP-5 size exclusion columns from GE-Biosciences. This method is used for separating the QD nanoparticles from cross-linkers of molecular weight inferior to 1 kDa, namely NHS–maleimide and NHS–PEG_{750}–OCH_3.
3. Purification of the QDs after cross-linker or PEG polymer grafting onto their surface is achieved using dialysis against 1× PBS using dialysis membranes with a cut-off of 12–14 kDa (Roth Zellutrans) (*see* **Note 8**) and a microdialysis set up suitable for volumes from 100 μL up to 500 μL (QuickSep™, available, for example, from Fisher Bioblock). This method is used for separating the QD nanoparticles from cross-linkers of molecular weight from 1 to 5 kDa, namely NHS–PEG_{3400}–maleimide and NHS–PEG_{7000}–maleimide.

2.4. Animals

All animal experiments are conducted in agreement with the "Principles of Laboratory Animal Care" (NIH publication no. 86-23, revised 1985).

1. Female NMRI nude mice, 5–6 weeks old, are purchased from Janvier.
2. For manipulation and imaging, mice are anesthetized using a gaseous anesthesia device (Minerve) with isoflurane/air 3.5% for induction and 1.5% thereafter.
3. 10^7 Ts/Apc cells (murine breast carcinoma) are used to prepare xenografted tumor-bearing mice. Cells are cultivated with RPMI 1640 medium supplemented with 10% fetal calf serum, 50 U/mL penicillin, 50 mg/mLstreptomycin, and 2.5×10^{-5} M 2-MCE and incubated at 37°C in a 5% CO_2 atmosphere.
4. IV or ID injections are made with 1-mL sterile insulin syringes (29 gauge) (Becton Dickinson).
5. The solutions to be injected are prepared in sterile 1× PBS buffer.

2.5. Imaging System

Several two-dimensional (2D) whole-body fluorescence imaging systems are now commercially available. The IVIS™ set up from Xenogen (http://www.xenogen.com/wt/page/imaging), the Maestro™ system from CRi (http://www.cri-inc.com/products/maestro.asp), the NightOWL™ system from Berthold (http://www.bertholdtech.com/ww/en/pub/home.cfm), and the Aequoria™ system from Hamamatsu (http://sales.hamamatsu.com/) can be quoted, among others.

A homemade system, adapted from the commercially available Aequoria™ system from Hamamatsu, is used. It is made up of a dark box equipped with a mouse body temperature controller and a gas anesthesia device. The excitation device is composed of ten LEDS emitting at 633 nm (adapted from the LuxiFlux™ device available from Hamamatsu) and equipped with interference band pass filters (633BP10 nm from Schott) for a light illumination power of 15 μW/cm^2. The filtered fluorescence signal (filter RG665 from Schott) is measured by a cooled CCD camera (Orca II BT 512 G, Hamamatsu), placed at 160 mm from the imaging field, with an exposure time that can be adjusted from 10 ms to a few seconds. For QD imaging, the typical integration time used is 500 ms, the gain of the camera is set at medium and the binning is 1 × 1. The Wasabi™ software (Hamamatsu) is used to drive the set up and for image processing.

3. Methods

3.1. Preparation of Quantum Dots

In this chapter, we describe in vivo imaging in nude mice using the following QDs: Qtracker™705 nontargeted QDs (Invitrogen), QD–PEG$_{2750}$–OCH$_3$, QD–PEG$_{2000}$–OH, QD–PEG$_{5400}$–OH, and QD–PEG$_{7000}$–OH (*see* **Note 4**), prepared according to the global synthetic scheme presented **Fig. 1** and the procedures detailed below in **Subheadings 3.1.2–3.1.5**. These procedures can be adapted for the QD functionalization by other thiol-bearing molecules, such as cysteine residues of peptides and proteins.

For mouse injection, the solutions are prepared in sterile 1× PBS, at the following concentrations:

- 2 μM for ID injection (10 μL or 20 pmol injected per animal).
- 200 nM for IV tail injection (200 μL or 40 pmol injected per animal).

3.1.1. Qtracker™ 705 Nontargeted QDs

1. For IV injection, the commercial 2 μM Qtracker™705 nontargeted QD solution (Invitrogen), is diluted 10 times using sterile 1× PBS buffer just before injection.
2. For ID injection, Microcon YM100 spin columns from Millipore are used for exchanging the borate buffer in which the nanoparticles are stored with sterile 1× PBS buffer. The spin columns are used according to the manufacturer's instructions (*see* **Note 9**), with reduced time and speed of spinning.

3.1.2. QD–PEG$_{2750}$–OCH$_3$

These nanoparticles are prepared using commercial QD–PEG–$_{2000}$–NH$_2$ particles (ITK705-amino [PEG] 8 μM in borate buffer, Invitrogen) as starting materials, and the global functionalization scheme of **Fig. 1**. Hereafter, the procedure is detailed for 125 μL of solution (1 nmol of QDs), but can be scaled up by adjusting reactant quantities at least up to 300 μL of solution.

1. The borate buffer in which the QDs are purchased is exchanged against 1× PBS buffer using a Microcon YM100 spin column (*see* **Note 10**). The Microcon spin columns are used according to the manufacturer's instructions (*see* **Note 9**). 125 μL of QD–PEG_{2000}–NH_2 particles in 1× PBS is collected (*see* **Note 11**).
2. To the 125 μL of QD–PEG_{2000}–NH_2 particles in 1× PBS, the 1 μmol solution of NHS–PEG_{750}–OCH_3 in anhydrous DMSO is added, and the nanoparticles are incubated 30 min to 3 h at room temperature in the dark.
3. QD–PEG_{2750}–OCH_3 conjugates are purified using NAP-5 size-exclusion columns, according to the manufacturer's instructions (*see* **Note 12**).
4. QD–PEG_{2750}–OCH_3 conjugates are concentrated and transferred in sterile 1× PBS buffer using Microcon YM100 spin column (*see* **Note 10**). The Microcon spin columns are used according to the manufacturer's instructions (*see* **Note 9**). 100–150 μL of QD–PEG_{2750}–OCH_3 particles in sterile 1× PBS is collected. The nanoparticles are stored at 4°C, preferentially in low-binding silicon microtubes (*see* **Note 13**).
5. The concentration of the solution is determined using visible absorbance measurements at a wavelength between 450 and 600 nm. The absorbance of the solution diluted 100 times is compared with that of the ITK705-amino (PEG) 8 μM solution in borate buffer, diluted 100 times in sterile 1× PBS buffer. Typical recovery yields of functionalized nanoparticles are 80%.
6. 2 μM Solution for ID injection (10 μL per injection) or 200 nM solution for IV injection (200 μL per injection) are prepared by diluting the nanoparticles in sterile 1× PBS buffer.

3.1.3. QD-PEG_{2000}-OH

1. QD–PEG_{2000}–maleimide conjugates are prepared using the same procedure as described in **Subheading 3.1.2**, **steps 1–4** (*see* **Note 14**), replacing NHS–PEG_{750}–OCH_3 with NHS–maleimide, and sterile 1× PBS buffer with the conjugation buffer (1× PBS + 1 mM EDTA) in **step 4**.
2. To the solution of QD–PEG_{2000}–maleimide in the conjugation buffer (1× PBS + 1 mM EDTA), 10 μL (1 μmol) of the 2-MCE solution is added. The nanoparticles are incubated 1–3 h at room temperature in the dark.
3. QD–PEG_{2000}–OH nanoparticles are purified using NAP-5 microcolumns. The nanoparticles are concentrated and exchanged in sterile 1× PBS buffer, quantitated by visible spectrophotometry, and prepared for injection as described in

steps 3–6 of **Subheading 3.1.2**. Typical recovery yields of functionalized nanoparticles are 60%.

3.1.4. QD–PEG$_{5400}$–OH

1. QD–PEG$_{5400}$–maleimide conjugates are prepared using the same procedure as described in **Subheading 3.1.3**, **steps 1–2**, replacing NHS–maleimide by NHS–PEG$_{3400}$–maleimide.
2. QD–PEG$_{5400}$–maleimide is purified using dialysis against 2,000 times their volume (typically 300–400 mL) of 1× PBS, at least three times 30 min.
3. QD–PEG$_{5400}$-maleimide conjugates are concentrated and transferred in the conjugation buffer (1× PBS + 1 mM EDTA) using a Microcon YM100 spin column (*see* **Notes 10** and **14**).
4. QD–PEG$_{5400}$–OH is obtained from QD–PEG$_{5400}$–maleimide using the same procedure as described in **Subheading 3.1.3**, **steps 2–3**. Typical recovery yields of functionalized nanoparticles are 60%.

3.1.5. QD–PEG$_{7000}$–OH

QD–PEG$_{7000}$–OH conjugates are prepared using the same procedure as described in **Subheading 3.1.4** replacing NHS–PEG$_{3400}$–maleimide by NHS–PEG$_{5000}$–maleimide. Typical recovery yields of functionalized nanoparticles are 60%.

3.2. Subcutaneous Tumor-Bearing Mice Preparation

Female NMRI nude mice, 5-weeks old, are purchased from Janvier (*see* **Notes 15–17**).

3.2.1. Cell Culture and Amplification

10^6 Ts/Apc cells (murine breast carcinoma) are seeded in a 25-cm^2 flask and cultivated with RPMI 1640 medium supplemented with 10% fetal calf serum, 50 U/mL penicillin, 50 mg/mL streptomycin, and 2.5×10^{-5} M 2-MCE and incubated at 37°C in a 5% CO_2 atmosphere.

When 80% of confluence is achieved, the medium is removed, 5 mL of 1× PBS buffer is used to rinse the cells, and 0.5 mL trypsin-EDTA (0.5 and 0.2 g/L, respectively) (Invitrogen) is introduced into the flask to peel off the cells. After 5 min, 5 mL of the culture medium is used to stop the trypsin action, the cell suspension is harvested in a 5-mL Falcon tube and centrifuged for 5 min at 300 × *g*. The supernatant is removed and the pellet is gently homogenized with 50 mL of culture medium and seeded in a 75-cm^2 flask. Afterward, the cells are diluted 1:4 twice a week following the same procedure.

3.2.2. Subcutaneous Tumor Cells Implantation

On the day of implantation, the medium is removed, 20 mL of 1× PBS buffer is used to rinse the cells, and 4 mL of trypsin-EDTA is introduced into the flask to peel off the cells. After 5 min, 20 mL of the culture medium is used to stop the trypsin action, the cell suspension is harvested in a 50-mL Falcon tube, and an aliquot is used to count the cell density. Fifty microliters of 0.4% trypan blue (Invitrogen) is added to 450 μL of the cell suspension, and the preparation is deposed on a Malassez cell for counting. Meanwhile, the cell suspension is centrifuged for 5 min at 300 × *g*. The supernatant is eliminated and the pellet is gently homogenized with 1× PBS buffer to obtain a final concentration of 5 × 10^6 cells/mL.

The subcutaneous tumor cells implantation is performed on anesthetized mice (*see* **Note 18**). Using 1-mL sterile insulin syringes (29 gauge), 200 μL of the cell suspension is implanted subcutaneously in the mouse right flank. Mice are observed until they wake up. Twelve days after implantation, the tumors reach 80–120 mm^3, which is a convenient size for in vivo tumor imaging.

3.3. In Vivo Fluorescence Imaging (See Note 19)

Mice are anaesthetized in an induction chamber (4% isoflurane in air) then placed in the imaging dark box on a warm plate (37.5°C) connected to an anesthesia mask (1.5% isoflurane in air) (*see* **Note 20**).

Fluorescence images are taken in three different mouse body positions, the camera seeing the front, the side, and the back of the mouse (*see* **Note 21**), with various exposure times from 20 to 750 ms (*see* **Notes 22** and **23**). For each mouse position, a black and white picture of the mouse is also acquired with a white light to allow the superposition with the fluorescence image.

Before the injection, fluorescence images are taken to estimate the mouse autofluorescence (*see* **Note 24**).

3.3.1. Intravenous Injection

Outside of the dark box, an anesthetized mouse is placed on a warm plate (37.5°C) connected to an anesthesia mask (1.5% isoflurane in air). Using 1-mL sterile insulin syringes (29 gauge), 200 μL of particles (40 pmol, 200 nM in 1× PBS buffer) is injected intravenously via the tail vein (*see* **Note 25**) and the mouse is imaged at each desired time point. The fluorescence images obtained after IV injection of the different QD nanoparticles described here are presented **Fig. 2**.

3.3.2. Intradermal Injection

Using 1-mL sterile insulin syringes (29 gauge), 10 μL of particles (20 pmol, 2 μM in 1× PBS buffer) is injected in the right paw and the mouse is imaged at each desired time point (*see* **Note 26**). The fluorescence images obtained after ID injection of the different QD nanoparticles described here are presented **Fig. 3**.

3.3.3. Image Processing

The Wasabi™ software is used for image processing. All of the fluorescence images are set with the same contrast scale, and when necessary for better localization of the signal, can be superimposed on the corresponding black and white picture of the mouse.

At the end of the experiment (or if tumors exceed 200 mm^3), mice are killed by cervical dislocation and a dissection can be performed to image isolated organs using the same fluorescence

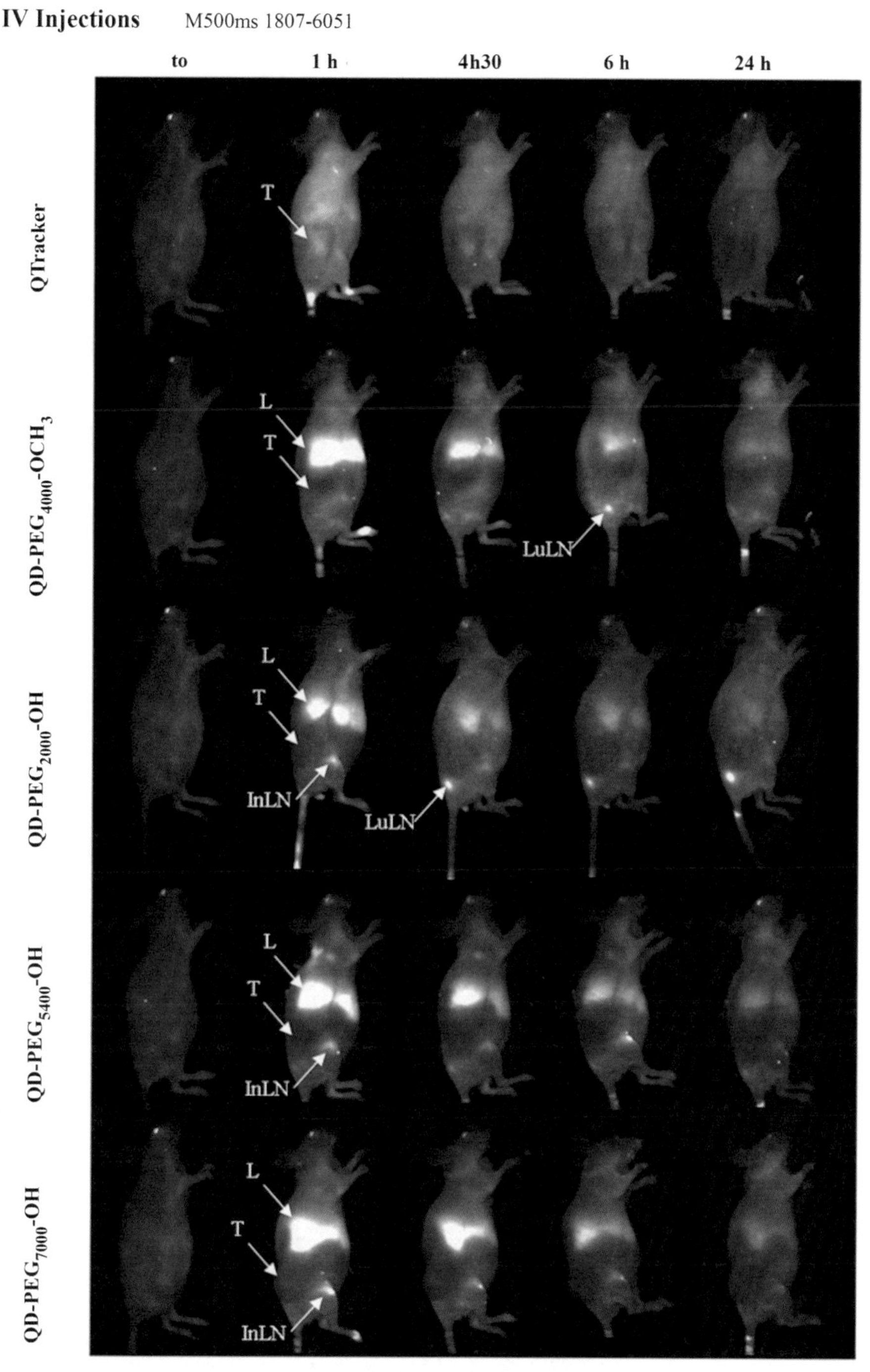

Fig. 2. Fluorescence images obtained after IV injection of 40 pmol of different QDs. The images have been taken using an exposure time of 500 ms and a medium gain of the camera. The contrast has been set between 1,807 and 6,051. *T* tumor; *L* liver; *LuLN* lumbar lymph node; *InLN* inguinal lymph node.

imaging set up. This can help in attributing the fluorescence observed during in vivo acquisitions to structures identified post mortem. Such an analysis is shown in **Fig. 4**.

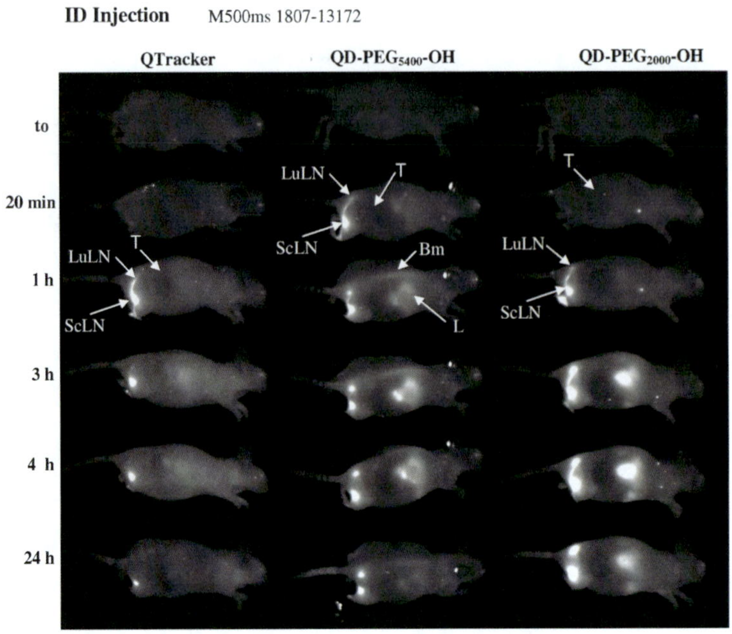

Fig. 3. Fluorescence images obtained after ID injection of 20 pmol of different QDs. The images have been taken using an exposure time of 500 ms and a medium gain of the camera. The contrast has been set between 1,807 and 13,172. ***T*** tumor; ***L*** liver; *LuLN* lumbar lymph node; *ScLN* sciatic lymph node; *Bm* bone marrow.

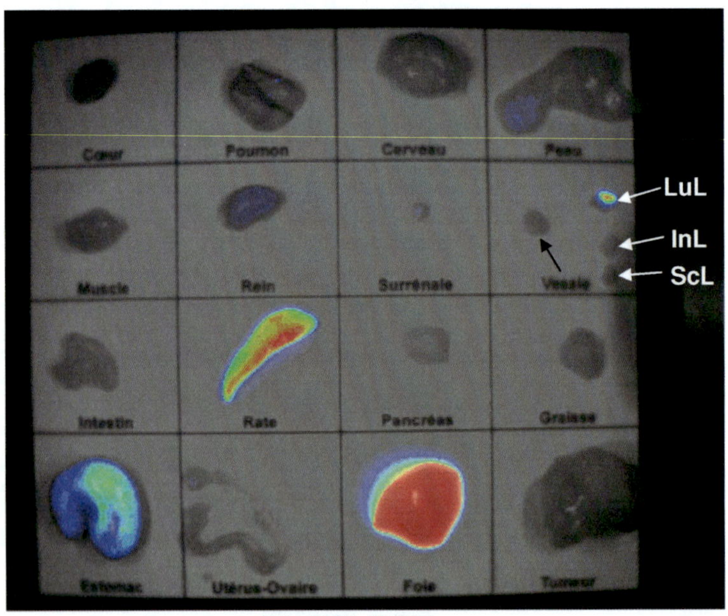

Fig. 4. Fluorescence image of isolated organs after dissection of a nude mice IV injected with 40 pmol of QD–PEG_{2000}–OH superimposed on the corresponding black and white picture. The fluorescence image has been taken using an exposure time of 1 s and a medium gain of the camera. The contrast has been set between 4,950 and 16,265. *LuLN* lumbar lymph node; *InLN* Inguinal lymph node; *ScLN* sciatic lymph node.

4. Notes

1. The core of QD nanoparticles is composed of heavy atoms (cadmium, telluride), which are known to be highly toxic when free in biological media. Even if any acute toxicity of QDs has not been observed in mouse or pig in vivo *(11–17,20)*, cellular toxicity has been reported in certain conditions *(21)*. Therefore, proper disposal of solutions and materials having contact with the nanoparticles should be ensured. We use the heavy metal disposal path for both liquid and material wastes. Personal protection equipment such as a lab coat, gloves, and glasses should be worn during manipulation.
2. The nanoparticles should never be frozen, otherwise they form aggregates, which cannot be further re-dispersed. The nanoparticles are stored at 4°C.
3. QDs are preferentially manipulated and stored using low-binding siliconized microtubes.
4. The subscript number refers to the molecular weight of the PEG chain.
5. NHS stands for the acid *N*-hydrosuccinimide ester group.
6. It might be necessary to heat the tube in the hand and gently vortex the tube for a few minutes to dissolve the product. The product might appear as a slightly pink-colored solution in case of NHS–PEG_{5000}–maleimide.
7. This solution is obtained by mixing 7 μL of pure 2-MCE and 993 μL of coupling buffer under an extraction hood. Note that 2-MCE is highly toxic for the environment and for the respiratory system, and should be manipulated under an extraction hood, and waste should be discarded with nonhalogenated organic solvents.
8. The membrane should be hydrated in the 1× PBS buffer for at least 30 min before use.
9. The spinning speed and time are reduced in comparison to the manufacturer's instructions to avoid drying of the nanoparticles onto the filter. For 125 μL of solution, 5,000 × *g* centrifugation during 5 min is used for concentrating the QDs onto the filter. After rinsing the nanoparticles with 100 μL of the new buffer, and another 5,000 × *g* 5 min centrifugation, the QDs are collected in the new buffer by turning the filter upside down in a new microtube and spinning for 1 min at 1,000 × *g* with 100 μL of the new buffer.

10. The Microcon YM100 spin columns (up to 500 μL) can be replaced by Amicon Ultra 15 centrifugal units (Millipore) for larger volumes.
11. If small aggregates are observed in the final solution, and if the borate buffer is not correctly eliminated by spinning, nanoparticles have probably aggregated. This typically happens whenever the nanoparticles have been frozen. They should be stored at 4°C, and never frozen.
12. For maximum efficiency, the NAP-5 purification columns should not dry out.
13. The solutions are preferentially used within the fortnight after their preparation. However, no differences are observed in in vivo biodistribution for nanoparticles used 3 months after their preparation.
14. The concentration step might be skipped if the volume of nanoparticles collected after the purification is approximately 400–500 μL or less. In this case, concentrated EDTA (100 mM, for example) should be added to obtain a final concentration of 1 mM EDTA in the buffer that will be used for the maleimide/thiol coupling.
15. Mice have to be delivered 1 week before the experiment to become acclimated to the environment and to recover from the stress of the transport.
16. Mice are maintained under specific pathogen-free conditions in a dedicated nude mice housing facility where temperature (25°C) and light (night/day cycles) are controlled.
17. Within the animal facilities, experimenters have to wear disposable sterile mobcaps, masks, coats, shoes, and gloves.
18. For the subcutaneous tumor cells implantation, in addition to the equipment mentioned above for the nude mice manipulation, the experimenter should wear protective glasses.
19. At least 1 h before the imaging experiment, the fluorescence camera and the 633-nm excitation device are switched on so that the camera can cool down to –70°C and the excitation device can stabilize its intensity. The mouse body temperature controller and the anesthesia device are switched on a few minutes before the imaging experiment and the RG665 filter is placed in front of the camera.
20. The advantage of gaseous anesthesia is that it can be repeated several times a day, without harm for the animal, which wakes up a few minutes after the gas is switched off.
21. Three different mouse body positions are needed for a good understanding of the biodistribution, because 2D FRI does not allow imaging of deep structures but detects mainly surface structures. To obtain three-dimensional (3D) imaging

of deep tissues, optical tomography associated with algorithms of image reconstruction is necessary.

22. The exposure time should be chosen so that the fluorescent signal uses the most of the camera dynamic range (65,535 grey levels for a 16-bit camera) and does not saturate the camera.
23. Because the same optical settings (filters, camera amplification, exposure time, binning, etc.) are required to compare images, several exposure times should be used for each time point.
24. Autofluorescence is mainly observed in the stomach and the guts of the mouse and, although it is weak in the near infrared range, it still needs to be estimated in case of a very low fluorescent signal after injection.
25. Just before the injection, the application of a warm compress on the tail can help for the IV injection by dilating the tail vein.
26. For better images, the ID injection site has to be masked by a black nonfluorescent tissue to distinguish the lymphatic network without saturating the image with the very fluorescent injection spot, especially during the 24 first hours after injection.

Acknowledgments

We thank Mélanie Guidetti and Toufic Jean Daou for their help in performing the experiments. This work was supported by the Commisariat à l'Energie Atomique (France), the European project EMIL (sixth PCRD-NOE contract no LSHC-CT-2004-503569), and the French Ministry of Research and Industry.

References

1. Licha, K., Olbrich, C. (2005). Optical imaging in drug discovery and diagnostic applications. *Adv. Drug Deliv. Rev.* **57**, 1087–1108.
2. Rao, J., Dragulescu-Andrasi, A., Yao, H. (2007). Fluorescence imaging in vivo: recent advances. *Curr. Opin. Biotechnol.* **18**, 17–25.
3. Gross, S., Piwnica-Worms, D. (2006). Molecular imaging strategies for drug discovery and development. *Curr. Opin. Chem. Biol.* **10**, 334–346.
4. Koo, V., Hamilton, P.W., Williamson, K. (2006). Non invasive in vivo imaging in small animal research. *Cell. Oncol.* **28**, 127–139.
5. Massoud, T.F., Gambhir, S.S. (2003). Molecular imaging in living subjects: seeing fundamental biological processes in a new light. *Genes Dev.* **17**, 545–580.
6. Weissleder, R. (2002). Scaling down imaging: molecular mapping of cancer in mice. *Nat. Rev.* **2**, 1–8.
7. Smith, A.M., Gao, X., Nie, S. (2004). Quantum dot nanocrystals for in vivo molecular and cellular imaging. *Photochem. Photobiol.* **80**, 377–385.
8. Michalet, X., Pinaud, F.F., Bentolila, L.A., Tsay, J.M., Doose, S., Li, J.J., Sundaresan, G.,

Wu, A.M., Gambhir, S.S., Weiss, S. (2005). Quantum dots for live cells, in vivo imaging, and diagnostics. *Science* **307**, 538–544.

9. Medintz, I.L., Uyeda, H.T., Goldman, E.R., Mattoussi, H. (2005). Quantum dot bioconjugates for imaging, labelling and sensing. *Nat. Mater.* **4**, 435–446.
10. Gao, X., Yang, L., Petros, J.A., Marshall, F.F., Simons, J.W., Nie, S. (2005). In vivo molecular and cellular imaging with quantum dots. *Curr. Opin. Biotechnol.* **16**, 63–72.
11. Ballou, B., Ernst, L.A., Andreko, S., Harper, T., Fitzpatrick, J.A.J., Waggoner, A.S., Bruchez, M.P. (2007). Sentinel lymph node imaging using quantum dots in mouse tumor models. *Bioconjugate Chem.* **18**, 389–396.
12. Kim, S., Lim, Y.T., Soltesz, E.G., De Grand, A.M., Lee, J., Nakayama, A., Parker, J.A., Mihaljevic, T., Laurence, R.G., Dor, D.M., Cohn, L.H., Bawendi, M.G., Frangioni, J.V. (2004). Near-infrared fluorescent type II quantum dots for sentinel lymph node mapping. *Nat. Biotech.* **22**, 93–97.
13. Kobayashi, H., Hama, Y., Koyama, Y., Barett, T., Regino, C., Urano, Y., Choyke, P.L. (2007). Simultaneous multicolor imaging of five different lymphatic basins using quantum dots. *Nano Lett.* 7, 1711–1716.
14. Zimmer, J.P. et al. (2006). Size series of small indium arsenide-zinc selenide core-shell nanocrystals and their application to in vivo imaging. *J. Am. Chem. Soc.* **128**, 2526–2527.
15. Akerman, M.E., Chan, W.C., Laakkonen, P., Bhatia, S.N., Ruoslathi, E. (2002). Nanocrystal targeting in vivo. *Proc. Nal Acad. Sci. U. S. A.* **99**, 12617–12621.
16. Cai, W., Shin, D.-W., Chen, K., Gheysens, O., Cao, Q., Wang, S.X., Gambhir, S.S., Chen, X. (2006). Peptide-labeled near-infrared quantum dots for imaging tumor vasculature in living subjects. *Nano Lett.* **6**, 669–676.
17. Gao, X., Cui, Y., Levenson, R.M., Chung, L.W., Nie, S. (2004). *In vivo* cancer targeting and imaging with semiconductor quantum dots. *Nat. Biotech.* **22**, 969–976.
18. Dubertret, B., Skourides, P., Norris, D.J., Noireaux, V., Brivanlou, A.H., Libchaber, A. (2002). In vivo imaging of quantum dots encapsulated in phospholipid micelles. *Science* **298**, 1759–1762.
19. So, M.K., Xu, C., Loening, A.M., Gambhir, S.S., Rao, J. (2006). Self-illuminating quantum dot conjugates for in vivo imaging. *Nat Biotech.* **24**, 339–343.
20. Ballou, B., Lagerholm, B.C., Ernst, L.A., Bruchez, M.P., Waggoner, A.S. (2004). Non invasive imaging of quantum dots in mice. *Bioconjugate Chem.* **15**, 79–86.
21. Hardman, R. (2006). A toxicologic review of quantum dots: toxicity depends on physicochemical and environmental factors. *Environ. Health Perspect.* **114**, 165–172.

Chapter 27

Semiconductor Fluorescent Quantum Dots: Efficient Biolabels in Cancer Diagnostics

Patricia M. A. Farias, Beate S. Santos, and Adriana Fontes

Summary

We present and discuss results and features related to the synthesis of water-soluble semiconductor quantum dots and their application as fluorescent biomarkers in cancer diagnostics. We have prepared and applied different core-shell quantum dots, such as cadmium telluride-cadmium sulfide, CdTe-CdS, and cadmium sulfide-cadmium hydroxide, $CdS/Cd(OH)_2$, in living healthy and neoplastic cells and tissues samples. The $CdS/Cd(OH)_2$ quantum dots presented the best results, maintaining high levels of luminescence as well as high photostability in cells and tissues. Labeled tissues and cells were analyzed by their resulting fluorescence, via conventional fluorescence microscopy or via laser scanning confocal microscopy. The procedure presented in this work was shown to be efficient as a potential tool for fast and precise cancer diagnostics.

Key words: Quantum dots, Cancer, Biological labeling, Diagnostics, Fluorescence, Colloids

1. Introduction

Research in the field of semiconductor nanostructures in the quantum confinement regime (quantum dots [QDs], quantum wells, quantum wires, and atomic and molecular clusters) has increased dramatically in the past two decades. These new versions of already well-known bulk materials show great potential for application in different research areas, ranging from microelectronics to fluorescent biolabels *(1–3)*. In the particular case of semiconductor QDs, their potential applications as biolabels is strongly related to the tuning of their optical properties, which is achieved by size, surface, and morphological control of the particles *(4, 5)*. This ability to tune the electronic band gap may be

James Weifu Lee and Robert S. Foote (eds.), *Micro and Nano Technologies in Bioanalysis*,
Methods in Molecular Biology, vol. 544
DOI 10.1007/978-1-59745-483-4_27, © Humana Press, a part of Springer Science+Business Media, LLC 2009

optimized for laser applications as well as for the improvement of solar cells *(6–8)* and for biolabeling purposes *(9–14)*.

Semiconductor QDs exhibit unique properties that are strongly related to the quantum confinement effect, which is also named quantum confinement regimen *(10)*. This intrinsic feature of semiconductor QDs can be observed by monitoring the optical properties of semiconductor nanocrystals, in which the diameter is smaller than 20 nm *(11)*. The size dependence of the emission wavelength of semiconductor QDs represents a direct consequence of this quantum effect, which can be described as follows: The quantum confinement regimen occurs when the QD radius is smaller than Bohr exciton´s radius, the QD electronic states become discretized, both in the valence band (VB) and in the conduction band (CB). Their values are not continuous, as they would be for the VB and the CB in a bulk semiconductor with the same composition of the QD. The quantization of electronic states, in which each state presents a very well defined electronic energy value, gives rise to an atom-like behavior as well as allows radiative electronic transitions in which the release of the electromagnetic energy related to every decay is directly proportional to the distance between the VB and CB states involved. As smaller is the QD, higher is the energy gap (Eg), which separates the VB and the CB. Thus, the tuning a QD emission wavelength is achieved by controlling the QD size *(10)*. In other words, this process generates excitons (electrons which give rise to a hole in their original state, when they are excited to an upper electronic state; excitons are also referred as "electron-hole pairs"). If compared to conventional organic dye molecules, QDs excited electrons can reach a set of higher energies. This explains why the QDs usually present broadband absorption spectra. When the exciton returns to a lower energy level, a narrow and symmetric emission spectrum is observed. The fluorescence lifetime is approximately 10–40 ns, which is significantly longer than the usual organic dyes, allowing images with greatly reduced levels of background noise in time-gated experiments *(13)*.

The fabrication of assemblies of perfect nanometer-sized materials, identically replicated in unlimited quantities, in such a state that they can be manipulated and used commercially, is an ultimate challenge of modern materials research, with outstanding fundamental and potential technological consequences. In this context, the use of soft chemistry synthesis remains a very important tool for achieving this goal *(9)*. Therefore, the synthesis and application of this class of materials is a subject of great interest, especially due to the fact that they can be obtained by easy and accessible preparation methodologies.

For biological labeling and diagnostic purposes, semiconductor QDs present many advantages when compared with conventional fluorescent organic dyes. They present high photostability *(14–16)*; broad excitation and narrow emission bands, allowing multiplexing of many QDs of different color in the same sample *(15–19)*; and low cytotoxicity *(17–20)*. It has been shown, since the first reports of this new class of biomarkers *(21, 22)*, that these fluorescent nanocrystals can be used to visualize, measure, and track individual molecular events using fluorescent and confocal microscopies, providing the ability to monitor dynamic molecular processes over extended periods (up to several minutes). All of these advantages, combined with the active chemical surface of these QDs in aqueous phase in conjunction with significant changes in cell properties caused by neoplastic mechanisms, make possible the use of QDs as a potential tool to expand conventional protocols used for cancer diagnostics.

Most of the semiconductor QDs have a *core-shell* structure. In the case of the aqueous $CdS/Cd(OH)_2$ QDs, which we focus on in this chapter, the cadmium sulfide (CdS) *core* is covered by a passivation layer of $Cd(OH)_2$. The *passivation shell* enhances the QD luminescence by reducing core surface defects. The passivation mechanism may be explained as follows: the shell layer prevents the excitons from spreading over the entire particle, forcing them to recombine while being spatially confined to the core. The resulting luminescence enhancement combined with a pronounced sharpening of the spectral bandwidth is an indication of the formation of the core-shell structure.

In order to make the QDs feasible for the use in biological systems, they need to be made biocompatible; this condition is achieved by functionalizing their surface with organic compounds. This process, called *functionalization*, results in a hybrid organic-inorganic nanomaterial that can be further modified with other capping species. The term *functionalization* (also known as *organic capping* and *ligand conjugation*) refers to the chemical modifications of the QD surface, which renders a biocompatible system feasible for binding with specific biomolecules that will target the QDs to specific sites in the biological systems. This process is thermodynamically favored because of the highly active chemical surfaces of the QDs, either presenting dangling bonds over the passivation shell or active binding groups in the stabilizing species that may be present in a colloidal system. **Figure 1** shows a diagrammatic representation of a passivated and functionalized QD.

It is well known that healthy and cancer cells (or tissues) present significant differences in their metabolic processes. Cell morphological and biochemical differentiation patterns are commonly used to characterize the presence or absence of cancer cells (or tissues) *(23, 24)*. For example, the detection of carbohydrate residues in human tissues above a normal value, using immunohistochemistry, is defined as a sign of abnormal metabolic processes associated with tumor cell growth *(25)*. On the other hand, for a nonspecific labeling process, the uptake rate of the biomarker itself can be used to quantify and differentiate healthy from cancer cells in in vitro experiments. This differentiation is

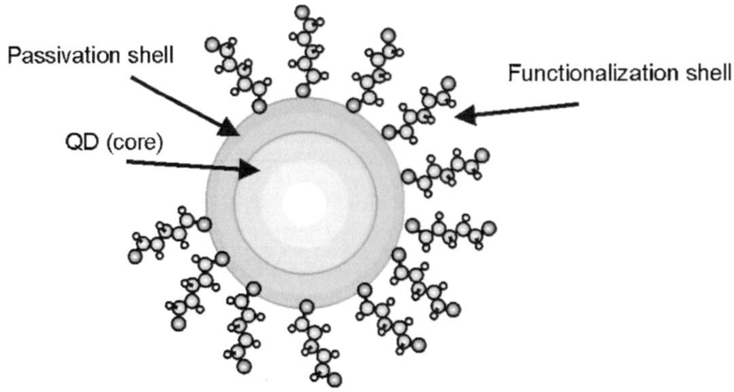

Fig. 1. Diagrammatic representation of a passivated and functionalized QD.

expected because of the faster metabolic rates observed for cancer cells, leading to enhanced growth and replication rates.

Here, we present and discuss some results of the application of water-soluble colloidal semiconductor QDs for diagnostic purposes in living cells. The fluorescence was used as a primary tool to explore and differentiate the labeling of the samples. Tissues and cells conjugated with QDs were analyzed by laser scanning confocal microscopy.

In this context, **Subheading 2** describes the materials used for synthesis, surface passivation, and functionalization of the QDs used in the procedures mentioned in **Subheading3**. **Subheading 2** also describes the materials used in the conjugation of the QDs with cells and tissues. In **Subheading 3**, the experimental methodology is presented. **Subheading 3.1** describes the methodology used in the synthesis, functionalization and labeling of $CdS/Cd(OH)_2$ nanocrystals. **Subheading3.2** presents some results obtained by conjugating $CdS/Cd(OH)_2$ highly fluorescent QDs to living healthy and neoplastic breast, glial, and cervical cell and tissue samples. In **Subheading 3.3**, some possible mechanisms of incorporation of the QDs by cancer cells and tissues are discussed. Concluding remarks are presented in **Subheading 3.4**. **Subheading 4 (Notes)** contains brief notes concerning relevant aspects that should be taken into account for biolabeling experiments using hydrophilic semiconductor QDs.

2. Materials

2.1. CdS Quantum Dot Synthesis

1. Ultrapure deionized water.
2. Sodium polyphosphate ($[(NaPO_3)]_9$, Sigma-Aldrich, 96%) as the stabilizing agent at 0.0051 g/mL.

3. Cadmium perchlorate ($Cd(ClO_4)_2$, Sigma-Aldrich, 99.99%) at 0.01 mol/L.
4. Sulfidric acid (H_2S) at 0.033 mol/L.

2.2. Passivation Of CdS Quantum Dots With $Cd(OH)_2$

1. Cadmium perchlorate ($Cd(ClO_4)_2$, Sigma-Aldrich, 99.99%) at 0.01 mol/L.
2. Sodium hydroxide (NaOH, Sigma-Aldrich, 99%) at 0.05 mol/L.

2.3. $CdS/Cd(OH)_2$ Quantum Dots Functionalization Using Glutaraldehyde

1. Glutaraldehyde at 0.01% v/v (Sigma-Aldrich, 99%).

2.4. Cell and Tissue Preparation

1. Saline solution (NaCl, Sigma-Aldrich, 99.99% at 0.9%).
2. DMEM: Dulbecco's modified Eagle medium.
3. Evans Blue solution (Sigma-Aldrich, 99.9% at 1%).
4. Normal and C6 neoplastic glial cells in culture medium at 10^6 cells/mL.
5. Fixed breast cancer tissue slices.
6. Fresh cervical epithelial cells.

2.5. Evaluation of Labeled Cells and Tissues

1. Conventional fluorescence microscopy and laser scanning confocal microscopy (Leica, AOBS). The images were obtained using high numerical aperture objectives. For the confocal microscopy, two wavelengths were used to excite the QDs present in the labeled samples: 488 and 543 nm.
2. Leica Lite and Carl Zeiss's LSM 510 software.

2.6. Optical and Structural Evaluation of the Quantum Dots

1. UV–visible absorption spectroscopy (Perkin-Elmer Lambda 6 spectrophotometer).
2. Transmission electron microscopy (Carl Zeiss TEM 902-80keV).
3. Image processing (Scion Image) was performed to enhance the morphological information on the nanoparticles.
4. The luminescence spectra of the CdS suspensions were obtained on a ISS K2 Multifrequency Phase Fluorimeter.

3. Methods

3.1. Synthesis, Functionalization, and Labeling

The CdS suspensions were prepared at room temperature by the addition of defined volumes of $H_2S(g)$ into aqueous solution containing Cd^{2+} ions (using $Cd(ClO_4)_2$ as the precursor) and sodium polyphosphate ($[(NaPO_3)]_9$) as the stabilizing agent, maintaining a Cd:S ratio of 1:1. After the synthesis of the QDs, their surface passivation was performed by depositing a $Cd(OH)_2$ layer as described elsewhere *(26)*. Basically, the pH of the original QD suspension is raised up to 10.5 and a solution of $Cd(ClO_4)_2$ is added dropwise to the suspension until its luminescence is increased to a stable level (~6 mL). The structural and optical characterization of the suspensions was performed using (a) X-ray diffraction, (b) electronic transmission microscopy (TEM), and (c) absorption, excitation, and emission spectroscopic analysis. The average size of the $CdS/Cd(OH)_2$ nanocrystals was 6 ± 1 nm. The excitation and emission spectra presented a broad excitation band and a narrow emission band of approximately 50 nm. At a pH of 7.2, the QDs were functionalized via a one-pot cross-linking glutaraldehyde (0.01% v/v) procedure *(19)*. Glutaraldehyde (Glut) is an organic functionalizing agent that was used to link the $CdS/Cd(OH)_2$ QDs with healthy and neoplastic living cells. Because glutaraldehyde is a homofunctional bidentade ligand, it establishes a hemi-acetal interaction with the outer shell of the QDs at the same time that it binds to cell surface proteins by Schiff's base interactions. No fixating action of the glutaraldehyde was observed in the analyzed samples with the concentration used for the experiments mentioned here. Labeled cells maintained their normal metabolic behavior in their respective culture medium for more than 5 days.

Healthy and neoplastic (C6) glial cells in separate culture medium wells were directly incubated with 5 μL of the aqueous suspension containing the glutaraldehyde-functionalized QDs, which corresponds to approximately 10^{10} particles/culture well. The healthy glial and the C6 cells were incubated with the QDs suspension for 1, 2, and 3 min in a prepared culture medium (DMEM) at room temperature (25°C) before their visualization *(27)*.

The fixed breast tissue slices were kept in 0.9% NaCl medium overnight and washed with fresh saline solution before the incubation process. The samples were labeled with $CdS/Cd(OH)_2$ after autofluorescence removal by chemical treatment using Evans Blue solution *(27)*.

Another procedure was performed for labeling the healthy and cancer cervical cells. For the cervical samples, a preliminary treatment for removing blood vestiges was performed by successively washing and centrifuging the samples (for 3 min at 1,000 rpm (67*g*)) with a diluted (5%) hydrogen peroxide aqueous solution (H_2O_2, Sigma–Aldrich). The blood vestiges (erythrocytes, platelets, and plasma) become a white-colored aggregate, which has a lower density than cervical tissue samples. These last cells remain in the bottom of the tube containing the sample, while the white aggregate migrates to the upper surface of the centrifuged sample.

The supernatant phase is then excluded from the sample and this process is repeated until the complete removal of blood vestiges.

Tissue slices and live cells stained with QDs were evaluated by fluorescence microscopy and laser scanning confocal microscopy. The acquisition parameters (pinhole, filters, beam-splitters, and also photomultiplier gain and offset) were maintained constant to allow the fluorescence intensity analysis to be a valid analytical tool. The images were reproduced at least three times for each cell/tissue sample and were further processed using appropriate software (described in **Subheading 2**). To confirm the presence of QDs inside the cells, some of the conjugated systems were also characterized by TEM. This type of measurement complements the fluorescence analysis, showing where the QDs are located after they are internalized by the cell.

3.2. Results

The application of QDs to the investigation of neoplastic processes, which may give rise to cancer, constitutes a topic of increasing interest, and many questions still await precise answers. In the pursuit of sensitive and quantitative methods to detect and diagnose cancer, nanotechnology has been identified as a field of great promise. Hydrophilic QDs, at physiological pH conditions, have the potential to expand conventional protocols used for cancer diagnostics, which need previous tissue/cell fixation, and extend the protocols to investigate neoplastic mechanisms in living cells and/or tissues in real time.

The cells showed no signs of damage after the conjugation procedure and maintained their integrity even after longer than 5 days of incubation time, demonstrating the low toxicity of the QDs for in vitro studies. **Figures 2** and **3** show confocal microscopy images and the corresponding fluorescence intensity maps for the time evolution (1–3 min incubation time) of the interaction between $CdS/Cd(OH)_2$-Glut QDs and healthy and neoplastic glial cells, respectively. In the fluorescence intensity maps, the black regions correspond to the absence of fluorescence, whereas the lighter gray to white regions correspond to areas of higher fluorescence intensities (this is valid for all of the fluorescence intensity maps shown in this chapter).

The analysis of the TEM images of the glial cells shows that the QDs accumulate near the nuclear membrane. **Figure 4** shows a representative TEM image of core-shell $CdS/Cd(OH)_2$ QDs functionalized with glutaraldehyde (QD-Glut) and conjugated in vitro with live human glioblastoma cells.

Figures 5 and **6** show confocal microscopy images as well as the fluorescence intensity maps for QDs labeling normal cervical cells and cervical intraepithelial neoplastic cells 3 (INC3), respectively. **Figure 7** shows a confocal microscopy image, fluorescence intensity map, and conventional transmission microscopy overlapped with fluorescence image of cervical cells presenting severe dyskaryosis, the last stage before cervical cancer.

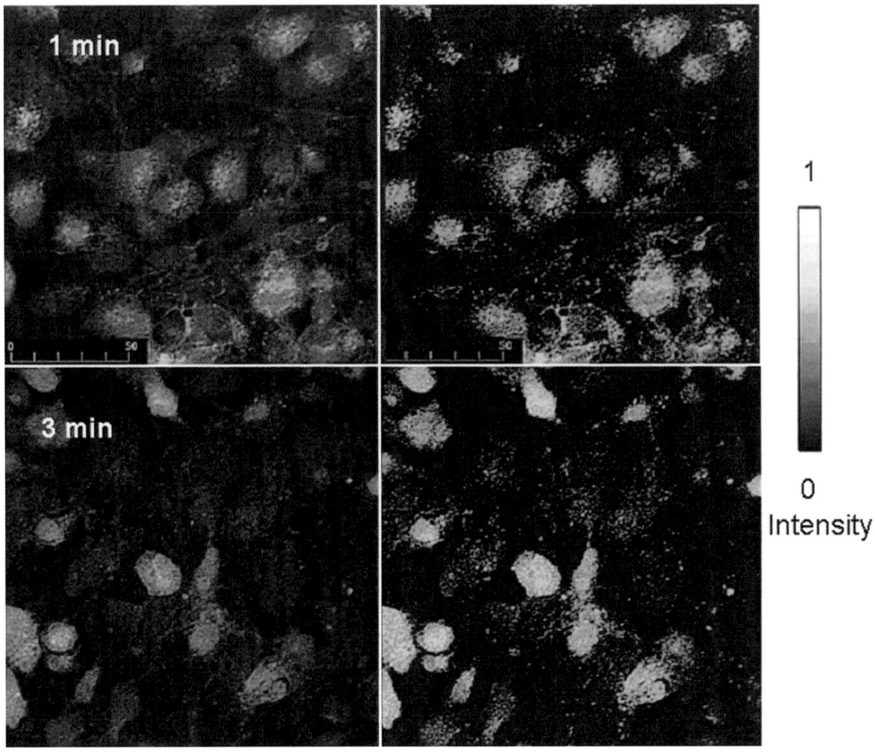

Fig. 2. Time evolution of the fluorescence pattern of healthy glial cells incubated with QDs-Glut. Confocal microscopy images (*left*) and the corresponding fluorescence intensity maps (*right*).

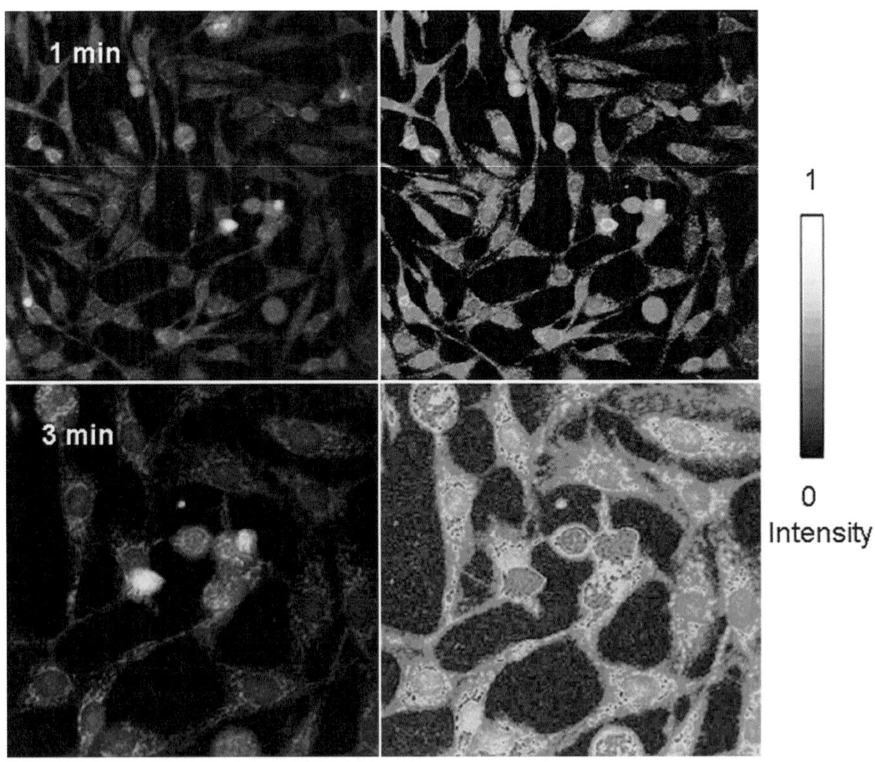

Fig. 3. Time evolution of the fluorescence pattern of neoplastic glial (glioblastoma) cells incubated with QDs-Glut. Confocal microscopy image (*left*) and the corresponding fluorescence intensity maps (*right*).

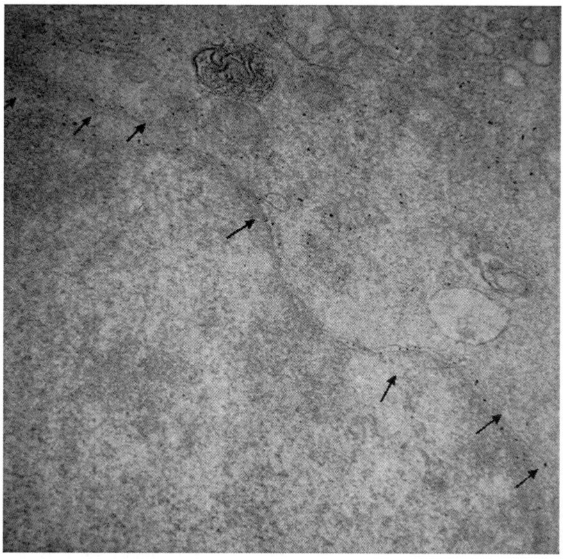

Fig. 4. Transmission electronic microscopy image of a glioblastoma labeled cell, in which the highest QDs concentration is nearby the nuclear envoltorium (*arrows*).

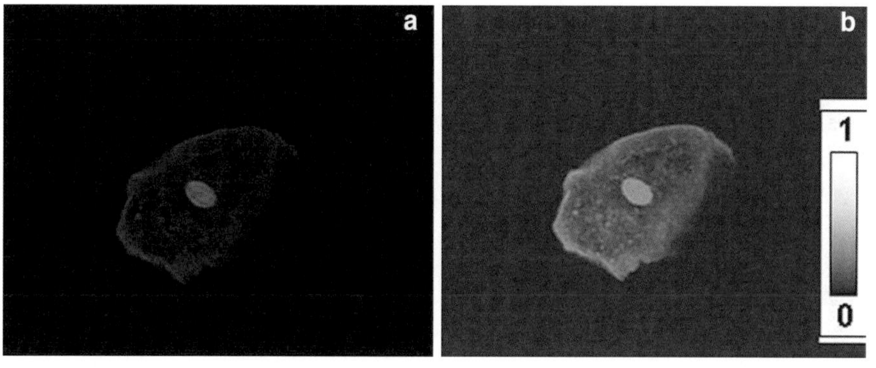

Fig. 5. (**a**) Confocal microscopy image of normal squamous cervical cell. (**b**) Corresponding fluorescence intensity map.

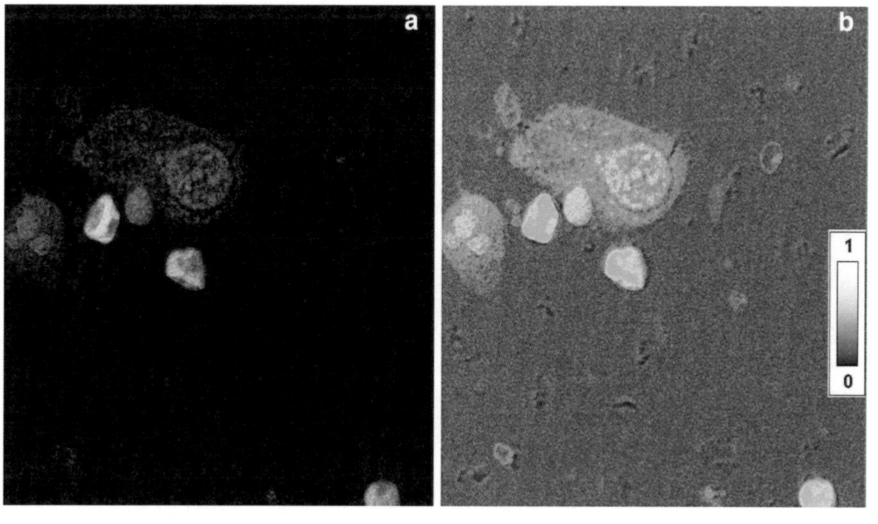

Fig. 6. Cervical intraepithelial neoplastic cells 3 (INC3). (**a**) Confocal microscopy image. (**b**) Fluorescence intensity map.

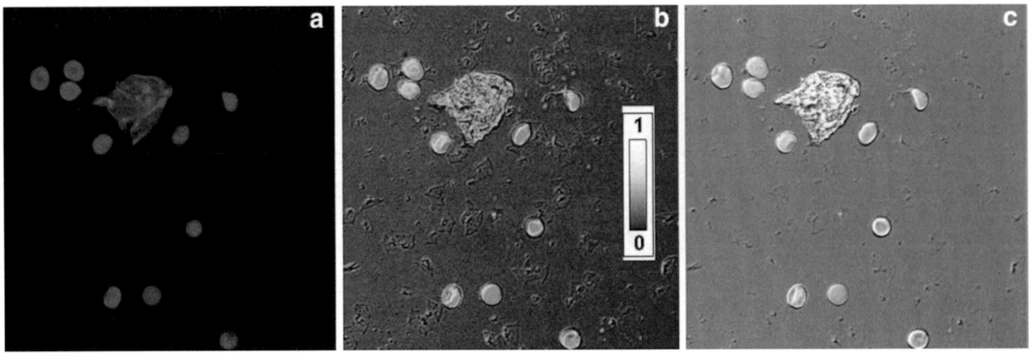

Fig. 7. Confocal microscopy image (**a**), fluorescence intensity map (**b**), and transmission microscopy overlapped with fluorescence image (**c**) of cervical cells presenting severe dyskaryosis.

Analyzing the images and the intensity maps, the time evolution of the interaction cells-QDs clearly reveals different labeling patterns as well as different fluorescence intensities. In general cancer cells present higher intensity levels of fluorescence, mainly because these cells are in higher and differentiated metabolic activity, also losing much of their membrane receptors specificity. One of the consequences is the fast endocytosis of QDs, a feature which is not observed in healthy cells. It also can be noticed that the QD-Glut easily interacts with both healthy and neoplastic cells.

3.3. Quantum Dot Uptake by Cancer Cells

Cancer is a complex disease caused by genetic instability and the accumulation of multiple molecular alterations, which are related to a wide variety of modifications in cell and tissue properties, such as increasing cell membrane permeability and hydraulic conductivity (which are, in general, significantly higher than in normal cells) and increasing the membrane channels and pores sizes. In tumor tissues, the blood vessels are leaky, also presenting larger pore sizes compared with normal tissues *(28)*. These features play an important role in the internalization of QDs by cancer cells and tissues. The results presented in the previous section clearly indicate that there are probably two different cell (or tissue)–QDs interaction mechanisms, which compete kinetically: (a) QD interaction with surface proteins, which occurs by the formation of Schiff's bases between amine terminals of these proteins and the terminal carboxyl group of the QD functionalizing agent (glutaraldehyde) and (b) QD uptake via endocytosis, which, depending on the cellular molecules involved in the internalization process, may be preferably named pinocytosis or phagocytosis. Although these last processes do differ in details, their common feature is that cells engulf the material to be incorporated. In the labeling procedures described in this chapter, the endocytosis mechanisms were prevalent in the interaction between cancer cells (or tissues) and QDs. The further analysis of the transmission electronic microscopy results indicated that the $CdS/Cd(OH)_2$ larger QDs (>6 nm) were predominantly localized in granular compartments around the nuclear region, whereas the smaller QDs (<6 nm) were internalized in the nuclear region. We also observed that, after cell division, the larger QDs can be found not only in the perinuclear region but also in cytosol regions. Beyond this, upon cell division, the ingested QDs are distributed between both cells *(27)*.

3.4. Concluding Remarks

Semiconductor QDs present a great potential in life sciences applications. Beyond their use as fluorescent labels and contrast agents *(29, 30)*, QDs also have a potential surgical utility by providing optical guidance that can result in reduction of cancer metastases *(31)*. Indicated references may provide relevant information about recent developments of such QDs applications.

The development of fast, cheap, and precise cancer diagnostic protocols can be achieved by using QD-based methodologies. In this chapter, we presented some features that can corroborate this conclusion. A simple methodology for the synthesis of water soluble QDs at physiological conditions, as well as simple labeling procedures were presented and the efficacy of QDs as a potential diagnostic tool was shown.

4. Notes

1. The QD surface has to be hydrophilic and present high fluorescence intensity.
2. The cells and tissues must be well preserved concerning their own morphological and physiologic characteristics.
3. The QDs/Glut as well as the QDs/cell ratios must be always taken into account to avoid undesirable features such as high background noise or insufficient material for labeling the sample.
4. The synthetic methods must be optimized to prepare high-quality QDs in affordable large-scale quantities.

References

1. Menezes, F. D., Brasil Jr., A. G., Moreira, W. L., Barbosa, L. C., Cesar, C. L., Ferreira, R., Farias, P. M. A. and Santos, B. S. (2005). CdTe/CdS core shell quantum dots for photonic applications, *Microelectron. J.* **36**, 989–991.
2. Alivisatos, A. (1996). Perspectives on the physical chemistry of semiconductor nanocrystals, *J. Phys. Chem.* **100**, 13226–13239.
3. Castro, V., Farias, P. M. A., Santos, B. S., Menezes, F. D., Ferreira, R., Fontes, A., Lima, P. R. M., Barjas-Castro, M. L. and Cesar, C. L. (2004). Quantum dots, efficient fluorescent markers for red cells, *Blood* **104**, 741a.
4. Brus, L. E., Bawendi, M., Wilson, W. L., Rothberg, L., Carroll, P. J., Jedju, T. M. and Steigerwald, M. L. (1991). Prymary photo-

physics of semiconductor crystallites, *Abstr. Pap. Am. Chem. Soc.* **201**, 409–414.

5. Kraus, R. M., Lagoudakis, P. G., Rogach, A. L., Talapin, D. V., Weller, H., Lupton, J. M. and Feldmann, J. (2007). Room-temperature exciton storage in elongated semiconductor nanocrystals, *Phys. Rev. Lett.* **98**, 0174011.
6. McCormick, C. (2007). Quantum dots and nanowires demonstrate potential for efficient solar cells, *MRS Bull.* **32**, 605–612.
7. Schaller, R. D. and Klimov, V. I. (2007). Optical absorption and ultrafast carrier dynamics characterization of CdSe quantum dots deposited on different morphologies of nanostructured TiO2 films, *Mat. Sci. Eng. C-Biomimetic Supramol. Syst.* **27**, 1514–1520.
8. Osovsky, R., Kloper, V., Kolny-Olesiak, J., Sashchiuk, A. and Lifshitz, E. (2007). Optical properties of CdTe nanocrystal quantum dots, grown in the presence of Cd-0 nanoparticles, *J. Phys. Chem C.* **111**, 10841–10847.
9. Pileni, M. P. (2000). II-VI semiconductors made by soft chemistry - Syntheses and optical properties, *Catal. Today.* **58**, 151–166.
10. Rossetti, R. and Brus, L. (1982). Electron-hole recombination emission as a probe of surface chemistry in aqueous CdS colloids, *J. Phys. Chem.* **86**, 4470–4472.
11. Bruchez, M. P. (2005). Turning all the lights on: quantum dots in cellular assays, *Curr. Opin. Chem. Biol.* 9, 533–537.
12. Michalet, X., Pinaud, F. F., Bentolila, L. A., Tsay, J. M., Doose, S. and Li, J. J. (2005). Quantum dots for live cells, in vivo imaging, and diagnostics, *Science* **307**, 538–544.
13. Alivisatos, A. P., Gu, W., Larabell, C. (2005). Quantum dots as cellular probes, *Annu. Rev. Biomed. Eng.* 7, 55–76.
14. Moreira, W. L., Fontes, A., Neves, A. A. R., Thomaz, A. A., Barbosa, L. C., Menezes, F. D., Farias, P. M. A., Santos, B. S. and Cesar, C. L. (2005). Synthesis and characterization of CdTe nanocrystals for biological labeling, *Genet. Eng. Opt. Probes Biomed. Appl. III.* **5704**, 240–248.
15. Rockenberger, J., Trögr, L., Rogach, A. L., Tischer, M., Grundmann, M., Eychmüller, A. and Weller, H. (1998). The contribution of particle core and surface to strain, disorder and vibrations in thiolcapped CdTe nanocrystals, *J. Chem. Phys.* **108**, 7807–7815.
16. Borchert, H., Talapin, D. V., Gaponik, N., McGinley, C., Adam, S., Lobo, A., Möller, T. and Weller, H. (2003). Relations between the photoluminescence efficiency of CdTe nanocrystals and their surface properties revealed by synchrotron XPS, *J. Phys. Chem. B.* **107**, 9662–9668.
17. Geho, N., Lahar, P., Gurnani, M., Huebschman, P., Herrmann, V., Espina, A., Shi, J., Wulfkuhle, H., Garner, E., Petricoin III, L. A., Liotta, K. and Rosenblatt, P. (2005). Pegylated, steptavidin-conjugated quantum dots are effective detection elements for reverse-phase protein microarrays, *Bioconjugate Chem.* **16**, 559–566.
18. Farias, P. M. A., Santos, B. S., Menezes, F. D., Ferreira, R., Barjas-Castro, M. L., Castro, V., Lima, P. R. M., Fontes, A. and Cesar, C. L. (2005). Investigation of red blood cell antigens with highly fluorescent and stable semiconductor quantum dots, *J. Biomed. Opt.* **10**, 0440231–0440234.
19. Farias, P. M. A., Santos, B. S., Menezes, F. D., Ferreira, R., Barjas-Castro, M. L., Castro, V., Lima, P. R. M., Fontes, A. and Cesar, C. L. (2005). Core-shell CdS/Cd(OH)2 quantum dots: Synthesis and bioconjugation to target red cells antigens, *J. Microsc.* **219**, 103–108.
20. Tsay, J. M., Trzoss, M., Shi, L. X., Kong, X. X., Selke, M., Jung, M. E. and Weiss, S. (2007). Singlet oxygen production by peptide-coated quantum dot-photosensitizer conjugates, *J. Am Chem. Soc.* **129**, 6865–6871.
21. Bruchez Jr., M., Morrone, M., Gin, P., Weiss, S. and Alivisatos, A. P. (1998). Semiconductor nanocrystals as fluorescent biological labels, *Science* **281**, 2013–2016.
22. Chan, W. C. W. and Nie, S. (1998). Quantum dot bioconjugates for ultrasensitive nonisotopic detection, *Science* **281**, 2016–2018.
23. Warburg, O. (1956). Origin of cancer cells, *Science* **123**, 309–314.
24. Wu, M., Neilson, A., Swift, A. L., Moran, R., Tamagnine, J., Parslow, D., Armistead, S., Lemire, K., Orrell, J., Teich, J., Chomicz, S. and Ferrick, D. A. (2007). Multiparameter metabolic analysis reveals a close link between attenuated mitochondrial bioenergetic function and enhanced glycolysis dependency in human tumor cells, *Am. J. Physiol. Cell Physiol.* **292**, C12–C136.
25. Shimma, Y. and Jigami, Y. (2004). Expression of human glycosyltransferase genes in yeast as a tool for enzymatic synthesis of sugar chain, *Glycoconjugate J.* **21**, 75–78.
26. Weller, H., Koch, U., Gutierrez, M. and Henglein, A. (1984). Photochemistry of colloidal metal sulfides. 7. Absorption and fluorescence of extremely small ZnS particles - The world of the neglected dimensions, *Berich. Bunsen-Gesellschaft-Physical Chem. Chem. Phys.*, **88**, 649–656.
27. Farias, P. M. A., Santos, B. S., Menezes, F. D., Ferreira, R., Fontes, A., Carvalho, H. F., Romão, L., Moura-Neto, V., Amaral, J. C. O.

F., Cesar, C. L., Figueiredo, R. C. B. Q. and Lorenzato, F. R. B. (2006). Quantum dots as fluorescent bio-labels in cancer diagnostic, *Phys. Stat. Sol. C.* **11**, 4001–4008.

28. Jain, R. K. (1999). Transport of molecules, particles, and cells in solid tumors, *Annu. Rev. Biomed. Eng.* **1**, 241–263.
29. Gerion, D., Herberg, J., Bok, R., Gjersing, E., Ramon, E., Maxwell, R., Kurhanewicz, J., Budinger, T. F., Gray, J. W., Shuman, M. A. and Chen, F. F. (2007). Paramagnetic silica-coated nanocrystals as an advanced MRI contrast agent, *J. Phys. Chem. C* **111**, 12542–12551.
30. Mulder, W. J. M., Griffioen, A. W., Strijkers, G. J., Cormode, D. P., Nicolay, K. and Fayad, Z. A. (2007). Magnetic and fluorescent nanoparticles for multimodality imaging, *Nanomedicine* **2**, 307–324.
31. Kim, S., Lim, Y. T., Soltesz, E. G., De Grand, A. M., Lee, J. and Nakayama, A. (2004). Near-infrared fluorescent type II quantum dots for sentinel lymph node mapping, *Nat. Biotechnol.* **22**, 93–97.

Chapter 28

The Monitoring and Affinity Purification of Proteins Using Dual Tags with Tetracysteine Motifs

Richard J. Giannone, Yie Liu, and Yisong Wang

Summary

Identification and characterization of protein-protein interaction networks is essential for the elucidation of biochemical mechanisms and cellular function. Affinity purification in combination with liquid chromatography-tandem mass spectrometry (LC-MS/MS) has emerged as a very powerful tactic for the identification of specific protein-protein interactions. In this chapter, we describe a comprehensive methodology that uses our recently developed dual-tag affinity purification system for the enrichment and identification of mammalian protein complexes. The protocol covers a series of separate but sequentially related techniques focused on the facile monitoring and purification of a dual-tagged protein of interest and its interacting partners via a system built with tetracysteine motifs and various combinations of affinity tags. Using human telomeric repeat binding factor 2 (TRF2) as an example, we demonstrate the power of the system in terms of bait protein recovery after dual-tag affinity purification, detection of bait protein subcellular localization and expression, and successful identification of known and potentially novel TRF2 interacting proteins. Although the protocol described here has been optimized for the identification and characterization of TRF2-associated proteins, it is, in principle, applicable to the study of any other mammalian protein complexes that may be of interest to the research community.

Key words: Dual-tags, TEV protease, Affinity purification, Freeze/thaw lysis, Tetracysteine motifs, FIAsH, Lumio, Subcellular localization, In-gel detection, Live cell monitoring, Fixed cell detection, TRF2, Mammalian protein complex, Mass spectrometry

1. Introduction

Among the plenitude of protein purification methods available, affinity purification is one of the most efficient and gentle discriminatory separation techniques for the capture of native protein complexes. Coupled with liquid chromatography (LC)

James Weifu Lee and Robert S. Foote (eds.), *Micro and Nano Technologies in Bioanalysis*,
Methods in Molecular Biology, vol. 544
DOI 10.1007/978-1-59745-483-4_28, © Humana Press, a part of Springer Science+Business Media, LLC 2009

and mass spectrometry (MS), the affinity purification of a bait protein of interest has allowed for the identification of specific interacting proteins at sub-picomolar levels *(1, 2)*. In spite of enormous success in yeast *(3, 4)*, this approach has encountered numerous challenges in terms of both specificity and efficiency of protein complex retrieval from mammalian cells. Despite the fact that a well-chosen affinity tag may allow a very high-degree of protein purification in a single step, it is impossible to reach purification homogeneity, which, in the area of protein complex elucidation, often leads to aberrant protein identifications and lengthy experimental verification. Although recent improvements to MS instrumentation have increased peptide detection sensitivities, problems inherent to purification specificity remain; a result that is largely caused by the impurity of the analyzed samples.

The development of the tandem affinity purification (TAP) tag was aimed at addressing the aforementioned specificity issue, with the technique relying on two separate, independent purification events to reduce the co-elution of nonspecific binding proteins. The TAP tag fusion construct encodes two IgG binding domains of *Staphylococcus aureus* protein A and a calmodulin binding peptide (CBP) domain separated by a tobacco etch virus (TEV) protease cleavage site *(2, 3)*, which allows for the TAP of a fusion protein under native conditions. The success of this technique has proven itself in the isolation and identification of protein complexes in yeast. Although several groups have applied TAP to examine the protein networks of mammalian cells *(5–10)*, the method offers an overall low recovery of bait and specific interacting proteins because of the sheer complexity of protein mixtures recovered from mammalian cellular extracts. With this drawback in mind, modifications to the original TAP construct and/or overall strategy have been introduced by several groups, each vying to augment the utility of TAP in the examination of mammalian systems *(11–15)*. Nevertheless, problems pertinent to any protein tagging strategy persist, such as insufficient exposure of the affinity tag, improper folding of the fusion proteins, steric exclusion of interacting proteins, and/or excessive overexpression of the fusion protein, all of which can inhibit the process of protein complex purification and identification.

To address some of the major hurdles inherent to protein tagging, detection, and purification in mammalian cells, we have developed a dual-tag affinity purification system that marries several useful features together: (a) Gateway®-compatible vectors for quick and simple cloning, (b) the option to regulate dual-tagged bait protein expression, (c) the incorporation of a tetracysteine motif (CCPGCC) for easy bait protein visualization in live, fixed, and lysed cells, (d) a second TEV protease recognition sequence to increase cleavage efficiency, and (e) various dual-tag constructs with different combinations of affinity tags to enhance bait protein compatibility *(16)*. Using human TRF2 as a test bed, this system demonstrated bait protein recoveries upward of 16% from as little as 1–7 × 10^7 cells upon dual-tag affinity purification while allowing us to easily monitor dual-tagged TRF2 expression and infer its conservation of function via proper localization to the telomere *(16)*. In this chapter, we focus on the description of protein complex purification as well as illustrate the usage of the tetracysteine motif to observe bait protein expression, purification, localization, and colocalization with other proteins.

2. Materials

2.1. Gateway Cloning and Cell Culture

1. Gene open-reading frame (ORF) representing the protein to be dual tagged, such as human TRF2 ORF *(16)*.
2. Gene-specific primers flanked by Gateway-compatible *attB* sequences (see Gateway protocol from Invitrogen, Carlsbad, CA). In this study, we used TRF2-specific primers (5′-GGG GAC AAG TTT GTA CAA AAA AGC AGG CTT GGC TGG TGG TGG TGG TT-3′ and 5′-GGG GAC CAC TTT GTA CAA GAA AGC TGG GTC TTA GTT CAT GCC AAG TCT TT-3)′.
3. Gateway entry clone/donor vector pDONR221 (Invitrogen). Store at –20°C
4. BP Clonase™ II (Invitrogen), Gateway BP recombination enzyme mix. Aliquot into single-use tubes and store at –80°C.
5. Dual-tag destination vectors (**Fig. 1a** and *(16)*). Store at –20°C.
6. LR Clonase™ II (Invitrogen), Gateway LR recombination enzyme mix. Aliquot into single-use tubes and store at –80°C.
7. T-REx compatible cell lines (already expressing the Tet repressor element, store in liquid N_2) or pcDNA™6/TR plasmid (store at –20°C) to create a new T-REx line (Invitrogen).
8. Blasticidin-S (Invitrogen) for antibiotic selection/maintenance of the Tet repressor element. Aliquot in several tubes and store at –20°C to avoid repeated freeze/thaw cycles. Once thawed, antibiotic is stable at 4°C for 2 weeks.
9. Cell culture-grade G418 disulfate salt (Sigma-Aldrich, St. Louis, MO) for selection/maintenance of cells with integrated dual-tag fusion protein(s). Store at 4°C.

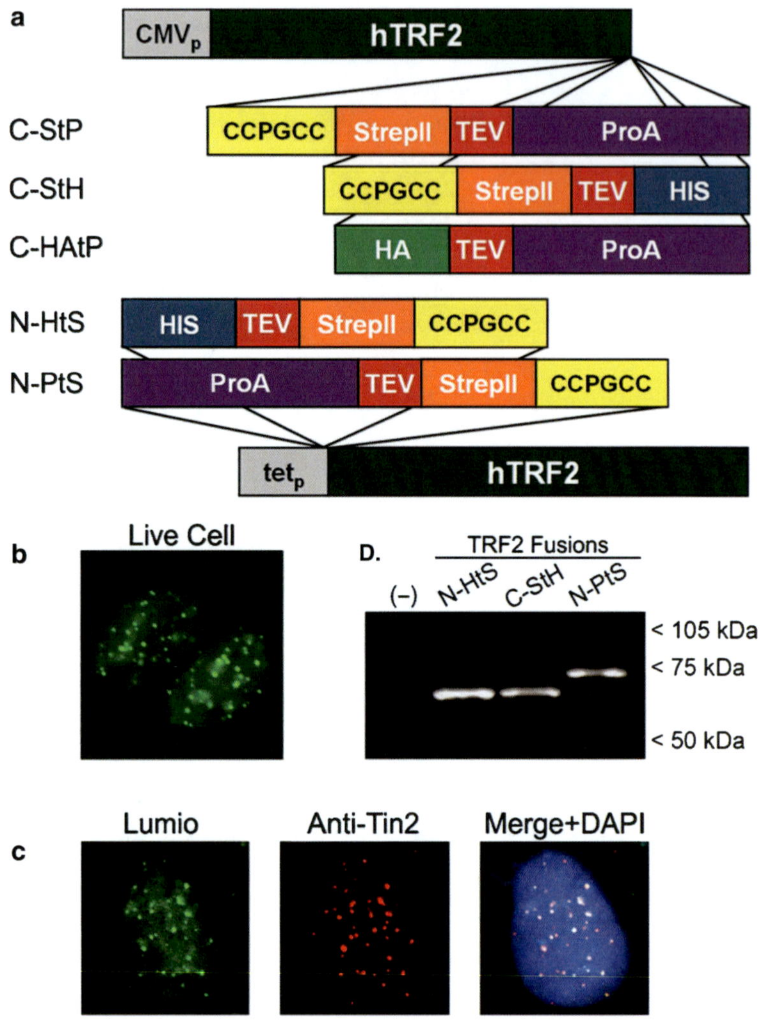

Fig. 1. Dual tags and Lumio-based visualization of dual-tagged TRF2. **(a)** Schematics of dual tags and their relative positions in TRF2 fusions. *C* C′-terminal; *N* N′-terminal; *CCPGCC* a tetracysteine motif; *S* Strep-Tactin binding peptides (StrepII tag); *t* tobacco etch virus (TEV) protease cleavage site; *P* IgG binding domain of protein A from *Staphylococcus aureus* (ProA tag); *H* 6x histidine (His tag); *HA* influenza hemagglutinin epitope (HA tag). (**b**) Live cell image of TRF2-C-StH punctate staining on telomeres revealed by Lumio. (**c**) Colocalization of transfected TRF2-C-StH with telomere specific protein Tin2 in U2OS cells after fixation. TRF2-C-StH (*green*) marked by Lumio green reagent, Tin2 (*red*) marked by anti-Tin2 antibody, chromatin (*blue*) stained by DAPI, and colocalization (*yellow*) represented in merged panel. (**d**) In-gel Lumio detection of TRF2 fusion protein expression and size, separated by SDS-PAGE (Reproduced partially from **ref. *16*** with permission from BioTechniques, Informa Life Sciences Group).

10. Cell culture-grade tetracycline hydrochloride (Sigma-Aldrich) to induce tetracycline-regulatable dual-tagged protein(s). Store at 4°C.
11. Cell culture medium supplemented with fetal bovine serum, consistent with culture conditions of desired cell line.

12. Trypsin (0.05%) with 1 mM ethylenediamine tetraacetic acid (EDTA) and phosphate-buffered saline (PBS) pH 7.4 for cell passage. Store at 4°C.

2.2. Detection of Dual-Tagged Proteins in Live Cells by Lumio

1. Lumio™ Green In-Cell Detection Kit (Invitrogen) containing 2 mM FlAsH-EDT$_2$ (Lumio Green Reagent) and Disperse Blue for background reduction.
2. 1,2-Ethanedithiol (EDT; Sigma-Aldrich) adjusted to 25 mM in dimethyl sulfoxide (DMSO). Warning: EDT is toxic and malodorous, dilute in fume hood. Store at room temperature (RT).
3. Dimethyl sulfoxide (Sigma-Aldrich). Store at RT in dark.
4. Dulbecco's phosphate-buffered saline (D-PBS) with Ca^{++} and Mg^{++} (Invitrogen).
5. Opti-MEM® Reduced Serum Media (Invitrogen). Light sensitive, store at 4°C.
6. Fluorescent microscope (inverted or vertical microscope with submersible lens [Zeiss, 63×/0.95W]) capable of excitation wavelength of 508 nm and equipped with a FITC filter set to visualize green fluorescence of dual-tag protein-bound Lumio.

2.3. Immunofluorescent Colocalization Studies with Lumio-Labeled Dual-Tagged Protein

1. Ice-cold PBS at pH 7.4 (does not need Ca^{++} and Mg^{++}). Make 10× stock by dissolving the following in 800 mL ddH$_2$O: 80 g NaCl, 2 g KCl, 14.4 g Na_2HPO_4, 2.4 g KH_2PO_4. Adjust to pH 7.4 and bring volume to 1 L with more ddH$_2$O.
2. Fixation: 2% paraformaldehyde (PFA; Sigma-Aldrich) in PBS. Store at 4°C.
3. Permeabilizer: 0.5% Nonidet P-40 (NP-40; Sigma-Aldrich) in PBS.
4. Blocking solution: 1% IgG-free bovine serum albumin (Sigma-Aldrich) in PBS.
5. Primary antibody against putative dual-tagged protein interacting partner, such as mouse anti-TIN2 antibody (Imgenex, San Diego, CA). Secondary antibody conjugated with a fluorophore that does not fluoresce green (because Lumio fluoresces in the green spectrum), such as Alexa-Fluor 594 donkey anti-mouse antibody (Invitrogen). Dilute both to manufacturer's recommended dilution in blocking solution.
6. VECTASHIELD Mounting Medium with 4′,6 diamidino-2-phenylindole (DAPI) (Vector Laboratories, Burlingame, CA).
7. Clear nail polish (any brand) to seal the cover slip with stained cells to the microscope slide.

8. Fluorescent microscope equipped with a filter sets to visualize green fluorescence of dual-tag protein-bound Lumio, fluorescence of secondary antibody bound to the putative interacting partner, and blue for DAPI (visualizes chromatin in nucleus).

2.4. In-Gel Lumio Detection of Dual-Tagged Proteins Separated by SDS-PAGE

1. Lysis buffer BF3-FT: 50 mM Tris–HCl, 150 mM NaCl, 50 mM NaH_2PO_4, 10 mM imidazole, 0.1% Nonidet™-P40 (NP-40), 10% glycerol, adjusted to pH 8.0. Store at 4°C. Before lysis, add the following fresh: 10 mM β-mercaptoethanol (BME), Complete Mini (EDTA-free) protease inhibitor cocktail (Roche, Switzerland), 1 mM phenylmethylsulphonyl fluoride (PMSF), and 50 μg/mL avidin (blocks biotinylated proteins to reduce nonspecific interaction with the Strep-Tactin affinity beads).
2. Bio-Rad Protein Assay (Hercules, CA) based on the Bradford Assay for protein concentration measurement.
3. Lumio™ Green Detection Kit (Invitrogen) containing Gel Sample Buffer, Lumio Reagent, and Lumio In-Gel Detection Enhancer.
4. Sodium dodecyl sulfate (SDS) polyacrylamide gel (purchased or made in house) at a percentage compatible with the dual-tagged protein. Standard electrophoresis apparatus and running buffers.
5. UV transluminator (UVP, Upland, CA) to visualize the dual-tagged protein in the gel.

2.5. Purification of Dual-Tagged Proteins with Affinity Columnsa

1. Affinity beads for purification by outer tag relative to the tagged proteins (**Fig. 1a**): Ni-NTA Superflow (Qiagen, Valencia, CA) for His-tag, immunoglobulin G (IgG) Sepharose™ 6 Fast Flow (GE Healthcare, Piscataway, NJ) for ProA-tag.
2. Poly-prep columns (Bio-Rad), empty, 10 mL maximum volume.
3. His-tag wash buffer (HTWB): 50 mM Tris–HCl, 50 mM NaH_2PO_4, 150 mM NaCl, 20 mM imidazole, 1 mM DTT, 0.1% NP-40, pH 8.0.
4. ProA-tag wash buffer (PTWB): same as HTWB but without imidazole and DTT.
5. TEV cleavage buffer (TCB): 50 mM Tris–HCl, 150 mM NaCl, 0.5 mM EDTA, 0.1% NP-40, 1 mM DTT, pH 8.0.
6. AcTEV protease (Invitrogen).
7. Affinity beads for purification by inner tag relative to the tagged protein (**Fig. 1a**): Strep-Tactin Superflow (IBA, St. Louis, MO) for StrepII-tag® purification; HA.11 antibody

(BabCo, Berkeley, CA) conjugated to ProA beads (Pierce, Rockford, IL) for HA-tag purification.

8. StrepII-tag wash buffer (STWB): 100 mM Tris-HCl, pH 8.0, 150 mM NaCl, 1 mM EDTA, 1 mM DTT. Note: compatible with HA-tag purification.
9. StrepII-tag elution buffer (STEB): 100 mM Tris-HCl, pH 8.0, 150 mM NaCl, 1 mM EDTA, 1 mM DTT, 20 mM desthiobiotin (IBA).
10. HA-tag elution buffer (HAEB): 80% acetonitrile, 0.025% formic acid in ddH_2O.

3. Methods

We generated five novel dual-tag affinity purification vectors, each with different combinations of affinity tags (two per construct, varying by composition, size, and terminal location) and promoters (constitutive expression [CMV_p]; C′-terminal tags; tetracycline-regulatable expression [Tet_p]; N′-terminal tags) to control expression of the bait proteins (**Fig. 1a**). We chose the Strep-Tactin-binding peptide (StrepII tag) in most of our dual-tag constructs because of its small (8-mer) size compared with the original streptavidin-binding peptide (Strep tag; 38-mer) in an effort to lessen its potential impact on the function of the tagged proteins or their specific interaction with other proteins. Other features include the incorporation of a second TEV protease recognition sequence to enhance cleavage efficiency and a tetracysteine motif (CCPGCC) to easily monitor bait protein expression and localization *(17)*. In addition, all of our dual-tags are constructed in Gateway®-compatible destination vectors, which allows for quick and simple cloning via site-specific recombination *(18)*. Once introduced into cells, dual-tagged bait proteins and their interacting partners can be visualized and purified (see below). The resultant complexes are preferably analyzed by Multidimensional Protein Identification Technology (MudPIT) *(19)* followed by tandem mass spectrometry (MS/MS), but may be analyzed by whatever means available to the researcher.

3.1. Gateway Cloning

1. Obtain the gene of interest (ORF of protein to be dual-tagged) to create a Gateway-compatible entry clone (see manufacturer's instructions). Two options are available: (a) construct Gateway-compatible primers that contain *attB* recombination sites flanking the 5′ and 3′ of the gene to be tagged; perform polymerase chain reaction (PCR) to amplify the ORF, such as TRF2, and perform a BP recombination

reaction to swap the PCR product into a Gateway DONR vector (pDONR221) or (b) buy gene of interest in a Gateway entry vector (if available, such as from ATCC, Manassas, VA).

2. Perform an LR recombination reaction with your entry clone and one of the dual-tag destination vectors (**Fig. 1a**), following the Gateway protocol.
3. Once verified (sequenced or digested with diagnostic restriction enzymes), transfect cell line of interest with the dual-tagged expression construct, and check for expression (*see* **Subheading 3.2** or **3.4**).
4. Dual-tagged vectors may be used for transient, constitutive expression of the protein of interest, but the ability to regulate expression (N′ terminal-based dual-tags) is better realized in tetracycline-regulatable stable cell lines (*see* **Note 1**).

3.2. Detection of Dual-Tagged Proteins in Live Cells by Lumio

1. Cells expressing the dual-tag construct (either transiently or stably) are grown to >50% confluence in a 35-mm cell culture plate (or on a coverslip if one wishes to eventually fix the cells and costain them with another probe (*see* **Subheading 3.3**) such as an antibody for immunofluorescence [IF] or a DNA probe for fluorescence in situ hybridization [FISH]) *(16)*. Proceed with Lumio staining as described below (modified version of protocol presented by Rudner et al. *(20)*).
2. Prepare the labeling solution by combining 1 μL of 2 mM FlAsH-EDT_2 (Lumio reagent) with 1 μL of 25 mM EDT and 1 μL DMSO. Incubate labeling solution for 15 min at RT in the dark.
3. Dilute the labeling solution to 200 μL with D-PBS (*see* **Note 2**) and incubate for 10 min at RT in the dark.
4. Wash cells expressing the dual-tagged protein of interest with warm PBS to remove serum proteins (which may cause nonspecific background staining) and apply 1.8 mL of warm PBS (with Ca^{++} and Mg^{++}) to the cells.
5. Apply the 200 μL labeling solution prepared above to the cells and place the culture dish back in the cell culture incubator (37°C with 5% CO_2) for 30 min (*see* **Note 3**).
6. Remove labeling solution (2 mL) from labeled cells and wash once with regular, prewarmed PBS (2 mL) and three times for 10 min each with PBS containing 350 μM EDT (2 mL for each wash). These three 10-min washes help remove Lumio reagent that is weakly bound to nonspecific proteins, resulting in a reduction of background fluorescence.
7. Apply 2 mL Opti-MEM reduced serum media to the cells when ready to visualize (*see* **Note 4**).

8. Visualize the labeled dual-tagged protein in live cells using a fluorescence microscope outfitted with an immersion lens (*see* **Note 5**) and a filter set that is compatible with FITC excitation and emission spectra (508 nm excitation, 528 nm emission). Acquire images using a charge-coupled device (CCD) camera attached to the microscope. An example of a Lumio-stained live cell expressing a TRF2-C-StH fusion protein is shown in **Fig. 1b**.
9. After live cell imaging, cells may fixed (if grown on coverslips) with 2% paraformaldehyde and labeled by IF or FISH (*see* **Subheading 3.3** or **ref.** *(16)*.

3.3. Colocalization Studies in Fixed Cells Pairing Lumio-Based Detection of Dual-Tagged Proteins with Immunofluorescent Detection of Putative Interacting Partners

1. Cells expressing the dual-tag construct are grown to >50% confluence on a microscope coverslip placed in a 35-mm culture dish.
2. Live cells expressing the dual-tagged protein are first stained with the Lumio reagent as described in **Subheading3.2** above (**steps 2–6**). If desired, the live cells may be imaged (*see* **Subheading 3.2**, **steps** 7–**8**) before fixation to check for proper expression, localization, and/or staining pattern of the dual-tagged fusion protein.
3. The following steps should be performed with the lights off because Lumio signal may decrease with light exposure.
4. Wash Lumio-stained cells three times with 2 mL of ice cold PBS to remove all traces of PBS-EDT or OPTI-MEM (*see* **Subheading 3.2**, **step 6** or 7, respectively).
5. Fix cells by applying 2 mL of a 2% paraformaldehyde solution (in PBS) to the culture dish (*see* **Note 6**). Incubate for 10 min at RT in the dark. Wash fixed cells three times with 2 mL of ice-cold PBS.
6. Permeabilize fixed cells with 2 mL of a 0.5% NP-40 solution (in PBS) for 10 min at RT. Wash cells three times with 2 mL of ice-cold PBS.
7. Block permeabilized cells with 2 mL of a 1% IgG-free BSA solution (in PBS) for 30 min at RT.
8. Prepare a primary antibody solution (against another protein of interest/putative interacting partner of the dual-tagged protein) containing the recommended antibody dilution for IF (see manufacturer's antibody information sheet). Primary antibody should be diluted in the blocking solution used in **step** 7. In this study, we used mouse anti-Tin2 (a telomere-specific protein) antibody with a dilution of 1:100 (5 μg/mL).
9. Apply 150 μL of primary antibody solution to the coverslip and incubate for 1 h at 37°C in the dark (or overnight at

4°C, *see* **Note 7**). Wash cells three times with 2 mL of ice-cold PBS.

10. Choose a secondary antibody that specifically reacts with the primary antibody. For IF, the secondary antibody must have a conjugated fluorophore that emits in the red spectrum (because the Lumio reagent in this study emits in the green spectrum). Dilute the secondary antibody in blocking solution to the manufacturer's recommended concentration. In our case, we used Alexa-Fluor 594 donkey anti-mouse at a 1:1,500 dilution.
11. Apply 150 μL of secondary antibody solution to the coverslip and incubate at 37°C for 1 h in the dark. Wash cells three times with 2 mL of ice-cold PBS.
12. Mount coverslip onto a microscope slide using Vectashield + DAPI and seal the edges with clear nail polish. Allow seal to dry completely (in the dark) before visualization.
13. Visualize labeled cells (Lumio-labeled dual-tagged protein, IF-labeled putative interacting partner) with a fluorescent microscope equipped with proper filters for Lumior (green), conjugated secondary antibody (red) and DAPI (blue), acquire separate images of each protein and DNA with a CCD microscope camera, and merge images to check for colocalization of the dual-tagged protein with its proposed interacting partner. An example of a dual-tagged TRF2 and Tin2 colocalization is shown in **Fig. 1c**.

3.4. In-Gel Lumio Detection of Dual-Tagged Proteins Separated by SDS-PAGE

1. Cells expressing the dual-tag construct are grown to >90% confluence (*see* **Note 8**) in 10-cm culture plates. Remove growth medium and wash once with 10 mL ice-cold PBS. Remove PBS and apply 2–3 mL of fresh ice-cold PBS. Using a rubber policeman, scrap cells off the culture plate and move cell/PBS solution to a 15-mL Falcon tube on ice. For a quick alternative if using loosely adherent cells such as 293T, *see* **Note 9**.
2. Cells are then pelleted in a 4°C centrifuge for 5 min at 500 × *g*. Remove supernatant and wash cells with exactly 10 mL of ice-cold PBS. Use this opportunity to estimate the packed cell volume (PCV) using the graduations on the pipette once the cells have been resuspended. Note the PCV. Centrifuge the cells again at 500 × *g* for 5 min at 4°C.
3. Remove the supernatant and apply twice the PCV of BF3-FT supplemented with fresh BME, PMSF, protease inhibitor, and avidin. Resuspend the cells with the pipette and place the tube on ice for 10 min.
4. Subject the crude lysate to three freeze/thaw cycles using liquid nitrogen (or ethanol/dry ice bath) to freeze and 4°C

water bath to thaw, breaking up clumps with a 1-mL pipette between each cycle.

5. Pass the lysate through a 25-gauge needle (*see* **Note 10**) to shear DNA and break up clumps. Take care to keep the lysate on ice (or in a cold room at 4°C). Centrifuge the Falcon tube with the processed lysate at 4,000 × *g* for 10 min at 4°C to remove cell debris, and transfer the crude lysate (supernatant) to a microfuge tube(s).
6. Preclear the lysate by centrifuging at 21,000 × *g* (or highest *g*-value available) for 30 min at 4°C. Transfer the supernatant to another clean microfuge tube and repeat the centrifugation. Again, transfer the supernatant to a clean microfuge tube and place on ice. Take care to avoid the pellet when transferring the supernatant at each step. The pellet contains insoluble proteins that can complicate and contaminate your analysis at later steps.
7. Determine the protein concentration by your preferred method, such as the Bradford Assay (*see* **Note 11**).
8. Prepare lysate for separation by SDS-polyacrylamide gel electrophoresis (PAGE) and detection using the Lumio Green Detection Kit as described by the manufacturer's protocol. Briefly, dilute 40 μg of protein lysate into 15 μL (total volume) of ddH_2O. Add 5 μL of 4× Lumio Gel Sample Buffer. Add 0.2 μL Lumio Green Detection Reagent to the 20-μL sample and incubate at 70°C for 10 min. Allow sample to cool (1–2 min) and add 2 μL Lumio In-Gel Detection Enhancer; incubate for 5 min at RT.
9. Pour an SDS-polyacrylamide gel with an acrylamide concentration compatible with the size of the dual-tagged protein. For example, a 10% gel (with 1 cm of a 5% stacking gel) will retain proteins >25 kDa if run for 1 h at 180 V. Load the sample and rainbow molecular weight marker (GE Healthcare Biosciences) onto the gel and run for 1 h at 180 V.
10. Remove the gel from the running apparatus, dry off the glass plates, and, using a standard office highlighter, mark the visible bands of the rainbow marker. These will fluoresce once exposed to the UV light of the transluminator.
11. Place the gel (still in the glass sandwich plate) on a UV transluminator and expose to UV at a wavelength of 302 nm (*see* **Note 12**). Using the ethidium bromide or SYBR® Green filter and a compatible camera, expose the gel for 4–10 s, adjusting brightness and contrast as necessary, and take a picture. A fluorescent band representing the dual-tagged protein should be visible (if it is above the Lumio detection limit of 1 pmol). Examples of SDS-PAGE separated,

Lumio-labeled dual-tagged TRF2 fusion proteins are shown in **Fig. 1d**.

12. If no band is visible, refer to the manufacturer's trouble-shooting section or perform a traditional Western blot, probing the membrane with an anti-StrepII tag primary antibody. See *(16)* for more details.
13. Once the lysate is verified to contain the dual-tagged protein, proceed to **Subheading 3.5** for purification of the dual-tagged protein and its interacting partners (*see* **Note 13**).

3.5. Purification of Dual-Tagged Proteins with Affinity Columns

1. Prepare precleared, concentrated lysate as described in **Subheading 3.4, steps 1–7**. Retain some lysate (at least 40 μg) to monitor purification progress by Lumio In-Gel Detection or Western blot if desired (recommended for initial attempts).
2. Aliquot beads (200 μL solid beads per 1.5 mL lysate; Ni-NTA for His-tag, IgG Sepharose™ 6 Fast Flow for ProA-tag) specific to the outer affinity tag (e.g., Ni-NTA for N-HtS-TRF2) into clean microfuge tube(s) and wash three times with 1 mL of lysis buffer BF3-FT (*see* **Note 14**), centrifuging at 350 × *g* between each wash to precipitate the beads.
3. Directly add prepared lysate to the washed beads and place sample(s) on a nutating platform for 2 h at 4°C. The dual-tagged protein and its interacting partners will be affinity purified by the interaction between the outer tag of the dual-tagged protein and the beads.
4. With a pipette, transfer the beads and supernatant to an empty, capped Poly-prep column. Collect the flow-through (lysate depleted of the dual-tag protein) by uncapping the column over a clean microfuge tube. If desired, retain the flow-through and load an amount equivalent to the lysate on a gel and perform a Western blot to check the efficiency of the pull-down (*see* **Note 15**).
5. Wash the beads in the column with three 10-mL additions of respective wash buffer (His-tag: HTWB; ProA-tag: PTWB) to remove nonspecifically bound proteins (*see* **Note 16**).
6. Equilibrate beads with one 10-mL addition of TEV cleavage buffer (TCB). Cap the column and add 1 mL of TCB to the beads. Resuspend beads with gentle pipetting and transfer bead/TCB solution to another clean microfuge tube.
7. Add 50 U (5 μL) of AcTEV protease and place sample(s) on a nutating platform for 1 h at RT to cleave/elute the dual-tagged bait protein and its associated partners from the beads (*see* **Note 17**).

8. Aliquot 200 μL solid beads (i.e., 400 μL of a 50% slurry) (Strep-Tactin beads for StrepII-tag, HA antibody conjugated to ProA beads for HA-tag) specific to the inner affinity tag (e.g., StrepII-tag for N-HtS-TRF2) into clean microfuge tube(s) and wash three times with 1 mL of TCB, centrifuging at 350 × *g* between each wash to precipitate the beads.
9. Centrifuge the TEV protease-treated sample at 350 × *g* to precipitate the beads, collect the supernatant (containing the freed dual-tagged protein), and place atop washed beads specific to the inner affinity tag (**step 8**). When pipetting the supernatant, be sure to avoid the precipitated beads. Place samples on a nutating platform for 2 h at 4°C.
10. With a pipette, transfer both beads and supernatant to a new, Poly-prep column. If desired, collect the flow-through to check the efficiency of the pull-down by Western blot. Wash the beads in the column to remove nonspecifically bound proteins with three 1-mL additions of ice cold STWB.
11. Cap the column and add 500 μL elution buffer (StrepII-tag: STEB; HA-tag: HAEB). Agitate the beads by holding the column and gently tapping the side. Incubate at RT for 5 min. Uncap the column and collect the flow-through (Elution 1). Repeat **step 11** two more times for a total of three elutions (1,500 μL) (*see* **Note 18**). Elution contains dual-tagged protein and interacting partners. An example depicting the recovery of a dual-tagged TRF2 protein is shown in **Fig. 2**.
12. As described in Giannone et al. *(16)*, the purified complexes are precipitated by trichloroacetic acid (TCA), digested with endoproteinase Lys-C and trypsin, and analyzed by two-dimensional (2D) LC-MS/MS. However, because the dual-tag purification process is performed in native conditions, eluted complexes may be analyzed by whatever means available to the researcher.

4. Notes

1. To regulate expression, the recipient cell line must be made T-REx compatible (or purchased from a company such as Invitrogen, if available). T-REx cell lines express a stably integrated tetracycline (Tet) repressor element (pcDNA™6/TR) that sits on the Tet operator sequence of N′-terminal dual tags and blocks expression until tetracycline is introduced to the culture media. The repressor element is very responsive

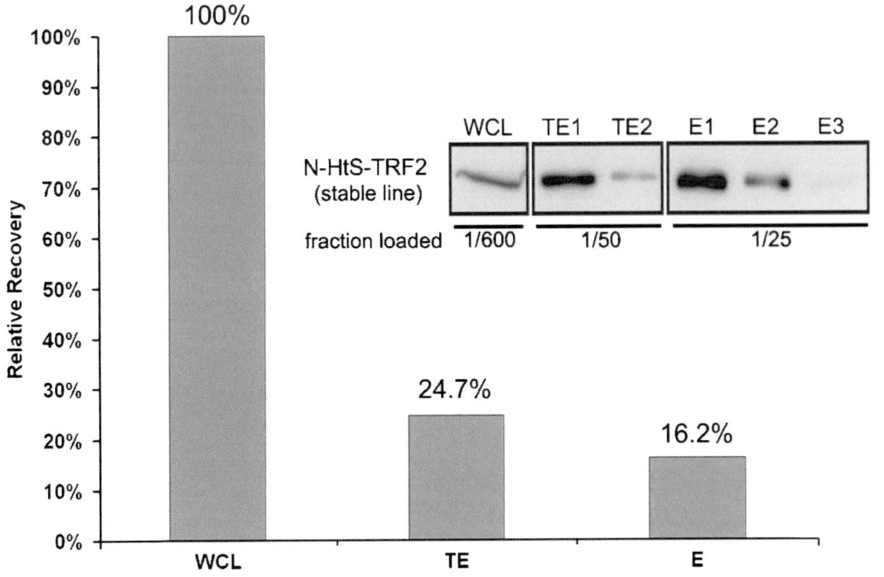

Fig. 2. Efficient purification of dual-tagged TRF2. Western blot (anti-StrepII tag) depicting the complete purification of stably expressed N-HtS-TRF2 in 293T T-REx cells, obtained by freeze/thaw lysis. *Bar graph* represents the estimated recovery of bait protein relative to input from the lysate. *WCL* whole-cell lysate; *TE* TEV protease-mediated elution(s) from Ni-NTA resin; *E* eluates from Strep-Tactin resin (Reproduced from **ref. *16*** with permission from BioTechniques, Informa Life Sciences Group).

to Tet concentration and thus dual-tag protein expression level can be titrated from levels as low as 10 ng/μL *(16)*. To create stable, tetracycline-regulatable systems, cells that have successfully integrated both the dual-tag construct and the Tet repressor element can be selected with G418 and Blasticidin-S, respectively. With our 293T cells, 800 μg/mL (selection) and 500 μg/mL (maintenance) of G418 and 5 μg/mL Blasticidin-S were used for selection/maintenance, but these must be adjusted accordingly depending on the cell line used.

2. The addition of calcium and magnesium in PBS helps maintain cell adhesion. Incubation for a period of time with regular PBS may lead to the cells lifting off the culture plate, especially cells that are loosely attached to begin with.
3. Thirty minutes is a good starting point, but this time should be optimized to your particular system because different cells and different dual-tagged proteins will vary in the amount of time needed to maximize signal while minimizing background.
4. The Lumio™ Green In-Cell Detection Kit provides a reagent called Disperse Blue that can be added to the Opti-MEM

before live cell visualization. Disperse Blue is a nonfluorescent, hydrophobic dye that reduces background fluorescence and may enhance detection of your dual-tagged protein.

5. Although not used in our laboratory, a better way to visualize the dual-tagged protein in live cells is to use an inverted fluorescent microscope.
6. Although 2% PFA was used in the study described in Giannone et al. *(16)*, it may not always be the best fixative. Other concentrations of PFA (a crosslinker) as well as ice-cold methanol (a dehydrator) should be tried if the putative interacting partner you are trying to visualize by IF is not compatible and/or exhibits no fluorescent signal.
7. If staining overnight, it is imperative that you not let the antibody solution evaporate. One option is to completely submerge the coverslip with antibody solution, but this can be expensive and a waste of reagent. Another option is to take a plastic sheet and cut out squares about the same size of, but preferably smaller than the coverslip. After applying 150 μL of your antibody solution to the coverslip, carefully place the plastic square on top. This will help prevent evaporation and also help spread the antibody solution across the coverslip. Also, use Parafilm to seal the plate containing your coverslip before overnight incubation.
8. If working with an inducible dual-tag construct, especially if it is a 293T-TREx stable line, grow the cells to approximately 70% confluence and then induce with tetracycline overnight. The cells should be fully confluent and ready to harvest the next morning.
9. Loosely adherent cells such as 293T can be removed easily with a pipette. To harvest these types of cells, remove the growth media and replace with ice-cold PBS. Aspirate the PBS with a pipette and dispense at full force, essentially blasting the cells off the plate. Repeat to remove all of the cells.
10. The lysate will most likely contain chromatin clumps that will clog the 25-gauge needle. If this is the case, remove the plunger on the syringe and use your 1-mL pipette to aspirate the lysate and dispense it into the syringe. Place the syringe over the lysate's original tube, replace the plunger, and force the lysate through the needle. Usually one pass is good enough before aspiration through the needle is attainable. Pass the lysate through the needle three times.
11. Performing this 2× PCV freeze/thaw lysis procedure on 293T cells, we usually obtain crude lysate concentration of approximately 18–20 μg/μL. This highly concentrated lysate improves bait capture during the first purification, especially with the low-affinity His-tag purification by Ni-NTA beads.

12. The actual excitation wavelength of Lumio is 508 nm and thus a more robust signal could be obtained using a laser-based scanner with an excitation maxima compatible with the stain.
13. If not proceeding to the dual-tag purification procedure outlined in **Subheading 3.5**, other quicker lysis procedures may be used. The main purpose of the freeze/thaw method is to obtain a very concentrated lysate, which, in our hands, was beneficial to the purification.
14. If performing the dual-tag purification with a ProA-tagged protein, try to avoid using reducing agents such as β-ME or DTT in the lysis buffer. These can reduce the disulfide bonds in the IgG molecules conjugated to the sepharose beads, leading to IgG contamination in the final eluate.
15. We have also successfully performed the whole purification procedure using a batch purification approach (without using columns). Although this method reduces sample handling (and potential protein loss on the membrane of the column), it is harder to wash the beads and care must be taken to avoid bead contamination in the final elution.
16. Notice that all buffers besides the lysis buffer avoid protease inhibitors. This is necessary because TEV protease is a required step in the purification. In addition, if the final detection step involves bottom-up mass spectrometry, residual protease inhibitors will inhibit trypsin.
17. An alternative method that sometimes increases recovery after TEV cleavage is to perform two consecutive 30-min TEV cleavage reactions, using 500 μL of TCB and 50 U of AcTEV protease for each incubation. In addition, if your protein of interest is particularly prone to degradation, you may perform the cleavage at 4°C, but incubation times should be increased (~3 h) to compensate for the loss of enzymatic activity.
18. In our experience, >90% of the dual-tag protein is obtained in elutions 1 and 2. Optimize accordingly.

Acknowledgments

We acknowledge Drs. Hayes McDonald and Gregory Hurst for help with the mass spectrometry analysis presented in our original manuscript and Ying Huang for technical support. The authors acknowledge the support of the Laboratory Directed Research and Development Program (LDRD) of Oak Ridge

National Laboratory, and the Office of Biological and Environmental Research, U.S. Department of Energy, under Contract DE-AC05–00OR22725 with UT-Battelle, LLC, and the DOE Genomics:GTL grants, respectively.

References

1. Aebersold, R. & Mann, M. (2003). Mass spectrometry-based proteomics. *Nature* **422**, 198–207.
2. Puig, O., Caspary, F., Rigaut, G., Rutz, B., Bouveret, E., Bragado-Nilsson, E., Wilm, M. & Seraphin, B. (2001). The tandem affinity purification (TAP) method: A general procedure of protein complex purification. *Methods* **24**, 218–229.
3. Rigaut, G., Shevchenko, A., Rutz, B., Wilm, M., Mann, M. & Seraphin, B. (1999). A generic protein purification method for protein complex characterization and proteome exploration. *Nat Biotechnol* **17**, 1030–1032.
4. Gavin, A. C., Bosche, M., Krause, R., Grandi, P., Marzioch, M., Bauer, A., Schultz, J., Rick, J. M., Michon, A. M., Cruciat, C. M., Remor, M., Hofert, C., Schelder, M., Brajenovic, M., Ruffner, H., Merino, A., Klein, K., Hudak, M., Dickson, D., Rudi, T., Gnau, V., Bauch, A., Bastuck, S., Huhse, B., Leutwein, C., Heurtier, M. A., Copley, R. R., Edelmann, A., Querfurth, E., Rybin, V., Drewes, G., Raida, M., Bouwmeester, T., Bork, P., Seraphin, B., Kuster, B., Neubauer, G. & Superti-Furga, G. (2002). Functional organization of the yeast proteome by systematic analysis of protein complexes. *Nature* **415**, 141–147.
5. Ahn, S. G., Kim, S. A., Yoon, J. H. & Vacratsis, P. (2005). Heat-shock cognate 70 is required for the activation of heat-shock factor 1 in mammalian cells. *Biochem J* **392**, 145–152.
6. Cox, D. M., Du, M., Guo, X., Siu, K. W. & McDermott, J. C. (2002). Tandem affinity purification of protein complexes from mammalian cells. *Biotechniques* **33**, 267–268, 270.
7. Davey, F., Hill, M., Falk, J., Sans, N. & Gunn-Moore, F. J. (2005). Synapse associated protein 102 is a novel binding partner to the cytoplasmic terminus of neurone-glial related cell adhesion molecule. *J Neurochem* **94**, 1243–1253.
8. Lesca, C., Germanier, M., Raynaud-Messina, B., Pichereaux, C., Etievant, C., Emond, S., Burlet-Schiltz, O., Monsarrat, B., Wright, M. & Defais, M. (2005). DNA damage induce gamma-tubulin-RAD51 nuclear complexes in mammalian cells. *Oncogene* **24**, 5165–5172.
9. Oshikawa, K., Matsumoto, M., Yada, M., Kamura, T., Hatakeyama, S. & Nakayama, K. I. (2003). Preferential interaction of TIP120A with Cul1 that is not modified by NEDD8 and not associated with Skp1. *Biochem Biophys Res Commun* **303**, 1209–1216.
10. Westermarck, J., Weiss, C., Saffrich, R., Kast, J., Musti, A. M., Wessely, M., Ansorge, W., Seraphin, B., Wilm, M., Valdez, B. C. & Bohmann, D. (2002). The DEXD/H-box RNA helicase RHII/Gu is a co-factor for c-Jun-activated transcription. *Embo J* **21**, 451–460.
11. Burckstummer, T., Bennett, K. L., Preradovic, A., Schutze, G., Hantschel, O., Superti-Furga, G. & Bauch, A. (2006). An efficient tandem affinity purification procedure for interaction proteomics in mammalian cells. *Nat Methods* **3**, 1013–1019.
12. Drakas, R., Prisco, M. & Baserga, R. (2005). A modified tandem affinity purification tag technique for the purification of protein complexes in mammalian cells. *Proteomics* **5**, 132–137.
13. Knuesel, M., Wan, Y., Xiao, Z., Holinger, E., Lowe, N., Wang, W. & Liu, X. (2003). Identification of novel protein-protein interactions using a versatile mammalian tandem affinity purification expression system. *Mol Cell Proteomics* **2**, 1225–1233.
14. Li, Q., Dai, X. Q., Shen, P. Y., Cantiello, H. F., Karpinski, E. & Chen, X. Z. (2004). A modified mammalian tandem affinity purification procedure to prepare functional polycystin-2 channel. *FEBS Lett* **576**, 231–236.
15. Zhou, D., Ren, J. X., Ryan, T. M., Higgins, N. P. & Townes, T. M. (2004). Rapid tagging of endogenous mouse genes by recombineering and ES cell complementation of tetraploid blastocysts. *Nucleic Acids Res* **32**, e128.
16. Giannone, R. J., McDonald, W. H., Hurst, G. B., Huang, Y., Wu, J., Liu, Y. & Wang, Y. (2007). Dual-tagging system for the affinity purification of mammalian protein complexes. *Biotechniques* **43**, 296–302.
17. Griffin, B. A., Adams, S. R. & Tsien, R. Y. (1998). Specific covalent labeling of recombinant protein molecules inside live cells. *Science* **281**, 269–272.

18. Hartley, J. L., Temple, G. F. & Brasch, M. A. (2000). DNA cloning using in vitro site-specific recombination. *Genome Res* **10**, 1788–1795.

19. McDonald, W. H., Ohi, R., Miyamoto, D. T., Mitchison, T. J. & Yates, J. R., III (2002). Comparison of three directly coupled HPLC MS/MS strategies for identification of proteins from complex mixtures: Single-dimension LC–MS/MS, 2-phase MudPIT, and 3-phase MudPIT. *Int. J. Mass Spectrom* **219**, 245–251.

20. Rudner, L., Nydegger, S., Coren, L. V., Nagashima, K., Thali, M. & Ott, D. E. (2005). Dynamic fluorescent imaging of human immunodeficiency virus type 1 gag in live cells by biarsenical labeling. *J Virol* **79**, 4055–4065.

Chapter 29

Use of Genomic DNA as Reference in DNA Microarrays

Yunfeng Yang

Summary

DNA microarray has become a mainstream technology to explore gene expression profiles, identify novel genes involved in a biological process of interest and predict their function, and determine biomarkers that are relevant to a given phenotype or disease. Typical two-channel microarray studies use an experimental design called the complementary DNA (cDNA) reference method, in which samples from test and control conditions are compared directly on a microarray slide. A substantial limitation of this strategy is that it is nearly impossible to compare data between experiments because the reference sample composition is subjected to changes at the level of experimental design and thereby not consistent from one experiment to another. Using genomic DNA as common reference will effectively overcome this limitation. This chapter describes detailed methods to prepare genomic DNA of high quality, label with fluorescent dye, co-hybridize with cDNA samples, and the subsequent data analyses. In addition, notes are provided to help the readers to obtain optimal results using the procedure.

Key words: DNA microarray, Genomic DNA, Universal reference, Microbe

1. Introduction

The advancement of sequencing technology has enabled the whole-genome sequencing at unprecedented speed. Subsequently, DNA microarray technology has been quickly adopted widely among the scientific community to explore gene expression profiling in the sequenced organisms *(1, 2)*. In two-channel DNA microarray experiments, both experimental and reference RNA samples are labeled with two different fluorescent dyes (typically Cy5 and Cy3), either directly or after reverse transcription into complementary DNA (cDNA) molecules. Then they are simultaneously hybridized with immobilized probes on microarray slides *(3)*. The ratios of signal intensities of the two fluorescently labeled

James Weifu Lee and Robert S. Foote (eds.), *Micro and Nano Technologies in Bioanalysis*,
Methods in Molecular Biology, vol. 544
DOI 10.1007/978-1-59745-483-4_29, © Humana Press, a part of Springer Science+Business Media, LLC 2009

cDNA representing the relative abundance of transcripts are calculated and interpreted for biological meanings. When a large number of samples needs to be compared, pair-wise comparisons across all samples are often desired. Simple math indicates that pairing all of the possible pairs for n samples yields a total of $n*(n-1)/2$ combinations, which is polynomially proportional to n. As a result, this approach is very costly and tedious for large numbers of samples. For example, a striking number of 4,950 microarrays are needed for comparing 100 samples. In addition, because feature geometry varies between DNA microarrays and there are differences of experimental design in different laboratories, it is nearly impossible to compare data across platforms, experiments, and laboratories. It has been desired for a long time to develop novel strategies to integrate data across multiple, initially unrelated studies between laboratories or over a long period of time to promote data sharing and integration.

A conceptually sound solution to the problems is to use "reference design," which adopts a common reference that is co-hybridized with each sample during microarray experiments. Typically, the ratio ($\gamma 1$) of signal intensities of cDNA over a common reference is compared with another ratio ($\gamma 2$) of signal intensities of cDNA over a common reference. The computed "ratio of ratios" ($\gamma 1/\gamma 2$) is mathematically equivalent to direct cDNA:cDNA comparisons. Only n microarrays are needed to calculate the ratios of any possible pairs of n samples. In contrast to the previous example, merely 100 microarrays are needed for 100 samples. Apparently, this strategy greatly reduces the costs and time incurred by direct ratiometric microarray experiments.

An excellent reference approach should fulfill the requirements of universality, reproducibility, and uniformity. That is, it should be universal across diverse microarrays in different laboratories, reproducible over a long period of time and by different researchers, and represent each gene at a uniform level. In practice, a commonly used reference is common RNA pools assembled from a number of different cell lines, tissues, and conditions, which are now commercially available for mouse and human samples (Stratagene). However, the RNA references fall well short of the aforementioned criteria. RNA is instable, and many RNAs could be underrepresented under a given condition. On the other hand, some RNAs are so prevalent that they saturate the hybridization to the DNA probes of the microarray. Although RNA pools are more comprehensive than a single source of RNA sample, it still partially represents the whole genome; and there is inherent biological variability among different RNA samples, making it difficult to reproduce faithfully for different preparations. As a result, data quality across multiple studies is inevitably compromised.

In light of these shortcomings, genomic DNA is thought to be a better choice of common reference *(4)*. It is easy and economic to prepare genomic DNA in large amounts; genomic DNA is stable and has a good shelf life; and it is independent of variations from one preparation to another, which is a desirable feature of a universal reference. In addition, genomic DNA represents entire genome completely and fairly uniformly, because the majority of genes are presented once in the prokaryotic genomes, or twice in most eukaryotic genomes. Several recent studies have proven that genomic DNA reference is indeed very effective and faithful to report gene expression profiles *(5–10)*. Furthermore, a comparative study between a genomic DNA reference and a universal RNA reference has reached the conclusion that genomic DNA is superior for routine use *(8)*.

Nevertheless, adopting genomic DNA as a reference also creates problems and challenges of its own. It is conceivable that this strategy enables the integration of disparate studies, but it brings in new variations. For example, spots with low signal intensity from labeled genomic DNA are prone to high standard errors for measurements, and spots with high intensity considerably interfere with the hybridization of cDNA samples to the probes via binding to cDNA or the probes, leading to low fidelity in the ratio of cDNA to genomic DNA. Meanwhile, the prevalent coexistence of exon and intron in large vertebrate genomes is also problematic. Because cDNA is free of introns, the presence of introns in the genomic DNA elevates noise and adversely affects absolute signal level. In addition, the existence of repetitive sequences in the genome also exerts unwanted effects on the signal quality. These complications are not present in small bacterial or fungal genomes, which have few or a limited number of intergenic regions and repetitive sequences in the genome. It has been observed that genomic DNA reference is very effective for bacteria, but is less effective for plants and higher animals *(5, 8)*.

This chapter aims to provide a detailed protocol to perform microarray experiments with a genomic DNA reference. It also includes an updated data analysis protocol that has been carefully tested in my group to improve data quality.

2. Materials

2.1. Preparation of Genomic DNA and Total RNA

1. Temperature-adjustable water bath.
2. 50 mM and 0.5 mM ethylenediamine tetraacetic acid (EDTA) in water.

3. 20 μg/μl lyticase (for yeast) (Sigma) or 100 μl of 20 μg/μl lysozyme (for gram-positive bacteria) (Fisher Scientific) in water.
4. Buffer S: 100 mM Tris-HCl, pH 8.0, 100 mM EDTA, pH 8.0, 1.5 M NaCl, 1% cetyltrimethyl-ammonium bromide (CTAB).
5. 10 μg/μl proteinase K (Invitrogen) in water.
6. 20% sodium dodecyl sulfate (SDS) solution in water (Ambion).
7. Phenol:chloroform:isoamyl alcohol (25:24:1) (Fisher Scientific).
8. Chloroform.
9. Isopropanol and ethanol.
10. 70% ethanol in water.
11. 3 M sodium acetate in water.
12. DNAse-free RNase A (1 μg/ml) (Ambion).
13. RNAse-free DNAse I (2 U/μl) (Ambion).
14. Trizol (Invitrogen).
15. RNeasy kit (Qiagen).
16. Rnase-free H_2O (Ambion).

2.2. DNA Labeling and Purification

1. Temperature-adjustable water bath.
2. SpeedVac Concentrator.
3. 2.5× random primer (Invitrogen) and 3 μg/μl random primer (Invitrogen).
4. dNTP mix for genomic DNA labeling: 5 mM dATP, 5 mM dGTP, 5 mM dCTP, and 2.5 mM dTTP. It should be stored at −20°C.
5. dNTP mix for cDNA labeling: 10 mM dATP, 10 mM dGTP, 10 mM dCTP, and 0.5 mM dTTP. It should be stored at −20°C.
6. Klenow fragment of DNA polymerase I (Invitrogen).
7. Cy3-dUTP and Cy5-dUTP (Amersham/Pharmacia).
8. 0.1 M Dithiothreitol (DTT) (Invitrogen).
9. RNAse inhibitor (Invitrogen).
10. SuperScript II H reverse transcriptase (Invitrogen).
11. Polymerase chain reaction (PCR) purification kit (Qiagen).

2.3. Microarray Experiments

1. Temperature-adjustable water bath.
2. Microarray slide (Corning).
3. Cover slip (Sigma).
4. Isopropanol.
5. Aerosol Whoosh-Duster (VWR).

6. Prehybridization solution: 40% formamide (VWR), 5× standard sodium citrate (SSC) (Ambion), 0.1% SDS, 0.1 mg/ml bovine serum albumin (BSA). Prepared in water.
7. Hybridization solution: 40% formamide, 5× SSC, 0.1% SDS, 0.1 mg/ml herring sperm DNA (Invitrogen). Prepared in water.
8. Wash buffer #1: 1× SSC and 0.2% SDS. Prepared in water.
9. Wash buffer #2: 0.1× SSC and 0.2% SDS. Prepared in water.
10. Wash buffer #3: 0.1× SSC. Prepared in water.

3. Methods

3.1. Genomic DNA Extraction and Purification

1. Grow up a large amount of biomass of bacterial, yeast, plant, and animal cells.
2. Harvest the cells by centrifugation at 14 krpm (equivalent of 22,000 g) for 1 min. Remove the supernatant.
3. For yeast and gram-positive bacteria: Resuspend the cells thoroughly in 3 ml of 50 mM EDTA. Add 100 μl of 20 μg/μl lyticase for yeast, or 100 μl of 20 μg/μl lysozyme for gram-positive bacteria to digest cell walls. Gently pipet to mix. Incubate at 37°C for 30–60 min. Centrifuge at 14 krpm for 1 min and remove the supernatant.
4. Resuspend the cells in Buffer S. Combine all of the cells in a total volume of 20 ml Buffer S.
5. Place the tubes at −70°C until frozen, and then transfer the tubes to a microwave oven. Heat the tubes up to boiling temperature. Repeat this step for four cycles. Cool down to room temperature.
6. Add 100 μl of fresh 10 μg/μl proteinase K. Mix thoroughly by vigorous vortexing. Make sure there are no clumps of cells, which impair DNA yields.
7. Add 2 ml of 20% SDS and mix the tubes gently by rotation. Then incubate the tubes in a 65–70°C water bath for 2 h or longer with gentle agitation. Cool down to room temperature.
8. Add 20 ml of phenol:chloroform:isoamyl alcohol (25:24:1). Gently mix the solution by inversion for 5 min, then centrifuge at 14 krpm for 2 min.
9. After centrifugation, the mixture separates into a red lower phenol–chloroform phase, an interphase mainly composed of proteins, and a colorless upper aqueous phase. DNA remains

exclusively in the upper phase. Carefully remove the upper phase liquid slowly and transfer it to a clean 50-ml centrifuge tube containing 20 ml chloroform. Leave the residual upper phase liquid in the original tube to avoid contaminating the DNA solution with proteins.

10. Gently mix the solution by inversion for 5 min, then centrifuge at 14 krpm for 2 min.
11. Carefully remove the upper phase liquid slowly and transfer it to a clean 50-ml centrifuge tube containing 20 ml isopropanol. Leave the residual upper phase liquid in the original tube to avoid contaminating the DNA solution with proteins.
12. Gently mix the solution by inversion until cotton-like precipitates appear. If DNA yield is low, place the samples at -20°C overnight.
13. Centrifuge the samples at 14 krpm for 30 min. DNA will precipitate as a pellet at the bottom of the tubes. Remove the supernatant and wash the pellet with 20 ml of 70% ethanol. Centrifuge the samples at 14 krpm for 20 min. Remove the supernatant.
14. Air-dry the DNA for 10 min. Add 500 μl of sterile water to dissolve the DNA, then add 20 μl of DNAse-free RNase A (5 μg/μl)and incubate at 37°C for 30 min.
15. Add 500 μl of chloroform, gently mix the solution by inversion, and centrifuge at 14 krpm for 2 min. Transfer the upper phase to a clean 2-ml tube containing 50 μl of 3 M sodium acetate and 1 ml of ethanol. Gently mix the solution by inversion.
16. Place the samples at –20°C for 1 hour. Centrifuge at 14 krpm for 30 min.
17. Remove the supernatant and wash the pellet with 1 ml of 70% ethanol. Centrifuge the samples at 14 krpm for 20 min. Remove the supernatant.
18. Air-dry the DNA for 10 min. Dissolve the DNA with an appropriate amount of sterile water (100–200 μl).
19. Examine the yield and quality of DNA by a nano-drop or electrophoresis on a 1% agarose gel with 1 μl of DNA sample. The absorbance of wavelength 260 nm/280 nm of pure DNA is 1.8. If DNA is not used for microarray immediately, it should be stored in a –20°C freezer, or at –80°C for longer storage.

3.2. Total RNA Extraction and Purification

1. Precipitate cells by centrifugation at room temperature at 14 krpm for 1 min. Remove the supernatant and store the pellet in a –80°C freezer or proceed immediately to the next step.

2. For yeast and gram-positive bacteria: Resuspend the cells thoroughly in 3 ml of 50 mM EDTA. Add 100 μl of 20 μg/μl lyticase for yeast, or 100 μl of 20 μg/μl lysozyme for gram-positive bacteria to digest cell walls. Gently pipet to mix. Incubate at 37°C for 30–60 min. Centrifuge at 14 krpm for 1 min and remove the supernatant.
3. Lyse cells in Trizol completely by repetitive pipetting or vortexing (*see* **Note 1**). Use 1 ml of the reagent per 5×10^6 of animal, plant, or yeast cells, or per 10^7 bacterial cells. Make sure there are no clumps of cells, which decrease RNA yields.
4. Incubate the homogenized samples at room temperature for 5 min to permit the complete dissociation of nucleoprotein complexes. Add 0.2 ml of chloroform per milliliter of Trizol reagent. Shake tubes vigorously by hand for 15 s and incubate them at room temperature for 2 min (*see* **Note 2**).
5. Centrifuge the samples at no more than 8 krpm (equivalent of 12,000 g) for 15 min at 4°C. After centrifugation, the mixture separates into a red lower phenol-chloroform phase, an interphase, and a colorless upper aqueous phase. RNA remains exclusively in the aqueous phase. The volume of the aqueous phase is approximately 60% of the volume of Trizol used for homogenization.
6. Make sure the working environment is RNAse-free from now on through the remaining steps of **Subheading 3.2** (*see* **Note 3**). Transfer the aqueous phase to a fresh tube. Precipitate the RNA from the aqueous phase by mixing with isopropyl alcohol. Use 0.5 ml of isopropanol per milliliter of Trizol used for the initial homogenization. Incubate samples at room temperature for 10 min and centrifuge at 8 krpm for 10 min at 4°C. The RNA precipitate, often invisible before centrifugation, forms a gel-like pellet at the bottom of the tube.
7. Remove the supernatant. Wash the RNA pellet with 1 ml of 75% ethanol per milliliter Trizol used for the initial homogenization. Mix the sample by vortexing and centrifuge at no more than 7 krpm (equivalent of 11,000 g) for 5 min at 4°C.
8. Air-dry the RNA pellet for 5–10 min (*see* **Note 4**). Dissolve 88 μl RNA in RNase-free water. Add 10 μl of 10× DNAse I buffer and 2 μl DNAse I. Mix gently by repetitive pipettings and incubate at 37°C for 20–30 min to remove residual DNA contaminant. Do not vortex because DNAse I is sensitive to physical denaturation.
9. Add 1 μl of 0.5 mM EDTA (pH 8.0), incubate at room temperature for 1 min, and then at 65°C for 10 min.
10. Use an RNeasy kit to further purify RNA. Add 350 μl Buffer RLT, which is included in the kit, to RNA samples, and mix thoroughly by repetitive pipettings. Add 250 μl of 100%

ethanol and mix thoroughly by repetitive pipettings. Proceed immediately to the next step (*see* **Note 5**).

11. Transfer the sample to an RNeasy minicolumn placed in a 2-ml collection tube (supplied by the manufacturer). Centrifuge for 15–30 s at 12 krpm (equivalent of 19,000 g), and discard the flow through.
12. Add 500 μl Buffer RPE, which is included in the kit, onto the RNeasy column. Centrifuge for 15–30 s at 12 krpm, and discard the flow through.
13. Repeat **step 11**.
14. Centrifuge the column for 1 min at 12 krpm to remove Buffer RPE completely.
15. Transfer the RNeasy column to a new 1.5-ml Eppendorf tube. Add 30–50 μl RNAse-free water directly onto the RNeasy silica-gel membrane. Centrifuge for 1 min at 12 krpm to elute purified RNA.
16. Examine the yield and quality of RNA by a nano-drop or electrophoresis on a 1% agarose gel with 1 μl RNA sample. The absorbance of wavelength 260 nm/280 nm of pure RNA is 2. If RNA is not used for microarray immediately, it should be stored in a –80°C freezer.

3.3. Genomic DNA Labeling

1. Prepare genomic DNA labeling by mixing 0.5 μg genomic DNA with 20 μl 2.5× Random Primer. Add nuclease-free water to bring the total volume to 35 μl. Vortex briefly.
2. Incubate at 100°C for 5 min. Chill on ice for 5 min.
3. Add 0.4 μl dNTPs for genomic DNA, 0.4 μl Cy3-dUTP, 12.7 μl nuclease-free water, and 1.5 μl Klenow enzyme. Limit light exposure because Cy3 is light sensitive. Vortex briefly.
4. Incubate at 37°C for 3 h. Transfer to 100°C for 3 min to inactivate the Klenow enzyme. Chill on ice. Proceed to **Subheading 3.5**.

3.4. cDNA Labeling

1. Prepare cDNA labeling by mixing 10 μg purified total RNA with 3.3 μl Random Primer (3 μg/μl). Add nuclease-free water to bring the total volume to 16.5 μl. Vortex briefly.
2. Incubate at 70°C for 10 min. Chill on ice for 5 min.
3. Add 1.5 μl dNTPs for cDNA labeling, 6 μl of 5× buffer, 3 μl of 0.1 M DTT, 1 μl RNAse Inhibitor, and 1 μl of 1 mM Cy5-dUTP. Limit light exposure because Cy5 is light sensitive. Vortex briefly.
4. Incubate at room temperature for 10 min. Add 1 μl reverse transcriptase. Vortex briefly.
5. Incubate at 42°C for 2 h. Transfer to 98°C for 2 min and then chill on ice. Proceed to **Subheading 3.5**.

3.5. Purification and Concentration of Labeled DNA

1. We use a Qiagene PCR purification kit to purify Cy3-labeled genomic DNA and Cy5-labeled cDNA. Add five volumes of Buffer PB, which is included in the kit, to the samples. Vortex briefly. Transfer the mixture to a spin column. Centrifuge for 30 s at 12 krpm, and discard the flow through.
2. Add 500 μl Buffer PE to the column, and centrifuge for 30 s at 12 krpm. Discard the flow through.
3. Repeat **step 2**.
4. Centrifuge the column for 1 min at 12 krpm to remove Buffer PE completely.
5. Transfer the column to a new 1.5-ml Eppendorf tube. Add 50 μl of 10 mM Tris–HCL pH 8.5 (Buffer EB) directly onto the RNeasy silica-gel membrane. Incubate for 2 min. Centrifuge at 12 krpm for 1 min to elute the purified DNA.
6. Optional: Examine the labeling efficiency of the product by a nano-drop.
7. Use a SpeedVac to centrifuge Cy3-labeled genomic DNA and Cy5-labeled cDNA under a vacuum until the samples are completely dried.
8. If the labeled genomic DNA is not used for microarray immediately, it should be stored at –20°C for up to a week. Otherwise, proceed to **Subheading 3.6**.

3.6. Microarray Prehybridization and Hybridization

1. Add 40 μl freshly prepared prehybridization buffer to microarray slides. Place a cover slip on top. Incubate at 45–50°C for 30 min.
2. Remove the cover slips and dip the slides in water for 2 min. Then dip the slides into isopropyl alcohol and remove the slides immediately. Dry the microarray with a Whoosh-Duster.
3. Add 100 μl of freshly prepared hybridization buffer to Cy5-labeled genomic DNA. Resuspend DNA by pipetting.
4. Add 20 μl of freshly prepared hybridization buffer to Cy3-labeled cDNA. Resuspend DNA by pipetting. Mix with 20 μl Cy5-labeled genomic DNA. Vortex briefly.
5. Transfer the mixture onto a microarray slide. Place a cover slip on top. Incubate at 45–50°C overnight.

3.7. Microarray Washing and Scanning

1. Remove the cover slip. Dip the microarray slide into freshly prepared washing buffer #1. Shake at room temperature for 7 min.
2. Transfer the microarray slide to freshly prepared washing buffer #2. Shake at room temperature for 7 min.
3. Transfer the microarray slide to freshly prepared washing buffer #3. Shake at room temperature for 40 s.
4. Dry the microarray slide with a Whoosh-Duster.

5. Use an appropriate microarray scanner to scan the slide with dual channels for Cy5 and Cy3, whose wavelengths are 635 nm and 532 nm, respectively. Store the raw images in the computer. Quantify the signals using appropriate software (ImaGene, AtlasImage, GenePix, etc.).

3.8. Microarray Data Analyses Using Software GeneSpring (Agilent Technologies, Inc.)

1. Load quantified files and their corresponding genome file into GeneSpring.
2. Select signal (cDNA) and reference (gDNA) files in pairs, click to add the pairs to the right panel, click "next", then click "yes" to create a new experiment. Assign a name and save the experiment (*see* **Note 6**).
3. Use "per spot and per chip LOWESS normalization" to define normalizations. Click "OK" If you want to use Floor, select Data transformation: set measurements <0.01–0.01, double click so that it appears on the right panel.
4. To define parameters, rename the first columns such as test 1, test 1, test 1, test 2, test 2, test 2, control 1, control 2, control 3. Save.
5. To define experiment interpretation, select mode as "Ratio" and measurement flagged as "present," select File name as "non-continuous" and save.
6. Select "replicates" to define error model. Save.
7. Select "filter by confidence" and "filter by number of replicates." Define minimal number of replicates (*see* **Note 7**). Save and name the generated list.
8. To view results, point to Spreadsheet on the View menu, and then click "show p-value". Click "copy to clipboard" on the Edit menu. Paste onto an Excel file to view as a spreadsheet.

4. Notes

1. When cells are taken out of the −80°C freezer, add Trizol immediately before the cells are completely thawed to prevent the degradation of messenger RNAs (mRNAs).
2. A speedy alternative method can be used to replace **steps 5–10** of **Subheading 3.2**, which can be described as:
   (a) Transfer the aqueous phase to a fresh tube. Slowly add an equal volume of 70% ethanol to the aqueous phase. Mix by repetitive pipettings.

(b) Use an RNeasy kit to purify RNA. Load the mixture onto an RNeasy mini column placed in a 2-ml collection tube (supplied by manufacturer). Centrifuge for 15–30 s at 12 krpm, and discard the flow through.

(c) Pipet 350 μl buffer RW1, which is included in the kit, into the RNeasy mini column, and centrifuge for 15 s at 12 krpm to wash. Discard the flow through.

(d) In a new tube, add 10 μl DNase I stock solution to 70 μl buffer RDD. Mix by gently inverting the tube.

(e) Pipet the DNase I incubation mix (80 μl) directly onto the RNeasy silica-gel membrane, and place at room temperature for 15 min. Digestion will be incomplete if part of the mix sticks to the walls of the O-ring of the column.

(f) Pipet 350 μl buffer RW1 into the RNeasy mini column, and centrifuge for 15 s at 12 krpm. Discard the flow through.

3. Because RNA is sensitive for degradation by RNAse, **steps 5–15** in **Subheading** 3.2 and all steps in **Subheading** 3.4 have to be carried out in an RNAse-free environment. Steps in **Subheadings 3.3–3.7** have to be performed under limited light exposure to prevent the degradation of light-sensitive dyes.
4. It is important not to let the RNA pellet dry completely because this will greatly decrease its solubility. Avoid drying RNA by centrifugation under vacuum (SpeedVac).
5. It is critical to proceed immediately from **steps 10** to **11** in **Subheading 3.2.** Long delay impairs RNA yield.
6. **Step 2** of **Subheading 3.8** determines the ratio of signal intensities of cDNA over the genomic DNA reference on the same slide, whereas **step 4** of **Subheading 3.8** is used to define test and control slides. In addition, replicates can also be defined at **step 4** of **Subheading 3.8.**
7. Changes in defining the minimal number of replicates significantly impact the microarray output. A high minimal number of replicates could filter out many spots of poor quality. However, such improvement comes along with the expense of losing some potentially biologically meaningful information.

Acknowledgments

The author is indebted to Jizhong Zhou and Soumitra Barua for their invaluable advice of the experimental procedure, and Liyou Wu and Mengxia Zhu for bioinformatics analyses. This work was supported by The United States Department of Energy under the Genomics:GTL Program through the *Shewanella* Federation, Microbial Genome Program and Natural and Accelerated Bioremediation Research Programs of the Office of Biological and Environmental Research, Office of Science.

References

1. Schena, M., Shalon, D., Davis, R.W. and Brown, P.O. (1995). *Science*, **270**, 467–470.
2. Shoemaker, D.D. and Linsley, P.S. (2002). *Curr Opin Microbiol*, **5**, 334–337.
3. Hegde, P., Qi, R., Abernathy, K., Gay, C., Dharap, S., Gaspard, R., et al. (2000). *Biotechniques*, **29**, 548–550
4. Eisen, M.B. and Brown, P.O. (1999). *Methods Enzymol*, **303**, 179–205.
5. Talaat, A.M., Howard, S.T., Hale, W.T., Lyons, R., Garner, H. and Johnston, S.A. (2002). *Nucleic Acids Res*, **30**, e104.
6. Bina, J., Zhu, J., Dziejman, M., Faruque, S., Calderwood, S. and Mekalanos, J. (2003). *Proc Natl Acad Sci U S A*, **100**, 2801–2806.
7. Belland, R.J., Zhong, G., Crane, D.D., Hogan, D., Sturdevant, D., Sharma, J., et al. (2003). *Proc Natl Acad Sci U S A*, **100**, 8478–8483.
8. Williams, B.A., Gwirtz, R.M. and Wold, B.J. (2004). *Nucleic Acids Res*, **32**, e81.
9. He, Q., Huang, K.H., He, Z., Alm, E.J., Fields, M.W., Hazen, T.C., et al. (2006). *Appl Environ Microbiol*, **72**, 4370–4381.
10. Yang, D.H., Barari, M., Arif, B.M. and Krell, P.J. (2007). *J Virol Methods*, **143**, 175–185.

Chapter 30

Single-Molecule Imaging of Fluorescent Proteins Expressed in Living Cells

Kayo Hibino, Michio Hiroshima, Masahiro Takahashi, and Yasushi Sako

Summary

This chapter focuses on single-molecule imaging (SMI) in living cells using green fluorescent protein (GFP) or its related fluorescent protein tags (GFPs). Use of GFPs is a convenient technique to achieve molecular imaging of most proteins in living cells. However, because of difficulties in preparing samples suitable for SMI and the instability of fluorescence signals, special care is required for SMI using GFPs in living cells. Techniques for vector preparation, protein expression, sample preparation, microscopy, and image processing for SMI of GFPs in living cells are discussed in this chapter, along with examples of imaging applications. Double labeling of single molecules and single-pair fluorescent resonance energy transfer (spFRET) are possible in living cells using GFP and YFP as fluorescent tags. The limitations of SMI using GFPs are also discussed.

Key words: GFP, Single-pair (sp) FRET, Total internal reflection (TIR) fluorescence microscopy, YFP

1. Introduction

Single-molecule imaging (SMI) is a technique used to visualize individual molecules of interest in aqueous solutions *(1, 2)*. In SMI, the molecules of interest need to be labeled with a fluorophore, and images of the molecules are acquired using a fluorescence microscope equipped with a high-sensitivity camera system. The positions, fluorescence spectra, and fluorescent intensities of each fluorophore conjugated to these molecules can be measured in single movie frames captured by SMI. SMI has already developed into a major technology used to study the structure and reaction of biological macromolecules *(3)*.

James Weifu Lee and Robert S. Foote (eds.), *Micro and Nano Technologies in Bioanalysis*,
Methods in Molecular Biology, vol. 544
DOI 10.1007/978-1-59745-483-4_30, © Humana Press, a part of Springer Science+Business Media, LLC 2009

The recent extension of the application of SMI into molecules within living cells allows us to obtain quantitative information on the dynamics and kinetics of unitary reactions in complicated reaction networks inside cells *(4, 5)*. Now, the many parameters of molecular reactions—including reaction rate constant, density (concentration) of molecules, transport velocity, diffusion coefficient, and cluster size distribution (construction of molecular complexes)—can be determined without disrupting cell structure integrity. Single molecules of green fluorescent protein (GFP) were first imaged in living cells in 2000 *(6)*, and the use of GFP and its related gene expression tags (GFPs) in modern cell biology is now indispensable for labeling proteins inside living cells, as well as for SMI. The techniques and applications of SMI in living cells using GFPs are described herein.

2. Materials

2.1. Preparation of Specimens

1. Cells and culture medium. Culture medium does not containing phenol red. Culture medium (without phenol red) containing 5 mM (Piperazine-1,4-bis(2-ethanesulfonic acid) (PIPES); pH 7.2–7.4).
2. Vectors of GFP (or other fluorescent proteins)-tagged proteins. Transfection reagents.
3. Glass coverslips (No. 1). 2% neutral detergent solution for glassware cleaning. Concentration H_2SO_4.
4. Bathtub-type sonicator.

2.2. Microscopy

1. Total internal reflection fluorescence microscope (TIR-FM) system (excitation lasers, fluorescence microscope, high-sensitivity camera, and image recording system).
2. Cambers for setting the coverslip on which cells are cultured (for example, Attofluor cell chamber [Molecular Probes] for 25-mm round coverslips).

2.3. Data Processing

1. Single-molecule tracking software (should be custom made).
2. General-purpose image processing software (such as Image J).

3. Methods

3.1. Preparation of Specimens

3.1.1. Choice of Fluorescent Protein Tag

The classic family of GFPs comprises BFP, CFP, GFP, YFP, and DsRed *(7)*. BFP and CFP are unsuitable for SMI because of the rapid photobleaching that occurs under strong excitation light to achieve SMI. Both GFP and YFP can be used for SMI, but GFP is better in terms of resistance to photobleaching and stability of fluorescence emission (**Fig. 1**). Among the many variants of GFPs, EGFP (Clontech) is the most ideal tag for SMI as far as we know. DsRed and other new variants of GFP can also be used for SMI, but very few examples of their application have been reported.

3.1.2. Preparation of Vectors for Protein Expression

Because of limited spatial resolution of the optical microscope (~250 nm), expression levels (density) of fluorescent molecules should be carefully controlled when used in SMI. Otherwise, the images of individual molecules will overlap or the out-of-focus background fluorescence will reduce the contrast of images. Commercial vectors for expression of GFPs contain very strong promoters. In addition, commercial transfection reagents are designed to obtain the most robust expression possible. Because of these attributes, the expression level of GFPs often exceeds the detection limits of SMI.

There are some methods to reduce the expression level of GFP-tagged proteins including: *[1]* truncation of the enhancer region

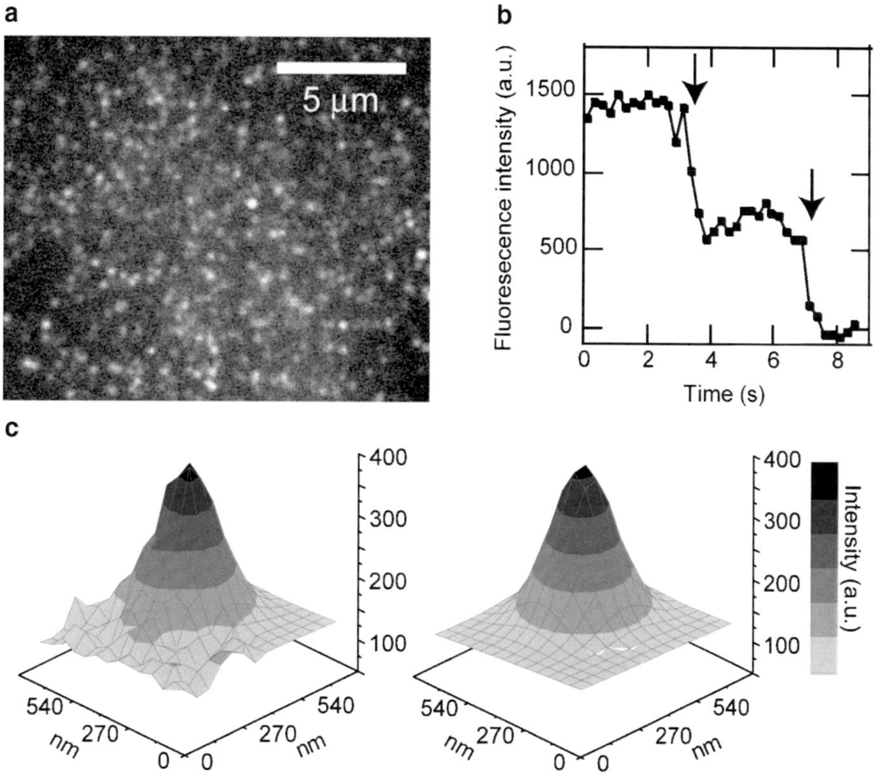

Fig. 1. Single-molecule imaging (SMI) of GFPs. (**a**) A single-molecule image of EGF receptor (EGFR) in a living CHO-K1 cell. The receptor was tagged with EGFP at the cytoplasmic C terminus. (**b**) Fluorescence intensity change of a single-molecule spot of EGFR-GFP. *Arrows* indicate photobleaching. This spot contained two EGFR-GFP molecules. (**c**) Two-dimensional fluorescence intensity profiles of a single EGFR-GFP molecule. Raw data (*right*) and data after fitting to the Gaussian distribution (*left*) are shown.

in the expression vectors; *[2]* engineering of the Kozak sequence, which acts as the ribosome-binding site *(8)*; and *[3]* shortening of the incubation time with vectors in the transfection procedure. We typically use Lipofectamine Plus (Clontech) as the transfection regent, with an incubation time of 1–3 h. The selection or cloning of cells with desirable expression levels is recommended not only for regulation of the protein expression but also for avoiding background contamination of the transfection regents.

GFPs tend to form homodimers at high concentrations. Therefore, dimerization of the probes should be avoided when making quantitative measurements. A point mutation, A206K, has been reported to decrease the affinities between GFPs *(9)*.

3.1.3. Cell Culture

Coverslips on which cells will be cultured should be cleaned thoroughly before using to reduce background signals.

1. Soak 50 coverslips (18 × 18 mm or 25-mm round) in 50 mL of 2% neutral glassware cleaning detergent for at least 1 day. Avoid overlapping of the coverslips.
2. Wash the coverslips in the detergent by sonication for 3 min using a bathtub-type sonicator.
3. Rinse the coverslips with ultrapure water. Change water three times, then sonicate in water for 3 min. Repeat this process at least two more times.
4. Transfer the coverslips into 50 mL of concentrated H_2SO_4 one by one (to avoid overlapping). This process is for oxidization of contaminants (*see* **Note 1**).
5. After at least 1 day, rinse the coverslips thoroughly with ultrapure water.
6. Keep the coverslips in ultrapure water after autoclaving or in 100% ethanol. They can be used longer than 1 month later.
7. Flame-sterilize the coverslips and soak them in serum-containing culture medium overnight before transferring cells onto them: otherwise, cells many not attach.

Cells can be cultured in their usual medium, but it is recommended that on the night before the experiments, the culture medium be changed to medium that contains no phenol red. In some experiments, the use of culture medium without serum is required. In such cases, we add bovine serum albumin (BSA) to the medium (1% final).

For microscopic observation, the culture medium (without phenol red) containing 5 mM PIPES (pH 7.2–7.4) is used to keep the pH neutral outside of the CO_2 incubator. If salines such as Hank's balanced salt solution (HBSS) can be used without loosing cellular activities, they provide good contrast in SMI, because some vitamins, serums, and BSA are slightly fluorescent.

3.2. Microscopy

3.2.1. Instruments

In light of recently developed high-sensitivity video cameras that have adequate sensitivity for SMI, the most important factor to consider with regard to microscopy for SMI is background reduction. For SMI of GFPs in living cells, the total internal reflection fluorescence microscope *(10)* is conventionally used to reduce background in out-of-focus regions of the specimens *(6)*. Low-angle oblique illumination microscopes and epi-illumination microscopes can be used for SMI in living cells using chemical fluorophores *(5)*; TIR-FM, however, provides the best contrast for GFPs, which are less ideal SMI probes compared with chemical fluorophores such as TMR or Cy3 (*see* **Note 2**). TIR-FM systems can be purchased from any major microscope company (*see* **Note 3**).

For excitation of GFP, the 488-nm line from an Ar ion laser or blue solid-state laser is suitable. YFP can be excited by the same laser as GFP or by the 514-nm line of an Ar ion laser. SMI requires a higher laser power for excitation compared with multiple-molecule imaging. In the conventional setup for TIR-FM, more than 10 mW of laser power after the objective is used to illuminate approximately 150 μm-ϕ of observation field (*see* **Note 4**).

At present, the best photodetector for SMI is a back-illumination cooled EM-CCD camera (Andor, Hamamatus, or Roper). An EB-CCD or chilled CCD camera can be used when coupled with image intensifiers using microchannel plates. Cooled CCD cameras can also be used at more than hundreds of milliseconds of exposure time.

3.2.2. Selection of the Optics

An oil-immersion objective with a high (>*1*.33) numerical aperture (NA) is required for TIR-FM. Most of the microscope companies have special objectives for TIR-FM with NAs of 1.45, 1.49, or 1.65.

Dichroic mirrors and excitation filters should be selected carefully to maximize fluorescence signal and to minimize background. A block diagram of our setup of optics for observation of GFP is shown in **Fig. 2a**.

Figure 2b shows an example of an optical setup for dual labeling using GFP and YFP. In this setup, signals in the GFP (short wavelength) channel come primarily from GFP; but in the YFP (long wavelength) channel, both GFP and YFP signals are observed because of the wide overlap between both excitation and emission spectra of GFP and YFP. Therefore, signals from GFP and YFP should be separated by calculation after the image acquisition by measuring the leak of the fluorescence signals of GFP and YFP into YFP and GFP channels, respectively, using specimens containing either GFP or YFP. Single-pair (sp) fluorescent resonance energy transfer (FRET) from GFP to YFP can be detected by using the same optics (**Fig. 3**).

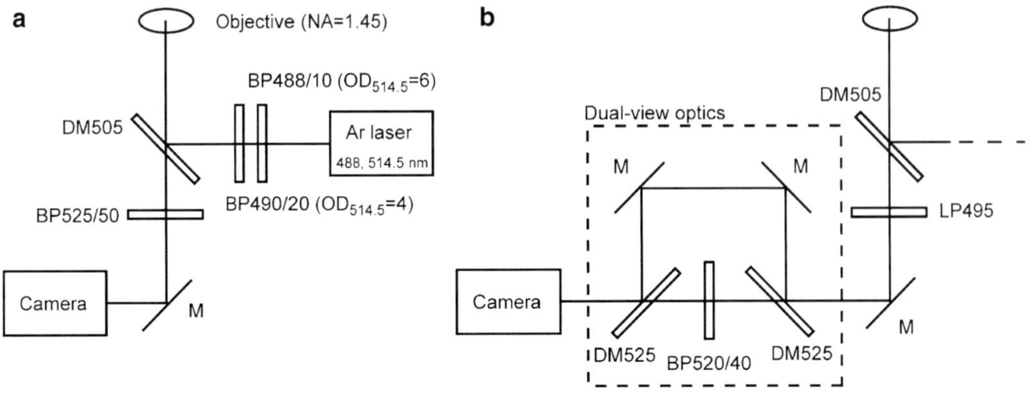

Fig. 2. Optics for SMI. *Block diagrams* of the optics for imaging GFP alone (**a**) and for simultaneous imaging of GFP and YFP (**b**) are shown (only optics for the emission side are shown in **b**). In **b**, fluorescence signals from GFP and YFP are separated using dual-view optics and projected to the same camera *(14)*. Filters and dichroic mirrors were purchased from Chroma Technology, Olympus, Omega Opical, and Semlock. Single-pair FRET from GFP to YFP can be detected using the same optics as shown in (**b**). BPX/Y: band-pass filter with center of transmission at X nm and full band-width of Y nm. DMX: dichroic mirror transmit wavelength longer than X nm. LPX: long-pass filter with cut-off wavelength of X nm. M: mirror.

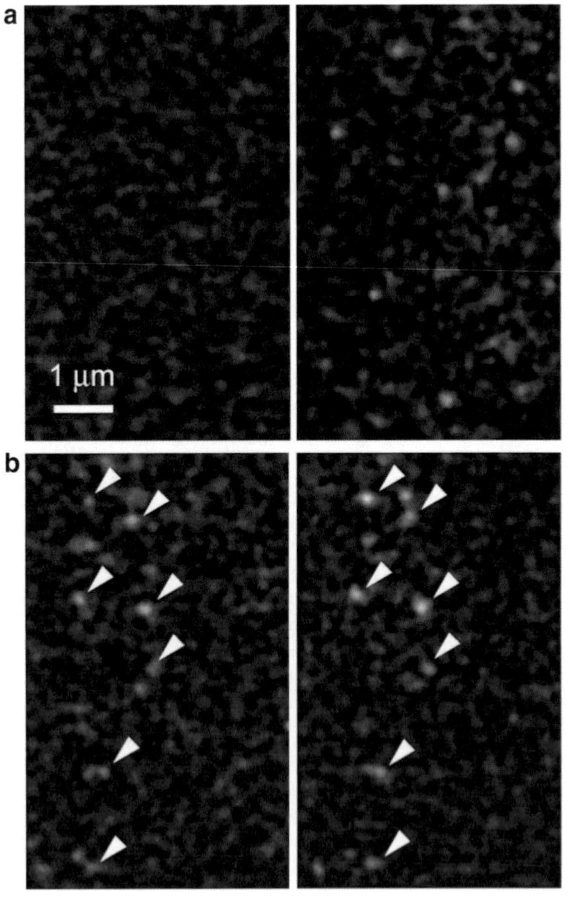

Fig. 3. Imaging of intramolecular spFRET. Single molecules of GFP-Raf1-YFP were observed in living HeLa cells using the optical setup shown in Fig. 2b. In this setup, signals in the GFP channel (*left*) primarily represent GFP fluorescence, but signals in the YFP channel (*right*) come from both GFP and YFP because of the large excitation and emission spectral overlap between GFP and YFP. Figures were acquired in the absence (**a**) and presence (**b**) of EGF. *Arrows* in (**b**) denote the same spots.

3.3. Data Processing

Determination of the position and fluorescence intensity of each fluorescent spot is the fundamental purpose of the data processing in SMI. Unfortunately, single-molecule signals are small and fluctuating (*see* **Note 5**). Because of heterogeneous backgrounds in both space and time, the conversion to binary images, which is a conventional image-processing technique for particle detection, is irrelevant for SMI in living cells.

In our laboratory, the image profiles of a single molecule are fit to a two-dimensional Gaussian distribution on an inclined plane (**Fig. 1c**). The center and the integral of the Gaussian distribution represent the position and the fluorescence intensity of the spots, respectively. This fitting is carried out using custom-made software.

3.4. Confirmation of Single-Molecule Detection

Several criteria are used to confirm single-molecule detection:

1. The spatial profile of single-molecule images must be the point-spread function of the optics (**Fig. 1c**).
2. Single molecules should emit nearly constant fluorescence radiation and photobleach in any single step (**Fig. 1b**).
3. Distributions of the fluorescence intensity (or step size of the photobleaching) of single molecules should have a single peak. The shape of the distribution should appear Gaussian (sometimes it appears log-normal).
4. Under continuous excitation, the number of fluorescent spots of single molecules in a cell decreases with time because of photobleaching; however, the fluorescence intensity of each spot should not change.

3.5. Applications of SMI of GFPs in Living Cells

3.5.1. Single-Molecule Tracking of the Movements of Ras

Ras is a small GTPase involved in the signaling pathways of cell proliferation and differentiation. Most Ras molecules localize to the cytoplasmic side of the plasma membrane. A fusion protein combining EGFP and human H-Ras (GFP-Ras) was expressed in HeLa cells and its movements were observed as single molecules *(11)*. Trajectories of GFP-Ras movement were obtained by connecting the center of the Gaussian distributions of single-molecule profiles frame by frame (**Fig. 4a**).

From the single-molecule trajectories, mean square displacements (MSDs) of the movements were calculated as a function of time, as follows:

$$\mathrm{MSD}(\Delta t) = \langle (X(j + \Delta t) - \mathrm{X}(j))^2 \rangle_j.$$

Here, $X(j)$ is the position of a fluorescent spots in the j-th frame and Δt is the sampling time *(12)*. The diffusion coefficient (D) can be calculated from the MSD. In two-dimensional random walks, such as unregulated movements of membrane proteins, $\mathrm{MSD}(\Delta t) = 4D\Delta t$.

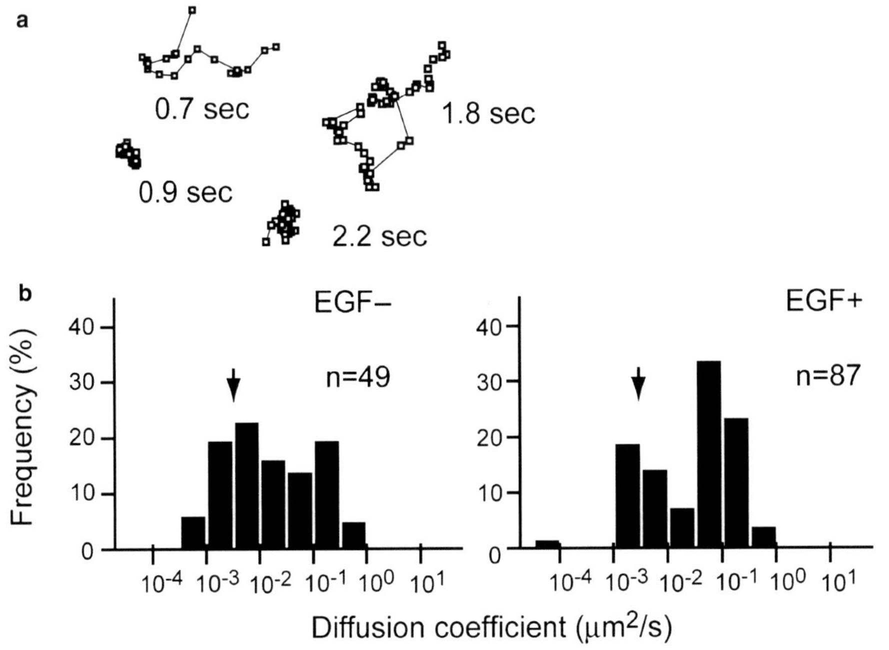

Fig. 4. Single-molecule tracking of GFP-Ras on the plasma membrane. GFP-Ras expressed in HeLa cells was observed in single molecules. Panel (**a**) shows typical trajectories of GFP-Ras observed at video rates of 33 ms/frame. Durations for observations are indicated. Diffusion coefficients for lateral movements of GFP-Ras on the plasma membrane were calculated for individual molecules in the time window of 0–150 ms (**b**). *Arrows* indicate the average of apparent diffusion coefficients for GFP molecules in fixed cells. Activation of Ras with EGF did not change the lateral movements.

Figure 4b shows the distribution of D for GFP-Ras in the time window of Δt = 0–150 ms. The distribution has two peaks; D for the smaller peak was similar to that for GFP-Ras in fixed cells. Therefore, there were two types of GFP-Ras molecules; one was randomly diffusing on the membrane surface and another was almost immobile. Activation of Ras changed these movements slightly.

3.5.2. Single-Molecule Kinetics of the Dissociation Between Ras and Raf1

Raf1 is a cytoplasmic serine/threonine kinase that recognizes activated Ras on the plasma membrane. The duration of GFP-Raf1 residence on the plasma membrane was measured for individual molecules. Because freely moving molecules in the cytoplasm cannot be detected as fluorescent spots because of rapid Brownian diffusion in solution, the duration between appearance and disappearance of GFP-Raf1 molecules on the plasma membrane represents the lifetime of interactions between Ras and Raf1.

Distribution of the lifetime, $f(t)$, contains information of dissociation kinetics between Ras and Raf1 *(13)*. Assuming a stochastic dissociation reaction with a reaction rate k, $f(t)$ is a simple exponential function, $f(t) = ke^{-kt}$. With regard to dissociation between inactive Ras and Raf1, $k = 2.6\ s^{-1}$. This reaction rate may be affected by relatively short photobleaching times of GFP measured in the same condition (3.7 s). Because it is highly probable that dissociation of Ras and Raf1 is independent of the GFP photobleaching, the true rate constant was estimated to be 2.3 s^{-1} (= 2.6 – 1/3.7).

3.5.3. Intramolecular FRET Observed in Single Molecules

It has been suggested recently that Raf has two conformations, closed and open, relating to Raf activation. In the closed conformation, its C-terminal kinase domain is thought to interact with the CRD domain in the middle of the molecule; in the open conformation, this interaction is lost and the Raf molecule is elongated. We attempted to detect this conformational change using intramolecular sp-FRET imaging.

GFP and YFP were tagged to the N- and C-terminus of Raf, respectively, and spFRET was imaged using the optics shown in **Fig. 2b**. In unstimulated cells, fluorescent spots were observed only in the YFP channel, indicating that FRET from GFP to YFP was very effective (**Fig. 3a**). In contrast, in cells stimulated with epidermal growth factor (EGF) to activate Raf, fluorescent spots were observed at the same positions in the GFP and YFP channels, indicating low FRET efficiency (**Fig. 3b**). These results suggest that Raf changes its conformation from closed to open depending on its activation.

3.6. Perspective

The determination of the quantitative parameters of reactions inside cells, without disrupting structural integrity, provides fundamental information pivotal to our understanding of reaction kinetics and dynamics in living cells. In addition, by detecting intramolecular spFRET, we can gain access to the molecular mechanisms of reactions inside cells. Thus, SMI is a realization of in situ biochemical and biophysical studies in living cells and will be an important experimental technique in molecular cell biology.

4. Notes

1. The treatment with H_2SO_4 can be replaced with sonication in 0.1N KOH for 1 h.
2. Adjust the laser beam carefully in TIR-FM. First, focus the microscope on the surface of the coverslip and set the incident angle of laser to 0°. At this time, the center of the image field should be illuminated with a parallel beam (with the smallest diameter at a far distance) aligning with the optic axis of

the objective. Then, increase the incident angle gradually. At the critical angle of TIR, a sudden increase of the fluorescent intensity will be observed.

3. All commercial TIR-FM systems use one-side illumination, which often causes inhomogeneous imaging. To avoid this problem, use multiple beams simultaneously from different directions or a circular beam for illumination *(5)*.
4. Lack of excitation power causes difficulty in single-molecule detection. Prepare lasers with the radiation power as high as possible. However, when an optical fiber is used to introduce the laser to the microscope, the excitation power that can be transferred through the fiber is limited.
5. The signal-to-noise ratio can be improved by using stronger excitation. However, because most of the fluorescent proteins are photobleached after emitting ~10^5 photons, there is a trade-off between higher signal and longer observation.

References

1. Funatsu, T., Harada, Y., Tokunaga, M., Saito, K., and Yanagida, T. (1995) Imaging of single fluorescent molecules and individual ATP turnover by single myosin molecules in aqueous solution. *Nature* **374**, 555–559.
2. Sase, I., Miyata, H., John, C. E. T., James, C. S., and Kinosita, K. Jr. (1995) Real time imaging of single fluorophores on moving actin with an epifluorescence microscope. *Biophys. J.* **69**, 323–328.
3. Cornish, P. and Ha, T. (2006) A survey of single molecule techniques in chemical biology. *ACS Chem. Biol.* **2**, 53–61.
4. Sako, Y. and Yanagida, T. (2003) Single-molecule visualization in cell biology. *Nature Rev. Mol. Cell Biol.* **4**, SS1-5.
5. Sako, Y. (2006) Imaging single molecules for systems biology. *Mol. Syst. Biol.* doi:10.1038/msb4100100.
6. Sako, Y., Hibino, K., Miyauchi, T., Miyamoto, Y., Ueda, M., and Yanagida, T. (2000) Single-molecule imaging of signaling molecules in living cells. *Single Mol.* **1**, 151–155.
7. Tsien, R. Y. (2005) Building and breeding molecules to spy on cells and tumors. *FEBS Lett.* **579**, 927–932.
8. Kozak, M. (1999) Initiation of translocation in prokaryotes and eukaryotes. *Gene* **234**,187–208.
9. Zacharias, D. A., Violin, J. D., Newton, A. C., and Tsien, R. Y. (2002) Partitioning of lipid-modified monomeric GFPs into membrane microdomains of live cells. *Science* **296**, 913–916.
10. Axelrod, D. (2001) Total internal reflection fluorescence microscopy in cell biology. *Traffic* **2**, 764–774.
11. Hibino, K., Watanabe, T., Kozuka, J., Iwane, A. H., Okada, T., Kataoka, T., Yanagida, T., and Sako, Y. (2003) Single- and multiple-molecule dynamics of the signaling from H-Ras to c-Raf1 visualized on the plasma membrane of living cells. *Chem. Phys. Chem.* **4**, 748–753.
12. Kusumi, A., Sako, Y., and Yamamoto, M. (1993) Confined lateral diffusion of membrane receptors as studied by single particle tracking (nanovid microscopy). Effects of calcium-induced differentiation in cultured epithelial cells. *Biophys. J.* **65**, 2021–2040.
13. Xie, S. (2001) Single-molecule approach to enzymology. *Single Mol.* **2**, 229–236.
14. Kinosita K., Ito, H., Ishiwata, S., Hirano, K., Nishizaka, T., and Hayakawa, T (1991) Dual-view microscopy with a single camera: real-time imaging of molecular orientations and calcium. *J. Cell Biol.* **115**, 67–73.

Chapter 31

MicroPET, MicroSPECT, and NIR Fluorescence Imaging of Biomolecules In Vivo

Zi-Bo Li and Xiaoyuan Chen

Summary

Molecular imaging is a newly merged multidisciplinary subject that requires contributions from biology, medical physics, and chemistry/radiochemistry. Integrin $\alpha_v\beta_3$, a cell adhesion molecule, plays pivotal roles in regulating tumor angiogenesis and the growth of new blood vessels. In this chapter, we use the cell adhesion molecule integrin $\alpha_v\beta_3$ as an example to demonstrate how one can synthesize appropriate arginine–glycine–aspartic acid (RGD) peptide-containing probes for visualizing and quantifying the receptor expression in vivo by means of microPET, microSPECT, and NIR fluorescence.

Key words: Tumor angiogenesis, Integrin $\alpha_v\beta_3$, RGD peptide, Molecular imaging, PET, SPECT, NIR fluorescence

1. Introduction

Cancer is the second leading cause of death in the United States (http://www.cdc.gov). Base on the report from American Cancer Society (ACS), approximately 1,444,920 new cancer cases are expected to be diagnosed and approximately 559,650 Americans are expected to die of cancer in 2007 (http://www.cancer.org). Early detection of cancer can greatly increase survival rates because it identifies cancer when it is most treatable, according to the National Cancer Institute (NCI). Many traditional medical imaging techniques, such as computed tomography (CT), magnetic resonance imaging (MRI), and ultrasound, have been routinely used for detecting cancers and monitoring the therapeutic

James Weifu Lee and Robert S. Foote (eds.), *Micro and Nano Technologies in Bioanalysis*,
Methods in Molecular Biology, vol. 544
DOI 10.1007/978-1-59745-483-4_31, © Humana Press, a part of Springer Science+Business Media, LLC 2009

effects of cancer intervention *(1, 2)*. The field of molecular imaging has flourished over the last decade, which may provide new ways to diagnose diseases and monitor therapies in patients.

1.1. Molecular Imaging

Molecular imaging, which is defined as "noninvasive, quantitative, and repetitive imaging of targeted macromolecules and biological processes in living organisms," originated from radiopharmacology and was established as a fast-growing interdisciplinary field. Unlike traditional imaging techniques, which primarily image differences in qualities such as densities or water content, molecular imaging has the unique ability to image very fine molecular changes within the area of interest. For example, by introducing molecular probes, molecular imaging can determine the expression of indicative molecular markers of the tumor development at different stages *(3–5)*. The ability to detect these molecular markers leads to an incredible number of exciting possibilities for biomedical applications, including early detection, treatment monitoring, and drug development.

1.2. Biomedical Imaging Modalities

Based on the molecular imaging probes and the means (instrument) by which to monitor these probes, molecular imaging could be divided into the following modalities: positron emission tomography (PET), single-photon emission computed tomography (SPECT), digital autoradiography, magnetic resonance imaging (MRI), optical bioluminescence/fluorescence, and ultrasound *(3)*. In this chapter, we demonstrate how to construct cyclic RGD peptide-based molecular probes for imaging integrin $\alpha_v\beta_3$ expression in vivo through microPET, microSPECT, and near-infrared (NIR) fluorescence.

1.2.1. Positron Emission Tomography (PET)

PET is a medical imaging technique in nuclear medicine that produces a three-dimensional image of functional processes in the body. The radioisotope emits a positron that annihilates with an electron, producing a pair of annihilation (gamma) photons moving in almost opposite directions. These photons will be detected in the scanning device and reconstructed to provide the imaging result. The sensitivity of PET is very high (10^{-11}–10^{-12} M), and there is no depth limitation for detecting tumor signal *(6, 7)*. Therefore, PET imaging has been a valuable technique in oncology, neurology, cardiology, and for studying various other diseases. The clinical PET systems usually have a spatial resolution of 5–7 mm, and high-resolution PET could have a spatial resolution of <2 mm. The positron-emitting radionuclides could be classified as biogenic elements (for example ^{11}C, ^{13}N, ^{15}O), radiohalogens (for example ^{18}F, ^{76}Br, ^{124}I), and radiometals (for example ^{64}Cu, ^{66}Ga, ^{68}Ga). PET usually provides information about the physiological function of the body because a biologically active tracer molecule is generally used.

1.2.2. Single-Photon Emission Computed Tomography (SPECT)

SPECT is a nuclear medicine tomographic imaging technique that provides three-dimensional (3-D) information based on gamma rays *(8, 9)*. Internal radiation is administered through a low mass amount of pharmaceutical labeled with a radioactive isotope, which is then inhaled, ingested, or injected. After the radio decay, γ-rays are emitted and detected by a γ-camera. The γ-camera can be used either in planar imaging to obtain two-dimensional (2-D) images, or in SPECT imaging to obtain 3-D images. The first object that an emitted γ-photon encounters after exiting the body is the collimator, which allows it to travel only along certain directions to reach the detector, to ensure that the position on the detector accurately represents the source of the γ-ray. Compared with PET, SPECT imaging has a very low detection efficiency ($<10^{-4}$ times the emitted number of γ-rays) because of the use of lead collimators to define the angle of incidence. However, the advantage of SPECT imaging is that it allows simultaneous imaging of multiple radionuclides because γ-rays emitted from different radioisotopes can be differentiated based on the energy. SPECT has been widely used in tumor imaging, infection (leukocyte) imaging, thyroid imaging, and bone imaging.

1.2.3. NIR Fluorescence

Besides PET and SPECT, there are also several nonradionuclide-based imaging techniques. In fluorescence imaging, excitation light illuminates the subject and the emission light is collected at a shifted wavelength. One advantage of optical imaging is that multiple probes with different spectral characters could potentially be used for multichannel imaging. However, because of the limited penetration and intense scattering of light in living objects, optical imaging will be only possible in humans in limited sites such as the tissues and lesions close to the surface of the skin, tissues accessible by endoscopy, and intraoperative visualization. Even though optical imaging may not be widely used in clinical settings, it is a relatively low-cost method suitable primarily for small-animal studies. Moreover, because the absorbance spectra for all biomolecules reach minima in the NIR (wavelength 700–900 nm) region, NIR provides a clear window for in vivo optical imaging *(10)*. Therefore, NIR fluorescent imaging has been used in rapid and cost-effective preclinical research in small-animal models before radionuclide-based imaging studies, which are more costly. Both charged-coupled device (CCD) cameras and fluorescence-mediated tomography have been developed to more efficiently detect the light emitted from the body *(11, 12)*. After exposing the subject with continuous wave or pulsed light from different sources, the emitted light is captured by multiple detectors arranged in a spatially defined order in an imaging chamber. These raw data would be processed mathematically for correction and the reconstructed tomographic image would be obtained.

1.3. Integrin $\alpha_v\beta_3$

Angiogenesis is an invasive process characterized by endothelial cell proliferation, modulation of the extracellular matrix, and cell adhesion/migration *(13)*. Among many angiogenic factors, the cell adhesion molecule integrin plays a key role. Integrins expressed on endothelial cells modulate cell migration and survival during tumor angiogenesis whereas integrins expressed on carcinoma cells potentiate metastasis by facilitating invasion and movement across blood vessels *(14)*. Integrins consist of two noncovalently bound transmembrane proteins: α and β subunits (in mammals, 18 α and 8 β subunits have been characterized) *(15)*. Among all of the integrins discovered to date, integrin $\alpha_v\beta_3$ is the most extensively studied, and is significantly upregulated on tumor vasculature but not on quiescent endothelium *(14, 16)*. Many monoclonal antibodies, cyclic arginine-glycine-aspartic acid (RGD) peptide antagonists, and peptidomimetics against integrin $\alpha_v\beta_3$ have been reported for anti-angiogenic cancer therapy *(17, 18)*. Recently, integrin $\alpha_v\beta_3$ targeted imaging has attracted considerable interest and many excellent reviews are available *(8, 19, 20)*. In this chapter, we demonstrate the general procedure for the construction of PET, SPECT, and NIR fluorescence imaging probes based on RGD peptides and their application in integrin $\alpha_v\beta_3$ imaging.

2. Materials

2.1. Reagents

1. ^{18}F-Fluoride (produced by PETtracer cyclotron in-house, GE Healthcare)
2. Trifluoroacetic acid (TFA) (Sigma-Aldrich); Potassium carbonate (anhydrous) (Aldrich).
3. Tetrabutylammonium hydroxide solution (40 wt% in H_2O) (Sigma-Aldrich).
4. *N*,*N*′-Diisopropylethylamine (DIPEA) (Sigma-Aldrich).
5. *N*,*N*,*N*′,*N*′-Tetramethyl-*O*-(*N*-succinimidyl) uronium tetrafluoroborate (TSTU) (Fluka).
6. *N*-Succinimidyl S-acetylthioacetate (SATA) (Pierce).
7. *N*-(2-aminoethyl)Maleimide trifluoroacetate salt (Fluka).
8. Tris(2-carboxyethyl) phosphine hydrochloride (TCEP HCl) (Pierce).
9. E[c(RGDyK)]$_2$ (Peptides International).
10. Mini-PEG-E[c(RGDyK)]$_2$ (PRGD2) (Peptides International).

11. *N*,*N*,*N*-Trimethyl-4-(ethoxy)carbonyl benzenaminium trifluoromethane Sulfonate (ABX).
12. *N*-Succinimidyl-4-fluorobenzoate (SFB) (ABX).
13. *N*-Hydroxysuccinimide (NHS) (Sigma).
14. Distilled, deionized water (Milli-Q®; >18 MΩ resistivity).
15. Acetonitrile (Sigma-Aldrich).
16. 0.1% TFA in water (Solution A, which has been rigorously degassed for high-performance liquid chromatography [HPLC] use).
17. 0.1% TFA in acetonitrile (Solution B, which has been rigorously degassed for HPLC use).
18. Phosphate-buffered saline (PBS), 7.4 (1×) (Invitrogen Corp.).
19. BupH borate buffer packs (Pierce).
20. Dimethyl sulfoxide (DMSO) (Acros).
21. Sodium hydroxide (NaOH) (Fluka).
22. Acetic acid, glacial (EMD).
23. Silicone oil (Fisher).
24. C18 Sep-Pak solid-phase extraction cartridge (Waters Corp.).
25. QMA cartridge (ABX).
26. Microcentrifuge tube (E&K Scientific).
27. Trisodium triphenylphosphine-3,3′3″-trisulfonate (TPPTS) (Sigma-Aldrich).
28. Tricine (Sigma-Aldrich).
29. $E\{E[c(RGDfK)]_2\}_2$ was prepared using the reported procedure *(21)*.
30. Sodium succinimidyl 6-(2-(2-sulfonatobenzaldehyde)hydrazono)-nicotinate (HYNIC-NHS) (ABX).
31. $Na^{99m}TcO_4$ (DuPont Pharma $^{99}Mo/^{99m}Tc$ generator).
32. Quantum dots (Qdot 705 ITK™ Amino [PEG] Quantum Dots; Invitrogen).
33. Cell binding buffer (20 mM Tris, 150 mM NaCl, 2 mM $CaCl_2$, 1 mM $MnCl_2$, 1 mM $MgCl_2$, 0.1% bovine serum albumin, pH 7.4).

2.2. Equipment

1. Digital heating block (VWR).
2. Rotary evaporator (Type: Antrieb VV-Mikro) (Heidolph).
3. Vydac protein and peptide column (5 μm, 250 ′ 10 mm) (Vydac).

4. Dionex 680 chromatography system with a UVD 340U absorbance detector and model 105S single-channel radiation detector (Carroll & Ramsey Associates). In a typical setup, the mobile phase starts from 95% solvent A (0.1% TFA in water) and 5% solvent B (0.1% TFA in acetonitrile [ACN]) at 0–2 min, to 35% solvent A and 65% solvent B at 32 min.
5. Synthesis module (GE Health, TRACERlab FXFN).
6. Filter multiscreen DV plates (96-well; pore size, 0.65 μm; Millipore).
7. microPET R4 scanner (Siemens Medical Solutions).
8. Dual head γ-camera (Siemens, E. CAM).
9. NAP-10 column (GE Healthcare).
10. Refrigerated microcentrifuge (Eppendorf).
11. Fluorometer (Fluoro Max-3; Jobin Yvon).
12. Glass-bottomed microwell dish (30-mm Petri dish, 14-mm Microwell; MatTek).
13. Inverted fluorescence microscope (Axiovert 200M; Zeiss).
14. Filter set (420/40 nm excitation, 705/40 nm emission, 475 DCXR dichroic; Chroma Technology).
15. Maestro imaging system (CRI, Inc.).

3. Methods

3.1. MicroPET of Tumor $\alpha_v\beta_3$ Integrin Expression

Imaging probes of integrin $\alpha_v\beta_3$ expression has important diagnostic and therapeutic applications. It can be used to visualize and quantify $\alpha_v\beta_3$ integrin expression levels, more appropriately select patients considered for anti-integrin $\alpha_v\beta_3$ treatment, and monitor treatment efficacy in $\alpha_v\beta_3$-positive patients. Multimeric cyclic RGD peptides are capable of improving the integrin $\alpha_v\beta_3$-binding affinity because of the polyvalency effect. Here we describe the detailed procedure of synthesizing a ^{18}F-labeled dimeric RGD peptide for PET of tumor $\alpha_v\beta_3$ expression *(22)*.

3.1.1. Preparation of ^{18}F-Labeled RGD Peptide (see Note 1)

We used a slightly different protocol from that of Vaidyanathan *(23)* for ^{18}F-SFB synthesis. The reaction is carried out using a commercially available synthesis module (GE TRACERlab FX_{FN}).

Synthesis of *N*-succinimidyl-4-^{18}F-fluorobenzoate (^{18}F-SFB)

1. Trap ^{18}F-F^- on a QMA cartridge first and wash it down with a solution of Kryptofix 222 and potassium carbonate (K_{222}: 15 mg/mL in 0.9 mL ACN; K_2CO_3: 3.0 mg/mL in 0.1 mL

H_2O). Evaporate the solvents by heating in the presence of a beam of helium gas. Add 1 mL acetonitrile and evaporate the solvent to obtain azeotropically dried ^{18}F-F^-.

2. Add 1 mL DMSO solution of *N,N,N*-trimethyl-4-(ethoxy) carbonyl benzenaminium trifluoromethane sulfonate (4 mg) to azeotropically dried ^{18}F-F^-. Heat the reaction mixture at 120 °C for 20 min to obtain 4-^{18}F-fluorobenzoic ethyl ester.
3. Hydrolyze this fluorinated ester with 0.5 mL of 0.2N NaOH solution to obtain 4-^{18}F-fluorobenzoic acid (^{18}F-FB). Neutralize the mixture with 10 mL water (18 MΩ, deionized) and 1 mL of 1N HCl.
4. After trapping on a C18 Sep-Pak cartridge, wash down the ^{18}F-FB with 1 mL acetonitrile and azeotropically dry the solution.
5. After azeotropic drying, add 15 mg *O*-(*N*-succinimidyl)-1-,1,3,3-tetramethyl-uronium tetrafluoroborate (TSTU) to ^{18}F-FB and incubate the mixture at 60 °C for 15 min in the presence of tetrabutylammonium hydroxide.
6. Add 30 mL of 0.1N HCl solution to the cooled solution and pass the diluted reaction mixture through the activated C18 Sep-Pak. Wash the cartridge with another 3 mL of water and elute ^{18}F-SFB to a 25-mL pear-shaped flask with 1 mL ACN.
7. After evaporating the solvent, redissolve the purified ^{18}F-SFB in DMSO for future conjugation with RGD peptide (*see* **Note 2**).

For our protocol, the total synthesis time of crude ^{18}F-SFB is approximately 90 min and the decay-corrected labeling yield is 60 ± 11% (n = 23).

Preparation of ^{18}F-FB-NH_2-PEG_3-RGD2 (^{18}F-FPRGD2)

1. Add purified ^{18}F-SFB to the DMSO solution of PRGD2 (200 µg, 0.12 µmol) and DIPEA (20 µL) (*see* **Note 3**).
2. Incubate the reaction mixture at 60 °C for 30 min and add 4 mL of water with 0.1% TFA to quench the reaction.
3. Inject the mixture onto the semipreparative HPLC and collect fractions containing ^{18}F-FPRGD2 (**Fig. 1**) under a typical setup condition. The flow rate is 5 mL/min and the UV absorbance is monitored at 218 nm. The HPLC retention time (R_t) for ^{18}F-FPRGD2 is 14.3 min.
4. Reconstitute the activity in normal saline and pass it through a 0.22-µm Millipore filter into a sterile multidose vial for in vivo experiments. The decay-corrected radiochemical yield based on ^{18}F-SFB is expected to be 30–60% (n = 10).

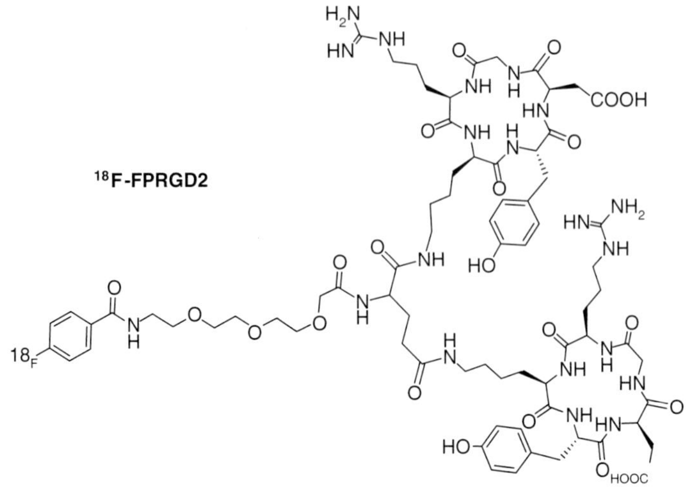

Fig. 1. Chemical structure of ^{18}F-FB-NH_2-PEG_3-RGD2 (^{18}F-FPRGD2).

5. The cold standard compound FPRGD2 is obtained from cold SFB through the same procedure.
6. Co-inject FPRGD2 cold standard and the purified ^{18}F-FPRGD2 to confirm the identity.

3.1.2. Cell Integrin Receptor-Binding Assay

To assess the integrin $\alpha_v\beta_3$-binding affinity of FPRGD2, displacement cell-binding assay, solid-phase receptor binding assay, or enzyme-linked immunosorbent assay (ELISA) need to be performed. The displacement cell-binding assay uses ^{125}I-Echistatin as the integrin $\alpha_v\beta_3$-specific radioligand. Experiments are performed with triplicate samples on U87MG human glioblastoma cells by the method described below (*see* **Note 4**).

1. U87MG cells are grown in Dulbecco's medium (Gibco) supplemented with 10% fetal bovine serum (FBS), 100 IU/mL penicillin, and 100 g/mL streptomycin (Invitrogen), at 37 °C in a humidified atmosphere containing 5% CO_2.
2. Harvest U87MG cells, wash them twice with PBS, and resuspend the cells (2×10^6 cells/mL) in binding buffer.
3. Add FPRGD2 peptide with increasing concentration (0–1,000 nM) to the filter multiscreen DV plates, followed by ^{125}I-Echistatin (0.025 μCi /well) and 1×10^5 cells (*see* **Note 5**).

4. Adjust the total volume to 200 μL and incubate the mixture for 2 h at room temperature.
5. Filter the plate through a multiscreen vacuum manifold and wash three times with cold binding buffer.
6. After drying, collect the hydrophilic polyvinylidenedifluoride (PVDF) filters and determine the radioactivity using a NaI(Tl) γ-counter (Perkin-Elmer).
7. Calculate the best-fit 50% inhibitory concentration (IC_{50}) values for the U87MG cells by fitting the data with nonlinear regression using GraphPad Prism (GraphPad Software, Inc.).

It is critical that the integrin $\alpha_v\beta_3$-binding affinity is maintained after chemical modification of RGD peptides. The binding of ^{125}I-Echistatin to U87MG cells should be inhibited by various concentrations of RGD peptide derivatives. A representative result is shown in **Fig. 2**.

3.1.3. In Vivo Tumor Imaging

1. Generate the U87MG tumor model by subcutaneous injection of 5 million cells into the left front flank of female athymic nude mice (Harlan). The mice are subjected to microPET studies when the tumor volume reaches 100–300 mm^3 (3–4 weeks after inoculation). Monitor the tumor size three times a week using a digital caliper following the equation $V = ab^2/2$.

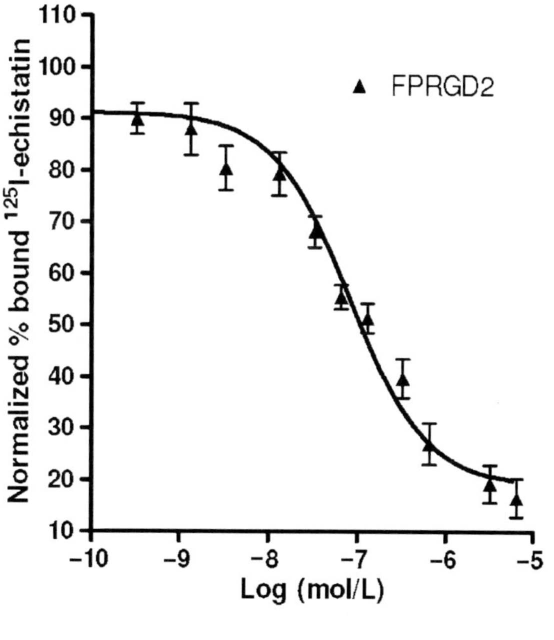

Fig. 2. In vitro inhibition of ^{125}I-Echistatin binding to integrin $\alpha_v\beta_3$ on human glioblastoma cell line U87MG by FPRGD2.

2. Inject approximately 3.7 MBq (100 μCi) of ^{18}F-FPRGD2 into a tumor-bearing mouse via the tail vein with the mouse under isoflurane anesthesia. Put the mouse back into the cage for future microPET scan.
3. For blocking experiments, inject tumor-bearing mice with 10 mg/kg mouse body weight of c(RGDyK). Inject approximately 3.7 MBq (100 μCi) of ^{18}F-FPRGD2 3 min later. Put the mice back into the cage for future microPET scan.
4. Set up the microPET system. Acquire 10-min static PET images. Reconstruct the images by a two-dimensional ordered-subsets expectation maximum (2D OSEM) algorithm and apply no attenuation or scatter correction.
5. Scan the animals at 1 h and 2 h after injection.

The U87MG human glioblastoma xenograft model has been well established to have high integrin expression. Representative microPET images are shown in **Fig. 3a**. ^{18}F-FPRGD2 delineates this integrin positive tumor. Because of the very low tracer uptake in most organs, especially in the abdominal region, ^{18}F-FPRGD2 is suitable for imaging integrin positive lesions in most areas except for the kidneys and the urinary bladder. This tracer is excreted predominantly through the renal route. The integrin $\alpha_v\beta_3$ specificity of ^{18}F-FPRGD2 in vivo is confirmed by a blocking experiment where the tracer is preinjected with c(RGDyK) (10 mg/kg). As can be seen from **Fig. 3b**, the U87MG tumor uptake in the presence of nonradiolabeled RGD peptide (0.5 ± 0.2%ID/g) is significantly lower than that without RGD blocking.

3.2. MicroSPECT of Tumor $\alpha_v\beta_3$ Integrin Expression

It has been demonstrated that integrin $\alpha_v\beta_3$ is an important molecular target for diagnosis and therapy of cancer. We have demonstrated that ^{99m}Tc-(HYNIC-RGDtetramer)(tricine)(TPPTS) displayed significant tumor accumulation with good contrast *(24)*. This high tumor uptake and fast renal excretion makes ^{99m}Tc(HYNIC-tetramer)(tricine)(TPPTS) a promising radiotracer

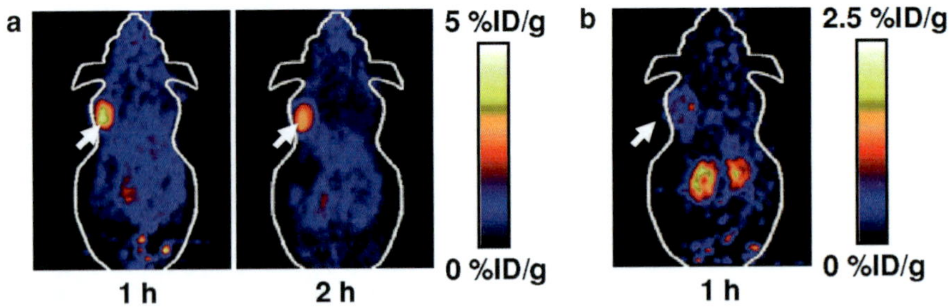

Fig. 3. (**a**) Coronal microPET images of U87MG tumor-bearing mice 1 and 2 h after intravenous injection of ^{18}F-FPRGD2. (**b**) Coronal microPET images of a U87MG tumor-bearing mouse 1 h after co-injection of ^{18}F-FPRGD2 and a blocking dose of c(RGDyK). *Arrows* indicate tumors in all cases.

for noninvasive imaging of integrin $\alpha_v\beta_3$-positive tumors by SPECT. The detailed labeling procedure is illustrated below.

3.2.1. Preparation of ^{99m}Tc-Labeled Cyclic RGD Tetramer

1. Dissolve NHS-HYNIC (4.17 mg, 10 μmol) in 300 μL DMSO and E{E[c(RGDfK)]$_2$}$_2$ (14.0 mg, 10.6 μmol) in 300 μL borate buffer (0.05 M, pH 8.5).
2. Mix the DMSO and buffer solutions and incubate the mixture at room temperature overnight, protect the reaction from light.
3. Inject the mixture onto the semipreparative HPLC and collect fractions containing HYNIC-RGDtetramer. Combine and lyophilize the fraction to give a pale yellow powder. The flow rate is 2.5 mL/min. The mobile phase is isocratic with 90% solvent A (0.1% acetic acid in water) and 10% solvent B (0.1% acetic acid in acetonitrile) at 0–5 min, followed by a gradient mobile phase going from 90% solvent A and 10% solvent B at 5 min to 60% solvent A and 40% solvent B at 20 min. The HPLC retention time (R_t) for HYNIC-RGDtetramer is 14 min.
4. Weigh 5 mg of TPPTS, 6.5 mg of tricine, 40 mg of mannitol, 38.5 mg of disodium succinate hexahydrate, and 12.7 mg of succinic acid and add to a vial. Add 1 mL water and adjust the pH to 4.8.
5. Add 0.2 mL of HYNIC-tetramer solution (100 μg/mL in water) and 0.3 mL of Na[^{99m}TcO$_4$] solution (370–1,850 MBq/mL) to the above mixture and heat the vial at 100 °C for 20–25 min in a lead-shielded water bath (*see* **Note 6**).
6. After heating, place the vial back into the lead pig and allow to stand at room temperature for 10 min.
7. Inject the mixture onto the semipreparative HPLC and collect fractions containing ^{99m}Tc(HYNIC-tetramer)(tricine)(TPPTS) (**Fig. 4**). Combine and rotary evaporate this solution to remove ACN and TFA. The flow rate is 1 mL/min. The mobile phase is isocratic with 90% solvent A (25 mM ammonium acetate buffer, pH 5.0) and 10% solvent B (acetonitrile) at 0–2 min, followed by a gradient mobile phase going from 10% solvent B at 2 min to 15% solvent B at 5 min, and to 20% solvent B at 20 min. The HPLC retention time (R_t) for ^{99m}Tc(HYNIC-tetramer)(tricine)(TPPTS) is 15.5 min (*see* **Note 7**).
8. Reconstitute the activity in normal saline for in vivo experiments.

3.2.2. Cell Integrin Receptor-Binding Assay

To assess the integrin $\alpha\beta_3$-binding affinity of tetramer derivatives, a displacement cell-binding assay is performed on MDA-MB-435 cells using ^{125}I-Echistatin as the integrin $\alpha_v\beta_3$-specific radioligand.

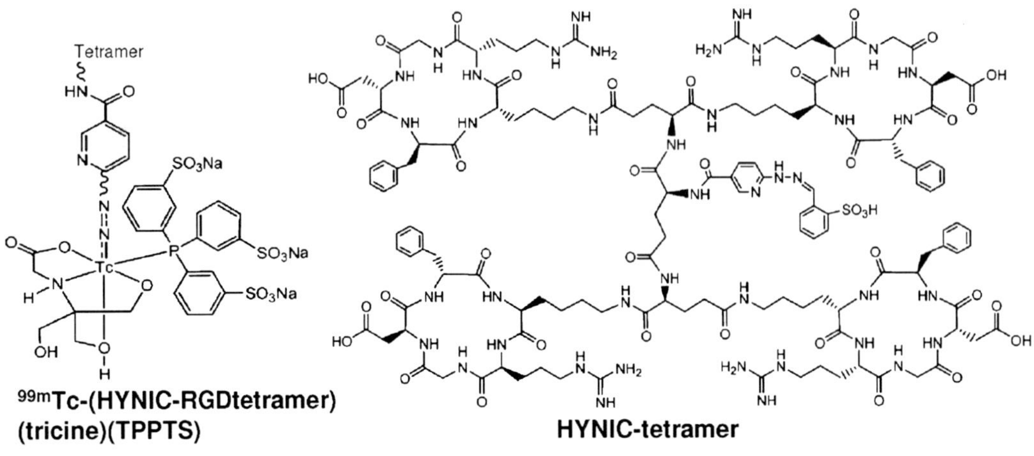

Fig. 4. Schematic structure of ^{99m}Tc(HYNIC-tetramer)(tricine)(TPPTS). Tetramer = E{E[c(RGDfK)]$_2$}$_2$, *TPPTS* trisodium triphenylphosphine-3,3′,3′-trisulfonate (Reproduced with permission from *(24)*. Copyright 2007 American Chemical Society).

MDA-MB-435 cells are grown in Gibco's Dulbecco's medium supplemented with 10% FBS, 100 IU/mL penicillin, and 100 μg/mL streptomycin (Invitrogen Co, Carlsbad, CA), at 37 °C in humidified atmosphere containing 5% CO_2. The detailed procedure is described in **Subheading 3.1.2**. A result from a typical competitive cell binding assay is shown in **Fig. 5**. There is no significant difference in the integrin $\alpha_v\beta_3$ binding affinity between their HYNIC conjugates within the experimental error. Attachment of HYNIC do not alter the binding affinity of the tetramer E{E[c(RGDfK)]$_2$}$_2$.

3.2.3. In Vivo Tumor Imaging

1. Implant 5 × 10^6 MDA-MB-435 human breast cancer cells into the mammary fat pad of female athymic nude (nu/nu) mice. When tumors reach 0.4–0.6 cm in mean diameter, the tumor-bearing mice are used for imaging studies.
2. Set up a γ-radiation camera with the following parameters: pulse height analyzer centered at 140 keV with a 20% window, 10-min acquisition time, 128 × 128 imaging matrix, 1,000 K acquisition count limit.
3. Anesthetize the animals using a rodent anesthesia system with isoflurane (2% isoflurane in 0.2 L/min of O_2 flow).
4. Inject 400 μCi of ^{99m}Tc(HYNIC-tetramer)(tricine)(TPPTS) solution into each mouse via the tail vein (total volume: 100–200 μL).
5. Scan the animal at 1, 2, and 4 h after injection.

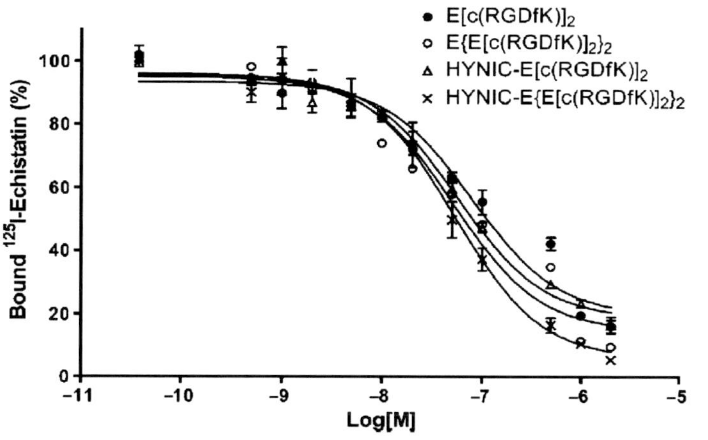

Fig. 5. In vitro inhibition of ^{125}I-Echistatin binding to $\alpha_v\beta_3$ integrin on MDA-MB-435 human breast cancer cells by E[c(RGDfK)]$_2$, E{E[c(RGDfK)]$_2$}$_2$, HYNIC-[c(RGDfK)]$_2$, and HYNIC-E{E[c(RGDfK)]$_2$}$_2$ (Reproduced with permission from *(24)*. Copyright 2007 American Chemical Society).

Radiolabeled HYNIC-tetramer is cleared rapidly from the circulation and the activity level in the abdominal region (particularly liver and lungs) is very low. This permits radionuclide imaging as early as 60 min after intravenous injection of mice. A representative planar radionuclide image is shown in **Fig. 6** and tumors were clearly visualized in SPECT images at all three time points.

3.3. NIRF Optical Imaging of $\alpha_v\beta_3$ Integrin Expression

In vivo fluorescent imaging allows visualization of biology in its intact and native physiological state. Because of the low absorbance of tissue chromophores, NIR light of wavelength 700–900 nm can propagate in the tissue for several centimeters. Therefore, imaging probes that absorb in the NIR region can be efficiently used to visualize and investigate in vivo molecular targets. The most common probes for in vivo fluorescence imaging are organic NIR fluorophores (such as polymethines) and semiconductor nanocrystals (quantum dots [QDs]). In particular, semiconductor QDs have generated significant interest for many biological applications because of their dramatically different properties compared with organic fluorophores. QDs have size- and composition-tunable fluorescence emission wavelength, symmetric and narrow emission spectra, high quantum yield, photostability, and broad absorption spectra, which allow the imaging of multicolor QDs by a single excitation with minimum signal overlap *(25)*. Because of their special properties, QDs have been widely used as in vitro probes in immunolabeling, cell tracking, in situ hybridization, and fluorescence resonance energy transfer (FRET) studies *(25–30)*. QDs have also been used for live animal targeting and imaging. Recently, in vivo targeting and imaging with QDs have

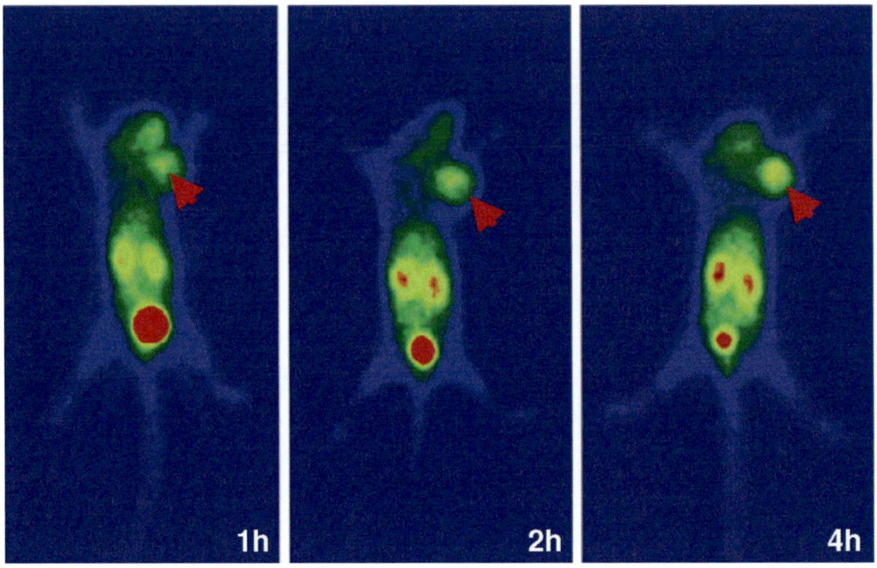

Fig. 6. Representative static SPECT images of the tumor-bearing mice administered with ~15 MBq of ^{99m}Tc(HYNIC-tetramer)(tricine)(TPPTS) and at 1, 2, and 4 h after injection. *Arrows* indicate the presence of tumors (Reproduced with permission from *(24)*. Copyright 2007 American Chemical Society).

been reported and reviewed *(31–38)*. As discussed above, to be more useful for in vivo optical imaging, NIR fluorescence (NIRF; 700–900 nm) probes are preferred because the absorbance of all biomolecules reaches a minimum in this region. In this section, we describe a step-by-step procedure for the preparation of a QD705-RGD conjugate and its subsequent use for cell staining and in vivo targeted imaging *(35)*.

3.3.1. Preparation of QD705-RGD Conjugate

Synthesis of Thiolated RGD Peptide

1. Dissolve SATA (6 μmol) in 100 μL of DMSO and add to the borate buffer (0.5 mL, pH 8.5) solution of 5 μmol c(RGDyK) peptide.
2. Incubate the reaction mixture at 25 °C for 60 min. Add 4 mL of water with 0.1% TFA to quench the reaction.
3. Inject the mixture into the semipreparative HPLC, and lyophilize the collected fractions containing SATA-c(RGDyK). In a typical setup condition, the HPLC retention time (R_t) for SATA-c(RGDyK) is 12.1 min.
4. Deprotect the peptide by adding 100 μL of 0.5 M hydroxylamine solution (pH 6.0) and purify the RGD-SH by semipreparative RP-HPLC. In a typical setup condition, The HPLC retention time (R_t) for SATA-c(RGDyK) is 10.7 min (*see* **Note 8**).

Synthesis of QD705-RGD Conjugate

1. Add 4-Maleimidobutyric acid *N*-succinimidyl ester (1 μmol) to 1 nmol QD705 solution in 300 μL borate buffer (0.05 M, pH 8.5) and incubate the mixture at room temperature for 1 h.
2. Purify the reaction mixture by a NAP-10 column with PBS. The fraction containing QD705 is collected with pH ~7.2.
3. Dissolve RGD-SH (1 μmol) in PBS buffer and add to the above purified mixture (*see* **Note 9**).
4. Incubate the mixture at room temperature for 1 h. The reaction mixture is purified by another NAP-10 column to provide the QD705-RGD conjugate.
5. Wash the obtained QD705-RGD fraction three times with PBS buffer through a Centricon (molecular cutoff: 100 K) to remove the residual unreacted RGD-SH.
6. Concentrate the QD705-RGD fractions to 1–4 μM for future in vivo imaging.

The chemistry of preparing this QD705-RGD conjugate is illustrated in **Fig. 7**.

3.3.2. Cell Integrin Receptor-Binding Assay

To assess the integrin $\alpha_v\beta_3$-binding affinity of QD705-RGD, a displacement cell-binding assay is performed using ^{125}I-Echistatin as the integrin $\alpha_v\beta_3$-specific radioligand. The detailed procedure

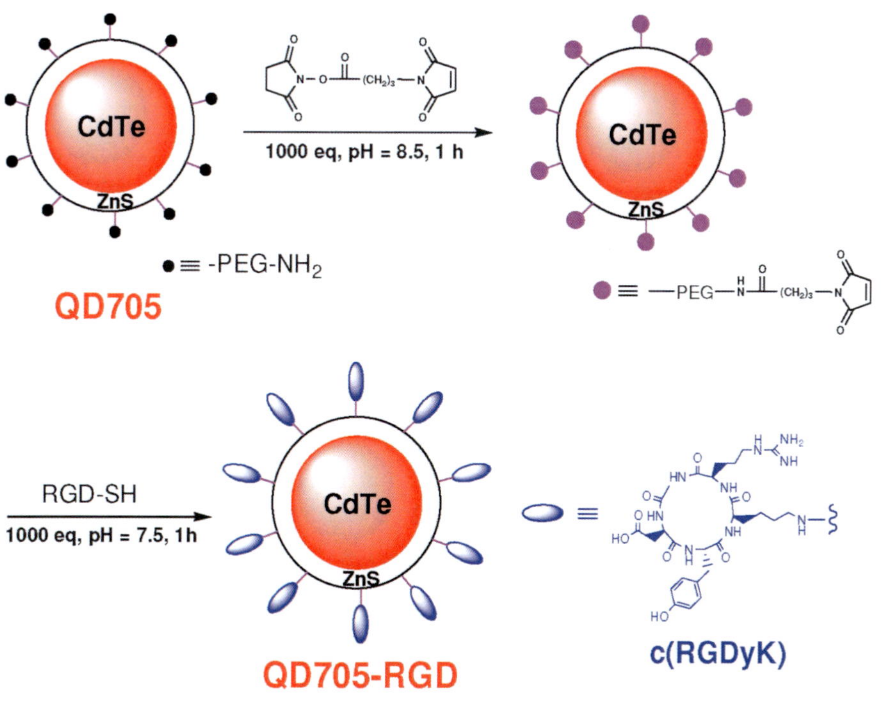

Fig. 7. Synthesis of QD705-RGD (Reproduced with permission from *(35)*. Copyright 2007 American Chemical Society).

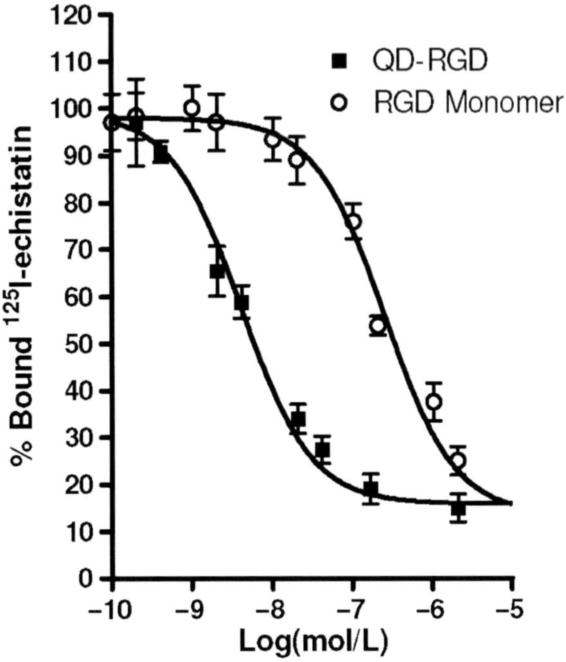

Fig. 8. In vitro inhibition of ^{125}I-Echistatin binding to integrin $\alpha_v\beta_3$ on human glioblastoma cell line U87MG by RGD peptide and QD705-RGD (*n* = 3, mean ?À SD).

is described in **Subheading 3.1.2**. A result from a typical competitive cell binding assay of QD705-RGD is shown in **Fig. 8**. The RGD monomer is also tested as a control.

3.3.3. Cell Staining

To determine the integrin $\alpha_v\beta_3$-binding specificity of QD705-RGD, MCF-7 cells (human breast cancer, integrin $\alpha_v\beta_3$ negative) and U87MG cells (human glioblastoma, integrin $\alpha_v\beta_3$ positive) are stained with 1 nM QD705, QD705-RGD, or QD705-RGD in the presence of 2 μM c(RGDyK) (*see* **Note 10**).

1. Culture MCF-7 cells in MEM supplemented with 10% (v/v) FBS at 37 °C. Culture U87MG cells in DMEM (low glucose) supplemented with 10% (v/v) FBS at 37 °C.
2. Pre-seed 0.1 × 10^5 U87MG cells or MCF-7 cells in a glass-bottomed microwell dish and incubate them overnight.
3. Aspirate the cell culture medium from the dish the next day and wash the cells two to three times with 1 mL of cell binding buffer (3- to 5-min each).
4. Add 150–300 μL of the QD705, QD705-RGD, or QD705-RGD plus 2 μM c(RGDyK) solution (1 nM, in cell binding buffer) to each dish and incubate the dishes in the incubator (37 °C) for approximately 30 min (*see* **Note 11**).
5. Aspirate the staining solution and wash the dish with cell binding buffer three times (3- to 5-min each).

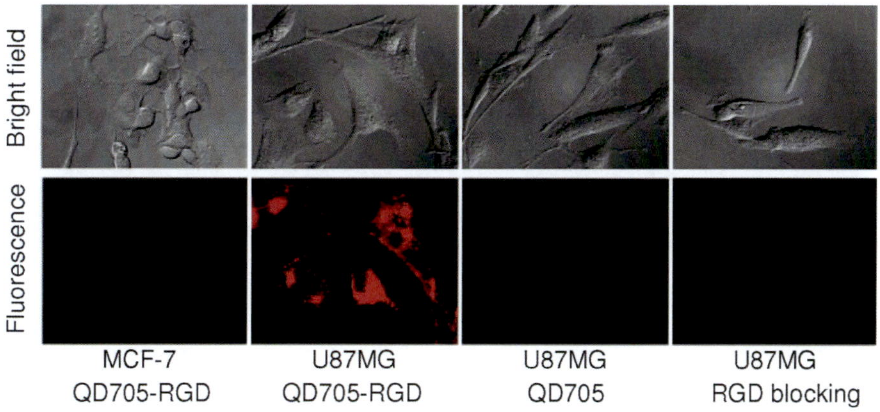

Fig. 9. In vitro staining of human breast cancer MCF-7 and human glioblastoma U87MG cells using QD705, QD705-RGD, and QD705-RGD plus RGD blocking. Magnification: 400?Á, 0.5-s exposure. All fluorescence images are acquired under the same condition and displayed under the same scale (Reproduced with permission from *(35)*. Copyright 2007 American Chemical Society).

6. Add 500–600 μL cell binding buffer to each dish and examine the cells under a microscope. Microscope filter set: excitation: 420/40 nm, emission 705/40 nm, dichrome 470 nm, 0.5 s exposure (*see* **Note 12**).

Representative brightfield and fluorescence images are shown in **Fig. 9**. QD705 only has minimal nonspecific binding on integrin $\alpha_v\beta_3$-positive U87MG cells, whereas QD705-RGD has significant integrin $\alpha_v\beta_3$-specific staining. QD705-RGD does not bind to integrin-negative MCF-7 cells.

3.3.4. In Vivo Tumor Imaging

1. Prepare the U87MG tumor bearing mice according to **Subheading 3.1.3**.
2. Setup Maestro with 590/30 nm excitation filter and 645 nm long-pass filter.
3. Inject ~200 pmol of QD705 or QD705-RGD conjugate into a tumor-bearing mouse via the tail vein.
4. Scan the animal at 1, 4, 6, and 27 h. Image acquisition time ranges from a few seconds to a few minutes per scan.
5. Spectral unmixing is performed by subtraction of the autofluorescence signal from the mixed signal of a mouse injected with QD705-RGD.

Representative Maestro fluorescence images are shown in **Fig. 10** after U87MG tumor-bearing mice are injected with 200 pmol QD705 or QD705-RGD. After spectral unmixing, the fluorescence signal resulting from QD705-RGD is clearly visible and the tumor fluorescence intensity reached maximum at approximately 6 h after injection with good contrast.

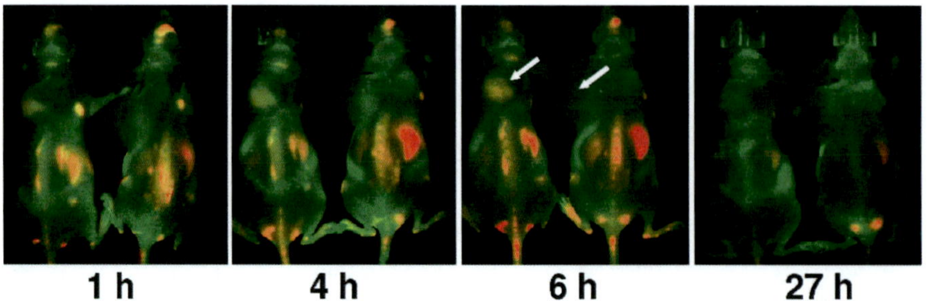

Fig. 10. In vivo NIR fluorescence imaging of U87MG tumor-bearing mice (left shoulder, indicated by *white arrows*) injected with 200 pmol of QD705-RGD (*left mouse*) and QD705 (*right mouse*), respectively (Reproduced with permission from *(35)*. Copyright 2007 American Chemical Society).

4. Notes

1. It is imperative to obtain appropriate training before handling radioactivity and to abide by all regulatory rules when handling radioactivity.
2. In **Subheading** "Synthesis of *N*-succinimidyl-4-^{18}F-Fluorobenzoate (^{18}F-SFB)," **step 7**, evaporation to complete dryness will make the radioactivity stick on the flask and it will be difficult to reconstruct the radioactivity, resulting in the loss of radioactivity.
3. In **Subheading** "Synthesis of QD705-RGD Conjugate," preparation of ^{18}F-FB-NH-PEG$_3$-RGD2 (^{18}F-FPRGD2), it is important to avoid water in the reaction mixture, otherwise the pH will be too high for the reaction.
4. In **Subheading 3.1.2**, the presence of certain metal ions (e.g., Mn^{2+} and Mg^{2+}) in the binding buffer is essential for integrin $\alpha_v\beta_3$ binding. Binding buffer without these ions will result in low count readings.
5. Newly purchased ^{125}I-Echistatin is recommended for more reliable/accurate results. Using ^{125}I-Echistatin after more than two half-lives (~4 months) will lead to low counts and irregular results.
6. ^{99m}Tc is a radionuclide emitting γ-radiation at 140 keV. Standard shielding and radionuclide-handling procedures must be used. Direct exposure to the radioactive dose should be kept to a minimum. Because ^{99m}Tc is a short-lived isotope ($t_{1/2}$ = 6.03 h), injected mice and the waste products for imaging do not represent any significant radiation hazard after 60 h (ten half-lives) of decay.
7. The ^{99m}Tc-(HYNIC-tetramer)(tricine)(TPPTS) labeling depends on maintaining the reducing environment. Therefore, freshly prepared reducing solution is suggested for the successful binding of ^{99m}Tc to the HYNIC-tetramer conjugate. All efforts to reduce air exposure should be taken.

8. In **Subheading 3.3.1** "Synthesis of *N*-succinimidyl-4-^{18}F-fluorobenzoate (^{18}F-SFB)," **step 4**, the pH of the reaction mixture is critical. Low pH of the mixture may result in incomplete deprotection and high pH will result in the formation of disulfide bonds because of oxidation. We also suggest storing RGD-SH under acidic conditions (pH 3–4) in water to prevent disulfide formation.
9. In **Subheading 3.3.1** "Preparation of ^{18}F-FB-NH_2-PEG_3-RGD2 (^{18}F-FPRGD2)," **step 3**, the pH of the reaction mixture is critical. TCEP.HCl (~1 mg or less) can also be added to reduce the disulfide that might be formed during the storage of RGD-SH.
10. In **Subheading 3.3.3**, if QD705-RGD has been stored for a while, we suggest centrifugation of the particle (3578 *g* for 10 min) to remove the aggregates before cell staining.
11. In **Subheading 3.3.3**, RGD blocking experiment, if the blocking RGD concentration is too high, it may cause detachment of the cells from the dish. 1 μM RGD peptide is sufficient to observe the blocking effect.
12. In **Subheading 3.3.3**, the fluorescence images should be acquired with the same microscope setup and they should be displayed at the same pseudocolor scale.

Acknowledgments

The authors thank Professor Shuang Liu, Dr. Yun Wu, Dr. Gang Niu, Dr. Zhanhong Wu, and Dr. Weibo Cai for developing the in vivo imaging probes discussed in this chapter. This work was supported by the National Institute of Biomedical Imaging and Bioengineering (NIBIB) (grant R21 EB001785), the National Cancer Institute (NCI) (grants R01 CA119053, R21 CA121842, R21 CA102123, P50 CA114747, U54 CA119367, and R24 CA93862), and the Department of Defense (DOD) (grants W81XWH-04-1-0697, W81XWH-06-1-0665, W81XWH-06-1-0042, and DAMD17-03-1-0143).

References

1. Gwyther, S.J. (2005) New imaging techniques in cancer management. *Ann Oncol.* **16** Suppl 2, ii63–70.
2. Rudin, M., and Weissleder, R. (2003) Molecular imaging in drug discovery and development. *Nat Rev Drug Discov.* **2**, 123–131.
3. Massoud, T.F., and Gambhir, S.S. (2003) Molecular imaging in living subjects: seeing fundamental biological processes in a new light. *Genes Dev.* **17**, 545–580.
4. Herschman, H.R. (2003) Molecular imaging: looking at problems, seeing solutions. *Science.* **302**, 605–608.

5. Gross, S., and Piwnica-Worms, D. (2005) Spying on cancer: molecular imaging in vivo with genetically encoded reporters. *Cancer Cell.* 7, 5–15.
6. Gambhir, S.S. (2002) Molecular imaging of cancer with positron emission tomography. *Nat Rev Cancer.* **2**, 683–693.
7. Sharma, V., Luker, G.D., and Piwnica-Worms, D. (2002) Molecular imaging of gene expression and protein function in vivo with PET and SPECT. *J Magn Reson Imaging.* **16**, 336–351.
8. Cai, W., Gambhir, S.S., and Chen, X. (2005) Multimodality tumor imaging targeting integrin αvβ3. *Bio Techniques.* **39**, s6–s17.
9. Peremans, K., Cornelissen, B., Van Den Bossche, B., Audenaert, K., and Van de Wiele, C. (2005) A review of small animal imaging planar and pinhole spect Gamma camera imaging. *Vet Radiol Ultrasound.* **46**, 162–170.
10. Frangioni, J.V. (2003) In vivo near-infrared fluorescence imaging. *Curr Opin Chem Biol.* **7**, 626–634.
11. Spibey, C.A., Jackson, P., and Herick, K. (2001) A unique charge-coupled device/xenon arc lamp based imaging system for the accurate detection and quantitation of multicolour fluorescence. *Electrophoresis.* **22**, 829–836.
12. Montet, X., Ntziachristos, V., Grimm, J., and Weissleder, R. (2005) Tomographic fluorescence mapping of tumor targets. *Cancer Res.* **65**, 6330–6336.
13. Brooks, P.C., Clark, R.A., and Cheresh, D.A. (1994) Requirement of vascular integrin $\alpha_v\beta_3$ for angiogenesis. *Science.* **264**, 569–571.
14. Hood, J.D., and Cheresh, D.A. (2002) Role of integrins in cell invasion and migration. *Nat Rev Cancer.* **2**, 91–100.
15. Hynes, R.O. (2002) Integrins: bidirectional, allosteric signaling machines. *Cell.* **110**, 673–687.
16. Xiong, J.P., Stehle, T., Diefenbach, B., Zhang, R., Dunker, R., Scott, D.L., Joachimiak, A., Goodman, S.L., and Arnaout, M.A. (2001) Crystal structure of the extracellular segment of integrin $\alpha_v\beta_3$. *Science.* **294**, 339–345.
17. Cai, W., and Chen, X. (2006) Anti-angiogenic cancer therapy based on integrin $\alpha_v\beta_3$ antagonism. *Anti-Cancer Agents in Medicinal Chemistry.* **6**, 407–428.
18. Xiong, J.P., Stehle, T., Zhang, R., Joachimiak, A., Frech, M., Goodman, S.L., and Arnaout, M.A. (2002) Crystal structure of the extracellular segment of integrin $\alpha_v\beta_3$ in complex with an Arg-Gly-Asp ligand. *Science.* **296**, 151–155.
19. Haubner, R. (2006) $\alpha_v\beta_3$-integrin imaging: a new approach to characterise angiogenesis? *Eur J Nucl Med Mol Imaging.* **33** Suppl 1, 54–63.
20. Liu, S. (2006) Radiolabeled multimeric cyclic RGD peptides as integrin $\alpha_v\beta_3$ targeted radiotracers for tumor imaging. *Mol Pharm.* **3**, 472–487.
21. Wu, Y., Zhang, X., Xiong, Z., Cheng, Z., Fisher, D.R., Liu, S., Gambhir, S.S., and Chen, X. (2005) microPET imaging of glioma integrin $\alpha_v\beta_3$ expression using 64Cu-labeled tetrameric RGD peptide. *J Nucl Med.* **46**, 1707–1718.
22. Wu, Z., Li, Z., Cai, W., He, L., Chin, F., li, F., and Chen, X. (2007) ^{18}F-labeled mini-PEG spacered RGD dimer (^{18}F-FPRGD2): synthesis and microPET imaging of $\alpha_v\beta_3$ integrin expression. *Eur J Nucl Med Mol Imaging.* **34**, 1823–1831.
23. Vaidyanathan, G., and Zalutsky, M.R. (2006) Synthesis of N-succinimidyl 4-[^{18}F]fluorobenzoate, an agent for labeling proteins and peptides with ^{18}F. *Nat Protoc.* **1**, 1655-1661.
24. Liu, S., Hsieh, W.Y., Jiang, Y., Kim, Y.S., Sreerama, S.G., Chen, X., Jia, B., and Wang, F. (2007) Evaluation of a ^{99m}Tc-labeled cyclic RGD tetramer for noninvasive imaging integrin $\alpha_v\beta_3$-positive breast cancer. *Bioconjug Chem.* **18**, 438–446.
25. Bruchez, M., Jr., Moronne, M., Gin, P., Weiss, S., and Alivisatos, A.P. (1998) Semiconductor nanocrystals as fluorescent biological labels. *Science.* **281**, 2013–2016.
26. Algar, W.R., and Krull, U.J. (2007) Towards multi-colour strategies for the detection of oligonucleotide hybridization using quantum dots as energy donors in fluorescence resonance energy transfer (FRET). *Anal Chim Acta.* **581**, 193–201.
27. Dahan, M., Levi, S., Luccardini, C., Rostaing, P., Riveau, B., and Triller, A. (2003) Diffusion dynamics of glycine receptors revealed by single-quantum dot tracking. *Science.* **302**, 442–445.
28. Kaul, Z., Yaguchi, T., Kaul, S.C., Hirano, T., Wadhwa, R., and Taira, K. (2003) Mortalin imaging in normal and cancer cells with quantum dot immuno-conjugates. *Cell Res.* **13**, 503–507.
29. Ness, J.M., Akhtar, R.S., Latham, C.B., and Roth, K.A. (2003) Combined tyramide signal amplification and quantum dots for sensitive and photostable immunofluorescence detection. *J Histochem Cytochem.* **51**, 981–987.
30. Pathak, S., Choi, S.K., Arnheim, N., and Thompson, M.E. (2001) Hydroxylated quantum dots

as luminescent probes for in situ hybridization. *J Am Chem Soc.* **123**, 4103–4104.

31. Akerman, M.E., Chan, W.C., Laakkonen, P., Bhatia, S.N., and Ruoslahti, E. (2002) Nanocrystal targeting in vivo. *Proc Natl Acad Sci U S A.* **99**, 12617–12621.
32. Ballou, B., Lagerholm, B.C., Ernst, L.A., Bruchez, M.P., and Waggoner, A.S. (2004) Noninvasive imaging of quantum dots in mice. *Bioconjug Chem.* **15**, 79–86.
33. Cai, W., Chen, K., Li, Z.B., Gambhir, S.S., and Chen, X. (2007) Dual-Function Probe for PET and Near-Infrared Fluorescence Imaging of Tumor Vasculature. *J Nucl Med.* **48**, 1862–1870.
34. Cai, W., and Chen, X. (2007) Nanoplatforms for targeted molecular imaging in living subjects. *Small.* **3**, 1840–1854.
35. Cai, W., Shin, D.W., Chen, K., Gheysens, O., Cao, Q., Wang, S.X., Gambhir, S.S., and Chen, X. (2006) Peptide-labeled near-infrared quantum dots for imaging tumor vasculature in living subjects. *Nano Lett.* **6**, 669–676.
36. Chan, W.C., and Nie, S. (1998) Quantum dot bioconjugates for ultrasensitive nonisotopic detection. *Science.* **281**, 2016–2018.
37. Dubertret, B., Skourides, P., Norris, D.J., Noireaux, V., Brivanlou, A.H., and Libchaber, A. (2002) In vivo imaging of quantum dots encapsulated in phospholipid micelles. *Science.* **298**, 1759–1762.
38. Li, Z.B., Cai, W., and Chen, X. (2007) Semiconductor quantum dots for in vivo imaging. *J Nanosci Nanotechnol.* **7**, 2567–2581.

Chapter 32

Ultrahigh Resolution Imaging of Biomolecules by Fluorescence Photoactivation Localization Microscopy

Samuel T. Hess, Travis J. Gould, Mudalige Gunewardene, Joerg Bewersdorf, and Michael D. Mason

Summary

Diffraction limits the biological structures that can be imaged by normal light microscopy. However, recently developed techniques are breaking the limits that diffraction poses and allowing imaging of biological samples at the molecular length scale. Fluorescence photoactivation localization microscopy (FPALM) and related methods can now image molecular distributions in fixed and living cells with measured resolution better than 30 nm. Based on localization of single photoactivatable molecules, FPALM uses repeated cycles of activation, localization, and photobleaching, combined with high-sensitivity fluorescence imaging, to identify and localize large numbers of molecules within a sample. Procedures and pitfalls for construction and use of such a microscope are discussed in detail. Representative images of cytosolic proteins, membrane proteins, and other structures, as well as examples of results during acquisition are shown. It is hoped that these details can be used to perform FPALM on a variety of biological samples, to significantly advance the understanding of biological systems.

Key words: FPALM, PALM, STORM, Hemagglutinin, Photoactivation, EosFP, Super-resolution, PA-GFP

1. Introduction

1.1. Resolution, Diffraction Barrier, and Point-Spread Function

Fluorescence light microscopy is one of the most frequently used imaging techniques in biological research *(1)*. However, despite extensive efforts over the past two centuries, the details of the structural organization and interaction of complex molecular assemblies have remained largely concealed. Diffraction of light ultimately blurs the tiny details of an observed object, effectively hiding the fine structures contained therein *(2)*. The smallest resolvable details are of the size described by the Rayleigh Criterion, which defines the resolution of a microscope as

James Weifu Lee and Robert S. Foote (eds.), *Micro and Nano Technologies in Bioanalysis*,
Methods in Molecular Biology, vol. 544
DOI 10.1007/978-1-59745-483-4_32,

$$R_0 = 0.61\lambda/\text{NA}, \tag{1}$$

where λ is the detection wavelength and NA is the numerical aperture of the objective lens (the product of the refractive index and the sine of the aperture angle of the lens). Two objects less than R_0 apart are difficult or impossible to distinguish. Although this is in principle still correctable by image processing for structures consisting only of a known (and small) number of point-like objects, it obscures objects that are more complex at size scales below R_0. More generally, the imaging characteristics of a fluorescence microscope (and other incoherent imaging systems that are spatially invariant over a field of view) can be described by the point-spread function (PSF) *(3–5)*. This quantity expresses the two- or three-dimensional spatial intensity distribution resulting from imaging a single point-like object. The size of the PSF hence represents the smallest observable feature in conventional imaging. If an emitting object is smaller than R_0, its image will be at least R_0 in size (including single molecules and atoms). A patterned sample featuring fine structures will appear nearly or completely homogeneous in its image at size scales below R_0.

We define a super-resolution microscopy method as one that allows imaging of features within the sample that are smaller than R_0. The PSF in fluorescence microscopy can be measured by imaging fluorescent particles significantly smaller than the PSF (beads for example) in two or three dimensions. The full-width-at-half-maximum (FWHM) of the PSF (d_F), is another common definition of the diffraction-limited resolution, but we will use R_0 as the definition of resolution. The parameter r_0, the $1/e^2$ radius of the PSF, is related to the FWHM by $d_F \sim 1.17r_0$, and is also related to s, the standard deviation of the PSF by $r_0 = 2s$.

1.2. Localization vs. Resolution

Although R_0 describes the smallest resolvable feature of complex structures that are imaged conventionally, finer details can be obtained if the structure is very simple and some information about its attributes is given. Most prominently, a single point-like object (i.e., an object much smaller than the PSF) well separated from any other structure in its neighborhood can be *localized*; that is, its position can be determined. Importantly, localization can be achieved with much higher accuracy than $\pm R_0$ *(6,7)*. Effectively, each detected photon emitted from an object constitutes a measurement of its position with uncertainty equal to s, the standard deviation of the PSF (s), which is approximately $0.37 \cdot R_0$. Repeated measurements of that position (from additional detected photons) will reduce the overall measurement uncertainty. The link between the resolution defined by the PSF (R_0) and the precision in determining the center of the PSF (σ_x) is analogous to the relationship between the standard deviation of a set of measurements (the detected photons distributed with standard deviation s, according to the PSF) and the standard deviation of the *mean*. Thus, in the absence of background and finite-sized pixel effects, the ratio between s and σ_x is $\sigma_x = s/\sqrt{N} \approx 0.39R_0/\sqrt{N}$, where N is the number of measurements (detected photons), and σ_x is the standard deviation in the position of the molecule. Localization precision can therefore be improved in an ideal system by as much as a factor of $\sqrt{N}$ compared with the resolution (see below for discussion of localization precision in the presence of background and finite pixel size).

1.3. Brief Review of Super-Resolution Methods

Although the diffraction barrier has limited the resolution of far-field (i.e., lens-based) light microscopy to ~250 nm, other techniques with better resolution have existed for more than half a century. They create high-resolution images by lowering the effective wavelength (e.g., electron and X-ray microscopy and tomography), by using near-field optics that are not governed by diffraction (e.g., scanning near-field optical microscopy or total internal reflection microscopy [TIRF]) or by avoiding optics completely (i.e., atomic force microscopy [AFM] and scanning tunneling microscopy).

On the other hand, light microscopy has many advantages when applied to biological systems, including the capability to image inside living biological specimens (necessitating far-field optics) in two and three dimensions, with single-molecule sensitivity, remarkable signal-to-noise and signal-to-background ratios, using a tremendous variety of fluorescent probes *(1,8)*. As a result, several concepts have been developed to attempt to achieve resolution beyond the classical diffraction limit. Confocal laser scanning microscopy enhances resolution by using diffraction-limited laser illumination and a detector aperture, resulting in a resolution improvement by up to $\sqrt{2}$, in addition to three-dimensional imaging capabilities. Mathematical image processing after acquisition (deconvolution) is able to enhance the resolution by amplifying high spatial frequency content in the image data. Depending on a the signal-to-noise ratio of the data, typically a resolution improvement of a factor of two can be achieved. This becomes especially powerful if combined with an optical scheme that enhances these spatial frequencies. For example, structured illumination microscopy *(9)* based on this principle has achieved close to 100 nm in lateral resolution. 4Pi microscopy *(10–14)* and I^5M *(15,16)* increase the effective aperture of the optics dramatically by using two opposing objective lenses. These lenses are optically combined in a coherent manner so that the lens assembly acts as if it would be a single lens with a strongly increased aperture, achieving ~100 nm resolution in the axial direction. All of these described methods however only *bend* the diffraction limit: the achievable resolution of approximately

80 nm (all methods combined) is still governed by diffraction and is still fundamentally limited.

The introduction of stimulated emission depletion (STED) microscopy in 1994 *(17)* demonstrated that the diffraction limit can be *broken*: in STED microscopy, excited fluorescent probe molecules are actively quenched by a second light beam by the physical process of stimulated emission before their spontaneous emission of fluorescence. Fluorescence emission can effectively be switched off in regions illuminated by the STED beam. Structuring the STED beam intensity profile, typically as a doughnut-shaped focus surrounding a standard excitation laser focus, restricts fluorescence emission to well-defined areas in the sample, the areas of low STED intensity. Although the STED beam itself is still diffraction limited, increasing the intensity *saturates* the depletion efficiency distribution: even areas close to the minima of the STED beam profile are now sufficiently bright to virtually switch off all fluorescence. This restricts the remaining fluorescence to smaller and smaller volumes—ultimately only the spots of zero STED intensity remain fluorescent. In practice, spatial resolutions of 16 nm, achieved with visible wavelengths, have been demonstrated by STED in a laser scanning microscope *(18)*. Recent publications have successfully demonstrated the applicability to biological research *(19,20)*. The concept of resolution enhancement well beyond the diffraction limit is not limited to the effect of stimulated emission or single laser focus-based microscopy, but can be used with any kind of probe molecule that shows reversible optical transitions between two distinguishable optical states and any structured illumination geometry *(21–23)*, as summarized by the RESOLFT concept *(24)*.

It is important to realize that super-resolution images in optical far-field microscopy can only be achieved by *sequential recording* in one form or another because the detection process for every emitter in the sample is diffraction limited. Reading out the complete ensemble of probe molecules at once would therefore lead to overlap between the images of molecules closer to one another than R_0, and render any structure smaller than R_0 unresolvable. STED and RESOLFT microscopy avoid this by spatially scanning the sample with an illumination pattern and recording a temporal data sequence. Nontemporal sequences, for example, based on different wavelengths *(25–27)*, are also possible.

The use of temporal sequences is also essential in a different area of modern microscopy, the field of particle tracking: single particles are imaged over time, and by localizing them in every image, particle trajectories can be created with position accuracies better than the diffraction limit even down to the ~1 nm range *(28)*. Such success suggested directly that the high precision achievable by localization could potentially be exploited to achieve high-resolution images of structures. Unfortunately, using the subsequent bleaching or statistical blinking of fluorescent molecules to generate sub-diffraction images is limited to only a few molecules within a diffraction-limited volume *(22,29,30)*.

Recently, we have solved this problem by developing fluorescence photoactivation localization microscopy (FPALM). In FPALM, single fluorescent molecules are actively switched between bright and dark states, allowing control of the number of visible probe molecules during a recording sequence. Complex structures are resolved by combining the localization information of a large number of different stochastic subsets of the ensemble of probe molecules into one data set. Simultaneous with our development, two other groups have developed very similar approaches called PALM and STORM *(31–33)*. Recently, another variant, PALMIRA, has been introduced *(34,35)*. Most of the methods described below can be applied directly to these techniques as well.

1.4. Principles and Theory

Normally, in fluorescence microscopy, a large number of fluorescent molecules are visible at a given time. Whenever two molecules are within R_0 of one another, they cannot be distinguished as separate individuals. As a result, normal fluorescence images are blurred by diffraction, and features smaller than the PSF are lost. In contrast, single fluorescent molecules can be localized with precision much better (smaller) than the diffraction-limited resolution. Localization of a molecule essentially amounts to measurement of its position; each photon detected from that molecule constitutes a measurement of its position, so a larger number of detected photons will result in improved localization precision. Quantitatively, the localization precision can be calculated using *(36)*:

$$\sigma_{xy}^2 = \frac{s^2 + q^2/12}{N} + \frac{8\pi s^4 b^2}{q^2 N^2}, \tag{2}$$

where σ_{xy} is the precision with which a fluorescent object can be localized in two dimensions, s is the standard deviation of the PSF (proportional to R_0), N is the total number of photons collected (not photons per pixel), q is the size of an image pixel within the sample space, and b is the background *noise* per pixel (not background intensity). For small pixel sizes and negligible background noise, **Eq. 2** reduces to $\sigma_{xy}^2 = s^2/N$.

FPALM enhances resolution by imaging only a sparse subset of the molecules in the sample at a given time, which allows each individual molecule to be identified and localized **(Fig. 1)**. To control the number of visible fluorescent probe molecules, two lasers are typically used (although a single laser can be used in some cases, see readout-induced activation below). The first laser (called the readout laser) typically illuminates the sample continuously, or at least most of the time, during acquisition, and

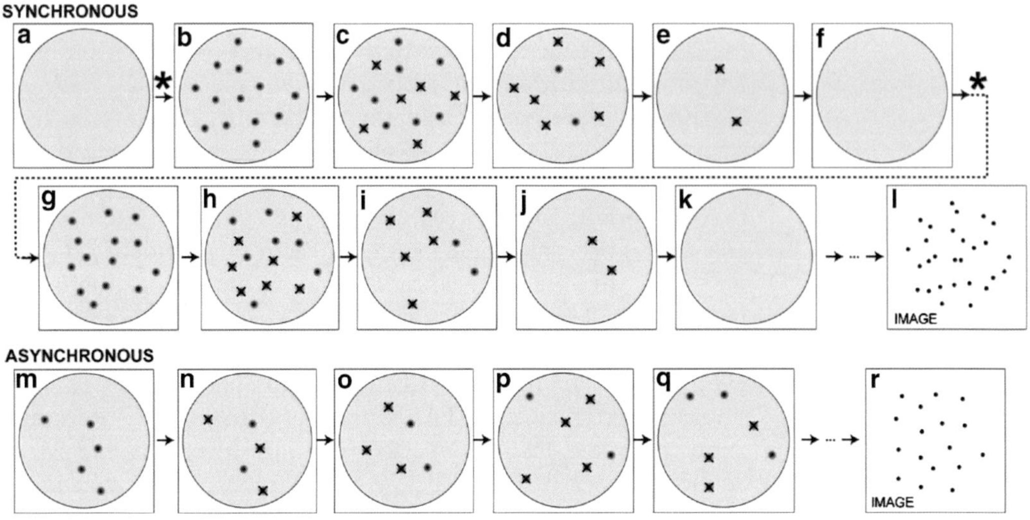

Fig. 1. Principle of fluorescence photoactivation localization microscopy (FPALM). Two methods are described, synchronous (**a–l**) and asynchronous (**m–r**). The *shaded squares* are successive simulated frames from an FPALM acquisition, in which a planar region of interest is being imaged with a widefield fluorescence microscope. *Fuzzy black dots* denote the diffraction-limited images of single fluorescent molecules, whereas *X*s denote single molecules that photobleach during the current frame. In synchronous FPALM (**a–k**), molecules are nonfluorescent ("inactive") and invisible initially. Activation using the appropriate laser wavelength (shown by a *large asterisk* between frames) causes inactive molecules to become active (meaning they are fluorescent when excited by a second laser, the readout beam). Here, irreversible activation is shown. Within the area illuminated by the readout beam (*shaded circular region* in each frame), any active molecule will fluoresce (*fuzzy black dots*) until it photobleaches. By adjusting the activation intensity to be low enough that only a few inactive molecules are activated per frame, the number of visible molecules is small at any given time, and molecules can be individually identified and localized. One cannot control which particular molecules activate in a given frame; rather, a stochastic subset of the inactive molecules absorbs a photon from the activation beam and becomes active. Once the brief activation pulse is complete, the activated molecules stay visible until they are photobleached by the readout beam (**b–e**). After most or all of the active molecules have bleached (e.g., frame **f**), an additional activation pulse is applied (between **f** and **g**), a different subset of molecules becomes active (frame **g**) and is imaged (**g–j**) until that subset bleaches (**k**). This cycle of activate, image, and photobleach is repeated until thousands or millions of molecules in the sample have been imaged. The plot of the positions of all localized molecules is the FPALM image (**l**). Asynchronous FPALM (**m–q**) means that activation, imaging, and photobleaching occur simultaneously. Initially active molecules (**m**) are imaged until they bleach, whereas new molecules activate either because of continuous illumination by the activation laser, or by readout-beam-induced activation. Similarly, the plot of the positions of all localized molecules is the FPALM image (**r**). Note that the *spots* shown in l and r are intentionally smaller and sharper than the imaged molecules in other frames because localization of molecules can be done more precisely than the diffraction-limited resolution. Because localization and molecular density are what limits the detail in FPALM, images can be obtained that depict features significantly smaller than the diffraction-limited resolution.

is used to excite any active molecules within the sample and cause them to fluoresce. Active molecules are defined as molecules that can be excited by the readout beam to produce fluorescence. Inactive molecules are (ideally) nonfluorescent under illumination by the readout beam. Inactive molecules can absorb photons at the activation wavelength (typically shorter than the readout wavelength) to become active.

The key difference in FPALM samples compared with normal fluorescent samples is that initially, almost all of the probe molecules are inactive (invisible) in FPALM, whereas, in normal fluorescent microscopy, the majority of probe molecules is visible (effectively in the active state). Whereas in normal fluorescence microscopy one would see bright fluorescence when illuminating the sample with the standard excitation wavelength, in FPALM, virtually no fluorescence is initially emitted from the majority of molecules. In FPALM, the molecules must first be photoactivated by the activation laser, and then they are visible under excitation by the *readout* laser. Thus, by carefully regulating the activation laser intensity, one can control the rate at which molecules become activated, and hence become visible under the readout laser. However, without some way to turn off currently active molecules, the number of active molecules would either grow (under activation) or stay constant (without activation). Eventually, after many molecules were activated, the total number of visible molecules would get so high that individual molecules would no longer be distinguishable from one another (much like raindrops on a windshield, which gradually coalesce if the windshield wiper is turned off). In FPALM, photobleaching provides the balancing factor that limits the total number of visible (active) molecules. Alternatively, if the fluorophore is reversibly photoactivatable, then active molecules can be switched off (deactivated) to reduce or limit the number of visible molecules. In either case, this mechanism to limit the number of active molecules is needed so that the distance between each molecule and its nearest neighbor stays at least as large as R_0, ideally much larger than R_0, so that each individual molecule can be localized.

Once the density of molecules can be controlled to be low enough for single-molecule localization, it is only a matter of imaging those molecules with a high-sensitivity camera as molecules are activated and then photobleached (or deactivated). Photobleaching typically occurs spontaneously in the presence of the readout laser, so that no explicit photobleaching step is required in practice (it happens on its own). From a time-lapse movie of many cycles (defined as activation, imaging, and photobleaching of molecules within a given field of view), one can determine the positions of a large number of (10^4–10^6) molecules. The uncertainty in the position of each molecule can also be determined experimentally from repeated imaging of the same molecule (after it has been activated, and before it photobleaches) or theoretically from ***Eq. 2*** using the measured number of photons detected from that molecule. The FPALM image is just the measured positions of all molecules, plotted together. Several methods for plotting these positions are discussed in more detail below (*see* **Subheading 3**).

1.5. Factors That Determine Localization Precision

Considering **Eq. 2**, which governs the localization precision of single molecules imaged by fluorescence microscopy *(36)*, several guidelines emerge for optimizing the localization precision (minimizing σ_{xy}). First, the number of detected photons (N)

should be maximized because both terms in **Eq. 2** decrease with increasing *N*. The left term corresponds to the contribution to σ_{xy} from shot noise and pixelation noise in the absence of background. Even in the absence of background, a molecule imaged with *N* detected photons will appear noisy and broad (the image will have a width of ~R_0). For localization, the image is typically analyzed by least-squares fitting the image with the known PSF, or a Gaussian approximation of the known PSF. The noisier the image of the molecule, the more uncertain its position will be. A larger number of detected photons will result in a less noisy image and a smaller uncertainty (σ_{xy}). The pixel size *q* also has a small influence: for large pixel sizes, σ_{xy} increases. In the extreme case of pixels much larger than r_0, the size of the pixel itself governs the localization uncertainty because it is not clear where in the pixel the molecule is located. In practice, however, the pixel size becomes negligible for $q<<s$. Because of other factors, such as readout noise and field of view, it is sometimes undesirable to decrease *q* dramatically below *s*.

The second term in **Eq. 2** accounts for localization uncertainties caused by background noise. Background signal can result from camera electronics; scattered light; autofluorescence from the specimen, mounting medium, or glass; nonspecific probe labeling; weak fluorescence of inactive probe molecules; and many other sources (*see* **Subheading 3**). This light detected from sources other than the molecule of interest leads to greater noise in the image, and degrades localization precision. Thus, the second term in **Eq. 2** contains the background *noise* per pixel (*b*), in photons (not the background signal level). A truly uniform (noiseless) offset (such as an offset in the black level of the camera that is identical for all pixels) can be subtracted and does not add to localization uncertainty. Minimizing the background noise and increasing the number of detected photons is therefore crucial in FPALM. Note that the dependence of the second term on *N* is as $1/N^2$, so the relative improvement as *N* increases is more significant for the second term compared with the first term. In principle, however, both of these terms can be arbitrarily decreased by increasing *N*, suggesting that localization with submolecular (<1 nm) precision is, in principle, possible. Experimentally, 1.5-nm localization precision has already been achieved *(6)*.

Although localization precision limits the quality of the image obtained by FPALM, the number of molecules detected also limits such quality. For example, a single molecule localized with 2-nm precision will not reveal much about a multimolecular structure of which it is a part. Rather, sufficient density is needed to provide comprehensive structural details *(37,38)*. The density of molecules must be high enough that a large number of molecules are localized within the structure of interest.

In super-resolution single-molecule localization methods such as FPALM, a new definition of resolution is needed that takes into account the localization precision and the molecular density (or sparseness). We refer to localization-based resolution to describe the smallest structure that can be imaged by a localization-based method. Thus, if r_L is the localization-based resolution of FPALM, r_L must be dependent on both σ_{xy} and r_{NN}, the nearest neighbor distance between molecules (which is dependent on the molecular density). One can therefore propose a relation between the FPALM resolution, σ_{xy} and r_{NN}:

$$r_L^2 = \sigma_{xy}^2 + r_{NN}^2. \quad (3)$$

If $\sigma_{xy} << r_{NN}$, this relation results in $r_L \sim r_{NN}$, the limit at which the density of molecules is so sparse that resolution is limited by that sparseness. On the other hand, if $r_{NN} << \sigma_{xy}$, the density does not limit the resolution, and $r_L \approx \sigma_{xy}$.

2. Materials

2.1. Photoactivatable Probes

FPALM in its original form requires the use of a fluorescent probe that is visible in small numbers of molecules at a time and whose density of visible molecules can be controlled. Typically, photoactivatable fluorescent probes satisfy these criteria. The choice of an appropriate probe is dependent on knowledge of its photophysical properties and the desired application. Probes with high photoactivation activation yields and low rates of spontaneous activation (relative to light-induced activation) are desirable for controlling the number of active molecules.

Particularly powerful probes for FPALM are the genetically encoded photoactivatable and photoswitchable proteins (PA-FPs) *(39)* derived from green fluorescent protein (GFP) *(40)* and other fluorescent proteins. Cells are typically transfected with a construct containing the gene for the PA-FP attached to the gene for the protein of interest. Because of the facility of genetic manipulations, probes can be tailored to specific biological applications by adjusting the promoter and PA-FP used, by making point mutations in the gene for the PA-FP, or by splicing various segments of genes to create chimeras. These changes can alter the expression level, absorption and emission spectra of the probe, its sensitivity to ion concentrations or other environmental parameters, the delay between transfection and expression (called the maturation time), and a number of other useful properties *(41)*.

Given the flexibility in genetically encoded markers, one can control the properties of PA-FPs to a large degree. What properties would be ideal? Several criteria are generally advantageous for any photoactivatable probe to be used in FPALM: (1) Large absorption coefficient at the activation and readout excitation wavelengths; (2) large fluorescence quantum yield in the active state; (3) little tendency for self-aggregation; (4) large quantum yield for activation; (5) negligible quantum yield for readout-induced activation, unless asynchronous FPALM is to be used (*see* **Fig.1**), in which case a small quantum yield is desirable; (6) small but finite quantum yield for photobleaching, because active molecules must emit a large number of photons, but must also eventually bleach or the density will become too high to allow individual identification and localization; (7) probes should have large contrast ratios, that is, the fluorescence from the inactive state must be weak in comparison with the active state because fluorescence from the inactive state contributes to the background noise *(42)*; (8) typically, for biological applications, the probe emission should be separable from known autofluorescence and other sources of background; (9) if possible, minimal sensitivity to other environmental variables is desirable (assuming the experimenter is not interested in probing those variables); and (10) simple probe photophysics is desirable, especially if the probe is relatively immune to fluorescence intermittency, which can significantly complicate interpretation of single-molecule fluorescence data in many situations *(43–49)*. *See also* **Note 1**.

Maximizing localization-based resolution demands maximizing the number of collected photons, which implies that probes with high fluorescence emission rates and large numbers of photons emitted before photobleaching are especially attractive candidates for FPALM applications. To control the number of active molecules requires that, under imaging conditions, the number of activated molecules per frame does not exceed the number of bleached molecules per time interval. This infers that the activation rate per molecule times the number of potentially activatable molecules must be smaller than or equal to the photobleaching rate (plus the deactivation rate in the case of reversible activation) times the number of active molecules *(42)*, *see also* **Eq. 7** below. *See* **Note 2**. Using a variety of probes, including PA-GFP *(50)*, PS-CFP2 *(51)* (available commercially through Evrogen), EosFP *(52)*, Dendra2 *(53)*, Kaede *(54)*, Dronpa *(55)*, and caged fluorescein, FPALM has been successfully performed. Each probe has particular advantages and disadvantages, detailed, for example, in several reviews and seminal papers *(39,52,53)*. Caged fluorescein is available commercially from Invitrogen (C-20050).

2.2. Filter Sets, Optics, and Microscopes

Optics. Optics should be chosen to maximize collection efficiency and resolution. Naturally, the objective lens is a crucial part of the microscope. The NA of the objective is typically between 1.2 (water immersion) and ≥1.4 (oil immersion) so that the standard resolution **(Eq. 1)** is already as high as possible (R_0 is minimized). Use of a high-NA objective also improves collection efficiency, which helps result in a large number of detected photons (N_{det}) and minimizes σ_{xy} (**Eq. 2**). Coverslip-corrected objectives are commonly used so that cells may be imaged in an inverted geometry. Objective magnification is also an important consideration, because the size of the image of each single molecule will depend on the magnification. Roughly speaking, pixels should be small enough that the image of a single molecule is more than one pixel in width. On the other hand, if pixels are made too small, readout noise and other background noise per pixel will begin to interfere with localization. Thompson et al. *(36)* find that the optimal pixel size (to give best localization precision) is given by $q / s = \sqrt[4]{96\pi b^2 / N}$. This pixel size q is the equivalent size of a pixel within the sample. That is, starting from the physical size of pixels on the camera chip, q_{chip} (16 μm, for example), the equivalent size of one pixel in the sample is given by $q=q_{chip}/M$, where M is the magnification. Larger values of q will result in difficulty localizing molecules because of the uncertainty about where, within the pixel, the molecule is localized. In other words, if the image of a single molecule does not span more than one pixel, it is not possible to determine where it is located more accurately than within that one pixel. If the image "spills over" into adjacent pixels, the relative amount of intensity in those adjacent pixels can be used to determine the location of the molecule. The relative intensities in adjacent pixels are the basis for localization of molecules with a precision of *better than one pixel*, despite the resolution being on the order of one pixel or larger. With the recent development of high-sensitivity cameras with on-chip electron multiplication (EM), the readout noise can be effectively reduced by a large factor. Because some of the background noise per pixel is reduced, smaller pixel sizes can be used with a reduced penalty from background noise, and localization uncertainty from the finite pixel size can be virtually eliminated.

Finally, the type of objective is also an important consideration. Although TIRF is ideal for imaging two-dimensional samples directly on a coverslip, for imaging structures *within* live cells or tissue, TIRF can be limiting. FPALM with normal excitation and detection allows a working distance of at least ~200 μm into the sample, allowing the focal plane to be positioned within the cell, rather than just on the bottom surface of the cell. Imaging cytoplasmic or nuclear proteins in a live cell, for example, would potentially be very difficult by TIRF, but has been successfully done using a conventional 1.2-NA water-immersion objective (**Figs. 6** and **7**). Furthermore, because drift of the sample relative to the objective is highly undesirable in FPALM, low-viscosity immersion fluids are convenient. Because biological samples are typically in aqueous buffer, use of a water-immersion lens to focus on cell features well above the coverslip (in an inverted geometry) can help reduce spherical aberrations compared with those expected using an oil-immersion lens at the same distance into the sample. However, the higher NA and the resulting higher resolution and light collection efficiency offered by oil-immersion lenses may be worth more in some applications than the other benefits of water-immersion lenses.

Filter Sets. The filter sets used in FPALM include: (1) laser cleanup filters and dichroic mirrors used to combine activation and readout laser beams (potentially more than one of each if multiple probes are being used) and direct them into the microscope, and (2) emission filters and dichroic mirrors used to direct the laser onto the sample, collect and direct fluorescence onto the detector, and attenuate laser and scattered light before it reaches the detector.

Considering first the case of a single-readout laser and a single activation laser, each with a distinct wavelength, a dichroic mirror is needed to combine the readout and activation beams (typically before the beams enter the microscope). Long-pass dichroic mirrors can be used, with the dichroic mirror placed in front of the activation laser (typically shorter wavelength than the readout beam). Ideally, the readout beam is first passed through a cleanup filter, then through the dichroic mirror, and the activation beam hits the dichroic mirror at the same spot as the readout beam and reflects. The angle of the dichroic mirror is then adjusted to make the two beams collinear. For example, an argon laser is used for readout and a Z488/10X cleanup filter (Chroma) is used to isolate the 488-nm beam, which then passes through a Z405RDC (Chroma long-pass filter with cutoff wavelength >405 nm). The 405-nm laser is reflected by the dichroic mirror and then the alignment is adjusted so that the two beams are collinear and travel into the microscope at the correct position and angle. For a larger number of beams, *see* **Note 3**.

Next, the dichroic mirror and emission filters for the microscope are considered. The dichroic mirror needs to reflect both the activation and readout beams while transmitting the fluorescence efficiently. The readout beam, which is usually on continuously and at high intensity, must not be allowed to be transmitted to the detector. Transmission coefficients of <0.1% at the laser wavelengths are recommended. The emission filter further attenuates the readout beam, and should have maximal transmission over the range of fluorescence wavelengths emitted by the species of interest. Emission filters produced by a novel sputtering method are allowing better transmission efficiencies (in some cases >90%), and feature attenuation of more than five orders of magnitude at the designed laser wavelength. The degree of transmission of the activation and readout lasers by the filters should be checked using a bare coverslip illuminated with the same intensity, dichroic mirror, emission filter, and camera settings being used for experiments. If significant light is observed at the detector, particularly light that has fringes, strong axial dependence, or the same color as either laser (suggesting inadvertent detection of laser light), the filter combination should be reconsidered.

Microscopes. For live-cell imaging, the typical FPALM geometry is an inverted fluorescence microscope with imaging performed by a sensitive camera. Although a variety of makes and models should suffice, the crucial quantities are: *(1)* access to a clear optical pathway between the exterior of the microscope, the dichroic mirror, and the objective back-aperture; *(2)* minimal lateral and focal-plane drift within the timescale of acquisition; and *(3)* ability to view and translate the sample conveniently. Additionally, simultaneous illumination of the sample with a transmitted lamp is helpful. Illumination of the sample with an arc-lamp is not necessary because the readout laser will play the role of exciting fluorescence.

Access to the dichroic mirror and objective back-aperture are important to allow for the shaping and positioning of the excitation and activation beams. A focusing lens is typically positioned on and aligned with the optical axis at one focal length from the objective back-aperture. The focused laser beam is centered within the objective back-aperture to yield a nearly Gaussian illumination profile at the sample, centered within the field. The activation laser can also be likewise aligned and centered. Beam expanders can be used to adjust the laser beam diameters before they are focused by the lens, to illuminate similarly sized areas within the sample. *See also* **Note 4**. For TIRF, published alignment procedures should be used *(32,56,57)*.

Microscope stability becomes increasingly crucial as acquisition times become long. Lateral drift of ~7 nm within 20 min *(58)* is acceptable for live-cell imaging, especially when acquisition times of ~10–30 s are possible. Clearly, as localization precision improves and FPALM is used for imaging with $\sigma_L < 10$ nm, lateral drift correction also needs to be improved. Scattering particles or fluorescent beads may be used as fiduciary markers to correct for drift *(32,33)*. Focal-plane (axial) drift can be compensated for manually, or automatic focus adjustment based on beam reflectance from the coverslip may be useful. When imaging two-dimensional samples, drift in the axial direction is not expected to be as disruptive to imaging if displacements are significantly less than the axial extent of the PSF (i.e., if drift is <<0.5 μm).

2.3. Detectors

As with all single molecule-based methods, where the available signal is low, careful selection of photon detectors is critical. Unlike confocal techniques that are more traditional, where images are obtained by raster scanning either the sample or a focused excitation beam, FPALM makes use of widefield illumination and collection, which mandates the use of an imaging device. The selection of an appropriate imaging device is arguably the most critical aspect of constructing an FPALM instrument. Although there are now a range of imaging technologies available, the two that dominate the commercial market are complementary metal oxide semiconductor (CMOS) and charge-coupled device (CCD)-based cameras.

Historically, CMOS cameras have been available at lower cost and with frame rates much larger than those available with CCD-based devices. CCD cameras, on the other hand, have consistently demonstrated superior quantum efficiencies, up to approximately 90% across much of the visible spectrum, as compared with CMOS cameras, which typically have quantum efficiencies peaking at approximately 15% in the middle of the visible spectrum. As a result, CMOS cameras have largely been used where abundant signal is available, and high-speed imaging is desirable, whereas CCD cameras have been used for low-signal time-integrated measurements. It would seem that the needs of FPALM lie squarely between the capabilities of these two technologies, requiring both high quantum efficiency (QE), and relatively large frame rates.

Recent advances in device design have improved the QE of CMOS devices and increased the frame rate of even the most sensitive CCD cameras. In fact, CMOS cameras are now available with reported quantum efficiencies in excess of 45%, and CCD cameras can be found with frame rates per second (fps) of greater than 10,000. One must proceed carefully, however, before selecting a device, because these improvements are often met at the cost of other performance characteristics. Generally, a wise approach is to select a camera only after considering the entire signal path, beginning with the source and intensity of emission, precise data collection method, and the analysis approach.

Frame Rate. For signal-starved applications, such as single-molecule imaging, the fluorescence emission rate and microscope collection efficiency place an upper limit on the frame rate. As described above, to achieve localization precision approximately 10 times better than the diffraction limit, the noise statistics of photon counting (shot noise), require that some minimum number of photons (*N*) be recorded for the individual molecule. Using PA-GFP as an example with illumination intensity of ~400 W/cm^2 at 496 nm and $N = 100$, this corresponds to single-frame exposure times of approximately 100 ms, which lies well within the capabilities of commercially available CMOS and CCD cameras. It is important, however, to note the difference between exposure time and frame rate when comparing device specifications. High-quality CCD cameras with large numbers of pixels and equipped for electronic gating are often referred to as "fast."

This often refers to their ability to rapidly change the active/inactive state of the CCD chip, not their ability to rapidly obtain sequential images. For these devices, the frame rate and the exposure time differ by the "dead time" associated with the time required to clear the CCD chip of charge, transfer the image information, and prepare it for the next exposure, which is a function of the number of pixels in the array. For FPALM applications in fixed samples, cameras with frame rates on the order of 20 fps or greater are desirable. For live-cell imaging, frame rates of >100 fps are desirable.

Detection Efficiency. The fill factor is a measure of the effective amount of dead space associated with each pixel (photosite) in an imaging array device. Although both CMOS and CCD devices make use of the photoactivity of metal oxide semiconductors, their pixel architecture differs significantly. CCD devices consist of a three-dimensional layered architecture where the photoactive semiconductor element covers the entire surface of the pixel, resulting in fill factors of nearly 100%, limited only by the finite space separating individual pixels. In contrast, CMOS devices are generated by the same methods used by the computer industry, where the photoactive element and the associated circuitry both lie in the same two-dimensional plane. Although this results in much lower processing costs, the on-chip circuitry can easily consume 50% or more of each photosite, resulting in a fill factor of less than 50%. This lower fill factor corresponds to fewer photons striking the active areas of the sensor, and consequently yields a lower detection efficiency. For single-molecule imaging, this characteristic alone typically precludes the use of CMOS devices for all but the brightest samples.

The intrinsic quantum efficiency (ϕ or QE) of a detector is usually defined as the fraction of absorbed photons (quanta of light) that are converted into an electrical signal. This simple definition only measures the electronic conversion efficiency, ignoring other properties such as the efficiency of the absorption process (as opposed to scattering), which is wavelength dependent. Thus, the use of an "external" (ϕ_E) or "total" (ϕ_T) QE is often adopted and describes the combination of all losses in the device, which can be measured using a well-characterized broadband light source.

The total QE takes into account fill factor, intrinsic QE, and electronic conversion efficiency of the device circuit (analog-to-digital). Although not typical, some manufacturers of inferior devices have in the past reported only the intrinsic QE of the CCD device, whereas the realizable efficiency is much lower. The highest quality back-thinned CCD devices exhibit quantum efficiencies around 90% for most of the visible spectrum, but drop off quickly to a few percent in either the infrared (IR) or ultraviolet (UV) wavelengths because of insufficient photon energy, or poor photon penetration, respectively. Some tunability in the wavelength dependence of detectors is available, and care should be taken to match the wavelength range with the highest QE (ϕ_E) to the emission spectrum of the (photoactivatable) fluorophores of interest.

Background and Noise Sources. When comparing candidate CCD devices for FPALM imaging, it is often helpful to consider separately those noise characteristics associated with the sensor (photon detection) and those associated with the device electronics (signal generation).

It is important to consider that the localization precision is a function of the total number of photons counted per molecule per frame (N), not per pixel **(Eq. 2)**. Returning to the PA-GFP example, if the collected photons associated with a single molecule (100 photons in 100 ms) are distributed over a 3 × 3 pixel area, and the PSF is approximately Gaussian, then the number of photons incident on the central pixel may be as many as 30–50, whereas those on the periphery may be as few as 6–10. For these peripheral pixels, fluctuations in the background (leading to background noise b per pixel) are expected to play a more significant role, and warrant discussion. The two most significant sources of background noise associated with CCD cameras are dark current noise and read noise.

Dark Current. Dark current is caused by impurities and defects in the bulk semiconductor (typically silicon for visible wavelengths) or at the various semiconductor-semiconductor oxide interfaces. Energetically, these impurities lie in the gap between the conduction and valence bands. Given sufficient ambient thermal energy, electrons can hop between these intermediate gap states and eventually migrate into the conduction band. The result is an increase in the electronic signal generated in the pixel, even in the absence of light. Because dark current is the result of a thermal process, an effective way to reduce dark current is to cool the CCD chip. For many years, cooling was accomplished with somewhat cumbersome liquid nitrogen cryostats, where dark currents lower than 0.0001 e–/pixel·s (Roper Scientific), can be achieved with temperatures of –120°C. The development of thermoelectric Peltier devices in recent years has increased dramatically, reducing the cost and dimensions of cooled CCD devices. Although the finite cooling power of these devices (typically $\Delta T < 100$°C) makes them somewhat less effective than liquid nitrogen cooling, dark currents of less than 1.0 e–/pixel·s (at –10°C) are routinely achieved, and, in high-performance devices, can be less than 0.01 e–/pixel·s (at –43°C).

For many image applications, especially those of intermediate integration times, the dark current can be treated as a systematic (noiseless) background offset (provided the temperature is held constant), and subtracted using a reference image acquired with the sensor shuttered (no light). For this approach, the dark frame integration time must match the data acquisition time.

Dark Current Noise (Dark Noise). Although the average number of dark counts can be subtracted from an image by acquiring a reference image, the noise associated with the dark counts cannot, and like photon shot noise, is equal to the square root of the number of dark counts. In devices in which the dark current is large, either intrinsically or as the result of an applied electronic gain, the dark noise can result in fluctuations comparable in magnitude to the expected single-molecule signal at each pixel (only tens of photons counted per read).

Because dark current is the result of a thermal process, and dark current noise scales as the square root of the dark current, an effective way to reduce dark current noise is to cool the CCD chip, as described above. Caution should be taken when comparing specifications for CCD cameras so as not to confuse dark current noise (given as root-mean-squared [rms]), with dark current (e–/pixel·s), either of which is reported for high-quality detectors.

Dark Noise Nonuniformity. Because of their small size (typically less than 25 × 25 μm), and because of challenges associated with device fabrication, each pixel in a CCD camera tends to have a slightly different density of defects or impurities. As a result, each pixel will exhibit a slightly different voltage offset, sensitivity to light, and rate of dark current production. This variation between pixels results in varying levels of dark noise, and is referred to as dark noise nonuniformity. As CCD arrays increase in size, and pixel dimensions decrease, the relative fluctuations in defect density are expected to become more significant. This nonuniformity imposes a variance in the background noise value (b) described in **Eq.2**, slightly increasing the localization uncertainty. The average dark noise and its variance can, in principle, be determined by sampling many dark frames, calculating the average dark noise for each pixel (B), and the statistical variance across all pixels (σ^2_B). In this case, b in **Eq.2** could be replaced by $B \pm \sigma_B$.

Read Noise. Although dark current noise is a function of charge accumulation in the photoactive element of the CCD, read noise arises from the process of extracting and converting the charge at each pixel first into an electrical signal (voltage), and second into a digital value (counts), using an analog-to-digital converter (ADC). In CCD arrays, this is accomplished in a line-by-line stepwise fashion, where pixel values in the first line are read and then reset to some offset value, then the entire array of values is shifted down one line, and the next line is read. This is repeated until the entire array has been read out, and the pixel values reset to the offset value. Unfortunately, the pixels are never quite cleared, leaving some unknown residual charge, which varies for each pixel and each shifting operation, giving rise to read noise (sometimes called reset noise). To improve CCD sensitivity, an electronic gain is typically used, which effectively multiplies the output voltage but can be related to the original charge (gain = number of charges per pixel/number of counts reported).

The read noise rises with increasing readout rate. If the CCD array is read slowly (tens of kpixels/s) the read noise can be relatively low (<2 e-rms). If the array is read quickly, as required for large arrays at high frame rates (>20 Mpixels/s), the read noise can be very high (>50 e-rms), overwhelming the already low (5–10) photon counts at the edge of the imaged PSF. This constraint arises from the charge amplifier in the CCD circuitry, which is required to convert the relatively small number of photoelectrons into a measurable voltage during the ADC process. For high-speed readout, large amplification (gain) is required. Unfortunately, the noise associated with the larger signal scales with the amplification. *See also* **Note 5**.

Electron-Multiplying CCDs (EMCCDs). Recently, a new class of CCDs has been released that offers both high frame rates (>30 fps) and extremely low readout noise. This is accomplished primarily by applying gain (amplification) during the line-shifting process, but before the analog-to-digital conversion process. As a result, these devices are referred to as "on-chip" or EMCCD cameras. By amplifying the signal on the chip before readout, the read noise associated with the amplification process in conventional CCDs is greatly reduced, and readout noise no longer limits sensitivity. Although the read noise is effectively eliminated in EMCCD cameras, there is an additional noise associated with the EM process.

Typically, the noise sources associated with a photonic measurement are a function of dark noise, readout noise, and photon counting or shot noise. The total noise characteristic of a device is then represented by adding all of these noise sources in quadrature:

$$\sigma_{\text{total}} = \sqrt{\sigma_{\text{dark}}^2 + \sigma_{\text{readout}}^2 + \sigma_{\text{photon}}^2}, \tag{4}$$

where the first two terms are characteristic of the detector. For noncooled CCD devices, the dark current can be very large, and dominate other noise sources, making these devices ill-suited for FPALM measurements. For either thermoelectrically or liquid nitrogen-cooled devices, readout noise is typically the limiting source of detector noise that can effectively limit the frame rate of the camera. EMCCD cameras are a relatively new technology that offers high sensitivity, rapid frame rate, and low device noise (readout and dark noise), and are recommended for most FPALM applications.

3. Methods

The typical FPALM setup is a fluorescence microscope with two lasers, high-NA objective lens, and a sensitive camera capable of imaging single fluorescent molecules (**Fig. 2**). The two lasers are aligned to be collinear and are focused by a lens into a small spot at the back-aperture of the objective lens to illuminate a fairly large region of the sample (from a few microns to 100 μm in size, roughly). Emitted fluorescence is collected by the objective and then focused by the microscope tube lens to form an image on the camera. Images are collected as a function of time, stored, and analyzed to determine the positions of molecules visible within the illuminated region of interest (ROI). The readout laser intensity is controlled to cause active fluorescent molecules to be clearly visible above the background, whereas the activation laser

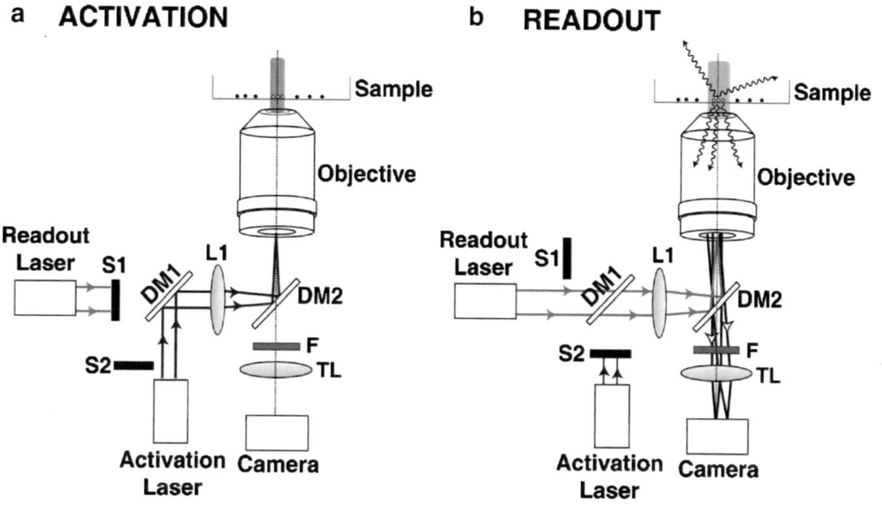

Fig. 2. Experimental setup. In FPALM, the experimental procedure entails cycles between (**a**) activation and (**b**) readout and photobleaching. (**a**) During activation, the activation laser is reflected by a dichroic mirror (DM1) into a lens (L1) that focuses the beam onto the objective back-aperture via a second dichroic mirror (DM2) inside the microscope filter cube. The focused beam emerges from the dichroic mirror as an approximately parallel beam that illuminates a circular area in the focal plane (sample). Photoactivatable molecules (*filled circles*) within the focal plane are activated with a low probability, such that just a few molecules become active per activation cycle (*open circles*). During activation, the readout laser is either blocked by a shutter (S1) or is allowed to illuminate the sample continuously. (**b**) During readout, the activation laser is blocked by a shutter (S2) and the readout beam passes through DM1 to be focused by L1 onto the objective back-aperture via DM2. The sample is then illuminated by the readout beam in a circular area (approximately Gaussian in profile), which causes any active molecules (*small open circles*) to emit fluorescence photons (*oscillating lines with arrowheads*). Some of those photons are collected by the same objective, filtered to remove laser and scattered light (*F*), then focused by the microscope tube lens (*TL*) to form an image on the electron-multiplying charge-coupled device (CCD; *camera*). Active molecules illuminated by the readout beam will eventually spontaneously photobleach, decreasing the number of visible molecules over time. Cycles of activation, readout/photobleaching are repeated until enough molecules have been imaged to obtain the desired information (*image*), or until all molecules have been photobleached.

intensity is chosen to be low enough that only a small number of (<100) molecules is activated by a single, short (~1–5 s) pulse of illumination. A detailed procedure is described below.

3.1. Alignment and Characterization of the Illumination Area

FPALM requires the collinear alignment of a readout laser beam and a (typically shorter wavelength) activation laser beam. These beams are then focused by a lens to a small spot at the center of the back-aperture of the microscope objective lens to produce an illumination area at the sample that is large enough to encompass the desired ROI, such as an entire cell. If a long-pass dichroic mirror is used to merge the two beams, alignment is most efficiently achieved by first aligning the straight-in (parallel) beam (typically the readout laser) into the center of the objective back-aperture, without the focusing lens in place (see L1 in **Fig. 2**). This lens, typically mounted near or just inside one of the input ports of the microscope, should then be aligned to focus the readout beam at the center of the objective back-aperture. The profile of the expanded beam area can then be viewed via the display of a CCD camera by focusing into a dilute solution of an appropriate fluorophore (see examples in **Fig. 3**). This solution should be dilute enough so as not to saturate the camera, and the emission range of the fluorophore should be chosen to be compatible with the filter sets being used. Collinear alignment of the activation laser beam is now easily accomplished by adjusting the dichroic mirror while monitoring the camera view. Alignment of the centers of both beams is recommended. However, although the beam centers should be aligned as closely as reasonably achievable, as long as the two profiles are well overlapping it will be possible to control the number of active molecules within the area illuminated by both the activation and readout beams. Images of the profile of both beams should now be obtained for later reference (see example in **Fig. 3**). The activation beam area may be smaller than the readout beam to maximize activation intensity. To create a nearly uniform illumination intensity within the ROI, the readout beam is typically spread over an area larger than the desired ROI. In addition, beam expanders can be used to match the diameters of the incoming activation and readout beams (before they strike the focusing lens), so that the areas illuminated by both beams will be similar at the sample.

Generally, illumination by the activation source will be intermittent, as is required to maintain a small number of (from ten to a few hundred) visible molecules within the ROI **(Fig. 3)**. Activation pulse duration is ideally regulated electronically (e.g., by computer) to allow a well-defined timing protocol or synchronization with various events such as camera frames, but it is also possible to manually control activation. Activation times vary strongly with the density of inactive molecules, but for PA-GFP an activation intensity of 1–5 W/cm^2 at 405 nm at the sample for

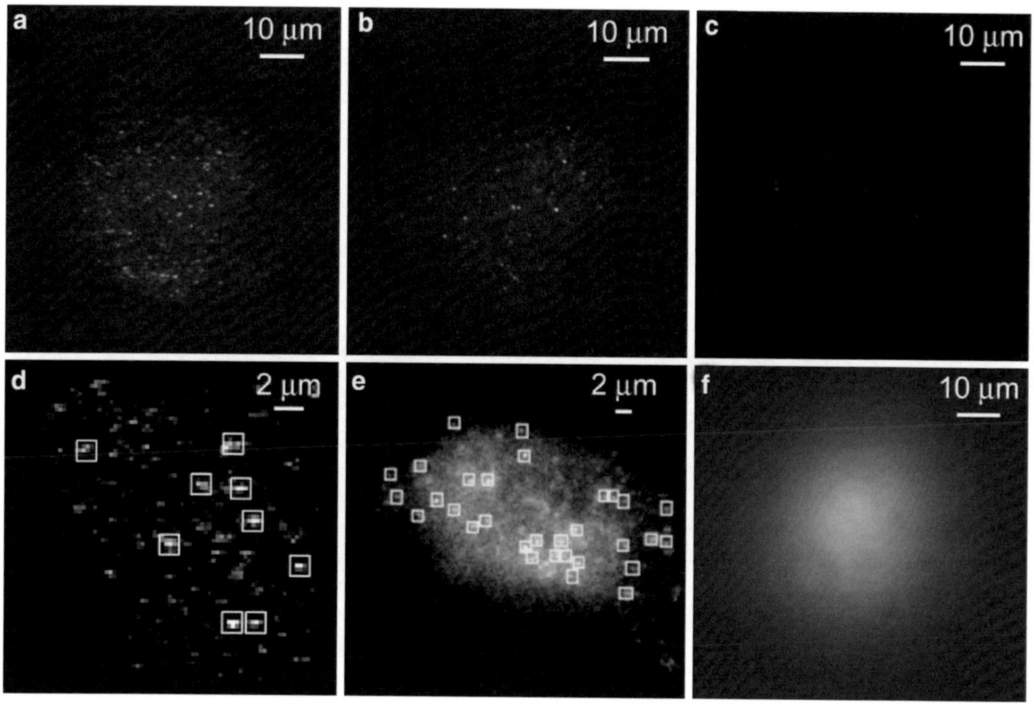

Fig. 3. Imaging single molecules. (**a**) Fluorescence background. Single fluorescent molecules, under illumination by 6 mW at 514 nm, are visible in an Olympus IX71 microscope with 60× 1.2-NA water-immersion objective. A 0.5-μL drop of HPLC water was placed on a 0.17-mm glass coverslip and allowed to dry in the air. The molecules visible within the circular area illuminated (~2,800 μm^2) are actually background from contaminants in the water. (**b**) Background was significantly decreased in water that had been treated by a UV lamp before use. A Cascade512B EMCCD camera was used for the imaging. Regions depicted are near the center of the drop. (**c**) Background can be reduced by photobleaching before acquisition begins. A similar drop of water with dilute fluorophore is shown after photobleaching for ~100 s. (**d**) Single molecules of EosFP in a fixed fibroblast illuminated at 532 nm using the same microscope and objective. Image analysis permits identification of single molecules (*white boxes*) and localization of those molecules. (**e**) Single molecules of PA-GFP-HA in a live fibroblast at room temperature were imaged by 496-nm excitation, identified (*white boxes*), and localized. (**f**) Profile of the 496-nm readout laser just before imaging the cell shown in (**e**). Images have been adjusted linearly for brightness and contrast.

1–3 s was sufficient to activate tens to hundreds of molecules in the sample per frame. For PA-GFP, readout laser-induced activation is also significant (in many cases no 405 nm illumination is needed at all), with readout intensity of 200–400 W/cm^2 at 496 nm *(42,58)*. Caged 5-carboxyfluorescein could be activated and imaged using similar wavelengths and powers as for PA-GFP **(Fig. 4)**. For EosFP or Dendra 2, 2–5 W/cm^2 at 405 nm for 1–2 s achieved satisfactory activation, and readout intensities of 200–1,200 W/cm^2 at 532 nm are typical. It is also convenient to have shutter control over the readout source (e.g., Thorlabs SH05 or FW102). In cases where having an expanded illumination area results in insufficient activation intensity, it may be necessary to have the lens near the back port of the microscope mount in a

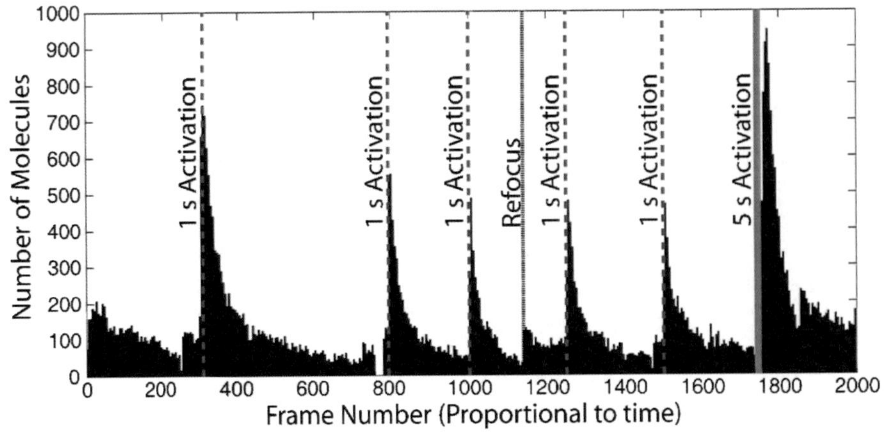

Fig. 4. Photoactivation of caged fluorescein. A sample of CMNB-caged 5-carboxyfluorescein (Invitrogen C-20050) was dried on a coverslip and imaged by FPALM. As pulses of the 405-nm activation laser were applied, the number of active molecules increased suddenly, and then photobleached gradually as the 496-nm readout laser continuously illuminated the sample. The number of localized molecules as a function of frame number (time) within the acquisition shows these increases and decreases as several short (1 s) activation pulses were applied (*dashed black line*), the objective was refocused (*dotted black line*), and a longer (5 s) photoactivation pulse was used (*solid gray line*).

motorized filter wheel so that the lens can be rotated out of the beam path in coordination with the activation pulse to produce a more intense (although much smaller) activation area.

3.2. Data Acquisition

1. *Single-Molecule Detection*. Single-molecule detection for FPALM requires high-intensity excitation, fluorophores capable of emitting a large number of photons (as quickly as possible) before photobleaching, and highly sensitive detection. Because fluorescence detection is approximately proportional to $(NA)^2$, where NA is the numerical aperture of the objective lens, a high-NA lens (NA > 1) is preferable. A variety of models have been used successfully for FPALM and methods similar to FPALM ***(31–34,42,58)***. Secondly, fluorophores should emit as many photons as possible within the acquisition time per frame (τ_F). For localization precision that is tenfold better than the diffraction-limited resolution, at least 100 detected photons are required (in the absence of background). Even larger numbers of photons are required if significant background is detected as well **(Fig. 3a, b, d, e)**, as dictated by **Eq. 2**. The rate of emitted photons per fluorophore (k_{em}) can be estimated (in the absence of fluorescence saturation) using:

$$k_{em} = \Phi_{FL}\sigma I, \quad (5)$$

where Φ_{FL} is the fluorescence quantum yield (dimensionless), σ is the excitation cross-section (units of area), and I is the readout excitation intensity (units of photons per second per unit area). The value of σ can be calculated from the extinction coefficient ε using $\sigma = \varepsilon \times 3.82 \times 10^{-21}$ cm^3 M *(59)*. The average excitation intensity can be estimated from the total laser power at the sample divided by the area over which that power is spread. Fluorescence saturation at high intensities results in a deviation from the linear relationship between intensity and emitted photons as given by **Eq. 5**. However, **Eq. 5** can still provide an upper bound on the number of photons that could possibly be emitted (the actual number will be equal to or less than this value on average). Furthermore, not every emitted photon is detected. Below saturation (at intensities typical of FPALM), the number of detected photons (N_{det}) per fluorophore per camera frame is given by:

$$N_{\text{det}} = \eta k_{\text{em}} \tau_{\text{F}} = \eta \Phi_{\text{FL}} \, \sigma \, I \tau_{\text{F}} \tag{6}$$

where typical detection efficiencies of the complete setup (η) are approximately 0.02–0.1 (2–10% of emitted photons are detected). Therefore, it is crucial to use as high an efficiency camera as possible, as high an NA as possible, and as optimal a filter combination as possible. Consider a fluorophore such as PA-GFP with a large quantum yield of $\Phi_{\text{FL}} = 0.79$ ***(39,50)*** and reasonable extinction coefficient $\varepsilon \sim 17.9 \times 10^3$ M^{-1} cm^{-1}, which gives $\sigma = 6.65 \times 10^{-17}$ cm^2. Suppose the illuminated area is 30 μm in radius, which gives an area of $\pi r^2 = 2{,}827$ μm^2. If the excitation laser (488 nm) power is 10 mW at the sample (energy per photon is $E = h\nu = hc/\lambda$, where h is Planck's constant, c is the speed of light, and λ the excitation wavelength), this results in 2.45×10^{16} photons/s of excitation light. Thus, $I = 8.68 \times 10^{20}$ photons/cm^2 s and $k_{\text{em}} = 4.69 \times 10^4$ emitted photons per second per molecule. With a detection efficiency $\eta = 0.03$, in a 50-ms frame, one will then detect $N_{\text{det}} \sim 70$ photons on average. So, to achieve at least 100 photons per frame, the acquisition time should be $\tau_{\text{F}} = 71$ ms. As long as the PA-GFP does not photobleach before these photons are emitted, these settings are roughly adequate. On the other hand, with higher readout laser intensities, faster frame rates can be achieved. Performance of other fluorophores can be estimated in the same way, using appropriate extinction coefficients, quantum yields, detection efficiencies (which depend on filters and fluorophore fluorescence emission spectra), excitation wavelengths, and powers.

2. *Control of the Number of Active Molecules.* Control of the number of active molecules is typically achieved by adjusting the relative excitation intensities of the activation and readout beams. Detailed calculations of the degree of control over the number of molecules are provided elsewhere ***(42)*** and summarized by:

$$N_{\text{A}} = \frac{(k_{\text{A}}\Phi_{\text{A}} + k_x \Phi_{\text{RA}} + k_0)}{(k_x \Phi_{\text{B}} + k_{\text{D}}\Phi_{\text{D}} + k_{\text{SD}})} N_{\text{I}}, \tag{7}$$

where k_A is the activation excitation rate, Φ_A is the activation quantum yield, k_0 is the spontaneous activation rate, Φ_{RA} is the readout-induced activation quantum yield, Φ_D is the light-dependent deactivation quantum yield, k_D is the light-dependent deactivation rate, k_{SD} is the spontaneous deactivation rate, k_x is the excitation rate from the readout laser, Φ_B is the photobleaching quantum yield, and N_I and N_A are the number of inactive molecules within the illuminated area, respectively. The value of N_A should be made small to allow localization. The rule of thumb is that the density of active (fluorescent) molecules should not exceed 0.1 μm^{-2}, or roughly <0.01 molecules within the area of the diffraction-limited PSF for standard diffraction-limited optics. *See* **Note 6**.

Labeling density can also be adjusted so that N_A is low enough to permit localization of individual molecules. For example, in samples labeled with a probe that shows significant readout-induced activation, as soon as the sample is illuminated, the number of active molecules will begin increasing. If that number exceeds the maximum density for localization, data analysis will become difficult. In this case, the sample can be illuminated (photobleached) continuously for a long enough time that the density of active molecules decreases to a low enough density to permit localization. Although this means that some of the molecules will be lost to photobleaching, some useful information can still be obtained about the structure of the sample once the density of active molecules has dropped. Alternatively, cells transfected with PA-FPs can be imaged sooner after transfection, which will generally yield a lower expression level of the PA-FP and help ensure that the density is low enough for FPALM imaging. A second (preferred) alternative is to use a PA-FP with a lower probability of readout-induced activation.

3. *Calibration Samples.* For confirmation of sub-resolution imaging, it is recommended to do FPALM imaging of a sample with known structure. For example, comparison of results with electron microscopy *(32,58)* or AFM *(42)* of the same sample can confirm accurate representation of the structures in the sample. Fabrication of some kind of structure with linear or symmetric structures of a known size is particularly helpful. A crystalline sample cut at an angle close to that of a crystal plane can be labeled with photoactivatable dye and imaged by FPALM and AFM. Even a scratched coverslip can provide small structures that can be used to check resolution. It is not recommended to rely exclusively on calculated resolution (e.g., **Eq. 1**), because many calculations predict the outcome in ideal cases, but do not take into account all factors that could compromise resolution. If the resolution measurement is based on theoretical calculations, at least the PSF FWHM should be measured by imaging fluorescent beads. Repeated

measurement of the position of the same molecule (for example, in fixed cell preparations, with settings that allow single molecules to remain fluorescent for several camera frames) can also provide a means of estimation of resolution, and thus calibration of the system.

Calibration of the number of photons detected by the camera is also a crucial step in characterization of an FPALM microscope. Measurement of the slope of variance versus mean intensity for many intensities can be used to determine the number of photons that is equivalent to a given pixel value. For photons, the expected relationship is that the variance in the number of detected photons from a region with a given intensity should equal the mean number of photons detected from that same region ***(60)***.

A scale of known size should be imaged under standard conditions to determine the magnification of the system and the resulting size of pixels within the image, so that dimensions of structures can be reliably measured. Drift of the microscope should also be characterized using immobilized fluorescent beads or other objects that are easily localized, but that will remain visible for as long as an acquisition could potentially last.

4. *Management of Fluorescent Background in Single-Molecule Experiments.* In single-molecule detection systems, it is important to minimize the fluorescence background from molecules other than the species of interest. Excessive background adds to noise and reduces localization precision **(Eq. 2)**. Thus, it is desirable to reduce it to a noninterfering level with respect to the fluorescent molecules of interest.

Usually the sources of background fluorescent molecules include: (1) coverslips or slides used to support the sample; (2) solutions used to embed, dilute, wash, or fix the sample, or reagents used on the objective or to clean the coverslips or slides; (3) autofluorescence from the biological system itself. Typically, high-purity water such as high-performance liquid chromatography (HPLC) water is used to embed or dilute samples and to clean objectives and coverslips. However, most commercially available HPLC water contains a significant amount of background fluorescence, considering the purity required for single-molecule fluorescence experiments. For example, an experiment using an inverted type microscope with a 60× 1.2-NA water-immersion objective with a CCD camera is configured to capture images with 1 pixel spanning an area of 0.267 μm × 0.267 μm in the object space. It is typical to illuminate and observe a circular area of ~100-pixel radius under this kind of microscope. The area observed is $A = \pi \times (100\ \text{pixels})^2 = \pi \times 100^2 \times (0.267\ \mu\text{m})^2 = 2{,}240\ \mu\text{m}^2 = 2.24 \times 10^{-3}\ \text{mm}^2$. If a drop of 0.5 μL of HPLC water is deposited on the coverslip with estimated radius of 1 mm and area of 3.14 mm^2, the illuminated area will correspond to a volume of roughly V_i = 0.5 μL·(2.24 × 10^{-3} mm^2/3.14 mm^2) = 3.57 × 10^{-4} μL, which will contain ~1.2 × 10^{16} water molecules. Assuming particularly clean water (with fluorescent contaminants only at the 1 part per trillion level), this yields ~1.2 × 10^4 fluorescent molecules within the area of observation, all of which will be somewhere on the coverslip surface after the drop evaporates. Thus, clearly, even the slightest amounts of fluorescent residues may have significant adverse effects.

Background can be categorized into two types, uniform and nonuniform. Uniform background is defined as a background distribution that does not spatially vary (apart from statistical noise), as might result from a weakly fluorescent contaminant within the buffer. The individual molecules contributing to a uniform background are typically not resolvable from one another.Such background, despite being spatially uniform, can still increase the noise in the detected signal because it consists of detected photons, which themselves exhibit shot noise ***(59,60)***. Background subtraction can eliminate the *average* background, but there will still be variations in the detected intensity because of photon shot noise. Thus, uniform background is to be minimized whenever possible.

Nonuniform background is even more difficult to compensate for, because its spatial dependence is frequently complex, owing to nonrandomly distributed fluorophores such as caused by cellular autofluorescence (see other sources below). It not only decreases the localization accuracy as uniform background, but can result in incorrectly localized positions and cause imaging artifacts.

Methods used to reduce the background levels in aqueous solutions are numerous. First, careful consideration of the initial water source is highly worthwhile. Water labeled to have minimal organic residue may still contain fluorescent contaminants, and the relative content of fluorescent contaminants may not scale directly with organic residue content. Opening containers within the lab allows dust and other atmospheric particles to enter the container and potentially deposit fluorescent contaminants into the solution. Samples opened repeatedly in even modestly dusty laboratory environments have been observed to degrade and become “hot” within weeks or even days. Creating aliquots of high-purity water can reduce contamination caused by repeated opening of containers, but plastic containers also leach fluorescent molecules into solution within days. The use of colored pipette tips can also contribute to background from contaminants. Because contamination is a priori very difficult to avoid, procedures that eliminate existing contaminants are useful. One procedure that is capable of significantly reducing background levels is described as follows: (1) A glass beaker cleaned with low-residue detergent and tap water should be rinsed thoroughly with tap water and then again with the purest water available. (2) Add 25–30 mL of the purest water available to the beaker. (3) Illuminate the water with a high-power (500-W) UV lamp for >20–30 min. **Figure 3a, b** compares untreated HPLC with UV-treated HPLC water. The image comparison was done after the HPLC water was cleansed using the above procedure. These observations suggest that background levels improved by at least ~100-fold. If the background levels are still too high, one may pre-illuminate the sample with the laser beams used for normal illumination (readout) in the microscope system. If possible, staining the sample with excess photoactivatable fluorescent dye before prebleaching will enable the postbleach levels of fluorophore of interest to exceed the background by a reasonable factor. Of course, excessive photobleaching may also lead to photodamage of the sample.

3.3. Biological Applications

1. *Labeling of Specific Structures.* Before beginning an FPALM experiment on a biological system of interest, the procedure for labeling the relevant structures should be considered. Genetically encoded photoactivatable fluorophores offer many advantages. Cells can be transfected 1–2 days before imaging, and imaged directly (living or fixed). Alternatively, antibodies conjugated to a photoactivatable fluorophore can be used as labels, as long as standard labeling procedures using that antibody can be carried out with minimal exposure of the sample to light at the activation wavelength. STORM-type switches can also be used to label DNA or to conjugate to antibodies *(31,33)*. The photoactivatable fluorophore density should be made as high as possible, under several constraints. First, too much label can disrupt the biological structures of interest. Such effects are always possible in fluorescence experiments, but need to be addressed with appropriate controls. Second, the density of active molecules needs to be low enough that visible molecules are individually distinct for localization. If the probe exhibits readout-induced activation or spontaneous activation, there will be visible (active) molecules even before a deliberate activation has been executed. If this number of unintentionally activated molecules is already close to the density limit or above the density limit, the sample will need to photobleached (i.e., illuminated with the readout laser) until the density is reduced below the necessary threshold. See methods below for estimating the density required for imaging at a particular resolution. Because photobleaching causes photodamage, for living samples, this excess initial density of molecules (and in the presence of readout-induced activation, a long-lasting excess density of molecules) can be a significant limitation. Although the density of fluorophore can be adjusted by labeling with less antibody, imaging sooner after transfection, or intentionally photobleaching the sample,

it is generally best to adjust the sample labeling first, establish a standard protocol, and then proceed with imaging.

2. *Identification of Sample Region.* When ready to begin imaging, the transmitted light (appropriately filtered to remove wavelengths in the activation range of the fluorophore being used, e.g., $\lambda < 500$ nm for PA-GFP and most other PA-FPs) from a microscope lamp can be used to locate cells or other sample features. The readout laser itself can also be used to locate an ROI by exploiting any preactivated molecules. For control over the density of active molecules during imaging, the sample ROI must be positioned within the overlapping area of the readout and activation beams. Sample regions should only be imaged if the presence of photoactivatable molecules can be observed (during illumination with the readout source) by eye or with the camera as discrete on/off fluorescence from single emitters and overall increase in fluorescence after a brief pulse (~1 s) of the activation beam. In addition, comparison of the observed density of molecules with densities observed in unlabeled samples is a crucial step to avoid accidental imaging of background (which will often consist of a hazy glow, and a few bright single fluorescent molecules). If cells have been transfected, and not every cell is transfected, visual comparison of bright-looking cells with mock-transfected cells of the same type (in a different well) can make what a transfected cell looks like more apparent. At the start of an acquisition, numerous fluorescent molecules may be visible during initial illumination by the readout beam because of spontaneous activation, inadvertent activation (e.g., by exposure to room light or ultraviolet sterilization lamps inside the cell incubator), or readout activation. As such, it may be necessary to photobleach the ROI before beginning an acquisition if so many molecules are emitting at once that single molecules cannot be distinguished by eye. Alternatively, when a large number of molecules are fluorescent, it can serve as an opportunity to observe and record the distribution of those molecules as they look by widefield microscopy (for comparison with any expectations for the distribution). However, to localize an efficient number of activated molecules, it is desirable to have an average separation between active molecules of ~$4R_0$ *(42)*. For example, for PA-GFP imaged by a 1.2-NA objective ($R_0 = 264$ nm), the optimal density would be approximately three activated molecules per 10-μm^2 area, although densities as high as 10 molecules per 10-μm^2 area would certainly be workable (occasional molecules that are too close together can either be analyzed very carefully or ignored). Once an appropriate density of activated (fluorescent) molecules has been achieved, this density can be controlled with intermittent pulses of the activation beam and a suitable continuous intensity of the readout beam

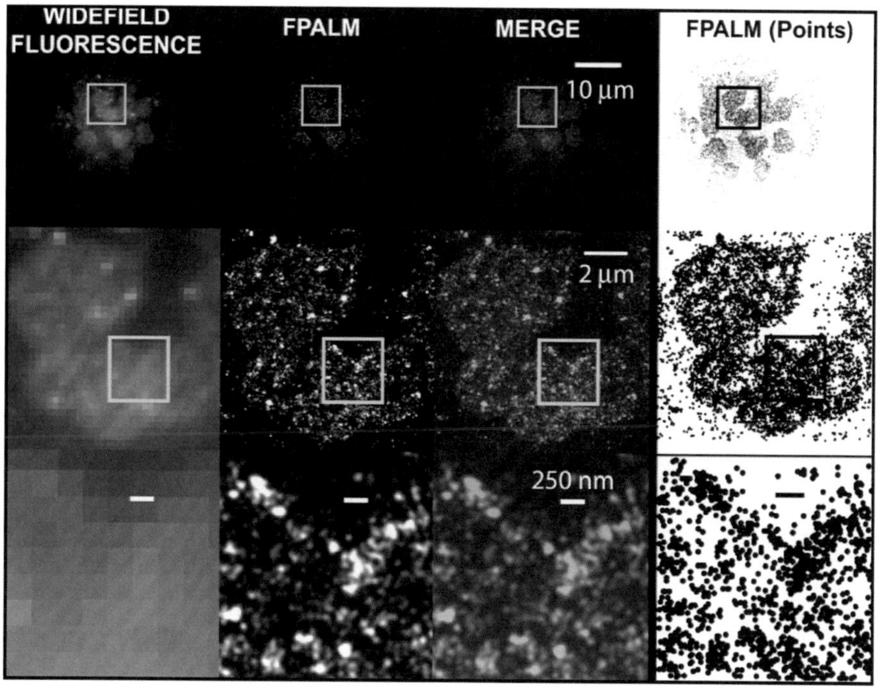

Fig. 5. FPALM images of caged fluorescein on a surface. Comparison between widefield fluorescence images and FPALM images of the same sample described in Fig. 4. (*left column*) The widefield fluorescence image shown was generated by averaging all widefield fluorescence images during the entire acquisition. The *second* and *third rows* show successively higher zooms of the *boxed region in the row above.* The FPALM images of the same region (*second column from left*) show significantly higher detail, especially at the highest zoom (*bottom row*), where the scale bar is 250 nm for all images shown. The *third column from the left* shows a merge of the FPALM and widefield fluorescence images. The *rightmost column* shows all localized molecules plotted as a *black point* with a fixed size, whereas the *second column* from the left shows an FPALM image rendered with each molecule plotted as a Gaussian with size equal to the calculated uncertainty in its position, and the intensity proportional to the number of photons detected from that molecule. Scale bars shown apply to all images within the same row. Images were adjusted linearly for brightness and contrast.

(see the activation pulse sequence shown in **Fig. 4** and resulting FPALM images in **Fig. 5**). Although an acquisition generally consists of continuous illumination by the readout beam, exact activation protocols will vary according to fluorophore(s) being used and readout beam intensity. Typically, ~1- to 5-s pulses of the activation beam are administered whenever the number of visible molecules is approximately fewer than ~0.1 μm^{-2} as viewed by eye on the camera display.

3. *Background and Autofluorescence in Cellular Environments.* Contributions to the accumulation of background signal can be generated from both external and internal sources. Before data acquisition, experimental considerations such as shielding the image beam path should always be taken to minimize external background light from reaching the camera(s). Common internal sources of background include fluorescence from inactive molecules (less significant for PA-FPs with higher contrast

ratios), out-of-focus active molecules (minimized by TIRF), the immersion liquid (which can be treated as detailed above), a dirty or dye-contaminated objective lens, autofluorescence from the coverslip (fused quartz is sometimes better than glass), and scattered laser light. Dark noise and read noise from the camera can also be an additional source of background noise, although these effects are usually negligible in cooled EMCCD cameras operated at high EM gain. Imaging cells or cellular structures introduces additional sources of background including autofluorescence, fluorescence from the growth media (including ingredients such as phenol red and serum), and residual transfection reagents. The presence of cellular autofluorescence from cellular structures can result in a significantly position-dependent background signal that decreases over time because of photobleaching. Background signal can also be reduced by efficiently filtering the image beam to exclude fluorescence outside the range of emission of the PA-FP.

4. *Optical Sectioning*. Standard FPALM imaging acquires a two-dimensional image from fluorescent molecules that can be localized within the focal plane of the objective. The greatly expanded area of and reduced collection efficiency for out-of-focus molecules cause out-of-focus molecules to be greatly reduced in intensity, and significantly larger (in apparent size) than in-focus molecules. Molecules that are significantly out of focus (more than approximately ±1 μm above or below the focal plane in the case of a 1.2-NA objective, detecting at 520 nm) will contribute to background levels, but will not be localized. Thus, even if a three-dimensional sample is imaged by FPALM, one really obtains two-dimensional information about the molecules that are within a slice approximately equal to the depth of field in thickness. Positioning of this focal plane is crucial, because single molecules may be visible at many different focal positions. Careful use of the microscope focus control or a calibrated *z*-stage will allow determination of the depth of the focal plane within the sample. Fluorescent beads attached to the coverslip can serve as a fiduciary mark for drift-correction of lateral coordinates and for reference to a known axial position. *See* **Note 7**.

3.4. Live-Cell Imaging

For live-cell imaging, one of the key parameters is the acquisition rate. Because structures can be moving quickly within a live cell, especially on the molecular level, it is crucial to maximize the acquisition frame rate. Many EMCCD cameras are now capable of imaging at >100 Hz. However, there are several considerations. First, how many molecules need to be localized, at what precision, and over what total area (A)? Suppose σ_{xy} = 40 nm resolution is sufficient, and the area of interest is 100 μm^2. One needs to estimate the number of localized molecules N_{loc} that is needed. Second, over what fraction of the area (f_A) does the species of interest reside? The necessary density of localized molecules needs to be high enough that sparseness of the probe does not limit resolution any more than σ_{xy}. If the probe is distributed within an area $f_A A$, then $N_{loc}/f_A A$ molecules per unit area will be observed within those structures, and the average distance between molecules will be approximately:

$$r_{ave} = (f_A A / N_{loc})^{1/2}.$$

To avoid having too sparse a distribution of probe, $r_{ave} < 40$ nm is needed. If the structure of interest covers ~20% of the area of interest, $f_A \times A = 20\ \mu m^2$ and $r_{ave} = 40$ nm, then $N_{loc} = F_A A/r^2_{ave}$ ~ $20\ \mu m^2/(0.04\ \mu m)^2 = 12{,}500$ at minimum. Assuming 20 molecules per frame can be localized ($1\ \mu m^{-2}$), 625 frames would be required, and at 10 ms/frame, an acquisition time per image of 6.25 s is needed. Thus, the structures of interest should move as little as possible (no more than 40 nm) within 6 s. For a higher frame rate (~500 Hz), the imaging time could approach 1 s. For 20-nm resolution of the same structure, four times more localized molecules would be needed, and thus the acquisition time would rise to ~25 s at 100 Hz.

In addition to the motion of the sample during the acquisition time, the motions of the individual molecules during a frame can cause considerable difficulty. For molecules diffusing in two dimensions (as in a membrane), the mean-squared displacement τ. If x^2 is larger than r^2_0, the $1/e^2$ radius of the PSF squared, then the image of the molecule will be blurred significantly by the motion of the molecule during a frame of duration τ. Thus, the frames must be kept short enough that $4D\tau << r^2_0$. Alternatively, this constraint limits the fastest diffusing molecules that can be reliably localized by FPALM in live cells to those with $D << r^2_0/4\tau$. For a frame rate of 100 Hz and 1.2-NA water-immersion lens at emission wavelength $\lambda = 520$ nm, $D << 1.4 \mu m^2/s$. For membrane proteins, such as hemagglutinin (HA) from influenza, diffusion coefficients can be small enough to satisfy this constraint. Recent work on FPALM imaging of HA *(58)* was successful largely because of the atypically small value of D for HA ($<0.1\ \mu m^2/s$). Clearly, faster acquisition rates are advantageous for several reasons.

3.5. Activation and Readout Procedures

Generally, the goal during FPALM acquisition is to image as many molecules of interest as possible, at a density low enough to localize them individually. The sample is placed on the stage, brought into focus, and a movie is acquired while activation and readout lasers are controlled in some pattern. In live cells, it is also crucial to acquire data in as little time as possible. Thus, the procedures for fixed- and live-cell imaging are slightly different.

In fixed cells, slower, synchronous (**Fig. 1**) readout can be performed. The readout laser is switched on and active molecules are imaged for tens to hundreds of frames, until most molecules have photobleached. Roughly, when the distance between visible molecules increases to more than a few microns within a field of interest, an activation pulse is applied. During each cycle, to reduce photobleaching of molecules by the readout beam, the readout beam can be blocked during activation. The intensity and duration of the activation pulse depend entirely on the density of inactive molecules, the quantum yield for activation, and the area illuminated. However, the goal is to activate as many molecules as possible without exceeding the density where molecules begin to overlap in the image or become closer than $4 \cdot R_0$ (~1 μm for green fluorescence). Because inactive molecules are invisible (making estimation of their concentration in a cell difficult), it usually takes some trial and error to determine, for a given sample, what length and intensity of pulse to use. For many samples, however, intensities of 0.1–10 W/cm^2 were sufficient to activate tens to hundreds of molecules of PA-GFP or EosFP. Tens to hundreds of cycles between activation and readout can be used, with total imaging times of minutes to tens of minutes being typical. **Fig. 6** shows an example of images of EosFP expressed in a fibroblast and imaged by FPALM.

A convenient procedure for high-speed live-cell FPALM, where time is precious, is to use asynchronous acquisition (*see* **Fig. 1**), where the readout laser illuminates the sample continuously and the camera images the sample continuously. This minimizes any delay caused by switching of optics or shutters. Activation laser pulses are applied whenever the density drops below ~0.1 μm^{-2} (very roughly) or low enough that the minimum distance between molecules is >>1 μm. Alternatively, for probes that do not show significant readout-induced activation, the activation laser may be allowed to illuminate the sample continuously as well, with an intensity that increases over time to induce the same number of activations per unit time even when the remaining density of inactive molecules has dropped (near the end of the acquisition). This solution may lead to increased background if the inactive form of the probe can be directly excited by the activation beam (as is the case for PA-GFP). However, the properties of PA-GFP are best suited to a third procedure for live-cell imaging. Because of the significant readout-induced activation in PA-GFP, acquisitions can be performed without activation laser pulses *(34,42,58)*. The readout beam performs all three functions: activation, readout (fluorescence excitation), and photobleaching. No shutters are required; the readout laser stays on continuously, the camera images continuously, and the entire data set can be obtained within seconds. The main problem with this method is that if the readout-induced activation is exploited, there is no way (so far)

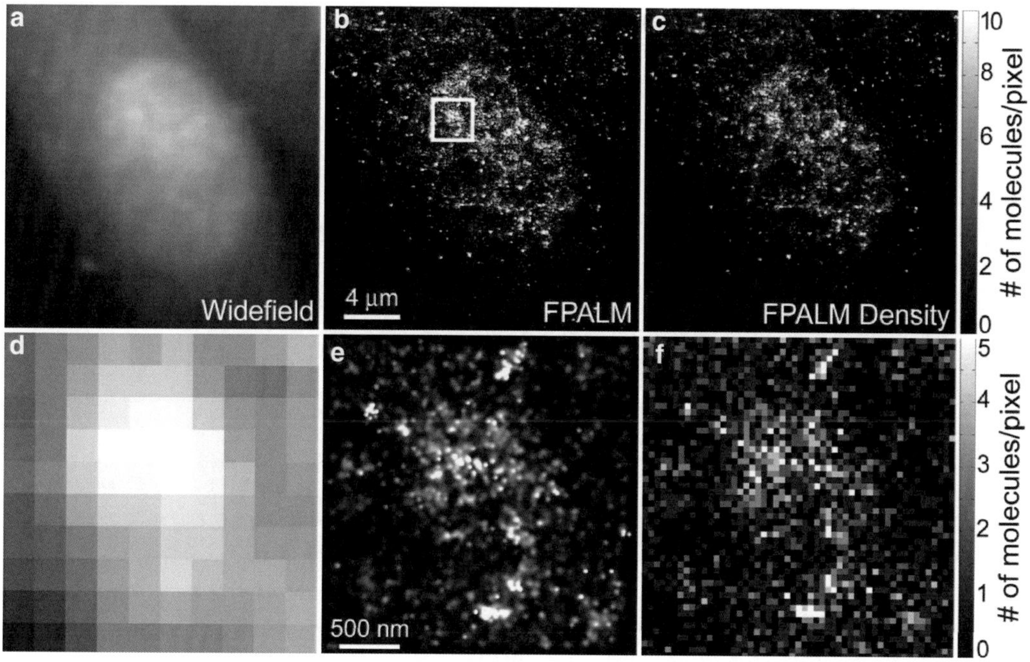

Fig. 6. Fixed cell imaging of EosFP. (**a**) Widefield image and (**b**) FPALM rendering (38,824 localized molecules) of a fixed fibroblast transfected with untargeted EosFP show comparison of FPALM and standard widefield fluorescence imaging. (**c**) Corresponding density plot of localized positions binned into 100-nm × 100-nm pixels. Note that the number of molecules per pixel does not distinguish between molecules localized more than once (in successive frames). (**d–e**) Zoom of *boxed region* in (**b**) shows significant improvement in resolution of (**e**) FPALM over (**d**) widefield fluorescence. (**f**) Corresponding localized molecule density within 50 nm × 50 nm pixels. Scale bar in (**b**) also applies to (**a**) and (**c**). Scale bar in (**e**) also applies to (**d**) and (**f**). Note that contrast was adjusted linearly in (**a**) and (**d**) for visualization.

to control the rate of activation, other than to change the readout beam intensity, which also affects the fluorescence emission and photobleaching rates. So, cell labeling density becomes even more important because if the density is too high, there is nothing to be done to decrease it except photobleach the given cell or try a different cell. On the other hand, many PA-GFP-labeled probes have been successfully imaged this way in times as little as 10–30 s per image (limited only by camera frame rate). **Fig. 7** shows an example of such a live-cell acquisition.

3.6. Data Analysis in Live and Fixed Cells

1. *Background Subtraction.* Before the position of a single molecule can be determined, a background subtraction is typically performed (background counts do not, in general, help to determine the position of an object, and add artificially to the apparent brightness of molecules and required threshold levels). The simplest method is the subtraction of a uniform baseline, in which a single (potentially time dependent, but spatially independent) value is subtracted from every pixel within the given image. This value is typically chosen as the average pixel value from a region in the image where there is

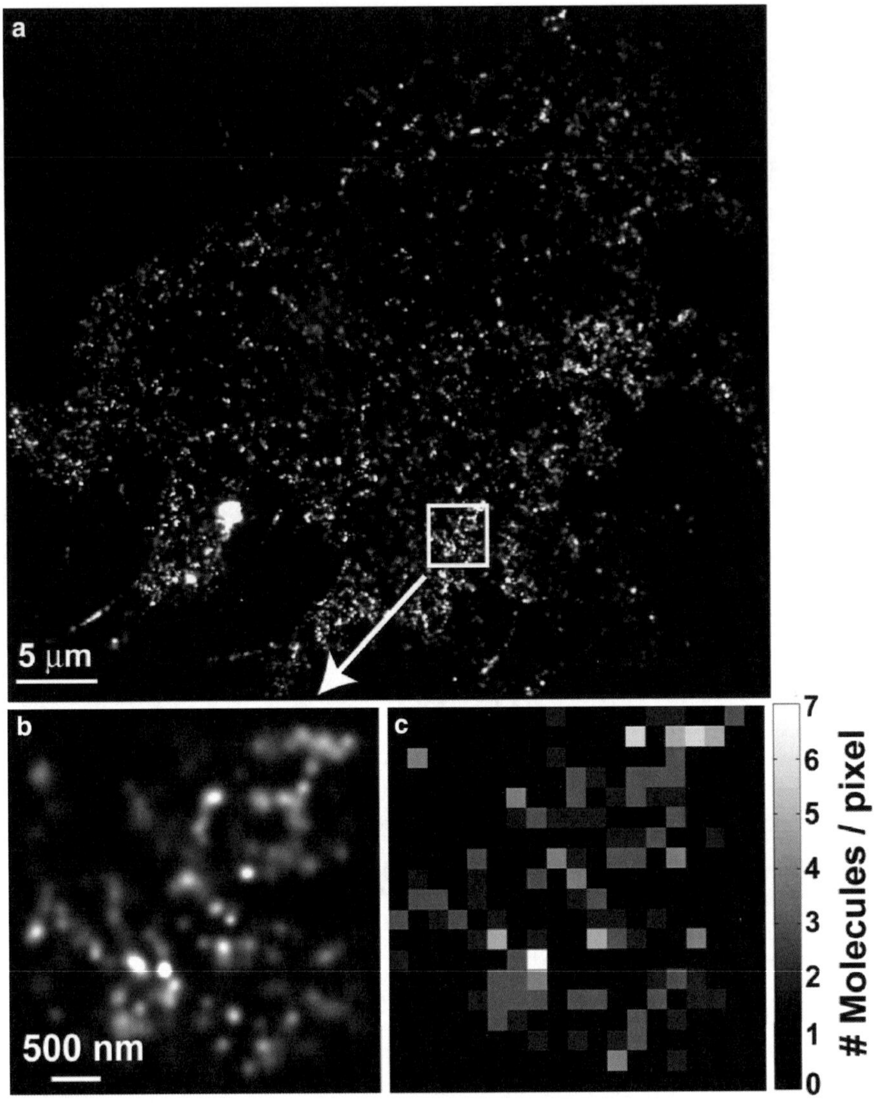

Fig. 7. Live-cell imaging of PA-GFP-HA. (**a**) FPALM rendering (11,937 localized molecules) of living fibroblast transfected with PA-GFP-HA and imaged at room temperature. (**b**) Zoom of *boxed region* in (**a**) shows structures resolved below the diffraction limit. (**c**) Corresponding density plot (shows number of molecules per unit area) of localized positions binned into 200-nm × 200-nm pixels. Scale bar in (**b**) also applies to (**c**).

no fluorescence, or is chosen from analysis of the image intensity histogram. This method is most appropriate for uniform distributions of background signal (such as dark counts from the camera) and can also be accounted for using a constant offset as a fitting parameter in the localization procedure.

In imaging applications that inevitably generate position- and time-dependent distributions of background signal (such as cell imaging), a nonuniform background subtraction is more appropriate, although it should be noted that any subtraction scheme will fail if the background signal is so high that the signal from a single emitting FP is indistinguishable from the background noise. One nonuniform method is to generate a time-averaged widefield image from all frames in the acquisition *(58)*. From each individual frame to be analyzed, the average widefield image (providing the position dependence of the background profile) is subtracted, weighted by (typically 95% of) the average intensity of that given frame (providing the time dependence of the background profile). This method requires the illumination profile within the area of interest to be as uniform as possible so that photobleaching occurs uniformly across the area.

Another example of nonuniform background subtraction is the differential image method in which the (n + 1)th frame of a time series acquisition is subtracted from the nth frame and positive intensity peaks that meet the single-molecule criteria are analyzed *(32)*. Additional consideration must be given for molecules that are visible in successive fames. In this manner, a series of N frames is sequentially analyzed backward starting with the (N− 1)th frame.

Yet another such method is the rolling ball algorithm in which the background subtraction is performed by "rolling" a sphere of a given radius (larger than the radius of the image of a single molecule) along the underside of the three-dimensional surface generated by the intensity of an image plotted as a function of x and y *(61)*. This method has the advantage that each frame of an acquisition has an independent background subtraction and an algorithm for its implementation is part of the commonly used imaging software ImageJ (Wayne Rasband, National Institute of Mental Health, Bethesda, MD, USA).

2. *Localization Algorithms.* The standard localization algorithm follows a series of steps, executed frame by frame: (1) identification of objects, (2) thresholding of objects, and (3) determination of coordinates. The first step is to search through the given image frame looking for any pixels above an intensity threshold, T_1. For this analysis, pixel values are calibrated to determine the number of detected photons corresponding to a given intensity (threshold). Any pixels above T_1 in intensity are marked as high pixels and tabulated. High pixels that are closer than some minimum radius (typically ~0.9 μm or ~$3.5R_0$) from any other high pixel are analyzed as a single object. Next, a box is created that contains the high pixel and the surrounding pixels (typically 5 × 5 pixels). The box should be large enough to contain the entire image of a single molecule (at least $4R_0$ in width) but not too much larger, because boxes should not contain more than one molecule. Each box is then thresholded. To pass, the box must have a minimum number of pixels above a second threshold (T_2), but must not have more than a maximum number of pixels above a third

threshold (T_3). Typically $T_2 \sim 0.6T_1$–$0.9T_1$, with a minimum of three pixels required to be above it (including the high pixel). The role of T_2 is to ensure that single (noisy) pixels are not analyzed as molecules. Typically, T_3 is the same as T_1, and the maximum number of pixels above it can be from 5 to 15. The role of T_3 is to eliminate objects (such as aggregates of fluorophore or fluorescent dust) that are too large and too bright to be single molecules. **Fig. 3d–e** shows examples of molecules identified and localized in fixed and living cells.

Boxes that have passed all three thresholds are then analyzed to determine the coordinates and intensity of the molecule within. A center-of-mass calculation provides an initial guess for the x–y coordinates of the molecule, which are used to initialize the least-squares fitting routine, which fits a two-dimensional Gaussian to the image of the molecule. The x and y coordinates, the amplitude, and a constant offset are used as the fitting parameters. The PSF width is usually fixed to a value determined experimentally, but can be allowed to float. Alternatively, instead of fitting the offset, the minimum intensity contained within that given box may be subtracted from all pixels in the box before fitting. The amplitude of the Gaussian fit is used to calculate the number of photons detected from that molecule, accounting for the finite size of the PSF by multiplying by the area of the PSF (in pixels, normalized to 1 at the peak).

3. *Rendering the Results.* FPALM images can be rendered by either of two methods: (1) unweighted, point-like plots of the positions of localized molecules, or (2) weighted plots of the positions, with each localized molecule plotted as a spot with a Gaussian profile, an intensity proportional to the number of photons detected from that molecule, and a radius equal to the calculated or experimentally determined localization-based resolution. Because the weighted plots take into account the position uncertainty and intensity of each molecule, the resulting image is in some ways a more realistic representation of a fluorescence image with ultrahigh resolution. Typically, all molecules localized within a particular area are rendered simultaneously, but in live cells or other time-dependent samples, time-dependent images may be rendered using subsets of molecules localized during various time periods. A threshold that includes only molecules within a particular range of intensities, or above a minimum intensity, can also be applied.

Figures 5–7 show examples of rendered FPALM images of caged 5-carboxyfluorescein on a coverslip, EosFP imaged in fixed fibroblasts, and PA-GFP-HA imaged in live fibroblasts at room temperature. The significant improvement in resolution is visible when comparing the FPALM images with normal widefield fluorescence images of the same samples. The density plots shown in **Figs. 6c**, **f**, and **7c** are another way to display the results, and show the number of molecules localized within each pixel of the image, color coded by intensity. Density plots are a good measure of the sparseness of the distribution, whereas the spots in the standard rendering (**Figs. 5, 6b, e,7a, b**) show the calculated localization precision for each rendered molecule.

4. Notes

1. For standard photoactivatable probes (nongenetically encoded), similar criteria apply, except that targeting is dependent on the physical properties of the probe, or on conjugation to an appropriate antibody or other biomolecule. Photoactivatable quantum dots would potentially be very powerful, because of their resistance to photobleaching, as long as they could be targeted to the desired structure.
2. If multiple probes are to be used, consideration must be taken to ensure that the emission of each probe will be spectrally separable using appropriate filter combinations. Alternatively, probes could be separated based on other properties. For example, differences in activation wavelength could be used even if the probes have similar emission spectra (in their active form). If one probe is pH or ion sensitive and the other is not, one probe could be imaged under conditions that temporarily quench the other.
3. To accommodate a larger number of excitation beams, several dichroic mirrors can be used to combine multiple beams together into a single path. As before, the positions of each dichroic mirror should be adjusted so that the beams meet at the same spot on the dichroic mirror, and then the angle of each dichroic mirror is adjusted so that the paths of the beams are collinear. If lasers are arranged sequentially from longest wavelength to shortest wavelength, long-pass dichroic mirrors can be used with successively shorter cutoff wavelengths. The cutoff wavelength for a given dichroic mirror needs to be in between the wavelength of the laser being reflected and the shortest wavelength of the lasers being transmitted.
4. Optionally, an engineered diffuser (Thorlabs ED1C20) can be used to shape the excitation beam and yield a circular (flat top) or square illumination profile at the sample.
5. In addition to the design of the CCD circuitry, the read noise can be influenced by intrinsic properties of the sensor. Unfortunately, read noise shows little temperature dependence, and as such, is often referred to as the fundamental limiting noise of a CCD device.

6. Note that the value of N_A may be limited to values larger than some minimum if either k_0 or Φ_{RA} is finite. If Φ_{RA} is too large, the number of active molecules may be difficult to control.
7. It is sometimes helpful to first find the coverslip surface, on which there are invariably some small numbers of fluorescent molecules initially visible in many areas, as a known *z*-position.

Acknowledgments

The authors thank J. Zimmerberg and P. Blank for useful discussions and loaned equipment, G. Patterson for PA-GFP protein and constructs, J. Wiedenmann and U. Nienhaus for EosFP protein and constructs, J. Gosse for helpful discussions, V. Verkhusha for Dendra2 constructs, T. Tripp for machining services, and J. Lozier for molecular biology services. This work was supported in part by National Institutes of Health (NIH) grant K25-AI65459, National Science Foundation (NSF) grant CHE-0722759, and University of Maine startup funds.

References

1. Pawley, J. B. (2006). *Handbook of Biological Confocal Microscopy*, 3rd ed., Springer, New York, NY.
2. Abbe, E. (1873). Beiträge zur Theorie des Mikroskops und der mikroskopischen Wahrnehmung, *Archive für mikroskopische Anatomie* **9**, 413–68.
3. Sandison, D. R., Piston, D. W., Williams, R. M. and Webb, W. W. (1995). Quantitative comparison of background rejection, signal-to-noise ratio, and resolution in confocal and full-field laser-scanning microscopes, *Appl Opt* **34**, 3576–88.
4. Gu, M. (1999). *Advanced Optical Imaging Theory*, Springer, Heidelberg.
5. Sandison, D. R. and Webb, W. W. (1994). Background rejection and signal-to-noise optimization in confocal and alternative fluorescence microscopes, *Appl Opt* **33**, 603–15.
6. Yildiz, A., Forkey, J. N., McKinney, S. A., Ha, T., Goldman, Y. E. and Selvin, P. R. (2003). Myosin V walks hand-over-hand: single fluorophore imaging with 1.5-nm localization, *Science* **300**, 2061–5.
7. Barak, L. S. and Webb, W. W. (1982). Diffusion of low density lipoprotein-receptor complex on human fibroblasts, *J Cell Biol* **95**, 846–52.
8. Lakowicz, J. R. (2006). *Principles of Fluorescence Spectroscopy*, 3rd ed., Springer, New York.
9. Gustafsson, M. G. (2000). Surpassing the lateral resolution limit by a factor of two using structured illumination microscopy., *J Microsc* **198**, 82–7.
10. Hell, S. and Steltzer, E. H. K. (1992). Properties of a 4Pi confocal fluorescence microscope, *J Opt Soc Am A* **9**, 2159–67.
11. Bewersdorf, J., Bennett, B. T. and Knight, K. L. (2006). H2AX chromatin structures and their response to DNA damage revealed by 4Pi microscopy, *Proc Natl Acad Sci U S A* **103**, 18137–42.
12. Egner, A., Jakobs, S. and Hell, S. W. (2002). Fast 100-nm resolution three-dimensional microscope reveals structural plasticity of mitochondria in live yeast, *Proc Natl Acad Sci U S A* **99**, 3370–5.
13. Gugel, H., Bewersdorf, J., Jakobs, S., Engelhardt, J., Storz, R. and Hell, S. W. (2004).

Cooperative 4Pi excitation and detection yields sevenfold sharper optical sections in live-cell microscopy, *Biophys J* **87**, 4146–52.

14. Bewersdorf, J., Schmidt, R. and Hell, S. W. (2006). Comparison of I5M and 4Pi-microscopy, *J Microsc* **222**, 105–17.
15. Gustafsson, M. G., Agard, D. A. and Sedat, J. W. (1995). Sevenfold improvement of axial resolution in 3D widefield microscopy using two objective lenses, *Proc SPIE* **2412**, 147–56.
16. Gustafsson, M. G., Agard, D. A. and Sedat, J. W. (1999). I5M: 3D widefield light microscopy with better than 100 nm axial resolution, *J Microsc* **195**, 10–6.
17. Hell, S. W. and Wichmann, J. (1994). Breaking the diffraction resolution limit by stimulated-emission – stimulated-emission-depletion fluorescence microscopy, *Opt Lett* **19**, 780–82.
18. Westphal, V. and Hell, S. W. (2005). Nanoscale resolution in the focal plane of an optical microscope, *Phys Rev Lett* **94**, 143903.
19. Kittel, R. J., Wichmann, C., Rasse, T. M., Fouquet, W., Schmidt, M., Schmid, A., Wagh, D. A., Pawlu, C., Kellner, R. R., Willig, K. I., Hell, S. W., Buchner, E., Heckmann, M. and Sigrist, S. J. (2006). Bruchpilot promotes active zone assembly, Ca^{2+} channel clustering, and vesicle release, *Science* **312**, 1051–4.
20. Terskikh, A., Fradkov, A., Ermakova, G., Zaraisky, A., Tan, P., Kajava, A. V., Zhao, X. N., Lukyanov, S., Matz, M., Kim, S., Weissman, I. and Siebert, P. (2000). "Fluorescent timer": protein that changes color with time, *Science* **290**, 1585–88.
21. Hell, S. W. and Kroug, M. (1995). Ground-state depletion fluorescence microscopy, a concept for breaking the diffraction resolution limit, *Appl Phys B* **60**, 495–97.
22. Lidke, K. A., Rieger, B., Jovin, T. M. and Heintzmann, R. (2005). Superresolution by localization of quantum dots using blinking statistics, *Opt Express* **13**, 7052–62.
23. Gustafsson, M. G. (2005). Nonlinear structured-illumination microscopy: wide-field fluorescence imaging with theoretically unlimited resolution, *Proc Natl Acad Sci U S A* **102**, 13081–6.
24. Hofmann, M., Eggeling, C., Jakobs, S. and Hell, S. W. (2005). Breaking the diffraction barrier in fluorescence microscopy at low light intensities by using reversibly photoswitchable proteins, *Proc Natl Acad Sci U S A* **102**, 17565–9.
25. Burns, D. H., Callis, J. B., Christian, G. D. and Davidson, E. R. (1985). Strategies for attaining superresolution using spectroscopic data as constraints, *Appl Opt* **24**, 154.
26. Hwang, J., Tamm, L. K., Bohm, C., Ramalingam, T. S., Betzig, E. and Edidin, M. (1995). Nanoscale complexity of phospholipid monolayers investigated by near-field scanning optical microscopy, *Science* **270**, 610–14.
27. Esa, A., Edelmann, P., Kreth, G., Trakhtenbrot, L., Amariglio, N., Rechavi, G., Hausmann, M. and Cremer, C. (2000). Three-dimensional spectral precision distance microscopy of chromatin nanostructures after triple-colour DNA labelling: a study of the BCR region on chromosome 22 and the Philadelphia chromosome, *J Microsc* **199**, 96–105.
28. Kural, C., Kim, H., Syed, S., Goshima, G., Gelfand, V. I. and Selvin, P. R. (2005). Kinesin and dynein move a peroxisome in vivo: a tug-of-war or coordinated movement?, *Science* **308**, 1469–72.
29. Qu, X., Wu, D., Mets, L. and Scherer, N. F. (2004). Nanometer-localized multiple single-molecule fluorescence microscopy, *Proc Natl Acad Sci U S A* **101**, 11298–303.
30. Gordon, M. P., Ha, T. and Selvin, P. R. (2004). Single-molecule high-resolution imaging with photobleaching, *Proc Natl Acad Sci U S A* **101**, 6462–5.
31. Bates, M., Huang, B., Dempsey, G. T. and Zhuang, X. (2007). Multicolor super-resolution imaging with photo-switchable fluorescent probes, *Science* **317**, 1749–53.
32. Betzig, E., Patterson, G. H., Sougrat, R., Lindwasser, O. W., Olenych, S., Bonifacino, J. S., Davidson, M. W., Lippincott-Schwartz, J. and Hess, H. F. (2006). Imaging intracellular fluorescent proteins at nanometer resolution, *Science* **313**, 1642–45.
33. Rust, M. J., Bates, M. and Zhuang, X. (2006). Sub-diffraction-limit imaging by stochastic optical reconstruction microscopy (STORM), *Nat Methods* **3**, 793–6.
34. Egner, A., Geisler, C., von Middendorff, C., Bock, H., Wenzel, D., Medda, R., Andresen, M., Stiel, A. C., Jakobs, S., Eggeling, C., Schonle, A. and Hell, S. W. (2007). Fluorescence nanoscopy in whole cells by asynchronous localization of photoswitching emitters, *Biophys J* **93**, 3285–90.
35. Folling, J., Belov, V., Kunetsky, R., Medda, R., Schonle, A., Egner, A., Eggeling, C., Bossi, M. and Hell, S. W. (2007). Photochromic rhodamines provide nanoscopy with optical sectioning, *Angew Chem Int Ed Engl* **46**, 6266–70.
36. Thompson, R. E., Larson, D. R. and Webb, W. W. (2002). Precise nanometer localization analysis for individual fluorescent probes, *Biophys J* **82**, 2775–83.

37. Lacoste, T. D., Michalet, X., Pinaud, F., Chemla, D. S., Alivisatos, A. P. and Weiss, S. (2000). Ultrahigh-resolution multicolor colocalization of single fluorescent probes, *Proc Natl Acad Sci U S A* **97**, 9461–6.
38. Michalet, X. and Weiss, S. (2006). Using photon statistics to boost microscopy resolution, *Proc Natl Acad Sci U S A* **103**, 4797–98.
39. Lukyanov, K. A., Chudakov, D. M., Lukyanov, S. and Verkhusha, V. V. (2005). Photoactivatable fluorescent proteins, *Nat Rev Mol Cell Biol* **6**, 885–91.
40. Tsien, R. Y. (1998). The green fluorescent protein, *Annu Rev Biochem* **67**, 509–44.
41. Giepmans, B. N., Adams, S. R., Ellisman, M. H. and Tsien, R. Y. (2006). The fluorescent toolbox for assessing protein location and function, *Science* **312**, 217–24.
42. Hess, S. T., Girirajan, T. P. and Mason, M. D. (2006). Ultra-high resolution imaging by fluorescence photoactivation localization microscopy, *Biophys J* **91**, 4258–72.
43. Hess, S. T., Sheets, E. D., Wagenknecht-Wiesner, A. and Heikal, A. A. (2003). Quantitative Analysis of the fluorescence properties of intrinsically fluorescent proteins in living cells, *Biophys J* **85**, 2566–80.
44. Heikal, A. A., Hess, S. T. and Webb, W. W. (2001). Multiphoton molecular spectroscopy and excited–state dynamics of enhanced green fluorescent protein (EGFP): acid-base specificity, *Chem Phys* **274**, 37–55.
45. Haupts, U., Maiti, S., Schwille, P. and Webb, W. W. (1998). Dynamics of fluorescence fluctuations in green fluorescent protein observed by fluorescence correlation spectroscopy, *Proc Natl Acad Sci U S A* **95**, 13573–78.
46. Hess, S. T., Heikal, A. A. and Webb, W. W. (2004). Fluorescence photoconversion kinetics in novel green fluorescent protein pH sensors, *J Phys Chem B* **108**, 10138–48.
47. Heikal, A. A., Hess, S. T., Baird, G. S., Tsien, R. Y. and Webb, W. W. (2000). Molecular spectroscopy and dynamics of intrinsically fluorescent proteins: Coral red (dsRed) and yellow (Citrine), *Proc Natl Acad Sci U S A* **97**, 11996–2001.
48. Schwille, P., Kummer, S., Heikal, A. A., Moerner, W. E. and Webb, W. W. (2000). Fluorescence correlation spectroscopy reveals fast optical excitation-driven intramolecular dynamics of yellow fluorescent proteins, *Proc Natl Acad Sci U S A* **97**, 151–56.
49. Brasselet, S., Peterman, E. J. G., Miyawaki, A. and Moerner, W. E. (2000). Single-molecule fluorescence resonant energy transfer in calcium concentration dependent cameleon, *J Phys Chem B* **104**, 3676–82.
50. Patterson, G. H. and Lippincott-Schwartz, J. (2002). A photoactivatable GFP for selective photolabeling of proteins and cells, *Science* **297**, 1873–77.
51. Chudakov, D. M., Verkhusha, V. V., Staroverov, D. B., Souslova, E. A., Lukyanov, S. and Lukyanov, K. A. (2004). Photoswitchable cyan fluorescent protein for protein tracking, *Nat Biotechnol* **22**, 1435–9.
52. Wiedenmann, J., Ivanchenko, S., Oswald, F., Schmitt, F., Rocker, C., Salih, A., Spindler, K. D. and Nienhaus, G. U. (2004). EosFP, a fluorescent marker protein with UV-inducible green-to-red fluorescence conversion, *Proc Natl Acad Sci U S A* **101**, 15905–10.
53. Gurskaya, N. G., Verkhusha, V. V., Shcheglov, A. S., Staroverov, D. B., Chepurnykh, T. V., Fradkov, A. F., Lukyanov, S. and Lukyanov, K. A. (2006). Engineering of a monomeric green-to-red photoactivatable fluorescent protein induced by blue light, *Nat Biotechnol* **24**, 461–65.
54. Ando, R., Hama, H., Yamamoto-Hino, M., Mizuno, H. and Miyawaki, A. (2002). An optical marker based on the UV-induced green-to-red photoconversion of a fluorescent protein, *Proc Natl Acad Sci U S A* **99**, 12651–56.
55. Ando, R., Mizuno, H. and Miyawaki, A. (2004). Regulated fast nucleocytoplasmic shuttling observed by reversible protein highlighting, *Science* **306**, 1370–73.
56. Axelrod, D. (2001) Total internal reflection fluorescence microscopy in cell biology, *Traffic* **2**, 764–74.
57. Axelrod, D., Thompson, N. L. and Burghardt, T. P. (1983). Total internal inflection fluorescent microscopy, *J Microsc* **129**, 19–28.
58. Hess, S. T., Gould, T. J., Gudheti, M. V., Maas, S. A., Mills, K. D. and Zimmerberg, J. (2007). Dynamic clustered distribution of hemagglutinin resolved at 40 nm in living cell membranes discriminates between raft theories, *Proc Natl Acad Sci U S A* **104**, 17370–75.
59. Lakowicz, J. R. (1983). *Principles of Fluorescence Spectroscopy*, Plenum, New York.
60. Bass, M. and Optical Society of America. (1995). *Handbook of Optics*, 2nd ed., McGraw-Hill, New York.
61. Sternberg, S. R. (1983). Biomedical image processing, *IEEE Comput* 22–34.

Part IV

Nanotechnologies for Biomolecular Delivery, Gene Therapy and Expression Control

Chapter 33

Real-Time Imaging of Gene Delivery and Expression with DNA Nanoparticle Technologies

Wenchao Sun and Assem G. Ziady

Summary

The construction of safe, efficient, and modifiable synthetic DNA nanoparticles is an emerging technology that has achieved important milestones of success in the past 5 years. Advances in chemical conjugation, purification, and controlled synthesis have allowed researchers to produce uniform and stable particles, whose physical characteristics can be well characterized and monitored. As a result of these improvements, DNA nanoparticles have now been cleared for clinical testing, and show good potential for human gene therapy. A very important recent development in the study of DNA nanoparticles is the use of small-animal imaging. Real-time imaging has become a valuable technique for tracking particle biodistribution and gene transfer efficacy. In this chapter, we discuss how bioluminescent, positron emission tomography, and magnetic resonance imaging can be used separately or in concert to study particle delivery, localization, and magnitude of gene expression in vivo.

Key words: Small-animal imaging, DNA nanoparticles, Bioluminescent imaging, Positron emission tomography, Magnetic resonance imaging

1. Introduction

1.1. DNA Nanoparticles

The basic design of DNA nanoparticles (DNP) consists of three elements: *modifying agents*, such as a ligand, coupled with a *polycation*, which is capable of condensing *DNA*. In initial reports, the efficacy of this nonviral method of gene delivery was low and inconsistent, especially in animals *(1)*. However, recent developments in the field have improved a number of factors including homogeneity and stability, allowing for the design of superior second- and third-generation particles that are capable of consistently

James Weifu Lee and Robert S. Foote (eds.), *Micro and Nano Technologies in Bioanalysis*,
Methods in Molecular Biology, vol. 544
DOI 10.1007/978-1-59745-483-4_33, © Humana Press, a part of Springer Science+Business Media, LLC 2009

efficient gene transfer in vivo *(2–4)*. Antigenicity is usually low for poly-L-amino acids *(5)* and DNA *(6)*, and the packaging capacity is high *(7)*, allowing for the inclusion of native promoters, enhancer sequences, and intronic sequences that augment gene expression. Some of the advantages of DNA nanoparticles include their ability to introduce DNA plasmids into specific cell types, when targeted *(8–14)*, and to access the nuclei of nondividing cells *(15)*, which is required for success in vivo. Selective targeting of cells is important for proper processing of therapeutic transgene products, and the ubiquitous expression of proteins otherwise not expressed in most tissues can be harmful, so indiscriminate gene transfer is undesirable. Furthermore, lower doses of targeted nanoparticles can achieve the efficacy of higher doses of nontargeted particles, because only a fraction of nontargeted particles reach the cell type of interest. Using lower doses lowers toxicity. Therefore, receptor ligands, ranging from simple sugars to complex antibodies, have been used to target DNA nanoparticles (DNPs) to specific cell types *(1)*.

Early generations of DNPs made use of polyK conjugates to ligands to condense DNA into spheroid charge-neutral particles measuring ~25 nm in diameter *(8–12)*. These DNPs are formed in a high-salt solution (~1 M NaCl) that prevents aggregation and promotes stability of the DNA condensates *(8–12)*. Modification of short chains of polyK with polyethylene glycol (PEG) stabilized (PS) these new-generation DNA nanoparticles (PS–DNPs) in physiological saline for up to 3 years at 4°C, and allowed production of highly concentrated formulations of compacted DNA (up to 12 mg/mL) *(9, 15)*. In addition to overcoming many of the barriers to good gene transfer, such as protecting DNA from nuclease degradation, accessing the nuclei of nondividing cells, and not interfering with transcription *(10, 16)*, PS–DNPs fail to elicit the inflammatory responses *(3)* activated by viral-based *(17)*, liposomal-based *(18)*, or polyethylenimine-based *(19)* vectors. In animals, depending on the route of administration, cells in the lung or liver can be targeted for reporter *(12)* or therapeutic *(4, 12)* gene delivery. Because polymers of L-amino acids fail to elicit an immune response *(5)*, DNA itself is a poor antigen, and PEG protects the complex from immune surveillance, PS–DNPs may be readministered without decrement in effect *(12)*. The safety and efficacy profile of PS–DNPs prompted the first clinical trial with cystic fibrosis transmembrane regulator (CFTR) polyplexes in cystic fibrosis (CF) patients *(4)*. Twelve CF patients received three escalating doses applied to the inferior turbinate. Vector was detected by polymerase chain reaction (PCR) at day 14 in nasal scrapings from all of the six patients who received the highest dose. Furthermore, partial nasal potential difference (NPD) correction was observed in 8 of 12 patients for up to 6 days. No adverse events were attributed to the treatment.

Table 1 summarizes the structural and functional characteristics of the different-generation DNPs.

1.2. Real-Time Animal Imaging of DNA Nanoparticles

Small-animal imaging (SAI) is an ideal way to regularly assess the delivery and efficacy of DNPs in vivo. We will discuss three non-invasive imaging modalities. The first is bioluminescent imaging (BLI), conducted, for example, with a Xenogen 200 charge-coupled device (CCD) that is cryogenically cooled to –105°C. This imager can efficiently detect chemiluminescence and/or fluorescence. The second is positron emission tomography (PET), conducted, for example, by a Concord Microsystems microPET R4 scanner. This rodent imaging system consists of four lutetium oxyorthosilicate (LSO) detector rings with an axial field of view (FOV) of 7.89 cm and is able to image a whole-body mouse in one bed position with a spatial resolution less than 2.0 mm at the center and a sensitivity of 900 cps/μCi. It acquires data in the list mode, which allows re-binning into desirable dynamic frames, and its rotating point source provides transmission scan for measured attenuation correction. Both the BLI and PET imaging instrumentation can be fitted with X-ray imaging to enhance localization accuracy. The third modality is small-animal magnetic resonance (MR) imaging (MRI), which may be performed with 7- and 9.4-T small-animal MRI and magnetic resonance spectroscopy systems. MRI has spatial resolution as high as 100 μm, but is less sensitive than BLI or PET imaging. All of these imaging techniques allow for serial and longitudinal imaging to be performed on the same living animal, enabling one to follow a single experimental subject over time and monitor both the localization of our trackable DNPs (PET and MRI) and their activity (BLI and PET).

To assess the activity of different formulations of DNPs, we use BLI to localize and quantitate firefly luciferase expression, or PET to image radiolabeled cell-permeable tracers (^{125}I-FIAU or ^{18}F-FHBG), which are specifically phosphorylated by HSV-1 tk *(20)*. Cells expressing this transgene sequester FIAU or FFHPG, allowing for localization by PET *(20)*. The use of BLI and PET in a confirmatory manner (two different imaging modalities of two different expression systems) reduces artifacts. For tracking formulations of DNPs, PET and/or MRI may be used. Because PET is more sensitive than MRI but has lower resolution, these techniques are complementary, and their combined use ensures that weak signals are detected, and strong signals are imaged with good resolution. A novel and versatile method for in vivo tracking is the conjugation of 1,4,7,10-tetraazacyclododecane-*N*,*N*′,*N*″,*N*‴ -tetraacetic acid (DOTA) to our DNPs. DOTA serves a dual role, because it can be chelated with metals for MRI (gadolinium) *(21)* or PET (indium) imaging *(22)*.

Macrocyclic metal-DOTA chelates are often administered to patients and animal models to create or enhance contrast in biomedical molecular imaging studies *(21)*. More recently,

Table 1
Characteristics and efficacy of 1st and 2nd generation DNA nanoparticles

	Polycation carrier								
Type (generation)	Formulation	Linkage between PEG and CK30	PolyK counter ion	Targeting ligand	Physical and chemical characteristics	In vitro gene delivery	In vivo gene delivery	Toxicity in vivo	References
DNA·polyK-SECR/PepL (1st)	PolyK 22.5 PolyK 100 PolyK 256 kDa	No PEG	Bromide	C105Y C1315	Formulated in high salt, unstable. Spheroidal particles. 17~25 nm in diameter	HuH7 and Hep G2 cells	Nasal/airway epithelia, and tissue macrophages	Low at amount <25 μg	*(1, 8–12)*
DNA·polyK-pIgR/Fab (1st)	PolyK 20 kDa	No PEG	Bromide	Anti-secretory component (human or rat) Fab fragment	Formulated in high salt, unstable. Toroid, ~25 nm in diameter	HT29.74 human colon carcinoma expressing pIgR and primary human tracheal epithelia	Rat airway epithelial cells	Low at amount <25 μg	*(1, 13)*
DNA·polyK-pIgR/scFv (1st)	PolyK 53 kDa	No PEG	TFA	Anti-human secretory component scFv	Formulated in high salt, unstable. Spheroidal or toroid, ~25 nm in diameter	MDCK cells expressing pIgR and primary human tracheal epithelia	Not reported	Low at amount <25 μg	*(1, 14)*
DNA·PEG10k-CS-CK30 (2nd)	m-PEG10k-maleimide + CK30	Thioether bond	TFA	Non	Formulated in water. Stable. Charge neutral. Short rods or ellipsoids in shape	HeLa cells. Transfect SY5Y cells and HuH7 cells only with Lipofectin	Airway epithelial cells of WT mice	None at amount <10 μg. Low at <100 μg	*(15)*

DNA·PEG10k-CS-CK30 (2nd)	m-PEG10k-maleimide + CK30	Thioether bond	Acetate	Non	Formulated in water. Stable. Slightly positively charged. Long rods in shape	HeLa cells	Airway epithelial cells of WT mice	None at <10 μg. Low at <100 μg	*(2–4, 7, 15)*
DNA·PEG10k-SS-CK30 (2nd)	m-PEG10k-OPSS + CK30	Disulfide-bond	Acetate	Non	Formulated in water. Stable. Slightly positively charged. Long rods in shape	HuH7 cells	Not reported	Not evaluated	N.R.
DNA·PEG10k-SS-CK30/PepL (2nd)	PEG10k (OPSS)2 + CK30	Disulfide-bond	Acetate	C105Y	Formulated in water. Stable. Slightly positively charged. Long rods in shape	HuH7 cells	Not reported	Not evaluated	N.R.

Non,; *WT*, Wild Type; *N.R.*, non Reported

metal-DOTA chelates have been conjugated to peptides to affect the in vivo pharmacokinetics of the metal–DOTA imaging agent over a wide range of spatial scales *(23, 24)*. Numerous other applications have been reported, including peptide-DOTA probes for multimodality imaging studies *(25)*. For the purposes of labeling DNPs, standard solid-phase peptide synthesis (SPPS) methods have been used to couple DOTA to the N terminus of peptides bound to a PEGA Rink amide resin *(21)*. An amine-derivatized DOTA (aminoDOTA) can couple to the carboxylates of an amino acid residue on resin. In this way, DOTA can be incorporated at any location within a peptide backbone *(21)*.

In addition to basic MR imaging of DOTA-conjugated DNPs with gadolinium (Gd), it is also possible to assess pH *(26)* and cellular internalization *(27)* using other metals. MRI signals can be quantified and so we will be able to determine what fraction of administered vector is routed to destruction in lysosomes, the suspected fate of many nonviral delivery agents. MR contrast agents that change the MR signal properties of surrounding water (e.g., the T1 relaxation rate) can be detected and calibrated to measure agent concentrations and pH effects *(26, 28)*. A pH-independent MR contrast agent, such as Gd, must also be evaluated, and the results from both agents must be ratiometrically compared to distinguish the pH measurement from the measurement of the concentrations of the agents.

Recently, researchers demonstrated that pH can be measured with a fundamentally new type of MRI contrast agent that is detected through a mechanism of PARAmagnetic Chemical Exchange Saturation Transfer (PARACEST) *(26, 29)*. The MR frequencies of a PARACEST agent's protons are shifted to very unique values because of their proximity to a paramagnetic lanthanide ion such as thallium (Tm), which can be chelated to DOTA. PARACEST agents contain protons that undergo rapid chemical exchange with protons of surrounding water molecules *(29)*. Selective saturation is applied at the specific frequency of the exchangeable protons, which reduces the detectable magnetization from these protons. Rapid chemical exchange with water causes a transfer of reduced detectable magnetization to the water signal, and continuous selective saturation and chemical exchange enhances the PARACEST effect, providing improved MRI sensitivity *(29)*. Importantly, chemical exchange between water and amide functional groups is dependent on pH, so that the PARACEST effect can be produced by a Tm-DOTA-peptide conjugate. The utility of the DOTA molecule for many PET and MR imaging approaches, the ease of coupling it to DNPs with no significant impact on size or function, and its low toxicity profile *(30, 31)* are all characteristics that have made it very desirable for use in real-time in vivo imaging of DNPs.

2. Materials

Material amounts outlined in this section reflect the quantities necessary for imaging a single 8- to 12-week-old mouse weighing ~25 g.

2.1. Synthesis of Molecular Conjugate

1. Spectrophotometer, such as Molecular Devices' SpectraMax M2.
2. Fourier transform infrared (FT-IR) spectrometer, such as ABB BOMEM's MB-104 spectrometer operating from 600 to 4,000 cm^{-1}.
3. High-performance liquid chromatography (HPLC) apparatus with a Grace Vydac OD-300 C-18 reverse-phase analytical column and a UV detector operating at 280 nm.
4. Binding resin with 1.0 mmol/g hydroxyl groups.
5. Wang resin (GenScript Corp.).
6. Fmoc-Cys-OH, Fmoc-Lys-OH, and Fmoc-amino acid of choice (as needed based on sequence of desired targeting ligand) are used as building amino acids, and O-benzotriazole-*N*,*N*,*N*,*N*-tetramethyl-uronium-hexafluoro-phosphate (HBTU, 0.5 mmol) and *N*-hydroxybenzotriazole (HOBt, 0.5 mmol) are used as coupling agents.
7. Freshly distilled TEA (140 μL, 1 mmol) added to 70 mL *N*-methylpyrolidone (NMP) and 20% piperidine in NMP.

2.2. Nontargeted DNP Construction

1. Polylysine: 1 cysteine residue followed by 30 lysine residues (CK_{30}), prepared as trifluoroacetate (TFA) salt using an automated solid-phase peptide synthesizer.
2. Polyethylene glycol (PEG): unifunctional methoxy-PEG-maleimide 10 kDa (mPEG10k-MAL), commercially available.
3. Ellman's assay reagent (Thermo Scientific Pierce Protein Research Products).
4. PE buffer: 50 mM sodium phosphate, pH 7.0, with 5 mM EDTA.

2.3. Targeted DNP Construction

1. **Items 1**, **3**, and **4** from **Subheading 2.1**.
2. Polyethylene glycol (PEG): bifunctional OPSS-PEG-OPSS 10 kDa ($PEG10k(OPSS)_2$), commercially available.
3. Targeting ligand containing cysteine residue with sulfhydryl group available for conjugation.

2.4. Metal Chelation with DOTA-Modified Peptides

1. $TmCl_3$, $GdCl_3$, or radioactive $InCl_3$ in equimolar concentration to DOTA used.
2. 0.5 N NaOH.

3. Arsenazo III is used to determine the amount of free metal in solution.
4. Matrix assisted laser desorption ionization (MALDI) mass spectrometer.

2.5. Plasmid and PS–DNP Preparation

1. *Escherichia coli* DH5α strain in competent state (Invitrogen Corp.).
2. Endotoxin free plasmid preparation kit, size as needed (Qiagen Inc.).
3. Restriction endonuclease enzymes, as needed, depending on the plasmid used.
4. 1.0% agarose gel electrophoresis.
5. Limulus amebocyte lysate (LAL) assay (Kinetic-QCL) for measuring endotoxin levels.
6. Synthesized molecular conjugate, capable of compacting DNA.
7. DNA plasmid encoding luciferase gene controlled by CMV promoter (*see* **Note 1**).
8. UV/Vis spectrophotometer (for example, the Molecular Devices SpectraMax).
9. Normal saline (0.9% NaCl).
10. Centrifugal concentrator: Vivaspin 20 3k molecular weight cutoff (MWCO) or 100k MWCO (Thermo Fisher)
11. Vibrax shaker set to 1,200 rpm (IKA-Vibrax-VXR, Janke & Kunkel).
12. Lyophilizer, as available.

2.6. PS–DNP Characterization

1. UV/Vis spectrophotometer (for example, the Molecular Devices SpectraMax).
2. Optical density (O.D.) data analysis software, such as the SoftMax pro software, that may be used to analyze multiple data points.
3. 400-Mesh electron microscope carbon grid.
4. 0.04% Uranyl acetate dissolved in HPLC-grade methanol.
5. An electron microscope, such as the JEOL-100C (Particle Sizing Systems, model NICOMP 380 ZLS), capable of 40,000× magnification.
6. Kinetic-QCL limulus amebocyte lysate (LAL) assay kit (Lonza Biosciences).

2.7. Animal Protocol

1. Small animal (<500 g). Usually, we use mice that are 8–12 weeks old. All groups are age, sex, and strain matched (*see* **Note 2**).
2. Anesthetic cocktail made up of 2.13 mg/mL xylazine, 0.36 mg/mL acepromazine, and 10.75 mg/mL ketamine.

3. Compacted DNA nanoparticles (nontargeted or targeted PS-DNPs).
4. One-microliter syringe with 27.5- or 30-gauge needle for intraperitoneal (IP) or intravenous (IV) injection, respectively.
5. For intratracheal (IT) administrations, a 22-gauge plastic catheter (Abbot Laboratories).

2.8. Small-Animal MRI

1. Magnetic resonance imager, such as the 9.4-T Bruker Biospin small-animal MRI scanner.
2. For cannulation of the tail vein, a 27- or 30-gauge cannula is needed.
3. Nuclear magnetic resonance spectrometer, such as the Varian Inova 600 MHz NMR spectrometer.

2.9. Bioluminescent Imaging

1. Bioluminescent imager, such as the Xenogen 200 instrument. The instrument should include a CCD camera that can sensitively detect photons.
2. Imaging software.
3. D-Luciferin purchased as a solid and dissolved at 30 mg/mL in phosphate-buffered saline (PBS) immediately before use.

2.10. Small-Animal PET Imaging

1. For X-ray images, X-SPECT scanner (Gamma Medica) is used.
2. Small-animal PET imager, such as the R4 microPET scanner (Siemens/Concord).
3. Dried ^{125}I-labeled 2′-fluoro-2′-deoxy-1-β-D-arabinofuranosyl-5-iodouracil ([^{125}I]-FIAU) or [^{18}F]-labeled 9-[4-fluoro-3-(hydroxymethyl)butyl]guanine ([^{18}F]-FHBG) are dissolved in PBS.
4. ^{111}Indium chloride ($InCl_3$), is obtained with an activity concentration of 15–25 mCi/mL.

3. Methods

3.1. Synthesis of the DNA Compacting Agent CK_{30}, and DOTA Functionalization

All peptides are synthesized using a Wang resin and follow Fmoc chemistry methods with HOBt and HBTU as coupling agents *(21)*.

1. It is necessary to test a number of parameters that affect synthesis yield when developing a new method. To test the efficiency of coupling each amino acid residue, use the Kaiser's ninhydrin test *(21)*. The loading efficiencies of the Fmoc compounds and the amine content of the resin are quantitatively analyzed by UV/Vis/fluorescence spectroscopy, and FT-IR spectroscopy is used to analyze the functional groups on the resin. Once these parameters are established for a given

scheme, the scale of peptide synthesis can be calculated for a given yield.

2. Add Fmoc-amino acid reagents in a 1:1 molar stoichiometry. Peptides are built from the C to N terminus while attached to resin. The protecting Fmoc group is removed after each reaction to allow for the following coupling step. Add base to prime and 20% piperidine in NMP to cleave the Fmoc groups on the resin.
3. For the production of CK_{30}, react Fmoc-Lys-OH through 30 cycles followed by the addition of 1 equivalent of Fmoc-Cys-OH that is reacted with resin-bound K_{30} amine to produce CK_{30}. DOTA can be added at any stage of this synthesis. Usually, C-DOTA-K_{30} is produced (*see* **Note 1**). This is accomplished by the addition of Fmoc-DOTA amine to the 30mer of lysine followed by the coupling of the Fmoc-Cys-OH to produce C-DOTA-K_{30}.
4. After the peptide synthesis, the Fmoc group of Fmoc-Cys-OH is removed and the peptide is cleaved from the resin with HF acid. This reaction can yield up to 95%. TFA is commonly used as the counter ion to lysine.
5. Purify the final product on a reverse phase column and characterize by MALDI mass spectrometry. Lyophilize the sample to remove any solvents, wash the product with diethylether, and dry under vacuum. Synthesis is schematically displayed in **Fig. 1**. Overall yield from starting materials is 60–75%.

3.2. Synthesis of Targeting Ligand and DOTA Functionalization

1. Following the same procedure outlined for **Subheading 3.1**, peptide targeting ligands with known sequences can be produced (*see* **Note 3**) as desired, with or without DOTA **(Fig. 1)**. Differently, HPLC is used to analyze and purify the peptide or peptide-DOTA amide final product.
2. Sequence integrity and order should be checked with liquid chromatography-mass spectrometry (LC–MS), because nonsense sequences will mostly likely fail to bind the desired molecular target.

3.3. Monofunctional PEGylation of DNA Compacting Agent

1. Prepare CK_{30} or C-DOTA-K_{30} solution (TFA salt at 2–6 mM in 15 mL of PE buffer) and equal molar concentration of mPEG–MAL solution in 15 mL of PE buffer (*see* **Note 4**).
2. Add the CK_{30} or C-DOTA-K_{30} solution dropwise to an equal volume of mPEG–MAL solution while mixing on a Vibrax shaker.
3. Incubate the reaction mixture in argon at room temperature overnight on a rotator.
4. Apply the reaction mixture on an ion exchange column, pool the fractions containing polymers based on absorbance at 220 nm.

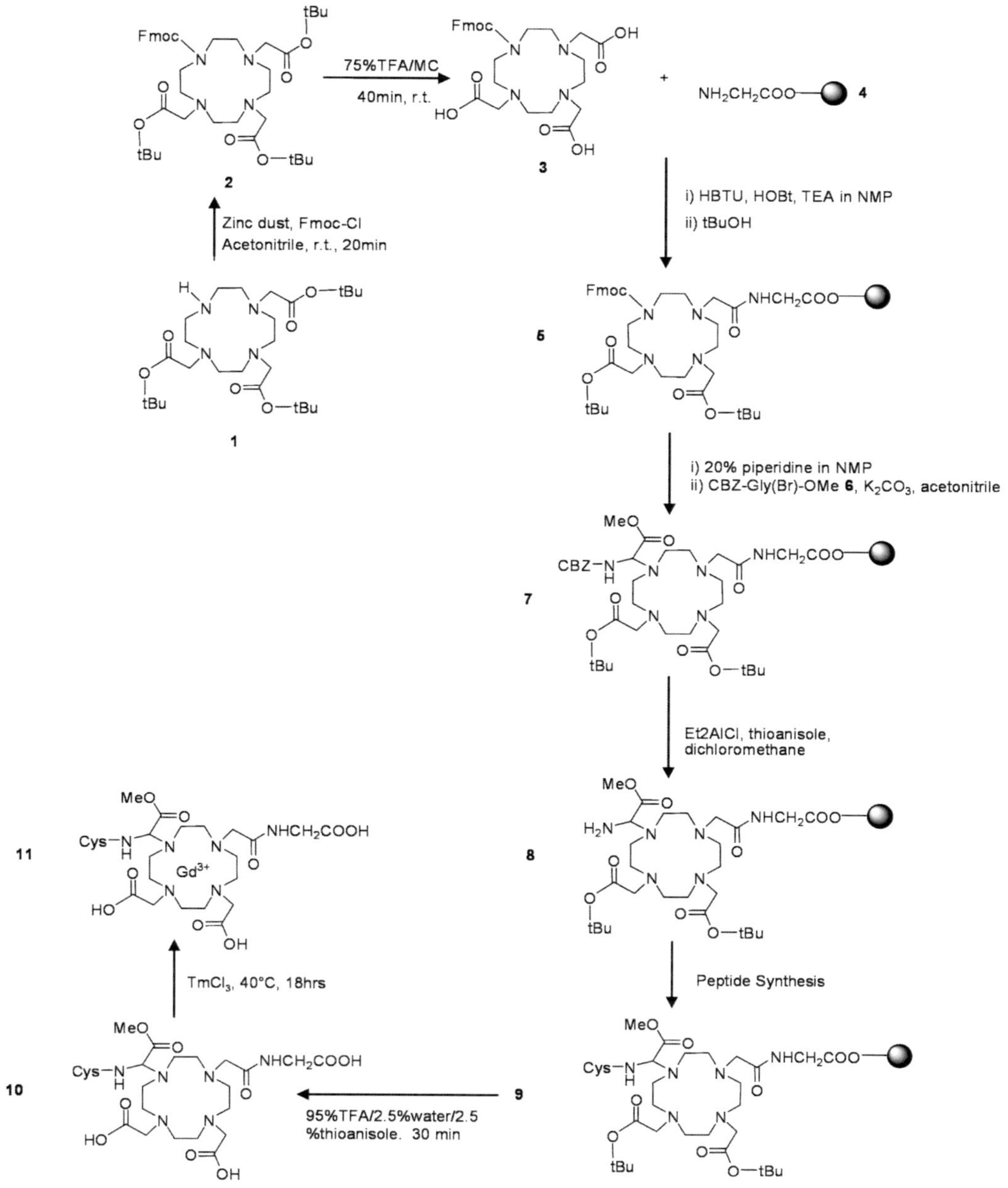

Fig. 1. Scheme for solid-phase synthesis of peptides with a DOTA moiety. Although this schematic incorporates a glycine residue between the DOTA and Wang resin for clarity, the SIPPEVKFNKPFVYLI peptide sequence was used in place of this glycine to produce the C-DOTA-105Y targeting ligand. To produce C-DOTA-K_{30}, a 30-mer of lysine was inserted in place of the glycine.

5. Solvent-exchange the pooled fractions with ammonium acetate using a centrifugal concentrator (3k MWCO).
6. Lyophilize the CK_{30}PEG10k, or C-DOTA-K_{30}PEG10k (acetate salt) and dissolve the powder in high-performance liquid

chromatography (HPLC)-grade water at a final concentration of 6.4 mg/mL.

3.4. Bifunctional PEGylation of DNA Compacting Agent

1. Prepare CK_{30} or C-DOTA-K_{30} solution (TFA salt at 2 mM in 2 mL of PE buffer) and four times molar concentration of PEG10k$(OPSS)_2$ solution (8 mM) in 2 mL of PE buffer (*see* **Notes 4** and **5**).
2. Add the CK_{30} or C-DOTA-K_{30} solution dropwise to an equal volume of PEG10k$(OPSS)_2$ solution while vortexing at room temperature.
3. Incubate the reaction mixture in argon at room temperature overnight on a rotator. Conjugation efficiency can be assessed by measuring the release of a byproduct of the reaction, pyridyl-2-thione, at 343-nm wavelength (moles of pyridyl-2-thione released are equivalent to moles of conjugated PEG).
4. Apply the reaction mixture on an ion-exchange column equilibrated with 50 mM sodium phosphate, pH 7.0, elute the CK_{30}PEG10kOPSS or C-DOTA-K_{30}PEG10kOPSS with a step gradient of NaCl in 50 mM sodium phosphate, pH 7.0, and collect the fraction containing CK_{30}PEG10kOPSS or C-DOTA-K_{30}PEG10kOPSS based on urea–acetic acid gel electrophoresis.
5. Desalt and solvent-exchange the fraction containing CK_{30}PEGOPSS or C-DOTA-K_{30}PEG10kOPSS with 20 mM ammonium acetate using a centrifugal concentrator (3k MWCO).
6. Lyophilize the CK_{30}PEG10kOPSS or C-DOTA-K_{30}PEG10kOPSS (as acetate salts) and dissolve the powder in HPLC-grade water at a final concentration of 6.4 mg/mL.

3.5. Complex Formation of Gd^{3+}, Tm^{3+}, or $^{111}In^{3+}$ with DOTA–Peptides

1. Dissolve synthesized peptides in water at pH 6.5 and 40°C, then add $TmCl_3$, $GdCl_3$, or radioactive $InCl_3$ in water dropwise for 1 h and adjust to pH 7.5 with 0.5 N NaOH. Stir the solution for 18 h at 40°C and adjust to pH 7.5 when the pH drops below 5.
2. The complete complex formation is evaluated by an Arsenazo III color test *(32)*. When the test shows negative results for free metal ions, cool the reaction mixture to room temperature. Adjust the pH to 9, remove any residual metal–hydroxide white precipitate by filtration, and freeze-dry the solution. Analyze product by MALDI mass spectrometry.

3.6. Plasmid Preparation and Compaction to Form PS–DNPs

1. DNA plasmids can be produced by common methods. Importantly, however, plasmid preparations must be endotoxin free to avoid eliciting inflammatory responses in vivo. Special consideration should be given to the design of expression constructs because plasmid size, CpG sequence content *(18)*, and promoter can all impact particle size *(7, 8)* and toxicity *(3)*.

2. For real-time imaging of luciferase expression with BLI, we use two plasmids encoding luciferase (pKCPIRlucBGH, KANr 5.3 kb), or luciferase and HSV-1 tk (VVN2069, APr 5.8 kb). The pKCPIRlucBGH plasmid contains a CMV enhancer, the CMV promoter, and the bovine growth hormone (BGH) polyadenylation signal *(2)*. The VVN2069 plasmid is a bicistronic construct that contains the CMV promoter. The advantage of this plasmid is that it allows dual imaging with microPET for HSV-1 tk expression, and BLI for luciferase expression.
3. In our experience, plasmids should be grown in *E. coli* DH5α, extracted, and purified from endotoxin. Identity of the plasmids is confirmed by restriction endonuclease digestion, and purity is established by 1.0% agarose gel electrophoresis.
4. Add 0.9 mL of DNA plasmids (0.222 mg/mL in HPLC-grade water) in 100-μL aliquots to a vortexing solution of 0.1 mL of CK_{30}PEG10k, C-DOTA-K_{30}PEG10k, CK_{30}PEG-10kOPPS, or C-DOTA-K_{30}PEG10kOPPS (6.4 mg/mL) at room temperature. The DNA concentration in the final solution is 0.2 mg/mL, and the end point ratio of positive to negative charges (NH_3^+/PO_4^-) is 2:1. Mixtures of the molecular conjugates (monofunctional and bifunctional PEG substituted) can also be used to compact DNA. The ratio of this mixture determines the number of available conjugation sites for targeting ligands. In this way, particles that are fully (~367 ligand modifications per 5.5-kb plasmid) or partially modified can be produced.
5. Filter the compacted DNA sample through a 0.22-μm polyethlylene sulfone (PES) membrane, solvent-exchange the filtered sample with normal saline, and concentrate the solution to 1 mg/mL with a centrifugal concentrator (100k MWCO).
6. For nontargeted PS–DNPs, characterize DNase I digestion followed by trypsinization and agarose gel electrophoresis, transmission electron microscopy (TEM), sedimentation assay, and turbidity assay (**Subheading 3.7**). For targeted PS–DNPs, go to **step 7**.
7. After compaction of DNA, react the available OPSS groups on the bifunctional PEG with C-ligand or C-DOTA–ligand to produce targeted PS–DNPs. Reaction completion can be assessed by pyridine-2-thione release by measurement of OD_{343} (*see* **Note 6**).

3.7. Characterization of PS–DNPs

1. Sedimentation analysis is a good measure for the level of precipitation of PS-DNPs. Centrifuge a minimum of 50 μL of compacted DNA at 6,000 rpm or 3,406 × *g* for 1 min. Measure the DNA concentration at OD_{260} before versus after centrifugation and determine the loss. Aggregated DNA

particles sediment and result in lower OD_{260} values. Acceptable samples for in vivo gene transfer give a recovery of 80–100% of the DNA after centrifugation.

2. Turbidity analysis is another measure of aggregation. Measure the OD of PS-DNPs at wavelengths 330 nm, 347 nm, 364 nm, 381 nm, 398 nm, and 415 nm. The slope (turbidity parameter) of the curve produced by plotting the log of the apparent absorbance versus the log of the wavelength correlates with the size and structure of the compacted DNA particles *(2–4, 7)*. Acceptable formulations have a turbidity parameter of ~–4, with an apparent absorbance ≥0.040 at 330 nm, ≤7% standard error of slope, and ≤10% variability of duplicates.
3. To examine the structure of PS-DNPs, use electron microscopy as previously described *(2–4, 7)*. Immediately after formulation, 10 μL of a 1:100 dilution of compacted DNA in saline is applied on a 400-mesh electron microscope carbon grid, washed with water, stained with 0.04% uranyl acetate (in methanol), washed in ethanol, and air dried. Examine samples with electron microscope (magnification: 40,000×). Acceptable formulations contain at least 80% of particles within the following size range: rods ≤300-nm long and ≤20-nm wide, toroids ≤100 nm in outer diameter.
4. It is important to test the charge of PS–DNPs (*see* **Note** 7). Zeta potential is evaluated using a dynamic light scattering with a run time of 3 min. Carboxylated latex microspheres are used to validate the zeta potential measurements.
5. To test the integrity of PS–DNPs and their ability to protect DNA against degradation, load samples on a 0.8% agarose gel after DNase I treatment, followed by trypsinization to remove poly K, or after a 2-h incubation in 75% mouse serum at 37°C followed by trypsinization.
6. Finally, measure the endotoxin levels in PS–DNP samples. Use the Kinetic-QCL limulus amebocyte lysate (LAL) assay. All reagents used in this assay are pyrogen-free. Naked or compacted DNA samples are diluted in LAL water containing 0.5% (v/v) Pyrosperse reagent to minimize false-positive results from the polycationic compacting polymer. Positive controls in each assay include diluted test samples that are spiked with endotoxin (0.5 EU/mL). In these controls, endotoxin activity is not diminished by addition of Pyrosperse.
   (a) Samples are aliquoted into a plate and preincubated for 10 min at 37°C before the addition of a colorimetric substrate.
   (b) Measure the rate of reaction at 405 nm and 37°C. Endotoxin content is calculated from the kinetic. The endotoxin levels of both naked and compacted plasmid DNA

are typically 9.3 EU/mg, or 40.9 EU/dose/kg, assuming 20-g mice, whereas levels for bacterial genomic DNA are on the order of 10^3–10^6 EU/mg.

3.8. Animal Protocol

1. All animal protocols should be approved by the Institutional Animal Care and Use Committee. Mice with identical genetic backgrounds and that are age and sex matched (*see* **Note 2**) should be used. Choice of animals is based on their suitability for each study based on previous experience.
2. PS–DNPs that meet acceptance criteria (*see* **Subheading 3.7**) should be kept on ice before administration to animals. Gene transfer and control treatments are administered by the intratracheal (IT), intranasal (IN), or intravenous (IV) route.
3. For IT administration, mice receive an intraperitoneal injection of 150 μL anesthetic cocktail. Once anesthetized, animals receive a tracheostomy and DNA, either compacted or naked, or saline alone, is administered as a bolus (volume = 25 μL of 1–4 mg/mL DNA) through a 22-gauge catheter (*see* **Note 8**). IT instillation is performed while mice are affixed to a surgical board at a 30% incline.
4. For IN administration, no anesthetic is necessary. Instill the treatment dose (in 25 μL) by aliquot to either nostril while avoiding the nasal septum (2.5 μL every 2–3 s). IN instillation is performed while mice are held upright and reclined slightly backward. To contain the dosing to the nose, animals receive the instillation while affixed to a surgical board at a 30% decline.
5. For IV administration, tail vein or retroorbital injection is used to deliver the treatment. On the day of the experiment, animals slated for injection with targeted or nontargeted PS-DNPs or controls are lightly anesthetized by isoflurane. Concentrate the treatment (dose as needed per experiment, usually ~100 μg with respect to DNA) into 50 μL, and slowly inject into the tail or retroorbital vein (*see* **Note 9**). Transgene expression is analyzed at different time intervals as needed using the specific methods below.

3.9. Small-Animal MRI

The addition of the DOTA molecule to CK_{30} or attached ligand allows for the tracking of PS-DNPs in vivo using MRI. Depending on the contrast agent used, different iterations are necessary for successful imaging (**Fig. 2**).

1. For Gd imaging using MRI relaxation:
   (a) The T1 relaxivity of a peptide–DOTA conjugate can be measured to assess the efficiency of the chelate to alter T1-weighted MRI contrast. T1 measurements of peptide–(Gd-DOTA) are conducted at 37°C and pH 7.0, with sample concentration ranging from 5 to 50 mM.

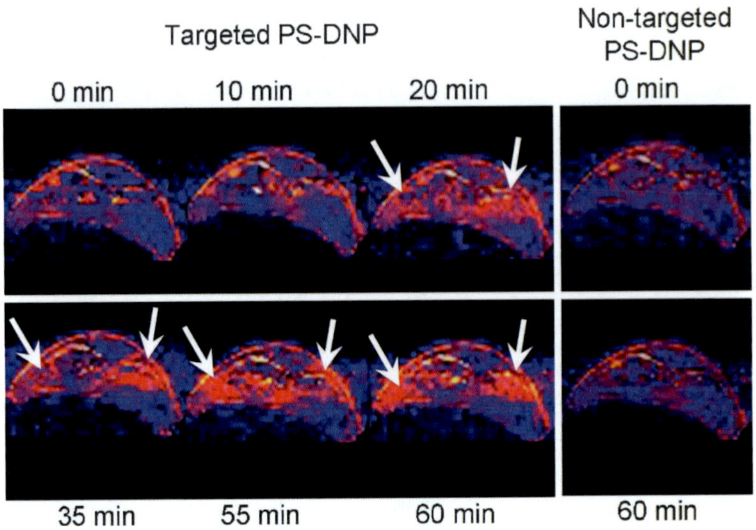

Fig. 2. MR imaging of Gd-loaded DOTA-modified targeted or nontargeted PS–DNPs. DOTA-modified PEG-CK_{30} was used to compact DNA into PS–DNPs. These particles were reacted with targeting ligand C105Y, loaded with Gd, and then injected IV into mice tightly restrained to the imaging stage of the MRI (reduces artifacts from breathing). Horizontal sections of mouse liver were immediately imaged and monitored for 1 h with repeated imaging every 5 min. The control group received Gd-DOTA nontargeted PS–DNPs.

(b) The T1 inversion-recovery experiment is conducted with a 600-MHz nuclear magnetic resonance (NMR) spectrometer for initial calibration. Animals are imaged before administration to obtain background measurements. For organ reference, a full body computed tomography (CT) scan may be obtained during each experiment.

(c) Cannulate the tail vein and inject the sample over a 30-s period. Collect MR images over 2 h for each tested experimental subject.

2. For Yb or Tm (using PARACEST MRI):

(a) PARACEST spectra are measured in a solution of 5% D_2O in water using a 600-MHz NMR spectrometer.

(b) A series of one-dimensional (1D) NMR spectra of the water signal are acquired with a selective saturation in 1 ppm increments between 100 and –100 ppm, and selective saturation is performed with a continuous wave pulse applied for 4 s at 3 T.

(c) Use identical conditions to acquire PARACEST MR images as in **step 1**, except do not use D_2O for in vivo studies. Sample heating should not occur under these radio frequency (RF) saturation conditions. MR images are collected for 2–4 h for each tested experimental subject.

3.10. BLI of Gene Expression

1. Imaging of cell or organ homogenates or whole animals can be performed. We sometimes remove tissues after in vivo imaging, homogenize, and image them to obtain photon emission numbers that are not attenuated as they are in vivo (*see* **Note 2**).
2. Animals are usually imaged 2 days after administration of PS–DNPs or controls **(Fig. 3)**, and can be repeatedly imaged at different intervals, as desired (*see* **Note 9**).
3. Animals are anesthetized under constant isoflurane throughout the imaging process. Inject 150 μL D-luciferin IP 10 min before imaging to allow for diffusion throughout the body.
4. The side of the mouse nearest the tissue targeted for gene delivery should be shaved to minimize photon attenuation by hair (*see* **Note 2**).
5. Mice are positioned on the imaging stage and images are acquired over a 6–15 min exposure time, as needed, than digitally filtered and processed to remove background and thermal noise.

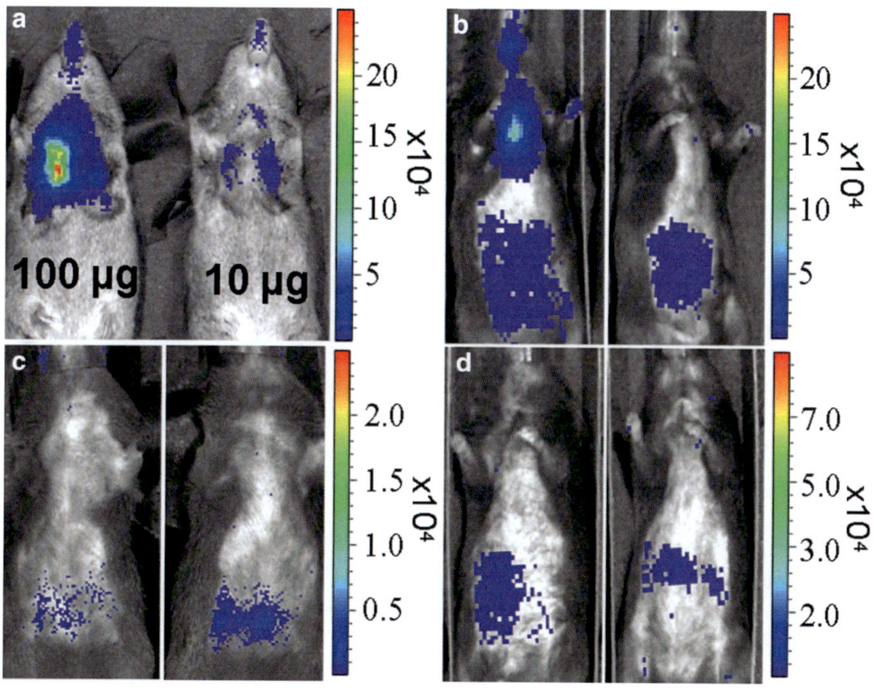

Fig. 3. Bioluminescent imaging of PS-DNP mediated luciferase gene expression. (**a**) FABP (fatty acid binding protein) CF mice dosed IT with either 100 or 10 μg nontargeted luciferase PS–DNP (unifunctional PEG) were imaged 4 days after administration. Panels (**b**)–(**d**) are C57BL/6j wild-type mice imaged 4 days after IV administration of targeted C105Y-PS-DNPs (**b**), nontargeted PS–DNPs (**c**), or targeted cystamine-modified PS–DNPs (**d**).

3.11. MicroPET Imaging of PS–DNPs Localization and HSV-1 tk Expression

MicroPET imaging can be used to sensitively track the localization of DOTA-modified PS-DNPs, or the expression of the transgene HSV-1 tk **(Fig. 4)**.

1. For tracking PS-DNPs:
   (a) All procedures including DOTA loading, PS–DNP preparation, and animal handling are performed in a specialized radioactive facility that is institution approved for $InCl_3$ administration to animals.
   (b) After C-DOTA–ligand and/or C-DOTA-K_{30} synthesis, peptides are loaded with (In) and free (In) is removed by precipitation in basic solution, and peptide solutions are exchanged to water.
   (c) Use either loaded C-DOTA-K_{30} or C-DOTA–ligand in PS–DNPs formulation to assess the fate of each of the components of the molecular conjugate (it may become different over time).
   (d) DNA of interest is compacted and a solution with ~150 mCi/100 μg PS–DNP/50 μL is administered via tail or retroorbital vein injection, as needed. Inject free $InCl_3$ in saline with identical activity as a control.
   (e) The mice are anesthetized by isoflurane throughout the entire procedure and in the imaging instruments. Animals are scanned using the CT component of the X-SPECT

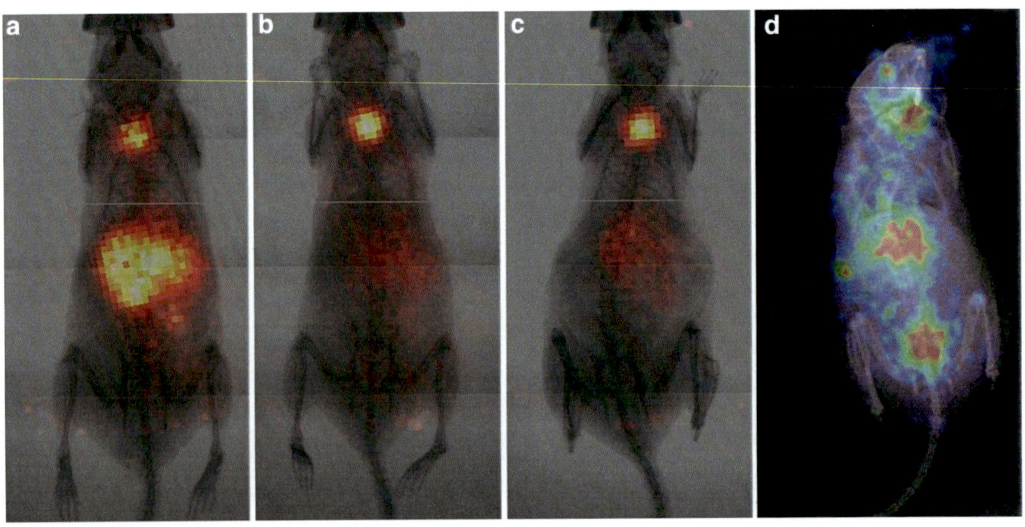

Fig. 4. MicroPET imaging of transgene expression and DNP biodistribution. C57BL/6j mice were dosed via the tail vein with 25 μg targeted or nontargeted HSV-1 tk DNPs, or naked HSV-1 tk DNA. Two days later, animals were injected IP with [^{125}I]FIAU substrate and imaged by PET 1 h later. (**a**) Targeted DNPs. (**b**) Nontargeted DNPs. (**c**) Saline injected control. For biodistribution, animals received ^{125}I-labeled targeted DNPs and were imaged over 24 h. (**d**) Six hours after administration.

scanner, and then placed in the R4 microPET scanner for a 20-min transmission scan using a point source. After the CT anatomical scan, PET scanning for the distribution of [^{111}In]-loaded PS–DNPs or free Indium is conducted. Approximately 250 μCi activity per animal (~25–30 g mouse) yields best signal-to-noise ratio images.

(f) Scanning can be repeated at time intervals as needed. Typically, free $InCl_3$ is eliminated from circulation rapidly (<2 h) via the excretory route, whereas chelated ^{111}In is excreted within 24 h.

2. For imaging transgene expression:

(a) Animals that receive PS–DNPs that contain the marker gene HSV-1 tk or controls are anesthetized with isoflurane 2 days after administration, and injected with either [^{125}I]-FIAU or [^{18}F]-FHBG shortly before imaging (*see* **Note 10**).

(b) Approximately 250 μCi of ^{18}F or ^{125}I-labeled PET tracer in 0.2 mL of sterile saline solution is injected IV via the tail vein, and CT and PET scans are obtained as in **step 1e**. Tracer administered to saline-injected animals serves as an imaging control. Animals are imaged for 1–3 h depending on the signal strength, and are imaged repeatedly, as needed, to follow expression in the same subject over time.

3. Image data are reconstructed using a two-dimensional (2D) ordered subset expectation maximization algorithm with a voxel size of 0.845 × 0.845 × 1.211 mm. Alignment between PET emission and CT images is done by registering the CT images with PET transmission using image registration software.

4. Various regions of interest (ROI) are defined according to study designs on tissues/organs as seen on the CT images that were aligned with the PET images. Regional data is defined as the sum of the measured radioactivity within a given region at a specific time point. For each ROI, data is decay corrected.

5. The time activity curve for each region can be obtained from the dynamic PET data. Relative uptake is calculated as the ratio of ROI-based uptakes between regions at earlier and later time frames.

4. Notes

1. Placing the DOTA or other modifying group within the polymer of lysines rather than at the ends interferes with DNA compaction by the polymer. Therefore, it is preferable to add the DOTA at or near the N or C terminals of the polycation.

2. The age and sex of mice has an impact on gene transfer efficacy. In our experience, younger mice are more susceptible to gene delivery by PS–DNPs. Animal tissues and coat color also have an impact on bioluminescent imaging efficiency. Bone, dark skin, and dark hair color attenuate light and diminish measured signals.
3. The addition of cysteine residues at the N terminus of components of the protein portion of DNA nanoparticles allows for the use of sulfhydryl chemistry for the conjugation of these components. Disulfide linkages lower the toxicity and increase the efficacy of DNA nanoparticles.
4. The CK_{30} solution should be prepared immediately before use. The concentration of active sulfhydryl group can be determined by Ellman's assay.
5. The reactivity of the OPSS group can be determined by monitoring the reaction of $PEG10k(OPSS)_2$ with excessive cysteamine based on absorbance at 343 nm, which comes from the released pyridine-2-thione.
6. The OPSS group's reaction with Cys is very efficient, so ~95% of available sites are usually reacted. In our experience, targeted particles retain the stability characteristics of nontargeted ones, which have been extensively reported *(2–4, 7)*.
7. The charge of DNPs can have a profound impact on uptake and toxicity *(33)*. We avoid producing particles that have a high positive charge to avoid activating the complement cascade in vivo. Typically, our particles have a slight negative charge.
8. DNA concentrations and speed of injection should be controlled to avoid nonspecific uptake of DNA after IV administration. Rapid (<10 s) venous injections of high concentrations of DNA (>300 μg) result in a phenomenon known as hydrodynamic shock (reviewed in **ref.** *(34)*), which delivers treatment to the liver without specific uptake.
9. In our experience, PS-DNPs containing DNA plasmids with viral promoters exhibit expression for up to 12 days, with a peak between 2 and 4 days after administration in vivo *(2, 3)*. Targeting the particles as well as using plasmids that contain mammalian promoter can significantly increase duration of expression.
10. FIAU and FHBG are both cell permeable-specific substrates of HSV-1 tk that are trapped within cells once phosphorylated. Therefore, the radiotracer accumulates in regions exhibiting HSV-1 tk expression.

Acknowledgments

The authors thank Junnan Chen and Samuel Shank for technical assistance. This work was funded by the Cystic Fibrosis Foundation and the State of Ohio. Dr. Ziady is an inventor on patents related to the technology discussed in this chapter. He also holds equity in Copernicus Therapeutics Inc., a nonviral gene therapy company, which has licensed these patents.

References

1. Ziady, A. G. and Davis, P. B. (2001). Receptor-directed molecular conjugates for gene transfer. *In: Methods in Molecular Medicine: Gene Therapy Protocols,* **69** (J. Morgan, ed.), Humana Press Inc
2. Ziady, A. G., Gedeon, C. R., Miller, T., Quan, W., Payne, J. M., Hyatt, S., Fink, T., Muhammad, O., Oette, S., Kowalczyk, T., Pasumarthy, M. K., Moen, R., Cooper, M. J., and Davis, P. B. (2003). Transfection of airway epithelium by stable PEGylated poly-L-lysine DNA nanoparticles in vivo. *Mol. Ther.* **8**, 936–947
3. Ziady, A. G., Gedeon, C. R., Muhammad, O., Stillwell, V., Oette, S., Fink, T., Quan, W., Kowalczyk, T., Hyatt, S. L., Peischl, A., Seng, J. E., Moen, R., Cooper, M. J., and Davis, P. B. (2003). Minimal toxicity of stabilized compacted DNA in the murine lung. *Mol. Ther.* **8**, 948–956
4. Konstan, M. W., Davis, P. B., Wagener, J. S., Hilliard, K. A., Stern, R. C., Milgram, L. J., Kowalczyk, T. H., Hyatt, S. L., Fink, T. L., Gedeon, C. R., Oette, S. M., Payne, J. M., Muhammad, O., Ziady, A. G., Moen, R. C., and Cooper, M. J. (2004). Compacted DNA nanoparticles administered to the nasal mucosa of cystic fibrosis subjects are safe and demonstrate partial to complete cystic fibrosis transmembrane regulator reconstitution. *Hum. Gene Ther.* **15**, 1255–1269
5. Maurer, P. H. (1962). Antigenicity of polypeptides (poly-alpha amino acids). *J. Immunol.* **88**, 330–345
6. Colter, J. S. and Ellem, K. A. (1961). Antigenicity of deoxyribonucleic acids from mouse liver and from the Ehrlich ascites tumour. *Nature* **190**, 550–551
7. Fink, T. L., Klepcyk, P. J., Oette, S. M., Gedeon, C. R., Hyatt, S. L., Kowalczyk, T. H., Moen, R. C., and Cooper, M. J. (2006). Plasmid size up to 20 kbp does not limit effective in vivo lung gene transfer using compacted DNA nanoparticles. *Gene Ther.* **13**, 1048–1051
8. Ziady, A., Perales, J. C., Ferkol, T., Gerken, T., Beegen, H., Perlmutter, D. H., and Davis, P. B. (1997). Gene transfer into hepatocyte cell lines via the serpin enzyme complex (SEC) receptor. *Am. J. Physiol.* 273 (*Gastrointest. Liver Physiol.* **36**,) G545–G552
9. Ziady, A. G., Ferkol, T., Gerken, T., Dawson, D. V., Perlmutter, D. H., and Davis, P. B. (1998). Ligand substitution of receptor targeted DNA complexes affects gene transfer into hepatoma cells. *Gene Ther.* **5**, 1685–1697
10. Ziady, A. G., Ferkol, T., Dawson, D. V., Perlmutter, D. H., and Davis, P. B. (1999). Chain length of the polymer portion of receptor-targeted DNA complexes modulates gene transfer both in vitro and in vivo. *J. Biol. Chem.* **274**, 4908–4916
11. Ziady, A. G., Kelley, T., Milliken, E., Ferkol, T., and Davis, P. B. (2002). Functional evidence of CFTR gene transfer in nasal epithelium of CF mice following apical application of SEC receptor targeted DNA complexes. *Mol. Ther.* **5**, 413–419
12. Ziady, A. G., Kim, J., Colla, J., and Davis, P.B. (2004). Defining strategies to extend duration of gene expression from targeted compacted DNA vectors. *Gene Ther.* **11** 1378–1390
13. Gupta, S., Eastman, J., Silski, C., Ferkol, T., and Davis, P.B. (2001). Single chain Fv: a ligand in receptor-mediated gene delivery. *Gene Ther.* **8**(8), 586–592
14. Ferkol, T., Perales, J.C., Eckman, E., Kaetzel, C.S., Hanson, R.W., and Davis, P.B. (1995). Gene transfer into the airway epithelium of animals by targeting the polymeric immunoglobulin receptor. *J. Clin. Invest.* **95**(2), 493–502
15. Liu, G., Li, D., Pasumarthy, M. K., Kowalczyk, T. H., Gedeon, C. R., Hyatt, S. L., Payne, J. M., Miller, T. J., Brunovskis, P., Fink, T. L., Muhammad, O., Moen, R. C., Hanson, R. W., and Cooper, M. J. (2003). Nanoparticles of compacted DNA transfect postmitotic cells. *J. Biol. Chem.* **278**, 32578–32586

16. Pollard, H., Remy, J. S., Loussouarn, G., Demolombe, S., Behr, J. P., and Escande, D. (1998). Polyethylenimine but not cationic lipids promotes transgene delivery to the nucleus in mammalian cells. *J. Biol. Chem.* **273**, 7507–7511

17. Raper, S. E., Chirmule, N., Lee, F. S., Wivel, N. A., Bagg, A., Gao, G. P., Wilson, J. M., and Batshaw, M. L. (2003). Fatal systemic inflammatory response syndrome in a ornithine transcarbamylase deficient patient following adenoviral gene transfer. *Mol. Genet. Metab.* **80**, 148–158

18. Yew, N. S., Wang, K. X., Przybylska, M., Bagley, R. G., Stedman, M., Marshall, J., Scheule, R. K., and Cheng, S. H. (1999). Contribution of plasmid DNA to inflammation in the lung after administration of cationic lipid: pDNA complexes. *Hum. Gene Ther.* **10**, 223–234

19. Chollet, P., Favrot, M. C., Hurbin, A., and Coll, J. L. (2002). Side-effects of a systemic injection of linear polyethylenimine-DNA complexes. *J. Gene Med.* **4**, 84–91

20. Tjuvajev, J. G., Doubrovin, M., Akhurst, T., Cai, S., Balatoni, J., Alauddin, M. M., Finn, R., Bornmann, W., Thaler, H., Conti, P. S., and Blasberg, R. G. (2002). Comparison of radiolabeled nucleoside probes (FIAU, FHBG, and FHPG) for PET imaging of HSV-1 tk gene expression. *J. Nucl. Med.* **43**, 1072–1083

21. Yoo, B. and Pagel, M. D. (2007). Peptidyl molecular imaging contrast agents using a new solid-phase peptide synthesis approach. *Bioconjug. Chem.* **18**, 903–911

22. Olafsen, T., Kenanova, V. E., Sundaresan, G., Anderson, A. L., Crow, D., Yazaki, P. J., Li, L., Press, M. F., Gambhir, S. S., Williams, L. E., Wong, J. Y., Raubitschek, A. A., Shively, J. E., and Wu, A. M. (2005). Optimizing radiolabeled engineered anti-p185HER2 antibody fragments for in vivo imaging. *Cancer Res.* **65**, 5907–5916

23. Prantner, A. M., Sharma, V., Garbow, J. R., and Piwnica-Worms, D. (2003). Synthesis and characterization of a Gd-DOTA-D-permeation peptide for magnetic resonance relaxation enhancement of intracellular targets. *Mol. Imaging.* **2**, 333–341

24. De Leon-Rodriguez, L. M., Ortiz, A., Weiner, A. L., Zhang, S., Kovacs, Z., Kodadek, T., and Sherry, A. D. (2002). Magnetic resonance imaging detects a specific peptide-protein binding event. *J. Am. Chem. Soc.* **124**, 3514–3515

25. Achilefu, S., Bloch, S., Markiewicz, M. A., Zhong, T., Ye, Y., Dorshow, R. B., Chance, B., and Liang, K. (2005). Synergistic effects of light-emitting probes and peptides for targeting and monitoring integrin expression. *Proc. Natl. Acad. Sci. U.S.A.* **102**, 7976–7981

26. Vinogradov, E., Zhang, S., Lubag, A., Balschi, J. A., Sherry, A. D., and Lenkinski, R. E. (2005). On-resonance low B1 pulses for imaging of the effects of PARACEST agents. *J. Magn. Reson.* **176**, 54–63

27. Herborn, C. U., Waldschuetz, R., Lauenstein, T. C., Goyen, M., Lauffer, R. B., Moeroey, T., Debatin, J. F., and Ruehm, S. G. (2002). Contrast-enhanced magnetic resonance imaging (MS-325) in a murine model of systemic lupus erythematosus. *Invest. Radiol.* **37**, 464–469

28. Ragunand, N., Howison, C., Sherry, A. D., Zhang, S., and Gillies, R. J. (2003). Renal and systemic pH imaging by contrast-enhanced MRI. *Magn. Reson. Med.* **49**, 249–257

29. Zhang, S., Merritt, M., Woessner, D. E., Lenkinski, R. E., and Sherry, A. D. (2003). PARACEST agents: modulating MRI contrast via water proton exchange. *Acc. Chem. Res.* **36**, 783–790

30. Bousquet, J. C., Saini, S., Stark, D. D., Hahn, P. F., Nigam, M., Wittenberg, J., Ferrucci, J. T. Jr. (1988). Gd-DOTA: characterization of a new paramagnetic complex. *Radiology* **166**, 693–698

31. Bourrinet, P., Martel, E., El Amrani, A. I., Champeroux, P., Richard, S., Fauchou, N., Le Coz, F., Drici, M., Bonnemain, B., and Gaillard, S. (2007). Cardiovascular safety of gadoterate meglumine (Gd-DOTA). *Invest. Radiol.* **42**, 63–77

32. Dadachova, E., Chappell, L. L., and Brechbiel, M. W. (1999). Spectrophotometric method for determination of bifunctional macrocyclic ligands in macrocyclic ligand-protein conjugates. *Nucl. Med. Biol.* **26**, 977–982

33. Thakor, D., Spigelman, I., Tabata, Y., and Nishimura, I. Subcutaneous peripheral injection of cationized gelatin/DNA polyplexes as a platform for non-viral gene transfer to sensory neurons. *Mol. Ther.* **15**(12), 2124–2131

34. Herweijer, H. and Wolff, J. A. (2007). Gene therapy progress and prospects: hydrodynamic gene delivery. *Gene Ther.* **14**, 99–107

Chapter 34

Nanoparticle-Mediated Gene Delivery

Sha Jin, John C. Leach, and Kaiming Ye

Summary

Nonviral gene delivery has been gaining considerable attention recently. Although the efficacy of DNA transfection, which is a major concern, is low in nonviral vector-mediated gene transfer compared with viral ones, nonviral vectors are relatively easy to prepare, less immunogenic and oncogenic, and have no potential of virus recombination and no limitation on the size of a transferred gene. The ability to incorporate genetic materials such as plasmid DNA, RNA, and siRNA into functionalized nanoparticles with little toxicity demonstrates a new era in pharmacotherapy for delivering genes selectively to tissues and cells. In this chapter, we highlight the basic concepts and applications of nonviral gene delivery using super paramagnetic iron oxide nanoparticles and functionalized silica nanoparticles. The experimental protocols related to these topics are described in the chapter.

Key words: Nonviral gene delivery, Magnetic nanoparticles, Silica nanoparticles, Gene delivery

1. Introduction

The development of nonviral vectors for safe and efficient gene delivery has been gaining considerable attention recently. Compared with viral vectors, nonviral vectors are relatively easy to prepare, less immunogenic and oncogenic, and have no potential of virus recombination and no limitation in the size of a gene that can be transferred. In addition, nonviral vectors can be vested readily to carry genetic materials to target cells through structural modification of the vectors. Of many nanoconstructs, nanoparticles (NPs) are an attractive vector for nonviral gene transfer. NPs have been successfully tested for both in vitro *(1, 2)* and in vivo gene delivery *(3, 4)*. In principle, NPs can be made to reach a

James Weifu Lee and Robert S. Foote (eds.), *Micro and Nano Technologies in Bioanalysis*,
Methods in Molecular Biology, vol. 544
DOI 10.1007/978-1-59745-483-4_34, © Humana Press, a part of Springer Science+Business Media, LLC 2009

target site by virtue of their size and charge *(5)*. The high surface area-to-volume ratio makes NPs ideal for nonviral gene transfer.

The size of NPs ranges from 10 to 200 nm. That is comparable to the size of a protein. The small size offers the potential of targeted gene delivery, allowing the penetration of tissues deep in place like solid tumors with a high level of specificity *(6)*. With the ultrafine size, NPs can readily interact with biomolecules on the cell surface or inside cells, and deliver genetic materials such as DNA, RNA, or small interfering RNA (siRNA) into target cells or tissues for gene expression *(7–10)*. Among a number of types of NPs, magnetic NPs (MNPs) are particularly attractive because MNP-mediated DNA delivery can use a magnetic field to guide DNA-carried NPs to target tissues and cells, and because the MNPs are safe for use in humans *(11, 12)*. MNPs have been exploited for monitoring and guiding NP-mediated gene delivery *(13, 14)*, and the transfection efficiency can be significantly enhanced with surface modifications of MNPs *(13, 15)*.

One of the easiest ways to load DNA onto NPs is to modify the surface of the NPs to a positive charge so that the NP–DNA complexes can simply be formed through electrostatic binding between the positively charged NPs and the negatively charged DNA. This mechanism has been widely used in liposomes and other polymer-mediated gene transfers *(3, 16–19)*. Nevertheless, these gene transfer vectors suffer from several disadvantages. For example, the reproducibility of polymeric vector synthesis is relatively low because of the polymers' high polydispersity. These polymers also do not tolerate heat, making autoclaving virtually impossible. In contrast, inorganic NPs such as inert-silica NPs have low polydispersity and low toxicity as well as high biocompatibility, making them ideal for gene delivery *(10)*. It has been found that silica NPs are resistant to bile salts and lipase encountered in the gastrointestinal tract, have physical stress during aerosolization, and can withstand autoclaving *(10, 20)*. We describe herein the characterization of the super paramagnetic iron oxide NPs (SPION)–DNA complex as functional NPs for gene delivery and the methods of functionalization of silica NPs for gene delivery.

1.1. Super Paramagnetic Iron Oxide Nanoparticles–DNA Complex Formation and Properties

The capabilities and possible applications of NPs are in a large part the result of the variety of NPs that can be synthesized in the laboratory. These physical and chemical differences offer new and innovative ideas of how to further use the NPs in gene delivery. One type of such particles is the SPION. SPIONs have recently been characterized and used to study membrane transport and targeted delivery applications using magnetic forces *(11)* to aid in delivery to the cells *(21)*. These SPIONs have also been used as contrast agents for imaging the bone marrow *(22)*, the liver *(23)*, and the lymph nodes using magnetic resonance imaging (MRI) technologies *(24)*. Here, we describe an approach to form

SPION–DNA complexes in hopes of using them for future gene therapy applications.

1.2. Silica Nanoparticle-Mediated Gene Delivery

Silica-based NP is another common material that allows easy and efficient surface modifications for gene delivery. Recently, a novel three-component DNA delivery system was developed by using dense silica NPs as transfection enhancers *(25, 26)*. The DNA transfection efficiency can be enhanced by fourfold to sevenfold at optimized transfection conditions, which includes the size and number of silica NPs and the types of commercially available transfection reagents, as well as the types of cell lines. Silica NP-mediated gene delivery has been shown to have a high efficiency, a low toxicity, and self-assembly properties for mediating nonviral gene therapy. Technically, hydrated, organically modified silica NPs can be synthesized by constructing dioctyl sodium sulfosuccinate (aerosol-OT)/DMSO/water microemulsions as a nonpolar core. The amino groups can then be immobilized on the surface of the NPs through synchronous hydrolysis of the triethoxyvinylsilane (TEVS) precursor and 3-aminopropyltriethoxysilane (APTES). The functionalized silica NPs will be capable of condensing DNA and extensively staining the cytoplasm of tumor cells in vitro *(2)*. It has been observed that the DNA can be released from silica NPs inside the cytoplasm and can then migrate into the nucleus for gene delivery. Thus, we describe herein the detailed experimental protocols of using silica NPs for gene transfer, based on these works *(2, 4, 25–27)*.

2. Materials

2.1. SPION–DNA Formation

1. SPION (Ocean Nanotech Inc., Fayetteville, AR).
2. 10 mM Tris–HCl buffer, pH 7.3.
3. DNA Plasmid pHSCR in 10 mM Tris-HCl buffer. Note: This plasmid encodes an enhanced green fluorescent protein (EGFP) expressed from a human cytomegalovirus immediate-early gene (CMV) promoter.
4. 1× TAE electrophoresis buffer: 40 mM Tris, 20 mM acetic acid, and 1 mM EDTA.
5. Ethidium bromide (1% solution).
6. 0.8% of agarose gel with 1% of ethidium bromide in 1× TAE buffer.
7. Gel XL Ultra Electrophoresis System (PGC Scientifics, San Diego, CA).
8. Bio-Rad 2D gel imaging system (ChemiDoc XRS, Hercules, CA).

9. Dulbecco's Modified Eagle's Medium (DMEM) (Gibco, Bethesda, MD).
10. 10% fetal bovine serum (FBS) (Hyclone, Ogden, UT).

2.2. Synthesis of Amino Group Functionalized Silica/DNA NPs

1. DNA Plasmid pHSCR.
2. Surfactant aerosol OT (AOT; 98%), triethoxyvinylsilane (TEVS; 97%), cosurfactant 1-butanol (99.8%), and 3-aminopropyltriethoxysilane (APTES; 99%) (Aldrich, St. Louis, MO).
3. Phosphate-buffered saline (PBS) (Gibco, Bethesda, MD).
4. 12- to 14-kDa cutoff cellulose membrane (Spectrum Laboratories, Inc., Rancho Dominguez, CA).
5. 0.2-μm membrane filters (Nalgene, Lima, OH).

2.3. Gene Delivery with Amino Group Functionalized Silica/DNA NPs

1. COS-7 cells (ATCC, Manassas, VA): African green monkey kidney cells derived from CV-1 simian cells transformed by an origin-defective mutant of SV40. The cells support the growth of recombinant SV40 viruses and should be handled at laboratory containment level 2.
2. 0.05% Trypsin/0.53 mM EDTA in HBSS buffer without sodium bicarbonate, calcium, and magnesium (Mediatech, Inc., Manassas, VA).
3. CO_2 incubator (model 2400, VWR Scientific, Bridgeport, NJ).
4. T-75 vented cap tissue culture-treated flasks and tissue culture-treated 6-well plates (Fisher Scientific, Pittsburgh, PA).
5. Tissue culture-treated 60-mm dishes (Fisher Scientific).
6. Dulbecco's phosphate-buffered saline (DPBS) without calcium and magnesium (Mediatech, Inc. VA).
7. Hemocytometer with a cover slip (Sigma, St. Louis, MO).
8. Tally counter (Fisher Scientific).
9. Microscope (Inverted microscope Olympus CK-40).
10. Inverted fluorescence microscope Olympus IX-71 equipped with Q-Imaging CCD camera and imaging processing software Slidebook.

2.4. Transfection by DNA/SuperFect/Silica NPs Three Component

1. COS-7 cells.
2. DMEM with 4.5 g/l glucose and sodium pyruvate without L-glutamine.
3. Fetal calf serum (FBS) (ATCC).
4. L-Glutamine (200 mM) (Mediatech, Inc.).
5. Penicillin-Streptomycin (10,000 IU/ml penicillin and 10,000 μg/ml streptomycin) solution (Mediatech, Inc.).
6. 0.05% Trypsin-EDTA (Mediatech, Inc.).

7. SuperFect transfection reagent (Qiagen Inc., Valencia, CA).
8. 0.225-μm-radius silica NPs (Polysciences, Warrington, PA).
9. Plasmid pHSCR (minimum DNA concentration: 0.1 μg/μl).
10. PBS buffer (Gibco, Bethesda, MD).
11. T-75 vented cap tissue culture treated flasks and 6-well plate (Fisher Scientific, Pittsburgh, PA).
12. Hemocytometer with cover slip.
13. Tally counter.
14. CO_2 incubator.
15. Inverted fluorescence microscope Olympus IX-71 equipped with Q-Imaging CCD camera and imaging processing software Slidebook.

3. Methods

3.1. Magnetic Nanoparticle Synthesis

The coating on the particles is based on electrostatic interaction through the use of a polymer with carboxyl groups to provide negative charges. The surface is then coated with polydiallyldimethylammonium chloride (PDDA) to provide positive charges on the surface of the SPIONs. These NPs are roughly 200 nm in diameter, as measured by scanning electron microscopy (data not shown).

3.2. Magnetic Nanoparticle–DNA Binding Assays

1. Heat the SPION solution to 75°C in a water bath for 10 min to inactivate any possible DNase contamination in the NP solution.
2. Cool the solution to room temperature.
3. Add varying amounts of the DNA plasmid pHSCR to the SPION solution.
4. Incubate solutions at room temperature for 30 min to allow for the formation of NP–DNA complexes.
5. Load the NP–DNA complexes into the agarose gel with ethidium bromide for visualization.
6. Run the electrophoresis at 100 V for 30 min.
7. Place the gel on top of a UV transilluminator (Fisher Scientific) so that the migration of the DNA–SPION complex within the gel can be observed through the UV transilluminator. A Bio-Rad 2D gel imaging system (ChemiDoc XRS) can be used to image the gels. An example of the results produced is presented in **Fig. 1** (*see* **Note 1**).

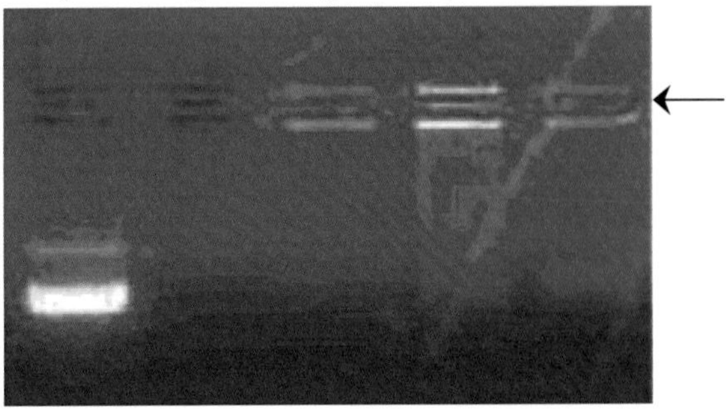

Fig. 1. Agarose gel electrophoresis of the magnetic NP–DNA complexes demonstrated that negatively charged DNA can be efficiently absorbed onto the positively charged magnetic NPs. DNA was mixed with SPION at different volume ratios. *Lane 1*: DNA standard. *Lane 2*: SPION in sterile water; *Lane 3*: 1:2.6 DNA to SPION; *Lane 4*: 1:2.5 DNA to SPION; and *Lane 5*: 1:2.4 DNA to SPION. The resistance of migration of the DNA within the agarose gel suggested the formation of SPION-DNA complexes because the SPIONs were positively charged. We observed partial migration of DNA in *lanes 4* and *5*, implying that part of the DNA was not bound to the SPION. This could be because of the insufficient amount of particles available for DNA binding. DNA was completely bound to the SPION when their ratio was set to 1:2.6.

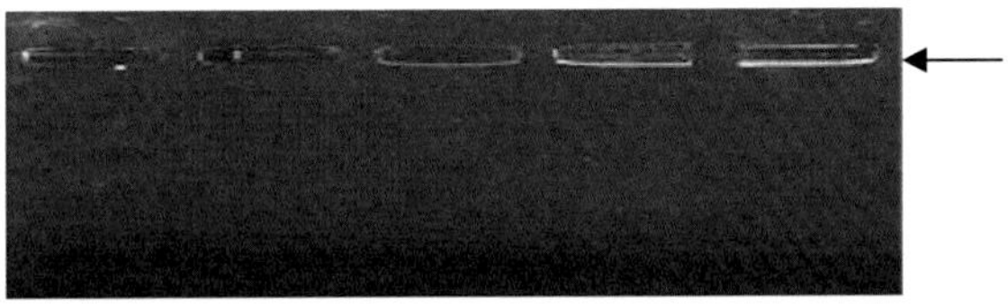

Fig. 2. DNA can be protected from degradation by forming a SPION–DNA complex. Varying amount of DNase was added to the NP–DNA complexes (1:2.6 DNA to SPION) and incubated for 10 min before the DNase was inactivated by heating to 75°C for 10 min. *Lane 1*: DNA alone with 0.2 U DNase; *Lane 2*: DNA alone with 0.05 U DNase; *Lane 3*: SPION–DNA complex with 0.2 U DNase; *Lane 4*: SPION–DNA complex with 0.05 U DNase; *Lane 5*: SPION–DNA complex in NaCl solution.

3.3. Determination of the Protection of NP–DNA Complex from DNA Degradation

1. Add 0.05, 0.1, 0.2, 0.5, and 1 U of DNase per sample (1 U/ml, New England Biolabs, Ipswich, MA) to uncomplexed plasmid DNA and the NP–DNA complex solutions containing 1 μg of DNA.
2. Allow 10 min for the DNase to act upon the complexes.
3. Inactivate the DNase by heating the solutions to 75°C in a water bath for 10 min.
4. Load the solutions into the agarose gel and run the electrophoresis at 100 V for 30 min.
5. Place the gel on the top of a UV transilluminator to observe the migration of DNA–SPION complex within the gel. A Bio-Rad 2D gel imaging system can be used to image the gels. An example of the results produced is presented in **Fig. 2** (*see* **Note 2**).

3.4. Determination of the Dissociation of the DNA from the Magnetic Nanoparticle

1. Mix the NP–DNA complex at the full-binding ratio (1:2.6 DNA to SPION) with media such as NaCl solution, DMEM, or DMEM with 10% heat-inactivated FBS. Incubate for approximately 10 min.
2. Load the above solution into an agarose gel and run the electrophoresis at 100 V for 30 min.
3. Detect the NP–DNA complexes, dissociated DNA, and NPs as described above.

   A typical example is presented in **Fig. 3**.

3.5. Synthesis of Amino Group Functionalized Silica/DNA NPs

The NPs are synthesized in the nonpolar core of AOT/DMSO/water micelles at room temperature as described below *(27)*:

1. Dissolve 0.44 g of surfactant AOT and 800 μl of 1-butanol in 20 ml of double-distilled water with vigorous magnetic stirring.
2. Add 200 μl of neat triethoxyvinylsilane to the resulting micellar solutions, and stir for approximately 1 h, or until the solution becomes clear.
3. Add 40 μl of neat 3-aminopropyltriethoxysilane to incorporated cationic amino groups on the surface of the silica NPs and stir the solutions for approximately 20 h. A blue–white translucency indicates the formation of NPs.
4. Remove surfactant AOT and cosurfactant 1-butanol completely by dialyzing the solutions against water in a 12- to 14-kDa cutoff cellulose membrane for 48 h.
5. Filter the dialyzed solutions through a 0.2-μm cutoff membrane filter.
6. Incubate 10^{14} of silica NPs with 135 μg of plasmid DNA expressing EGFP for 30 min.
7. Suspend the resulting silica NPs–DNA complex in the PBS for DNA transfection.

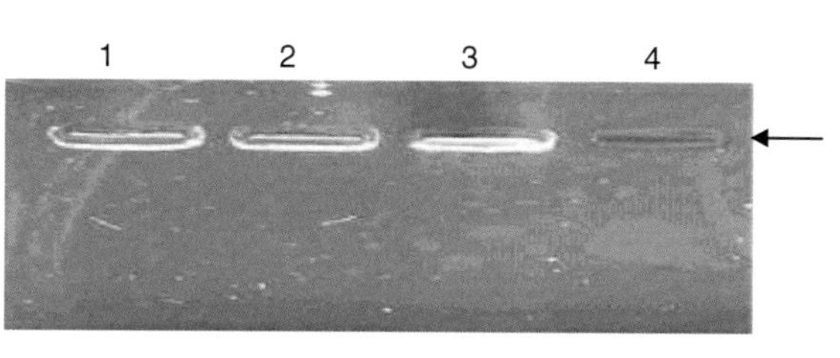

Fig. 3. Magnetic NP–DNA complex media dissociation. The NP–DNA complexes at a ratio of 1:2.6 DNA to SPION were subjected to DMEM, DMEM with 10% heat-inactivated FBS, and NaCl solution. *Lane 1*: complex in DMEM; *Lane 2*: complex in DMEM with serum; *Lane 3*: complex in NaCl solution; and *Lane 4*: DNA alone with 1 U of DNase. The complexes did not dissociate upon introduction of the physiological media.

3.6. Gene Delivery with Amino Group Functionalized Silica/DNA NPs

1. Prepare the cell culture medium by adding L-glutamine, penicillin–streptomycin, and FBS to the DMEM medium (the final concentrations of L-glutamine, penicillin–streptomycin, and FBS are 2 mM, 100 IU/ml-10 μg/ml, and 10%, respectively.) The L-glutamine-containing cell culture medium can be stored at 4°C for up to 1 month.
2. Grow the COS-7 cells in a T-75 vented cap flask with 12 ml of cell culture medium in a CO_2 incubator at 37°C and 5% CO_2.
3. Once the cells reach 60–80% of confluence, remove the medium from the cell culture by aspiration.
4. Rinse the cells with 10 ml of DPBS to remove the residue serum, calcium, and magnesium (serum needs to be removed because it will inhibit trypsin activity, and the removal of calcium and magnesium will be beneficial to the detachment of the cells from the flask).
5. Add 2 ml of 0.05% trypsin–EDTA to the flask and incubate in CO_2 incubator for 3 min at 37°C.
6. Examine the cells under a phase-contrast inverted microscope using a 10× low magnification object to check on the detachment of the cells from the flask (rounding up of the cells indicates initiation of the detachment).
7. Add 10 ml of the prewarmed cell culture medium to the flask and incubate in a CO_2 incubator for another 10 min (serum contained in the medium will stop the trypsinization by neutralizing the trypsin).
8. Examine the cells again under the inverted microscope and make sure all of the cells detach from the flask.
9. Centrifuge the cells at 200 × *g*, 4°C for 10 min.
10. Discard the cell culture medium carefully.
11. Resuspend the cells in prewarmed cell culture medium.
12. Count the cells using a hemocytometer.
13. Transfer 7.5×10^3 cells/ml to a 60-mm tissue culture dish in 5 ml of cell culture medium.
14. Incubate the cells in a CO_2 incubator at 37°C and 5% CO_2 overnight.
15. Rinse the cells with DPBS, followed by the addition of 5 ml of prewarmed growth medium to the dish.
16. Add 50 μl of silica/DNA NPs obtained from **Subheading 3.5** to the cells and swirl the dish gently.
17. Incubate the NP–DNA-treated cells at 37°C and 5% CO_2 for 2 h.
18. Exchange with fresh cell culture medium and then incubate the cells in a CO_2 incubator.

19. Examine the expression of EGFP 24–48 h after transfection using an inverted fluorescence microscope equipped with a narrow-band GFP filter set (Chroma, Brattleboro, VT) by exciting at 480 nm and detecting the emission of the fluorescence at 510 nm (the narrow-band GFP filter eliminates autofluorescence).

3.7. Transfection by Three Components of DNA/SuperFect/ Silica NPs

The DNA transfection efficiency can be significantly enhanced by combining the silica NPs with existing transfection kits such as the SuperFect DNA transfection kit from Qiagen (*see* **Note 3**). Here is the protocol:

1. Seed 4 × 10^5 cells/well in each 6-well plate in 3 ml of growth medium the day before transfection, as described in **Subheading 3.6**.
2. On the day of transfection, mix 2 μg of plasmid DNAs (in sterile DPBS buffer) with the cell culture medium without serum and antibiotics to a total volume of 100 μl. Mix and spin down the solution for a few seconds to remove the drops from the top of the tube.
3. Add 10 μl of SuperFect Transfection Reagent to the DNA solution. Mix by pipetting up and down five times or by briefly vortexing for 10 s.
4. Incubate the samples for 5–10 min at room temperature to allow Superfect-DNA complex formation.
5. Incubate the above Superfect-DNA complex with 3 × 10^8 silica NPs/ml for another 5–10 min. The concentration of beads required for a given cell type should be empirically determined.
6. Gently aspirate the cell culture medium from the 6-well plate and wash the cells once with 2 ml DPBS.
7. Add 0.6 ml of the culture medium to the reaction tube containing the DNA/SuperFect/silica NPs complexes.
8. Mix by pipetting up and down twice and transfer the total volume to the cells in the 6-well plate.
9. Gently swirl the plate to ensure uniform distribution of the complexes.
10. Incubate cells with the complexes for 2 h at 37°C and 5% CO_2.
11. Gently remove the medium containing the complexes from the cells, and wash the cells once with 2 ml DPBS.
12. Add 2 ml of fresh culture medium.
13. Examine the expression of EGFP as described in **Subheading 3.6**.

4. Notes

1. The DNA electrophoresis demonstrated that SPION-DNA complexes could bind the negatively charged plasmid DNA because the SPION–DNA complexes were retained at the baseline in the well of the agarose gel, as revealed in **Fig. 1.** Partial migration of DNA was observed in lanes 4 and 5 of **Fig. 1**, implying that part of the DNA was not bound to the SPION. This can be because of the insufficient amount of NPs available for DNA binding. DNA is completely bound to the SPION when the ratio was set to 1:2.6.
2. It is important to determine whether the DNA can be protected from degradation by forming a SPION–DNA complex, because the DNA will be exposed to many types of enzymes when delivered in vivo. Without appropriate protection, DNA will decompose rapidly on the way to the targeted tissues or cells. This can be ascertained by testing whether the DNA enclosed in the MNP–DNA complex is decomposed when exposed to DNase. It is found that the DNA can indeed be protected from degradation through the formation of the MNP–DNA complex. As shown in **Fig. 2**, the SPION–DNA complexes digested by DNase are still visible in sample wells under fluorescence (Lanes 3 and 4), whereas the lanes containing only DNA and DNase were not (Lanes 1 and 2), because the free plasmid DNAs are degraded completely and cannot be visualized in the agarose gel. These results demonstrate the protection of DNA by the SPION, suggesting that the enzymatic digestion of DNA can be inhibited by these NP-DNA complexes. This may be caused by the smaller accessibility of the enzyme to the DNAs that are entrapped inside the MNPs. This raises another interesting point, that the MNP needs to be further improved to make the DNA more difficult to access for the DNase so that the DNA can be better protected from degradation in vivo.
3. It has been shown that the combination of silica NP with transfection kits can increase DNA transfection efficiency by fourfold to sevenfold depending on the size and number of silica NPs and the types of commercially available transfection reagents, as well as the types of cell lines *(26)*.

References

1. Tan, W., et al., Bionanotechnology based on silica nanoparticles. *Med Res Rev*, 2004. **24**(5): p. 621–38.
2. Roy, I., et al., Optical tracking of organically modified silica nanoparticles as DNA carriers: a nonviral, nanomedicine approach for gene delivery. *Proc Natl Acad Sci U S A*, 2005. **102**(2): p. 279–84.
3. Singh, M., et al., Cationic microparticles: A potent delivery system for DNA vaccines. *Proc Natl Acad Sci U S A*, 2000. **97**(2): p. 811–6.

4. Bharali, D.J., et al., Organically modified silica nanoparticles: a nonviral vector for in vivo gene delivery and expression in the brain. *Proc Natl Acad Sci U S A*, 2005. **102**(32): p. 11539–44.
5. Nomura, T., et al., Effect of particle size and charge on the disposition of lipid carriers after intratumoral injection into tissue-isolated tumors. *Pharm Res*, 1998. **15**(1): p. 128–32.
6. Cuenca, A.G., et al., Emerging implications of nanotechnology on cancer diagnostics and therapeutics. *Cancer*, 2006. **107**(3): p. 459–66.
7. Kaul, G. and M. Amiji, Cellular interactions and in vitro DNA transfection studies with poly(ethylene glycol)-modified gelatin nanoparticles. *J Pharm Sci*, 2005. **94**(1): p. 184–98.
8. Kaul, G. and M. Amiji, Tumor-targeted gene delivery using poly(ethylene glycol)-modified gelatin nanoparticles: in vitro and in vivo studies. *Pharm Res*, 2005. **22**(6): p. 951–61.
9. Kneuer, C., et al., A nonviral DNA delivery system based on surface modified silica-nanoparticles can efficiently transfect cells in vitro. *Bioconjug Chem*, 2000. **11**(6): p. 926–32.
10. Kneuer, C., et al., Silica nanoparticles modified with aminosilanes as carriers for plasmid DNA. *Int J Pharm*, 2000. **196**(2): p. 257–61.
11. Mondalek, F.G., et al., The permeability of SPION over an artificial three-layer membrane is enhanced by external magnetic field. *J Nanobiotechnology*, 2006. **4**: p. 4.
12. Jin, S. and K. Ye, Nanoparticle-mediated drug delivery and gene therapy. *Biotechnol Prog*, 2007. **23**(1): p. 32–41.
13. Morishita, N., et al., Magnetic nanoparticles with surface modification enhanced gene delivery of HVJ-E vector. *Biochem Biophys Res Commun*, 2005. **334**(4): p. 1121–6.
14. Prow, T., et al., Construction, gene delivery, and expression of DNA tethered nanoparticles. *Mol Vis*, 2006. **12**: p. 606–15.
15. Pan, B., et al., Dendrimer-modified magnetic nanoparticles enhance efficiency of gene delivery system. *Cancer Res*, 2007. **67**(17): p. 8156–8163.
16. Reszka, R., Zhu, J.H., Weber, F., Liposome mediated transfer of marker and cytokine genes into rat and human Glioblastoma cells in vitro and in vivo. *J Lipsome Res.*, 1995. **5**: p. 149–154.
17. Junghans, M., J. Kreuter, and A. Zimmer, Antisense delivery using protamine-oligonucleotide particles. *Nucleic Acids Res*, 2000. **28**(10): p. E45.
18. Schwab, G., et al., Antisense oligonucleotides adsorbed to polyalkylcyanoacrylate nanoparticles specifically inhibit mutated Ha-ras-mediated cell proliferation and tumorigenicity in nude mice. *Proc Natl Acad Sci U S A*, 1994. **91**(22): p. 10460–4.
19. Erbacher, P., et al., Chitosan-based vector/DNA complexes for gene delivery: biophysical characteristics and transfection ability. *Pharm Res*, 1998. **15**(9): p. 1332–9.
20. He, X., Wang, K., Tan, W., Liu, B., Liu, X., Huang, S., Li, D., He, C., Li, J., A novel gene carrier based on amino-modified silica nanoparticles. *Chinese Science Bulletin*, 2003. **48**(3): p. 223–228.
21. Barnes, A.L., et al., Magnetic characterization of superparamagnetic nanoparticles pulled through model membranes. *Biomagn Res Technol*, 2007. **5**: p. 1.
22. Fukuda, Y., et al., Superparamagnetic iron oxide (SPIO) MRI contrast agent for bone marrow imaging: differentiating bone metastasis and osteomyelitis. *Magn Reson Med Sci*, 2006. 5(4): p. 191–6.
23. Savranoglu, P., et al., The role of SPIO-enhanced MRI in the detection of malignant liver lesions. *Clin Imaging*, 2006. **30**(6): p. 377–81.
24. Mack, M.G., et al., Superparamagnetic iron oxide-enhanced MR imaging of head and neck lymph nodes. *Radiology*, 2002. **222**(1): p. 239–44.
25. Luo, D. and W.M. Saltzman, Enhancement of transfection by physical concentration of DNA at the cell surface. *Nat Biotechnol*, 2000. **18**(8): p. 893–5.
26. Luo, D., et al., A self-assembled, modular DNA delivery system mediated by silica nanoparticles. *J Control Release*, 2004. **95**(2): p. 333–41.
27. Roy, I., et al., Ceramic-based nanoparticles entrapping water-insoluble photosensitizing anticancer drugs: a novel drug-carrier system for photodynamic therapy. *J Am Chem Soc*, 2003. **125**(26): p. 7860–5.

Chapter 35

Magnetic Nanoparticles for Local Drug Delivery Using Magnetic Implants

Rodrigo Fernández-Pacheco, J. Gabriel Valdivia, and M. Ricardo Ibarra

Summary

This chapter is a brief description of the state of the art of the field of targeted drug delivery using magnetic implants. It describes the advantages and drawbacks of the use of internal magnets to concentrate magnetic nanoparticles near tumor locations, and the different approaches to this task performed in vitro and in vivo reviewed in literature are presented.

Key words: Magnetic Implants, Magnetic nanoparticles, Drug delivery

1. Introduction

The targeted delivery of drugs is an elegant and noninvasive strategy to treat certain diseases. Magnetic drug delivery uses magnetic carriers that can be loaded with the drugs and directed toward a desired zone, such as tumor cells or blood clots *(1)*, by means of a magnetic field gradient. The drugs are adsorbed on the surface or embedded inside these nanosystems, and likely released only at the target organ. In this manner, systemic effects caused by the administration of highly toxic drugs are minimized.

Two aspects have to be taken into account when tackling the problem of local drug delivery: the first one is the choice of adequate magnetic carriers. Magnetic nanoparticles are submicron moieties that, because of their unique properties in size and structure and their huge surface-to-volume ratio are perfect candidates for the targeted delivery of drugs. Because of their

James Weifu Lee and Robert S. Foote (eds.), *Micro and Nano Technologies in Bioanalysis*,
Methods in Molecular Biology, vol. 544
DOI 10.1007/978-1-59745-483-4_35, © Humana Press, a part of Springer Science+Business Media, LLC 2009

biocompatibility and magnetic properties, iron oxide nanoparticles are the preferred material for most biomedical applications. Iron oxide nanoparticles consist of a magnetite (Fe_3O_4) or maghemite (γ-Fe_2O_3) core encapsulated in an organic *(2, 3)* or inorganic matrix *(4)*. An alternative is the use of magnetic nanocapsules, such as magnetic liposomes, where the drugs are confined inside a cavity surrounded by an inorganic *(5)* or a polymeric membrane *(6)*.

The second issue is how to direct the particles to the target. The classic approach has been the use of an external source of magnetism, such as an electromagnet, applied on the desired area to generate a magnetic field gradient and thus localize and concentrate magnetic particles. However, the use of external magnets presents serious limitations. The homogeneity of the magnetic field on the target zone generates very weak field gradients, unable to concentrate the carriers in an effective way. Because of the weakness of magnetic forces, which have to overcome hydrodynamic forces in the blood stream, the administration method is limited to an artery close to the tumor. Consequently, it can only be applied to the treatment of superficial organs. Moreover, well-defined magnetic field geometries are a must, depending on the location, to perform an efficient magnetic drug delivery *(7–9)*.

To avoid these inconveniences, the insertion of small magnetic implants directly into the affected zone, or the combination of both external and internal sources of magnetism, appear to be promising alternatives. A small stent or a magnetic wire is inserted near the diseased area, and an external source of magnetism is applied, magnetizing the implant. An internal local magnetic field is thus created, which is more effective than fields produced by external magnets.

For a proper discussion on the advantages and drawbacks of internal versus external magnetic fields, the basic principles of magnetic attraction have to be understood *(10, 11)*. To exert a force at a distance, a magnetic field gradient is required. A uniform field gives rise to a torque, but no translational action. The magnetic force acting on a point-like magnetic dipole **m** is:

$$F_{\mathrm{m}} = (m \cdot \nabla)B. \tag{1}$$

If **B** is applied along the z direction and the medium is isotropic, then:

$$F_{\mathrm{m}} = \mathrm{F}_{\mathrm{m}}^{z} \cdot \hat{z} = m_z \cdot \frac{\partial B_z}{\partial z} \hat{z}. \tag{2}$$

This is the physical basis for drug delivery using drug-loaded magnetic carriers. The use of external magnets is limited by the requirement of using very strong magnets to retain the carriers at the desired zone. On the contrary, the insertion of a magnetic implant enhances the attraction of the particles toward the targeted site, because a ferromagnetic material placed under the influence of a magnetic field produces an additional, localized magnetic field, and hence a greater force in the vicinity of its surroundings *(12)*.

2. Magnetic Particles

Within the vast field of nanoparticulate systems, magnetic nanoparticles attract special attention because of their unique properties. These are nanostructures that contain a metal such as iron, nickel, or cobalt, or some of their oxides. Because of their response to an applied magnetic field, magnetic nanoparticles can be used as magnetic carriers in a number of biomedical applications, both in vitro or in vivo, including the release of a specific drug or molecule (magnetic drug delivery) at a desired target site inside an organism.

There are several requirements for the use of magnetic nanoparticles in targeted drug delivery: their size has to be small enough to avoid the action of gravitational forces that would make them precipitate in solution. Moreover, they have to be small enough to pass through the smallest capillaries in the body, and, at the same time, they have to be big enough to overcome the strength of blood flow *(13)*. A balance between the two factors is then needed. Any material used in vivo must be sterile and biocompatible as well, therefore, the use of toxic metals such as nickel, cobalt, neodymium, or samarium is preferably avoided. Iron and iron oxides, especially magnetite (Fe_3O_4) and maghemite (γ-Fe_2O_3), are by far the nanosystems most used as magnetic carriers *(14–18)*. Finally, magnetic nanoparticles must have high magnetic susceptibilities to respond rapidly and in an effective way to the action of a magnet. Bigger particles generally have higher saturation magnetization and magnetic susceptibility than smaller particles, so a compromise is again required. The stability of magnetic particles suspensions is vital, and is not a trivial issue. The stability depends enormously on the synthesis process, hence an appropriate preparation and isolation of the materials is a must *(19)*.

With the aim of producing the most efficient delivery system, in the last years, there has been a big effort in the synthesis of core-shell particles. These consist of a magnetic core encapsulated in an inorganic (i.e., carbon or silica) or polymeric coating. The coating renders the particles biocompatible, prevents them from oxidation in suspension, isolates them from each other, avoids the formation of magnetic aggregates, and serves as a support for the chemical bonding or physical adsorption of biomolecules on the surface of the particle.

Depending on their size and their hydrophilicity or hydrophobicity, nanoparticles without a coating after systemic administration are most likely to be removed from circulation in blood, captured by the reticuloendothelial system (RES). Different proteins of the blood serum, known as opsonins, bind to the surface of a nanoparticle, enhancing the uptake of the particle by phagocytes. Therefore, particles that undergo opsonization are eventually cleared to the macrophages of the liver, spleen, and bone marrow *(20)*. Biodegradable coatings, such as dextran *(21)*, and nonbiodegradable organic coatings can be used to reduce the uptake by the macrophages of the RES. These molecules contribute to avoid the degradation of the core and the release of toxic material to the media. They provide a hydrophilic surface that contributes to the stabilization of particles in aqueous suspensions, as in the case of in vivo applications. Diverse functional groups can be grafted, or specific drugs adsorbed on this surface *(22)*. Poly(ethylene glycol) (PEG) is currently the most used molecule to prevent the action of the RES. The attachment of PEG to the nanoparticle surfaces provides a hydrophilic surface that serves to delay recognition and depletion by the macrophages. Despite this stealth effect, there are several drawbacks implied in the use of PEG, for example, the branching of the PEG could affect the diffusion of the drug in magnetic drug delivery *(23, 24)*.

Inorganic coatings, such as carbon or silica, are also good candidates for attaining suspensions of particles that are suitable and long-term stable. **Figure 1** shows a ultrahigh-resolution transmission electron microscopy (UHRTEM, Titan 80–300) image of an iron nanoparticle encapsulated in carbon. Both materials are biocompatible and nontoxic *(25)*. In addition, amorphous silica is a heat-resisting material, with a low specific gravity, high surface area, and good mechanical strength. The small pore size of silica can give rise to a very selective interaction with the adsorbed molecules depending on their size, shape, and chemical characteristics *(26, 27)*.

This coating often plays a key role in the stabilization of the particles in suspension. First of all, it avoids leaching of the core, that is, the loss of magnetic material. In addition, the appearance of charges on the surface can stabilize liquid suspensions with a high saline concentration or slightly acid or alkaline pH. For example, the isoelectric point of silica is pH = 2–3, so, at physiological pH = 7.4, the surface of silica will be negatively charged, inducing electrostatic repulsion between particles to help avoid aggregation. A low ionic force, i.e., a low concentration

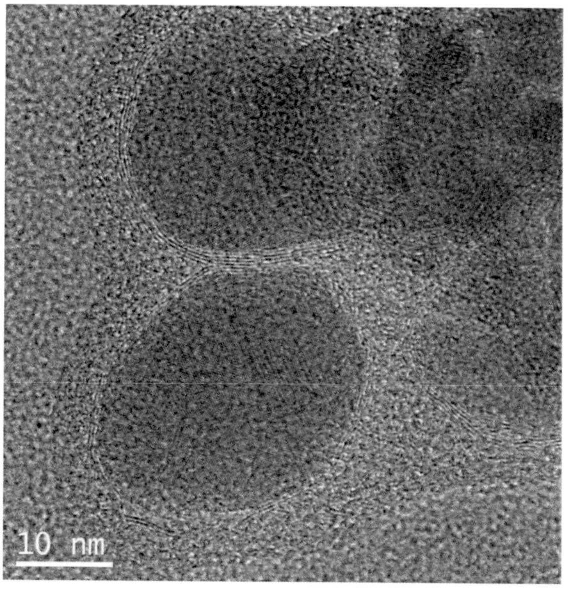

Fig. 1. Ultrahigh-resolution transmission electron microscopy (UHRTEM Titan 80–300) image of an iron nanoparticle encapsulated in carbon. Atomic planes of the metallic core and the graphitic coating are clearly visible.

of salts in solution will also help to keep particles separated from each other to yield stable suspensions *(22)*.

Carbon and silica have hydroxyl (OH) groups on the surface that can be substituted with more reactive chemical chains that will be capable to bind covalently to the amine (NH_2) or carboxyl (COOH) terminal groups of a protein, for example. The number of hydroxyl groups on the carbon surface is very low in comparison with silica, and much less reactive. This fact makes carbon less attractive for chemical functionalization but, on the other hand, it renders its surface highly hydrophobic, that way it easily adsorbs any organic molecule in solution. This is very important for the spontaneous adsorption and release of chemotherapy drugs in the case of drug delivery, and at the same time it makes carbonous particles a perfect target for the RES, because, due to the hydrophobic nature of their surface, they are more likely to undergo opsonization processes. Therefore, the objective is double: on one hand, to introduce on the particles some chemical groups that avoid the adsorption of opsonin, and therefore the action of the immunological defences of the body; and on the other hand, the hydrophobic nature of carbon must be preserved so it can still be able to efficiently adsorb a drug.

3. Magnetic Field Targeting

Several strategies have been described to achieve an effective targeted accumulation of the drug carriers, and only a very small number of the strategies have been devoted to the theoretical or experimental development of magnetic implants. The classic approach has been to use a single source of magnetism to magnetize the carriers and attract them to a specific site inside the organism *(8, 28, 29)*. This, however, presents many disadvantages, especially the impossibility of generating strong magnetic fields that permit a high capture efficiency of the carriers at the targeted site.

To overcome this, Yellen et al. suggested the use of two independent magnetic sources. A small-size magnetic mesh, consisting of a 3 × 3 array of magnetic wires, in combination with long-range externally applied fields were used to magnetize and localize the carriers in an in vitro experiment *(30)*. They proved that fields in the range of 0.1 T should be used to reach magnetic saturation of both the particles and the implant. Fields greater than 0.1 T would have little additional effect on particle capture. Therefore, their results demonstrate that superparamagnetic nanoparticles at high concentrations can be captured at flow conditions consistent with the dimensions and flow velocity occurring in the coronary artery in the human blood.

Ritter and co-workers studied the feasibility of using a ferromagnetic wire implanted under the skin and next to the carotid artery for the collection of magnetic carriers using external magnets. A complete two-dimensional (2D) mathematical analysis of the hydrodynamic and magnetic forces involved in the process was performed *(31)*. Based on these calculations, they designed a coiled ferromagnetic wire stent to simulate an intravascular stent in vitro. They used magnetite nanoparticles embedded in poly(styrene/divinylbenzene) matrixes as carriers, and an external permanent magnet to magnetize the wire stent. Changes of the capture efficiency by varying the applied magnetic field, the particle concentration, and the fluid velocity were studied *(32)*. They concluded that an increase in the fluid velocity and the concentration of particles had positive effects on the efficiency of the capture of the magnetic carriers, because of the increase in the hydrodynamic force of the fluid and the formation of small aggregates. On the other hand, an increase in the applied magnetic field strength yielded a very slight improvement in the capture efficiency, because not very high fields were required for the saturation of particles and implants.

The implantation of small permanent magnets by minimally invasive surgery is an alternative to these procedures. Despite the preliminary achievements in in vitro models, very little research has been performed in experimental animal models. First in vivo experiments of localization of magnetic nanoparticles with implanted permanent magnets have already been performed in animals *(33, 34)*. Neodymium–iron–boron permanent magnets were coated with gold to render them biocompatible and inserted by minimally invasive surgery into the main organs of New Zealand rabbits. The left kidney was used as a test organ and the right kidney was punctured as a control (**Fig. 2**). Stable suspensions of carbon-coated iron nanoparticles were then injected intravenously, and the capability of the magnet to attract the magnetic particles was tested. The laparoscopic insertion of a small magnetic implant inside the renal parenchyma is an unprecedented, effective, and easily reproducible method to deliver magnetic carriers to a target organ. The tolerance of well-coated, biocompatible magnets was excellent, and their positions can be easily tracked after operation by means of abdominal ultrasound scanning or by means of radiography. The implant creates a magnetic field gradient of 30 T/m in a region of 1 cm around the zone of insertion. **Figure 3** shows the simulation of the magnetic field created by a Nd-Fe-B cylindrical permanent magnet. The dimensions of the magnet are 2-mm high and 4-mm diameter. The lines in the figure are the magnetic field lines; the north and south poles of the magnet are situated in the vertical axis. The maximum number of field lines is found in the vicinity of the edge of the magnet, which means that maximum variation of field is achieved in this region. Because magnetic field gradient, and not magnetic field, is what really drives particles or objects subject to magnetism to move toward the magnet, maximum attraction of the particles shall be found near the edges of the magnet.

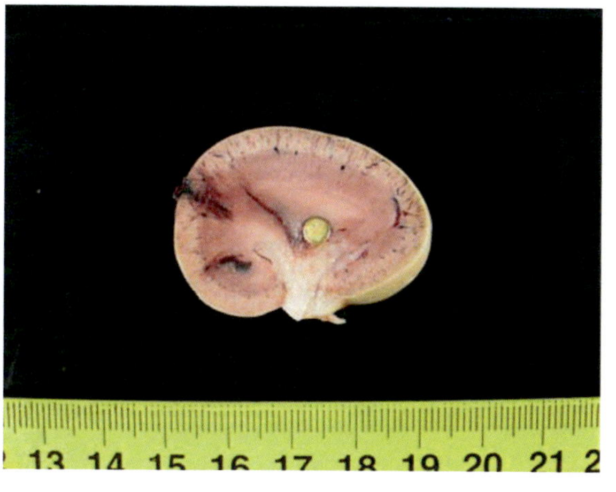

Fig. 2. Longitudinal section of a rabbit's left kidney with a magnet inserted in the middle. Several hemorrhagic trajectories caused by the insertion of the magnetic implant can be observed.

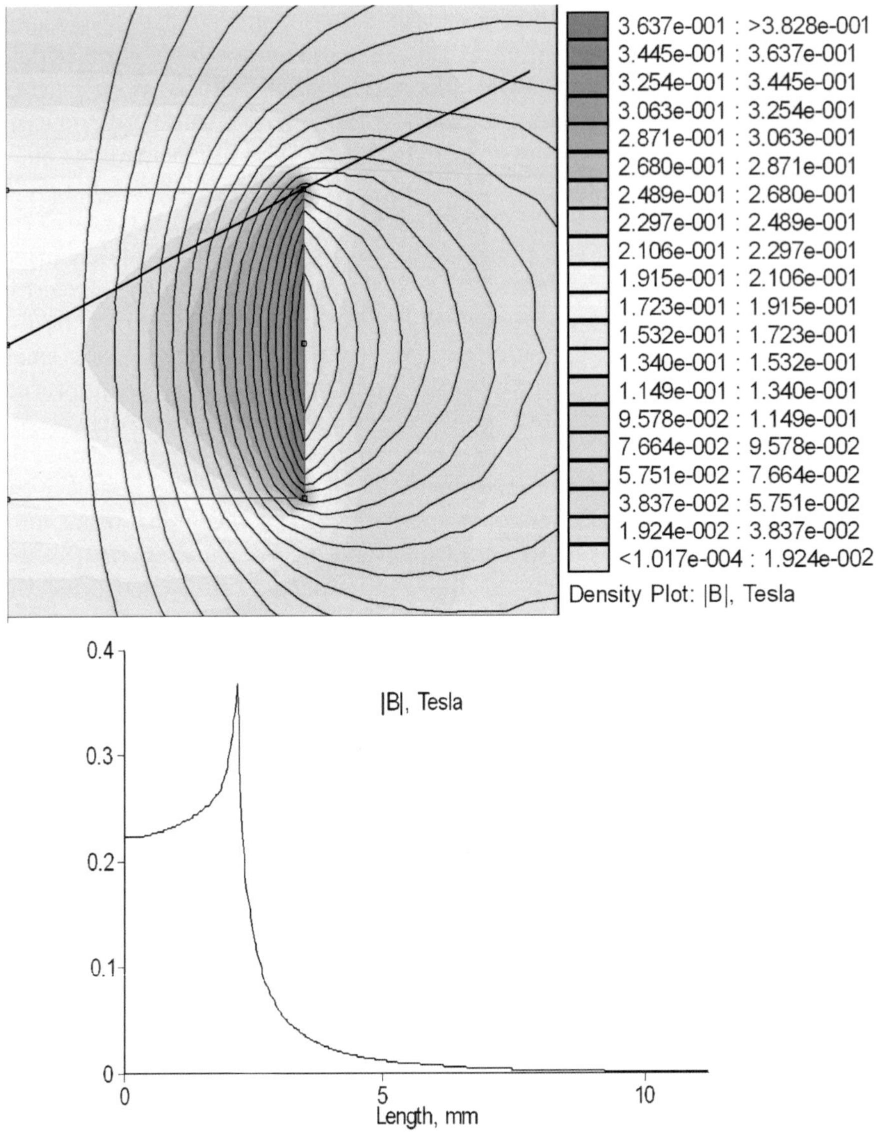

Fig. 3. (**a**) Simulation of the magnetic field created by a cylindrical permanent magnet. The figure represents half of a magnet with cylindrical symmetry. (**b**) Variation of magnetic field with distance from the middle to the vortex of the magnet.

Histopathology studies showed the distribution inside the organism and the short-term and medium-term tolerance of nanoparticles inside the main organs of the body. Particles were attracted under the influence of the internally created magnetic field gradient, and were retained near the magnet. Future work will involve drug-loaded particles, which will permit a slow and controlled diffusion of the chemotherapy agent in the desired zone.

From histological studies, a larger particle concentration in the implanted kidney was observed in comparison with the

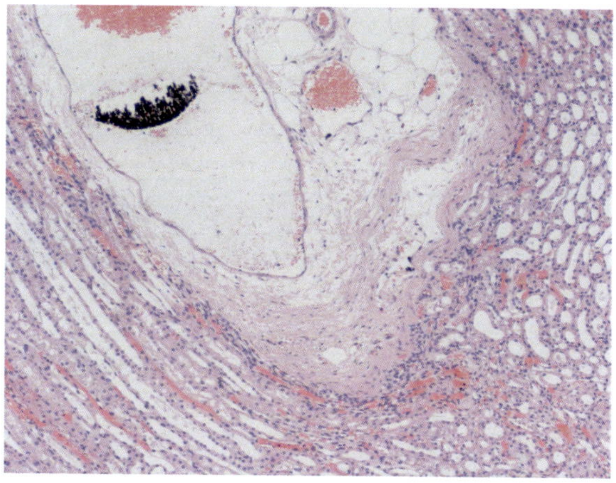

Fig. 4. Particles accumulate in the vessel of a kidney in which a magnetic implant has been inserted. The particles remain there and tend to align in the direction of the magnetic field gradient. Hematoxylin and eosin, ×100.

nonimplanted kidney (**Fig. 4**). Although it is difficult to quantify the results, a significant quantity of the particles was concentrated in the kidney in which a magnet was inserted. This is an indication of the capability of the magnet to attract particles specifically to the target organ. It is the key issue to the method, and the main advantage of using internal magnets over the use of external magnetic fields.

A relatively important amount of particles was found to be concentrated in the main filters of the body, the liver and spleen, whereas a much lower quantity of particles was found in the lungs of some animals. Despite the inert surface of the material, the size of the particles was large enough to be recognized by the macrophages of these organs, which captured and retained them, eventually excreting them after several days. This fact could be a drawback for the purpose of delivering chemicals to specific targets in the body but, on the other hand, it could also be helpful for other applications, such as treatment of liver or spleen tumors, or for the treatment of leishmania *(35)*. The functionalization of the nanoparticles' surface with chemical groups that specifically repel macrophages seems a profitable solution.

Acknowledgments

Financial support by the Spanish Ministry of Science (through projects NAN2004-09270-CO3-03, and CONSOLIDER NANOBIOMED) and the Department of Science, Technology and University of the Government of Aragón is acknowledged. We also acknowledge Alba García and Dr. José A. García de Jalón from the Animal Pathology Department at the University of Zaragoza for histopathology images. Dr. Clara Marquina, Dr. Manuel Arruebo, Dr. Valeria Grazu, Dr. Jesús M. de La Fuente, Dr. Gerardo Goya, Dr. Américo Viloria, Dr. Teresa Higuera, Javier Gómez-Arrúe, Alicia Laborda, Ángel García, María Moros, Laura Asín, Teobaldo Torres, and Sara Puertas also played key parts in this research.

References

1. Cha, Y., Chen, L., Askew, T., Veal, B., Hull, J. (2007). Manipulation of magnetic carriers for drug delivery using pulsed-current high Tc superconductors. *J. Magn. Magn. Mater.* **311**, 312–317.
2. Alexiou, C., Arnold, W., Klein, R. J., Parak, F., Hulin, P., Bergemann, C., Erhardt, W., Wagenpfel, S., Lübbe, A. (2000). Locoregional cancer treatment with magnetic drug targeting. *Cancer Res.* **60**, 6641–6648.
3. Aurich, K., Schwalbe, M., Clement, J. H., Weitschies, W., Buske, N. (2007). Polyaspartate coated magnetite nanoparticles for biomedical applications. *J. Magn. Magn. Mater.* **311**, 1–5.
4. Fernández-Pacheco, R., Arruebo, M., Marquina, C. I., Ibarra, M. R., Arbiol, J., Santamaría, J. (2006). Highly magnetic silica-coated iron nanoparticles prepared by the arc-discharge method. *Nanotechnology* **17**, 1188–1192.
5. Arruebo, M., Galán, M., Navascués, N., Téllez, C., Marquina, C., Ibarra, M. R., Santamaría, J. (2006). Development of magnetic nanostructured silica-based materials as potential vectors for drug-delivery applications. *Chem. Mater.* **18**, 1911–1919.
6. Shinkai, M., Suzuki, M., Iijima, S., Kobayashi, T. (1994). Antibody-conjugated magnetoliposomes for targeting cancer cells and their application in hyperthermia. *Biotechnol. Appl. Biochem.* **21**, 125–137.
7. Shinkai, M. (2002). Functional magnetic particles for medical application. *J. Biosci. Bioeng.* **94**, 606–613.
8. Lübbe, A., Alexiou, C., Bergemann, C. (2001). Clinical applications of magnetic drug targeting. *J. Surgical Res.* **95**, 200–206.
9. Alexiou, C., Jurgons, R., Schmid, R., Hilpert, A., Bergemann, C., Parak, F., Iro, H. (2005). In vitro and in vivo investigations of targeted chemotherapy with magnetic nanoparticles. *J. Magn. Magn. Mater.* **293**, 389–393.
10. Pankhurst, Q. A., Connolly, J., Jones, S. K., Dobson, J. (2003). Applications of magnetic nanoparticles in biomedicine. *J. Phys. D: Appl. Phys.* **36**, R167–R181.
11. Goya, G. F., Grazú, V., Ibarra, M. R. (2007). Magnetic nanoparticles for cancer therapy. *Curr. Nanosc.* **4**, 1–16.
12. Ritter, J. A., Ebner, A. D., Daniel, K. D., Stewart, K. L. (2004). Application of high gradient magnetic separation principles to magnetic drug targeting. *J. Magn. Magn. Mater.* **280**, 184–201.
13. Kuznetsov, A. A., Filippov, V. I., Kuznetsov, O. A., Gerlivanov, V. G., Dobrinsky, E. K., Malashin, S. I. (1999). New ferro-carbon adsorbents for magnetically guided transport of anti-cancer drugs. *J. Magn. Magn. Mater.* **194**, 22–30.
14. Jain, T. K., Morales, M. A., Sahoo, S. K., Leslie-Pelecky, D. L., Labhasetwar, V. (2005). Iron oxide nanoparticles for sustained delivery of anticancer agents. *Mol. Pharm.* 2, **194**–205.
15. Finotelli, P. V., Morales, M. A., Rocha-Leão, M. H., Baggio-Saitovitch, E. M., Rossi, A. M. (2004). Magnetic studies of iron (III) nanoparticles in alginate polymer for drug delivery applications. *Mater. Sci. Eng.: C* **24**, 625–629.
16. Neuberger, T., Schopf, B., Hofmann, H., Hofmann, M., von Rechenberg, B. (2005). Superparamagnetic nanoparticles for biomedical applications: Possibilities and limitations of a new drug delivery system. *J. Magn. Magn. Mater.* **293**, 483–496.
17. Gupta, A. K., Gupta, M. (2005). Synthesis and surface engineering of iron oxide nanoparticles for biomedical applications. *Biomaterials* **26**, 3995–4021.
18. Li, L., Chen, D., Zhang, Y., Deng, Z., Ren, X., Meng, X., Tang, F., Ren, J., Zhang, L. (2007) Magnetic and fluorescent multifunctional chitosan nanoparticles as a smart

drug delivery system. *Nanotechnology* **18**, 405102–405107.

19. Muller, R. H., Keck, C. M. (2004). Challenges and solutions for the delivery of biotech drugs – a review of drug nanocrystal technology and lipid nanoparticles. *J. Biotech.* **113**, 151–170.
20. Berry, C. C., Curtis, A. S. G. (2003). Functionalisation of magnetic nanoparticles for applications in biomedicine. *J. Phys. D: Appl. Phys.* **36**, R198–R206.
21. Zhang, J. L., Srivastava, R. S., Misra, R. D. K. (2007). Core-shell magnetite nanoparticles surface encapsulated with smart stimuli-responsive polymer: synthesis, characterization, and LCST of viable drug-targeting delivery system. *Langmuir* **23**, 6342–6351.
22. Arruebo, M., Fernández-Pacheco, R., Ibarra, M. R., Santamaría, J. (2007). Magnetic Nanoparticles for Drug Delivery. *Nanotoday* **2**, 22–32.
23. Xu, Z. P., Hua Zeng, Q., Lu, G. Q., Bing Yu, A. (2006). Inorganic nanoparticles as carriers for efficient cellular delivery. *Chem. Eng. Sci.* **61**, 1027–1040.
24. Liu, Y. L., Hsu, C. Y., Wang, M. L., Chen, H. S. (2003). A novel approach of chemical functionalization on nano-scaled silica particles. *Nanotechnology* **14**, 813–819.
25. Spange, S. (2000). Silica surface modification by cationic polymerization and carbenium intermediates. *Prog. Polym. Sci.* **25**, 781–849.
26. Moghimi, S. M. (2002). Chemical camouflage of nanospheres with a poorly reactive surface: towards development of stealth and target-specific nanocarriers. *Biochim. Biophys. Acta* **1590**, 131–139.
27. Gref, R., Domb, A., Quellec, P., Blunck, T., Muller, R. H., Verbavatz, J. M., Langer, R. (1995). The controlled intravenous delivery of drugs using PEG-coated sterically stabilized nanospheres. *Adv. Drug Deliv. Rev.* **16**, 215–233.
28. Rudge, S., Peterson, C., Vessely, C. (2001). Adsorption and desorption of chemotherapeutic drugs from a magnetically targeted carrier (MTC). *J. Contr. Release* **74**, 335–340.
29. Flores, G. A., Liu, J. (2002). In vitro blockage of a simulated vascular system using magnetorheological fluids as a cancer therapy. *Eur. Cell. Mater.* **3**, 9–11.
30. Yellen, B. B., Forbes, Z. G., Halverson, D. S., Fridman, G., Barbee, K. A., Chorny, M., Levy, R., Friedman, G. (2005). Targeted drug delivery to magnetic implants for therapeutic applications. *J. Magn. Magn. Mater.* **293**, 647–654.
31. Avilés, M. O., Ebner, A. D., Chen, H., Rosengart, A. J., Kaminski, M. D., Ritter, J. A. (2005). Theoretical analysis of a transdermal ferromagnetic implant for retention of magnetic drug carrier particles. *J. Magn. Magn. Mater.* **293**, 605–615.
32. Avilés, M. O., Chen, H., Ebner, A. D., Rosengart, A. J., Kaminski, M. D., Ritter, J. A. (2007). In vitro study of ferromagnetic stents for implant assisted-magnetic drug targeting. *J. Magn. Magn. Mater.* **311**, 306–311.
33. Fernández-Pacheco, R., Ibarra, M. R., Valdivia, J. G., Marquina, C. I., Serrate, D., Romero, M. S., Gutiérrez, M., Arbiol, J. (2005). Carbon Coated Magnetic nanoparticles for local drug delivery using magnetic implants. Technical Proceedings of the 2005 NSTI Nanotechnology Conference and Trade Show, Vol 1. NSTI Publications, Danville (USA), pp.144–147.
34. Fernández-Pacheco, R., Marquina, C. I., Valdivia, J. G., Gutiérrez, M., Romero, M. S., Cornudella, R., Laborda, A., Viloria, A., Higuera, T., García, A., García de Jalôn, J. A., Ibarra, M. R. (2007). Magnetic nanoparticles for local drug delivery using magnetic implants. *J. Magn. Magn. Mater.* **311**, 318–322.
35. Chellat, F., Merhi, Y., Moreau, A., Yahia, L. (2005). Therapeutic potential of nanoparticulate systems for macrophage targeting. *Biomaterials* **26**, 7260–7275.

Chapter 36

Functionalized Magnetic Nanoparticles as an In Vivo Delivery System

Shu Taira, Shinji Moritake, Takahiro Hatanaka,
Yuko Ichiyanagi, and Mitsutoshi Setou

Summary

We developed extremely small functionalized magnetic nanoparticles (MNPs) for use as an in vivo delivery system for pharmaceuticals and biomolecules. We functionalized the MNPs (d = 3 nm) by silanization of amino groups on the particles with (3-aminopropyl)triethoxysilane for subsequent cross-linking with pharmaceuticals and biomolecules. The MNPs were successfully introduced into living cells without any further modification, such as the use of cationic residues, to enhance endocytic internalization. The particles could be incorporated into the subcutaneous tissue of a mouse's ear through the skin of the ear and could be localized by application of an external magnetic field.

We also developed a cell-specific delivery system that makes use of MNPs (d = 3 nm) conjugated with folic acid and a coumarin fluorophore for recognition by folate receptors on the cell surface. The modified MNPs were internalized by human pharyngeal cancer cells (KB cells) after an incubation time that was short compared with the time required for internalization of MNPs without folic acid. Cellular recognition of MNPs may lead to the development of other cell-specific delivery systems.

These functionalized MNPs are expected to be useful as a new drug delivery tool.

Key words: Nanotechnology, Magnetic particle, Delivery system, Folate receptor, Cell recognition, Silanization

1. Introduction

1.1. Recent Developments in Nanotechnology

Recently, the use of nanoparticles (NPs) in the fields of nanomachines *(1–4)*, imaging methods *(5–7)*, biosensors *(8–10)*, diagnostics *(11)*, and drug delivery systems *(12–14)* has been reported. NPs with diameters in the 100- to 1,000-nm range are widely used as carriers for macromolecules (such as plasmid DNA

James Weifu Lee and Robert S. Foote (eds.), *Micro and Nano Technologies in Bioanalysis*,
Methods in Molecular Biology, vol. 544
DOI 10.1007/978-1-59745-483-4_36, © Humana Press, a part of Springer Science+Business Media, LLC 2009

(15), small interfering RNA *(16)*, peptides *(17)*, and genes *(18)*) and small molecules (such as corticosteroids *(19)* and alkaloids *(20)*), with the goal of improving the therapeutic efficacy of cellular drug delivery systems. NPs are also used in bioengineering applications *(1, 8, 21)* and in the field of fundamental physics *(22)*. NP surfaces can be conjugated with various chemical compounds, including impermeable biomolecules, which can then be carried into living cells by the NPs. Cationic residues that facilitate the endocytic introduction of the carriers are sometimes necessary *(23, 24)*. So far, however, this method has been used only for nonspecific endocytosis processes; that is, the destination of the particles is not strictly controlled. If NPsould be modified to recognize specific cell types, controlling the destination of a drug using particles modified with a cell-selective moiety in addition to the drug might be feasible *(13, 25, 26)*. Previously reported methods for the preparation of such modified NPs are complicated in that several steps are required. In addition, the NPs are more than several tens of nanometers *(27, 28)* in diameter, which reduces the introduction efficiency.

We prepared magnetic nanoparticles (MNPs) surrounded by amorphous SiO_2 by mixing aqueous solutions of 3*d* transition-metal chlorides ($M\mathrm{Cl}_2 \cdot n\mathrm{H}_2\mathrm{O}$) and sodium metasilicate nonahydrate ($Na_2SiO_3 \cdot 9H_2O$) to obtain monodisperse nanoparticles with diameters of less than 10 nm in a single step *(29–34)*. We also prepared MNPs functionalized with surface amino groups by means of a silanization procedure to covalently bond target molecules to the particles without increasing the particle diameter. Because the particles are extremely small, there is no barrier effect of the plasma membrane against the particles; therefore, the particles can be introduced into the cells without a cationic coating. We found that MNPs can be accumulated at high concentrations in a mouse's ear under the influence of an external magnetic field (**Fig. 1a**) *(12)*.

We also synthesized functionalized MNPs designed to recognize certain cell types for specific delivery. The folate receptor (FR) is overexpressed on the surfaces of human tumor cells *(13, 35, 36)*, and folic acid (FA) has been used as a ligand to target tumor cells. When the γ-carboxyl group of FA is linked to a drug or an imaging reagent, the linked compound binds strongly to the FR ($K_D \sim 10^{-10}$), and receptor-mediated endocytosis occurs smoothly *(26–28, 37)*. Therefore, in a single step, we synthesized MNPs covalently modified with FA on the particle surfaces. We foundhat the FA-conjugated MNPs would bind to FR on human pharyngeal cancer cells (KB cells) and be selectively and preferentially introduced to the target cells *(14)* (**Fig. 1b**).

1.2. Preparation and Characterization of Functionalized MNPs

γ-Fe_2O_3 particles *(29)* with diameters ranging from 1.3 to 3 nm were used to prepare functionalized MNPs by means of a silanization procedure (**Fig. 2a**). The silanized particles bear surface amino groups, which can be used to covalently attach various

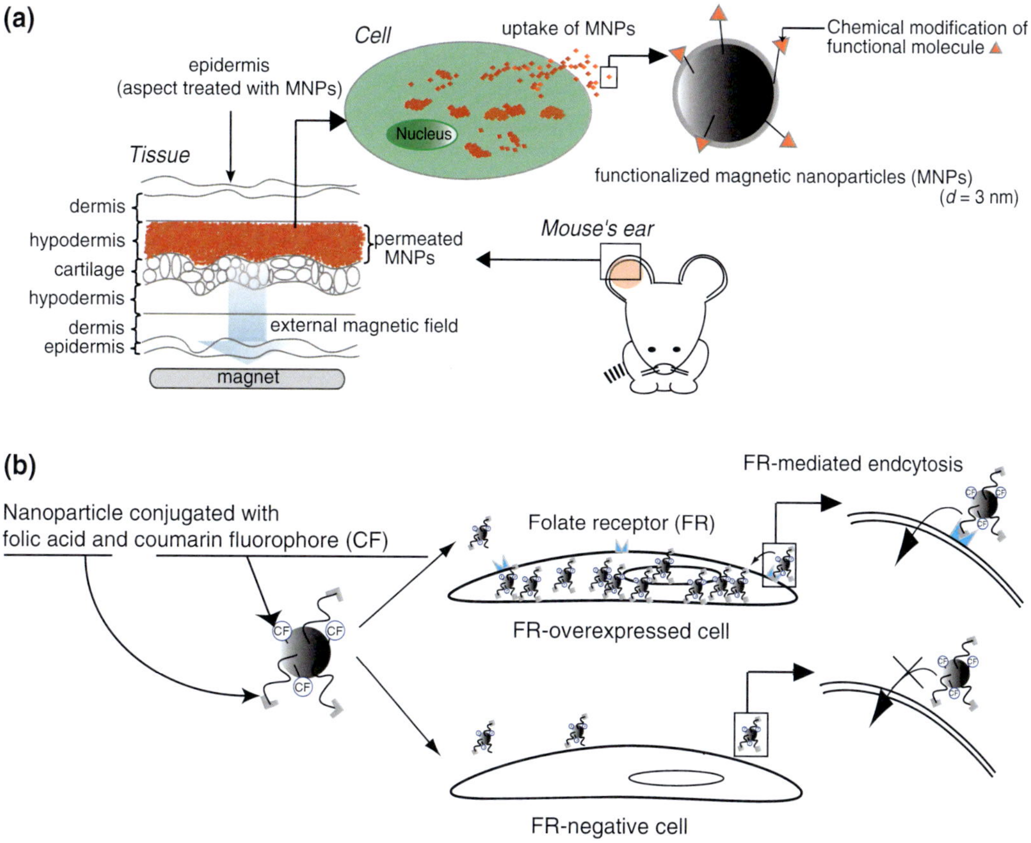

Fig. 1. Schematic illustration of (**a**) functionalized magnetic nanoparticles (MNPs) as an in vivo delivery system and (**b**) of cell-specific uptake of modified MNPs.

materials to the particle surface. The presence of the amino groups on the amino-MNPs was confirmed by Fourier transform infrared spectroscopy (FT-IR). The typical O-H peak (3,660–2,991 cm^{-1}), the C-H peak (2,972–2,842 cm^{-1}), and the C-N peak (1,583–1,481 cm^{-1}) were detected in the spectrum of the amino-MNPs **(Fig. 2b**-I), whereas only the O-H (3,660–2,991 cm^{-1}) peak was observed in the spectrum of the unfunctionalized MNPs **(Fig. 2b**-II).

The X-ray diffraction (XRD) pattern of the amino-MNPs **(Fig. 2c**-I) indicated that the functionalized particles retained the spinel γ-Fe_2O_3 structure of the unfunctionalized MNPs **(Fig. 2c**-II). The broad peaks at $2\theta = 35°$ and $60°$ were confirmed to be the (311) reflection and the (440) reflection of γ-Fe_2O_3, respectively; and the peak below $2\theta = 30°$ was determined to have originated from amorphous SiO_2 *(38)*. The diameters of the amino-MNPs and the MNPs were independently estimated at 3 nm from the half-width of the XRD peaks for the amino-MNPs and MNPs *(31)*. The transmission electron microscopy (TEM) images showed no significant difference between the shape of the

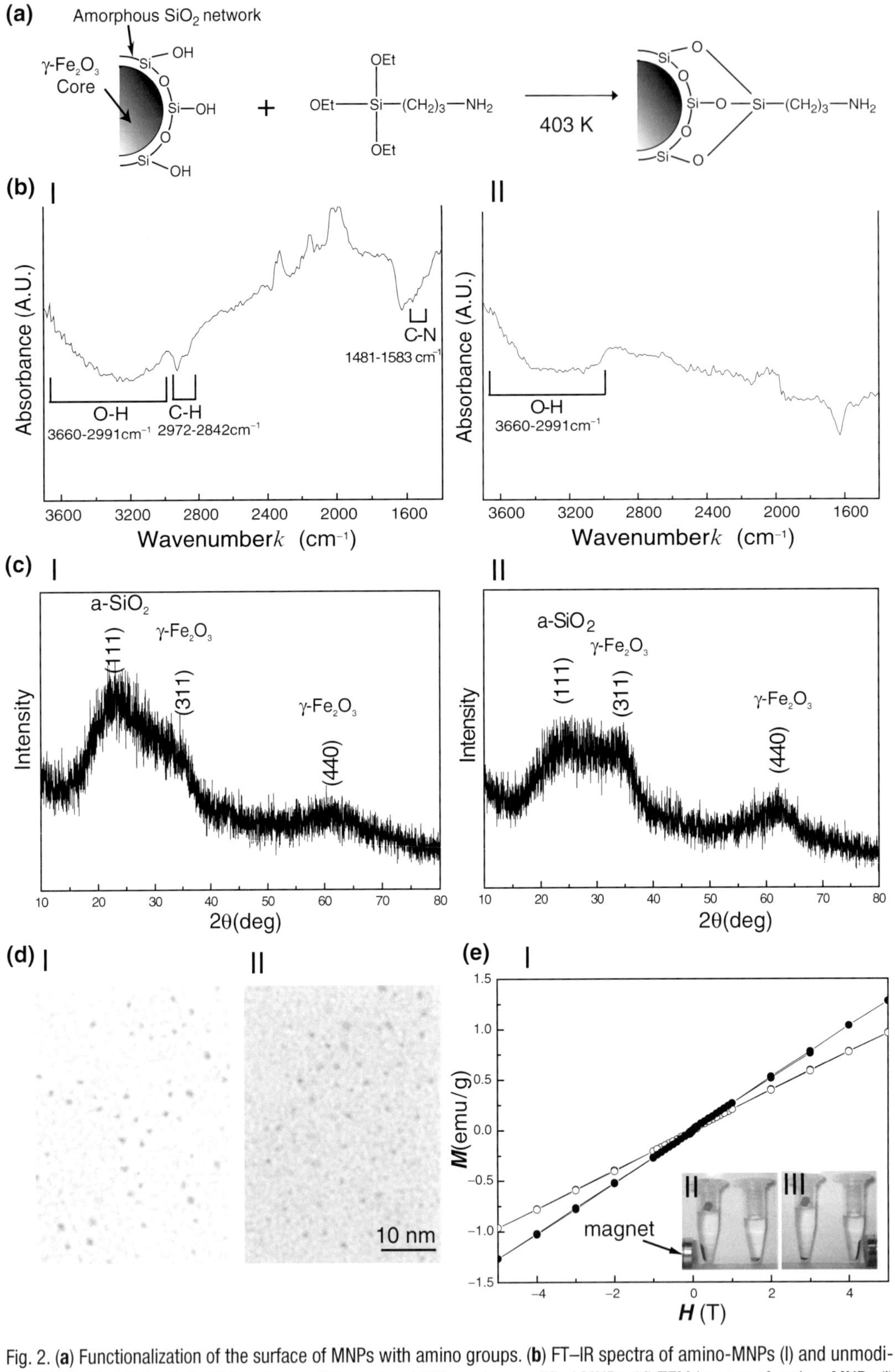

Fig. 2. (**a**) Functionalization of the surface of MNPs with amino groups. (**b**) FT–IR spectra of amino-MNPs (I) and unmodified MNPs (II). (**c**) The CuK$_{\alpha}$ XRD patterns of the amino-MNPs and unmodified MNPs. (**d**) TEM images of amino–MNPs (I) and unmodified MNPs (II). (**e**) *M–H* curves for amino-MNPs (*filled circle*) and unmodified MNPs (*open circle*) (I) and attraction of MNPs by an external magnetic field (340 mT) (II and III). Both photographs show amino–MNPs (*left*, a blue sticker was posted on an Eppendorf tube) and MNPs (*right*).

amino-MNPs (**Fig. 2d**-I) and that of the unfunctionalized MNPs (**Fig. 2d**-II). The number-average diameter of the amino-MNPs was determined to be approximately 3 nm, which is in good agreement with the XRD experimental value. The data obtained with a superconducting quantum interference device (SQUID) magnetometer indicated that the magnetism of both the amino-MNPs and the unfunctionalized MNPs increased linearly with increasing magnetic field; that is, both types of particles showed paramagnetism (**Fig. 2e**-I). The MNP precipitates could be easily attracted from water solution by means of an external magnetic field, regardless of whether or not they were functionalized (**Fig. 2e**-II, III). These results indicate that the magnetism was basically unchanged by functionalization with the amino groups.

1.3. Cellular Uptake of Functionalized MNPs

We were able to introduce the MNPs into living cells without cationic coating on the particle surface. We investigated the intracellular space of rat kangaroo kidney epithelium cells (PtK2 cells) by TEM (accelerating voltage, 100 kV) to confirm the introduction of the MNPs into the cells. The TEM image of the cells into which MNPs had been introduced showed high-density spots (**Fig. 3a**, dashed red circles), thought to be aggregated MNPs. In the control TEM image (**Fig. 3b**), no large high-density spots (d = 200 nm) were observed, which indicates that no particularly large particles were introduced by endocytosis. These results suggest that the nanoparticles without a cationic coating could be introduced into cells and that particle size may markedly influence cellular uptake. Both TEM images showed small dots (indicated by yellow arrows). On the basis of the shape of the dots and the massive number of them, we think that they may be ribosomes. Microtubules (a cytoskeletal filament) with an outer diameter of 25 nm were observed (indicated by dotted rectangles), which indicates that MNPs had little effect on the cytoskeletal ultrastructure.

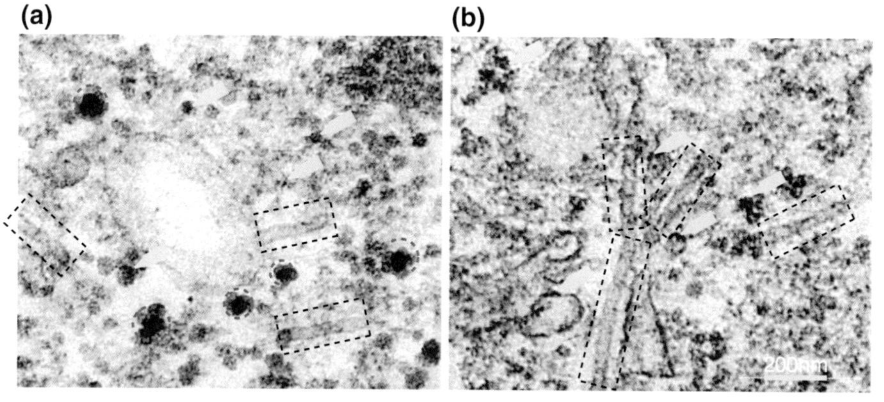

Fig. 3. (**a**) TEM images of a PtK2 cell into which MNPs and (**b**) control particles (d = 200 nm) were introduced. The *red dashed circles* indicate aggregated MNPs, the *yellow arrows* indicate ribosomes, and the *dotted rectangles* indicate microtubules.

The colloidal stability and surface electric charge of the rhodamine-labeled MNPs were evaluated. As expected, no migration toward either the anode or the cathode occurred after 10 min at 100 mV, which indicates that the residual amino groups on the particles were successfully quenched.

1.4. Localization of MNPs in Mouse Ear by Means of an External Magnetic Field

We addressed the question of whether MNPs could permeate the epidermis into subcutaneous tissue under the influence of a magnetic field **(Fig. 4a)**. We assumed that the magnetization of the MNPs was determined by the surface flux density (240 mT) and the *M–H* curves **(Fig. 2e)**, and therefore we estimated

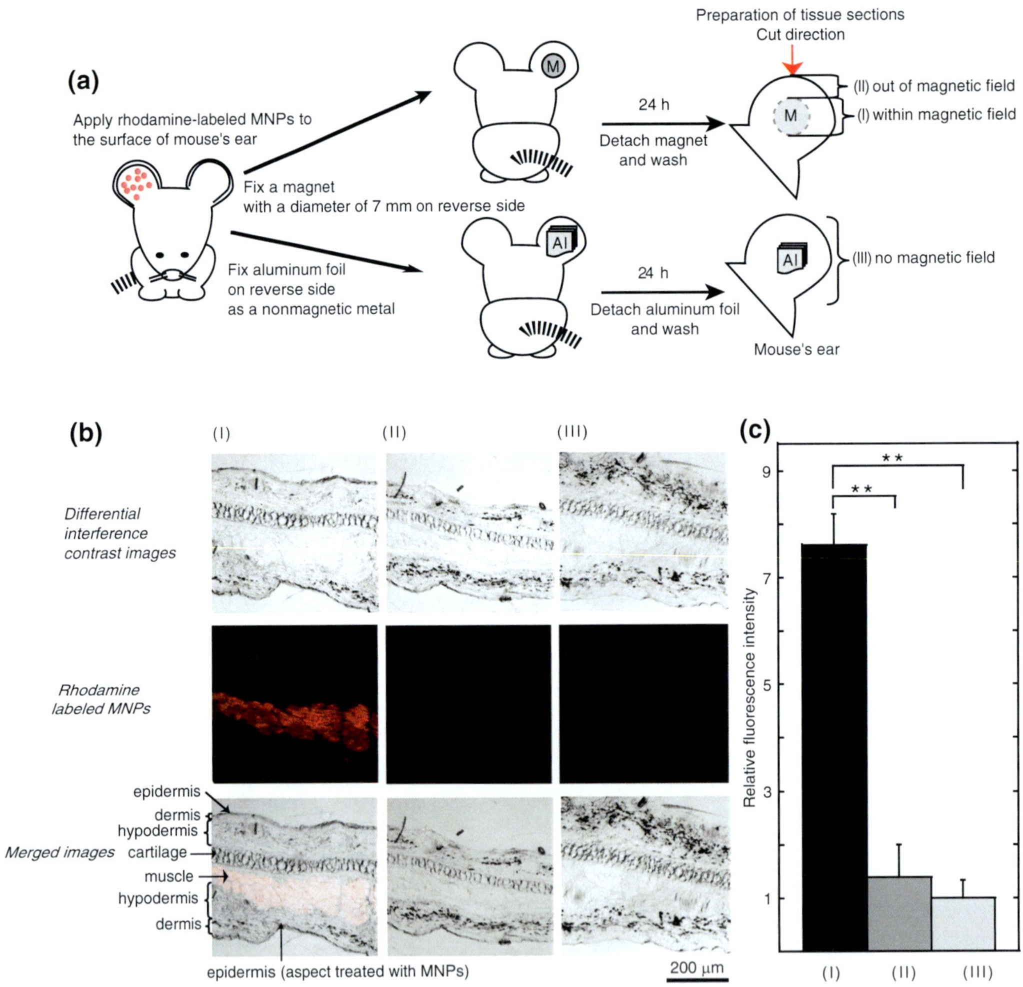

Fig. 4. (**a**) Preparation of mouse external ear section for external magnetic field specific internalization of MNPs. (**b**) Confocal laser scanning microscopy images of mouse external ear sections: differential interference contrast images (*upper panel*), rhodamine fluorescence images (*middle panel*), and merged images (*lower panel*). Localization of the MNPs (d = 3 nm) was investigated in the presence (**b**-I) and absence (**b**-II) of an external magnetic field and without the permanent magnet (aluminum foil control) (**b**-III). (**c**) The relative fluorescence intensity was determined by dividing the sum of all pixel intensities. The values presented are mean ± s.e.m. (N = 13). **P < 0.01 with Student's t test.

the magnetization of the particles to be 5.2×10^{-2} emu/g. We applied rhodamine (Rh)-labeled MNPs to the surface of a mouse's ear and then fixed a magnet to the reverse side of the ear. Using a confocal laser scanning microscope, we obtained differential interference contrast images and rhodamine fluorescence images of the mouse's external ear sections (**Fig. 4b**). We evaluated the effect of the magnetic field by comparing the part of ear to which the magnet was directly attached (**Fig. 4b-I**) with the part that was not directly covered by the magnet (**Fig. 4b-II**). In the former case, strong fluorescence was observed, and the merged image clearly shows that the Rh-MNPs were localized at the hypodermis. In contrast, in the area outside the external magnetic field, no fluorescence was observed (**Fig. 4b**-II), which indicates that no introduction or localization of MNPs occurred outside the area affected by the external magnetic field. In addition, no fluorescence was detected in a control experiment using aluminum foil (a nonmagnetic metal; **Fig. 4b**-III). The relative fluorescence intensities in each section are shown in **Fig. 4c**. The ear sections to which the magnet was directly attached showed intensity (**Fig. 4c**-I) that was fivefold and sevenfold, respectively, the intensity of the section that was not directly covered by the magnet (**Fig. 4c**-II) and of the section covered by the aluminum foil (**Fig. 4c**-III).

1.5. Preparation and Characterization of Folic Acid-Conjugated MNPs

The amino-MNPs were modified with FA and a coumarin fluorophore (CF) (**Fig. 5a**). The FT-IR spectra of the FA–CF–MNPs showed a peak at 1,419 cm^{-1}(**Fig. 5b**-I) corresponding to the *p*-amino benzoic acid moiety of FA *(39)*(**Fig. 5b**-III), and no peaks for the amino-MNPs were observed (**Fig. 5b**-II). The IR peaks of the CF moiety could not be identified, because they overlapped the FA peaks. The presence of the CF moiety on the MNPs was confirmed by fluorescence microscopy. We also evaluated the colloidal stability and surface electric charge of the FA–CF–MNPs and CF–MNPs. FA–CF–MNPs migrated toward the anode. For CF–MNPs, no migration toward either the anode or the cathode occurred after 10 min at 100 mV (**Fig. 5c**). These results indicated that FA–CF–MNPs have anionic properties caused by the presence of the FA groups, which have an α-carboxyl group.

Transmission electron microscopy showed that the shape of the FA–CF–MNPs (**Fig. 5d**) did not differ markedly from the shape of the unmodified MNPs. The number-average diameter of the FA–CF–MNPs was determined to be approximately 3 nm.

1.6. Cellular Uptake of FA-Conjugated MNPs

We evaluated the cellular uptake of the FA–CF–MNPs in three dimensions. After KB cells had been cultured for 3 h, fluorescence was observed from the cells treated with the MNPs. The lateral image clearly shows that the aggregated particles were

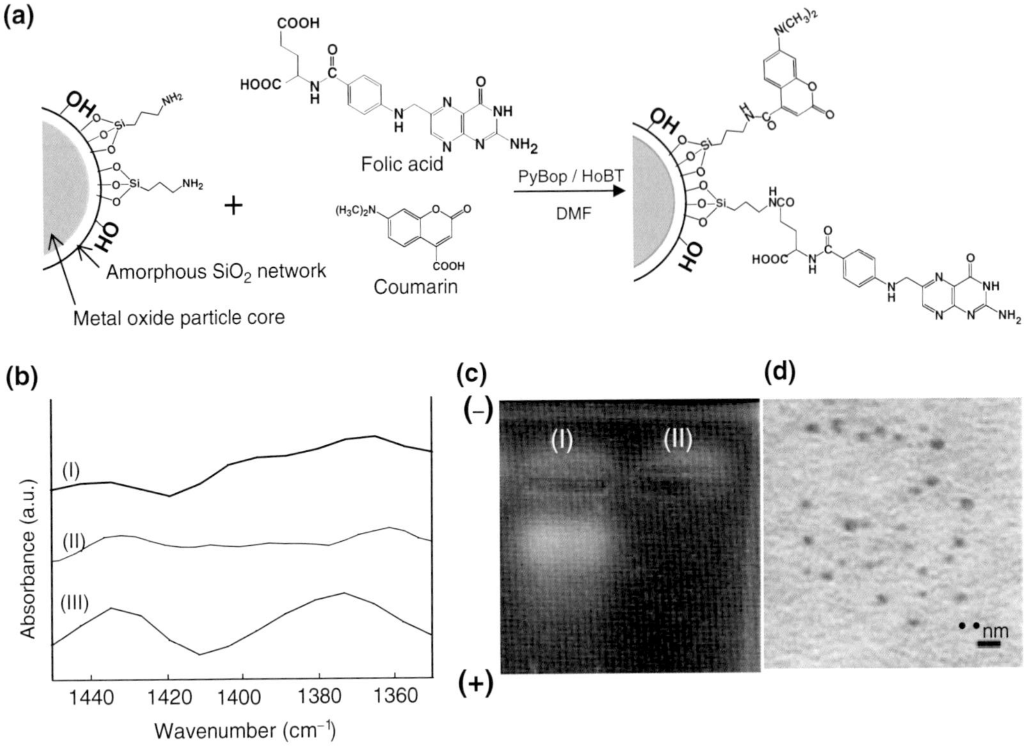

Fig. 5. (**a**) Preparation of folic acid (FA)–coumarin (CF)–conjugated MNPs. (**b**) FT–IR spectra of (I) FA–CF-conjugated MNPs, (II) amino-MNPs, and (III) FA. (**c**) Gel electrophoretic analyses of (I) FA–CF-conjugated MNPs and (II) CF-conjugated MNPs. After electric migration, the gel was exposed to UV light to detect the CF.

present inside the cells **(Fig. 6a**-I). In contrast, no fluorescence was observed from inside the cells treated with CF–MNPs **(Fig. 6a**-II), although weak fluorescence was observed from the cell surface and hollow region, respectively, owing to nonspecific adsorption of the particles. Untreated KB cells showed no fluorescence **(Fig. 6a**-III), which indicates that the fluorescence observed from the KB cells treated with the FA–CF–MNPs was not autofluorescence originating from the cells themselves. Normalized fluorescence intensities (F.I.)/μm^2 were estimated. All fluorescence intensities were normalized with respect to the fluorescence intensity of the untreated control cells **(Fig. 6b)**. The ratio of the fluorescence intensity of the FA–CF–MNPs internalized in the cells to the intensity of the cells treated with CF–MNPs to the intensity of the untreated cells was 158:2:1. In addition, no fluorescence was observed in cells treated with FA–CF–MNPs and excess free FA at the same time. As another control experiment, FA–CF–MNPs were added to rat kangaroo kidney epithelium (PtK2) cells. No fluorescence was observed from either the inside cells or the superficial cells treated with FA–CF–MNPs. These results indicate that FA–CF–MNPs were internalized in the KB cells via FR-mediated endocytosis (ligand-receptor interaction).

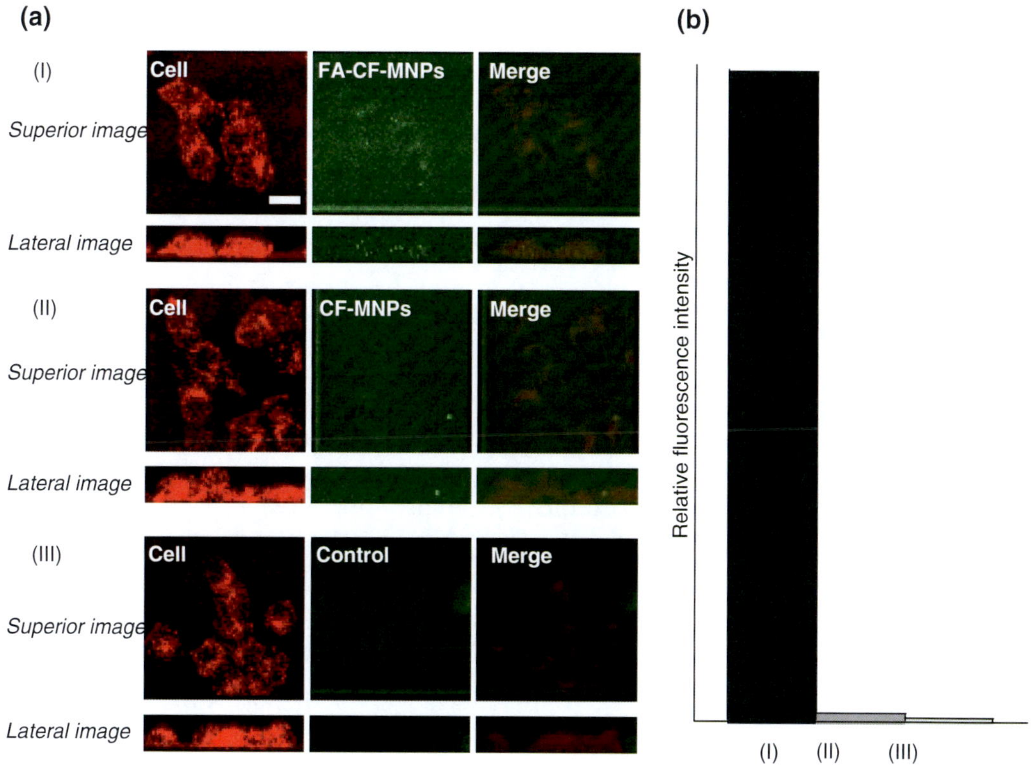

Fig. 6. (**a**) Confocal laser scanning microscopy images of (I) KB cells treated with FA–CF-conjugated MNPs, (II) KB cells treated with CF-conjugated MNPs, and (III) untreated KB cells as a control. Dil fluorescence images (cell) (*left panels*), coumarin fluorescence images (MNPs) (*middle panels*), and merged images (*right panels*). (**b**) Relative fluorescence intensities determined by dividing the sum of all pixel intensities of coumarin. (I) FA–FC-conjugated MNPs internalized in the cells, (II) cells treated with FC-conjugated MNPs, and (III) untreated cells. Scale bar = 20 μm.

In summary, we studied the feasibility of in vivo introduction of functionalized MNPs. The particles could be incorporated into subcutaneous tissue of a mouse's ear by means of a magnetic field. We demonstrated that FA–CF–MNPs could recognize a specific cell type and be internalized in tumor cells. We may be able to take advantage of these properties to deliver therapeutic agents to diseased tissue by application of a magnetic field and thus achieve selective delivery at the cellular level using our MNPs.

2. Materials

1. $FeCl_2 \cdot 4H_2O$ (Wako Pure Chemical, Osaka, Japan).
2. $Na_2SiO_3 \cdot 9H_2O$ (Junsei Chemical, Japan).
3. Tube furnace (Isuzu, Japan).

4. (3-Aminopropyl)triethoxysilane (γ-APTES; Shin-Etsu Chemical, Tokyo, Japan).
5. Ultrapure water (≥18.2 Ω, Milli-Q SP, Millipore, Tokyo, Japan).
6. Fourier transform infrared spectrophotometer (FT–IR; FT-720, Horiba, Kyoto, Japan).
7. Transmission electron microscope (TEM; JEM-1230, JEOL, Tokyo, Japan).
8. CuK_α X-ray powder diffractometer (λ = 0.154 nm, Geigerflex, Rigaku, Tokyo, Japan).
9. Superconducting quantum interference device (SQUID) magnetometer (Quantum Design, San Diego, CA, USA).
10. 5- and 6-Carboxytetramethyl rhodamine succinimidyl ester (Sigma-Aldrich, St. Louis, MO, USA).
11. FluoreSphere beads, standard, which show hydrophobic and nonelectric charge (*d* = 200 nm, Invitrogen, Carlsbad, CA, USA).
12. Rat kangaroo kidney epithelium cells (PtK2 cells).
13. Phosphate-buffered saline (PBS; 0.15 M NaCl, 5 mM phosphate from KH_2PO_4 and K_2HPO_4, pH 7.0).
14. Phosphate buffer (PB; 0.1 M phosphate from KH_2PO_4 and K_2HPO_4, pH 7.4).
15. Ultramicrotome (Reichert-Nissei Ultracut N, Nissei Sangyo Co., Tokyo, Japan).
16. Transmission electron micrograph (Tecnai G2 Sphera, FEI, Hillsboro, OR, USA).
17. Confocal laser scanning microscopy (LSM5, Carl Zeiss, Germany).
18. C57BL/6 mice, 44 weeks, female and male.
19. Permanent magnet (*d* = 7 mm, surface flux density 240 mT, Niroku Seisakusyo, Shiga, Japan).
20. Heparin–saline solution (10,000 U/L).
21. Fixation solution (4% paraformaldehyde in 0.1 M PB) (*see* **Note 1**).
22. 50 mM glycine in 0.1 M PB.
23. 10, 20, and 30% sucrose in 0.1 M PB.
24. OCT Compound (Tissue-Tek).
25. Cryostat (CM3050S, Leica).
26. Confocal laser scanning microscopy (LSM5 PASCAL, Carl Zeiss Co.).
27. 1-Hydroxy-*1H*-benzotriazole (HoBT) (Nacalai Tesque, Tokyo, Japan).

28. Benzotriazol-1-yl-oxytripyrrolidinophosphonium hexafluorophosphate (PyBop) (Nacalai Tesque, Tokyo, Japan).
29. *N*-Ethyl-*N*-(1-methylethyl)-2-propanamine (DIEA) (Nacalai Tesque, Tokyo, Japan).
30. 7-Dimethylaminocoumarin-4-acetic acid fluorophore (CF) (AnaSpec, San Jose, CA, USA).
31. Sodium borohydride ($NaBH_4$) (Sigma-Aldrich).
32. Human pharyngeal cancer cells (KB cells).
33. KB cell culture medium (174 mL MEM [Earle's salts, without L-glutamine, without phenol red], 20 mL of 10% FBS, 1% nonessential amino acids [NEAA], 1% L-glutamine, 1% penicillin/streptomycin).
34. PtK2 cell culture medium (174 mL MEM, 20 mL 10% FBS, 1% NEAA, 1% L-glutamine, 1% gentamicin, 0.1% pyruvic acid).
35. Folic acid (Wako Pure Chemical, Osaka, Japan).
36. Carbocyanine fluorescent dyes (DiI) (Takara Bio, Ootsu, Japan).

3. Methods

3.1. Preparation of MNPs

1. γ-Fe_2O_3 MNPs surrounded by amorphous SiO_2 (a-SiO_2) were prepared by mixing aqueous solutions of $FeCl_2{\cdot}4H_2O$ (10 mM, dissolved in 500 mL distilled water) and $Na_2SiO_3{\cdot}9H_2O$ (10 mM, dissolved in 500 mL of distilled water).
2. After 10 min of stirring, the resulting precipitates were centrifuged and washed three times with 1,000 mL distilled water and dried at 353 K in a water bath.
3. The dried precipitates were crushed in a porcelain mortar and then annealed in air for 7 h at 873 K in a tube furnace (*see* **Note 2**).

3.2. Functionalization of MNPs

1. For functionalization of the MNP surfaces with amino groups, 20 mL of γ-APTES was added to the solid MNPs (20 mg), and the reaction mixture was stirred at reflux temperature (403 K).
2. After 20 h at reflux, the mixture was cooled and then washed three times with ultrapure water and ethanol and dried at 353 K in an oven.

3.3. Characterization of Amino-MNPs

1. The presence of the amino groups was confirmed by FT-IR.
2. The morphology of the MNPs was investigated by TEM (accelerating voltage 100 KV).

3. The XRD patterns of the unfunctionalized MNPs and amino-MNPs were measured at ambient temperature to confirm particle structure and diameter.

4. Magnetization was measured with a SQUID magnetometer under an external field between –5 and 5 T at 300 K.

3.4. Preparation of Rhodamine-Labeled MNPs

1. To confirm the ability of the amino-MNPs to bind molecules to their surfaces and to visualize the introduction of MNPs into tissue, rhodamine (Rh)-labeled MNPs were prepared as a follows. A 0.1 M dimethyl sulfoxide solution of 5- and 6-carboxytetramethyl rhodamine succinimidyl ester (final concentration, 10 mM) was added to an aqueous solution of the amino-MNPs, and the total volume of the solution was brought to ~500 μL.

2. The labeling reaction was allowed to proceed at 310 K for 10 min and was then quenched for 30 min at room temperature by the addition of formaldehyde (final concentration, 0.1 M).

3. After reduction of the Schiff base by $NaBH_4$ (final concentration, 0.1 M), the resulting precipitates were washed several times with ultrapure water.

4. The colloidal stability and surface electric charge of the Rh–MNPs were investigated by electrophoresis of the Rh–MNPs in 1% agarose gel (100 V, 10 min). Other chemicals were used in the all experiments without further purification.

3.5. Cellular Uptake of Rhodamine-Labeled MNPs

1. The Rh–MNP suspension (final number concentration, 1.4 × 10^{12}mL^{-1}) and large control particles (d = ~200 nm, FluoreSphere beads, standard; final number concentration, 4.8 × 10^{7} mL^{-1}) with culture medium were independently added to PtK2 cells in culture medium and incubated for 24 h.

2. After the culture dishes were washed with PBS, PtK2 cells were fixed with 2.5% glutaric aldehyde in 0.1 M PB for 15 min at room temperature and then washed with PB. Osmium (0.1%) in 0.1 M PB was added to the cells at 277 K. The cells were rinsed with distilled water, dehydrated with ethanol, and embedded in epoxy resin (Epon 812).

3. Ultrathin sections were cut with an ultramicrotome, collected on copper slot grids, stained with uranyl acetate/lead citrate (5:1, w:w), and then observed by TEM.

3.6. Localization of Rhodamine-Labeled MNPs in Mouse Ear by Means of an External Magnetic Field

1. C57BL/6 mice were used in this study. All animal experiments were approved by the Ethical Committee for Experimental Animals of our institute and were performed according to international guidelines and the guidelines of the committee.

2. The Rh–MNPs were mixed with Vaseline in a 1:1 (v/v) ratio to minimize desorption from the surface of the ear. The Rh–MNP

suspension was applied on the face aspect of one ear (*see* **Note 3**), and the permanent magnet was fixed on the reverse side of the ear. A piece of aluminum foil folded multiple times was fixed similarly for use as a nonmagnetic metal control.

3. After 24 h, the magnet and the aluminum were detached.
4. The ears treated with Rh–MNPs were washed with ultrapure water and ethanol.
5. Perfusion fixation was performed on the mice (*see* **Note 4**).
6. After sucrose substitution (*see* **Note 5**), the ears were embedded in OCT Compound (*see* **Note 6**).
7. The mouse ear sections were prepared by cryostat and observed by confocal laser scanning microscopy.

3.7. Synthesis of FA–CF–MNPs

1. A vessel containing a suspension of amino-MNPs (25 mg) in dimethylformamide (DMF) was cooled in an ice bath. HoBT (23 mM), PyBop (23 mM), and DIEA (2.3 mM) were added to the particle suspension.
2. A mixture of 12.5 mg each of FA and 7-dimethylaminocoumarin-4-acetic acid fluorophore (CF) (230 μM) in dry DMF was added to the above amino-MNPs suspension dropwise over a period of 10 min, and then the reaction was allowed to proceed for 20 h in the dark.
3. After the reaction was complete, the crude suspension was washed three times with dry DMF and then quenched for 30 min at room temperature with formaldehyde (final concentration, 0.1 M) in methanol. After reduction of the Schiff base with $NaBH_4$ (final concentration, 0.1 M), the resulting precipitates were washed several times with ultrapure water.

3.8. Characterization of FA–CF–MNPs

1. The presence of the FA was confirmed by FT–IR.
2. The morphology of the MNPs was investigated by TEM.
3. The colloidal stability and surface electric charge of the FA–CF–MNPs and CF–MNPs were investigated by electrophoresis of the MNPs on 0.75% agarose gel (100 V, 10 min).

3.9. Cellular Uptake of FA–CF–MNPs

1. Human pharyngeal cancer cells (KB cells) were used as a model tumor cell line, and rat kangaroo kidney epithelium (PtK2) cells were used as a control. Approximately 2.0×10^4 KB cells or PtK2 cells were initially cultured in dishes with 2 mL of medium and were then incubated for 48 h.
2. Either FA–CF–MNPs or CF–MNPs were added to the cell culture dishes for 2 h to initiate cellular uptake of the MNPs.
3. To confirm the shape of the KB and PtK2 cells, 0.8 μg/mL DiI was added to the dishes after 1 h to stain the cell surfaces.
4. Cellular uptake of MNPs was investigated by confocal laser scanning microscopy.

5. To confirm specific introduction of FA–CF–MNPs, a competitive uptake inhibition assay was examined. FA–CF–MNPs were co-incubated with free folic acid (1.1 mM) and KB cells.

4. Notes

1. The 8% paraformaldehyde solution was prepared as follows. Ultrapure water (800 mL) was heated to 80°C in a fume hood. Paraformaldehyde (80 g) was added, and the mixture was stirred for 1 min. NaOH (4 M, 0.5–1 mL) was slowly added until the solution cleared. Ultrapure water was added to bring the final volume to 1 L. The solution was cooled and filtered through a funnel lined with filter paper. Equal amounts of the stock 8% paraformaldehyde solution and 0.2 M PB were combined.
2. The annealing sequence was as follows. The furnace temperature was raised from room temperature to 873 K over a period of 6 h and maintained at 873 K for 7 h; then the furnace was switched off and allowed to cool.
3. The MNPs (suspended in Vaseline) were spread on the mouse's ears with a brush.
4. Perfusion fixation was performed as follows.
   (a) The heparin–saline solution was maintained at 37?C with a water bath, and the fixation solution was cooled over ice.
   (b) The perfusion pump was set up, and a perfusion needle (catheter) was attached. The heparin-saline solution was run through the tube of the pump, the flux was adjusted (7 mL/min), and then the pump was stopped.
   (c) A narcotized mouse was placed on a rack, and an incision the length of the diaphragm was made with sharp scissors through the abdomen. The connective tissue at the bottom of diaphragm was cut to allow access to the rib cage.
   (d) With large scissors, blunt side down, the ribs were cut just to the left of the rib cage midline.
   (e) While the heart was held steady (it should still have been beating), taking care not to extend the needle too far, the catheter was inserted directly into the protrusion of the left ventricle so that the catheter extended straight up ~2 mm. The needle was secured in position by clamping it in place. The pump was started and

adjusted to achieve a slow, steady flow (~7 mL/min) of heparin-saline solution.

(f) A cut was made in the right auricle with sharp scissors, making sure that the solution was flowing freely. If the solution was not flowing freely or was coming from mouse's nostrils or mouth, the needle was repositioned.

(g) When blood had been cleared from the body (heparin-saline solution ~100 mL per mouse), perfusion with the fixation solution (~100 mL) was begun; care was taken not to introduce air. Spontaneous movement ("formalin dance") and lightened color of the liver were good indicators of clearing.

(h) The mouse's ear was cut and submersed in fixation solution for 2–3 h on ice.

5. Sucrose substitution was performed as follows (all processes were performed in a 4°C room).

   (a) The mouse's ear was placed in 50 mM glycine in 0.1 M PB and shaken for 2 h.

   (b) The ear was placed in 10% sucrose in 0.1 M PB overnight.

   (c) The ear was placed successively in 20% and 30% sucrose in 0.1 M PB overnight.

6. The mouse ears were embedded in OCT Compound.

   (a) A cubic box was formed from aluminum foil, and a little OCT Compound was poured into the box.

   (b) The ear was removed from the 30% sucrose in 0.1 M PB and wiped.

   (c) The ear was placed in the box, and enough OCT Compound to cover the ear was added.

   (d) A boat was made from aluminum foil and placed on liquid nitrogen; the box containing the ear was placed on the boat to rapidly freeze the ear.

References

1. Du, Y.-Z., Hiratsuka, Y., Taira, S., Eguchi, M., Uyeda, T.Q.P., Yumoto, N., and Kodaka, M. (2005). Motor protein nano-biomachine powered by self-supplying ATP. *Chem. Commun.* **16**, 2080–2082
2. Taira, S., Du, Y.-Z., Hiratsuka, Y., Uyeda, T.Q.P., Yumoto, N., and Kodaka, M. (2007). Loading and unloading of molecular cargo by DNA-conjugated microtubule. *Biotechnol. Bioeng.* **99**, 734–739
3. Hiratsuka, Y., Tada, T., Oiwa, K., Kanayama, T., and Uyeda, Q.P.T. (2001). Controlling the direction of kinesin-driven microtubule movements along microlithographic tracks. *Biophys. J.* **81**, 1555–1561
4. Eelkema, R., Pollard, M.M., Vicario, J., Katsonis, N., Ramon, B.S., Bastiaansen, C.W.M., Broer, D.J., and Feringa, B.L. (2006). A molecular motor in a liquid-crystal film uses light to turn items thousands of times larger than itself. *Nature* **440**, 163–163
5. Taira, S., Sugiura, Y., Moritake, S., Shimma, S., Ichiyanagi, Y., and Setou, M. (2007). Nanoparticle-assisted laser desorption/ionization

for ultra resolution imaging mass spectrometry. Proceedings of 55th ASMS conference on Mass Spectrometry and Allied Topics

6. Eghtedari, M., Oraevsky, A., Copland, J.A., Kotov, N.A., Conjusteau, A., and Motamedi, M. (2007). High sensitivity of in vivo detection of gold nanorods using a laser optoacoustic imaging system. *Nano Lett.* 7, 1914–1918
7. Wang, X. and Liu, C. (2005) Multifuntional probe array for nano patterning and imaging. *Nano Lett.* **5**, 1867–1872
8. Taira, S., Du, Y.Z., and Kodaka, M. (2006). Trap and release of oligonucleotide using pH-responsive amphoteric particle prepared by interfacial polymerization in W/O miniemulsion system. *Biotechnol. Bioeng.* **93**, 396–400
9. Taira, S., Du, Y.-Z., Hiratsuka, Y., Konishi, K., Kubo, T., Uyeda, T.Q.P., Yumoto, N., and Kodaka, M. (2006). Selective detection and transport of fully matched DNA by DNA-loaded microtubule and kinesin motor protein. **95**, 533–538
10. Du, Y.Z., Tomohiro, T., Zang, G., Nakamura, K., and Kodaka, M. (2004). Biotinylatedand enzyme-immobilized carrier prepared by hetero-bifunctional latex beads. *Chem. Commun.* **5**, 616–617
11. Huh, Y.M., Jun, Y.W., Song, H.T., Kim, S., Choi, J.S., Lee, J.H., Yoon, S., Kim, K.S., Shin, J.S., Suh, J.S., and Cheon, J. (2005). In vivo magnetic resonance detection of cancer by using multifunctional magnetic nanocrystals. *J. Am. Chem. Soc.* **127**, 12387–12391
12. Moritake, S., Taira, S., Ichiyanagi, Y., Morone, N., Song, S.Y., Hatanaka, T., Yuasa, S., and Setou, M. (2007). Functionalized nano magnetic particles for an in vivo delivery system. *J. Nanosci. Nanotechnol.* 7, 937–944
13. Bae, Y., Jang, W.D., Nishiyama, N., Fukushima, S., and Kataoka, K. (2005). Multifunctional polymeric micelles with folate-mediated cancer cell targeting and pH-triggered drug releasing properties for active intracellular drug delivery. *Mol. Biosyst.* **1**, 242–250
14. Taira, S., Hatanaka, T., Moritake, S., Kai, Y., Ichiyanagi, Y., and Setou, M. (2007). Cellular recognition of functionalized with folic acid nanoparticles. *e-J. Surf. Sci. Nanotechol.* **5**, 23–28
15. Sun, X., Duan, Y.R., He, Q., Lu, J., and Zhang, Z.R. (2005). PELGE nanoparticles as new carriers for the delivery of plasmid DNA. *Chem. Pharm. Bull.* **53**, 599–603
16. Kakizawa, Y., Furukawa, S., and Kataoka, K. (2004). Block copolymer-coated calcium phosphate. nanoparticles sensing intracellular environment for oligodeoxynucleotide and siRNA delivery. *J. Control. Release* **97**, 345–356
17. Grenha, A., Seijo, B., and Remunan-Lopez, C. (2005). Microencapsulated chitosan nanoparticles for lung protein delivery. *Eur. J. Pharm. Sci.* **25**, 427–437
18. RaviKumar, M.N.V., Mohapatra, S.S., Kong, X., Jena, P.K., Bakowsky, U., and Lehrd, C.M. (2004). Cationic poly(lactide-co-glycolide) nanoparticles as efficient in vivo gene transfection agents. *J. Nanosci. Nanotechnol.* **4**, 990–994
19. Murakami, T., Ajima, K., Miyawaki, J., Yudasaka, M., Iijima, S., and Shiba, K. (2004) Drug-loaded carbon nanohorns: adsorption and release of dexamethasone in vitro. *Mol. Pharm.* 1, **399**–405
20. Hasegawa, M., Ohno, H., Tanaka, H., Hatakeyama, M., Kawaguchi, H., Takahashi, T., and Handa, H. (2005). Affinity identification of δ opioid receptors using latex nanoparticles. *Bioorg. Med. Chem. Lett.* **16**, 158–161
21. Ohtsu, Y., Ohba, R., Imamura, Y., Kobayashi, M., Hatori, H., Zenkoh, T., Hatakeyama, M., Manabe, T., Hino, M., Yamaguchi, Y., Kataoka, K., Kawaguchi, H., Watanabe, H., and Handa, H. (2005). Selective ligand purification using high-performance affinity beads. *Anal. Biochem.* **338**, 245–252
22. Komiyama, M., Kobayashi, J., and Morita, S. (1990). Structure of platinum ultrafine particles in Pt/C catalyst observed by scanning tunneling microscopy. *J. Vac. Sci. Technol. A* **8**, 608–609
23. Song, H.T., Choi, J.S., Huh, Y.M., Kim, S., Jun, Y.W., Suh, J.S., and Cheon, J. (2005). Surface modulation of magnetic nanocrystals in the development of highly efficient magnetic resonance probes for intracellular labeling. *J. Am. Chem. Soc.* **127**, 9992–9993
24. Lewin, M., Carlesso, N., Tung, C.H., Tang, X.W., Cory, D., Scadden, D.T., and Weissleder, R. (2000). Tat peptide-derivatized magnetic nanoparticles allow in vivo tracking and recovery of progenitor cells. *Nat. Biotechnol.* **18**, 410–414
25. Hatanaka, T., Haramura, M., Fei, Y.J., Miyauchi, S., Bridges, C.C., Ganapathy, P.S., Smith, S.B., Ganapathy, V., and Ganapathy, M.E. (2004). Transport of amino acid-based prodrugs by the Na^{+} and Cl^{-}coupled amino acid transporter $ATB^{0,+}$ and expression of the transporter in tissues amenable for drug delivery. *J. Pharmacol. Exp. Ther.* **308**, 1138–1147
26. Antony, A.C. (1992). The biological chemistry of folate receptors. *Blood.* **79**, 2807–2820
27. Stella, B., Arpicco, S., Peracchia, M.T., Desmaele, D., Hoebeke, J., Renoir, M., D'angelo, J., Cattel, L., and Couvreur, P. (2000). Design

of folic acid-conjugated nanoparticles for drug targeting *J. Pharm. Sci.* 89, 1452–1464
28. Lee, R.J. and Low, P.S. (1994). Delivery of liposomes into cultured KB cells via folate receptor- mediated endocytosis. *J. Biol. Chem.* **269**, 3198–3204
29. Ichiyanagi, Y., Uozumi, T., and Kimishima, Y. (2001). Magnetic properties of Fe_2O_3 nanoparticles. *Trans. Mater. Res. Soc. Jpn.* **26**, 1097–1100
30. Ichiyanagi, Y. and Kimishima, Y. (2002). Structural, magnetic and thermal characterizations of Fe_2O_3 nanoparticle systems. *J. Therm. Anal. Calorim.* **69**, 919–923
31. Ichiyanagi, Y., Wakabayashi, N., Yamazaki, J., Yamada, S., Kimishima, Y., Komatsu, E., and Tajima, H. (2003). Magnetic properties of NiO nanoparticles. *Physica.* B **329**, 862–863
32. Ichiyanagi, Y., Kondoh, H., Yokoyama, T., Okamoto, K., Nagai, K., and Ohta, T. (2003). X-ray absorption fine structure study on the $Ni(OH)_2$ monolayer nanostructures. *Chem. Phys. Lett.* **379**, 345–350
33. Ichiyanagi, Y., Kimishima, Y., and Yamada, S. (2004). Magnetic study on Co_3O_4 nanoparticles. *J. Magn. Magn. Mater.* **272**, 1245–1246
34. Ichiyanagi, Y., Uehashi, T., and Yamada, S. (2004) Magnetic properties of Ni–Zn ferrite nanoparticles. *Phys. Stat. Sol.* **12**, 3485–3488
35. Weitman, S.D., Lark, R.H., Coney, L.R., Fort, D.W., Frasca, V., Zurawski V.R., Jr., and Kamen, B.A. (1992). Distribution of the folate receptor GP38 in normal and malignant cell lines and tissues. *Cancer Res.* **52**, 3396–3401
36. Ross, J.F., Chaudhuri P.K., and Ratnam, M. (1994). Differential regulation of folate receptor isoforms in normal and malignant tissue in vivo and in established cell lines. *Cancer* **73**, 2432–2443
37. Reddy, J.A. and Low, P.S. (1998). Folate-mediated targeting of therapeutic and imaging agents to cancers. *Crit. Rev. Ther. Drug Carrier Syst.* **15**, 587–627
38. Ichiyanagi, Y., and Kimishima, Y. (1996). Magnetic properties of Lanthanoid-VO3. *Jpn. J. Appl. Phys.* **35**, 2140–2144
39. Sun, C., Sze, R., and Zhang, M. (2006). Folic acid-PEG conjugated superparamagnetic nanoparticles for targeted cellular uptake and detection by MRI. *J. Biomed. Mater. Res A.* **78**, 550–557

Chapter 37

Formulation/Preparation of Functionalized Nanoparticles for In Vivo Targeted Drug Delivery

Frank Gu, Robert Langer, and Omid C. Farokhzad

Summary

Targeted cancer therapy allows the delivery of therapeutic agents to cancer cells without incurring undesirable side effects on the neighboring healthy tissues. Over the past decade, there has been an increasing interest in the development of advanced cancer therapeutics using targeted nanoparticles. Here we describe the preparation of drug-encapsulated nanoparticles formulated with biocompatible and biodegradable poly(D,L-lactic-co-glycolic acid)-*block*-poly(ethylene glycol) (PLGA-*b*-PEG) copolymer and surface functionalized with the A10 2-fluoropyrimidine ribonucleic acid aptamers that recognize the extracellular domain of prostate-specific membrane antigen (PSMA), a well-characterized antigen expressed on the surface of prostate cancer cells. We show that the self-assembled nanoparticles can selectively bind to PSMA-targeted prostate cancer cells in vitro and in vivo. This formulation method may contribute to the development of highly selective and effective cancer therapeutic and diagnostic devices.

Key words: Aptamers, Nanoparticles, Chemotherapy, Targeted drug delivery, Bioconjugated chemistry

1. Introduction

Nanomaterials have unique physicochemical properties, such as large surface area-to-mass ratios and high surface reactivity, which are different from bulk materials of the same composition. These unique physical properties allow the materials to interact with the human body on the molecular scale with a high degree of specificity. The application of nanotechnology in medicine, also known as nanomedicine, involves the use of precisely engineered nanomaterials for medical diagnosis and therapeutic treatments *(1)*. One of the most exciting research topics in nanomedicine is targeted drug delivery. By combining molecular targeting capabilities and controlled drug

James Weifu Lee and Robert S. Foote (eds.), *Micro and Nano Technologies in Bioanalysis*,
Methods in Molecular Biology, vol. 544
DOI 10.1007/978-1-59745-483-4_37, © Humana Press, a part of Springer Science+Business Media, LLC 2009

release properties, targeted drug delivery offers the possibility of achieving precision-guided drug delivery to individual diseased cells with minimal side effects on neighboring healthy cells *(2, 3)*.

In this chapter, we describe the method for preparing prostate cancer-targeted nanoparticles (NPs) *(4–7)*. We used an A10 2′-fluoropyrimidine ribonucleic acid aptamer (Apt) *(8)*, which binds to prostate-specific membrane antigen (PSMA) on the surface of prostate cancer (PCa) cells, as a model hydrophilic targeting molecule; the poly(D,L-lactide-co-glycolide) (PLGA) as a model controlled release polymer system; and polyethylene glycol (PEG) as a model hydrophilic polymer with antibiofouling properties, to develop a proof-of-concept NP-Apt that is potentially suitable for selectively targeting PSMA PCa cells in vitro and in vivo.

2. Materials

2.1. Polymer Conjugation

1. All chemical reagents used in this study were cell culture grade (purity >95%) and were purchased from Sigma-Aldrich, St. Louis, MO, USA, unless otherwise noted.
2. PSMA A10 2′-fluoropyrimidine RNA Apt (sequence: 5′-NH_2-spacer GGG/AGG/ACG/AUG/CGG/AUC/AGC/CAU/GUU/UAC/GUC/ACU/CCU/UGU/CAA/UCC/UCA/UCG/GCiT-3′ with 2′-fluoro pyrimidines, a 5′-amino group attached by a hexaethyleneglycol spacer and a 3′-inverted T cap) was custom synthesized by RNA-TEC (Leuven, Belgium). The aptamers were stored as lyophilized powder at −80°C.
3. Heterobifunctional PEG (amine-PEG-carboxylate) (MW = 34,00 g/mol) (Nektar Therapeutics, San Carlos, CA, USA) was stored in the dark at −20°C.
4. Poly(D,L-lactide-co-glycolide) (PLGA) (Lactel Absorbable Polymers, Pelham, AL, USA) with terminal carboxylate groups (PLGA–carboxylate) was stored at −20°C.
5. Conjugation cross-linkers: 1-ethyl-3-(3-dimethylaminopropyl)-carbodiimide (EDC) was stored in the dark at −20°C, and *N*-hydroxysuccinimide (NHS) was stored at 4°C.
6. *N,N*-diisopropylethylamine (DIEA) was stored in the dark at room temperature.
7. Solvents: Dichloromethane (DCM), ethyl ether, acetonitrile, and methanol were molecular biology grade (>99% in purity or higher).
8. Amicon ultracentrifugation tubes with molecular weight cut-off of 100,000 Da (Millipore, Billerica, MA, USA).
9. Washing solution: Anhydrous ethyl ether and methanol (50/50, v/v).

2.2. Cell Culture

1. Prostate cancer cell lines LNCaP and PC3 were both purchased from ATCC (Manassas, VA, USA).
2. LNCaP cells were cultured in RPMI-1640 (ATCC) supplemented with 100 U/mL aqueous penicillin G, 100 g/mL streptomycin, and 10% fetal bovine serum (FBS).
3. PC3 cells were cultured in F-12K (ATCC) supplemented with 100 U/mL aqueous penicillin G, 100 g/mL streptomycin, and 10% FBS.
4. Phenol-red-reduced OptiMEM media (Invitrogen, Carlsbad, CA, USA).

2.3. Immunohistochemistry for Tracking Nanoparticle Endocytosis

1. PLGA-*b*-PEG (50 mg/mL in acetonitrile [ACN]).
2. Fixative: 4% formaldehyde in phosphate-buffered saline (PBS) (freshly prepared).
3. Blocking solution: 1% bovine serum albumin (BSA) in PBS.
4. Blocking and permeabilization solution: 0.1% Triton-X100 in blocking solution.
5. Alexa phalloidin (5 U/mL) (Invitrogen).
6. 4′,6-diamidino-2-phenylindole (DAPI): 0.1 mg/mL.
7. Vectashield mounting media kit (Vector Laboratories, Burlingame, CA, USA).

2.4. Tumor Preparation In Vivo

1. LNCaP cells were cultured in T-175 flasks (BD Biosciences, Franklin Lakes, NJ, USA).
2. Matrigel (BD Biosciences) was stored at –20°C.
3. 8-week-old balb/c nude mice (Charles River Laboratories, Wilmington, MA, USA) (*see* **Note 1**).

3. Methods

3.1. Polymer Conjugation Chemistry

1. 5 g of PLGA–carboxylate (0.28 mmol) was dissolved in 10–20 mL DCM (*see* **Note 2**).
2. NHS (135 mg, 1.1 mmol) and EDC (230 mg, 1.2 mmol) were dissolved in 2 mL DCM (*see* **Note 3**).
3. PLGA–carboxylate was converted to PLGA–NHS by adding the EDC/NHS solution prepared in **step 2** to the PLGA–carboxylate solution.
4. PLGA–NHS was precipitated with 10 mL ethyl ether/methanol washing solvent to remove residual NHS and EDC.
5. The precipitated PLGA–NHS was collected by centrifugation at 4,000 × *g* for 20 min.

6. Washing and centrifugation (**step 4** and **5**) were repeated two times.
7. The PLGA–NHS pellet was dried under vacuum for 30 min to remove the residual ether and methanol.
8. After drying under vacuum, PLGA–NHS (1 g, 0.059 mmol) was dissolved in DCM (4 mL) followed by addition of amine–PEG–carboxylate (250 mg, 0.074 mmol) and DIEA (28 mg, 0.22 mmol).
9. The resulting PLGA-*b*-PEG block copolymer was precipitated with ether/methanol washing solvent and washed with the same solvent to remove unreacted PEG.
10. The resulting purified PLGA-*b*-PEG block copolymer was dried under vacuum and used for NP preparation without further treatment (*see* **Note 4**).
11. The composition of PLGA-*b*-PEG was characterized using a 400 MHz ^{1}H nuclear magnetic resonance (Bruker, Billerica, MA, USA). The nuclear magnetic resonance (NMR) characterization sample was prepared by dissolving 5 mg of the PLGA-*b*-PEG diblock copolymer in 1 mL of deuterated chloroform ($CDCl_3$). An example of a PLGA-*b*-PEG NMR spectrum is shown in **Fig. 1**.

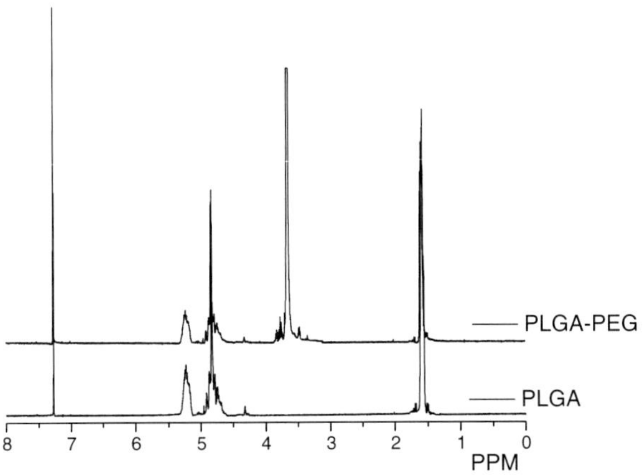

Fig. 1. PLGA-*b*-PEG characterization using nuclear magnetic resonance (NMR). The presence of PLGA and PEG were visualized in using a 400 MHz ^{1}H NMR ppm 5.2 (m, ((OCH(CH3)C(O)OCH2C(O))n–(CH2CH2O)m), 4.8 (m, ((OCH(CH3)C(O)OCH2C(O))n–(CH2CH2O)m), 3.7 (s, ((OCH(CH3)C(O)OCH2C(O))n–(CH2CH2O)m), 1.6 (d, ((OCH(CH3)C(O)OCH2C(O))n–(CH2CH2O)m) (reproduced from **ref.** *(7)* with permission from PNAS, Copyright (2008) National Academy of Sciences, U.S.A.).

3.2. Nanoparticle Preparation Methods

3.2.1. Nanoprecipitation for Encapsulating of Hydrophobic Compounds

1. PLGA-*b*-PEG (10 mg/mL) and docetaxel (0.1 mg/mL) were dissolved in acetonitrile.
2. The PLGA-*b*-PEG and docetaxel mixture was added drop wise to three to five volumes of stirring water (*see* **Note 5**), giving a final polymer concentration of 3.3 mg/mL.
3. The NPs were stirred for 2 h, and the remaining organic solvent was removed in a rotary evaporator at reduced pressure.
4. The NPs were concentrated using Amicon ultracentrifugation at 4,000 × *g* for 15 min and washed with deionized water and reconstituted in PBS.
5. The particle size and size distribution can be measured by dynamic light scattering (Brookhaven Instruments Corporation 90 plus particle sizer, 676-nm laser) at 25°C and at a scattering angle of 90° at a concentration of approximately 1 mg NP/mL water (*see* **Note 6**).

3.2.2. Double Emulsion (w/o/w) for Encapsulating Hydrophilic Compounds

1. An aqueous solution of rhodamine-labeled dextran (2.5 mg/mL, 0.4 mL) was emulsified in 2 mL PLGA-*b*-PEG dissolved in DCM (50 mg/mL) using a probe sonicator (Fisher Scientific, Pittsburgh, PA, USA) at 20 W for 45 s.
2. The emulsion was then transferred to an aqueous solution of PVA (0.1%, w/v, 50 mL), and sonicated at 20 W for 1 min.
3. The w/o/w emulsion formed was gently stirred at room temperature for 2 h or until the evaporation of the organic phase was complete.
4. The nanoparticles were then recovered using Amicon ultracentrifugation as described in **step 4** of **Subheading 3.2.1** (*see* **Note 6**).
5. The particle size and size distribution can be measured as described in **step 5** of **Subheading 3.2.1**.

3.2.3. NP–Apt Conjugation

1. PLGA-*b*-PEG NPs (10 mg/mL) were suspended in DNase- and RNase-free water, and were mixed with EDC (400 mM) and NHS (200 mM) for 20 min.
2. NPs were then washed three times in DNase- and RNase-free water using Amicon ultracentrifugation tubes.
3. The resulting NHS-activated NPs were reacted with PSMA A10-Apt (1 mg/mL) for 2 h.
4. The NP–Apt bioconjugates were washed 3 times as described in **Subheading 3.2.3**, **step 2**.
5. NP–Apt bioconjugates were denatured at 90°C and allowed to assume binding conformation during snap-cooling on ice.
6. The NP suspensions were kept at 4°C until use.

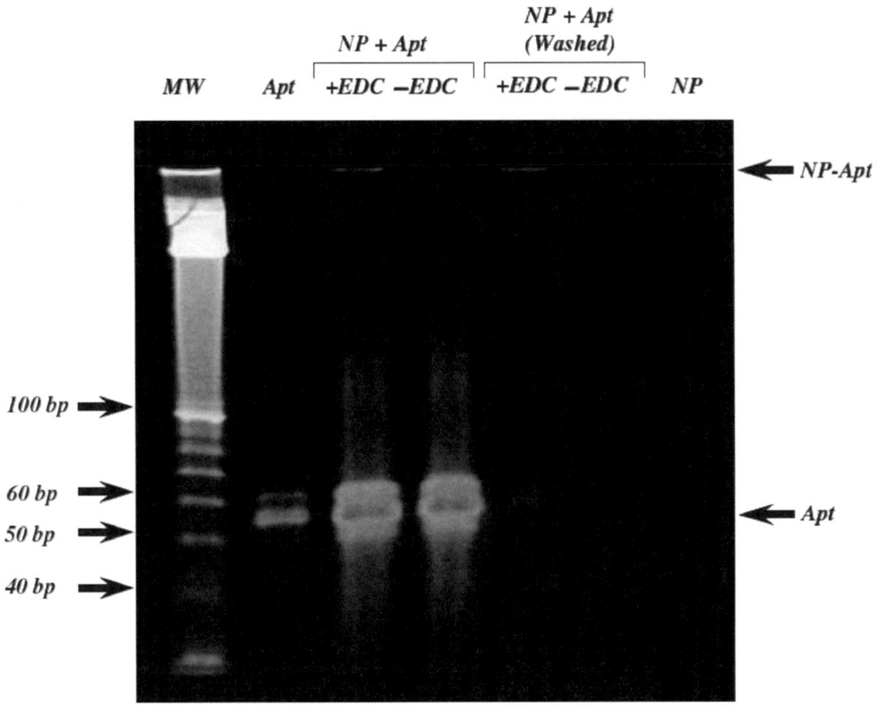

Fig. 2. Confirmation of NP–Apt conjugation. The A10 PSMA aptamer (Apt) was incubated with PLGA–*b*–PEG NP in the absence (–) or presence (+) of EDC and the reactions were resolved on a 10% TBE–urea PAGE directly, or after washing to remove any unconjugated Apt. The bands corresponding to the A10 PSMA Apt and NP–Apt are indicated by *arrows*. The molecular weight (MW) DNA marker and free aptamer served as standards for a 57-base pair band on the gel and are shown on the *left* (reproduced from **ref.** *(4)* with permission from Elsevier).

7. NP–Apt conjugation was confirmed using 10% TBE–urea polyacrylamide gel electrophoresis (PAGE). Samples containing nanoparticles surface functionalized with aptamers (NP+Apt), native aptamers (Apt), and nanoparticles without surface modification (NP) were loaded in PAGE. A sample gel image is shown in **Fig. 2**.

3.3. Targeted Nanoparticle Uptake In Vitro

1. PCa LNCaP andPC3 cells were grown in 8-well chamber slides in RPMI 1640 and Ham's F12K medium, respectively, supplemented with 100 U/mL aqueous penicillin G, 100 μg/mL streptomycin, and 10% fetal bovine serum at concentrations to allow 70% confluence (i.e., LNCaP: 40,000 cells/cm^2).
2. LNCaP and PC3 cells were washed with prewarmed PBS and incubated with phenol red-reduced OptiMEM media for 30 min at 37°C.
3. Cells were incubated with 50 μg of NP–Apt prepared as described in **Subheading 3.2.2** and **Subheading 3.2.3** for 15 min to 1 h at 37°C.
4. Cells were washed with prewarmed PBS three times.
5. Cells were fixed with 4% paraformaldehyde for 20 min, followed by washing with PBS.

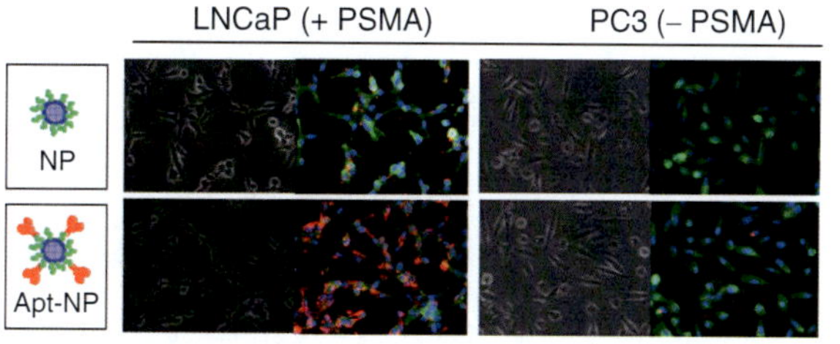

Fig. 3. Binding of NP–Apt bioconjugates to prostate epithelial cells. LNCaP cells and PC3 cells were grown on chamber slides and incubated in culture medium with rhodamine-labeled dextran-encapsulated pegylated NP (*red*), or rhodamine-labeled dextran-encapsulated pegylated NP–Apt bioconjugates (*red*). Cells were washed in PBS three times, fixed, and permeabilized, stained with 4′,6-diamidino-2-phenylindole (nuclei, *blue*) and Alexa-Fluor phalloidin (cytoskeleton, *green*), washed, and analyzed by light transmission or fluorescent microscopy (reproduced from **ref.** *(6)* with permission from American Association for Cancer Research).

6. Cells were counterstained with 4′,6-diamidino-2-phenylindole (DAPI) and Alexa-Fluor phalloidin.
7. The cell culture chamber slides were then mounted and visualized by fluorescent microscopy (*see* **Note 7**).
8. Where indicated, the number of nanoparticle aptamer bioconjugates or control nanoparticles attached to individual LNCaP or PC3 cells was quantified by fluorescent microscopy under oil immersion at 100× magnification (*see* **Note 8**). A sample figure of targeted NP uptake by LNCaP and PC cells is shown in **Fig. 3**.

3.4. Efficacy of Tumor Reduction In Vivo

1. The NPs were traced by encapsulating docetaxel using the nanoprecipitation method explained in **steps 1–5** of **Subheading 3.2.1**.
2. The NP formulations were suspended in 200 μL PBS before administration.
3. LNCaP cells were cultured in RPMI-1640 medium supplemented with 10% fetal bovine serum, 100 U/mL penicillin G, and 100 μg/mL streptomycin.
4. Mice were anesthetized by intraperitoneal injection of Avertin (200 mg/kg body weight), and dosed with 3 million LNCaP cells suspended in 600 μL of 1:1 (v/v) media and Matrigel (*see* **Note 9**).
5. LNCaP tumors were induced in 8-week-old balb/c nude mice (Charles River Laboratories).
6. Mice were injected subcutaneously in the right flank with 3×10^6 LNCaP cells suspended in a 1:1 mixture of media and Matrigel (BD Biosciences).
7. Tumor-targeting studies were carried out after the mice developed ~100 mg tumors (*see* **Note 10**).
8. Mice were randomly divided into different groups, minimizing tumor size variations between groups.

9. Mice were anesthetized by intraperitoneal injection of Avertin (200 mg/kg body weight), and dosed with NP formulations via intratumoral injection.
10. After dosing, the mice were monitored for weight and implanted tumor size change daily for 2 weeks and every 3 days thereafter.
11. If body weight loss (BWL) persisted beyond 20% of predosing weight, the animals were euthanized.
12. The length and width of the tumors were measured by digital calipers, calculating tumor volume by the following formula: $(\text{width}^2 \times \text{length})/2$.
13. Mice were monitored for a maximum of 109 days, until the tumor was completely regressed or until the tumor volume exceeded 800 mm^3, for which the mice were euthanized for excessive tumor load.
14. For animals that were euthanized because of tumor load or BWL, the tumor size at the time of euthanasia was used for the purpose of mean tumor size calculation. Initial volume of the tumors averaged 328 mm^3. Tumor efficacy study results are shown in **Fig. 4**.

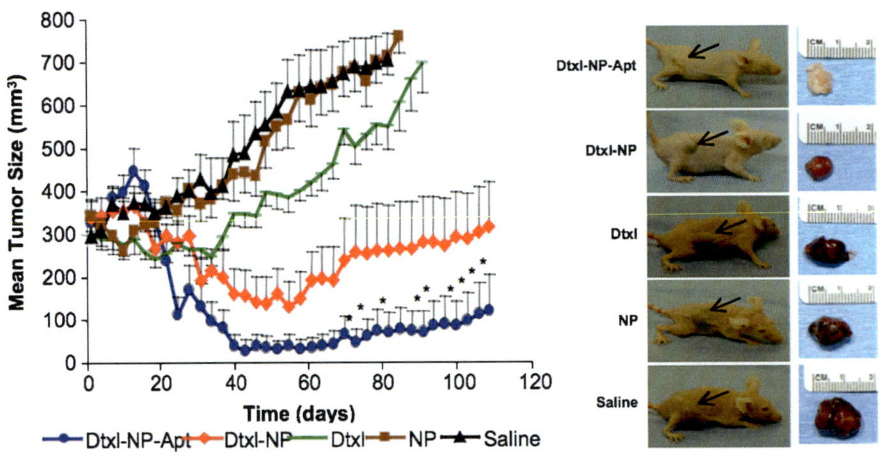

Fig. 4. Comparative efficacy study in LNCaP subcutaneous xenograft nude mouse model of PCa. (**a**) The comparative efficacy study of single intratumoral injection (day 0) of (i) saline (*black*); (ii) pegylated PLGA NP without drug (NP, *brown*); (iii) emulsified Dtxl (Dtxl, *green*), 40 mg/kg; (iv) Dtxl–encapsulated NPs (Dtxl–NP, *red*), 40 mg/kg; or (v) Dtxl–encapsulated NP–Apt bioconjugates (Dtxl–NP–Apt, *blue*), 40 mg/kg was evaluated over 109 days and demonstrated that targeted NPs are significantly more efficacious in tumor reduction as compared with other groups. Data represent mean ± SEM of seven mice per group. Data points labeled with "*" for the Dtxl–NP–Apt group were statistically significant compared with all other groups by analysis of variance (ANOVA) at a 95% confidence interval. (**b**) A representative mouse at the end point for each group is shown (*left*) alongside images of excised tumors (*right*). For the Dtxl–NP–Apt group, which achieved complete tumor regression, the scar tissue and underlying skin at the site of injection are shown. *Black arrows* point to the position of the implanted tumor on each mouse (reproduced from **ref.** *(5)* with permission from PNAS, Copyright (2008) National Academy of Sciences, U.S.A.).

4. Notes

1. All animal studies were carried out under the supervision of the Massachusetts Institute of Technology's Division of Comparative Medicine and in compliance with the National Institutes of Health's Principles of Laboratory Animal Care.
2. The PLGA viscosity can influence the rate of PLGA-*b*-PEG conjugation. For high-viscosity PLGA, dilute PLGA in DCM to 0.1–0.25 g/mL before adding EDC/NHS.
3. For maximum conjugation efficiency, dissolve EDC/NHS in DCM immediately before adding to PLGA–carboxylate.
4. To achieve more efficient polymer conjugation, use a high-power vacuum pump to rapidly evaporate residual solvents in the polymer formulation.
5. To avoid nanoparticle aggregation, the acetonitrile:water volume should be greater than 2:1.
6. To maintain NP colloidal stability, always formulate NPs in pure water, then reconstitute NPs in PBS or other desired media.
7. To preserve the imaging quality, we recommend imaging the mounted slides right away, or at most within 5 days after mounting, in which case slides should be stored in the dark at −20°C until imaging.
8. The fluorescent dyes encapsulated in NPs are released in a time-dependent manner. Always prepare a fresh batch of NPs for the in vitro release study to maximize the amount of dyes encapsulated in the NPs.
9. To obtain fast growing tumors, always reconstitute LNCaP cells in full growth media containing serum before mixing with Matrigel.
10. For maximum LNCaP growth, ensure all media are phenol free.

Acknowledgment

This work was supported by National Institutes of Health grants CA119349 and EB003647, and by a Koch-Prostate Cancer Foundation Award in Nanotherapeutics. FG was supported by a Postdoctoral Fellowship from the Canadian Natural Sciences and Engineering Research Council.

References

1. Gu, F., Karnik, R., Wang, A., Alexis, F., Levy-Nissenbaum, E., Hong, S., Langer, R., & Farokhzad, O. C. (2007). Targeted nanoparticles for cancer therapy. *Nano Today* 2, 14–21
2. Langer, R. (1998). Drug delivery and targeting. *Nature* **392**, 5–10
3. Langer, R. & Tirrell, D. A. (2004). Designing materials for biology and medicine. *Nature* **428**, 487–492
4. Cheng, J., Teply, B. A., Sherifi, I., Sung, J., Luther, G., Gu, F. X., Levy-Nissenbaum, E., Radovic-Moreno, A. F., Langer, R., & Farokhzad, O. C. (2007). Formulation of functionalized PLGA-PEG nanoparticles for in vivo targeted drug delivery. *Biomaterials* **28**, 869–876
5. Farokhzad, O. C., Cheng, J., Teply, B. A., Sherifi, I., Jon, S., Kantoff, P. W., Richie, J. P., & Langer, R. (2006). Targeted nanoparticle-aptamer bioconjugates for cancer chemotherapy in vivo. *Proc. Natl. Acad. Sci. U.S.A.* **103**, 6315–6320
6. Farokhzad, O. C., Jon, S., Khademhosseini, A., Tran, T. N., Lavan, D. A., & Langer, R. (2004). Nanoparticle-aptamer bioconjugates: a new approach for targeting prostate cancer cells. *Cancer Res.* **64**, 7668–7672
7. Gu, F., Zhang, L., Teply, B., Mann, N., Wang, A., Radovic-Moreno, A. F., Langer, R., & Farokhzad, O. (2008). Precise engineering of targeted nanoparticles by using self-assembled biointegrated block copolymers. *Proc Natl Acad Sci U. S. A.* **105**, 2586–2591
8. Lupold, S. E., Hicke, B. J., Lin, Y., & Coffey, D. S. (2002). Identification and characterization of nuclease-stabilized RNA molecules that bind human prostate cancer cells via the prostate-specific membrane antigen. *Cancer Res.* **62**, 4029–4033

Chapter 38

Detection of mRNA in Single Living Cells Using AFM Nanoprobes

Hironori Uehara, Atsushi Ikai, and Toshiya Osada

Summary

Examining messenger RNA (mRNA) expression is useful for the determination of cell and tissue conditions. Many methods of determining mRNA expression require total RNA extraction or cell fixation. These processes cause difficulties in examining mRNA expression in single living cells without causing cell death. We think that analysis of specific mRNA expression in single living cells will become important in cell biology. In this chapter, we present a method to examine mRNA expression of single living cells without killing the cells. The single-cell nanoprobe (SCN) method uses atomic force microscopy (AFM) to extract mRNA. We also present examples of β-actin mRNA detection and multiple mRNA detection from single living cells.

Key words: Single cell nanoprobe, Atomic force microscopy, mRNA, RT–PCR, Real-time quantitative PCR, multiple PCR1

1. Introduction

The principal features of the single-cell nanoprobe (SCN) method *(1–4)* are outlined in **Fig. 1**. We directly pick up cell components containing messenger RNA (mRNA) using an atomic force microscopy (AFM) probe by physical adsorption, and analyze it by real-time polymerase chain reaction (RT–PCR), nested polymerase chain reaction (PCR), and quantitative PCR. Although simple in principal, carrying out the experiment has many difficulties, because of the small amount of mRNA able to be adsorbed by an AFM probe. Because of the small amount of mRNA, we need to prepare the sample and instruments by

James Weifu Lee and Robert S. Foote (eds.), *Micro and Nano Technologies in Bioanalysis*,
Methods in Molecular Biology, vol. 544
DOI 10.1007/978-1-59745-483-4_38, © Humana Press, a part of Springer Science+Business Media, LLC 2009

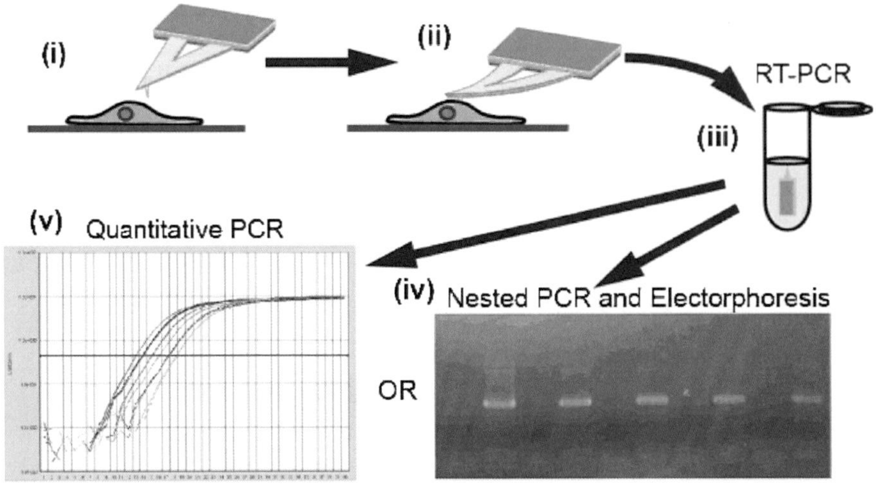

Fig. 1. Scheme of the SCN method. (i) Place the top of an AFM probe over the target position of the cell using inverted phase-contrast microscopy. (ii) Lower the AFM probe and insert it into the cell. (iii) After 30–45 s, lift up the AFM probe, put it into a tube containing RT–PCR solution, and perform RT–PCR. (iv) Using the RT–PCR solution as a template, perform nested PCR and detect target mRNA expression by electrophoresis after ethidium bromide staining. (v) Alternatively, mRNA expression is quantified by using real-time quantitative PCR.

special protocols to remove all contamination. We must also establish PCR conditions of amplification on very small amounts of mRNA.

In this chapter, we explain the SCN method using an example of β-actin mRNA detection from rat vomeronasal organ fibroblast-like cells (VNO). In addition, we present multiple mRNA detection as an advanced technique.

2. Materials

2.1. Cell Culture and Cell Preparation

1. Dulbecco's Modified Eagle Medium Nutrient Mixture (DMEM/F12) (GIBCO, Invitrogen Corp.) supplemented with 50 U/mL penicillin/50 μg/mL streptomycin (GIBCO), and 10% fetal bovine serum (FBS) (GIBCO) (*see* **Note 1**).
2. Solution of trypsin (0.25%) and ethylenediamine tetraacetic acid (EDTA) (1 mM) from GIBCO.
3. Cover glasses, 0.12- to 0.17-mm thick, from Matunami (Japan).
4. Phosphate buffered saline (PBS) (–), pH 7.4: 137 mM sodium chloride, 8.1 mM disodium hydrogen phosphate, 2.68 mM potassium chloride, 1.47 mM potassium dihydrogen phosphate.
5. Collagen (CELLGEN type I-AC, 3% solution) from KOKEN (Japan). At use, dilute 50 times with PBS (–).

2.2. The Single-Cell Nanoprobe Method

1. Atomic force microscope (NVB-100, Olympus, Inc.).
2. AFM probe (NP, Digital Instruments, Canada). Its spring constant should be low (~0.1–0.2 N/m). The AFM probe should be preserved in low humidity.
3. Water should be fresh MilliQ water.
4. 70% ethanol is made by diluting 99.5% ethanol with MilliQ water.
5. DNAZap from Ambion.

2.3. RT–PCR, Nested PCR, Real–Time Quantitative PCR, cDNA Preparation, and Electrophoresis

1. We used the one-step RT–PCR kit (Qiagen) for RT–PCR, HotStarTaq® Master Mix Kit (Qiagen) for nested PCR and SYBR Green I RT–PCR Mastermix (Qiagen) for real-time quantitative PCR. We used RNaseOUT Ribonuclease Inhibitor from Invitrogen.
2. All primers are HPLC grade, purchased from Operon Biotechnology.

Primer sequences for RT–PCR are:
Rat β-actin:
5′ primer: 5′-TTGTAACCAACTGGGACGATATGG-3′
3′ primer: 5′-GATCTTGATCTTCATGGTGCTAGG-3′
Rat fibronectin 1 (FN1):
5′ primer: 5′-GAAAGGCAACCAGCAGAGTC-3′
3′ primer: 5′-AAGTGACCCGCATAGTGTCC-3′
Rat glyceraldehyde-3-phosphate dehydrogenase (GAPDH):
5′ primer: 5′-TGCCACTCAGAAGACTGTGG-3′
3′ primer: 5′-TGTGAGGGAGATGCTCAGTG-3′
Rat c-jun:
5′ primer: 5′-ACCCCCACTCAGTTCTTGTG-3′
3′ primer: 5′-CTTGATCCGCTCCTGAGACT-3′
Rat c-myc:
5′ primer: 5′-ACGGCCTTCTCTTCTTCCTC-3′
3′ primer: 5′-CTCGCCGTTTCCTCAGTAAG-3′
Primers sequences for nested PCR are:
Rat β-actin:
5′ primer: 5′-AAGATTTGGCACCACACTTTCTAC-3′
3′ primer: 5′-ACACTTCATGATGGAATTGAATGT-3′
Rat FN1:
5′ primer: 5′-CGAGGTGACAGAGACCACAA-3′
3′ primer: 5′-AACTCTGATCGGCATGAACC-3′
Rat GAPDH:
5′ primer: 5′-AAGGTCATCCCAGAGCTGAA-3′
3′ primer: 5′-ATGTAGGCCATGAGGTCCAC-3′
Rat c-jun:
5′ primer: 5′-CAAGAACGTGACCGACGAG-3′
3′ primer: 5′-AGTTGCTGAGGTTGGCGTAG-3′
Rat c-myc:
5′ primer: 5′- CAACGTCTTGGAACGTCAGA-3′
3′ primer: 5′- TCATCTGCTTGAACGGACAG-3′

Primers sequences for quantitative PCR of the β-actin gene are:
Rat β-actin:
5′ primer: 5′-GTAGCCATCCAGGCTGTGTT-3′
3′ primer: 5′-CCCTCATAGATGGGCACAGT-3′

3. For total RNA purification, RNeasy® Micro Kit from Qiagen or PURESCRIPT® RNA Isolation Kit from Gentra systems is used following kit protocols.
4. Tris–Borate–EDTA Buffer (TBE buffer): 0.089 M Tris, 0.089 M borate acid, 0.002 M EDTA.
5. Wizard® SV Gel and PCR Clean-Up System from Promega is used to extract cDNA from agar gel following the kit protocols.

3. Methods

In the single-cell nanoprobe method, it is difficult to detect target cDNA after only one round of RT–PCR, because the amount of mRNA picked up is very small. To fully detect cDNA, we performed a second round of PCR using the first round RT–PCR reaction product as a template. All PCR experiments were performed according to the manufacturer's kit protocols. **Subheadings 3.1–3.6** explain the β-actin mRNA detection. In **Subheading 3.7**, we show an example experiment of multiple mRNA detection.

3.1. Cell Preparation on Cut Cover Glass for SCN Method

Figure 2 summarizes cell preparation on cover glass for SCN. Cell culture medium containing FBS usually results in failure of the SCN method because of contamination. To remove FBS before the experiment, we prepared cell samples as follows:

1. Cut cover glass to 8 × 3-mm rectangles with a glass cutter. Discard cracked or damaged glass covers.
2. Ultrasonicate the cut cover glasses in a 1.5-mL tube containing 1 mL of 1 M sodium hydrate for 15 min.
3. Wash the ultrasonicated cover glasses with 3 mL of MilliQ water in a 35-mm culture dish three times; using clean tweezers to transfer cover glasses to clean dishes containing MilliQ water for each wash.
4. Wash the cover glasses with 99.5% ethanol three times, moving them to a clean culture dish between each wash.
5. Air-dry the washed cover glasses on a clean bench.

From this step on, all experimental steps should be performed on a clean bench.

6. Immerse the dry cover glasses in 0.06% collagen/PBS for 15 min.

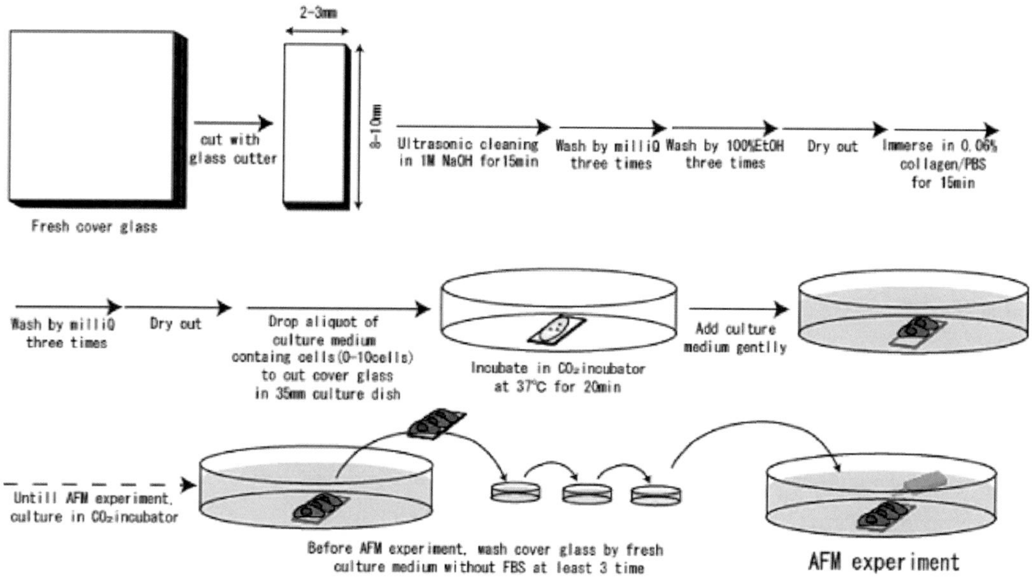

Fig. 2. Preparation of cells on cover glass.

7. Wash the cover glasses with MilliQ water three times.
8. Air-dry the cover glasses on a clean bench.
9. Store the dry cover glasses in 35-mm tissue culture dish, for up to 1 day.

Next, prepare cells to put onto the collagen-coated cover glasses. Cells approaching 80–90% confluent in 25-cm^2 flask are detached by incubation with 3 mL of trypsin/EDTA solution at 37°C for 5 min. Transfer the cells into a 15-mL centrifuge tube, and add an equivalent volume of 10% FBS DMEM/F12 to inactivate the trypsin.

10. Centrifuge the 15-mL tube at 700 × *g* for 10 min.
11. Discard the supernatant and gently suspend the precipitated cells in 1 mL of 10% FBS DMEM/F12.
12. Dilute the suspension to 0–10 cells/one drop (almost 5 μL) with 10% FBS DMEM/F12 (*see* **Note 2**).
13. Place up to six collagen-coated cover glasses in one 35-mm culture dish, using clean tweezers.
14. Apply 5 μL of diluted cell solution on the cover glass; avoid cell solution contamination of the culture dish.
15. Incubate dishes in a CO_2 incubator (5%) at 37°C for 20 min. Some cells can attach to the collagen-coated cover glass weakly.
16. Add 3 mL of 10% FBS DMEM/F12 to the dish containing the cover glasses.
17. Incubate the dishes in a CO_2 incubator for at least 2 days before use.

3.2. Clean AFM and Instruments

1. Clean all of the AFM instrument parts exposed to the sample chamber with 70% ethanol.
2. Rinse the AFM instrument parts with DNAZap following the manufacturer's protocol.
3. The parts that are directly dipped into sample culture medium are the most dangerous for contaminations of DNA, RNA, and DNAZap in routine AFM experiments. Wash the AFM probe holder and AFM head (both are directly dipped into the sample culture medium) with MilliQ water ten times or more, then incubate in MilliQ water until use.
4. Wash the other instruments (tweezers, holder substrate, etc.) with MilliQ water followed by a brief vigorous wash in 100% ethanol. Store on a clean bench until use.

3.3. The Single-Cell Nanoprobe Method for β-Actin mRNA Detection

Before starting the SCN method, it is necessary to prepare the RT–PCR solution and store it on ice or refrigerate. RT–PCR is described in **Subheading 3.5.**

1. Remove the AFM head from the MilliQ water.
2. Put an AFM probe into 3 mL of 99.5% ethanol in a 35-mm dish using clean tweezers, and wash the probe briefly.
3. Transfer the AFM probe to 3 mL of MilliQ water in a 35-mm dish, and wash it briefly two times.
4. After washing the AFM probe holder by spraying it with 99.5% ethanol, place the AFM probe on the AFM probe holder, and place the complex on the AFM head.
5. Dip the AFM head in 5 mL of DMEM/F12 (not containing FBS) in a 35-mm dish for 1 min.
6. Remove the AFM probe from the AFM probe holder, and put it into the PCR tube containing RT–PCR solution. This will be negative control 1 (*see* **Note 3**).
7. Remove the culture cells on cover glasses from the CO_2 incubator.
8. Check the number and condition of the cells (*see* **Note 4**).
9. Using clean tweezers, gently transfer the cover glasses to a fresh 35-mm dish containing DMEM/F12. Repeat this step five times using new media and dishes. Wash the tweezers with 99.5% ethanol each time.
10. Place the 35-mm dish containing sample under the AFM head.
11. Perform **steps 2–4**, and dip the AFM head into sample medium for 1 min without inserting the AFM probe into a cell. Carefully lift it up, and put the AFM probe into a PCR tube as in **step 6**. This will be negative control 2 (*see* **Note 3**).

12. Perform **steps 2–4**. While observing with the inverted phase contrast microscope, lower the AFM tip and insert it into the target position of a cell by step motor and hold for 30–45 s (*see* **Notes 5** and **6**). Lift the AFM tip up and put it into a PCR tube as in **step 6**. This will be the sample for mRNA detection.
13. Repeat **steps 11** and **12**. Negative control 2 and mRNA detection should be done alternatively, and it is ideal to change culture medium every 30 min.
14. Perform RT–PCR (described in **Subheading 3.5**).

3.4. β-Actin cDNA Preparation for Standard of Quantitative PCR

1. Total RNA is prepared by extracting from 80 to 90% confluent cells in a 25-cm² flask using RNeasy® Micro Kit (Qiagen) and is quantified with an absorption spectrophotometer.
2. Perform RT–PCR for 35 cycles using 50 ng total RNA as a template and RT–PCR primers for β-actin mRNA following the kit manufacturer's protocol. Conditions for RT–PCR are set to 50°C for 30 min, 95°C for 15 min, and 35 cycles of 94°C for 15 s, 55°C for 30 s, and 72°C for 45 s, followed by 72°C for 10 min.
3. Electrophoresis is performed in 1% (w/w) agar gel/TBE buffer at 100 mV for 30 min.
4. The gel is stained in 1 μg/mL ethidium bromide/TBE for 15 min.
5. Extract cDNA from agar gel using Wizard®SV Gel and PCR Clean-Up System (Promega).
6. Determine the cDNA concentration by absorption spectrophotometer.

3.5. RT–PCR and Nested PCR for the Single-Cell Nanoprobe Method of β-Actin mRNA

1. Prepare the RT–PCR solution following the kit protocol. We recommend each primer concentration be 0.2 μM in the β-actin case for the SCN method. Each volume of RT–PCR solution is 50 μL.
2. Perform the SCN method (*see* **Subheading 3.3**).
3. After placing the AFM probe in each tube, perform RT–PCR for 30 cycles. RT–PCR cycle conditions are 50°C for 30 min, 95°C for 15 min, and 30 cycles of 94°C for 15 s, 55°C for 30 s, and 72°C for 45 s, and 72°C for 10 min.
4. Using 1 μL of the RT–PCR reaction buffer as a template, perform nested PCR for 35 cycles. Nested PCR cycle conditions are 95°C for 15 min, and 35 cycles of 94°C for 15 s, 55°C for 30 s, and 72°C for 45 s, and 72°C for 10 min. Analyze the PCR reaction buffer by electrophoresis (refer to **Subheading 3.4**).

3.6. Quantitative PCR for the Single-Cell Nanoprobe Method of β-Actin mRNA

1. Prepare standard samples before quantitative PCR. To make the standards, perform RT–PCR using a concentration series of complementary DNA (cDNA) (prepared in **Subheading3.4**) at fourfold intervals (1,000; 250; 64; 16; and 4 molecules of cDNA) using the same conditions for RT–PCR as in the single-cell nanoprobe method.
2. Perform quantitative PCR using SYBR Green I PCR Mastermix (Qiagen) and the Applied Biosystems Prism 7000 sequence detection system as a thermal cycler. PCR conditions are 95°C for 15 min, and 40 cycles of 94°C for 15 s, 55°C for 30 s, and 72°C for 45 s. Analyze the results with ABI Prism 7000 SDS software. **Figure 3** shows examples of β-actin mRNA detection from several parts of a single living cell.

3.7. Multiple mRNA Detection (β-Actin, GAPDH, FN1, c-jun, and c-myc)

Here, we show multiple mRNA detection by the SCN method. **Figure 4a** summarizes the strategy to detect multiple mRNAs by RT–PCR after nested PCR, and **Fig. 4b** explains the cell conditions of this experiment. We show two examples of the results in **Fig. 4c, d**.

1. Prepare the cells on cover glasses according to **Subheading 3.1**.
2. After 2 days of cultivation in 10% FBS DMEM/F12, transfer the cells to DMEM/F12 without FBS.
3. Incubate the cells for 1 day. In this step, the cells transfer into the stationary (G0) phase.
4. Prepare the RT–PCR solution containing each primer. The primer concentration for β-actin, GAPDH, FN1, c-jun, and c-myc is 0.2 μM.
5. Transfer the cells to 10% FBS DMEM/F12 containing 10 μg/mL cycloheximide to stimulate the cells.
6. Incubate the cells in a CO_2 incubator for 2 h.

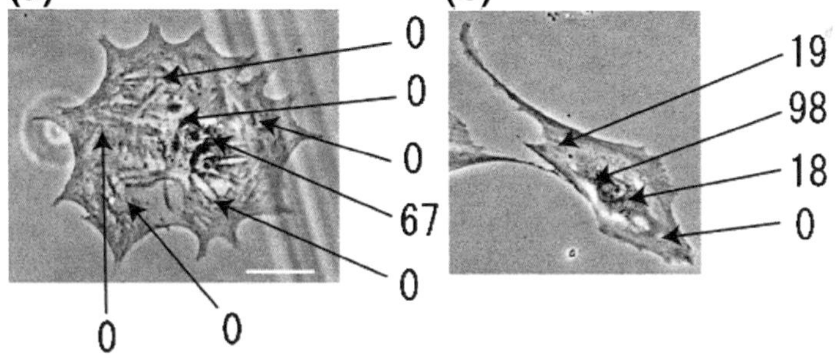

Fig. 3. β-actin mRNA detection by the SCN method. *Arrows* indicate points of AFM probe insertion. Each number indicates the β-actin mRNA quantity obtained from the insertion point, as determined by real-time quantitative PCR. Scale bar is 55 μm.

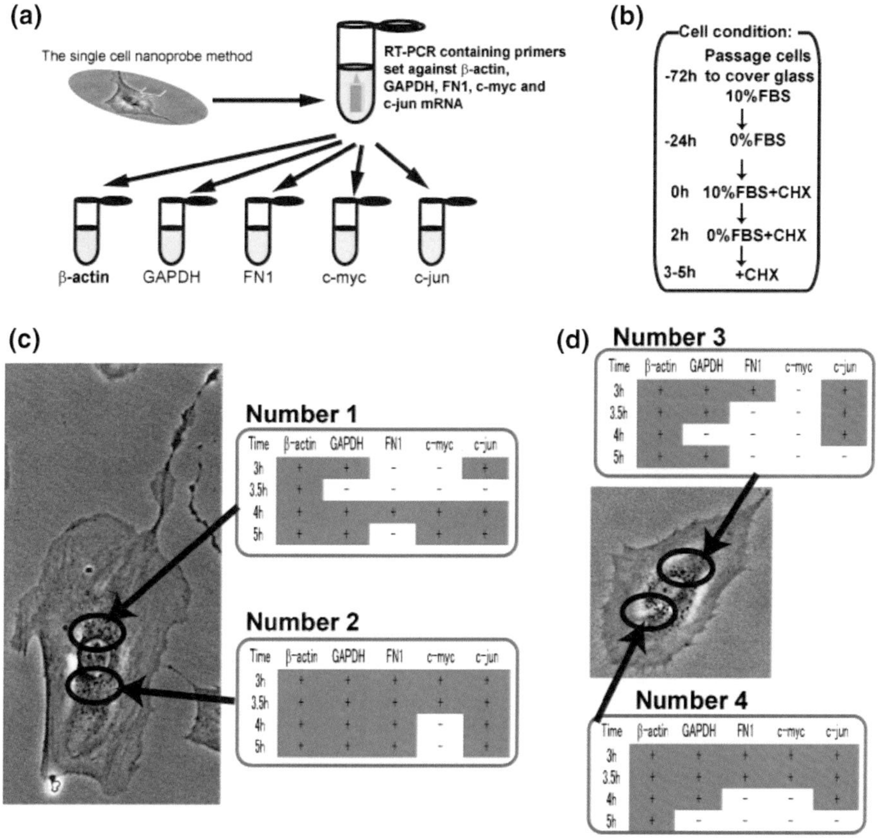

Fig. 4. Five kinds of mRNA detection by the SCN method. (**a**) Strategy to detect five kinds of mRNA. (**b**) Conditions of this experiment. (**c** and **d**) Results of SCN detection. We picked up cell components from two different positions with two different AFM probes, alternatively (Numbers 1–4). "+" and "–" indicate mRNA detection and nondetection, respectively.

7. Transfer the cells to DMEM/F12 containing 10 μg/mL cycloheximide without FBS.
8. Incubate the cells in a CO_2 incubator for 1 h.
9. Perform the single-cell nanoprobe method using DMEM/F12 containing 10 μg/mL cycloheximide. In this experiment, we obtained cell components from two positions with two different AFM probes on each target time (3 h, 3.5 h, 4 h, and 5 h).
10. Perform RT–PCR according to **Subheading 3.6**. These primers are designed to have the same Tm value.
11. Using 1 μL of RT–PCR sample as a template, perform nested PCR for each gene. Each primer concentration is 0.6 μM. Nested PCR cycle conditions are 95°C for 15 min, and 35 cycles of 94°C for 15 s, 55°C for 30 s, and 72°C for 45 s, and 72°C for 10 min.

12. Electrophoresis and detection. **Figure 4c, d** shows examples of multiple mRNA detections. Cell components were obtained from two positions (numbers 1 and 2 from cell (**c**), numbers 3 and 4 from cell (**d**) on the time course.

4. Notes

1. In all steps, DMEM/F12 contains antibiotics.
2. Drop 5 μL of mixture in a 35-mm dish and count the approximate number of cells.
3. Negative control 1 checks DNA or RNA contamination from AFM instruments or regents. Negative control 2 is used to check sample contamination.
4. The number of cell should be less than ten, and it is important that cells are located in the center of the cover glass. Cells on the edge of the glass may be damaged by tweezers.
5. Inserting an AFM tip into a VNO cell's nucleus caused cell death in our experiments.
6. The general loading force to insert is estimated to be 150 nN, however we succeeded in detecting β-actin mRNA in force curve mode, at a final force of 2 nN.

References

1. Osada, T., Uehara, H., Kim, H., and Ikai, A. (2003). mRNA analysis of single living cells. *J. Nanobiotechnology*, **1**, 2
2. Osada, T., Uehara, H., Kim, H., and Ikai, A. (2004). Clinical laboratory implications of single living cell mRNA analysis, in *Advances in Clinical Chemistry*, Vol. 38 (Makowski, G., ed.), pp. 239–257. Academic Press, London
3. Uehara, H., Osada, T., and Ikai, A. (2004). Quantitative measurement of mRNA at different loci within an individual living cell. *Ultramicroscopy*, **100**, 197–201
4. Uehara, H., Kunitomi Y. Osada T., and Ikai A. (2007). mRNA detection of individual cells with the single cell nanoprobe method compared with in situ hybridization. *J. Nanobiotechnology*, **5**, 7

Chapter 39

Reverse Transfection Using Gold Nanoparticles

Shigeru Yamada, Satoshi Fujita, Eiichiro Uchimura, Masato Miyake, and Jun Miyake

Summary

Reverse transfection from a solid surface has the potential to deliver genes into various types of cell and tissue more effectively than conventional methods of transfection. We present a method for reverse transfection using a gold colloid (GC) as a nanoscaffold by generating nanoclusters of the DNA/reagent complex on a glass surface, which could then be used for the regulation of the particle size of the complex and delivery of DNA into nuclei. With this method, we have found that the conjugation of gold nanoparticles (20 nm in particle size) to the pEGFP-N1/Jet–PEI complex resulted in an increase in the intensity of fluorescence of enhanced green fluorescent protein (EGFP) (based on the efficiency of transfection) from human mesenchymal stem cells (hMSCs), as compared with the control without GC. In this manner, we constructed a method for reverse transfection using GC to deliver genes into the cells effectively.

Key words: Human mesenchymal stem cells, Reverse transfection, Gold colloid; Particle size, Solid surface

1. Introduction

Technical efforts are being made to develop methods for transfection of cell microarrays from a solid surface. The miniaturization and simplification provided by cell microarrays are available for high-throughput analysis and the evaluation for various types of cells *(1–8)*. Sabatini et al. were the first to report a method for reverse transfection. They mixed nucleic acids (plasmid DNA or RNA) with gelatin to transfer genes exclusively to cells in

James Weifu Lee and Robert S. Foote (eds.), *Micro and Nano Technologies in Bioanalysis*,
Methods in Molecular Biology, vol. 544
DOI 10.1007/978-1-59745-483-4_39, © Humana Press, a part of Springer Science+Business Media, LLC 2009

microarrays *(7)*. Ochiya et al. also reported a method for reverse transfection that used atelocollagen as a carrier *(2, 4)*. In our laboratory, we have used fibronectin to generalize the process of retention of nucleic acids and to promote gene transfer, and we have successfully transfected a variety of cell lines (e.g., HEK293, HeLa, NIH3T3, HepG2, and human mesenchymal stem cells [hMSCs]) *(8)*. Moreover, during application of our method to PC12 cells, we found that collagen IV functioned even better than fibronectin in our system *(5)*. Kato et al. developed a method for the reverse transfection of nonadherent cells, in which plasmid DNA was deposited in a defined area on *b*iocompatible *a*nchor for *m*embrane (BAM)-modified glass slides and introduced into nonadherent K562 cells that has also been immobilized on such slides *(3)*.

Although methods for reverse transfection have been developed, as described above, it is necessary to regulate the solid-phase condition if we are adequately to understand the mechanism and to increase the efficiency. For this, Shen et al. co-precipitated DNA with inorganic minerals onto cell culture surfaces. Gene transfer from the engineered surfaces was as efficient as an optimized commercial lipid-based transfection reagent; in addition, the degree of gene transfer was adjustable by varying the mineral composition *(9)*.

Recently, some researchers have suggested the effectiveness of colloidal gold particles as a nanoscaffold for the introduction of plasmid DNA, in the presence of polyethylenimine (PEI), into mammalian cells, such as Chinese hamster ovary (CHO) cells and monkey kidney (COS-7) cells *(10)*. Furthermore, some reports suggest that the particle size of the complex can affect the efficiency of gene transfer to cells in conventional methods of transfection *(11, 12)*. Although the control of the solid-phase surface had not been achieved in the reverse transfection, it might be able to promote the uptake of the DNA/reagent complex by regulating the particle size of the complex printed on the solid surface.

2. Materials

1. Human MSCs (Cambrex BioScience, Walkersville, MD, USA).
2. 0.25% trypsin solution (Nacalai Tesque, Kyoto, Japan).
3. Dulbecco's modified Eagle medium (DMEM; Nacalai Tesque).
4. Dulbecco's phosphate-buffered saline (DPBS; Nacalai Tesque).

5. Fetal bovine serum (FBS; Invitrogen, Carlsbad, CA, USA).
6. Kanamycin (Invitrogen).
7. Penicillin-streptomycin (Nacalai Tesque).
8. Poly-L-lysine (PLL)-precoated glass slides (Matsunami glass, Osaka, Japan).
9. Bovine plasma lyophilized fibronectin (Nacalai Tesque).
10. Fourfold concentrated Jet-PEI (30 mM; Q-Bio Gene, Irvine, CA, USA).
11. Four types of gold colloids (GC; 0.01% original solution; Sigma, St. Louis, MO, USA) with various particle sizes (10, 20, 50, and 100 nm).
12. pEGFP-N1 vector (1 mg/mL; BD Biosciences Clontech, Mountain View, CA, USA).
13. High-precision ink-jet microarrayer (Cartesian Technologies, Irvine, CA, USA).
14. Array scanner (ArrayWorks, Applied Precision LLC, Issaquah, WA, USA).

3. Methods

Figure 1 shows a schematic representation of reverse transfection. In standard methods, the gene delivery complex in solution phase undergoes several processes, such as diffusion in solution, entry

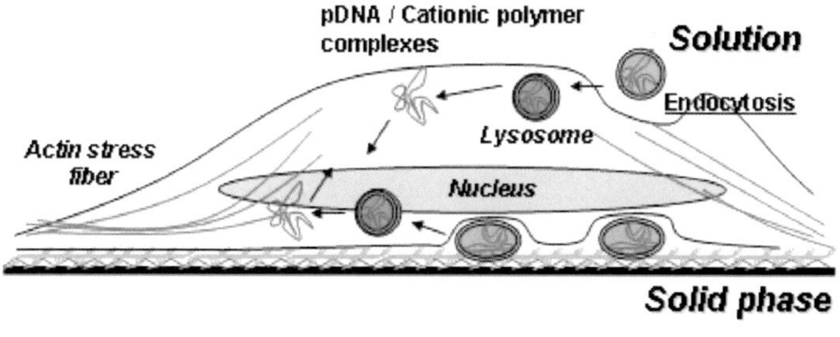

Fig. 1. Transfection in solution or from a solid surface. Gene delivery in the solution phase involves several steps, such as diffusion in solution, uptake by cells, and transfer to and incorporation into the nucleus. In the case of transfection from a solid surface, the increase in the projection area of the nucleus and the flattening of the cell should minimize the cytosolic distance between the cell membrane and the nucleus, further decreasing the intracellular distance that the DNA has to travel via diffusion. This might also significantly minimize the time during which incorporated DNA can be degraded by nucleases. In this solid-phase system, regulation of the area of the DNA/reagent complex that is printed on the solid surface should enhance the efficiency of uptake into cells.

into the cells, transfer to the nuclear membrane via lysosomes, and incorporation into the nucleus *(13, 14)*. During transfection from a solid surface, the increase in the projection area of the nucleus and the flattening of the cell are expected to minimize the cytosolic distance between the cell membrane and the nucleus, which decreases the intracellular distance that the DNA has to travel via diffusion *(5, 8, 15)*. These conditions should significantly minimize the time during which the incorporated DNA can be degraded by nucleases *(16)*.

To improve the efficiency of reverse transfection, we attempted to promote the uptake of the DNA/reagent complex by regulating the solid-surface conditions. Some reports suggest that the particle size of the complex can affect the efficiency of gene delivery into the cells *(11, 12)*. We postulated that we might be able to promote the uptake of the DNA/reagent complex by regulating the area on which the complex was printed on the solid surface. We used negatively charged particles of gold colloid (GC) as a nanoscaffold to achieve electrostatic interactions with the positive charged Jet-PEI reagent (**Fig. 2**). We anticipated that nanoclusters of the net positively charged PEI/GC would be formed that we could then use for the condensation of DNA and transfer of the DNA to the cells.

3.1. Culture of hMSCs

1. Culture the hMSCs in DMEM containing 10% FBS, 0.1 mg/mL kanamycin, and 0.05 mg/mL penicillin–streptomycin at 37°C in an atmosphere of 5% CO_2 in air.
2. Passage cells when they approach confluence with trypsin to provide new maintenance cultures on 100-mm-diameter dishes (*see* **Note 1**).
3. Use the cells after fewer than five passages to avoid phenotypic changes.

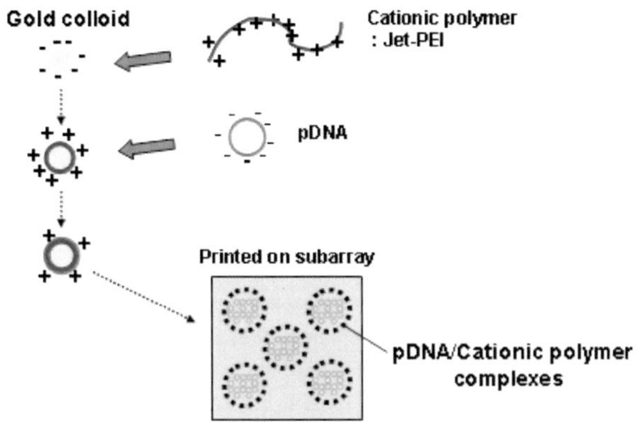

Fig. 2. Reverse transfection with a nanoscaffold. Use of particles of gold colloid (GC) as a nanoscaffold for formation of net positively charged PEI/GC complexes, which can be used the condensation of DNA and gene transfer to the cells.

3.2. Coating of Glass Slides with Fibronectin

1. Use PLL-precoated glass slides (*see* **Note 2**), with 48 squares (subarrays) of 9 mm^2 each, separated by a 25-μm-thick layer of Teflon.
2. Dissolve 10 mg bovine plasma lyophilized fibronectin in deionized distilled water (DDW) to a final concentration of 1 mg/mL (*see* **Note 3**).
3. Apply 1 mL of fibronectin solution on the glass slide, covering the entire subarrays.
4. Incubate for 1 h at room temperature.
5. Remove the solution from the glass slide by vacuum aspirate.
6. Wash the glass slide once with DDW (*see* **Note 4**).
7. Dry up the glass slide for 1 h under the clean bench.

3.3. Reverse Transfection with Gold Colloid

1. Dilute fourfold concentrated Jet-PEI with 0.006% GC solution **(Table 1)** (*see* **Note 5**).
2. Incubate for 1 h at room temperature to allow formation of Jet-PEI/GC complexes.
3. Mix the pEGFP-N1 vector with DMEM according to **Table 2** with pipetting.

Table 1
Gold colloid solution (0.006%; diluent: DMEM)

Particle size (nm)	Concentration (particles/mL)	Surface area (μm^2/mL)
10	3.4×10^{12}	10.67×10^{8}
20	4.2×10^{11}	5.27×10^{8}
50	2.7×10^{10}	2.12×10^{8}
100	3.4×10^{9}	1.07×10^{8}

Table 2
Preparation conditions for mixture

	N/P = 5	N/P = 10[a]	N/P = 20
DMEM (serum free)	10.5	7.5	1.5
Plasmid DNA (1 mg/mL)	1.5	1.5	1.5
JetPEI (1×) concentration	3	6	12
Fibronectin (4 mg/mL)	5.0	5.0	5.0
Final volume	20.0	20.0	20.0

[a]The most effective condition for hMSCs

4. Mix the solution of Jet-PEI/GC with the pEGFP-N1 solution at different N/P ratios (ratios of cations to anions in the pEGFP-N1/Jet–PEI complex) according to **Table 2** with pipetting (*see* **Note 6**).
5. Incubate for 20 min at room temperature to allow formation of pEGFP-N1/Jet-PEI/GC complexes.
6. Again mix the solution of pEGFP-N1/Jet–PEI/GC with pipetting (*see* **Note 7**).
7. Print the transfection complex on modified glass slides either manually, using a 10-μL pipette tip or by a high-precision ink-jet microarrayer in a regular pattern (five spots) on each 9-mm^2 subarray.
8. Dry the glass slide under the clean bench.
9. Seed the cells (400,000 cells/array) on each subarray set in a 100-mm-diameter dish (*see* **Note 8**).
10. Incubate for 30 min at 37°C in an atmosphere of 5% CO_2 in air (*see* **Note 9**).
11. Pour the culture media onto the entire surface of the subarrays.
12. Incubate the cells for 48 h under the culture conditions.

3.4. Fluorescence Imaging

1. Wash the glass slide once with DPBS.
2. Place the coverslip on the glass surface to eliminate background fluorescence of media components.
3. Determine the fluorescence caused by EGFP expressed in the cells on the glass slide with an array scanner.
4. Quantify the fluorescence intensities by a quantification program (IMAGENE).

4. Notes

1. Do not allow the cells to become confluent. Passage the cells at least twice per week at 80% confluence or less.
2. Coated with PLL, the sticking of printed complexes onto the substrate is improved.
3. Do not vortex. Mix by inverting the tube several times.
4. Do not wash extensively. Wash once gently.
5. Sonicate the tube containing the GC solution for 20–30 s, because GC aggregates easily. In the case of hMSCs, we investigated four types of GC with various particle sizes. GC = 20 nm is most suitable for reverse transfection.

6. Mix with vigorous pipetting ten or more times. In the case of hMSCs, the most effective N/P ratio is 10. If you use another type of cell, the mixture should be prepared with the suggested N/P ratio.
7. Mix with vigorous pipetting ten or more times.
8. The density of the cell seeding on subarrays is different depending on the cell type. Therefore, optimize the seeding density according to the type of cell.
9. The incubation time is dependent on cell precipitation and attachment on the substrate. Therefore, optimize the incubation time according to the type of cell.

Acknowledgments

This study was carried out as a part of "The Project for Development of Analysis Technology for Gene Functions with Cell Arrays," which was entrusted by the New Energy and Industrial Technology Development Organization (NEDO).

References

1. Bailey, S. N., Wu, R. Z., and Sabatini, D. M. (2002). Applications of transfected cell microarrays in high-throughput drug discovery. *Drug Discov. Today* 7, S113–S118
2. Honma, K., Ochiya, T., Nagahara, S., Sano, A., Yamamoto, H., Hirai, K., Aso, Y., and Terada, M. (2001). Atelocollagen-based gene transfer in cells allows high-throughput screening of gene functions. *Biochem. Biophys. Res. Commun.* **289**, 1075–1081
3. Kato, K., Umezawa, K., Miyake, M., Miyake, J., and Nagamune, T. (2004). Transfection microarray of nonadherent cells on an oleyl poly(ethylene glycol) ether-modified glass slide. *Biotechniques* **37**, 444–452
4. Ochiya, T., Takahama, Y., Nagahara, S., Sumita, Y., Hisada, A., Itoh, H., Nagai, Y., and Terada, M. (1999). New delivery system for plasmid DNA in vivo using atelocollagen as a carrier material: the Minipellet. *Nat. Med.* **5**, 707–710
5. Uchimura, E., Yamada, S., Uebersax, L., Yoshikawa, T., Matsumoto, K., Kishi, M., Funeriu, D. P., Miyake, M., and Miyake, J. (2005). On-chip transfection of PC12 cells based on the rational understanding of the role of ECM molecules: efficient, non-viral transfection of PC12 cells using collagen IV. *Neurosci. Lett.* **378**, 40–43
6. Wheeler, D. B., Carpenter, A. E., and Sabatini, D. M. (2005). Cell microarrays and RNA interference chip away at gene function. *Nat. Genet.* **37**(Suppl.), S25–S30
7. Wu, R. Z., Bailey, S. N., and Sabatini, D. M. (2002). Cell-biological applications of transfected-cell microarrays. *Trends Cell Biol.* **12**, 485–488
8. Yoshikawa, T., Uchimura, E., Kishi, M., Funeriu, D. P., Miyake, M., and Miyake, J. (2004). Transfection microarray of human mesenchymal stem cells and on-chip siRNA gene knockdown. *J. Control. Release* **96**, 227–232
9. Shen, H., Tan, J., and Saltzman, W. M. (2004). Surface-mediated gene transfer from

nanocomposites of controlled texture. *Nat. Mater.* **3**, 569–574

10. Thomas, M. and Klibanov, A. M. (2003). Conjugation to gold nanoparticles enhances polyethylenimine's transfer of plasmid DNA into mammalian cells. *Proc. Natl. Acad. Sci. U.S.A.* **100**, 9138–9143
11. Morimoto, K., Nishikawa, M., Kawakami, S., Nakano, T., Hattori, Y., Fumoto, S., Yamashita, F., and Hashida, M. (2003). Molecular weight-dependent gene transfection activity of unmodified and galactosylated polyethyleneimine on hepatoma cells and mouse liver. *Mol. Ther.* 7, 254–261
12. Ogris, M., Steinlein, P., Carotta, S., Brunner, S., and Wagner, E. (2001). DNA/polyethylenimine transfection particles: influence of ligands, polymer size, and PEGylation on internalization and gene expression. *AAPS PharmSci.* **3**, E21
13. Briane, D., Lesage, D., Cao, A., Coudert, R., Lievre, N., Salzmann, J. L., and Taillandier, E. (2002). Cellular pathway of plasmids vectorized by cholesterol-based cationic liposomes. *J. Histochem. Cytochem.* **50**, 983–991
14. Friend, D. S., Papahadjopoulos, D., and Debs, R. J. (1996). Endocytosis and intracellular processing accompanying transfection mediated by cationic liposomes. *Biochim. Biophys. Acta* **1278**, 41–50
15. Itano, N., Okamoto, S., Zhang, D., Lipton, S. A., and Ruoslahti, E. (2003). Cell spreading controls endoplasmic and nuclear calcium: a physical gene regulation pathway from the cell surface to the nucleus. *Proc. Natl. Acad. Sci. U.S.A.* **100**, 5181–5186
16. Godbey, W. T., Wu, K. K., and Mikos, A. G. (1999). Poly(ethylenimine) and its role in gene delivery. *J. Control. Release.* **60**, 149–160

Chapter 40

Custom-Designed Molecular Scissors for Site-Specific Manipulation of the Plant and Mammalian Genomes

Karthikeyan Kandavelou and Srinivasan Chandrasegaran

Summary

Zinc finger nucleases (ZFNs) are custom-designed molecular scissors, engineered to cut at specific DNA sequences. ZFNs combine the zinc finger proteins (ZFPs) with the nonspecific cleavage domain of the FokI restriction enzyme. The DNA-binding specificity of ZFNs can be easily altered experimentally. This easy manipulation of the ZFN recognition specificity enables one to deliver a targeted double-strand break (DSB) to a genome. The targeted DSB stimulates local gene targeting by several orders of magnitude at that specific cut site via homologous recombination (HR). Thus, ZFNs have become an important experimental tool to make site-specific and permanent alterations to genomes of not only plants and mammals but also of many other organisms. Engineering of custom ZFNs involves many steps. The first step is to identify a ZFN site at or near the chosen chromosomal target within the genome to which ZFNs will bind and cut. The second step is to design and/or select various ZFP combinations that will bind to the chosen target site with high specificity and affinity. The DNA coding sequence for the designed ZFPs are then assembled by polymerase chain reaction (PCR) using oligonucleotides. The third step is to fuse the ZFP constructs to the FokI cleavage domain. The ZFNs are then expressed as proteins by using the rabbit reticulocyte in vitro transcription/translation system and the protein products assayed for their DNA cleavage specificity.

Key words: Gene therapy, Nonviral vectors, Zinc finger nucleases, Gene targeting, Genome engineering, Site-specific modification, Targeted mutagenesis, Gene correction, Homologous recombination, Nonhomologous end joining

1. Introduction

Cells use the universal process of homologous recombination (HR) to maintain their genomic integrity, particularly in the repair of a double-strand break (DSB), which otherwise would be lethal. DSB repair of a damaged chromosome by HR is a highly

James Weifu Lee and Robert S. Foote (eds.), *Micro and Nano Technologies in Bioanalysis*,
Methods in Molecular Biology, vol. 544
DOI 10.1007/978-1-59745-483-4_40,

accurate form of repair, which works via the copy and paste mechanism, using the homologous DNA segment from the undamaged chromosomal partner as a template. Gene targeting—the process of replacing a gene by HR—uses an extrachromosomal fragment of donor DNA and invokes the cell's HR for sequence exchange. Gene targeting is not a very efficient process in plant and mammalian cells; approximately only one in a million cells provided with excess donor sequence undergo the desired gene modification *(1–3)*. However, when a defined chromosomal break is introduced, HR is induced at that site in a large fraction of cells in a population *(3)*. There is then a need to create custom molecular scissors for precision genome surgery to greatly enhance site-specific manipulation of plant and mammalian cells, including human cells.

Zinc finger nucleases (ZFNs) that combine the nonspecific cleavage domain (N) of FokI endonuclease with zinc finger proteins (ZFPs) offer a general way to deliver a site-specific DSB to the genome, and stimulate local HR in cells by several orders of magnitude. The Cys_2His_2 ZF motifs bind DNA by inserting an α-helix into the major groove of the double helix *(4, 5)*. Each finger primarily binds to a triplet within the DNA substrate. Key amino acids at positions −1, 2, 3, and 6 relative to the start of the α-helix contribute most of the sequence-specific interactions to the ZF motifs *(4, 5)*. These amino acids can be changed while maintaining the remaining amino acids as a consensus backbone to generate ZFPs with different sequence specificities *(6, 7)*. The ZFP also has the additional advantage that greater specificity can be achieved by adding more ZF motifs (a maximum of six ZF domains) to the ZFPs *(8–10)*. Thus, ZF DNA-binding motifs, because of their modular nature and modular structure, offer an attractive framework for designing ZFNs with tailor-made sequence specificities *(1, 11, 12)*. Several 3- and 4-finger ZFPs, each recognizing a 9- or 12-bp sequence, respectively, have been fused to the nonspecific cleavage domain of FokI to form ZFNs. The cleavage specificity of ZFNs correlates directly with the binding specificity of the corresponding ZFPs that are used to make them *(13, 14)*. Binding of two 3- or 4-finger ZFN monomers (each recognizing a 9- or 12-bp inverted site) is necessary because dimerization of the FokI cleavage domain is required to produce a DSB. Thus, a pair of ZFNs effectively has an 18- or 24-bp recognition site, which is long enough to specify a unique genomic location in plants and mammals. Reports from several labs including ours have shown that 3- and 4-finger ZFNs find and cleave their chromosomal target in cells; and, as expected, they induce local HR to repair the DSB. In the absence of homology-directed repair via HR (for example, if both alleles of a gene are damaged), cells repair the DSB by simple ligation via nonhomologous end joining (NHEJ); repair by NHEJ is mutagenic.

Thus, custom ZFNs can be used with and without homologous donor DNA sequences to induce "directed" mutations, by HR and NHEJ, respectively *(1, 2)*. Designer ZFNs have become valuable molecular scissors to perform precision genome surgery for various biological and biomedical applications. The ability to target a DSB to a specific genomic locus and stimulate HR by several orders of magnitude at that local site has great potential not only in genome engineering that is targeted manipulation of the mammalian *(15, 16)* and plant genomes *(17, 18)*, but also to treat human diseases as a form of gene therapy.

Routine and facile production of custom ZFNs and their rapid characterization for sequence-specific DNA binding and cleavage properties in vitro is a prerequisite for ZFN-mediated gene targeting *(19, 20)*. We describe here the protocols that are needed to design (**Fig. 1**) and rapidly generate custom ZFNs (**Figs. 2** and **3**), and the protocol to rapidly characterize their properties by cell-free ZFN cleavage assays using the rabbit reticulocyte in vitro transcription/translation (IVTT) system. The cognate sites for the engineered ZFNs are encoded in a plasmid, which is then used as a substrate to monitor sequence-specific cleavage activity of the custom ZFNs (**Fig. 4**).

2. Materials

2.1. PCR Generation of ZFP

1. User-designed overlapping synthetic oligonucleotides and end primers.
2. Amplitaq Gold (ABI).

mTYR (exon 1 sequence)

```
2581  TTCCAGATCT CTGATGGCCA TTTTCCTCGA GCCTGTGCCT CCTCTAAGAA CTTGTTGGCA
2641  AAAGAATGCT GCCCACCATG GATGGGTGAT GGGAGTCCCT GCGGCCAGCT TTCAGGCAGA
2701  GGTTCCTGCC AGGATATCCT TCTGTCCAGT GCACCATCTG GACCTCAGTT CCCCTTCAAA
2761  GGGGTGGATG ACCGTGAGTC CTGGCCCTCT GTGTTTTATA ATAGGACCTG CCAGTGCTCA
2821  GGCAACTTCA TGGGTTTCAA CTGCGGAAAC TGTAAGTTTG GATTTGGGGG CCCAAATTGT
2881  ACAGAGAAGC GAGTCTTGAT TAGAAGAAAC ATTTTTGATT TGAGTGTCTC CGAAAAGAAT
2941  AAGTTCTTTT CTTACCTCAC TTTAGCAAAA CATACTATCA GCTCAGTCTA TGTCATCCCC
3001  ACAGGCACCT ATGGCCAAAT GAACAATGGG TCAACACCCA TGTTTAATGA TATCAACATC
3061  TACGACCTCT TTGTATGGAT GCATTACTAT GTGTCAAGGG ACACACTGCT TGGGGGCTCT
```

Fig. 1. Selection of ZFN target sites within the nucleotide sequences of mouse tyrosinase (mTYR) gene. The nucleotide sequence of the mTYR exon 1 is shown. The best targets are inverted sequences of the form $(NNC)_{3\ or\ 4}\cdots(GNN)_{3\ or\ 4}$ separated by 5 or 6 bp. The ZFN designs for the chosen targets that have been constructed and characterized for their DNA binding and cleavage properties are *shaded*. Other potential target sites for ZFN designs are *boxed*. The site of point mutation within the tyrosinase gene responsible for transition from pigmented (*black*) to nonpigmented (albino) mice is shown in *bold*.

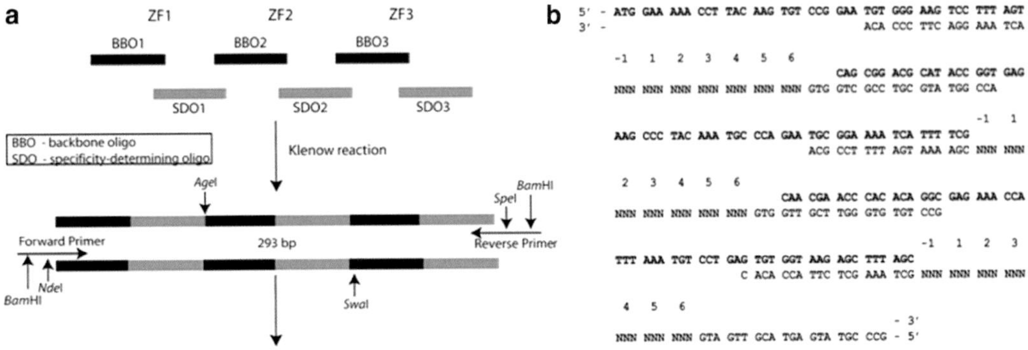

Fig. 2. Assembly of 3-finger ZFPs by using PCR. (**a**) The genes for the ZFPs are first assembled using the overlapping BBOs and SDOs (60-mers) in a Klenow reaction, which is then amplified by PCR using the outside forward and reverse primers, which are flanked by unique restriction sites (NdeI and SpeI sites, respectively) to facilitate cloning. BBO1, BBO2, and BBO3 correspond to the consensus backbone oligonucleotides whereas SDO1, SDO2, and SDO3 correspond to specificity-determining oligonucleotides for ZF1, ZF2, and ZF3, respectively. (**b**) Scheme for assembling the 3-finger ZFPs via the oligonucleotide assembly strategy using the consensus framework residues and the chosen contact amino acid residues at positions –1, +1, +2, +3, +4, +5, and +6 of the α-helix, which confer specificity to each of the ZFs. The indicated top strand (*bold*) and bottom strand oligonucleotides overlap and will be assembled using PCR. The bottom strand oligonucleotides are depicted as having NNN, which code for the contact residues that confer specificity to each ZF.

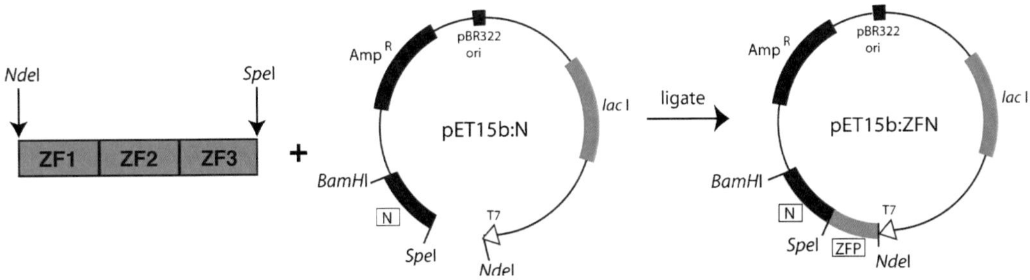

Fig. 3. Converting ZFPs into ZFNs. The NdeI/SpeI-cut ZFPs are ligated into the pET15b:N, the plasmid containing the FokI cleavage domain to form pET15b:ZFN.

3. 10 mM dNTP Mix (Invitrogen).
4. 1% Agarose gel.
5. Polymerase chain reaction (PCR) machine.
6. 1-kb DNA ladder (Invitrogen).

2.2. Cloning of PCR Fragments into a pUC18 Plasmid

1. Amplified PCR fragments (*see* **Subheading 3.2**).
2. BamHI-cut dephosphorylated pUC18 plasmid DNA.
3. T4 DNA ligase (Invitrogen or NEB).
4. Electroporation cuvettes (Bio-Rad).
5. Competent DH5α cells (Invitrogen).
6. SOC medium and LB medium.
7. Standard LB plates.

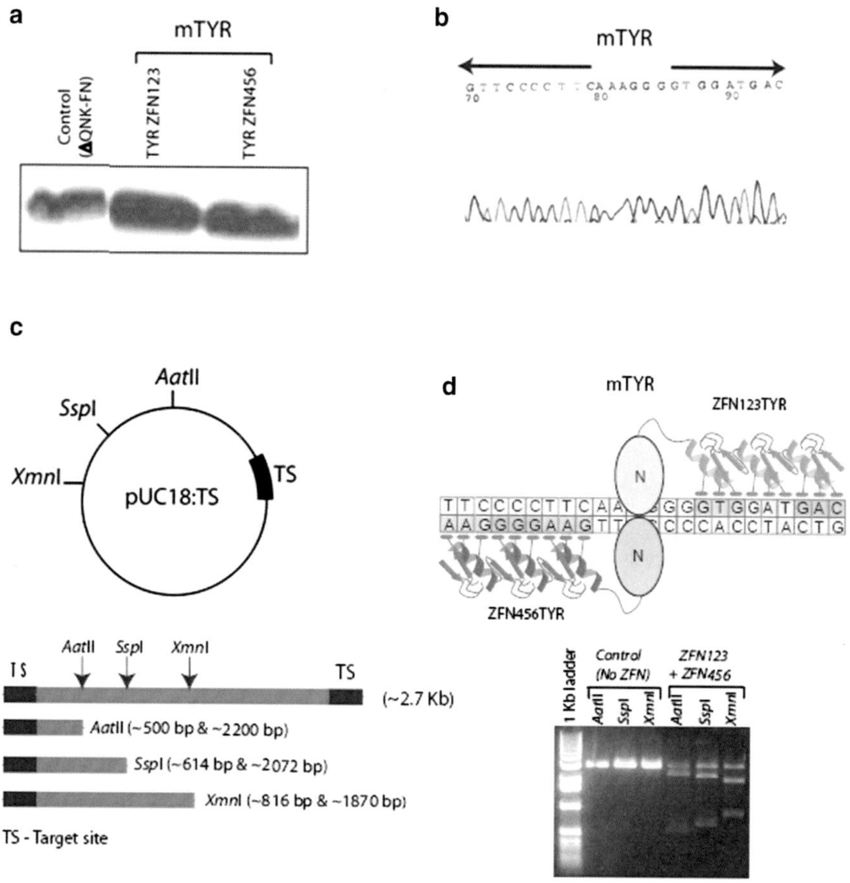

Fig. 4. Cell-free ZFN cleavage assays using the IVTT system. (**a**) Western blot profile of the fusion proteins made using the in vitro transcription-translation (IVTT) system. This yields sufficient fusion protein for rapid characterization of the cleavage specificity of the custom-designed ZFNs. (**b**) Nucleotide sequences of the ZFN target sites (TS) for mTYR, the gene encoded in the plasmid substrates (pUC18: TS) for use in the cleavage reactions. (**c**) Schematic representation of the plasmid substrates (pUC18: TS) encoding the ZFN target site of the mTYR gene at the multiple cloning site of pUC18. Four unique restriction enzyme sites, namely AatII, SspI, and XmnI, within the plasmid substrates are indicated. The expected sizes of the fragments upon cleavage by ZFNs, followed by AatII, SspI, or XmnI, respectively, are shown. (**d**) Agarose gel profile of engineered ZFN cleavage of mTYR plasmid substrates. The plasmid substrate was digested by the ZFNs, followed by one of the restriction enzymes, namely AatII, SspI, or XmnI. The particular restriction enzyme used in the reactions after the corresponding ZFN digestion is indicated on *top of each lane*. Plasmid substrates digested with the control IVTT product (which contained no ZFN plasmids), followed by one of the enzymes, AatII, SspI, or XmnI, respectively, for each are also shown. The 1-kb ladder marker is included in the gel profile.

8. Carbenicillin (Sigma).
9. X-gal (5-bromo-4-chloro-3-indolyl-β-D-galactoside): 40 mg/ml stock solution stored at −20°C.
10. IPTG (isopropylthio-β-D-galactoside): 1 M stock solution stored at −20°C.
11. Plasmid Mini purification kit (Qiagen).

12. Gel purification kit (Qiagen).
13. PCR purification kit (Qiagen).
14. Restriction Enzyme BamHI (NEB).

2.3. Conversion of ZFPS into ZFNS

1. pET15b:ZFN plasmid construct containing the FokI cleavage domain (available from our lab on request).
2. Plasmids containing various ZFP designs (*see* **Subheading 3.2**).
3. Restriction enzymes NdeI and SpeI (NEB).
4. T4 DNA ligase (Invitrogen or NEB).
5. 100 mM DTT (Invitrogen).
6. 100 mM ATP (Invitrogen).
7. RR1 electro-competent cells.

2.4. Cell-Free ZFNS Cleavage Assay Using the IVTT System

1. pUC18 plasmid containing the ZFN binding sites for use as substrates.
2. pET15b plasmid containing ZFN constructs.
3. TnTT7 Quick-Coupled Transcription/Translation kit (Promega).
4. 5 mM $ZnCl_2$.
5. Separating buffer (4×): 1.5 M Tris-HCl, pH 8.7, 0.4% sodium dodecyl sulfate (SDS). Store at room temperature.
6. Stacking buffer (4×): 0.5 M Tris-HCl, pH 6.8, 0.4% SDS. Store at room temperature.
7. Thirty percent acrylamide/bis solution (37.5:1 with 2.6% C) (Bio-Rad).
8. *N,N,N,N′*-Tetramethyl-ethylenediamine (TEMED) (Bio-Rad).
9. 10× Tris-glycine-SDS running buffer (Bio-Rad).
10. 10× Tris-glycine transfer buffer (Bio-Rad).
11. Ammonium persulfate: Prepare 10% solution in water and immediately freeze in single use (200 μl) aliquots at –20°C.
12. Water-saturated isobutanol. Shake equal volumes of water and isobutanol in a glass bottle and allow it to separate. Use the top layer. Store at room temperature.
13. Prestained molecular weight markers (Invitrogen or NEB).
14. Supported nitrocellulose membrane from Hybond ECL, chromatography paper from Whatman.
15. Tris-buffered saline with Tween (TBS-T): Prepare 10× stock with 1.37 M NaCl, 27 mM KCl, 250 mM Tris-HCl, pH 7.4, 1% Tween-20. Dilute 100 ml with 900 ml water for use.
16. Blocking buffer: 5% (w/v) nonfat dry milk in TBS-T.
17. FokI polyclonal antibody (available from our lab).

18. FokI restriction enzyme (NEB).
19. Secondary antibody: Antirabbit IgG conjugated to horseradish peroxidase (Amersham Biosciences).
20. Enhanced chemiluminescent (ECL) Western blotting detection reagents (Amersham Biosciences) and Bio-Max ML film (Kodak).

3. Methods

Engineering custom-designed ZFNs for an endogenous chromosomal gene target in mammalian cells entails the following steps: *(1)* Identify target sequences of the form $(NNC)_{3\text{ or }4}$...$(GNN)_{3\text{ or }4}$ separated by 5 or 6 bp within the gene of interest, which make for excellent targets. *(2)* Design or select ZFPs that recognize a chosen target site. *(3)* Convert the engineered ZFPs into ZFNs. *(4)* Characterize rapidly their in vitro cleavage specificity, which is essential before any cell culture studies can be performed using the designed ZFNs.

3.1. Protocol to Identify ZFN Targets Within the Gene of Interest

1. Search for ZFN binding targets within the desired gene. Inverted sequences of the form $(NNC)_{3\text{or }4}$...$(GNN)_{3\text{or }4}$ separated by 5 or 6 bp make for excellent targets. The efficiency of ZFN-mediated gene targeting falls off rapidly with increasing spacer length beyond 6 bp. Furthermore, for inverted ZFN target sites with greater than 6-bp separation, a 15-amino acid $(Gly_4Ser)_3$ linker needs to be inserted between the ZFP DNA-binding domain and the FokI cleavage domain for effective double-strand cleavage *(21)*. The ZFN target sequences could be within a hundred base pairs away from the mutation site for gene conversion (*see* **Notes 1** and **2**).
2. The Barbas lab has posted a website (http://zincfingertools.org) *(22)* that could be used to search for target sites in the desired gene. A similar program is available at the Zinc Finger Consortium website (http://www.zincfingers.org). An example of the ZFN target sequences within the mouse tyrosinase gene (mTYR) is shown in **Fig. 1**.

3.2. Protocol to Create Genes that Code for the Desired ZFPs

1. Our ZFPs were designed based on the previously described zinc-finger-framework consensus sequence derived from 131 ZF sequence motifs *(6)*. We designed 3-finger ZFPs that recognize a specific 9-bp sequence within the chosen genes as follows: *(1)* by using the consensus framework backbone sequence for each and every finger within the ZFPs of the three invariant amino acid backbone oligos (BBO1, BBO2,

and BBO3); *(2)* by varying the contact residues at positions −1, +1, +2, +3, +4, +5, and +6 of the α-helix within each ZF motif of the ZFPs using three specificity-determining oligos (SDO1, SDO2, and SDO3); the amino acid residues that confer DNA binding site specificity to each ZF motif were chosen from previously available DNA triplet recognition data for ZFPs in the literature *(23–26)* and wherever possible taking into account the positional data of each ZF motif in the context of its neighboring fingers *(27)* (*see* **Note 3**).

2. The overlapping oligonucleotide assembly strategy was used to construct the 3-finger ZFPs (**Fig. 2**). They were first assembled by Klenow reaction using the BBOs and SDOs. The assembled 3-finger ZFPs were then amplified by PCR using the forward primer (flanked by NdeI/BamHI site) and reverse primer (flanked by SpeI/BamHI site) to facilitate cloning of the engineered ZFPs into the desired target plasmid.

Component	Amount (ml)	Final concentration
0.83 ml of each 2 μM oligo	5.0	0.2 μM
1.25 mM dNTPs	1.0	0.125 mM
5× Second-strand buffer	2.0	1×
100 mM DTT	1.0	10 mM
5,000 U/ml Klenow	1.0	0.5 U/μl

3. Set up the Klenow reaction for each of the ZFPs with the following:

The Klenow reaction is carried out at 37°C for 30 min and then at room temperature for another 30 min.

Component	Amount (μl)	Final concentration
Klenow reaction product	2.5	
10 mM dNTPs	2.0	200 mM
10 μM Forward primer (NdeI)	2.5	250 nM
10 μM Reverse primer (SpeI)	2.5	250 nM
10× PCR buffer	10.0	1×
10 U/μl Amplitaq Gold Enzyme	1.0	0.1 U/μl
dH_2O	79.5	

4. Set up the PCR using the product from the Klenow reaction as the template:

The PCR is run for 30 cycles of amplification, where each cycle is programmed for 94°C for 30 s, 55°C for 30 s, and 72°C for 1 min.

5. PCR-amplified ZFPs are run on 1% agarose gel to ensure the correct size for the DNA fragment. The PCR product is then purified by using a Qiagen PCR purification kit.

Component	Amount (μl)	Final concentration
ZFP DNA fragment	10.0	1 μg/μl
10× NEB Buffer 2	2.0	1×
20,000 U/ml BamHI	2.0	1.0 U/μl
dH_2O	6.0	

6. The purified DNA fragment is subjected to digestion with BamHI:

The digestion reaction is performed at 37°C for 12–16 h.

7. The BamHI-digested DNA fragment is run on 1% gel and the 300-bp ZFP DNA fragment is gel purified using a Qiagen gel purification kit.

Component	Amount (μl)	Final concentration
BamHI-digested DNA	2.0	0.2 μg/μl
BamHI-cut pUC18 plasmid	1.0	0.1 μg/μl
1 U/μl T4 DNA ligase	1.0	0.05 U/μl
5× Ligase buffer	4.0	1×
dH_2O	12.0	

8. The purified fragment is then ligated to the BamHI-cut and dephosphorylated pUC18 plasmid:

The ligation reaction is incubated at 16°C overnight. The ligation product is extracted once with an equal volume of phenol-chloroform and twice with chloroform. The DNA is then precipitated with 100% ethanol at −80°C for 1 h. The precipitate is spun down in a microcentrifuge and washed with 70% ethanol. The pellet is air-dried and reconstituted in 10 μl of dH_2O.

9. One microliter of the purified ligation product is used to electroporate DH5α cells, the SOC medium is added immediately, and the cells are incubated at 37°C for 1 h to express the antibiotic resistance gene. LB plates containing 100 μg/ml of carbenicillin are spread with 40 μl of 40 mg/ml X-gal and 40 μl of 100 mM IPTG and incubated at 37°C until the cells are ready to be plated. The cells are then plated on the LB plates and incubated at 37°C overnight.
10. The next day, 12 white colonies are picked and grown in LB medium containing 100 μg/ml of carbenicillin overnight. The following day, plasmid purification is done using Qiagen tip20. The plasmids are digested with BamHI and the products are run on 1% agarose gel to ensure the size of the ZFP fragment. The plasmids containing the ZFP fragments are also sequenced.

3.3. Protocol to Convert ZFPS into ZFNS

The pUC18 plasmid DNA containing the ZFPs coding sequence is digested with NdeI/SpeI to release the DNA fragment encoding the ZFPs, which are then ligated into the NdeI/SpeI-cleaved pET15b:ZFN vector *(27)*, thereby replacing the existing ZFPs with the newly created ZFPs. These constructs link the designed consensus framework based ZFPs to the C-terminal 196 amino acids of FokI restriction enzyme, which constitutes the FokI cleavage domain. The ZFN fusions are of the form NH_3^+–ZF1–ZF2–ZF3–FokI (*N*)–CO_2^-. When the separations between the inverted ZFN target sites are 5 or 6 bp, which are optimal for efficient cleavage, no linker is included between the ZFPs and the FokI cleavage domain; however, for ZFN target sites with greater than 6-bp separation, the ZFP is connected to the FokI cleavage domain through a $(Gly_4Ser)_3$ linker. Furthermore, during the initial cloning of the engineered ZFNs into the bacterial cells, clones carrying the ZFN constructs are made more viable by increasing the constitutive levels of the DNA ligase within these cells *(13, 14)* (*see* **Note 4**).

Component	Amount (μl)	Final concentration
pUC18 plasmids containing ZFPs	10.0	1.0 μg/μl
10× NEB Buffer 2	2.0	1×
10× BSA	2.0	1×
20,000 U/ml NdeI	1.0	1.0 U/μl
10,000 U/ml SpeI	2.0	1.0 U/μl
dH_2O	3.0	

1. The pUC18 plasmids, those that are determined to be error-free by sequencing, are subjected to digestion with NdeI and SpeI. The reaction is performed at 37°C overnight. The digested products are run on 1% agarose gel and the 300 bp ZFP fragments are gel purified using a Qiagen gel purification kit.

Component	Amount (μl)	Final concentration
NdeI/SpeI-digested ZFP DNA	2.0	0.1 μg/μl
NdeI/SpeI-cut pET15b:N plasmid	1.0	0.05 μg/μl
5× Ligase buffer	4.0	1×
T4 DNA ligase	1.0	0.05 U/μl
dH_2O	12.0	

2. The gel-purified fragment is ligated to NdeI/SpeI-digested pET15b:N plasmid containing the FokI cleavage domain: The ligation reaction is incubated at 16°C overnight. The ligation product is extracted once with an equal volume of phenol-chloroform, and twice with chloroform. The DNA is then precipitated with 100% ethanol and incubated at –80°C for 1 h. The precipitate is spun down in microcentrifuge, and the precipitate washed with 70% ethanol. The precipitate is air-dried and then resuspended in 10 μl of dH_2O.
3. One microliter of the ligation product is electroporated into RR1 cells. During the initial cloning of the engineered ZFNs into the bacterial cells, clones carrying the ZFN constructs are made more viable by increasing the constitutive levels of the DNA ligase. The cells are plated on LB plates containing 100 μg/ml of carbenicillin and incubated at 37°C overnight. Twelve colonies are picked and grown overnight in LB medium containing carbenicillin. The next day, plasmid purification is done using Qiagen Tip20. Approximately 1 μg of plasmid is digested with NdeI and SpeI to confirm the presence of ZFPs in the recombinant plasmid. All of the above said steps are done to convert each pair of the ZFPs pair into corresponding ZFNs (**Fig. 3**).

3.4. Protocol for In Vitro Expression of ZFNS Using the IVTT System

The modified in vitro transcription-translation (IVTT) assay *(28)* was used to rapidly screen for the sequence specific cleavage by the engineered ZFNs. This protocol uses the rabbit reticulocyte IVTT system that yields a sufficient amount of the fusion protein product in the crude extract to study sequence-specific cleavage

Component	Amount (µl)	Final concentration
TNT Quick Master Mix	40.0	
1 mM Methionine	1.0	0.2 mM
T7 TNT PCR enhancer	1.0	
5 mM $ZnCl_2$	1.0	0.1 mM
1–2 µg pET15b:ZFN	2.5–5.0	0.02–0.04 µg/µl
dH_2O	4.5–2.0	

of the plasmid substrate encoding the ZFN target site. Each of the ZFNs is expressed from the pET15b:ZFN plasmids using the manufacturer's protocol. Assemble the reaction components in a 1.5-ml centrifuge tube with the following components:
Perform IVTT reactions for ZFN123 and ZFN456 in separate tubes. Incubate at 30°C for approximately 60 min. Analyze the reaction products by Western blot.

3.5. SDS–PAGE-Expressed ZFNS Using the IVTT System

1. Prepare a 1.5-mm thick, 10% gel by mixing 7.5 ml of 4× separating buffer with 10 ml acrylamide/bis solution, 12.5 ml water, 100 µl ammonium persulfate solution, and 20 µl TEMED. Pour the gel, leaving space for a stacking gel, and overlay with water-saturated isobutanol. This helps to smooth out the surface of the gel. The gel should polymerize in approximately 30 min.
2. Pour off the isobutanol and rinse the top of the gel twice with water.
3. Prepare the stacking gel by mixing 2.5 ml of 4× stacking buffer with 1.3 ml acrylamide/bis solution, 6.1 ml water, 50 µl ammonium persulfate solution, and 10 µl TEMED. Then pour the stacking gel and insert the comb. The stacking gel should polymerize in approximately 15–30 min.
4. Prepare the running buffer (Tris–glycine–SDS) by diluting 100 ml of the 10× running buffer with 900 ml of water in a measuring cylinder. Mix thoroughly.
5. Once the stacking gel has set, carefully remove the comb. Add the SDS running buffer to the upper and lower chambers of the gel unit and load 5–10 µl of each sample in a well. The samples are prepared by adding 2 µl of 5× SDS gel loading dye mixed with 8 µl of IVTT reaction product and heated at 60°C for 10 min. Include one well for prestained molecular weight marker and another well to run a positive control with the FokI enzyme.

6. Complete the assembly of the gel unit and connect to a power supply. The gel can be run during the day (in ~5 h) at 20 mA through the stacking gel and 30–40 mA through the separating gel.

3.6. Western Blot of ZFN Proteins Generated Using the IVTT System

1. The samples that have been separated by SDS-PAGE are transferred to supported nitrocellulose membranes by electrophoresis. A tray of transfer buffer is prepared that is large enough to lay out a transfer cassette with two pieces of foam and with two sheets of 3MM filter paper submerged on one side. A sheet of the nitrocellulose cut just larger than the size of the separating gel is laid on the surface of a separate tray of distilled water to allow the membrane to wet by capillary action. The membrane is then submerged in the setup buffer on top of the 3MM paper.
2. The SDS gel unit is disconnected from the power supply and disassembled. The stacking gel is removed and discarded and one corner cut from the separating gel to guide the loaded samples. The separating gel is then laid on top of the nitrocellulose membrane. Two further sheets of 3MM paper are wetted in the setup buffer and carefully laid on top of the gel, ensuring that no bubbles are trapped in the resulting sandwich. The second wet foam sheet is laid on top and the transfer cassette closed. The cassette is placed into the transfer tank such that the nitrocellulose membrane is between the gel and the anode. It is important to properly ensure this orientation or the proteins will be lost from the gel. The setup is placed in the cold room. The tank is covered and connected to the power supply. Transfers can be accomplished at 70 V for 2 h. Once the transfer is complete, the cassette is taken out of the tank and carefully disassembled, with the top sponge and sheets of 3MM paper removed. The gel is left in place on top of the nitrocellulose and these are laid on a glass plate so that the shape of the gel (including the cut corner for orientation) can be cut into the membrane using a razor blade. The gel and excess nitrocellulose can then be discarded. The colored molecular weight markers should be clearly visible on the membrane.
3. The nitrocellulose is then incubated in 50 ml blocking buffer (5% milk in TBS-T) for 1 h at room temperature on a rocking platform.
4. The blocking buffer is discarded and the membrane quickly rinsed before addition of a 1:2,000 dilution of the polyclonal FokI antibody in 5% milk for 1 h at room temperature on a rocking platform.
5. The primary antibody is then removed and the membrane washed three times for 5 min each with 50 ml TBS-T.

6. The secondary antibody is freshly prepared for each experiment as a 1:20,000-fold dilution in 5% milk and added to the membrane for 60 min at room temperature on a rocking platform.
7. The secondary antibody is discarded and the membrane washed six times for 10 min each with TBS-T.
8. During the final wash, 10-ml aliquots of each portion of the ECL reagent are warmed separately to room temperature and the remaining steps are done in a dark room. Once the final wash is removed from the blot, the ECL reagents are mixed together and then immediately added to the blot, which is then rocked gently by hand for 1 min to ensure even coverage.
9. The blot is removed from the ECL reagents, blotted with Kim-Wipes, and then placed between Saran wrap cut to appropriate size of the membrane. The Saran wrap containing the membrane is then placed in an X-ray film cassette with film for a suitable exposure time, typically a few minutes. An example of the results produced is shown in **Fig. 4**.

Component	Amount (μl)	Final concentration
Plasmid substrate (pUC18 encoding the targeted gene or ZFN sites)	2.0	0.005–0.01 μg/μl
10× NEB buffer 4	20.0	1×
ZFN123 IVTT extract	2.5	
ZFN456 IVTT extract	2.5	
dH_2O	173.0	

3.7. Protocol for the Cell-Free ZFNS Cleavage Assay Using the IVTT Extract

1. The chosen ZFN target sites are cloned into pUC18 plasmid, which serves as the substrate *(14, 21)* for ZFN digestion. The ZFN digestion reaction is set as follows:

The control for the reaction is set up with the IVTT reaction product obtained without adding for ZFN expression plasmid. The reaction is incubated at 30°C for 1 h.

2. The digest is extracted with phenol-chloroform and then precipitated with ethanol; the precipitate is air-dried and resuspended in 100 μl of autoclaved water. The digest is analyzed using a 1% agarose gel. Similarly, reactions using other restriction enzymes (AatII or ScaI or XmnI, respectively) can also be performed.

Component	Amount (μl)	Final concentration
ZFN-digested substrate plasmid	10.0	0.5 μg/μl
5,000 U/ml SspI	1.0	0.25 U/μl
10× NEB buffer 2	2.0	1×
dH_2O	7.0	

The digest is performed at 37°C overnight. One microliter of RNAse (5 mg/ml) is then added to the reaction mix and incubated for another 60 min at 37°C. The digest is analyzed using a 1% agarose gel. Similarly, reactions using other restriction enzymes (AatII or ScaI or XmnI, respectively) can also be performed. An example of this gel profile is shown in **Fig. 4** (for trouble shooting, *see* **Note 5**).

3.8. Applications of ZFN-Mediated Gene Targeting: Site-Specific and Permanent Modification of Plant and Mammalian Genomes, Including the Human Genome

Several laboratories have shown that designed 3- and 4-finger ZFNs find and cleave their chromosomal targets in a variety of cell substrates, which include frog oocytes *(29)*, *Drosophila (30, 31)*, plant cells *(17, 18)*, *Caenorhabditis elegans(32)*, and human cells *(15, 16, 33)*; and, as expected, they induce local HR at the site of cleavage to repair the DSB. In the absence of HR (for example, if both alleles of a gene are damaged), cells repair the DSB by simple ligation with the addition or deletion of some sequence (via NHEJ). Therefore, repair via NHEJ is mutagenic by nature. NHEJ-mediated repair has been shown for DSBs induced by ZFNs in *Drosophila (30, 31)* and *Arabidopsis(18)*. Thus, to induce targeted gene modifications in a variety of organisms and cells, ZFNs can be used with or without a homologous template DNA sequence (involving HR and NHEJ, respectively). Urnov et al. *(16)* have used designed 4-finger ZFNs in human cells to target an endogenous target site within the IL2Rγ gene (which causes the human X-linked disease, severe combined immune deficiency [SCID]). They achieved highly efficient permanent modification of the IL2Rγ gene in the K562 cell line—a remarkable gene-modification efficiency of 18% of K562 cells had the desired gene replacement without additional cell selection, and one third of these were altered on both X chromosomes. They obtained similar results using primary human T lymphocytes. Thus, the designer ZFNs have become powerful molecular tools to deliver a targeted genomic DSB to cells and stimulate local HR in human cells with an exogenously provided DNA template *(15–17, 30, 33–35)*. The high targeting efficiency attests to the power of the ZFN-evoked HR for site-specific engineering of the human genome and raises the possibility of developing ZFN-based strategies *(1)* for laboratory research of gene-modified human cells, to complement what can be done to study gene function in knock-out and knock-in models of mice and other species, *(2)* for targeted genome engineering of plant species, and *(3)* potentially for human therapeutics as a form of gene therapy in the future *(1, 16, 33, 35–38)* (*see* **Notes 4** and **6**).

3.9. Summary

ZFN technology has shown great promise and has the obvious potential to have a tremendous impact on both biotechnology and genetic medicine. ZFN technology has been referred to as a "game changing innovation" specifically for agriculture because it is now possible to target native genes in plants using custom ZFNs with exceptional precision. ZFNs have made it possible to target a preselected site and enable site-specific transgene integration in plants. It is only a matter of time before the ZFN technology is used toward trait generation and trait stacking across different crop plants. Thus, the ZFN technology platform has significant potential to contribute to the development of precision traits and to enable efficient and reproducible generation of combinations or stacks of multiple traits by targeted insertion of new traits into plants.

ZFN-based strategies as a form of gene therapy, in particular by modifying human stem cells ex vivo, may provide a new paradigm in genetic medicine for treating monogenic human diseases *(2, 39)* by correcting the causative genetic defect at its origins. Many of the difficulties associated with gene therapy are likely to be overcome if one could insert the corrected version of the mutation at the precise location of the genetic defect within the human genome. The current gene therapy vectors lack the requisite sequence specificity necessary for targeted correction of the defective site within the human genome. ZFN-based strategies for gene correction of human stem cells may provide a viable option to treat monogenic human diseases in the future. The ZFN technology is still at its infancy and it has the potential to fulfill its promise for human therapeutics, in particular of curing monogenic diseases over the next decade.

4. Notes

1. Inverted sequences of the form $(NNC)_{3or\ 4}$…$(GNN)_{3or\ 4}$ separated by 5 or 6 bp make for the best ZFN target sites. The ZF motif designs for GNN and ANN triplets are the best studied and well characterized; and therefore, they are highly preferred over the ZFN sites containing CNN and TNN triplets.

2. Whenever possible, select 4-finger ZFN sites, because they are expected to be more sequence specific and likely to show higher affinity for their cognate sites compared with the 3-finger sites, everything else being equal. The 4-finger ZFNs are also likely to be more selective in binding to their targets and less toxic to cells.
3. The binding of the ZF motif to its cognate triplet occurs in the context of neighboring fingers, which could affect the specificity and affinity of the ZFPs. Therefore, designing and tethering of ZF motifs to form a ZFP, and hence linking the ZFP to the FokI cleavage domain to form the ZFN, for a chosen target site do not always yield ZFNs with the desired sequence specificity and affinity. More often than not, one needs to optimize the sequence-specific binding and affinity of the designed ZFPs by tinkering with ZF motif designs within a designed ZFP by substituting with other ZF motif designs that are available for a particular triplet, as was done by Urnov et al. *(16)*.
4. The requirement for dimerization of the FokI cleavage domain restricts cleavage by a pair of ZFNs to long sequences, but it also introduces a potential problem, because this dimer interaction does not select for the heterodimer species, which could lead to unwanted off-target cleavage elsewhere in the genome of cells. Two different homodimers could result from the two individual ZFNs with different sequence specificities, and, more often than not, they do occur. The homodimers, although they are not relevant for gene modification, potentially could and often do affect the safety and efficacy of ZFN-mediated gene targeting of cells. Two recent articles have addressed this issue of ZFN cytotoxicity by redesigning the ZFN dimer interface to inhibit homodimerization, thereby, greatly reduce the off-target cleavage by ZFNs, and hence their cytotoxicity *(40,41)*. We strongly recommend using these improved FokI cleavage domains in pairs of ZFNs to reduce cytotoxicity.
5. The TnTT7 Quick-Coupled Transcription/Translation System also contains nonspecific nucleases that could degrade DNA. If present during ZFN digestion of plasmid substrate encoding the ZFN sites, they could degrade DNA fragments resulting from substrate cleavage. The transcribed/translated ZFN gene reaction mixture using IVTT will also contain messenger RNA (mRNA) and transfer RNAs (tRNAs), which could appear as background during agarose gel electrophoresis. RNaseA could help in their removal. ZFNs could also remain bound to substrate and to the product fragments resulting from ZFN cleavage, giving rise to gel-

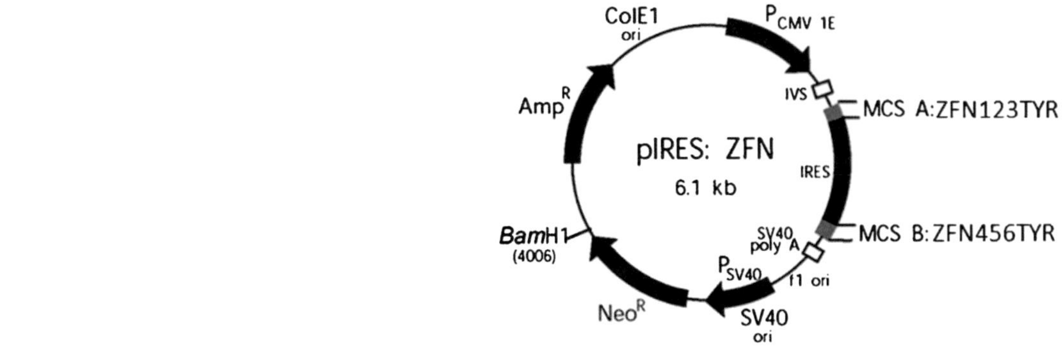

Fig. 5. Structure of pIRES: ZFNs. The pair of ZFNs that bind to the mTYR site, are cloned into a single pIRES plasmid at two different multiple cloning sites (MCS) downstream of the CMV promoter for use in the co-transfection of mouse melanocytes along with the exogenous donor DNA during ZFN-mediated gene targeting experiments.

shifted bands during agarose gel electrophoresis. Incubation of the reaction mixture with proteinase K might be needed for the removal of ZFNs from the resulting product fragments.

6. The pair of custom-designed ZFNs that target the mTYR gene are then cloned into a single pIRES plasmid at two different multiple cloning sites (MCS) downstream of the cytomegalovirus (CMV) promoter used in the co-transfection of mouse melanocytes along with the exogenous donor DNA during ZFN-evoked gene targeting experiments using cell cultures (**Fig. 5**). Alternatively, the pair of ZFNs with different sequence specificities could also be cloned individually into the pIRES plasmid and then both plasmids simultaneously introduced into the cells along with the donor DNA.

Acknowledgments

We thank Dr. Sundar Durai for drawing the figures. The research on ZFNs in our lab has been supported by various grants from National Institutes of Health, USA, during the past 13 years; it is currently being funded by the research grant GM077291 from NIGMS/NIH. Our work on ZFN-mediated gene targeting in human stem cells is partially supported by a grant from the Maryland Stem Cell Research Fund.

References

1. Kandavelou, K., M. Mani, S. Durai, and S. Chandrasegaran. (2005). "Magic" scissors for genome surgery. *Nat Biotechnol* **23**:686–687.
2. Wu, J., K. Kandavelou, and S. Chandrasegaran. (2007). Custom-designed zinc finger nucleases: what is next? *Cell Mol Life Sci* **64**:2933–2944.
3. Vasquez, K.M., K. Marburger, Z. Intody, and J.H. Wilson. (2001). Manipulating the mammalian genome by homologous recombination. *Proc Natl Acad Sci U S A* **98**:8403–8410.
4. Kim, C.A., and J.M Berg. (1996). A 2.2 Å resolution crystal structure of a designed zinc finger protein bound to DNA. *Nat Struct Biol* **3**:940–945.
5. Pavletich, N.P., and C.O. Pabo. (1991). Zinc finger-DNA recognition: crystal structure of a Zif268-DNA complex at 2.1 Å. *Science* **252**:809–817.
6. Desjarlais, J.R., and J.M. Berg. (1993). Use of a zinc-finger consensus sequence framework and specificity rules to design specific DNA binding proteins. *Proc Natl Acad Sci U S A* **90**:2256–2260.
7. Shi, Y., and J.M. Berg. (1995). A direct comparison of the properties of natural and designed zinc-finger proteins. *Chem Biol* **2**:83–89.
8. Beerli, R.R., D.J. Segal, B. Dreier, and C.F. Barbas, III. (1998). Toward controlling gene expression at will: specific regulation of the erbB-2/HER-2 promoter by using polydactyl zinc finger proteins constructed from modular building blocks. *Proc Natl Acad Sci U S A* **95**:14628–14633.
9. Kim, J.S., and C.O. Pabo. (1998). Getting a handhold on DNA: design of poly-zinc finger proteins with femtomolar dissociation constants. *Proc Natl Acad Sci U S A* **95**:2812–2817.
10. Liu, Q., D.J. Segal, J.B. Ghiara, and C.F. Barbas, III. (1997). Design of polydactyl zinc-finger proteins for unique addressing within complex genomes. *Proc Natl Acad Sci U S A* **94**:5525–5530.
11. Chandrasegaran, S., and J. Smith. (1999). Chimeric restriction enzymes: what is next? *Biol Chem* **380**:841–848.
12. Kandavelou, K., Mani, M., Durai, S., and Chandrasegaran, S. (2004). Engineering and applications of chimeric nucleases. Springer, Berlin. pp. 413–414.
13. Kim, Y.G., J. Cha, and S. Chandrasegaran. (1996). Hybrid restriction enzymes: zinc finger fusions to Fok I cleavage domain. *Proc Natl Acad Sci U S A* **93**:1156–1160.
14. Smith, J., J.M. Berg, and S. Chandrasegaran. (1999). A detailed study of the substrate specificity of a chimeric restriction enzyme. *Nucleic Acids Res* **27**:674–681.
15. Porteus, M.H., and D. Baltimore. (2003). Chimeric nucleases stimulate gene targeting in human cells. *Science* **300**:763.
16. Urnov, F.D., J.C. Miller, Y.L. Lee, C.M. Beausejour, J.M. Rock, S. Augustus, A.C. Jamieson, M.H. Porteus, P.D. Gregory, and M.C. Holmes. (2005). Highly efficient endogenous human gene correction using designed zinc-finger nucleases. *Nature* **435**:646–651.
17. Wright, D.A., J.A. Townsend, R.J. Winfrey, Jr., P.A. Irwin, J. Rajagopal, P.M. Lonosky, B.D. Hall, M.D. Jondle, and D.F. Voytas. (2005). High-frequency homologous recombination in plants mediated by zinc-finger nucleases. *Plant J* **44**:693–705.
18. Lloyd, A., C.L. Plaisier, D. Carroll, and G.N. Drews. (2005). Targeted mutagenesis using zinc-finger nucleases in Arabidopsis. *Proc Natl Acad Sci U S A* **102**:2232–2237.
19. Carroll, D., J.J. Morton, K.J. Beumer, and D.J. Segal. (2006). Design, construction and in vitro testing of zinc finger nucleases. *Nat Protoc* **1**:1329–1341.
20. Wright, D.A., S. Thibodeau-Beganny, J.D. Sander, R.J. Winfrey, A.S. Hirsh, M. Eichtinger, F. Fu, M.H. Porteus, D. Dobbs, D.F. Voytas, and J.K. Joung. (2006). Standardized reagents and protocols for engineering zinc finger nucleases by modular assembly. *Nat Protoc* **1**:1637–1652.
21. Smith, J., M. Bibikova, F.G. Whitby, A.R. Reddy, S. Chandrasegaran, and D. Carroll. (2000). Requirements for double-strand cleavage by chimeric restriction enzymes with zinc finger DNA-recognition domains. *Nucleic Acids Res* **28**:3361–3369.
22. Mandell, J.G., and C.F. Barbas, III. (2006). Zinc finger tools: custom DNA-binding domains for transcription factors and nucleases. *Nucleic Acids Res* **34**:W516–W523.
23. Liu, P.Q., E.J. Rebar, L. Zhang, Q. Liu, A.C. Jamieson, Y. Liang, H. Qi, P.X. Li, B. Chen, M.C. Mendel, X. Zhong, Y.L. Lee, S.P. Eisenberg, S.K. Spratt, C.C. Case, and A.P. Wolffe. (2001). Regulation of an

endogenous locus using a panel of designed zinc finger proteins targeted to accessible chromatin regions. Activation of vascular endothelial growth factor A. *J Biol Chem* **276**:11323–11334.

24. Liu, Q., Z. Xia, X. Zhong, and C.C. Case. (2002). Validated zinc finger protein designs for all 16 GNN DNA triplet targets. *J Biol Chem* **277**:3850–3856.
25. Dreier, B., R.R. Beerli, D.J. Segal, J.D. Flippin, and C.F. Barbas, III. (2001). Development of zinc finger domains for recognition of the 5′-ANN-3′ family of DNA sequences and their use in the construction of artificial transcription factors. *J Biol Chem* **276**:29466–29478.
26. Zhang, L., S.K. Spratt, Q. Liu, B. Johnstone, H. Qi, E.E. Raschke, A.C. Jamieson, E.J. Rebar, A.P. Wolffe, and C.C. Case. (2000). Synthetic zinc finger transcription factor action at an endogenous chromosomal site. Activation of the human erythropoietin gene. *J Biol Chem* **275**:33850–33860.
27. Mani, M., K. Kandavelou, F.J. Dy, S. Durai, and S. Chandrasegaran. (2005). Design, engineering, and characterization of zinc finger nucleases. *Biochem Biophys Res Commun* **335**:447–457.
28. Ruminy, P., C. Derambure, S. Chandrasegaran, and J.P. Salier. (2001). Long-range identification of hepatocyte nuclear factor-3 (FoxA) high and low-affinity binding sites with a chimeric nuclease. *J Mol Biol* **310**:523–535.
29. Bibikova, M., D. Carroll, D.J. Segal, J.K. Trautman, J. Smith, Y.G. Kim, and S. Chandrasegaran. (2001). Stimulation of homologous recombination through targeted cleavage by chimeric nucleases. *Mol Cell Biol* **21**:289–297.
30. Bibikova, M., K. Beumer, J.K. Trautman, and D. Carroll. (2003). Enhancing gene targeting with designed zinc finger nucleases. *Science* **300**:764.
31. Bibikova, M., M. Golic, K.G. Golic, and D. Carroll. (2002). Targeted chromosomal cleavage and mutagenesis in *Drosophila* using zinc-finger nucleases. *Genetics* **161**: 1169–1175.
32. Morton, J., M.W. Davis, E.M. Jorgensen, and D. Carroll. (2006). Induction and repair of zinc-finger nuclease-targeted double-strand breaks in *Caenorhabditis elegans* somatic cells. *Proc Natl Acad Sci U S A* **103**:16370–16375.
33. Alwin, S., M.B. Gere, E. Guhl, K. Effertz, C.F. Barbas, III, D.J. Segal, M.D. Weitzman, and T. Cathomen. (2005). Custom zinc-finger nucleases for use in human cells. *Mol Ther* **12**:610–617.
34. Beumer, K., G. Bhattacharyya, M. Bibikova, J.K. Trautman, and D. Carroll. (2006). Efficient gene targeting in *Drosophila* with zinc-finger nucleases. *Genetics* **172**:2391–2403.
35. Porteus, M.H. (2006). Mammalian gene targeting with designed zinc finger nucleases. *Mol Ther* **13**:438–446.
36. Moehle, E.A., J.M. Rock, Y.L. Lee, Y. Jouvenot, R.C. DeKelver, P.D. Gregory, F.D. Urnov, and M.C. Holmes. (2007). Targeted gene addition into a specified location in the human genome using designed zinc finger nucleases. *Proc Natl Acad Sci U S A* **104**:3055–3060.
37. Porteus, M.H., and D. Carroll. (2005). Gene targeting using zinc finger nucleases. *Nat Biotechnol* **23**:967–973.
38. Wilson, J.H. (2003). Pointing fingers at the limiting step in gene targeting. *Nat Biotechnol* 21:759–760.
39. Porteus, M.H., J.P. Connelly, and S.M. Pruett. (2006). A look to future directions in gene therapy research for monogenic diseases. *PLoS Genet* **2**:e133.
40. Miller, J.C., M.C. Holmes, J. Wang, D.Y. Guschin, Y.L. Lee, I. Rupniewski, C.M. Beausejour, A.J. Waite, N.S. Wang, K.A. Kim, P.D. Gregory, C.O. Pabo, and E.J. Rebar. (2007). An improved zinc-finger nuclease architecture for highly specific genome editing. *Nat Biotechnol* **25** :778–785.
41. Szczepek, M., V. Brondani, J. Buchel, L. Serrano, D.J. Segal, and T. Cathomen. (2007). Structure-based redesign of the dimerization interface reduces the toxicity of zinc-finger nucleases. *Nat Biotechnol* **25**:786–793.

Chapter 41

Determining DNA Sequence Specificity of Natural and Artificial Transcription Factors by Cognate Site Identifier Analysis

Mary S. Ozers, Christopher L. Warren, and Aseem Z. Ansari

Summary

Artificial transcription factors (ATFs) are designed to mimic natural transcription factors in the control of gene expression and are comprised of domains for DNA binding and gene regulation. ATF domains are modular, interchangeable, and can be composed of protein-based or nonpeptidic moieties, yielding DNA-interacting regulatory molecules that can either activate or inhibit transcription. Sequence-specific targeting is a key determinant in ATF activity, and DNA-binding domains such as natural zinc fingers and synthetic polyamides have emerged as useful DNA targeting molecules. Defining the comprehensive DNA binding specificity of these targeting molecules for accurate manipulations of the genome can be achieved using cognate site identifier DNA microarrays to explore the entire sequence space of binding sites. Design of ATFs that regulate gene expression with temporal control will generate important molecular tools to probe cell- and tissue-specific gene regulation and to function as potential therapeutic agents.

Key words: Artificial transcription factor, Polyamide, DNA-binding domain, Transcriptional activation domain, Cognate site identifier arrays

1. Introduction

Natural transcription factors are nuclear proteins that bind to specific DNA sequences localized at a gene or set of genes and recruit enzymes to modify chromatin structure and/or synthesize messenger RNA (mRNA) (**Fig. 1a**). Transcription factors are responsible for choreographing the complex cascade of signaling events that give rise to specific cell types and tissues. Sequencing of the human genome has indicated that nearly 6% of the genome (~2,000 genes) encodes transcription factors, yet the genome-wide role of these factors in controlling gene expression has

James Weifu Lee and Robert S. Foote (eds.), *Micro and Nano Technologies in Bioanalysis*,
Methods in Molecular Biology, vol. 544
DOI 10.1007/978-1-59745-483-4_41, © Humana Press, a part of Springer Science+Business Media, LLC 2009

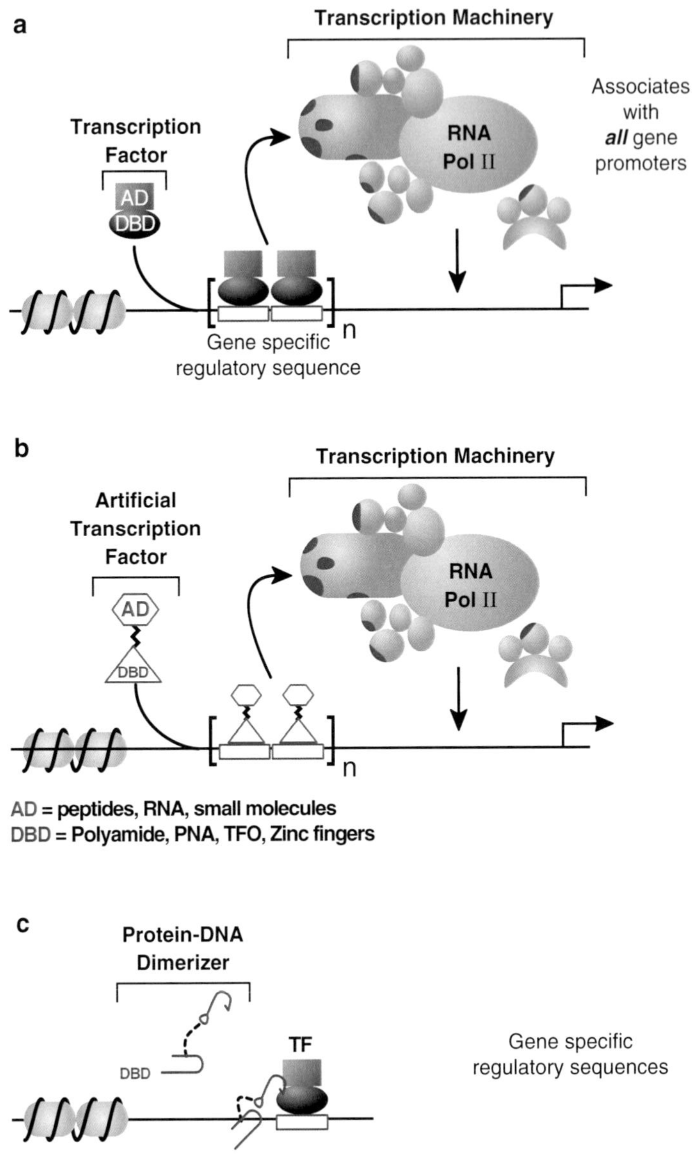

Fig. 1. Role of ATFs and protein–DNA dimerizers in transcription. (**a**) Transcription factors are modular proteins composed of a DNA-binding domain (DBD) that recognizes gene-specific regulatory sequences and an activation domain (AD) that recruits RNA polymerase II and/or associated proteins. (**b**) Artificial transcription factors (ATFs) can be created using transcriptional activation domains such as VP16 or small molecules and DBDs such as protein-based DNA-binding modules or synthetic molecules including polyamides, triplex-forming oligonucleotides (TFOs), or peptide nucleic acids (PNAs). (**c**) Protein–DNA dimerizers contain a DBD linked to a molecular "hook" such as a short peptide or small molecule, which facilitates binding of a natural transcription factor at an adjacent DNA site.

been well characterized for only a small subset *(1–6)*. Transcription factors are capable of activating or repressing gene expression, and their aberrant activity has been linked to an array of disorders including cancer, obesity, diabetes, and inflammation.

Artificial transcription factors (ATFs) that can be engineered to target specific genomic sites and modulate gene activity will be useful as novel therapeutics for disease treatment and for mechanistic studies of gene expression.

ATFs mimic the modular design of natural transcription factors and are comprised of a DNA-binding domain (DBD), a regulatory domain, and, in some cases, a linker region **(Fig. 1b)** *(7, 8)*. ATFs can also target a natural transcription factor to an adjacent DNA-binding site by use of a molecular "hook" such as a peptide or small molecule that specifically recruits the transcription factor **(Fig. 1c)**. Early design of ATFs combined the DBD of one protein with the regulatory domain of another, resulting in chimeric proteins with altered DNA binding specificity. Common activating domains include the herpes simplex viral transactivator VP16 or a 20-residue peptide known as amphipathic helix (AH), with the sequence PEFPGIELQELQELQALLQQ *(9)*. Despite decades of research to uncover a specific structure or sequence associated with transcriptional regulatory domains, the only preferences that emerged were acid-rich, glutamine-rich, or proline-rich activation domains and alanine-rich or positively charged repression domains *(10–14)*. These studies indicated that regulatory domain determinants such as structural folds or specific motifs were not clearly defined, whereas the DBDs could be more accurately described by structure, and their binding specificity could be mimicked by natural or synthetic molecules.

DBDs direct specific interaction with DNA sites to deliver the regulatory domain to a target gene or set of genes. Protein-based DBDs are often categorized by their structural fold such as a helix-turn-helix or coordination molecules such as a zinc finger. Zinc fingers of the type Cys_2His_2 are ~30-amino acid domains that fold into two β-strands and one α-helix, with the N-terminal residues of the α-helix making specific base contacts in the major groove of DNA. Cys_2His_2 zinc fingers are useful as ATF DBDs because they can be tailored to recognize any 3-bp DNA site with precise specificity and high affinity *(15–17)*. Although empirical rules for zinc finger recognition of all DNA 9-mers theoretically exist, some zinc finger domains make base contacts outside of their three target nucleotides, preferring a 4-bp rather than a 3-bp site. Computational approaches, including sequential and bipartite selection strategies, and use of predictive design tools have improved zinc finger design *(15, 17)* but also highlight the need to define and refine DBD binding specificity. Some limitations of using protein-based ATFs for disease therapy include the low efficiency of cellular delivery, poor nuclear localization, significant potential for antigenicity, and protein degradation *(8)*.

Synthetic DBDs were designed to address some of these limitations of cellular uptake and to provide a chemical approach to selectively alter gene expression; examples include polyamides, triplex-forming oligonucleotides (TFOs), and peptide nucleic acids (PNAs). Polyamides bind in the minor groove of DNA and are composed of *N*-methylpyrrole (Py) and N-methylimidazole (Im) amino acids or derivatives such as hydroxypyrrole (Hp). The advantage of polyamides is that the combination of the Py and Im aromatic rings in a side-by-side arrangement can be exploited, according to a set of pairing rules, to generate a DBD that targets a specific DNA sequence (**Fig. 2**). Important aspects of DNA binding include the specificity of DNA recognition and the binding affinity or efficacy for a particular site. For some DNA targets, polyamides possess improved specificity and affinity if polyamide sequence recognition rules are extended to include polyamide pairings such as ImPy and PyPy (i.e., two pairs of rings) to target Watson-Crick base pairs *(19)*. Especially promising in ATF design, polyamides are permeable to mammalian cells and can affect gene expression. Characterization of polyamide binding in colon cancer cells indicated that expression of particular genes was affected but some of the expected signaling pathways were not down-regulated, although it was unclear whether this occurred because of altered DNA recognition, chromatin structure, or

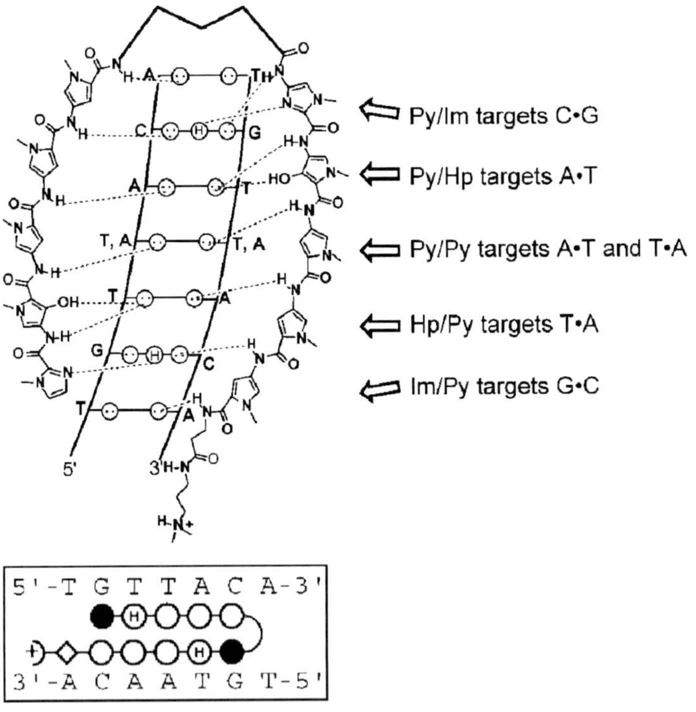

Fig. 2. Polyamide pairing rules. (**a**) Polyamides target DNA based on a set of pairing rules. An Im/Py ring pair recognizes G·C; Py/Py pair targets A·T or T·A; and Py/Hp binds to A·T *(18)*. (**b**) Polyamides can be represented by the following *abbreviations* and *symbols*: Im, *N*-methylimidazole, *filled circle*; Py, *N*-methylpyrrole, *open circle*; Hp, hydroxypyrrole, *open circle* with H; β, β-alanine, *diamond*; Dp, dimethylaminopripylamide, *half circle* with *positive charge*; γ or turn, γ-aminobutyric acid.

interplay with other transcription factor signaling pathways *(20)*. Thus, sequence-specific targeting and refining this DNA recognition remain critical areas of ATF development.

The specific binding of ATFs can be evaluated using cognate site identifier (CSI) arrays bearing every permutation of an 8- to 10-bp DNA that is double stranded and B-form **(Fig. 3)** *(21)*. DNA is synthesized on the array using maskless array synthesizer (MAS) technology with every feature containing multiple copies of a particular sequence *(22)*. DNA-binding molecules are applied to the array and are detected using fluorescence by direct labeling of the molecule of interest or with a fluorescent antibody. CSI analysis was applied to determine the binding specificity of a Cy3-labeled polyamide, PA1 **(Fig. 4a)**, which was engineered to target the sequence 5′-WWGWWCWW-3′ (W = A or T). Most features were not bound by PA1, as shown by the peak centered at zero, whereas a subset of features displayed high intensity **(Fig. 4b)**. The binding motif was very similar to the expected binding sequence for PA1 **(Fig. 4c)**. Further CSI analysis revealed subtle and unobvious contributions of the core 6 bp of the PA1-binding site to DNA binding specificity, highlighting the usefulness of CSI arrays in determining DNA specificity *(21)*.

DNA binding specificity and affinity can also be engineered using cooperative complexes. The *Drosophila* transcription factors, Hox and Extradenticle (Exd), bind DNA with lower specificity and affinity as individual proteins, but their binding specificity and affinity dramatically increase as a cooperatively bound complex *(23)*. Polyamides have been used in the creation of bifunctional molecules, known as protein-DNA dimerizers, comprised of

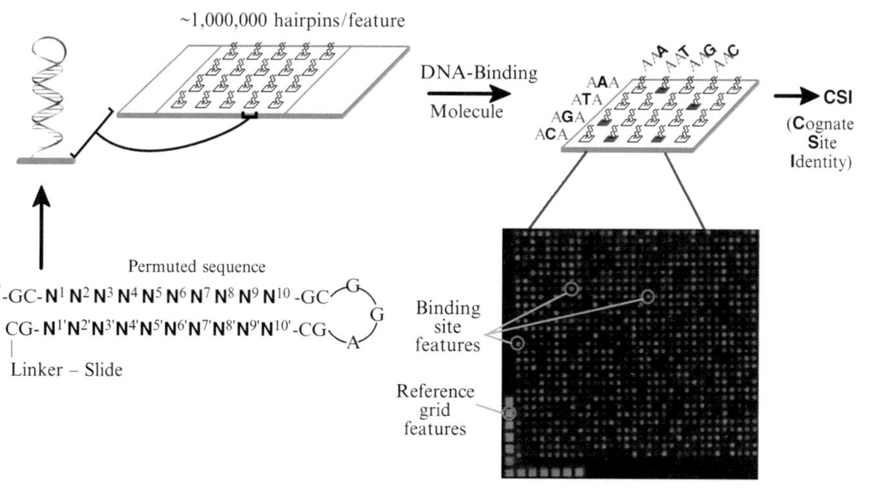

Fig. 3. Illustration of a CSI microarray. Every permutation of a 10 bp (N^1–N^{10}) sequence is displayed in a hairpin probe containing a GC stem and GGA turn. The array is incubated with a fluorescently labeled DNA-binding molecule. The fluorescent features are identified and used to determine DNA-binding preferences.

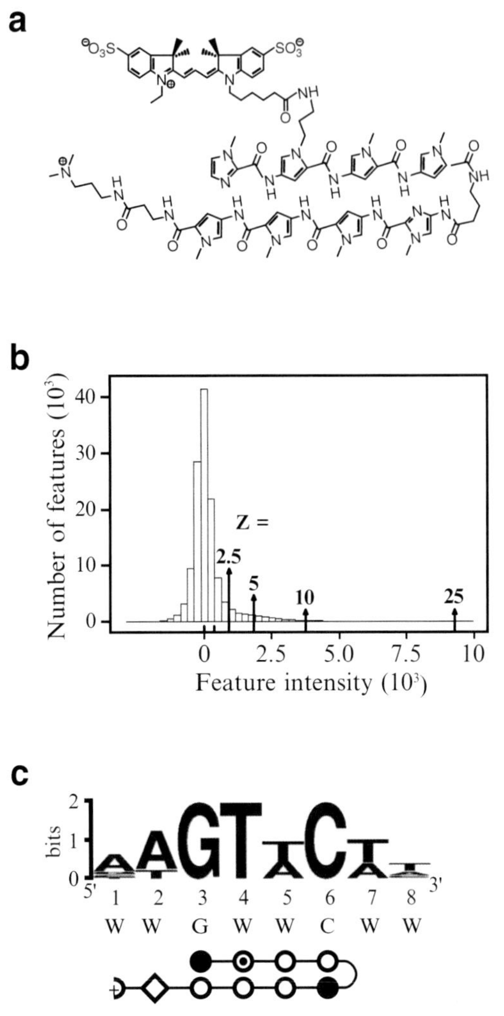

Fig. 4. Binding of PA1. (**a**) Structure of Cy3-conjugated polyamide PA1 (ImPy*PyPy-γ-ImPyPyPy-β-Dp). (**b**) Histogram of averaged intensities of all replicate features. *Z* scores (see Methods) are noted. (**c**) Logo obtained from top *Z*-score bin of 25. *Abbreviations:* Py*, *N*-methylpyrrole ring with Cy3 dye attached, *open circle* with *inner dot*; W = A or T.

a protein-interacting molecule linked to the polyamide DBD **(Fig. 1c)** *(24, 25)*. The protein-interacting molecule, also referred to as a "hook," can stabilize or enhance the binding of a natural transcription factor to an adjacent DNA site. A polyamide bearing the Hox peptide YPWM has been shown to recruit Exd in a cooperative complex on DNA, mimicking the developmental regulator Hox-Exd heterodimeric complex **(Fig. 5a)** *(25)*. The length of the linker between the protein-interacting moiety and the polyamide DBD determines the effectiveness of the protein-DNA dimerizer in recruiting the natural transcription factor

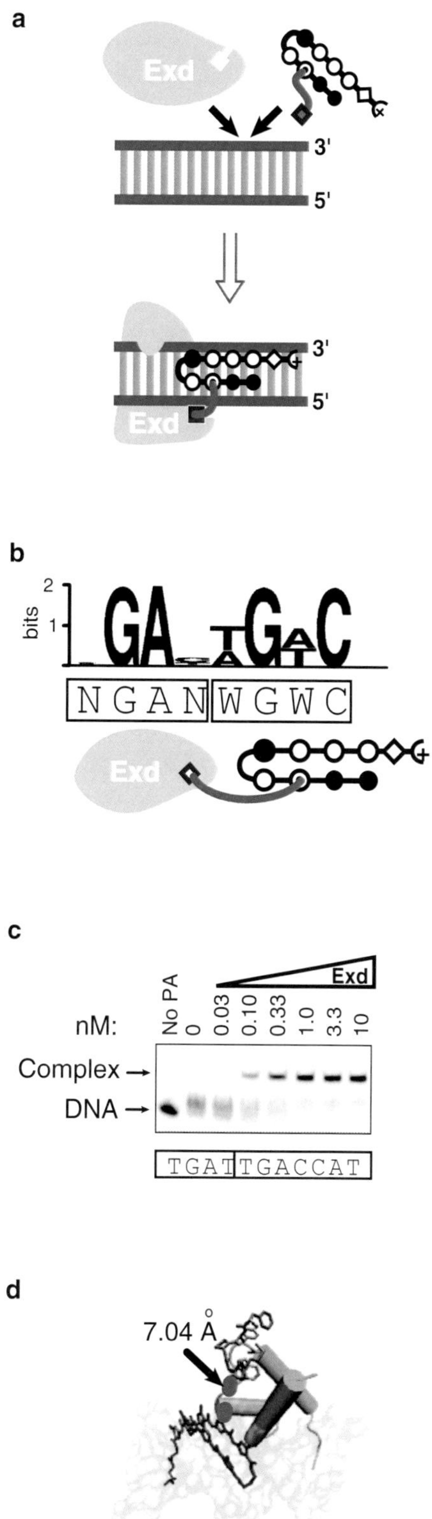

Fig. 5. Polyamide–Exd cooperative complex. (**a**) Schematic of polyamide–Exd cooperative complex on DNA. (**b**) Logo *(26–28)* obtained from CSI analysis of polyamide–Exd binding. *Boxed* sequence displays binding sites for Exd and polyamide. (**c**) The polyamide-peptide conjugate at 50 nM was incubated with increasing concentrations of Exd (nM). *Arrows* indicate free DNA (*lower arrow*) and DNA–polyamide complex (*upper arrow*). *Boxed* sequence denotes the Exd and polyamide-binding sites. (**d**) Molecular modeling *(29,30)* of Exd (*green*) and polyamide (*blue*) bound to consensus DNA site, with peptide hook (*brown*) and linker (*red*). Models were obtained by aligning crystal structures of the DNA complexed with Exd or hairpin polyamide (Protein Data Bank files 1B8I and 1M18).

(24, 31). The binding specificity of the polyamide-Exd complex was comprehensively defined using CSI arrays **(Fig. 5b)** and verified by electrophoretic mobility shift assays (EMSAs) **(Fig. 5c)** and molecular modeling **(Fig. 5d)**.

This review chapter focuses on methods to precisely and comprehensively define the DNA targeting specificity of ATFs, or any DNA-binding molecule or protein, using CSI microarrays. Biochemical/biophysical tools to validate the accuracy of the CSI platform are also described.

2. Materials

2.1. Polyamide Synthesis

1. Succinimidyl ester of Cy3 dye (Amersham, Piscataway, NJ) for dye labeling.
2. Phenylacetamidomethyl (PAM) resin (Peptides International, Louisville, KY).
3. Chemical reagents: dimethylaminopropylamine, trifluoroacetic acid (TFA), acetonitrile, methanol, toluene, dimethylformamide (DMF), diisopropylethylamine (DIEA).
4. N-methylpyrrole and N-methylimidazole.

2.2. DNA Array Synthesis

1. ArrayIt SuperClean microscope slides (TeleChem, Inc., Sunnyvale, CA).
2. UV-protected desiccator (Secador™) and Drierite desiccant (Fisher Scientific, Pittsburgh, PA).
3. Microarraying facility with MAS technology.
4. *N*-(3-Triethoxysilylpropyl)-4-hydroxy-butyramide (Gelest, Morrisville, PA).
5. Slide buffer: 95% ethanol, 0.1% glacial acetic acid (*see* **Note 1**).
6. Silane buffer: 95% ethanol, 0.1% glacial acetic acid, 1.6% *N*-(3-triethoxysilylpropyl)-4-hydroxy-butyramide.
7. Acetone.
8. Metal rack slide holder (Wheaton, Millville, NJ) and glass container (Pyrex or comparable) large enough to hold metal slide holder; prewashed with acetone.
9. Standard lab oven with vacuum drying capability (Lab-Line/Barnstead International, Dubuque, IA).
10. 200-Proof ethanol.
11. Forceps.
12. Adjustable speed platform shaker (Fisher Scientific).
13. Glass slide holder, ~100 mL size (Fisher Scientific).
14. Ethylenediamine (Fisher Scientific).

15. 50-mL Conical tubes.
16. Methanol (Fisher Scientific).
17. 7 M Urea in 1× phosphate-buffered saline (PBS).
18. 5× PBS: 685 mM NaCl, 13.5 mM KCl, 50 mM Na_2HPO_4, 9 mM KH_2PO_4. Adjust to pH 7.4 with HCl if necessary. Filter-sterilize or autoclave and store at room temperature.
19. Nonstringent wash buffer (0.9 M NaCl, 60 mM NaH_2PO_4, 7.6 mM EDTA, 0.01% v/v Tween-20).
20. Final wash buffer (NimbleGen Systems, Inc., Madison, WI).
21. ArrayIt array dryer (TeleChem, Inc., Sunnyvale, CA) (*see* **Note 2**).
22. Axon 4000B 5-μm microarray scanner (Molecular Devices, Sunnyvale, CA), or comparable, connected to an IBM-compatible computer with 1.6-GHz dual-core processor or faster; Windows XP or Vista (32-bit) operating system; 1-GB RAM; 40-GB hard drive.

2.3. Measurement of Microarray Signal Intensities

1. NimbleScan 2.4 or GenePix® Pro 6.0 microarray analysis software.
2. Excel software (Microsoft).
3. Access software (Microsoft).

2.4. Data Analysis

1. MEME/MAST System Motif Discovery and Search (http://meme.sdsc.edu/meme/intro.html).

2.5. Polyamide Binding by CSI Arrays

1. Hybridization buffer (100 mM MES, 1 M NaCl, 20 mM EDTA pH 7.5, 0.01% v/v Tween-20).
2. Secure-seal™ hybridization chamber (Grace Bio-Labs, Bend, OR).
3. Blocking buffer (2.5% nonfat dried milk in dH_2O).

2.6. Biochemical/Biophysical Validation of CSI Arrays

2.6.1. Electrophoretic Mobility Shift Assays

1. 10 μM DNA stock (Integrated DNA Technologies, Coralville, IA, or comparable), in dH_2O or TE buffer.
2. TE buffer (10 mM Tris-Cl pH 7.5, 1 mM EDTA).
3. 5× forward buffer (Invitrogen, Carlsbad, CA).
4. Kinase T4 (Invitrogen).
5. γ^{32}P-ATP (6,000 Ci/mmoL, PerkinElmer, Waltham, MA).
6. 0.6-mL, 1.5-mL, and 2-mL Eppendorf tubes.
7. G25 spin column (GE Healthcare, Piscataway, NJ).
8. Binding buffer (150 mM potassium glutamate, 50 mM HEPES pH 7.5, 2 mM DTT, 100 ng/μL bovine serum albumin (BSA), 10% DMSO, 10% glycerol).
9. 10% Acrylamide/3% glycerol gel.

10. 1× TBE: 90 mM Tris, 64.6 mM boric acid, 2.5 mM EDTA pH 8.3; can be prepared as 10× stock, kept at room temperature, and diluted one part in nine parts dH_2O.
11. Geiger counter.
12. Gel electrophoresis voltage box (Pharmacia/GE Healthcare).
13. Typhoon PhosphorImager System (GE Healthcare) or comparable.
14. ImageQuant software (GE Healthcare).

2.6.2. Fluorescence Polarization

1. Cy3-labeled polyamide (synthesized as in **Subheading 3.3.1**).
2. Double-stranded DNA (Integrated DNA Technologies or comparable) in dH_2O or TE buffer.
3. Instrument capable of fluorescence polarization (Tecan, Inc., Durham, NC or comparable).

2.6.3. Nuclease Protection Assay/Footprinting

1. Footprinting binding buffer: 10 mM KCl, 10 mM Tris-HCl (pH 7.5), 2 mM $CaCl_2$, 2 mM $MgCl_2$, 5% glycerol, 100 ng/μL BSA.
2. DNaseI (Invitrogen; diluted 1:50,000 before use).
3. Stop buffer: 10 mM EDTA, 10 mM NaOH, 80% formamide, 0.01% xylene cyanol, and 0.01% bromphenol blue.
4. 8% Acrylamide/7 M urea gel.
5. Software for nonlinear regression such as SigmaPlot (Systat Software, San Jose, CA) or Prism (GraphPad, San Diego, CA).

3. Methods

3.1. Polyamide Synthesis of Cy3-Conjugated PA1 (Fig. 4)

1. Use orthogonally protected N-methylpyrrole or N-methylimidazole building blocks in standard Boc-based solid-phase synthesis (*see* **Note 3**).
2. Cleave the polyamide from PAM resin (100 mg) by treatment with 1 mL dimethylaminopropylamine to remove the phthalimide protecting group and make the free base accessible.
3. Dilute the crude cleavage mixture with 0.1% TFA (aq) and acetonitrile to a final volume of 5 mL and load onto a preconditioned solid-phase extraction column (C_{18}-bonded phase).
4. Wash the column with a 4:1 (v:v) solution of 0.1% TFA (aq) and acetonitrile.
5. Elute the product with methanol. Remove solvents by azeotropic distillation from toluene.

6. The resulting product is the aminopropyl precursor of PA1 and should be a slightly yellow solid. Verify the identity and purity of the product using analytical high-performance liquid chromatography (HPLC) and matrix-assisted laser desorption/ionization time of flight mass spectrometry (MALDI-TOF MS).
7. Dissolve 0.5 μmol of the intermediate free base in 0.45 mL anhydrous DMF and 0.050 mL DIEA. Add 1 mg of pre-packaged amine-reactive Cy3 fluorophore and agitate in the absence of light, at ambient temperature, for 4 h.
8. Purify the crude product by preparative HPLC using C_{18}-bonded phase silica with 0.1% TFA and acetonitrile as mobile phases and confirm the purity and identity of product by analytical HPLC and MALDI-TOF MS.
9. Store polyamides in small aliquots at –80°C. Do not subject to multiple freeze-thaw cycles.

3.2. DNA Array Synthesis

3.2.1. Slide Derivatization

1. Immerse the slides arranged in the metal slide holder in silane buffer for 4 h with gentle agitation.
2. Wash slides twice in stock slide buffer for 20 min each wash with gentle agitation (*see* **Note 4**).
3. Bake slides in metal rack in oven for 1 h at 120°C with no vacuum. Then bake overnight at 120°C with vacuum applied (*see* **Note 5**).

3.2.2. Microarray Synthesis

1. Synthesize arrays using MAS technology *(22)*. Covalently attach homopolymer (T5) linkers to monohydroxysilane glass slides. Synthesize oligonucleotides on the homopolymers to create a high-density DNA microarray (*see* **Notes 5** and **6**).
2. Deprotect arrays by incubating in 50% ethanol, 50% ethylenediamine for 2 h at room temperature, protected from light. Rinse the slide in a 50-mL conical tube of dH_2O for 30 s and then in a 50-mL conical tube of methanol for 30 s.
3. As an alternative, DNA arrays may be obtained from NimbleGen Systems, Inc. (Madison, WI) and are ready for hairpin induction **(step 4)**.
4. To induce hairpins, incubate the slide in a 50-mL conical tube of 7 M urea prepared in 1× PBS for 30 min in a 65°C water bath, with a back-and-forth shake of the tube every 10 min. Next, immerse the slide in preheated 1× PBS and incubate in a 65°C water bath for 15 min with one back-and-forth shake during the incubation (*see* **Note 7**).
5. Incubate slide in nonstringent wash buffer for 5–10 min at room temperature.
6. Wash the array in final wash buffer for 10–20 s.

7. Dry the array for 20 s with an ArrayIt slide centrifuge (*see* **Note 2**).
8. Scan the microarray to check for low background with an Axon 4000B, ScanArray 5000 (GSI Lumionics, Billerica, MA) or comparable 5-μM scanner. Settings on the Axon 4000B at 532 nm (Cy3) are initially set at a photomultiplier tube (PMT) gain setting of 470 and 100% lamp power. Pixel size is 5, lines to average is 1, and focus position is 0.
9. Examine data with GenePix Pro version 6.0 or comparable for background.

3.3. Polyamide Binding by CSI Arrays

3.3.1. Polyamide Binding

1. Attach a hybridization chamber to the array. Rinse the chamber on the array by filling it with dH_2O and then withdrawing all dH_2O using a pipettor. Add 2.5% nonfat dried milk to fill the chamber and then withdraw enough such that the chamber is slightly more than half full (~200 μL). Incubate for 1.5 h at room temperature with rotation (approximately three to four rotations per minute).
2. Wash the hybridization chamber twice with hybridization buffer.
3. Dilute the PA1 to 5 nM in hybridization buffer and add to the hybridization chamber for incubation (1–16 h) with rotation (*see* **Note 8**).
4. Remove the PA1 solution and wash the array with hybridization buffer.
5. Wash the array with final wash buffer for 10 s and dry.
6. Scan the microarray with a GenePix 4000B scanner (Molecular Devices). Settings on GenePix 4000B at 532 nm for Cy3 are initially set at a PMT gain setting of 470 and 100% lamp power. Pixel size is 5, lines to average is 1, and focus position is 0. Settings at 635 nm for Cy5 are initially set at a PMT of 710 and 100% lamp power. If features are saturated (=65,535 intensity), the PMT gain setting should be decreased until no saturated features are detected. Save data as a single-channel image TIF for data extraction (*see* **Note 9**).
7. Extract data using NimbleScan 2.1 (or comparable extraction software such as GenePix) and open the output file in an appropriate spreadsheet, such as Microsoft Excel. Data can be further sorted using Microsoft Access.

3.4. Data Analysis/ Normalization

1. Perform global mean normalization for each of three to four replicates to verify that the mean intensity of the replicate arrays is similar and to correct differences in array brightness from experimental variation *(32–34)*.
2. Perform local mean normalization to correct artifacts caused by uneven distribution and spatial abnormalities using a Loess function spanning 0.5–0.02 as necessary *(35)*.

3. Determine and remove outliers between replicate features using a Q test at 90% confidence *(36, 37)*.
4. Average the intensities of duplicate features from the same array.
5. Perform quantile-normalization of the replicates to eliminate any possible nonlinearity between arrays *(38)*.
6. Average the intensities of replicate features (different arrays) to give a single intensity for every sequence.
7. Subtract the center of the histogram peak of the averaged features to correct for background.
8. Calculate Z scores as |signal|/standard deviation to determine the signal-to-noise ratio and to indicate the probability (P-value) that a given sequence is preferentially bound by the DNA-interacting molecule *(39)*.
9. Graph the feature intensity (x-axis) versus the number of features (y-axis). If specific binding is observed, a peak around zero intensity (no binding) will be present with a right-handed tail indicating features with higher intensity.
10. Use MEME and WebLogo *(26)* or other algorithms to determine the specific DNA sequence recognized by PA1.

3.5. Biochemical/ Biophysical Validation of CSI Arrays

3.5.1. Electrophoretic Mobility Shift Assays

1. Incubate 1 μL of 10 μM double-stranded or hairpin DNA with 4 μL of 5× forward buffer, 1 μL kinase T4, 9 μL dH_2O, and 5 μL γ^{32}P-ATP in a 0.6-mL microfuge tube for 1 h at 37°C.
2. To purify radioactive probe using G25 spin column, centrifuge the spin column from the supplier in a 2-mL centrifuge tube at 735 × g for 75 s in a standard microcentrifuge and dispose of liquid. Add the radioactive reaction to the G25 spin column, place the spin column in a 1.5-mL microfuge tube, and centrifuge for 1 min at 735 × g. Quantitate 1 μL of eluted radioactive DNA probe using a Geiger counter.
3. Mix 50 nM polyamide-peptide conjugates with a ^{32}P-DNA dilution (150–500 cpm/lane) in binding buffer for 30 min at 4°C. Titrate in partner transcription factor (example is Exd at final concentrations of 0.033, 0.1, 0.33, 10, 33, and 100 nM) for a final reaction volume of 20 μL, and incubate for 1 h at 4°C.
4. Pour 10% acrylamide/3% glycerol gel in 1× TBE and prerun the electrophoresis for 20 min.
5. Load 15 μL of reaction onto gel and run for 2–3 h at 220 V. Place the gel on Whatman filter paper, cover with plastic wrap, and dry. Expose the dried gel overnight on a phosphorimager screen and visualize the bound DNA versus free DNA using a Typhoon phosphorimager or comparable instrument (*see* **Note 10**).

3.5.2. Fluorescence Polarization

Depending on the fluorescence polarization instrument, final volumes for each sample will be 20–100 μL.

1. Label polyamide with FITC, as described in **Subheading 3.1**.
2. Titrate DNA from 0.1 nM to 1 μM.
3. Add 1 or 5 nM FITC-polyamide in binding buffer to each sample. The concentration of FITC-polyamide may need to be optimized.
4. Measure fluorescence polarization at the appropriate wavelength (fluorescein excitation: 485 nm; emission: 530 nm).

3.5.3. Nuclease Protection Assay/Footprinting

1. Use standard PCR with one ^{32}P-labeled primer and one unlabeled primer to prepare labeled double-stranded DNA.
2. Prepare several dilutions of polyamide in footprinting binding buffer.
3. Incubate polyamide dilutions with 10,000 cpm of ^{32}P-DNA for 1 h at room temperature, in a final volume of 10 μL.
4. Add 1 μL DNaseI and incubate for 1 min.
5. Add 10 μL stop buffer and heat to 95°C for 5 min.
6. Chill samples on ice and then load immediately on an 8% acrylamide/7 M urea gel.
7. Run gel at 2,000 V in 0.5× TBE buffer until the bromophenol blue is run off the gel.
8. Dry gel, expose to phosphorimager screen overnight, and visualize using a Typhoon imager.
9. Run experiment in triplicate and determine binding constants using ImageQuant and appropriate software for nonlinear regression.

4. Notes

1. All solutions are prepared in Nanopure distilled water (dH_2O) with a resistivity of 18 MΩ cm.
2. As an alternative, arrays may be dried using an argon gas stream applied in a back-and-forth motion evenly across the top and bottom of the array.
3. All polyamide synthesis work was done in collaboration with Dervan and colleagues, and the reader is referred to their publications for additional details of synthesis *(40, 41)*.
4. Clean slides are critical for efficient array synthesis. All buffers should be made fresh, and all containers should be cleaned with acetone and rinsed with dH_2O before use.

5. Store derivatized slides and synthesized DNA arrays in a UV-protected desiccator, when not being used.
6. Each array can be synthesized with a distinct "reference" sequence synthesized at the edges for quality control and to align the grid for data extraction.
7. Hairpin percentage formation can be assessed by including two distinct features on the arrays: one that forms a hairpin (5′-CGC-TTAGTTCA-CGC-TCCT-GCG-TGAACTAA-GCG-3′) and a single-stranded version (5′-CGC-TTAG-TTCACGC-3′). Cy3-labeled 5′-GCG-TGAACTAA-GCG-3′ was added to the array at 50 nM and annealed to both the hairpin and single-stranded features. The fluorescence intensity of the hairpin sequence was divided by the fluorescence intensity of the single-stranded sequence, averaged, and background-subtracted, yielding a hairpin formation of 95.6%.
8. Protein binding to the array can be done using similar techniques as for PA1. We recommend storing protein aliquots at −80°C and minimizing freeze-thaw cycles. Binding buffer and protein concentration will need to be optimized for each protein. Other blocking agents such as 5–10% fetal bovine serum (FBS) or 1–5% BSA may be tested. Protein incubation with the array may also be stabilized at 4°C, versus room temperature.
9. If the PMT is reduced too much to eliminate signal saturation, this may indicate that DNA saturation is occurring. The concentration of fluorescent DNA-binding molecule may need to be reduced to address this issue.
10. To validate the DNA microarray results, a subset of DNA sequences that yielded various fluorescent intensities on the DNA microarray can be tested with polyamide/ATF titrations by EMSA to determine affinity values. The association constant (K_a) values form a linear relationship with the fluorescence intensities obtained by CSI arrays. This linear relationship can be used to estimate the affinity of the polyamide for any DNA sequence on the array. Other techniques such as fluorescence polarization may also be used to determine the affinity of the ATF-DNA interaction.

Acknowledgments

The authors thank Karl Hauschild and Clayton Carlson for reviewing the manuscript and Laura Vanderploeg for help with figures. This work was supported by the National Institutes of Health grant R01 GM069420 to (AZA), University of Wisconsin

Innovation and Economic Development Research Program (AZA and MSO), National Foundation—March of Dimes (AZA), US Department of Agriculture—Hatch/McIntire/Stennis grant (AZA), the Greater Milwaukee Foundation—Shaw Scientist Award (AZA), and Computation and Informatics in Biology and Medicine Training Grant T15LM007359 (CLW).

References

1. Ren B., Robert F., Wyrick J.J., Aparicio O., Jennings E.G., Simon I., Zeitlinger J., Schreiber J., Hannett N., Kanin E., Volkert T.L., Wilson C.J., Bell S.P., and Young R.A. (2000). Genome-wide location and function of DNA binding proteins. *Science* **290**, 2306–2309.
2. Venter J.C., et al. (2001). The sequence of the human genome. *Science* **291**, 1304–1351.
3. Wells J., Graveel C.R., Bartley S.M., Madore S.J., and Farnham P.J. (2002). The identification of E2F1-specific target genes. *Proc. Natl Acad. Sci. U. S. A.* **99**, 3890–3895.
4. Horak C.E., Mahajan M.C., Luscombe N.M., Gerstein M., Weissman S.M., and Snyder M. (2002). GATA-1 binding sites mapped in the beta-globin locus by using mammalian ChIP-chip analysis. *Proc. Natl Acad. Sci. U. S. A.* **99**, 2924–2929.
5. Martone R., Euskirchen G., Bertone P., Hartman S., Royce T.E., Luscombe N.M., Rinn J.L., Nelson F.K., Miller P., Gerstein M., Weissman S., and Snyder M. (2003). Distribution of NF-kappaB-binding sites across human chromosome 22. *Proc. Natl Acad. Sci. U. S. A.* **100**, 12247–12252.
6. Wei C.L., Wu Q., Vega V.B., Chiu K.P., Ng P., Zhang T., Shahab A., Yong H.C., Fu Y., Weng Z., Liu J., Zhao X.D., Chew J.L., Lee Y.L., Kuznetsov V.A., Sung W.K., Miller L.D., Lim B., Liu E.T., Yu Q., Ng H.H., and Ruan Y. (2006). A global map of p53 transcription-factor binding sites in the human genome. *Cell* **124**, 207–219.
7. Ansari A.Z., and Mapp A.K. (2002). Modular design of artificial transcription factors. *Curr. Opin. Chem. Biol.* **6**, 765–772.
8. Mapp A.K., and Ansari A.Z. (2007). A TAD further: exogenous control of gene activation. *ACS Chem. Biol.* **2**, 62–75.
9. Mapp A.K., Ansari A.Z., Ptashne M., and Dervan P.B. (2000). Activation of gene expression by small molecule transcription factors. *Proc. Natl Acad. Sci. U. S. A.* **97**, 3930–3935.
10. Ptashne M., and Gann A. (2001). Genes & Signals. New York: Cold Spring Harbor Laboratory.
11. Gerber H.P., Seipel K., Georgiev O., Hofferer M., Hug M., Rusconi S., and Schaffner W. (1994). Transcriptional activation modulated by homopolymeric glutamine and proline stretches. *Science* **263**, 808–811.
12. Courey A.J., Holtzman D.A., Jackson S.P., and Tjian R. (1989). Synergistic activation by the glutamine-rich domains of human transcription factor Sp1. *Cell* **59**, 827–836.
13. Mermod N., O'Neill E.A., Kelly T.J., and Tjian R. (1989). The proline-rich transcriptional activator of CTF/NF-I is distinct from the replication and DNA binding domain. *Cell* **58**, 741–753.
14. Saha S., Brickman J.M., Lehming N., and Ptashne M. (1993). New eukaryotic transcriptional repressors. *Nature* **363**, 648–652.
15. Mandell J.G., and Barbas C.F., III (2006) Zinc finger tools: custom DNA-binding domains for transcription factors and nucleases. *Nucleic Acids Res.* **34**, W516–W523.
16. Beerli R.R., and Barbas C.F., III (2002). Engineering polydactyl zinc-finger transcription factors. *Nat. Biotech.* **20**, 135–141.
17. Isalan M., Klug A., and Choo Y. (2001). A rapid, generally applicable method to engineer zinc fingers illustrated by targeting the HIV-1 promoter. *Nat. Biotech.* **19**, 656–660.
18. Dervan P.B., and Edelson B.S. (2003). Recognition of the DNA minor groove by pyrrole-imidazole polyamides. *Curr. Opin. Struct. Biol.* **13**, 284–299.
19. Buchmueller K.L., Staples A.M., Howard C.M., Horick S.M., Uthe P.B., Le N.M., Cox K.K., Nguyen B., Pacheco K.A., Wilson W.D., and Lee M. (2005). Extending the language of DNA molecular recognition by polyamides: unexpected influence of imidazole and pyrrole arrangement on binding affinity and specificity. *J. Am. Chem. Soc.* **127**, 742–750.
20. Supekova L., Pezacki J.P., Su A.I., Loweth C.J., Riedl R., Geierstanger B., Schultz P.G., and Wemmer D.E. (2002). Genomic effects of polyamide/DNA interactions on mRNA expression. *Chem. Biol.* **9**, 821–827.

21. Warren C.L., Kratochvil N.C., Hauschild K.E., Foister S., Brezinski M.L., Dervan P.B., Phillips G.N., Jr., and Ansari A.Z. (2006). Defining the sequence-recognition profile of DNA-binding molecules. *Proc.Natl Acad. Sci.U. S. A.* **103**, 867–872.
22. Singh-Gasson S., Green R.D., Yue Y., Nelson C., Blattner F., Sussman M.R., and Cerrina F. (1999). Maskless fabrication of light-directed oligonucleotide microarrays using a digital micromirror array. *Nat. Biotech.* **17**, 974–978.
23. Mann R.S., and Chan S.K. (1996). Extra specificity from extradenticle: the partnership between HOX and PBX/EXD homeodomain proteins. *Trends Genet.* **12**, 258–262.
24. Hauschild K.E., Metzler R.E., Arndt H.D., Moretti R., Raffaelle M., Dervan P.B., and Ansari A.Z. (2005). Temperature–sensitive protein–DNA dimerizers. *Proc. Natl Acad. Sci. U. S. A.* **102**, 5008–5013.
25. Arndt H.D., Hauschild K.E., Sullivan D.P., Lake K., Dervan P.B., and Ansari A.Z. (2003). Toward artificial developmental regulators. *J. Am. Chem. Soc.* **125**, 13322–13323.
26. Schneider T.D., and Stephens R.M. (1990). Sequence logos: a new way to display consensus sequences. *Nucleic Acids Res.* **18**, 6097–6100.
27. Bailey T.L., and Elkan C. (1994). Fitting a mixture model by expectation maximization to discover motifs in biopolymers. *Proc. Int. Conf. Intell. Syst. Mol. Biol.* **2**, 28–36.
28. Liu X.S., Brutlag D.L., and Liu J.S. (2002). An algorithm for finding protein-DNA binding sites with applications to chromatin-immunoprecipitation microarray experiments. *Nat. Biotech.* **20**, 835–839.
29. Guex N., and Peitsch M.C. (1997). SWISS-MODEL and the Swiss-PdbViewer: an environment for comparative protein modeling. *Electrophoresis* **18**, 2714–2723.
30. Humphrey W., Dalke A., and Schulten K. (1996). VMD: visual molecular dynamics. *J. Mol. Graph.* **14**, 33–38, 27–38.
31. Stafford R.L., and Dervan P.B. (2007). The reach of linear protein-DNA dimerizers. *J. Am. Chem. Soc.* **129**, 14026–14033.
32. Smyth G.K., and Speed T. (2003). Normalization of cDNA microarray data. *Methods* **31**, 265–273.
33. Quackenbush J. (2002). Microarray data normalization and transformation. *Nat. Gen.* **32** Suppl, 496–501.
34. Yang Y.H., Dudoit S., Luu P., Lin D.M., Peng V., Ngai J., and Speed T.P. (2002). Normalization for cDNA microarray data: a robust composite method addressing single and multiple slide systematic variation. *Nucleic Acids Res.* **30**, e15.
35. Colantuoni C., Henry G., Zeger S., and Pevsner J. (2002). Local mean normalization of microarray element signal intensities across an array surface: quality control and correction of spatially systematic artifacts. *BioTechniques* **32**, 1316–1320.
36. Dixon W. (1950). Analysis of extreme values. *Ann. Math. Stat.* **21**, 488–506.
37. Rorabacher D. (1991) Statistical treatment for rejection of deviant values: critical values of Dixon Q parameter and related subrange ratios at the 95 percent confidence level. *Anal. Chem.* **83**, 139–146.
38. Bolstad B.M., Irizarry R.A., Astrand M., and Speed T.P. (2003). A comparison of normalization methods for high density oligonucleotide array data based on variance and bias. *Bioinformatics* **19**, 185–193.
39. Adbi H. (2007). Encyclopedia of Measurement and Statistics. Thousand Oaks, CA: Sage.
40. Wurtz N.R., Turner J.M., Baird E.E., and Dervan P.B. (2001). Fmoc solid phase synthesis of polyamides containing pyrrole and imidazole amino acids. *Org. Lett.* **3**, 1201–1203.
41. White S., Szewczyk J.W., Turner J.M., Baird E.E., and Dervan P.B. (1998). Recognition of the four Watson-Crick base pairs in the DNA minor groove by synthetic ligands. *Nature* **391**, 468–471.

INDEX

A

B

C

D

E

O

P

S

T

FSC
www.fsc.org
MIX
Papier | Fördert
gute Waldnutzung
FSC® C083411

Kolibri 360 GmbH
Ferdinand-Jühlke-Straße 7
99095 Erfurt, Deutschland
produktsicherheit@kolibri360.de